T0335500

Mathematical Foundations of Imaging, Tomography and Wavefield Inversion

Inverse problems are of interest and importance across many branches of physics, mathematics, engineering and medical imaging. In this text, the foundations of imaging and wavefield inversion are presented in a clear and systematic way. The necessary theory is gradually developed throughout the book, progressing from simple wave-equation-based models to vector wave models. By combining theory with numerous MATLAB-based examples, the author promotes a complete understanding of the material and establishes a basis for real-world applications.

Key topics of discussion include the derivation of solutions to source radiation and scattering problems using Green-function techniques and eigenfunction expansions; the propagation and scattering of waves in homogeneous and inhomogeneous backgrounds; and the concepts of field time reversal and field back propagation and the key role that they play in imaging and inverse scattering.

Bridging the gap between mathematics and physics, this multidisciplinary book will appeal to graduate students working in established areas of inverse scattering and to researchers developing new computational imaging modalities. Additional resources, including solutions to end-of-chapter problems and MATLAB codes for all the examples presented in the book, are available online at www.cambridge.org/9780521119740.

Anthony J. Devaney is Distinguished Professor of Engineering at Northeastern University, Boston and he has worked in the general area of inverse problems for more than 40 years. He did his Ph.D. at the Institute of Optics at the University of Rochester and his thesis was supervised by Professor Emil Wolf. Professor Devaney has experience in geophysics inverse problems and inverse problems related to radar, optical and acoustic imaging. His patent on diffraction tomography (covered in Chapter 8 of this book) was selected as one of the top 50 patents from Schlumberger Doll Research (the principal basic research center of the Schlumberger Corporation) over its first 50 years of existence.

Mathematical Foundations of Imaging, Tomography and Wavefield Inversion

ANTHONY J. DEVANEY
Northeastern University, Boston

Shaftesbury Road, Cambridge CB2 8EA, United Kingdom

One Liberty Plaza, 20th Floor, New York, NY 10006, USA

477 Williamstown Road, Port Melbourne, VIC 3207, Australia

314–321, 3rd Floor, Plot 3, Splendor Forum, Jasola District Centre, New Delhi – 110025, India

103 Penang Road, #05–06/07, Visioncrest Commercial, Singapore 238467

Cambridge University Press is part of Cambridge University Press & Assessment, a department of the University of Cambridge.

We share the University's mission to contribute to society through the pursuit of education, learning and research at the highest international levels of excellence.

www.cambridge.org
Information on this title: www.cambridge.org/9780521119740

First published 2012

A catalogue record for this publication is available from the British Library

Library of Congress Cataloging-in-Publication data
Devaney, Anthony J.
Mathematical foundations of imaging, tomography and wavefield inversion / Anthony J. Devaney, Northeastern University, Boston.
pages cm
ISBN 978-0-521-11974-0 (hardback)
1. Wave equation. 2. Inverse problems (Differential equations) I. Title.
QC174.26.W28D382 2012
515′.357–dc23
2012000073

ISBN 978-0-521-11974-0 Hardback

Additional resources for this publication at www.cambridge.org/9780521119740

Contents

Preface

I started this book roughly 20 years ago with the intention of producing a finished product within a year or so. But reality in the form of government research grants and "publish or perish" soon set in and so now, at long last, I have finally finished. The final product has of course changed significantly over these intervening years, both in content and in breadth. My original plan was to put together a six- or seven-chapter treatise on basic "Fourier-based" coherent imaging and diffraction tomography complete with Matlab codes implementing the imaging and inversion algorithms presented in the text. The current book certainly includes this material, but also includes a host of other material such as the chapter on time-reversal imaging and the four chapters on the propagation and scattering of waves in homogeneous and inhomogeneous backgrounds. More importantly, the "Fourier-based" inversion schemes originally used to develop much of coherent imaging and linearized inverse scattering (diffraction tomography) have been replaced by the much more powerful singular value decomposition (SVD). This approach allows virtually all of the linearized inverse problems associated with the wave and Helmholtz equation both in homogeneous and in inhomogeneous backgrounds to be treated in a uniform "turn the crank" manner.

My work on imaging and wavefield inversion began as a graduate student under Professor Emil Wolf at the University of Rochester. Originally I had intended to pursue my Ph.D. in quantum optics, but had my plans changed significantly by an off-hand remark by Professor Wolf during one of our meetings. We were discussing the classical theory of imaging by lenses, at which point he asked the question "what exactly is an image?" The answer to that seemingly simple question set us off on a road that included non-radiating sources, non-scattering scatterers, and other bizarre objects that the mathematician would recognize as being members of the null space of the mapping from object to "image." While the purely non-radiating sources and non-scattering scatterers are in the null space of the mapping from object to image, there are other strange objects that I have chosen to call "essentially" non-radiating sources (or scatterers). These objects are not in the null space but are very close to it, having the property that they only radiate (or scatter) evanescent waves outside of their support and are the cause for instability of inverse problems related to the wave and Helmholtz equations. I have tried to couple these physical interpretations of non-uniqueness and instability to the purely mathematical view of these properties throughout the book. Indeed, the melding of physics with mathematics is one of my major goals in this book.

The general areas of imaging and inverse scattering are multidisciplinary in that they require a strong foundation in physics, mathematics, and signal processing. I have tried to include the necessary background in all three areas, but assume that the reader is already proficient in complex-variable theory and linear algebra at the senior undergraduate/first-year graduate level and has at least a rudimentary familiarity with the wave and Helmholtz equations in a homogeneous medium such as free space. I have also tried to emphasize the underlying physics of the various topics covered in the book but, unfortunately, at the expense of mathematical rigor. This is especially true in the development of time-independent scattering theory in Chapters 6 and 9, which follow the purely formal approach used in non-relativistic quantum scattering (collision) theory.

The vast majority of the book treats scalar wave theory, with only the last chapter devoted to vector waves in the form of the electromagnetic (EM) field. The reasons for this are that all of the essential ingredients of coherent imaging and inverse scattering are already contained in the scalar theory and that the vector theory, at least for the EM field, can be reduced to three or fewer coupled scalar wave problems. Indeed, by using the so-called Whittaker or Debye representations presented in Chapter 11, EM inverse source and scattering problems for planar or spherical geometries can be reduced to two uncoupled scalar wave problems that are treated exactly in the manner presented in earlier chapters of the book. I have also, for the most part, restricted the treatment of the various inverse problems to *linearized* formulations of the corresponding forward problems. The exceptions to this are the inverse source problem which, by its nature, is a linear problem and one of the formulations of inverse scattering from conducting surfaces in Chapter 7.

The goal of this book is to present the mathematical (and physical) *foundations* of imaging and wavefield inversion rather than to push specific inversion schemes or algorithms or to present detailed results of the use of such algorithms on real data. To this end, I have concentrated on simple yet representative Matlab-based examples that are easily understood and directly related to the theoretical development presented in the book. The myriad details that attend any actual application of these algorithms to real data are not presented. Such details include the methods required to retrieve the phase of an optical field in an optical-imaging or inverse-scattering algorithm and the need to align, usually through the use of digital filters, the outputs from antenna or transducer arrays in ultrasound or EM inverse-scattering or time-reversal imaging experiments.

Finally, a word about the references cited in the book. Originally I intended to include as complete a list as possible of the majority of papers and books by workers in the general field of inverse scattering and wavefield inversion. I soon found the list growing beyond bound and was forced to limit the list to those references that I felt to be directly related to the material presented in the book. The book is mostly about *linearized* formulations of inverse scattering and, thus, I have left out an enormous number of references, especially within the mathematics community, to exact non-linear approaches to inverse scattering. I have also left out virtually all references to applications since the book is about the underlying *theory* of linearized inverse scattering and is not concerned with applications of this theory in various fields such as optics, acoustics, etc. I apologize to the many researchers who may feel slighted by not being included in the bibliography or not being suitably referenced.

I would like to thank my former professor and good friend and colleague over the past (can it be 40?) years Emil Wolf. Much of the material in the book can be traced back to my Ph.D. thesis and to joint papers by Emil and myself. I would also like to thank my colleague of many years' standing Dr. George Sherman and the dozens of current and former students and colleagues who collaborated on the development and application of the material presented in the book. Special thanks go to my former friend and colleague Alan Witten, who died unexpectedly in 2005. Alan, who was professor of geophysics at the University of Oklahoma, used acoustic diffraction tomography to help find and unearth seismosaurus, the longest dinosaur yet discovered (see *NY Times* "New X-Ray Technique Helps Dinosaur Hunters," Science Section, Dec. 12, 1989), and whose work was, at least partially, the motivation for the opening scenes in the original *Jurassic Park* movie. I would also like to thank Dr. Arje Nachman of the AFOSR and Dr. Richard Albanese, director of the mathematical products division at the Brooks Air Force Base in San Antonio, for financial and inspirational support over the past 20 years. Finally, I must thank Simon Capelin and the wonderful staff at Cambridge University Press. Simon first met me about the book in 1990 in my company office in downtown Boston to discuss the project that I promised would be finished in less than a year.

1 Radiation and initial-value problems for the wave equation

1.1 The radiation problem

We consider the radiation of waves from a real-valued space- and time-varying source $q(\mathbf{r}, t)$ embedded in an infinite, homogeneous, isotropic and non-dispersive and non-attenuating medium such as free space. The real-valued radiated wavefield satisfies the inhomogeneous scalar wave equation

$$\left[\nabla^2 - \frac{1}{c^2}\frac{\partial^2}{\partial t^2}\right] u(\mathbf{r}, t) = q(\mathbf{r}, t) \tag{1.1}$$

throughout all of space and for all time, where c is the constant velocity of wave propagation in the background medium. We will assume that the source q is compactly supported in the space-time region $\{\mathcal{S}_0 | \mathbf{r} \in \tau_0,\ t \in [0, T_0]\}$, where τ_0 is its spatial volume and $[0, T_0]$ the interval of time over which the source is turned on. We also assume that the source possesses finite energy (is square-integrable in $\mathcal{S}_0$)

$$\mathcal{E}_q = \int_0^{T_0} dt \int_{\tau_0} d^3r |q(\mathbf{r}, t)|^2 < \infty, \tag{1.2}$$

although we will sometimes have to enlarge the class of sources to include Dirac delta functions, which are not square-integrable, but these cases are special and will be dealt with on an individual basis as required.

The reader may wonder why we have assumed that the source radiates only over a finite time period $[0, T_0]$ as opposed to being allowed to radiate over the semi-infinite interval $[0, \infty)$. The main reason is that it simplifies the mathematics without being a real restriction on the theory and results that we will obtain. In particular, although the time interval over which the source radiates is finite, it can be arbitrarily large so that this source model can apply to any real source to arbitrary accuracy. Moreover, most of our results will be valid in the limit $T_0 \to \infty$ so that the assumption places little or no restriction on our theoretical development.

The solution to the inhomogeneous wave equation Eq. (1.1) is not unique. In particular, it is clear that we can add to u any solution to the *homogeneous wave equation*

$$\left[\nabla^2 - \frac{1}{c^2}\frac{\partial^2}{\partial t^2}\right] \delta u(\mathbf{r}, t) = 0 \tag{1.3}$$

and still obtain a solution to Eq. (1.1). In order to obtain a unique solution it is necessary to specify *initial conditions* in the form of the *Cauchy conditions*

$$u(\mathbf{r},t)|_{t=0} = u_0(\mathbf{r}), \qquad \frac{\partial}{\partial t}u(\mathbf{r},t)|_{t=0} = u_0'(\mathbf{r}), \tag{1.4}$$

where u_0 and u_0' are arbitrary (real-valued) functions of position $\mathbf{r}$. We will show below that the inhomogeneous wave equation Eq. (1.1) together with Cauchy conditions at $t = 0$ suffice to uniquely determine the field u. The appropriate initial conditions required of the physically meaningful solution to Eq. (1.1) are derived from the requirement of *causality*; i.e., we seek the *particular solution* to the wave equation $u_+(\mathbf{r},t)$ that is causally related to the source; i.e., that vanishes prior to the turn-on time of the source ($t = 0$). The required causality of the field u_+ is equivalent to requiring that this field satisfy *homogeneous Cauchy conditions* at $t = 0$; i.e.,

$$u_+(\mathbf{r},t)|_{t=0} = 0, \qquad \frac{\partial}{\partial t}u_+(\mathbf{r},t)|_{t=0} = 0,$$

where we have denoted the causal solution to Eq. (1.1) with the subscript +. The problem of solving the inhomogeneous wave equation Eq. (1.1) under the condition of causality (or, equivalently, homogeneous Cauchy conditions at $t = 0$) is called the *radiation problem*. The problem of solving the *homogeneous* wave equation Eq. (1.3) subject to arbitrary inhomogeneous Cauchy conditions at $t = 0$ is called the *initial-value problem*. We will treat both problems in this chapter.

1.1.1 Fourier integral representations

We will make frequent use of Fourier integral representations of space and time dependent functions and fields throughout the book. We will assume throughout that any function (or field) $f(\mathbf{r},t)$ possesses a temporal Fourier transform defined by

$$F(\mathbf{r},\omega) = \int_{-\infty}^{\infty} dt\, f(\mathbf{r},t)e^{i\omega t}, \tag{1.5a}$$

and that the temporal transform can be further transformed via a spatial Fourier transform of the form

$$\tilde{F}(\mathbf{K},\omega) = \int d^3r\, F(\mathbf{r},\omega)e^{-i\mathbf{K}\cdot\mathbf{r}}, \tag{1.5b}$$

where the integration in Eq. (1.5b) is carried out over all of R^3. We further assume that each of the transforms can be inverted to yield Fourier integral representations given by

$$f(\mathbf{r},t) = \frac{1}{2\pi}\int_{-\infty}^{\infty} d\omega\, F(\mathbf{r},\omega)e^{-i\omega t}, \tag{1.6a}$$

$$F(\mathbf{r},\omega) = \frac{1}{(2\pi)^3}\int d^3K\, \tilde{F}(\mathbf{K},\omega)e^{i\mathbf{K}\cdot\mathbf{r}}. \tag{1.6b}$$

We can, of course, combine the temporal and spatial Fourier integral representations into a single *space-time* representation of the form

$$f(\mathbf{r},t) = \frac{1}{(2\pi)^4}\int_{-\infty}^{\infty} d\omega \int d^3K\,\tilde{F}(\mathbf{K},\omega)e^{i(\mathbf{K}\cdot\mathbf{r}-\omega t)}, \tag{1.7a}$$

where

$$\tilde{F}(\mathbf{K},\omega) = \int_{-\infty}^{\infty} dt \int d^3r\,f(\mathbf{r},t)e^{-i(\mathbf{K}\cdot\mathbf{r}-\omega t)}. \tag{1.7b}$$

As in the presentation given above, we will generally denote functions of space and time (space-time) by lower-case letters and their temporal (time) transforms by upper-case letters. The spatial Fourier transforms of the latter transforms are then represented by an upper-case letter with a tilde on top. Thus, we have the progression

$$f(\mathbf{r},t) \longleftrightarrow F(\mathbf{r},\omega) \Longleftrightarrow \tilde{F}(\mathbf{K},\omega), \tag{1.8}$$

where the double arrow $\Longleftrightarrow$ denotes the Fourier-transform operation. We will explicitly display the limits on one-dimensional integrals as in the temporal and inverse temporal transforms given above but will not explicitly display the limits on multi-dimensional integrals unless their integration domains are finite.

The classical theory of the Fourier integral requires that the functions $f(\mathbf{r},t)$, $F(\mathbf{r},\omega)$ and $\tilde{F}(\mathbf{K},\omega)$ are all absolutely integrable and that the integrals be interpreted as Lebesgue integrals for the above set of equations to hold. If, in addition, the functions decay sufficiently fast at infinity they will possess the important property that *multiplication by the frequency variable in the frequency domain corresponds to differentiation in the time or space domain.* The modern theory of the Fourier integral is based on distribution theory (the theory of generalized functions) and avoids all of the analysis and issues of the classical theory as well as enlarging the class of functions that can be transformed to include discontinuous and non-differentiable (generalized) functions. Within the context of distribution theory any generalized function $f(\mathbf{r},t)$ will possess transforms and inverse transforms as given above and partial derivatives that are related to their transforms via the equations

$$\frac{\partial^n}{\partial t^n}f(\mathbf{r},t) \Longleftrightarrow (-i\omega)^n F(\mathbf{r},\omega), \qquad \frac{\partial^n}{\partial x^n}F(\mathbf{r},\omega) \Longleftrightarrow (iK_x)^n\tilde{F}(\mathbf{K},\omega), \tag{1.9}$$

where n is any positive integer, x is any of the Cartesian components of $\mathbf{r}$, and the double arrow $\Longleftrightarrow$ denotes the Fourier-transform operation.

Important examples of generalized functions are the Dirac delta functions which are defined according to the equations[1]

[1] We will not employ different symbols for one-, two- and three-dimensional delta functions unless their dimensionality is not clear from their argument.

$$\phi(0) = \int_{-\infty}^{\infty} dt\, \delta(t)\phi(t), \qquad \chi(\mathbf{0}) = \int d^3r\, \delta(\mathbf{r})\chi(\mathbf{r}), \tag{1.10}$$

where $\phi(t)$ and $\chi(\mathbf{r})$ are any well-behaved ordinary functions of t and $\mathbf{r}$, respectively. The Dirac delta functions do not have meaning within the framework of classical function theory and must be interpreted within the framework of distribution theory, where the "integrals" in the above definitions are taken to be inner products defined on a suitable space of "testing functions." Although $\delta(t)$ and $\delta(\mathbf{r})$ are not ordinary functions, they can be formally manipulated and treated as such as long as at the end of a calculation they appear in integrals with ordinary functions that can then be given meaning through Eqs. (1.10). In this connection, we note that the Fourier transforms of the delta functions are given by

$$1 = \int_{-\infty}^{\infty} dt\, \delta(t)e^{i\omega t}, \qquad 1 = \int d^3r\, \delta(\mathbf{r})e^{-i\mathbf{K}\cdot\mathbf{r}}, \tag{1.11a}$$

which follows from Eqs. (1.10) on taking $\phi(t) = \exp(i\omega t)$ and $\chi(\mathbf{r}) = \exp(-i\mathbf{K}\cdot\mathbf{r})$. The delta functions then admit Fourier-integral representations given by

$$\delta(t) = \frac{1}{2\pi}\int_{-\infty}^{\infty} d\omega\, e^{-i\omega t}, \qquad \delta(\mathbf{r}) = \frac{1}{(2\pi)^3}\int d^3K\, e^{i\mathbf{K}\cdot\mathbf{r}}. \tag{1.11b}$$

We will interpret the Fourier integral throughout this book within the context of distribution theory, which amounts to using the transforms and inverse transforms in a purely formal way without any regard for the properties of the functions being transformed or inverse transformed. In most cases the results we obtain will hold within the classical theory of the transform but will be obtained using much less effort than would be required using the classical theory. In some cases the results cannot be obtained using classical theory but have a perfectly acceptable interpretation within distribution theory as, for example, will be the case when we compute the Green function of the wave equation in the following section. We will not detour into a review of distribution theory but will present certain results from the theory when needed. We refer the interested reader to the books on the subject listed at the end of the chapter.

Example 1.1 As a simple example of the use of the Fourier transform we consider the initial-value problem for the one-dimensional homogeneous wave equation

$$\left[\frac{\partial^2}{\partial z^2} - \frac{1}{c^2}\frac{\partial^2}{\partial t^2}\right] u(z,t) = 0,$$

together with the Cauchy conditions

$$u(z,t)|_{t=0} = u_0(z), \qquad \frac{\partial}{\partial t}u(z,t)|_{t=0} = u_0'(z).$$

It is easy to verify that the general solution is given by

$$u(z,t) = f(z-ct) + g(z+ct), \tag{1.12}$$

where f and g are arbitrary functions that have derivatives up to second order. These functions are uniquely determined from the Cauchy conditions via the equations

$$f(z)+g(z)=u_0(z), \qquad -\frac{\partial}{\partial z}f(z)+\frac{\partial}{\partial z}g(z)=\frac{1}{c}u_0'(z).$$

The above coupled set of equations is easily solved using the Fourier transform. In particular, on Fourier transforming both sides of the above equations we obtain the result

$$\tilde{f}(K)+\tilde{g}(K)=\tilde{u}_0(K), \qquad -iK\tilde{f}(K)+iK\tilde{g}(K)=\frac{1}{c}\tilde{u}_0'(K),$$

where

$$\tilde{f}(K)=\int_{-\infty}^{\infty} dz f(z)e^{-iKz}$$

and similarly for the other transformed quantities. We conclude that

$$\tilde{f}(K)=\frac{1}{2}\left[\tilde{u}_0(K)+\frac{i}{cK}\tilde{u}_0'(K)\right],$$
$$\tilde{g}(K)=\frac{1}{2}\left[\tilde{u}_0(K)-\frac{i}{cK}\tilde{u}_0'(K)\right],$$

so that the solution to the Cauchy initial-value problem is given by Eq. (1.12) with

$$f(z)=\frac{1}{2\pi}\int_{-\infty}^{\infty} dK\, \overbrace{\frac{1}{2}\left[\tilde{u}_0(K)+\frac{i}{cK}\tilde{u}_0'(K)\right]}^{\tilde{f}(K)} e^{iKz},$$
$$g(z)=\frac{1}{2\pi}\int_{-\infty}^{\infty} dK\, \overbrace{\frac{1}{2}\left[\tilde{u}_0(K)-\frac{i}{cK}\tilde{u}_0'(K)\right]}^{\tilde{g}(K)} e^{iKz}.$$

Example 1.2 In many cases treated here and in later chapters the temporal and/or spatial Fourier transforms are not only ordinary (as opposed to generalized) functions but also analytic functions of the transform variables. As an example, consider the temporal transform $Q(\mathbf{r},\omega)$ of the source $q(\mathbf{r},t)$, which is assumed to be square-integrable and supported in the finite time interval $[0,T_0]$. This transform is an entire analytic function of ω throughout the entire complex-ω plane. To show this we set $f(\mathbf{r},t)=q(\mathbf{r},t)$ in Eq. (1.5a) and expand the exponential in a Taylor series centered at the origin. We then interchange the orders of summation and integration (which is allowable since the integral has finite limits and the series converges uniformly) to obtain the result

$$Q(\mathbf{r},\omega)=\sum_{n=0}^{\infty} A_n\omega^n, \tag{1.13}$$

where $Q(\mathbf{r},\omega)$ is the temporal transform of $q(\mathbf{r},t)$ and

$$A_n=\frac{(i)^n}{n!}\int_0^{T_0} dt\, t^n q(\mathbf{r},t).$$

Since the time-dependent source $q(\mathbf{r},t)$ is square-integrable and compactly supported in $[0,T_0]$, it must be at least piecewise continuous within this interval so that

$$|A_n| \le \frac{\max|q(\mathbf{r},t)|}{n!}\int_0^{T_0} dt\, t^n = \frac{\max|q|T_0^{(n+1)}}{(n+1)!}.$$

It then follows that the Taylor series of $Q(\mathbf{r},\omega)$ given in Eq. (1.13) is term by term smaller than the series

$$G(\omega) = \sum_{n=0}^{\infty} \frac{\max|q|T_0^{(n+1)}}{(n+1)!}\omega^n. \tag{1.14}$$

But the latter series has, by the ratio test, an infinite radius of absolute convergence, i.e.,

$$\frac{\left|\frac{\mathrm{Max}|q|T_0^{(n+2)}}{(n+2)!}\omega^{n+1}\right|}{\left|\frac{\mathrm{Max}|q|T_0^{(n+1)}}{(n+1)!}\omega^{n}\right|} = T_0\frac{|\omega|}{n+2} \to 0, \quad n\to\infty, \forall\omega.$$

It then follows by the comparison test that the series Eq. (1.13) also converges absolutely for all ω and, hence, is entire analytic.

1.2 Green functions

The commonly used expressions "the Green's function" and "a Green's function" represent an atrocity to the English language. I doubt that those who use them ever refer to "a Shakespeare's sonnet." (Rohrlich, 1965)

We define a *Green function* of the wave equation to be any (real) solution to the wave equation Eq. (1.1) for the special case in which the source $q(\mathbf{r},t) = \delta(\mathbf{r}-\mathbf{r}')\delta(t-t')$, where $\delta(\cdot)$ is the Dirac delta function and $\mathbf{r}'$ and t' are considered to be free parameters that can assume any values in space-time. A Green function to the wave equation thus satisfies the equation

$$\left[\nabla^2 - \frac{1}{c^2}\frac{\partial^2}{\partial t^2}\right] g(\mathbf{r},\mathbf{r}',t,t') = \delta(\mathbf{r}-\mathbf{r}')\delta(t-t'). \tag{1.15}$$

It is seen that a Green function $g(\mathbf{r},\mathbf{r}',t,t')$ is simply the field radiated by an impulsive source located at the space-time point $\mathbf{r}',t'$. It follows from the homogeneity of infinite free space and time that any physically meaningful Green function must then be a function only[2] of the difference vector $\mathbf{R} = \mathbf{r}-\mathbf{r}'$ and the time difference[3] $\tau = t - t'$; i.e.,

[2] This follows from the fact that any two space-time points in infinite homogeneous isotropic space-time are indistinguishable; i.e, since there are no physical boundaries or inhomogeneities all space-time points are equivalent so that solutions to the wave equation must be translationally invariant in space-time.

[3] We will also use the Greek symbol τ to denote regions of space throughout the book; e.g., τ_0 as the space support of a source $q(\mathbf{r},t)$ to the wave equation. No confusion should arise, however, since the meaning of the symbol will be clear from the context.

$$g(\mathbf{r}, \mathbf{r}', t, t') = g(\mathbf{r} - \mathbf{r}', t - t') = g(\mathbf{R}, \tau).$$

We can thus replace Eq. (1.15) by the simpler equation

$$\left[\nabla^2 - \frac{1}{c^2}\frac{\partial^2}{\partial \tau^2}\right] g(\mathbf{R}, \tau) = \delta(\mathbf{R})\delta(\tau), \tag{1.16}$$

where the Laplacian operator ∇^2 is taken with respect to the components of the **R** vector.

A Green function, like any solution to the inhomogeneous wave equation, is not unique. In particular, given any Green function $g(\mathbf{R}, \tau)$ we can obtain a new Green function by adding *any* function $\delta g(\mathbf{R}, \tau)$ that satisfies the homogeneous wave equation Eq. (1.3). The different Green functions obtained in this way will satisfy the same defining equation Eq. (1.16) but different *initial conditions*. As discussed earlier, the choices of initial conditions that result in a unique solution of the inhomogeneous wave equation are known to be the value of the field and its first time derivative at some initial time (the Cauchy conditions). A time-domain Green function $g(\mathbf{R}, \tau)$ to the wave equation in infinite free space is thus uniquely determined by Cauchy conditions at $\tau = 0$.

To compute a Green function we represent it in the space-time Fourier integral representation given in Eqs. (1.7) and make use of Eqs. (1.9). We then find that

$$\left[\nabla^2 - \frac{1}{c^2}\frac{\partial^2}{\partial \tau^2}\right] g(\mathbf{R}, \tau) \Longleftrightarrow \left[-K^2 + \frac{\omega^2}{c^2}\right] \tilde{G}(\mathbf{K}, \omega),$$

where $\Longleftrightarrow$ denotes a four-dimensional Fourier transformation and $K^2 = \mathbf{K}\cdot\mathbf{K}$. The space-time Fourier transform of Eq. (1.16) is then found to be

$$[-K^2 + k^2]\tilde{G}(\mathbf{K}, \omega) = 1,$$

where $k = \omega/c$ is the "wavenumber" of the background medium and we have used the fact that the transforms of the delta functions in Eq. (1.16) are each unity. On using the inverse space-time Fourier transform we then obtain

$$g(\mathbf{R}, \tau) = \frac{1}{(2\pi)^4}\int_{-\infty}^{\infty} d\omega \int d^3K \, \frac{e^{i(\mathbf{K}\cdot\mathbf{R} - \omega\tau)}}{-K^2 + k^2}. \tag{1.17}$$

1.2.1 Retarded and advanced Green functions

As discussed above, a Green function is not unique and, indeed, the Fourier integral representation Eq. (1.17) does not uniquely define a Green function due to the improper nature of the integral resulting from the poles of the integrand at $k = \omega/c = \pm K$. In order to give meaning to the integral it is necessary to deform the ω contour to avoid these poles, which is possible in a number of ways, with each scheme yielding a different Green function. The Green function of most physical interest is the *causal* Green function, which vanishes for negative time $\tau < 0$ and results in homogeneous Cauchy conditions at $\tau = 0$. The causal Green function, which is referred to as the *retarded Green function* for reasons discussed below, results from deforming the ω integration contour to lie above both poles, leaving the upper half-plane (u.h.p.) free of singularities. If $\tau < 0$ the contour can be closed in the u.h.p., from which it follows from Cauchy's integral theorem that the integral vanishes and the Green function is causal. If $\tau > 0$ we can close the ω integration contour in the lower half-plane (l.h.p.) to obtain

$$g_+(\mathbf{R}, \tau) = \frac{c^2}{(2\pi)^4} \int d^3K \, e^{i\mathbf{K}\cdot\mathbf{R}} \int_C d\omega \, \frac{e^{-i\omega\tau}}{\omega^2 - c^2K^2}, \quad \tau > 0, \tag{1.18}$$

where C is the causal contour that lies above both poles and is closed over an infinite semicircle in the l.h.p., and where we have used the subscript $+$ to denote the causal Green function. We can now evaluate the integral using residue calculus to find that

$$g_+(\mathbf{R}, \tau) = -\frac{c}{(2\pi)^3} \int d^3K \, e^{i\mathbf{K}\cdot\mathbf{R}} \frac{\sin(cK\tau)}{K}, \quad \tau > 0. \tag{1.19}$$

To finish the calculation we transform to spherical polar integration variables in Eq. (1.19) with the polar axis aligned along the direction of $\mathbf{R}$. We then have that $\mathbf{K} \cdot \mathbf{R} = KR\cos\theta$, with θ the polar angle of $\mathbf{K}$. The integration over the azimuthal angle in Eq. (1.19) can then be performed, and we obtain

$$\begin{aligned} g_+(\mathbf{R}, \tau) &= -\frac{c}{(2\pi)^2} \int_0^\infty K\,dK \, \sin(cK\tau) \int_0^\pi e^{iKR\cos\theta} \sin\theta \, d\theta \\ &= -\frac{c}{(2\pi)^2 R} \int_{-\infty}^\infty dK \, \sin(cK\tau)\sin(KR), \quad \tau > 0. \end{aligned}$$

The final step is to expand the sine functions using Euler's identity and use the Fourier-integral representation of the delta function given in Eq. (1.11b). We obtain after some minor algebra

$$g_+(\mathbf{R}, \tau) = -\frac{1}{4\pi} \frac{\delta(\tau - R/c)}{R}, \tag{1.20a}$$

and $g_+(\mathbf{R}, \tau) = 0$ for $\tau < 0$.

As mentioned above, the causal time-domain Green function defined in Eq. (1.20a) is generally known as the *retarded Green function*. The motivation for this nomenclature is that $g_+(\mathbf{r} - \mathbf{r}', t - t')$ represents the field radiated from an impulsive source located at the space-time point $\mathbf{r}'$, t' and this field is not observed at the space point $\mathbf{r}$ until the time $\tau = R/c \Rightarrow t = t' + |\mathbf{r} - \mathbf{r}'|/c$; i.e., the observation time is *retarded* by the distance between the two field points divided by the velocity c of the background medium.

Another time-domain Green function of interest is the acausal or *advanced* Green function $g_-(\mathbf{R}, \tau)$. This Green function results from taking the ω contour of integration in the Fourier-integral representation Eq. (1.17) to lie below the two poles. In this case the l.h.p. is free of singularities and the integral vanishes if $\tau > 0$, resulting in the acausal Green function. If $\tau < 0$ we can close the contour in the u.h.p. and, following steps almost identical to those used to compute $g_+(\mathbf{R}, \tau)$, we obtain

$$g_-(\mathbf{R},\tau) = -\frac{1}{4\pi}\frac{\delta(\tau + R/c)}{R}, \tag{1.20b}$$

and $g_-(\mathbf{R},\tau) = 0$ for $\tau > 0$. The term *advanced* Green function comes from the property that $g_-(\mathbf{r}-\mathbf{r}', t-t')$ is observed at the space point $\mathbf{r}$ at the time $\tau = -R/c \Rightarrow t = t' - |\mathbf{r}-\mathbf{r}'|/c$; i.e., the observation time is *before* the pulse is emitted: it is *advanced* by the distance between the two field points divided by the velocity c of the background medium. Although the advanced Green function does not arise naturally in the solution of any physical problem, it does play an important role in the class of *inverse problems* treated in later chapters.

1.2.2 Frequency-domain Green functions

Any of the time-domain Green functions $g(\mathbf{R},\tau)$ can be represented according to Eq. (1.6a) in terms of a *frequency-domain Green function* $G(\mathbf{R},\omega)$ with

$$G(\mathbf{R},\omega) = \int_{-\infty}^{\infty} d\tau\, g(\mathbf{R},\tau)e^{i\omega\tau}, \qquad g(\mathbf{R},\tau) = \frac{1}{2\pi}\int_{-\infty}^{\infty} d\omega\, G(\mathbf{R},\omega)e^{-i\omega\tau}. \tag{1.21}$$

The frequency-domain Green functions are solutions to a partial differential equation, called the reduced wave equation or *Helmholtz equation*, that results from performing a temporal Fourier transform of the wave equation Eq. (1.15) satisfied by the time-domain Green functions $g(\mathbf{R},\tau)$. On making use of Eqs. (1.9) we conclude that

$$\left[\nabla^2 - \frac{1}{c^2}\frac{\partial^2}{\partial\tau^2}\right] g(\mathbf{R},\tau) \Longleftrightarrow \left[\nabla + \frac{\omega^2}{c^2}\right] G(\mathbf{R},\omega),$$

which then yields the inhomogeneous Helmholtz equation

$$[\nabla^2 + k^2]G(\mathbf{R},\omega) = \delta(\mathbf{R}). \tag{1.22}$$

Like the inhomogeneous wave equation satisfied by the time-domain Green function Eq. (1.16), the inhomogeneous Helmholtz equation Eq. (1.22) does not possess a unique solution until an appropriate *boundary condition* is appended. The requirement of causality in the time domain yields a boundary condition in the frequency domain known as the *Sommerfeld radiation condition* (SRC) (Sommerfeld, 1967) and a Green function denoted by $G_+(\mathbf{R},\omega)$ that is generally referred to as the "outgoing-wave" Green function for reasons to be discussed below. The outgoing-wave Green function can be computed from the Helmholtz equation Eq. (1.22) using a Fourier-based scheme entirely parallel to that used to compute the retarded Green function (cf. Example 1.3 below and our derivation of G_+ presented in the next chapter). Alternatively, it can be obtained by simply taking the temporal Fourier transform of the causal (retarded) Green function found above. Using the latter procedure we obtain

$$G_+(\mathbf{R},\omega) = -\frac{1}{4\pi}\frac{e^{ikR}}{R}, \tag{1.23a}$$

with $k = \omega/c$.

The justification for the use of the name "outgoing-wave" Green function for G_+ is apparent when we examine the Fourier-integral representation of the retarded Green function in terms of G_+:

$$g_+(\mathbf{R},\tau) = \frac{1}{2\pi}\int_{-\infty}^{\infty} d\omega \left\{ -\frac{1}{4\pi}\frac{e^{-i\omega(\tau-R/c)}}{R} \right\}.$$

In particular, we can regard this representation as a superposition of elemental spherical time-harmonic waves

$$u_+(\mathbf{R},\tau) = -\frac{1}{4\pi}\frac{e^{ik(R-c\tau)}}{R},$$

which expand *outward* from the origin $\mathbf{R} = 0$ with increasing time τ. This can be visualized by keeping the phase factor $R - c\tau$ fixed and seeing that increasing τ requires that the distance R must increase also in order for the phase to remain constant. An "incoming wave" on the other hand would be of the form

$$u_-(\mathbf{R},\tau) = -\frac{1}{4\pi}\frac{e^{ik(R+c\tau)}}{R},$$

and would have the property that the surfaces of constant phase contract *inward* toward the origin with increasing τ.

The frequency-domain Green function corresponding to the advanced Green function $g_-(\mathbf{r},\tau)$ is the "incoming-wave" Green function $G_-(\mathbf{R},\omega)$. On taking a temporal Fourier transform of g_- we find that

$$G_-(\mathbf{R},\omega) = -\frac{1}{4\pi}\frac{e^{-ikR}}{R} = G_+^*(\mathbf{R},\omega), \tag{1.23b}$$

where the wavenumber k is assumed to be real-valued. It is easy to verify using an argument similar to that employed above for G_+ that G_- is associated with incoming waves, thus justifying its name as the *incoming-wave* Green function. The Green functions g_- and G_- are associated with the important operations of field *time reversal* and *back propagation*, as we will see later in this chapter.

Example 1.3 The time-domain Green function for the one-dimensional wave equation satisfies the equation

$$\left[\frac{\partial^2}{\partial z^2} - \frac{1}{c^2}\frac{\partial^2}{\partial t^2}\right] g(z,t) = \delta(z)\delta(t). \tag{1.24}$$

Fourier transforming Eq. (1.24) leads to the one-dimensional Helmholtz equation

$$\left[\frac{\partial^2}{\partial z^2} + k^2\right] G(z, \omega) = \delta(z), \tag{1.25}$$

which under a spatial Fourier transformation yields

$$\tilde{G}(K, \omega) = \frac{1}{-K^2 + k^2}.$$

The Green function to the one-dimensional Helmholtz equation thus admits the Fourier-integral representation

$$G(z, \omega) = \frac{1}{2\pi} \int_{-\infty}^{\infty} dK \, \frac{e^{iKz}}{-K^2 + k^2}. \tag{1.26}$$

The Green function is not unique due to the improper nature of the integral caused by the poles of the integrand at $K = \pm k$. The outgoing-wave Green function $G_+(z, \omega)$ is obtained by requiring $k = \omega/c$ to have a positive imaginary part. This conclusion follows from the fact that we required the ω contour of integration defining the causal time-domain Green function to lie in the u.h.p., which translates to $\Im k > 0$. The poles in the integrand of Eq. (1.26) occur at $K = \pm k$. One pole lies in the u.h.p. and the other lies in the l.h.p. Since the integrand goes to zero exponentially fast in the u.h.p. if $z > 0$ and in the l.h.p. if $z < 0$, the integration contours can be closed for both cases and the integral can be computed using residue calculus. We obtain the result

$$G_+(z, \omega) = -\frac{i}{2k} e^{ik|z|}. \tag{1.27}$$

We can directly verify that the Green function defined in Eq. (1.27) satisfies the defining equation Eq. (1.25) by direct differentiation. In particular, we have that

$$\frac{\partial}{\partial z} G_+(z, \omega) = \frac{1}{2} e^{ik|z|} \operatorname{Sgn}(z); \qquad \frac{\partial^2}{\partial z^2} G_+(z, \omega) = \frac{ik}{2} e^{ik|z|} \operatorname{Sgn}^2(z) + \frac{1}{2} e^{ik|z|} \frac{\partial}{\partial z} \operatorname{Sgn}(z),$$

where

$$\operatorname{Sgn}(z) = \begin{cases} 1 & \text{if } z > 0 \\ -1 & \text{if } z < 0 \end{cases}$$

is the sign function. By using the fact that

$$\operatorname{Sgn}(z) = 2\Theta(z) - 1,$$

where Θ is the step function, we conclude that

$$\frac{\partial^2}{\partial z^2} G_+(z, \omega) + k^2 G_+(z, \omega) = \frac{ik}{2} e^{ik|z|} + e^{ik|z|} \delta(z) - \frac{ik}{2} e^{ik|z|} = \delta(z)$$

as required.

Example 1.4 The retarded (causal) time-domain Green function for the one-dimensional wave equation can be easily obtained from the frequency-domain Green function G_+ defined in Eq. (1.27) of Example 1.3 by performing an inverse temporal Fourier transform of G_+. We find that

$$g_+(z,\tau) = \frac{1}{2\pi}\int_{-\infty+i\epsilon}^{\infty+i\epsilon} d\omega \overbrace{\left\{-\frac{i}{2k}e^{ik|z|}\right\}}^{G_+(z,\omega)} e^{-i\omega\tau}$$

$$= \frac{-ic}{4\pi}\int_{-\infty+i\epsilon}^{\infty+i\epsilon} d\omega \frac{e^{i(\omega/c)|z|}}{\omega} e^{-i\omega\tau}. \tag{1.28}$$

We have explicitly indicated that the ω integral in Eq. (1.28) be taken along a contour lying in the upper half of the complex-ω plane required by causality by including the term $+i\epsilon$, $\epsilon > 0$ in the limits of the integral. The integrand has a single pole at $\omega = 0$ and goes to zero exponentially fast in the u.h.p. if $|z| > c\tau$ and in the l.h.p. if $|z| < \tau$. Since the only pole is on the real-ω axis, Cauchy's integral theorem yields the result

$$g_+(z,\tau) = -\frac{c}{2}\Theta[c\tau - |z|], \tag{1.29}$$

where Θ is the step function

$$\Theta(x) = \begin{cases} 0 & \text{if } x < 0, \\ 1 & \text{if } x > 0. \end{cases}$$

The Green function $g_+(z - z', t - t')$ defined in Eq. (1.29) is clearly causal (vanishes for $t < t'$) but is very different than the three-dimensional retarded Green function defined in Eq. (1.20a). In the three-dimensional case the Green function is a delta function concentrated on the "light cone" $|\mathbf{r} - \mathbf{r}'| = c(t - t')$, while in the one-dimensional case it is a step function that begins at the retarded time $t = t' + |z - z'|/c$. Note also that we have used contour integration to obtain the Green function in the one-dimensional case and used the definition of the Dirac delta function in the three-dimensional case. This was necessary since the singularity of this quantity is higher in three dimensions than it is in one dimension.

1.3 Green-function solutions to the radiation problem

The *radiation problem* consists of obtaining the causal solution to the inhomogeneous wave equation Eq. (1.1) in terms of the source term $q(\mathbf{r}, t)$. To obtain this solution we use standard Green-function techniques applied to the pair of equations

$$\left[\nabla_{r'}^2 - \frac{1}{c^2}\frac{\partial^2}{\partial t'^2}\right] g(\mathbf{r} - \mathbf{r}', t - t') = \delta(\mathbf{r} - \mathbf{r}')\delta(t - t'), \tag{1.30a}$$

$$\left[\nabla_{r'}^2 - \frac{1}{c^2}\frac{\partial^2}{\partial t'^2}\right] u_+(\mathbf{r}', t') = q(\mathbf{r}', t'), \tag{1.30b}$$

where the subscript on the Laplacian operator indicates that it operates on the primed coordinate vector and g is an unspecified Green function to the wave equation. Here, $\mathbf{r}$ and t play the role of fixed parameters that are allowed to assume any values over some region $\mathcal{R} = \{\mathbf{r} \in \tau, t \in (t_0, t_1)\}$ of space-time within which we seek a solution to the radiation problem. We emphasize that the Green function g as well as the space-time region $\mathcal{R}$ are as yet unspecified and totally arbitrary.

By "standard Green-function techniques" we mean the standard procedure for converting the coupled partial differential equations given in Eqs. (1.30) into an integral representation for the radiated field in terms of the source term and field boundary conditions on a bounding surface to the space region τ in which we seek a solution to Eq. (1.30b) as well as Cauchy conditions at the two times t_0 and t_1. This "standard procedure" consists of multiplying Eq. (1.30a) by u_+ and Eq. (1.30b) by g and subtracting one of the two resulting equations from the other to obtain

$$\overbrace{u_+(\mathbf{r}',t')\nabla_{r'}^2 g(\mathbf{r}-\mathbf{r}',t-t') - g(\mathbf{r}-\mathbf{r}',t-t')\nabla_{r'}^2 u_+(\mathbf{r}',t')}^{I_1(\mathbf{r},\mathbf{r}',t,t')}$$
$$\overbrace{-\frac{1}{c^2}\left\{u_+(\mathbf{r}',t')\frac{\partial^2}{\partial t'^2}g(\mathbf{r}-\mathbf{r}',t-t') - g(\mathbf{r}-\mathbf{r}',t-t')\frac{\partial^2}{\partial t'^2}u_+(\mathbf{r}',t')\right\}}^{I_2(\mathbf{r},\mathbf{r}',t,t')}$$
$$= \overbrace{u_+(\mathbf{r}',t')\delta(\mathbf{r}-\mathbf{r}')\delta(t-t') - g(\mathbf{r}-\mathbf{r}',t-t')q(\mathbf{r}',t')}^{I_3(\mathbf{r},\mathbf{r}',t,t')}.$$

We now integrate both sides of the above equation over the space-time region $\mathcal{R} - \{\mathbf{r}' \in \tau,\ t' \in (t_0, t_1)\}$ to obtain the result

$$\overbrace{\int_{t_0}^{t_1} dt' \int_\tau d^3r'\, I_1}^{\chi_1(\mathbf{r},t)} + \overbrace{\int_{t_0}^{t_1} dt' \int_\tau d^3r'\, I_2}^{\chi_2(\mathbf{r},t)} = \overbrace{\int_{t_0}^{t_1} dt' \int_\tau d^3r'\, I_3}^{\chi_3(\mathbf{r},t)}, \tag{1.31}$$

where we have simplified the notation by no longer explicitly displaying the arguments of the three quantities $I_j(\mathbf{r},\mathbf{r}',t,t')$, $j = 1,2,3$. We will follow the same procedure in the following, where we will drop the arguments of all field quantities under integral signs unless there is the possibility of confusion.

By using Green's theorem the first term, χ_1, can be written in the form

$$\begin{aligned}\chi_1(\mathbf{r},t) &= \int_{t_0}^{t_1} dt' \int_\tau d^3r'\, \nabla\cdot[u_+\nabla g - g\nabla u_+]\\ &= \int_{t_0}^{t_1} dt' \int_{\partial\tau} dS' \left[u_+\frac{\partial}{\partial n'}g - g\frac{\partial}{\partial n'}u_+\right],\end{aligned} \tag{1.32a}$$

where $\partial\tau$ is the bounding surface to the volume τ and the partial derivatives are with respect to the outward directed normal to $\partial\tau$. The second term, χ_2, becomes

$$\begin{aligned}\chi_2(\mathbf{r},t) &= -\frac{1}{c^2}\int_{t_0}^{t_1} dt' \int_\tau d^3r'\, \frac{\partial}{\partial t'}\left[u_+\frac{\partial}{\partial t'}g - g\frac{\partial}{\partial t'}u_+\right]\\ &= -\frac{1}{c^2}\int_\tau d^3r' \left[u_+\frac{\partial}{\partial t'}g - g\frac{\partial}{\partial t'}u_+\right]\Bigg|_{t'=t_0}^{t_1}.\end{aligned} \tag{1.32b}$$

Finally, the third term, χ_3, reduces to

$$\chi_3(\mathbf{r},t) = \begin{cases} u_+ - \int_{t_0}^{t_1} dt' \int_\tau d^3r'\, gq & \text{if } \mathbf{r} \in \tau,\ t \in (t_0, t_1), \\ -\int_{t_0}^{t_1} dt' \int_\tau d^3r'\, gq & \text{otherwise.} \end{cases} \tag{1.32c}$$

We now examine the three quantities $\chi_j, j = 1, 2, 3$, defined in Eqs. (1.32) for different selections of the Green function g and for the space volume τ and time parameters t_0 and t_1. For any given selection of these quantities we will compute the corresponding field using the coupling equation Eq. (1.31).

1.3.1 The primary field solution

We first compute an expression for the causal field $u_+(\mathbf{r}, t)$ in terms of the source $q(\mathbf{r}, t)$ that is valid for all time t and over all of infinite space. We call this the *primary field solution.* To obtain this solution we select the Green function g to be the causal Green function g_+ and examine the three quantities $\chi_j, j = 1, 2, 3$, defined in Eqs. (1.32) in the limits where $\tau \to \infty$, $t_0 \to -\infty$, and $t_1 \to +\infty$. The simplest term is χ_2 defined in Eq. (1.32b). Because the field u_+ is required to be causal and, hence, vanishes for negative time the contribution from $t' = t_0 = -\infty$ must vanish. Moreover, the contribution from $t' = t_1$ also must vanish in the limit $t_1 \to \infty$ due to the causality of the Green function g_+. Thus, $\chi_2 = 0$. The quantity χ_1 can also be shown to vanish in the limit where $\tau \to \infty$. In particular, recalling that the Green function $g_+(\mathbf{r} - \mathbf{r}', t - t') = 0$ unless $|\mathbf{r} - \mathbf{r}'| = c(t - t')$ and that the field u_+ is causal, it follows that the contribution from the integral over the surface $\partial\tau$ in Eq. (1.32a) must vanish in the limit where the radius of the surface tends to infinity while keeping t_1 finite but arbitrarily large. Since $\chi_1 + \chi_2 = \chi_3$ we conclude that χ_3 must also vanish in the limit $\tau \to \infty$, $t_0 \to -\infty$, $t_1 \to +\infty$. Since $\mathbf{r} \in \tau$ and $t \in (t_0, t_1)$ we thus arrive at our final result:

$$u_+(\mathbf{r}, t) = \int_0^{T_0} dt' \int_{\tau_0} d^3r'\, g_+(\mathbf{r} - \mathbf{r}', t - t') q(\mathbf{r}', t'), \tag{1.33}$$

which holds over all space and for all time. By substituting the expression for the retarded Green function given in Eq. (1.20a) we can also write the field in the form

$$u_+(\mathbf{r}, t) = \frac{-1}{4\pi} \int_{\tau_0} d^3r'\, \frac{q(\mathbf{r}', t - |\mathbf{r} - \mathbf{r}'|/c)}{|\mathbf{r} - \mathbf{r}'|}.$$

The process of radiation as governed by Eq. (1.33) consists of a superposition of expanding spherical waves $g_+(\mathbf{r} - \mathbf{r}', t - t')$, each weighted by the amplitude of the source $q(\mathbf{r}', t')$ at the source point $\mathbf{r}', t'$. The contribution to the total field from any one source point is a delta function on the spherical surface (called the "light-cone")

$$|\mathbf{r} - \mathbf{r}'| = c(t - t')$$

and, hence, expands outward with the velocity c from each spatial source point $\mathbf{r}'$. For the case in which the source volume τ_0 is a sphere centered at the origin $\mathbf{r} = 0$ and having a radius a it then follows from the above considerations that at any given time t the field is identically zero outside a sphere $V_+(t)$ that is also centered at the origin and has a radius $R_+(t) = a + ct$. Moreover, since the source ceases to radiate at time $t = T_0$, the radiated field also vanishes for all time $t > T_0 + a/c$ everywhere *inside* a different sphere $V_-(t)$ that is also centered at the origin and has a radius $R_-(t) = c(t - T_0) - a$. The picture of radiation that can be inferred from Eq. (1.33) is of a "ball of energy" that expands outward from the source volume τ_0 and eventually is contained within the annular region between the two expanding concentric spheres $V_-(t)$ and $V_+(t)$. This picture of the radiation process is useful to have in mind when dealing with the time-domain radiated field.

Example 1.5 We again consider the one-dimensional wave equation

$$\left[\frac{\partial^2}{\partial z^2} - \frac{1}{c^2}\frac{\partial^2}{\partial t^2}\right] u_+(z,t) = q(z,t),$$

where the source is compactly supported in some finite space-time region $\mathcal{S}_0|z \in [-a_0, a_0],\ t \in [0, T_0]$ and the field is required to be causal. By following the sequence of steps employed in the three-dimensional case we find that

$$\begin{aligned} & u_+(z',t')\frac{\partial^2}{\partial z'^2}g_+(z-z',t-t') - g_+(z-z',t-t')\frac{\partial^2}{\partial z'^2}u_+(z',t') \\ & -\frac{1}{c^2}\left[u_+(z',t')\frac{\partial^2}{\partial t'^2}g_+(z-z',t-t') - g_+(z-z',t-t')\frac{\partial^2}{\partial t'^2}u_+(z',t')\right] \\ & = u_+(z',t')\delta(z-z')\delta(t-t') - g_+(z-z',t-t')q(z',t'). \end{aligned}$$

On integrating both sides of the above equation over a space-time region $-Z < z' < Z, -T < t' < T$, where $Z > a_0$ and $T > T_0$, we obtain the result

$$\begin{aligned} & \int_{-T}^{T} dt'\left[u_+(z',t')\frac{\partial}{\partial z'}g_+(z-z',t-t') - g_+(z-z',t-t')\frac{\partial}{\partial z'}u_+(z',t')\right]\Bigg|_{-Z}^{Z} \\ & -\frac{1}{c^2}\int_{-Z}^{Z} dz'\left[u_+(z',t')\frac{\partial}{\partial t'}g_+(z-z',t-t') - g_+(z-z',t-t')\frac{\partial}{\partial t'}u_+(z',t')\right]\Bigg|_{-T}^{T} \\ & = u_+(z,t) - \int_{-T}^{T} dt' \int_{-Z}^{Z} dz'\, g_+(z-z',t-t')q(z',t'). \end{aligned}$$

If we now assume that $t \in (-T, T)$ and $z \in (-Z, Z)$ the second line of the above equation vanishes on account of the causality of the field and Green function. The first line can also be shown to vanish due to the fact that, as shown in Example 1.4, the causal Green function vanishes if $|z - z'| > c(t - t')$ and this will be the case in the limit where we take $Z \to \infty$. In the limit $T \to \infty$ and $Z \to \infty$ we thus arrive at the result

$$\begin{aligned} u_+(z,t) &= \int_0^{T_0} dt' \int_{-a_0}^{a_0} dz'\, g_+(z-z',t-t')q(z',t') \\ &= -\frac{c}{2}\int_0^{T_0} dt' \int_{-a_0}^{a_0} dz'\, \Theta[c(t-t') - |z-z'|]q(z',t') \\ &= -\frac{c}{2}\int_{-a_0}^{a_0} dz' \int_0^{t-|z-z'|/c} dt'\, q(z',t'). \end{aligned}$$

Example 1.6 The primary field solution obtained above can also be used to represent fields radiated by time-periodic sources. As an example we consider a unit-amplitude point source traveling at constant velocity in a circle confined to the plane $z = 0$:

$$q(\mathbf{r}, t) = \frac{\delta(\rho - a)}{a}\delta(z)\delta(\phi - \nu_\phi t),$$

where a is the radius of the circular orbit and ν_ϕ the angular velocity in radians per second of the point source, and ρ and ϕ are the cylindrical coordinates on the plane $z = 0$. The motion is time-periodic with period $T_\phi = 2\pi/\nu_\phi$, and the source can thus be expanded into the Fourier series

$$q(\mathbf{r}, t) = \frac{\delta(\rho - a)}{2\pi a}\delta(z) \sum_{n=-\infty}^{\infty} e^{in\phi} e^{-in\nu_\phi t}. \tag{1.34a}$$

The temporal Fourier transform of the source is readily computed and we obtain

$$Q(\mathbf{r}, \omega) = \frac{\delta(\rho - a)}{a}\delta(z) \sum_{n=-\infty}^{\infty} e^{in\phi}\,\delta(\omega - n\nu_\phi). \tag{1.34b}$$

It follows from Eq. (1.34b) and the linearity of the radiation process that the radiated field will consist of a Fourier series with frequencies $\omega_n = n\nu_\phi$. In particular, we find using the primary field solution that

$$u_+(\mathbf{r}, t) = \frac{1}{2\pi} \sum_{n=-\infty}^{\infty} U_n(\mathbf{r}, \omega_n) e^{-i\omega_n t}, \tag{1.35a}$$

where

$$U_n(\mathbf{r}, \omega_n) = -\frac{1}{4\pi}\int_0^{2\pi} d\phi'\, e^{in\phi'} \frac{e^{ik_n R}}{R}, \tag{1.35b}$$

where $R = \sqrt{(x - a\cos\phi')^2 + (y - a\sin\phi') + z^2}$ and $k_n = \omega_n/c$, with c being the velocity of the background medium.

1.3.2 Representation of the radiated field in terms of boundary values via the Kirchhoff–Helmholtz theorem

The solution to the radiation problem obtained in the previous section represents the radiated field directly in terms of the source $q(\mathbf{r}, t)$. If we now take τ to be a finite region that contains the source region τ_0 and has closed boundary $\partial\tau$ and restrict our attention to space points lying outside τ, it is possible to represent the radiated field in terms of the field and its normal derivative evaluated on the closed boundary surface $\partial\tau$. We again look for a causal solution to the inhomogeneous wave equation Eq. (1.30b) over a time interval (t_0, t_1) with $t_0 \to -\infty$ and $t_1 \to +\infty$ and take the Green function g in Eq. (1.30a) to be the causal (retarded) Green function g_+.

We are now seeking a solution in the region $\tau^\perp$ (the complement to τ) so that the space integrals in Eqs. (1.32) are over $\tau^\perp$, which is bounded by $\partial\tau$ and a closed sphere Σ_∞ at infinity. If we then use an argument identical to that employed above in deriving the primary field solution, we conclude from the set of Eqs. (1.32) that $\chi_2 = 0$ and that the contribution to χ_1 from the integral over the bounding surface Σ_∞ also vanishes in the limit $t_0 \to -\infty$, $t_1 \to +\infty$. If we then make use of the primary field solution given in Eq. (1.33), we find that the field throughout the *exterior* region $\tau^\perp$ can be expressed in the form

$$u_+(\mathbf{r}, t) = \int_{-\infty}^{\infty} dt' \int_{\partial\tau} dS' \left[g_+ \frac{\partial}{\partial n'} u_+ - u_+ \frac{\partial}{\partial n'} g_+ \right], \quad \mathbf{r} \in \tau^\perp, \tag{1.36a}$$

where the partial derivatives $\partial/\partial n'$ are directed *outward* from the interior region τ into $\tau^\perp$. We also find, if the field point $\mathbf{r}$ is located in the interior region τ, that the field and its normal derivative over $\partial\tau$ are coupled via the homogeneous integral equation

$$\int_{-\infty}^{\infty} dt' \int_{\partial\tau} dS' \left[g_+ \frac{\partial}{\partial n'} u_+ - u_+ \frac{\partial}{\partial n'} g_+ \right] = 0, \quad \mathbf{r} \in \tau, \tag{1.36b}$$

where, as in Eq. (1.36a), the partial derivatives $\partial/\partial n'$ are directed *outward* from τ into the exterior region $\tau^\perp$ (Fig. 1.1).

Equations (1.36) together constitute a time-domain version of the *Kirchhoff–Helmholtz Theorem*. Equation (1.36a) is an expression for the radiated field valid everywhere outside any bounding surface $\partial\tau$ to the source region τ_0 in terms of the field and its (outward directed) normal derivative over $\partial\tau$ and is sometimes referred to as the *first Helmholtz*

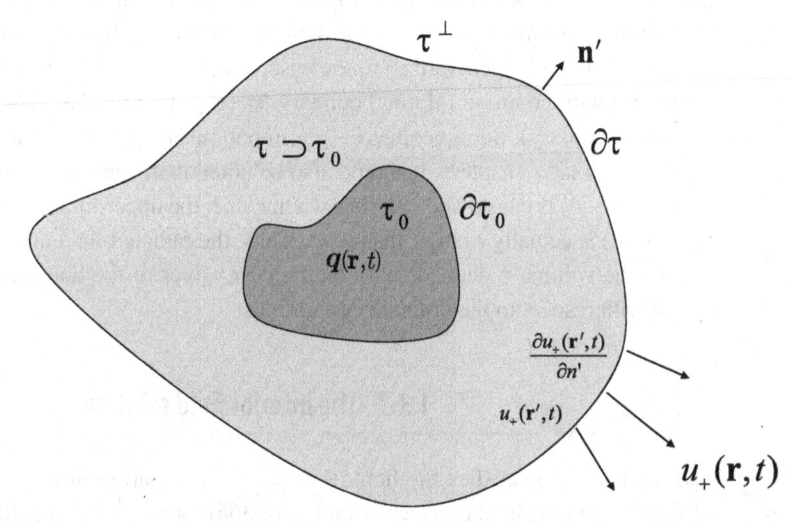

Fig. 1.1 Computing the radiated field in terms of boundary values over a closed surface $\partial\tau$ to a space region $\tau \supseteq \tau_0$ that contains the source. The radiated field throughout the region $\tau^\perp$ exterior to τ can be represented in terms of the boundary value of the field and its normal derivative specified everywhere on $\partial\tau$. These two quantities are not independent but are coupled via the homogeneous integral equation Eq. (1.36b).

identity. Equation (1.36b), referred to as the *second Helmholtz identity*, is an identity that must be satisfied at all space points lying in the interior region τ. This second Helmholtz identity is, in fact, a homogeneous integral equation that must be satisfied by the field and normal derivative on $\partial\tau$ and indicates that these two quantities are not independent and that, in principle, one can be determined from the other. Thus, in fact, Eq. (1.36a) is an *over-determined* solution to a boundary value problem of the wave equation in the region $\tau^{\perp}$ lying outside $\partial\tau$. A properly posed solution to this boundary-value problem would involve only the field or its normal derivative (or a linear combination of the two) on $\partial\tau$, not both.

The above development assumes that the source is located within the finite space volume τ and that the field satisfies the homogeneous wave equation throughout the exterior complement region $\tau^{\perp}$. However, a similar development can be employed to obtain a solution of the radiation problem within a finite region τ when the source is located in the infinite exterior $\tau^{\perp}$. Indeed, it is not difficult to show that in such a case the above two Helmholtz identities still hold where, however, the two regions τ and $\tau^{\perp}$ are now interchanged and the normal derivatives are now directed out of $\tau^{\perp}$ (the region containing the source) and into τ (the region in which the field satisfies the homogeneous wave equation).

Although the Kirchhoff–Helmholtz-type representation of the radiated field given in Eq. (1.36a), as well as its companion form for a source located in $\tau^{\perp}$, is over-determined in that it involves both the field and its normal derivative over $\partial\tau$, this representation is extremely useful in theoretical studies of the radiation problem and will be employed extensively in this and later chapters of the book. Properly posed solutions to boundary-value problems of the wave equation will be treated in the frequency domain in Chapter 2.

Finally, we mention that throughout the above development we have tacitly assumed that the region τ was finite and simply connected and that the complement region $\tau^{\perp}$ was infinite and multiply connected. However, it is not difficult to show that the entire development can be generalized to the case in which both regions are infinite and simply connected with common infinite boundary $\partial\tau$ (e.g., two infinite half-spaces separated by a planar boundary). Such geometries are important in applications and will be employed frequently in later chapters. It should also be noted that, since the retarded Green function $g_+(\mathbf{r}-\mathbf{r}', t-t')$ is causal and vanishes when $t' > t$, the upper limit in the time integration in Eqs. (1.36) is actually t rather than $+\infty$. Thus, the radiated field at any time t everywhere outside the volume τ depends only on the *past* values of the boundary conditions; i.e., is causal with respect to the boundary conditions.

1.3.3 The interior field solution

The field $u_+(\mathbf{r}, t)$ satisfies the homogeneous wave equation in the region exterior to the surface $\partial\tau$ and the field representation Eq. (1.36a) for the radiated field within this exterior region is a formal solution of an *exterior boundary-value problem* for the homogeneous wave equation; i.e., it solves the homogeneous wave equation in the region $\tau^{\perp}$ exterior to $\partial\tau$ in terms of boundary conditions on $\partial\tau$. If the time t exceeds the turn-off time T_0 of the source then the field within the interior region τ that contains the source also satisfies the homogeneous wave equation and can be represented as the formal solution of a boundary value problem for this region in terms of the field and its normal derivative over $\partial\tau$.

To obtain this field representation we use Eqs. (1.32) with the Green function g equal to the acausal Green function g_- and again select τ to be finite with closed boundary $\partial\tau$ and take $t_0 \to -\infty$, $t_1 \to +\infty$. The term χ_2 vanishes since the field u_+ is causal and will vanish over all of τ at $t' = \pm\infty$, and we obtain the result

$$u_+ = \int_0^{T_0} dt' \int_{\tau_0} d^3r'\, g_- q + \int_{-\infty}^{\infty} dt' \int_{\partial\tau} dS' \left[u_+ \frac{\partial}{\partial n'} g_- - g_- \frac{\partial}{\partial n'} u_+ \right], \tag{1.37a}$$

if $\mathbf{r} \in \tau$, and

$$0 = \int_0^{T_0} dt' \int_{\tau_0} d^3r'\, g_- q + \int_{-\infty}^{\infty} dt' \int_{\partial\tau} dS' \left[u_+ \frac{\partial}{\partial n'} g_- - g_- \frac{\partial}{\partial n'} u_+ \right], \tag{1.37b}$$

if $\mathbf{r} \in \tau^\perp$, where the partial derivatives in the above equations are directed outward from the interior region τ into the exterior $\tau^\perp$.

Now, if the observation time t exceeds the turn-off time T_0 of the source, the first term on the r.h.s. of Eqs. (1.37) must vanish since the advanced Green function vanishes when its argument $t - t' + |\mathbf{r} - \mathbf{r}'|$ is positive, which will be the case $\forall t' \in [0, T_0]$ if $t > T_0$. In this case Eq. (1.37a) reduces to

$$u_+ = \int_{-\infty}^{\infty} dt' \int_{\partial\tau} dS' \left[u_+ \frac{\partial}{\partial n'} g_- - g_- \frac{\partial}{\partial n'} u_+ \right], \quad \mathbf{r} \in \tau,\ t > T_0. \tag{1.38a}$$

For field points lying in the exterior region $\tau^\perp$ and for times $t > T_0$ Eq. (1.37b) reduces to

$$0 = \int_{-\infty}^{\infty} dt' \int_{\partial\tau} dS' \left[u_+ \frac{\partial}{\partial n'} g_- - g_- \frac{\partial}{\partial n'} u_+ \right], \quad \mathbf{r} \in \tau^\perp,\ t > T_0. \tag{1.38b}$$

As was the case for the solution of the exterior boundary-value problem given in Eq. (1.36), the solution to the interior boundary-value problem given in Eq. (1.38a) is over-determined because of the second Helmholtz identity Eq. (1.36b) and, thus, can be expressed entirely in terms of the field *or* its normal derivative on $\partial\tau$. We emphasize that this over-determination is due to the fact that the field u_+ is *radiated* by the source $q(\mathbf{r}, t) \in \tau_0$ that is bounded by the surface $\partial\tau$. Thus the wavefield over $\partial\tau$ will consist only of "outgoing" waves that constrain the value of the field and its normal derivative to satisfy the second Helmholtz identity over any bounding surface to the source. The fact that we are able to express the field interior to the bounding surface $\partial\tau$ in terms of field boundary values on $\partial\tau$ when $t > T_0$ is a consequence of the fact that the source $q(\mathbf{r}, t)$ vanishes if $t > T_0$ so that the field satisfies the *homogeneous* wave equation for all such times. The solution Eq. (1.38a) is, in fact, a formal solution of the *interior boundary-value problem* of the homogeneous wave equation; i.e., it solves the homogeneous wave equation throughout the interior of $\partial\tau$ in terms of boundary conditions on $\partial\tau$.

1.4 The initial-value problem for the wave equation

We can follow the same general procedure as was employed in the previous section to solve the radiation problem to solve the *homogeneous wave equation* subject to inhomogeneous Cauchy initial conditions. In particular, suppose that a field $u(\mathbf{r}, t)$ satisfies the homogeneous wave equation Eq. (1.3) throughout all of space and for all time $t > t_0$ and satisfies the inhomogeneous Cauchy conditions

$$u(\mathbf{r}, t)|_{t=t_0} = u_{t_0}(\mathbf{r}), \qquad \frac{\partial}{\partial t} u(\mathbf{r}, t)|_{t=t_0} = u'_{t_0}(\mathbf{r}) \tag{1.39}$$

at the initial time $t = t_0$. By following the same steps as led to the set of equations Eqs. (1.32) we find that for all space points interior to an arbitrary volume τ and for all times in the interval $[t_0, t_1]$

$$u(\mathbf{r}, t) = \int_{t_0}^{t_1} dt' \int_{\partial\tau} dS' \left[u(\mathbf{r}', t') \frac{\partial}{\partial n'} g(\mathbf{r} - \mathbf{r}', t - t') - g(\mathbf{r} - \mathbf{r}', t - t') \frac{\partial}{\partial n'} u(\mathbf{r}', t') \right]$$
$$- \frac{1}{c^2} \int_{\tau} d^3r' \left\{ u(\mathbf{r}', t') \frac{\partial}{\partial t'} g(\mathbf{r} - \mathbf{r}', t - t') - g(\mathbf{r} - \mathbf{r}', t - t') \frac{\partial}{\partial t'} u(\mathbf{r}', t') \right\} \Big|_{t'=t_0}^{t_1},$$

where g is *any* Green function to the wave equation. Since we are free to choose the Green function, we will select it to be the retarded Green function g_+. It then follows using arguments identical to those used in computing the solution to the radiation problem in the preceding section that in the limit where $\tau \to \infty$ and $t_1 \to \infty$ the contributions to u from the surface integral over $\partial\tau$ and from the volume integral at the final time $t' = t_1$ vanish and we obtain

$$u(\mathbf{r}, t) = \frac{1}{c^2} \int d^3r' \left[u_{t_0}(\mathbf{r}') \frac{\partial}{\partial t_0} g_+(\mathbf{r} - \mathbf{r}', t - t_0) - g_+(\mathbf{r} - \mathbf{r}', t - t_0) u'_{t_0}(\mathbf{r}') \right], \tag{1.40a}$$

which is valid over all of space and for all times $t > t_0$.

Equation (1.40a) gives the solution to the Cauchy initial-value problem over all of space and for all time $t > t_0$. If the homogeneous wave equation holds also for times $t < t_0$ it is also possible to compute its solution for these times using the same inhomogeneous Cauchy conditions as employed in Eq. (1.40a). In particular, by again following steps similar to those used in obtaining Eq. (1.40a) we find that

$$u(\mathbf{r}, t) = -\frac{1}{c^2} \int d^3r' \left[u_{t_0}(\mathbf{r}') \frac{\partial}{\partial t_0} g_-(\mathbf{r} - \mathbf{r}', t - t_0) - g_-(\mathbf{r} - \mathbf{r}', t - t_0) u'_{t_0}(\mathbf{r}') \right], \tag{1.40b}$$

which now holds over all of space and for all times $t < t_0$ and where g_- is the *advanced Green function* defined in Eq. (1.20b). The fact that the fields defined in Eqs. (1.40) satisfy the homogeneous wave equation is trivial to verify. The fact that they both satisfy the same Cauchy conditions Eq. (1.39) is a bit more difficult to prove, although a proof based on the plane-wave expansion of these fields is quite simple and will be established in Example 3.1 of Chapter 3.

The above results hold for *any* field that satisfies the homogeneous wave equation and given Cauchy conditions at $t = t_0$. If $t > T_0$, the field u_+ radiated by a source that turns off at T_0 satisfies the homogeneous wave equation so that these results also apply to u_+ where $t_0 > T_0$ and Eq. (1.40a) applies for $t > t_0$ and Eq. (1.40b) for $T_0 < t < t_0$. For $t < T_0$ the field so computed will not equal the actual radiated field (which satisfies the *inhomogeneous wave equation* if $t < T_0$) but, as we will find later, is a *time-reversed* version of this field that satisfies the homogeneous wave equation everywhere.

We note that, since the retarded Green function $g_+(\mathbf{r} - \mathbf{r}', t - t_0)$ is causal and vanishes for $t < t_0$ and the advanced Green function $g_-(\mathbf{r} - \mathbf{r}', t - t_0)$ is acausal and vanishes for $t > t_0$, we can combine Eqs. (1.40a) and (1.40b) into a single equation that is valid for all time t:

$$u(\mathbf{r}, t) = \frac{1}{c^2} \int d^3r' \left[u_{t_0}(\mathbf{r}') \frac{\partial}{\partial t_0} g_f(\mathbf{r} - \mathbf{r}', t - t_0) - g_f(\mathbf{r} - \mathbf{r}', t - t_0) u'_{t_0}(\mathbf{r}') \right], \quad (1.41)$$

where

$$g_f(\mathbf{R}, \tau) = g_+(\mathbf{R}, \tau) - g_-(\mathbf{R}, \tau).$$

We will call the quantity $g_f(\mathbf{R}, \tau)$ the *free-field propagator*, since it governs the evolution of the solution of the initial-value problem for the wave equation (the "free field") from its Cauchy conditions at some initial time.

1.4.1 Uniqueness

The uniqueness of the solution to the inhomogeneous wave equation subject to specified Cauchy conditions at $t = 0$ is easy to establish using the solution to the Cauchy initial-value problem given above. In particular, assume that the same source generates two fields $u^{(1)}$ and $u^{(2)}$ that satisfy the inhomogeneous wave equation Eq. (1.1) and the same homogeneous or inhomogeneous Cauchy conditions at $t = 0$. The difference field

$$u_+(\mathbf{r}, t) = u^{(2)}(\mathbf{r}, t) - u^{(1)}(\mathbf{r}, t)$$

then satisfies the homogeneous wave equation and homogeneous Cauchy conditions at $t = 0$. Taking $t_0 = 0$ in the solution to the initial-value problem given in Eq. (1.40a), it follows that $u = 0$, $\forall t > 0$. In a similar manner, taking $t_0 = 0$ in the solution to the initial-value problem for negative $t - t_0$ given in Eq. (1.40b), we conclude that $u = 0$, $\forall t < 0$. Since $u = 0$ for $t = 0$ because of the Cauchy conditions, we conclude that $u = 0$, which establishes the uniqueness. It follows from this, of course, that the solution to the radiation problem is unique.

1.4.2 Field back propagation

The expression for the radiated field in terms of the field and its normal derivative over $\partial\tau$ given in Eq. (1.36a) performs the process of *field forward propagation*; i.e., it "propagates" the field outward from a boundary $\partial\tau$ that completely encloses the source region τ_0 to points that are further removed from the source than is the boundary surface. Similarly, the solution of the initial-value problem Eq. (1.40a) for Cauchy data acquired at some time t_0 exceeding the turn-off time T_0 of the source forward propagates the field from the Cauchy data to future times $t > t_0$. It is also possible to "back propagate" the field from the boundary $\partial\tau$ to field points lying within τ and from the Cauchy data to times $t < t_0$ via Eqs. (1.38a) and (1.40b), respectively. The process of forward propagation from either boundary-value data over the surface $\partial\tau$ or Cauchy conditions at $t_0 > T_0$ is exact; i.e., forward propagation exactly reproduces the field radiated by the source. On the other hand, the process of field back propagation as implemented from boundary-value data via Eq. (1.38a) or Cauchy data via Eq. (1.40b) is approximate and will exactly reproduce the field $u_+(\mathbf{r}, t)$ only for times exceeding the turn-off time of the source. Although back propagation as described here reproduces the field exactly only for times $t > T_0$, it generates an approximation to the exact field for all times and is an extremely important operation that will be used throughout this book. We will describe a theoretically exact (but mathematically unstable) form of back propagation from boundary-value data in Chapter 4 but will, in practice, employ approximate forms of back propagation such as described here for virtually all of the inverse problems that we treat in this book.

1.5 Frequency-domain solution of the radiation problem

We can express the radiated field $u_+(\mathbf{r}, t)$ in the frequency domain by making use of the Fourier-integral representation of the causal Green function g_+ given in Eq. (1.21). In particular, on setting $\mathbf{R} = \mathbf{r} - \mathbf{r}'$ and $\tau = t - t'$ in that equation and substituting the result into Eq. (1.33) we obtain

$$u_+(\mathbf{r}, t) = \int_0^{T_0} dt' \int_{\tau_0} d^3r' \overbrace{\frac{1}{2\pi}\int_{-\infty}^{\infty} d\omega\, G_+(\mathbf{r} - \mathbf{r}', \omega) e^{-i\omega(t-t')}}^{G_+(\mathbf{r}-\mathbf{r}', t-t')} q(\mathbf{r}', t')$$

$$= \frac{1}{2\pi}\int_{-\infty}^{\infty} d\omega \overbrace{\int_{\tau_0} d^3r'\, G_+(\mathbf{r} - \mathbf{r}', \omega) Q(\mathbf{r}', \omega)}^{U_+(\mathbf{r},\omega)}\, e^{-i\omega t}, \tag{1.42}$$

where

$$Q(\mathbf{r}, \omega) = \int_0^{T_0} dt'\, q(\mathbf{r}, t) e^{i\omega t}$$

is the temporal Fourier transform of the source and

$$U_+(\mathbf{r}, \omega) = \int_{\tau_0} d^3r'\, G_+(\mathbf{r} - \mathbf{r}', \omega) Q(\mathbf{r}', \omega) \tag{1.43a}$$

is the frequency-domain representation (temporal Fourier transform) of the radiated field $u_+(\mathbf{r}, t)$. By making use of the defining equation for a Green function to the Helmholtz equation Eq. (1.22) we find that the frequency-domain field U_+ satisfies the inhomogeneous Helmholtz equation

$$[\nabla^2 + k^2]U_+(\mathbf{r}, \omega) = Q(\mathbf{r}, \omega). \tag{1.43b}$$

1.5.1 The radiation pattern and the Sommerfeld radiation condition

By setting

$$|\mathbf{r} - \mathbf{r}'| = \sqrt{r^2 + r'^2 - 2\mathbf{r} \cdot \mathbf{r}'} \sim r - \mathbf{s} \cdot \mathbf{r}' \quad \text{as} \quad r \to \infty$$

in the expression Eq. (1.23a) for the outgoing-wave Green function we find that

$$G_+(\mathbf{r} - \mathbf{r}', \omega) \sim -\frac{1}{4\pi} e^{-ik\mathbf{s} \cdot \mathbf{r}'} \frac{e^{ikr}}{r} \quad \text{as} \quad r \to \infty, \tag{1.44}$$

where $\mathbf{s} = \mathbf{r}/r$ is the unit vector along the $\mathbf{r}$ direction having direction cosines $\sin\theta\cos\phi$, $\sin\theta\sin\phi$, and $\cos\theta$, with (r, θ, ϕ) being the spherical polar coordinates of the vector $\mathbf{r}$. If we now substitute Eq. (1.44) into Eq. (1.43a) we obtain the following *far-field* approximation to $U_+(\mathbf{r}, \omega)$:

$$U_+(\mathbf{r}, \omega) \sim \int_{\tau_0} d^3r' \overbrace{\left[-\frac{1}{4\pi} e^{-ik\mathbf{s} \cdot \mathbf{r}'} \frac{e^{ikr}}{r}\right]}^{G_+(\mathbf{r}-\mathbf{r}', \omega)} Q(\mathbf{r}', \omega),$$

which can be expressed in the form

$$U_+(\mathbf{r}, \omega) \sim f(\mathbf{s}, \omega) \frac{e^{ikr}}{r} \quad \text{as} \quad r \to \infty, \tag{1.45a}$$

where

$$f(\mathbf{s}, \omega) = -\frac{1}{4\pi} \int_{\tau_0} d^3r \, Q(\mathbf{r}, \omega) e^{-ik\mathbf{s} \cdot \mathbf{r}} \tag{1.45b}$$

is known as the *radiation pattern* of the source. It can be seen from Eq. (1.45a) that the radiated field at large distances from the source is an outgoing spherical wave with a complex angularly dependent amplitude given by the radiation pattern.

The spatial Fourier transform of the source $Q(\mathbf{r}, \omega)$ is defined by

$$\tilde{Q}(\mathbf{K}, \omega) = \int_{\tau_0} d^3r' \, Q(\mathbf{r}', \omega) e^{-i\mathbf{K} \cdot \mathbf{r}'}. \tag{1.46}$$

It then follows from Eq. (1.45b) that the radiation pattern is proportional to the spatial Fourier transform of the (frequency-domain) source *evaluated on the surface of the sphere* $\mathbf{K} = k\mathbf{s}$; i.e.,

$$f(\mathbf{s},\omega) = -\frac{1}{4\pi}\tilde{Q}(\mathbf{K},\omega)|_{\mathbf{K}=k\mathbf{s}}. \tag{1.47}$$

It can also, of course, be interpreted as being the space-time transform $\tilde{q}(\mathbf{K},\omega)$ of the time-dependent source $q(\mathbf{r},t)$ over the same surface in Fourier space.

The surface $\mathbf{K} = k\mathbf{s}$ is a special case of the more general *Ewald sphere* that plays a fundamental and dominant role in inverse-scattering theory treated later in the book. For now, we simply note that, due to our assumption that the source is at least piecewise continuous in the finite volume τ_0, the spatial Fourier transform $\tilde{Q}(\mathbf{K},\omega)$ is an entire analytic function of the three Cartesian components of the $\mathbf{K}$ vector[4] so that its boundary value $\tilde{Q}(k\mathbf{s},\omega)$ and, hence, the radiation pattern $f(\mathbf{s},\omega)$ are entire analytic functions of the unit vector $\mathbf{s}$; i.e., are entire functions of the real- *and* complex-valued direction cosines $\sin\theta\cos\phi, \sin\theta\sin\phi, \cos\theta$ of $\mathbf{s}$ as well as the real and complex polar and azimuthal angles θ,ϕ defining this unit vector. Note that the *observable* radiation pattern will be associated with real direction cosines and real polar and azimuthal angles but, due to the analyticity, this boundary value can be continued from the real unit sphere into the complex unit sphere where $\mathbf{s}$ becomes complex but still has unit length[5] $\sqrt{\mathbf{s}\cdot\mathbf{s}} = 1$.

The asymptotic expression Eq. (1.45a) is one form of the famed *Sommerfeld radiation condition* (SRC). An alternative, and the most often quoted, form of the SRC is given by

$$\lim_{r\to\infty} r\left[\frac{\partial U_+(\mathbf{r},\omega)}{\partial r} - ikU_+(\mathbf{r},\omega)\right] \to 0, \tag{1.48}$$

with a similar expression holding for G_+. The equivalence of the two forms of the SRC is easily established.

The time-domain radiation pattern

The time-domain far-field approximation is obtained by taking the inverse Fourier transform of the frequency far-field approximation Eq. (1.45a). We obtain

$$u_+(\mathbf{r},t) \sim \frac{1}{2\pi}\int_{-\infty}^{\infty} d\omega\, f(\mathbf{s},\omega)\frac{e^{i(\omega/c)r}}{r}e^{-i\omega t} = \frac{F(\mathbf{s},t-r/c)}{r}, \tag{1.49}$$

where again $\mathbf{r} = r\mathbf{s}$ and

$$F(\mathbf{s},t) = \frac{1}{2\pi}\int_{-\infty}^{\infty} d\omega\, f(\mathbf{s},\omega)e^{-i\omega t} \tag{1.50}$$

is known as the *time-domain radiation pattern* (Shlivinski *et al.*, 1997). Equation (1.49) is a representation of the time-domain far field in terms of a superposition of spherical wave

[4] The proof follows similar lines to those employed in Example 1.2 to establish that the temporal transform of a source supported in a finite time interval is an entire analytic function of the transform variable ω.

[5] Note that the "length" is *not* $|\mathbf{s}| = \sqrt{\mathbf{s}^*\mathbf{s}}$ but is the square root of the product of the complex unit vector with itself.

pulses that propagate *outward* from the origin to infinity with a propagation velocity of c. Note that because of this interpretation of the time-domain far field the term "outgoing-wave field" is sometimes applied to the frequency- and time-domain fields as well as to the frequency-domain Green function G_+.

The time-domain radiation pattern $F(\mathbf{s}, t)$ can be expressed directly in terms of the time-domain source $q(\mathbf{r}, t)$ by substituting Eq. (1.45b) into Eq. (1.50). We obtain

$$F(\mathbf{s}, t) = \frac{1}{2\pi} \int_{-\infty}^{\infty} d\omega \overbrace{\left\{ -\frac{1}{4\pi} \int_{\tau_0} d^3r' \, Q(\mathbf{r}', \omega) e^{-ik\mathbf{s}\cdot\mathbf{r}'} \right\}}^{f(\mathbf{s},\omega)} e^{-i\omega t}$$
$$= -\frac{1}{4\pi} \int_{\tau_0} d^3r' \, q\left(\mathbf{r}', t + \frac{\mathbf{s}\cdot\mathbf{r}'}{c}\right). \qquad (1.51)$$

It is interesting to note that the time-domain radiation pattern need not be causal even though the source q is causal. Indeed, source points for which $\mathbf{s}\cdot\mathbf{r}' > 0$ generate a contribution to F even for negative time. However, the time-domain radiation pattern is not directly observed and the observed quantity (the far field) defined in Eq. (1.49) is causally related to the source.

1.6 Radiated power and energy

The power radiated by the source $q(\mathbf{r}, t)$ can be computed from the *energy flux vector* $\mathbf{P}(\mathbf{r}, t)$ defined to be the energy flow per unit time through a differential surface dS located at $\mathbf{r}$ and having unit normal $\mathbf{n}$. For the scalar wavefields under consideration here this quantity is given by the expression

$$\mathbf{P}(\mathbf{r}, t) = -\kappa \frac{\partial u_+(\mathbf{r}, t)}{\partial t} \nabla u_+(\mathbf{r}, t), \qquad (1.52a)$$

where κ is a real and positive constant that depends on the nature of the radiation and is, for example, equal to the fluid density for the case of acoustic waves in compressible fluids. The total power $P(t)$ radiated out of a region τ containing the source region τ_0 and having a closed boundary surface $\partial\tau$ is then given by the integral

$$P_{\partial\tau}(t) = \int_{\partial\tau} dS \, \mathbf{n}\cdot\mathbf{P}(\mathbf{r}, t) = -\kappa \int_{\partial\tau} dS \frac{\partial u_+(\mathbf{r}, t)}{\partial t} \frac{\partial u_+(\mathbf{r}, t)}{\partial n}, \qquad (1.52b)$$

where the normal derivative is directed out of the interior region τ and into the exterior of $\partial\tau$. The total *energy* radiated out of $\partial\tau$ by the source is the integral of the radiated power $P_{\partial\tau}(t)$ over all of time. This quantity is then found to be given by

$$E_q = \int_{-\infty}^{\infty} dt \, P_{\partial\tau}(t) = -\kappa \int_{\partial\tau} dS \int_{-\infty}^{\infty} dt \frac{\partial u_+(\mathbf{r}, t)}{\partial t} \frac{\partial u_+(\mathbf{r}, t)}{\partial n}. \qquad (1.52c)$$

We will show below in Theorem 1.1 that the total radiated energy E_q is independent of the particular bounding surface $\partial\tau$ used in its computation, which allows us to label this quantity with q alone and not $\partial\tau$.

The total energy E_q radiated by the source is an important quantity that is used frequently in applications. This quantity has an especially simple form and interpretation in the frequency domain. In particular, on making use of Eq. (1.9) we have that

$$\frac{\partial u_+(\mathbf{r},t)}{\partial t} \Leftrightarrow -i\omega U_+(\mathbf{r},\omega), \qquad \frac{\partial u_+(\mathbf{r},t)}{\partial n} \Leftrightarrow \frac{\partial U_+(\mathbf{r},\omega)}{\partial n},$$

where, as before, the double arrow denotes a Fourier transform. It then follows from Parseval's theorem that

$$\int_{-\infty}^{\infty} dt \, \frac{\partial u_+(\mathbf{r},t)}{\partial t}\frac{\partial u_+(\mathbf{r},t)}{\partial n} = \frac{1}{2\pi}\int_{-\infty}^{\infty} d\omega \left[i\omega U_+^*(\mathbf{r},\omega)\frac{\partial U_+(\mathbf{r},\omega)}{\partial n}\right].$$

If we now use the above result in Eq. (1.52c) we find that

$$\begin{aligned} E_q &= -i\frac{\kappa}{2\pi}\int_{-\infty}^{\infty} d\omega\,\omega \int_{\partial\tau} dS\, U_+^*(\mathbf{r},\omega)\frac{\partial U_+(\mathbf{r},\omega)}{\partial n} \\ &= -i\frac{\kappa}{2\pi}\int_{0}^{\infty} d\omega\,\omega \int_{\partial\tau} dS\left[U_+^*(\mathbf{r},\omega)\frac{\partial U_+(\mathbf{r},\omega)}{\partial n} - U_+(\mathbf{r},\omega)\frac{\partial U_+^*(\mathbf{r},\omega)}{\partial n}\right] \\ &= \frac{1}{2\pi}\int_0^{\infty} d\omega \left[2\kappa\omega\Im\int_{\partial\tau} dS\, U_+^*(\mathbf{r},\omega)\frac{\partial U_+(\mathbf{r},\omega)}{\partial n}\right], \end{aligned}$$

where we have made use of the fact that $U_+(\mathbf{r},-\omega) = U_+^*(\mathbf{r},\omega)$.

The quantity

$$E_Q(\omega) = 2\kappa\omega\Im\int_{\partial\tau} dS\, U_+^*(\mathbf{r},\omega)\frac{\partial U_+(\mathbf{r},\omega)}{\partial n} \tag{1.53}$$

can be interpreted as being the *energy spectra* (energy per unit frequency) of the radiated field, with the total radiated energy given by

$$E_q = \frac{1}{2\pi}\int_0^{\infty} d\omega\, E_Q(\omega).$$

Important properties of the energy spectra E_Q and radiated energy E_q are that they are independent of the shape and location of the surface $\partial\tau$ over which they are computed so long as this surface completely surrounds the source spatial volume τ_0. In particular, we have the following theorem.

Theorem 1.1 (Radiant energy theorem) *The energy spectra $E_Q(\omega)$ and radiated energy E_q are independent of the surface $\partial\tau$ as long as $\partial\tau$ completely surrounds the source spatial volume τ_0. Moreover, the energy spectra can be expressed in terms of the frequency-domain radiation pattern $f(\mathbf{s},\omega)$ via the integral*

$$E_Q(\omega) = \frac{2\kappa\omega^2}{c}\int_{4\pi} d\Omega_s |f(\mathbf{s},\omega)|^2, \tag{1.54}$$

where $\mathbf{s}$ is a unit vector and the integration is over 4π steradians.

We first show that $E_Q(\omega)$ is independent of the surface $\partial\tau$ as long as the surface completely surrounds the source. Let $\Sigma_<$ and $\Sigma_>$ be any two surfaces that completely surround the source and such that $\Sigma_>$ lies entirely outside (i.e., encloses) $\Sigma_<$. Since these two surfaces both lie outside τ_0 and the wavenumber k is strictly real, we have that

$$[\nabla^2 + k^2]U_+(\mathbf{r},\omega) = 0, \qquad [\nabla^2 + k^2]U_+^*(\mathbf{r},\omega) = 0,$$

throughout the region $\delta\tau$ enclosed between these two surfaces. It then follows that

$$\int_{\delta\tau} d^3r\{U_+^*(\mathbf{r},\omega)\nabla^2 U_+(\mathbf{r},\omega) - U_+(\mathbf{r},\omega)\nabla^2 U_+^*(\mathbf{r},\omega)\} = 0.$$

On using Green's theorem we then obtain the result

$$\int_{\Sigma_>} dS\left[U_+^*\frac{\partial U_+}{\partial n} - U_+\frac{\partial U_+^*}{\partial n}\right] = \int_{\Sigma_<} dS\left[U_+^*\frac{\partial U_+}{\partial n} - U_+\frac{\partial U_+^*}{\partial n}\right],$$

which proves the first part of the theorem.

To derive Eq. (1.54) we take $\partial\tau = \Sigma_\infty$, where Σ_∞ is the surface of a sphere centered within the source region and having radius $R \to \infty$. The radiated field over Σ_∞ is then given by Eq. (1.45a) so that

$$\begin{aligned} E_Q(\omega) = -i\kappa\omega \int_{\Sigma_\infty} R^2\, d\Omega_s \Bigg[& f^*(\mathbf{s},\omega)\frac{e^{-ikR}}{R} f(\mathbf{s},\omega)\frac{ike^{ikR}}{R} \\ & - f(\mathbf{s},\omega)\frac{e^{ikR}}{R} f^*(\mathbf{s},\omega)\frac{-ike^{-ikR}}{R}\Bigg] \\ = 2\kappa\omega k \int_{4\pi} d\Omega_s |f(\mathbf{s},\omega)|^2, & \end{aligned}$$

where we have used the result that $dS = R^2\, d\Omega$.

Finally, we note that we can express the energy spectra and total radiated energy directly in terms of the source by making use of Eq. (1.47). In particular, we find that

$$E_Q(\omega) = \frac{2\kappa\omega^2}{c}\int_{4\pi} d\Omega_s \overbrace{\left|\frac{\tilde{Q}(k\mathbf{s},\omega)}{-4\pi}\right|^2}^{|f(\mathbf{s},\omega)|^2} = \frac{\kappa\omega^2}{8\pi^2 c}\int_{4\pi} d\Omega_s |\tilde{Q}(k\mathbf{s},\omega)|^2. \tag{1.55}$$

It is, of course, also possible to express the energy spectra directly in terms of the source $Q(\mathbf{r},\omega)$ by making use of Eqs. (1.46) in Eq. (1.55) (cf. Example 2.6 of Chapter 2).

1.7 Non-radiating sources

There exist non-trivial sources $q_{\rm nr}(\mathbf{r},t)$ that are compactly supported in $\{\mathcal{S}_0|\mathbf{r} \in \tau_0,\ t \in [0, T_0]\}$ and that have the interesting (and surprising) property that they generate fields $u_+(\mathbf{r},t)$ that *vanish identically outside the space-time region* $\mathcal{S}_0$. Such sources are called *non-radiating sources* (NR sources) and will be designated using the subscript "nr" throughout the book. The NR sources were first investigated by physicists interested in developing microscopic models for stable classical atoms and molecules (Bohm and Weinstein, 1949; Goedecke, 1964). The general idea was that if the atom or molecule didn't radiate a field it would not lose energy and, hence, would not suffer the inevitable collapse associated with classical models for atoms and molecules. Our interest in this book in the class of NR sources is more pragmatic. In particular, we will find in the following chapters that these sources and their scattering equivalents, the *non-scattering potentials*, cause the inverse source and scattering problems to have non-unique solutions. For now we will show that such sources do indeed exist (at least mathematically) and will present a recipe for generating examples of such sources.

An NR source is simple to construct. In particular, consider a function $\pi(\mathbf{r}, t)$ of position and time that is compactly supported in $\{\mathcal{S}_0 | \mathbf{r} \in \tau_0, t \in [0, T_0]\}$ and possesses continuous first partial derivatives throughout the closed domain $\mathcal{S}_0$ but is otherwise arbitrary. We construct an NR source by simple application of the D'Alembertian operator:

$$q_{\rm nr}(\mathbf{r}, t) = \left[\nabla^2 - \frac{1}{c^2}\frac{\partial^2}{\partial t^2}\right]\pi(\mathbf{r}, t). \tag{1.56}$$

The associated causal NR field is given by Eq. (1.33):

$$u_{\rm nr}(\mathbf{r}, t) = \int_0^{T_0} dt' \int_{\tau_0} d^3r'\, g_+(\mathbf{r} - \mathbf{r}', t - t')\overbrace{\left\{\left[\nabla_{r'}^2 - \frac{1}{c^2}\frac{\partial^2}{\partial t'^2}\right]\pi(\mathbf{r}', t')\right\}}^{q_{\rm nr}(\mathbf{r}', t')}$$

$$= \int_0^{T_0} dt' \int_{\tau_0} d^3r' \overbrace{\left[\nabla_{r'}^2 - \frac{1}{c^2}\frac{\partial^2}{\partial t'^2}\right] g_+(\mathbf{r} - \mathbf{r}', t - t')}^{\delta(\mathbf{r}-\mathbf{r}')\delta(t-t')}\pi(\mathbf{r}', t') = \pi(\mathbf{r}, t),$$

where we have twice integrated by parts and dropped the surface terms due to the assumption that π has continuous first partials throughout $\mathcal{S}_0$. The field $u_{\rm nr}$ vanishes outside $\mathcal{S}_0$, which then establishes that the source defined via Eq. (1.56) is an NR source and, moreover, that the field it generates is the compactly supported function $\pi(\mathbf{r}, t)$.

The quantity $\pi(\mathbf{r}, t)$ appearing in Eq. (1.56) must be supported in $\mathcal{S}_0$ and possess continuous first partial derivatives but is otherwise arbitrary. Thus, by means of the prescription Eq. (1.56) it is possible to generate a countably infinite number of NR sources by different choices of the "generating function" $\pi(\mathbf{r}, t)$. Moreover, each such source will be at least piecewise continuous and will possess finite energy[6] since the generating function π has continuous partials so that $q_{\rm nr}$ is at least piecewise continuous in $\mathcal{S}_0$.[7]

[6] By "energy" we mean here the source energy as defined in Eq. (1.2). This "energy" is simply the squared L^2 norm of the source and should not be confused with the radiated energy of the source introduced in Section 1.6.

[7] A larger class of NR sources can be constructed by taking limits of infinite convergent sequences of the piecewise continuous NR sources constructed via Eq. (1.56). However, such sources will not be discussed or employed in this book since we will limit our attention to sources that are, at worst, piecewise continuous.

From the discussion presented in Section 1.3 for the radiation process we can describe the radiation process for an NR source. In particular, during the time period $[0, T_0]$ the field $u_{nr}(\mathbf{r}, t)$ generated by the NR source evolves in time completely contained within the source spatial region τ_0. However, as soon as the source turn-off time T_0 occurs the field completely vanishes! There is no attendant "ball of energy" that expands outward as time increases and, indeed, there is no trace whatsoever that a source ever existed. This rather perplexing conclusion is one reason why many scientists believe that NR sources do not exist in the real world and are, at best, a mathematical oddity. However, to date there is no evidence to rule out their physical (as opposed to "mathematical") existence and some reason to believe (Goedecke, 1964) that they play an important but not fully understood role in nature.

1.7.1 Non-radiating sources in the frequency domain

A time-dependent NR source generates a radiated field $u_{nr}(\mathbf{r}, t)$ that vanishes for all time t everywhere outside the source spatial volume τ_0. This then requires that its Fourier transform $U_{nr}(\mathbf{r}, \omega)$ vanish everywhere outside τ_0 at *all* frequencies ω. It is worthwhile to weaken this requirement when dealing with frequency-domain fields and define an NR source in the frequency domain according to the following definition.

Definition 1.1 (Frequency-domain non-radiating sources) Let $Q(\mathbf{r}, \omega)$ be a finite energy source compactly supported in the spatial volume τ_0 that radiates an outgoing-wave (causal) field $U_+(\mathbf{r}, \omega)$ according to Eq. (1.43a). Then this source is a *non-radiating source* (NR source) at any given frequency ω if and only if its radiated field $U_+(\mathbf{r}, \omega)$ vanishes outside its spatial support τ_0 at that frequency.

We note that we have defined the NR source in the frequency domain in a frequency-by-frequency manner. A consequence of this is that an NR time-domain source will be NR in the frequency domain but not, necessarily, vice versa. In order to have complete equivalence one has to apply the frequency-domain NR requirement at all frequencies over which the source is defined.

The frequency-domain representation of piecewise-continuous time-domain NR sources can be constructed via a formula that is obtained by straightforward Fourier transformation of Eq. (1.56). Alternatively, this representation can be obtained directly in the frequency domain using a procedure that is completely parallel to the procedure that yielded the time-domain formula (Devaney and Wolf, 1973). In particular, in analogy to the steps leading up to Eq. (1.56), we consider a function $\Pi(\mathbf{r}, \omega)$ of position and frequency that is compactly supported in the spatial volume τ_0 at any given frequency ω and possesses continuous partial spatial derivatives throughout this volume but is otherwise arbitrary. We construct an NR source by simple application of the Helmholtz operator:

$$Q_{nr}(\mathbf{r}, \omega) = [\nabla^2 + k^2]\Pi(\mathbf{r}, \omega). \tag{1.57}$$

The associated frequency-domain NR field is found using Eq. (1.43a):

$$U_{nr}(\mathbf{r},\omega) = \int_{\tau_0} d^3r'\, G_+(\mathbf{r}-\mathbf{r}',\omega)\overbrace{\{[\nabla^2_{r'}+k^2]\Pi(\mathbf{r}',\omega)\}}^{Q_{nr}(\mathbf{r}',\omega)}$$

$$= \int_{\tau_0} d^3r'\, \overbrace{[\nabla^2_{r'}+k^2]G_+(\mathbf{r}-\mathbf{r}',\omega)}^{\delta(\mathbf{r}-\mathbf{r}')}\,\Pi(\mathbf{r}',\omega) = \Pi(\mathbf{r},\omega),$$

where we have twice integrated by parts and dropped the surface terms due to the assumption that Π has continuous partials throughout τ_0. The field U_{nr} vanishes outside τ_0, which then establishes that the source defined via Eq. (1.57) is an NR source at frequency ω and, moreover, that the field it generates is the compactly supported function $\Pi(\mathbf{r},\omega)$.

We thus conclude that whereas the time-domain NR sources are generated by applying the D'Alembertian operator to a time-dependent generating function that is compactly supported with continuous first partial derivatives in space and time, the frequency-domain NR sources are generated by applying the Helmholtz operator to a time-independent generating function that is compactly supported with continuous first partial derivatives in space alone. The remarks that were made concerning more general (non-piecewise-continuous) NR sources in the time domain apply also in the frequency domain. In particular, it is possible to generate more general NR frequency-domain sources using limits of infinite sequences of the piecewise-continuous NR sources constructed via Eq. (1.57). As discussed earlier for the time-dependent case, we will not make use of such sources in this book.

The energy spectra of an NR source

A compactly supported NR source generates zero field outside its space-time support and, hence, must radiate zero energy. This means, of course, that the energy spectra $E_{Q_{nr}}(\omega)$ must vanish at any frequency ω at which it is NR. It then follows from Eq. (1.55) that

$$\tilde{Q}_{nr}(k\mathbf{s},\omega) = \tilde{Q}_{nr}(\mathbf{K},\omega)|_{\mathbf{K}=k\mathbf{s}} = 0 \tag{1.58}$$

is a necessary condition for a compactly supported source to be NR at frequency $\omega = ck$. In fact, *the vanishing of the energy spectra and the boundary value of the source spatial Fourier transform according to Eq. (1.58) are also sufficient conditions for such a source to be NR.*[8] In particular, we have the following theorem.

Theorem 1.2 *A necessary and sufficient condition for a piecewise continuous source Q compactly supported in τ_0 to be NR at any given frequency ω is that its energy spectra vanish at that frequency; i.e.,*

$$E_Q(\omega) = \frac{\kappa\omega^2}{8\pi^2 c}\int_{4\pi} d\Omega_s |\tilde{Q}(k\mathbf{s},\omega)|^2 = 0. \tag{1.59}$$

[8] However, this is true only for compactly supported sources whose spatial Fourier transforms are entire functions of the spatial frequency variable $\mathbf{K}$. For example, a source supported in an infinite rectangular "strip" can radiate a field that possesses zero radiated field energy but that need not vanish identically outside this strip (see Section 4.11.1 of Chapter 4).

To prove the theorem we note that the energy spectra will vanish if and only if

$$\tilde{Q}(k\mathbf{s}, \omega) = 0 \rightarrow \tilde{Q}(\mathbf{K}, \omega)|_{\mathbf{K}=k\mathbf{s}} = 0 \rightarrow \tilde{Q}(\mathbf{K}, \omega)|_{K^2=k^2} = 0. \quad (1.60)$$

Moreover, if the source is compactly supported and piecewise continuous its spatial Fourier transform $\tilde{Q}(\mathbf{K}, \omega)$ must be an entire analytic function of $\mathbf{K}$ and, according to Eq. (1.60), must have zeros at $K = \pm k$. The transform must then admit the representation

$$\tilde{Q}(\mathbf{K}, \omega) = [-K^2 + k^2]\tilde{\Pi}(\mathbf{K}, \omega), \quad (1.61)$$

where $\tilde{\Pi}$ is an entire analytic function of $\mathbf{K}$ whose inverse Fourier transform $\Pi(\mathbf{r}, \omega)$ is twice differentiable and compactly supported in τ_0. If we now take the inverse Fourier transform of Eq. (1.61) we obtain the result

$$Q(\mathbf{r}, \omega) = \frac{1}{(2\pi)^3}\int d^3K[-K^2 + k^2]\tilde{\Pi}(\mathbf{K}, \omega)e^{i\mathbf{K}\cdot\mathbf{r}} = [\nabla^2 + k^2]\Pi(\mathbf{r}, \omega), \quad (1.62)$$

which establishes the theorem.

Example 1.7 As an example of a frequency-domain NR source consider a one-dimensional source $Q(z, \omega)$ that is zero except at $z = -z_0$ and $z = +z_0$ with $z_0 > 0$, where it is a delta function with weights $A_-(\omega)$ and $A_+(\omega)$, respectively:

$$Q(z, \omega) = A_-(\omega)\delta(z + z_0) + A_+(\omega)\delta(z - z_0).$$

The field radiated by this source is given by

$$u_+(z, \omega) = -\frac{i}{2k}\int dz'[A_-(\omega)\delta(z + z_0) + A_+(\omega)\delta(z - z_0)]e^{ik|z-z'|}$$
$$= -\frac{i}{2k}[A_-(\omega)e^{ik|z+z_0|} + A_+(\omega)e^{ik|z-z_0|}].$$

We conclude that the field will vanish for all $z > z_0$ if

$$A_-(\omega)e^{ikz_0} + A_+(\omega)e^{-ikz_0} = 0,$$

and for all $z < -z_0$ if

$$A_-(\omega)e^{-ikz_0} + A_+(\omega)e^{+ikz_0} = 0.$$

The source will thus be NR if both of the above equations hold, and this will happen so long as $A_- = (-1)^{N+1}A_+$ and $2kz_0 = N\pi$, where N is any integer.

1.7.2 A source decomposition theorem

An important theorem that we will use in our treatment of the *inverse source problem* (ISP) in Chapter 5 establishes that an arbitrary square-integrable source in either the time or frequency domain can be uniquely decomposed into the sum of an NR source and a second component that is orthogonal to the NR source and satisfies the homogeneous wave equation in the time domain and the homogeneous Helmholtz equation in the frequency domain. Here we will present and prove the theorem in the time domain, leaving the proof in the frequency domain as an exercise for the reader at the end of the chapter.

Theorem 1.3 *Let $q(\mathbf{r},t)$ be a square-integrable source compactly supported within the space-time region $\{\mathcal{S}_0|\mathbf{r}\in\tau_0,\ t\in[0,T_0]\}$. Then this source can be uniquely decomposed into an NR component $q_{\rm nr}(\mathbf{r},t)$ and a second component $\hat{q}(\mathbf{r},t)$ such that*

$$\int_{\mathcal{S}_0} dt\, d^3r\, q_{\rm nr}(\mathbf{r},t)\hat{q}(\mathbf{r},t)=0, \tag{1.63a}$$

$$\left[\nabla_r^2-\frac{1}{c^2}\frac{\partial^2}{\partial t^2}\right]\hat{q}(\mathbf{r},t)=0, \tag{1.63b}$$

$$q_{\rm nr}(\mathbf{r},t)=\left[\nabla_r^2-\frac{1}{c^2}\frac{\partial^2}{\partial t^2}\right]\pi(\mathbf{r},t), \tag{1.63c}$$

where $\pi(\mathbf{r},t)$ is a square-integrable function supported in $\mathcal{S}_0$ that possesses continuous first partial derivatives.

We have already shown at the beginning of the section that any NR source can be represented according to Eq. (1.63c). To prove the remainder of the theorem we make use of the fact that the set of all functions $\pi(\mathbf{r},t)$ that are compactly supported in $\mathcal{S}_0$ and have continuous first partial derivatives span the Hilbert space[9] $\mathcal{H}_q$ of all square-integrable functions supported in $\mathcal{S}_0$. It then follows that the NR sources constructed from this set via Eq. (1.63c) span a subspace $\eta\subset\mathcal{H}_q$ comprised of all sources that generate zero field outside of $\mathcal{S}_0$. We can identify η as being the null space of the transformation from the source to the field at all space-time points exterior to the source region $\mathcal{S}_0$. The entire Hilbert space $\mathcal{H}_q$ can be decomposed into η and its orthogonal complement $\eta^\perp$ composed of all sources orthogonal to the NR sources. It then follows that any source $q\in\mathcal{H}_q$ admits the unique decomposition

$$q(\mathbf{r},t)=q_{\rm nr}(\mathbf{r},t)+\hat{q}(\mathbf{r},t),$$

where $q_{nr}\in\eta$ and $\hat{q}\in\eta^\perp$ satisfy Eq. (1.63a).

We now have only to prove that the sources $\hat{q}\in\eta^\perp$ satisfy the homogeneous wave equation. This follows from the orthogonality condition Eq. (1.63a) and the NR source representation Eq. (1.63c). In particular, we have that

$$\int_{\mathcal{S}_0} dt\, d^3r\, \overbrace{q_{\rm nr}(\mathbf{r},t)}^{\mathcal{D}\pi(\mathbf{r},t)}\hat{q}(\mathbf{r},t)=\int_{\mathcal{S}_0} dt\, d^3r\,\pi(\mathbf{r},t)\mathcal{D}\hat{q}(\mathbf{r},t)=0, \tag{1.64}$$

where

$$\mathcal{D}=\nabla_r^2-\frac{1}{c^2}\frac{\partial^2}{\partial t^2}$$

is the D'Alembertian operator, and we have twice integrated by parts and used the fact that the functions $\pi(\mathbf{r},t)$ have continuous first-order partials. But Eq. (1.64) must hold for all $\pi(\mathbf{r},t)\in\mathcal{H}_q$ and, since these functions span $\mathcal{H}_q$, it then follows that $\mathcal{D}\hat{q}(\mathbf{r},t)=0$ and the proof of the theorem is complete.

[9] For a brief review of Hilbert spaces see Chapter 5.

1.7.3 Essentially non-radiating sources

A frequency-domain NR source $Q_{\rm nr}(\mathbf{r},\omega)$ generates *zero field* outside the source's space volume τ_0 at the given frequency ω. It is also possible for a source to generate a non-zero field outside τ_0 that has an energy spectrum $E_Q(\omega)$ that is *essentially* zero at one or more frequencies ω and, hence, radiates negligible power and energy over this set of frequencies. We call such sources *essentially non-radiating sources* (essentially NR sources).

We showed in Section 1.6 in Theorem 1.1 that the energy spectra $E_Q(\omega)$ of a source can be expressed in terms of the radiation pattern via Eq. (1.54) so that an obvious requirement for an essentially NR source is that

$$E_Q(\omega) = \frac{2\kappa\omega^2}{c}\int_{4\pi} d\Omega |f(\mathbf{s},\omega)|^2 < \epsilon(\omega), \tag{1.65}$$

where $\epsilon(\omega)$ is some small parameter that is used to characterize the essentially NR condition and may depend on the frequency ω. Although the condition Eq. (1.65) defining an essentially NR source seems somewhat arbitrary, we will now show that there is a "natural" choice for the parameter $\epsilon(\omega)$ that leads to a useful and meaningful criterion for essentially NR sources.

We begin with Eq. (1.45b) relating the frequency-domain radiation pattern to the frequency-domain representation of the source. We can expand the plane wave $\exp(-ik\mathbf{s}\cdot\mathbf{r})$ in the series (see Example 3.4 in Chapter 3)

$$e^{-ik\mathbf{s}\cdot\mathbf{r}} = 4\pi\sum_{l=0}^{\infty}\sum_{m=-l}^{l}(-i)^l j_l(kr)Y_l^m(\mathbf{s})Y_l^{m*}(\hat{\mathbf{r}}),$$

where j_l is the spherical Bessel function of order l and Y_l^m the spherical harmonic of degree l and order m (see Chapter 3). In this equation and throughout the book the unit vectors $\mathbf{s}$ and $\hat{\mathbf{r}}$ are employed as shorthand notation to denote the polar and azimuthal angles of the unit propagation vector and field point $\mathbf{r}$ in the arguments of the spherical harmonics. If we use the above expansion in Eq. (1.45b) we find that

$$f(\mathbf{s},\omega) = \sum_{l=0}^{\infty}\sum_{m=-l}^{l}(-1)^{l+1}i^l q_l^m(\omega)Y_l^m(\mathbf{s}), \tag{1.66a}$$

where

$$q_l^m(\omega) = \int_{\tau_0} d^3r\, Q(\mathbf{r},\omega)j_l(kr)Y_l^{m*}(\hat{\mathbf{r}}). \tag{1.66b}$$

The energy spectrum is obtained by substituting Eq. (1.66a) into Eq. (1.65) and, on using the orthonormality of the spherical harmonics, we conclude that

$$E_Q(\omega) = \frac{2\kappa\omega^2}{c}\sum_{l=0}^{\infty}\sum_{m=-l}^{l}|q_l^m(\omega)|^2. \tag{1.66c}$$

Now it is clear that for the energy spectrum to be finite the expansion coefficients $q_l^m(\omega)$ must tend to zero as $l \to \infty$. However, these coefficients will actually tend to zero much faster than is required by simple convergence considerations due to the constraint that the source is compactly supported in the space volume τ_0. To see this we note that q_l^m as defined in Eq. (1.66b) can be regarded as an L^2 inner product between the source Q and the functions $j_l Y_l^m$ so that, by application of the Schwarz inequality, we conclude that

$$|q_l^m(\omega)|^2 \leq \mathcal{E}_Q \int_0^{a_0} r^2\, dr |j_l(kr)|^2 = \mathcal{E}_Q \overbrace{\frac{a_0^3}{2}[j_l^2(ka_0) - j_{l-1}(ka_0) j_{l+1}(ka_0)]}^{\mu_l^2(ka_0)}, \tag{1.67a}$$

where

$$\mathcal{E}_Q = \int_{\tau_0} d^3 r |Q(\mathbf{r}, \omega)|^2 \tag{1.67b}$$

is the L^2-norm square of the source (source energy) and a_0 the radius of the smallest bounding sphere to τ_0.

The quantities

$$\mu_l^2(ka_0) = \frac{a_0^3}{2}[j_l^2(ka_0) - j_{l-1}(ka_0) j_{l+1}(ka_0)] \tag{1.68}$$

appearing in Eq. (1.67a) are important parameters that will re-occur many times throughout this book. These parameters are exponentially decreasing functions of their index l when $l > ka_0$ and, indeed, $l_0 = ka_0$ can be considered to be a cutoff value beyond which μ_l^2 are effectively zero. This is illustrated in Fig. 1.2, which shows plots of μ_l^2 and of the log (base 10) of μ_l^2 plotted as a function of the index l for various values of $l_0 = ka_0$. It then follows that the expansion coefficients q_l^m tend to zero exponentially fast for $l > ka_0$ so that *any source for which the expansion coefficients q_l^m vanish for $l < ka_0$ will radiate negligible energy; i.e., will be "essentially NR."*

Non-radiating sources and essentially NR sources play important roles in the ISP, which will be treated in Chapter 5. The pure NR sources will be shown to be responsible for the non-uniqueness of solutions to the ISP, while the essentially NR sources are responsible for the ill-posedness of this problem. The inequality Eq. (1.67a) which leads to essentially NR sources also limits the angular resolution of the radiation pattern radiated by any finite-norm source. This conclusion follows directly from the expansion of the radiation pattern given in Eq. (1.66a). In particular, because of the exponential decay of the source multipole moments $q_l^m(\omega)$ for $l > ka_0$ it follows that this expansion is effectively terminated at $l = ka_0$. Moreover, the spherical harmonics $Y_l^m(\mathbf{s})$ are periodic functions of the polar and azimuthal angles of the unit vector $\mathbf{s}$ with minimum angular periods of $2\pi/l$ radians for fixed index l. It then follows that any finite-norm source cannot possess a radiation pattern having angular periods smaller than $2\pi/(ka_0) = \lambda/a_0$ radians, which thus sets a limit on the achievable resolution of this class of source that depends only on the wavelength and the source radius.

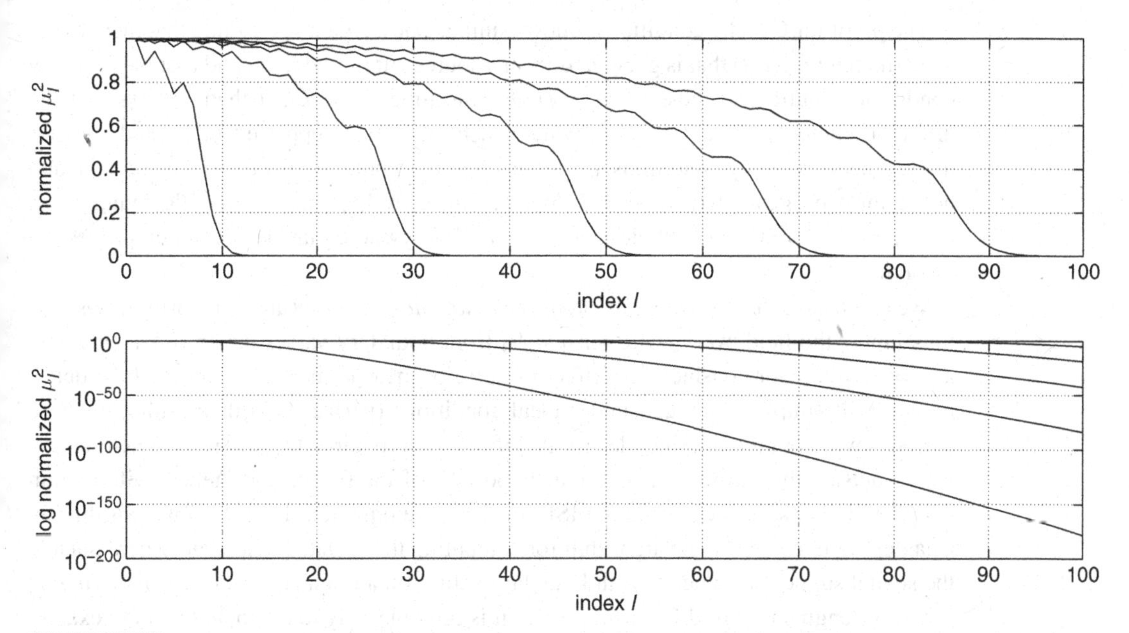

Fig. 1.2 (Top) Plots of μ_l^2 normalized by their peak value for values of $ka_0 = 10, 30, 50, 70$ and 90 plotted as a function of the index l. (Bottom) Plots of the log to base 10 of the μ_l^2 plotted as a function of the Index l for the same set of values of ka_0.

1.7.4 The field uniqueness theorem

The following theorem plays an important role in the ISP, which will be treated in Chapter 5.

Theorem 1.4 (Field uniqueness theorem) *Let $u_+(\mathbf{r}, t)$ be the field radiated by a causal source compactly supported in the space-time domain $\{\mathcal{S}_0 | \mathbf{r} \in \tau_0,\ t \in [0, T_0]\}$. Then this field is uniquely determined over all space-time points lying outside $\mathcal{S}_0$ by its boundary values (field and normal derivative) over any closed surface $\partial\tau$ that completely surrounds τ_0 or by Cauchy data specified at any time t_0 exceeding the turn-off time T_0 of the source.*

We first prove the theorem for boundary-value data over $\partial\tau$. Let us assume that there exists also a second field $\hat{u}_+(\mathbf{r}, t)$ that is generated from a source supported in $\mathcal{S}_0$ and assumes boundary conditions over the arbitrary bounding surface $\partial\tau$ identical to those for u_+. It then follows that the difference field $\delta u = u_+ - \hat{u}_+$ is generated by a source supported in $\mathcal{S}_0$ and has zero boundary values on $\partial\tau$. According to the first Helmholtz identity Eq. (1.36a) δu must then vanish identically outside the bounding surface $\partial\tau$ and, hence, must be generated by an NR source compactly supported in the interior spatial region $\tau \supseteq \tau_0$ bounded by the surface $\partial\tau$. But, by hypothesis, δu is generated by a source supported in $\mathcal{S}_0$. We thus conclude that $\delta u = u_+ - \hat{u}_+$ is generated by an NR source that is compactly supported in $\mathcal{S}_0$ so that $u_+ = \hat{u}_+$ everywhere outside $\mathcal{S}_0$.

The proof for Cauchy conditions follows similar lines. Again we assume that there exists a second field $\hat{u}_+(\mathbf{r}, t)$ that is generated from a source supported in $\mathcal{S}_0$ and assumes Cauchy conditions identical to those of $u_+(\mathbf{r}, t)$ at some time $t_0 > T_0$. It then follows that the difference field $\delta u = u_+ - \hat{u}_+$ is generated by a source supported in $\mathcal{S}_0$ and satisfies homogeneous Cauchy conditions at t_0. This field then must vanish for all time $t > t_0$ and hence must be generated by an NR source. Since, by hypothesis, this NR source must be supported in $\mathcal{S}_0$ we conclude that $u_+ = \hat{u}_+$ everywhere outside $\mathcal{S}_0$, which proves the theorem.

We emphasize that the above theorem is a *uniqueness theorem* and neither indicates *how* to compute the field everywhere outside $\mathcal{S}_0$ from boundary values on $\partial\tau$ or Cauchy data at $t = t_0 > T_0$ nor considers the effect of noise or measurement error on the field determination. It simply states that under ideal conditions (perfect field data) either of these data sets will contain, in principle, all the information required to compute the field everywhere outside the (known) space-time support $\mathcal{S}_0$ of the source and, hence, will contain complete information regarding the ISP for the wave equation. In fact, as we will find in Chapter 4, there is no stable algorithm for computing the radiated field everywhere outside the spatial support volume τ_0 from boundary values on a boundary $\partial\tau$ that is more than a few wavelengths removed from τ_0. Rather, it is possible only to compute this field exactly and stably for space points $\mathbf{r}$ that lie outside the measurement boundary $\partial\tau$ or approximately and stably for space points $\mathbf{r}$ that lie inside this boundary. We also note that the theorem requires that both the field and the normal derivative be specified on $\partial\tau$ to insure uniqueness. However, only one of the two quantities is actually required, since the two data sets are connected via the second Helmholtz identity Eq. (1.36b) and either one can, in principle, be computed from the other.

1.8 Surface sources

Up to this point we have considered only three-dimensional (3D) (volume) sources $q(\mathbf{r}, t)$ that are distributed throughout a 3D space volume τ_0 and radiate a field that consists of a superposition of outgoing-wave spherical waves each weighted by the source amplitude at the various source points $\mathbf{r} \in \tau_0$. It is possible to generalize this model to include two-dimensional (2D) sources that are distributed over a surface $\partial\tau_0$ and that radiate a field that consists of a superposition of outgoing spherical waves centered at the various source points $\mathbf{r}_0 \in \partial\tau_0$. The most general model of this type radiates a field according to the formula

$$u_+(\mathbf{r}, t) = \int_0^\infty dt' \int_{\partial\tau_0} dS_0 \left[q_s(\mathbf{r}_0, t') g_+ - q_d(\mathbf{r}_0, t') \frac{\partial}{\partial n_0} g_+ \right], \tag{1.69}$$

where $\mathbf{r}_0 \in \partial\tau_0$ denotes a point on the surface $\partial\tau_0$, which can be closed with finite interior τ_0 and infinite exterior $\tau_0^\perp$ or the two regions can be infinite with common boundary $\partial\tau_0$. The normal derivative in the above equation can be selected to be directed out of the region τ_0 and into the $\tau_0^\perp$ or vice versa. The selection is arbitrary but is sometimes physically motivated, as we will see in our treatment of the connection of time reversal to field back propagation in Chapter 4. Here, and elsewhere unless stated otherwise, we will assume for the sake of definiteness that it is directed outward from the region τ_0 into the region $\tau_0^\perp$. The field $u_+(\mathbf{r},t)$ radiated by the source pair is clearly causal and satisfies the homogeneous wave equation over all space-time points that lie away from the boundary $\partial\tau_0$; i.e., within both τ_0 and $\tau_0^\perp$.

The source component q_s is termed the "singlet" component since it radiates the simple monopole spherical wave g_+, while q_d is termed the "doublet" component since it radiates the dipole $(\partial/\partial n_0)g_+$. The two components are generally required to radiate different fields into the two spaces τ_0 and $\tau_0^\perp$. In cases where $\partial\tau_0$ is a *separable surface* (see Chapter 3) defined by generalized coordinates (ξ_1,ξ_2,ξ_3) with, say, $\xi_3=\xi_{30}$ equal to a constant defining the surface, the corresponding surface source can be expressed as a 3D source distribution in the form

$$q(\mathbf{r},t)=q_s(\xi_1,\xi_2,t)\frac{\delta(\xi_3-\xi_{30})}{h_3(\mathbf{r}_0)}+q_d(\xi_1,\xi_2,t)\frac{h_1(\mathbf{r}_0)h_2(\mathbf{r}_0)}{h_1(\mathbf{r})h_2(\mathbf{r})}\frac{\frac{\partial}{\partial\xi_3}\delta(\xi_3-\xi_{30})}{h_3(\mathbf{r})},\tag{1.70}$$

where $\delta(\cdot)$ is the Dirac delta function and h_1,h_2,h_3 are the scale factors for the coordinate system. The field radiated by q is then given by the primary field representation Eq. (1.33) with the source given by Eq. (1.70), which reduces to Eq. (1.69).

The singlet–doublet sources are completely arbitrary and, in particular, are *not, necessarily, equal to the boundary values of the field u_+ and its normal derivative on $\partial\tau_0$.* Indeed, although Eq. (1.69) resembles the Kirchhoff–Helmholtz representation of the field radiated into $\tau_0^\perp$ by a volume source supported within τ_0 and given in Eq. (1.36a), the above field representation is much more general and reduces to the Kirchhoff–Helmholtz representation only in the special case that the two source components are such that u_+ satisfies the second Helmholtz identity Eq. (1.36b) throughout the region τ_0; i.e., vanishes within τ_0. In general this identity will not be satisfied and the singlet–doublet pair will radiate a non-zero field both into τ_0 and into $\tau_0^\perp$.

To prove the above assertion that Eq. (1.69) reduces to the Kirchhoff–Helmholtz representation Eq. (1.36a) if the field u_+ vanishes throughout τ_0 we make use of the following limiting values (Green, 1969) of the field u_+:

$$\lim_{\mathbf{r}\to\mathbf{r}_0}u_+(\mathbf{r},t)=u_{+\mathrm{P}}(\mathbf{r}_0,t)\pm\frac{1}{2}q_d(\mathbf{r}_0,t),\tag{1.71a}$$

$$\lim_{\mathbf{r}\to\mathbf{r}_0}\frac{\partial}{\partial n}u_+(\mathbf{r},t)=\frac{\partial}{\partial n_0}u_{+\mathrm{P}}(\mathbf{r}_0,t)\pm\frac{1}{2}q_s(\mathbf{r}_0,t),\tag{1.71b}$$

where the + sign is used when the limit is taken from $\tau_0^\perp$ and the minus sign when the limit is taken from τ_0, and the subscript P stands for the principal value of the quantity being subscripted. Now, if u_+ satisfies the second Helmholtz identity and, hence, vanishes throughout τ_0 then Eqs. (1.71) require that

$$u_{+\mathrm{P}}(\mathbf{r}_0,t)-\frac{1}{2}q_{\mathrm{d}}(\mathbf{r}_0,t)=0\Rightarrow u_{+\mathrm{P}}(\mathbf{r}_0,t)=\frac{1}{2}q_{\mathrm{d}}(\mathbf{r}_0,t),$$

$$\frac{\partial}{\partial n_0}u_{+\mathrm{P}}(\mathbf{r}_0,t)-\frac{1}{2}q_{\mathrm{s}}(\mathbf{r}_0,t)=0\Rightarrow \frac{\partial}{\partial n_0}u_{+\mathrm{P}}(\mathbf{r}_0,t)=\frac{1}{2}q_{\mathrm{s}}(\mathbf{r}_0,t),$$

which, when used in Eqs. (1.71), yield

$$\lim_{\mathbf{r}\to\mathbf{r}_0}u_+(\mathbf{r},t)=q_{\mathrm{d}}(\mathbf{r}_0,t),\qquad \lim_{\mathbf{r}\to\mathbf{r}_0}\frac{\partial}{\partial n}u_+(\mathbf{r},t)=q_{\mathrm{s}}(\mathbf{r}_0,t),\tag{1.72}$$

where the limits are taken from the region $\tau_0^{\perp}$ onto the boundary $\partial\tau_0$. The Kirchhoff–Helmholtz representation Eq. (1.36a) then results from Eq. (1.69) upon making the above substitutions for the singlet–doublet pair. It then follows that in this case the field u_+ can be radiated by a 3D volume source located within the region τ_0 and the surface sources $q_{\mathrm{s}}(\mathbf{r}_0,t)=\partial u_+(\mathbf{r}_0,t)/\partial n_0$ and $q_{\mathrm{d}}(\mathbf{r}_0,t)=u_+(\mathbf{r}_0,t)$ are sometimes referred to as "secondary sources" that result from the "primary" 3D source.

1.8.1 Non-radiating surface sources

When the singlet–doublet pair $q_{\mathrm{s}},q_{\mathrm{d}}$ constitutes a secondary source corresponding to the boundary values of a field u_+ radiated by a 3D source supported within the region τ_0 then the second Helmholtz identity requires that the field u_+ radiated by this pair as defined via Eq. (1.69) must vanish identically throughout this region. We can interpret this as meaning that *the singlet–doublet pair $q_{\mathrm{s}}=\partial u_+/\partial n$, $q_{\mathrm{d}}=u_+$, where u_+ is radiated by a 3D source supported in τ_0, constitutes an NR surface source relative to the same region; i.e., relative to the region τ_0.* Similarly, it is not difficult to show that if the singlet–doublet pair results from the boundary values of a field radiated by a source supported entirely within the region $\tau_0^{\perp}$ then the field radiated by this pair must vanish everywhere throughout $\tau_0^{\perp}$ and, hence, must be an NR surface source relative to $\tau_0^{\perp}$. These conclusions are simply a consequence of the second Helmholtz identity written for fields radiated by 3D sources confined to either τ_0 or $\tau_0^{\perp}$ and are illustrated in Fig. 1.3.

We have just seen that all secondary surface sources that result from the boundary values of a field radiated by 3D sources located within τ_0 or $\tau_0^{\perp}$ are NR into those regions; e.g., if the singlet–doublet pair results from a 3D source located within τ_0 then this surface source is NR into τ_0 and similarly for $\tau_0^{\perp}$. Moreover, we will show in the following chapter that *all NR surface sources can be so constructed.* This then provides a general algorithm for generating any NR surface source in analogy to the general algorithm presented in Eq. (1.56) for volume NR sources. In particular, if $\psi(\mathbf{r},t)$ is a field that is radiated by a source $q(\mathbf{r},t)$ that is supported entirely within τ_0 or its complement $\tau_0^{\perp}$ then, for any point $\mathbf{r}_0$ lying on the boundary $\partial\tau_0$,

$$q_{\mathrm{snr}}(\mathbf{r}_0,t)=\frac{\partial}{\partial n_0}\psi(\mathbf{r}_0,t)=-\frac{1}{4\pi}\int_{\tau_0}d^3r'\,\frac{\partial}{\partial n_0}\frac{q(\mathbf{r}',t-|\mathbf{r}_0-\mathbf{r}'|/c)}{|\mathbf{r}_0-\mathbf{r}'|},$$

$$q_{\mathrm{dnr}}(\mathbf{r}_0,t)=\psi(\mathbf{r}_0,t)=-\frac{1}{4\pi}\int_{\tau_0}d^3r'\,\frac{q(\mathbf{r}',t-|\mathbf{r}_0-\mathbf{r}'|/c)}{|\mathbf{r}_0-\mathbf{r}'|}$$

and the singlet–doublet pair is NR into the region in which the source is located.

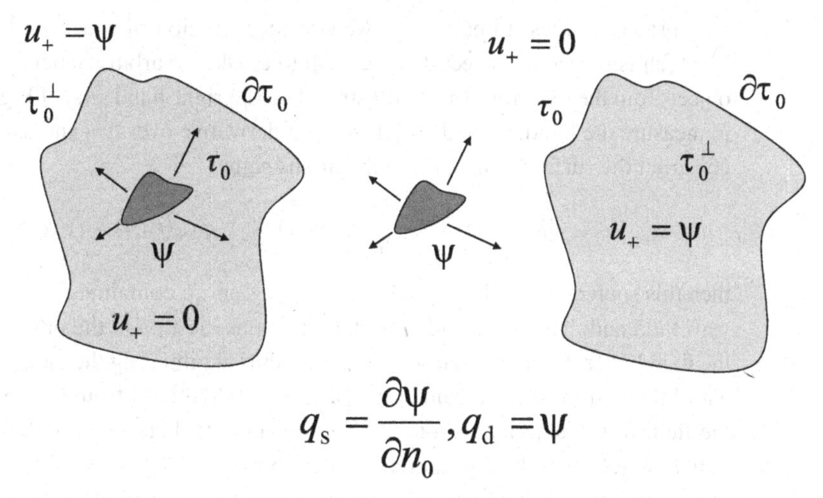

Fig. 1.3 Illustrating the construction of NR surface sources. On the left an NR pair for the interior region τ_0 is produced by the field radiated by a primary 3D source located in τ_0 and on the right an NR pair for the exterior region $\tau_0^\perp$ is produced by a primary 3D source located in $\tau_0^\perp$.

The natural question which arises is that of whether there exist surface sources that are NR simultaneously in both τ_0 and $\tau_0^\perp$. In fact, no such source can exist. To see this we consider the case of a surface source that is NR in τ_0. Then the field u_+ radiated by this singlet–doublet pair must vanish throughout τ_0 and, hence, must satisfy Eq. (1.72) with the limits taken from within $\tau_0^\perp$. But if the field is also NR within $\tau_0^\perp$ then these two limits must be zero and, hence, this equation requires that both q_s and q_d vanish, which then establishes the result.

Finally, we mention that besides purely NR surface sources it is also possible to have essentially NR surface sources that radiate a field that possesses an energy spectrum that is essentially zero at one or more temporal frequencies. Such sources are secondary surface sources generated by the boundary values of essentially NR 3D volume sources located within τ_0 or $\tau_0^\perp$ and will radiate an identical field to that of the essentially NR primary source throughout the complement of the region in which this primary source is located. Since these surface sources are secondary sources they must satisfy the second Helmholtz identity and are thus purely NR throughout the region in which the primary essentially NR 3D source is located.

1.8.2 Active object cloaking

An interesting and potentially very important application is that of "object cloaking" whereby an object can be made invisible to electromagnetic or other types of wavefields by the use of specially designed embedding materials (so-called "meta-materials") or by surrounding the object by an active antenna system (Miller, 2006). That this is theoretically possible with the use of a surrounding surface source implemented in the form of an active antenna system is easily established using the results obtained above concerning

NR surface sources. In particular, we consider a region of space τ_0 illustrated in Fig. 1.3 in which is placed an object that we wish to cloak. An arbitrary field ψ is incident on this object from the exterior of τ_0 as illustrated on the right-hand side of Fig. 1.3. If we are able to measure the incident field and its normal derivative over $\partial\tau_0$ and use this information to construct the surface source (note the minus signs)

$$q_{\mathrm{snr}}(\mathbf{r}_0, t) = -\frac{\partial}{\partial n_0}\psi(\mathbf{r}_0, t), \qquad q_{\mathrm{dnr}}(\mathbf{r}_0, t) = -\psi(\mathbf{r}_0, t) \tag{1.73}$$

then this source will be NR into the exterior region $\tau_0^\perp$ containing the source of the incident wavefield and, hence, will not alter this field outside τ_0. On the other hand, it will radiate the field $-\psi(\mathbf{r}, t)$ into the interior region τ_0, thus annihilating the incident field and creating a null field within τ_0 that cannot generate a scattered field from the object that could alter the field in the exterior region $\tau_0^\perp$. The object will then be totally cloaked (invisible) to radiation generated by any source exterior to τ_0.

The practical difficulties in achieving active field cloaking as described above are that (i) both the incident field and its normal derivative would need to be measured and (ii) the active cloaking antenna system needs to include both singlet and doublet sources. The first requirement can, in principle, be eliminated since the normal derivative of the incident field over $\partial\tau_0$ can, in principle, be determined from knowledge of the incident field (and vice versa) via the second Helmholtz identity implemented in the form of a fast digital algorithm, which is certainly possible for separable surfaces (see Chapter 3). On the other hand, the second requirement is not so easily eliminated. However, it is possible to implement such a cloaking antenna system using two concentric dipole arrays that together approximate a general surface source or, alternatively, using directional antenna elements that radiate only over a limited angular range.

We should note that if both the field and its normal derivative over $\partial\tau_0$ are measured then it is not necessary to isolate the incident wavefield component of the total field over $\partial\tau_0$ in constructing the sources in Eq. (1.73). The reason for this is that even if the cloaked object generates a scattered field its boundary values over $\partial\tau_0$ will constitute a surface NR source relative to the interior region τ_0 and, hence, will not change the cloaking field within τ_0. The overall cloaking scheme will thus be insensitive (stable) to changes in the object or cloaking surface $\partial\tau_0$.

Further reading

The key literature on the radiation and propagation of waves and on much of the material presented in this book is the two-volume work by Morse and Feshbach (Morse and Feshbach, 1953). Other excellent texts include the books by Stratton (Stratton, 1941), Jackson (Jackson, 1998), Courant and Hilbert (Courant and Hilbert, 1966), Born and Wolf (Born and Wolf, 1999) and Chew (Chew, 1990) and, from a more engineering perspective, the book by Balanis (Balanis, 1989). A complete account of the initial-value problem is contained in the classic text by Hadamard (Hadamard, 1952). Very readable accounts of distribution theory include Zemanian (1965) and Richards and Youn (1995), while an excellent account of classical Fourier methods as applied to partial differential equations is given in Weinberger (1995). The book by Arsac (Arsac, 1984) contains a combined treatment of Fourier transforms and distribution theory written from a physicist's point of view. A simple but complete treatment of complex-variable theory at the level required in this book is given in Arfken and Weber (2001). An excellent overall review of the history of NR sources can be found in Gbur (2003). The theory of surface sources presented in Section 1.8 relied heavily on the book by Baker and Copson (Baker and Copson, 1950).

Problems

1.1 Determine the relationship between the two Cauchy conditions for the one-dimensional (1D) wave equation such that (a) only a wavefield propagating in the $+z$ direction is present and (b) only a wave propagating in the $-z$ direction is present. Are there any non-trivial Cauchy conditions that result in a zero field?

1.2 (a) Compute the temporal Fourier transform of the "Rect" function

$$\mathrm{Rect}(t) = \begin{cases} 1 & -T_0 \le t \le +T_0, \\ 0 & \text{else.} \end{cases}$$

(b) Use the Cauchy–Riemann equations to prove that the transform that you computed is an entire analytic function of the frequency variable ω.

1.3 Perform the steps leading from the Fourier-integral representation of the causal Green function in Eq. (1.18) to its final form given in Eq. (1.20a).

1.4 Prove using Cauchy's integral theorem that the difference between the causal (retarded) and acausal (advanced) Green functions satisfies the homogeneous wave equation and, hence, is not a Green function.

1.5 Compute the frequency-domain outgoing- and incoming-wave Green functions $G_+(\mathbf{R}, \omega)$ and $G_-(\mathbf{R}, \omega)$ by performing spatial Fourier inversions of $\tilde{G}(\mathbf{K}, \omega)$.

1.6 Compute the 1D incoming-wave Green function $G_-(z, \omega)$ from Eq. (1.26) of Example 1.3.

1.7 Directly verify by differentiation that the 1D causal Green function given in Eq. (1.29) of Example 1.4 satisfies the defining equation Eq. (1.24) of Example 1.3.

1.8 Verify by direct differentiation that the difference between the causal and acausal 1D Green functions to the wave equation satisfies the homogeneous wave equation.

1.9 Verify that the interior field representations given in Eqs. (1.37) remain valid with g_- replaced by g_+; i.e., show that the two new equations are also correct.

1.10 Derive Eqs. (1.34a) and (1.34b) in Example 1.6.

1.11 Derive the expression for the radiated field given in Eq. (1.35a) of Example 1.6.

1.12 Determine the Cauchy conditions satisfied by the free-space propagator $g_f(\mathbf{R}, \tau)$ at $\tau = 0$ directly from its definition as the difference between the retarded and advanced Green functions to the wave equation.

1.13 Verify that the solution to the initial-value problem given in Eq. (1.41) reduces to the Cauchy initial conditions found in Problem 1.12.

1.14 Derive the time-domain Porter–Bojarski integral equation from the interior field solution Eq. (1.37a):

$$\int_0^{T_0} dt' \int_{\tau_0} d^3r'\, g_f(\mathbf{r}-\mathbf{r}', t-t') q(\mathbf{r}', t') = \phi(\mathbf{r}, t), \quad \mathbf{r} \in \tau,$$

where

$$g_f(\mathbf{R}, \tau) = g_+(\mathbf{R}, \tau) - g_-(\mathbf{R}, \tau)$$

is the free-field propagator and

$$\phi(\mathbf{r}, t) = \int_{-\infty}^{\infty} dt' \int_{\partial\tau} dS' \left[u_+ \frac{\partial}{\partial n'} g_- - g_- \frac{\partial}{\partial n'} u_+ \right].$$

1.15 Compute the frequency-domain solution and radiation pattern for the 1D wave equation from the time-domain solution found in Example 1.5.

1.16 Compute the frequency- and time-domain radiation patterns for the time-periodic source considered in Example 1.6.

1.17 Transform the SRC as defined in Eq. (1.48) into the time domain and interpret the result.

1.18 Prove that the far-field approximation given in Eq. (1.49) is causal.

1.19 Express the energy spectra $E_Q(\omega)$ directly in terms of the source $Q(\mathbf{r}, \omega)$.

1.20 Derive the general expression for an NR source for the 1D wave equation.

1.21 Show that the solution to the 1D radiation problem can be expressed entirely in terms of $\tilde{q}(\pm k, \omega)$ everywhere outside the source region. Using this solution, show that the field everywhere outside the source region is uniquely determined by the value of the field at any two points $z_1 < -a_0$ and $z_2 > a_0$, where $[-a_0, a_0]$ is the space support for the source. Give an expression for the field in terms of the field amplitude at these two points.

1.22 Derive the equation satisfied by a frequency-domain NR source Eq. (1.57) directly from the equation satisfied by the time-domain NR source Eq. (1.56).

1.23 Construct an NR source using the classical testing function of distribution theory

$$\Pi(\mathbf{r}) = \begin{cases} 0 & r \geq a_0, \\ \exp[1/(r^2 - a_0^2)] & r < a_0. \end{cases}$$

1.24 Determine whether the rotating point source considered in Example 1.6 can ever be NR at one or more temporal frequencies.

1.25 Determine whether the rotating point source considered in Example 1.6 can ever be essentially NR. Construct an essentially NR source by high-pass filtering the Fourier-series expansion of the delta function $\delta(\phi - v_\phi t)$.

1.26 Use the second Helmholtz identity to verify that the cloaking field within the interior τ_0 generated from the surface source given in Eq. (1.73) is not modified when the incident field is replaced by the total field (incident plus scattered). Discuss why this modification (using total rather than incident field measurements) requires that both the field and its normal derivative be separately measured.

2 Radiation and boundary-value problems in the frequency domain

We return to the problem of computing the field $u_+(\mathbf{r}, t)$ radiated by a real-valued space- and time-varying source $q(\mathbf{r}, t)$ embedded in an infinite homogeneous medium such as free space. As in Chapter 1 we will assume here that the time-dependent source $q(\mathbf{r}, t)$ is compactly supported in the space-time region $\{\mathcal{S}_0 | \mathbf{r} \in \tau_0,\ t \in [0, T_0]\}$, where τ_0 is its spatial volume and $[0, T_0]$ the interval of time over which the source is turned on. In the case in which the medium is non-dispersive the radiated wavefield satisfies the inhomogeneous scalar wave equation Eq. (1.1). More generally, if the background medium is dispersive it is necessary to replace the second time derivative in this equation by an integral (convolutional) operator, so that the wave equation is actually an integral-differential equation. In this chapter we will treat the radiation problem in the frequency domain so that this complication is avoided and our results apply both to dispersive and to non-dispersive backgrounds.

In addition to treating the radiation problem we also treat the classical boundary-value problem for the scalar wave Helmholtz equation in a (possibly dispersive) uniform background medium. Special attention is devoted to the famous Rayleigh–Sommerfeld boundary-value problem, which consists of computing a radiated field throughout a half-space that is exterior to the source region τ_0 from Dirichlet or Neumann conditions prescribed over an infinite bounding plane to the source. The inverse problem of (approximately) computing a radiated field throughout an interior half-space that includes the source region from Dirichlet or Neumann conditions on an infinite bounding plane is also addressed. This problem, which is a form of field *back propagation*, is important in a number of imaging and wavefield-inversion applications and will be addressed more completely in Chapter 4.

2.1 Frequency-domain formulation of the radiation problem

The key to treating the radiation problem in the frequency domain is the observation that even if the background medium is dispersive the Fourier transforms $Q(\mathbf{r}, \omega)$ and $U(\mathbf{r}, \omega)$ of the time-dependent source and field are still related by the inhomogeneous Helmholtz equation Eq. (1.43b) and the frequency-domain Green functions satisfy the Helmholtz equation Eq. (1.22) with a delta-function source. The only difference is that the wavenumber k is no longer simply ω/c but will be a product of ω/c with an analytic function of ω (see the discussion below). The frequency- and time-domain quantities are still related via a temporal Fourier transform and the time-dependent radiated field u_+ is required to be causally related to the source and, hence, must vanish for all negative time. This then guarantees that its temporal transform $U_+(\mathbf{r}, \omega)$ is analytic throughout the upper half of the complex-ω plane and vanishes as $|\omega| \to \infty$ in that half-plane. The causal solution to the radiation problem and the causal time-dependent Green function are thus obtained, as we showed in Chapter 1, by employing an ω contour of integration in their temporal Fourier-integral representations that lies above the real-ω axis. The "outgoing-wave Green function" $G_+(\mathbf{R}, \omega)$ that yields the causal time-domain Green function g_+ is then obtained as the solution to the Helmholtz equation Eq. (1.22) with $\Im\omega > 0$ and is still given by Eq. (1.23a), where, however, the wavenumber will depend on the frequency variable ω. The radiated field is still given in terms of G_+ by Eq. (1.43a) of Section 1.5, again with $k = k(\omega)$. Indeed, virtually all of the frequency-domain results obtained in the previous chapter can be extended to dispersive media by simply allowing k to be an analytic function of ω that satisfies certain conditions imposed by causality as discussed below.

2.1.1 Analytic-signal representation of time-domain fields

Unless stated otherwise, we will assume throughout this chapter that the temporal frequency ω is a strictly real quantity. It is also sometimes useful to restrict ω to being positive. This is certainly possible in the case of *real-valued* time-dependent fields and sources since their time Fourier transforms satisfy the reciprocity condition

$$\tilde{F}(\mathbf{r}, -\omega) = \tilde{F}^*(\mathbf{r}, \omega), \tag{2.1}$$

where $F(\mathbf{r}, \omega)$ is the time Fourier transform of some arbitrary *real-valued* function $f(\mathbf{r}, t)$, the superscript asterisk $*$ denotes the complex conjugate and we have assumed that ω is real-valued. Because of the reciprocity condition we can restrict our attention to only positive frequencies and recover the negative-frequency components (if necessary) using Eq. (2.1).

In fact, it is not even necessary to use the reciprocity relationship Eq. (2.1) to determine the associated, real-valued time-domain quantities. In particular, we can make use of the *analytic signal*

$$f^{(+)}(\mathbf{r}, t) = \frac{1}{2\pi} \int_0^\infty d\omega \, \tilde{F}(\mathbf{r}, \omega) e^{-i\omega t},$$

which can be computed using only positive-frequency components of $\tilde{F}$. If we use the reciprocity relationship Eq. (2.1) it is easy to show that

$$f(\mathbf{r}, t) = \frac{1}{2\pi} \int_{-\infty}^\infty d\omega \, \tilde{F}(\mathbf{r}, \omega) e^{-i\omega t} = 2\Re f^{(+)}(\mathbf{r}, t) \tag{2.2}$$

so that only the analytic signal $f^{(+)}$ need be computed and the real (physically meaningful) quantity f can then be obtained directly in the time domain using Eq. (2.2).

Although we will not make extensive use of the analytic-signal representation in this book, this representation is often used in the signal-processing, physics and engineering literature since it simplifies time-domain analysis. However, in this book we will work primarily in the frequency domain, where the analytic-signal representation is not needed.

2.1.2 The Helmholtz equation

Fourier transformation of the wave equation Eq. (1.1) yields the *reduced scalar wave equation* (scalar Helmholtz equation)

$$[\nabla^2 + k^2]U(\mathbf{r}, \omega) = Q(\mathbf{r}, \omega), \tag{2.3}$$

where $k = \omega/c$ is the *wavenumber*. Although we have derived the Helmholtz equation directly from the wave equation, this equation is, as we mentioned above, much more general than the wave equation and should be considered in its own right rather than simply as the frequency-domain counterpart of the wave equation. For example, although the wave equation applies only to a non-dispersive and non-attenuating background medium, the Helmholtz equation Eq. (2.3) describes the radiation of waves in a dispersive medium characterized by an index of refraction $n(\omega)$ and wavenumber $k(\omega) = (\omega/c)n(\omega)$. *We thus take the Helmholtz equation to be the fundamental equation governing radiation and wave propagation in a dispersive medium characterized by a complex background wavenumber* $k(\omega)$.

The index of refraction $n(\omega)$ will be complex and analytic in the u.h.p. due to the required causality of the medium but will have singularities in the form of branch points in the l.h.p. It then follows that the particular solution U_+ to the Helmholtz equation that corresponds to a causal time-dependent field $u_+(\mathbf{r}, t)$ will also be analytic throughout the upper half of the complex-ω plane. Moreover, conservation-of-energy arguments also require that the imaginary part of the index of refraction $\Im n(\omega)$ be non-negative, corresponding to the requirement that the background medium can only absorb energy from a propagating wave rather than add energy to such a wave. Thus, we will assume throughout this and following chapters that $\Im k(\omega) = \Im[(\omega/c)n(\omega)] \geq 0$, at least along the real-ω axis and throughout the upper half of the complex-ω plane.

2.1.3 Lorentz dispersive medium

A *Lorentz model* for the complex index of refraction of a dispersive medium is often used in theoretical studies of wave propagation and scattering in such media. The simplest example of such a model is provided by the equation (Oughstun and Sherman, 1997)

$$n(\omega) = \left(1 - \frac{b^2}{\omega^2 - \omega_0^2 + 2\delta i\omega}\right)^{1/2},$$

where ω_0 is the resonance frequency of the material and δ and b are positive real constants that are specific to the particular medium that is being modeled. We note from the above equation that, when ω is real, $n(-\omega) = n^*(\omega)$, which is a general feature of the complex index of refraction of dispersive media. The above equation can be rewritten in the form

$$n(\omega) = \left(\frac{(\omega - \omega'_+)(\omega - \omega'_-)}{(\omega - \omega_+)(\omega - \omega_-)}\right)^{1/2}, \tag{2.4}$$

where

$$\omega'_{\pm} = \pm\sqrt{\omega_1^2 - \delta^2} - \delta i,$$
$$\omega_{\pm} = \pm\sqrt{\omega_0^2 - \delta^2} - \delta i,$$

with $\omega_1^2 = \omega_0^2 + b^2$. It is clear from Eq. (2.4) that the index of refraction is free of singularities in the upper half of the complex-ω plane but has singularities in the form of four branch points at $\omega'_{\pm}$ and $\omega_{\pm}$ in the lower half of the complex-ω plane. From the definitions of these quantities it can be seen that these branch points are symmetrically placed along a line parallel to the real-ω axis in the lower half of the complex-ω plane as illustrated in Fig. 2.1. The branch cuts can then be selected as indicated in that figure. We will use this simple Lorentz model later in our discussion of the time-domain Green function of a dispersive medium in Section 2.3.1.

We will also have cause to employ the analytic continuation of the complex conjugate of $n(\omega)$ from its boundary value on the real-ω axis. This quantity is algebraically given by $n^*(\omega^*)$, so it follows from Eq. (2.4) that for Lorentz media this quantity is given by

$$n^*(\omega^*) = \left(\frac{(\omega - \omega'^*_{+})(\omega - \omega'^*_{-})}{(\omega - \omega^*_{+})(\omega - \omega^*_{-})}\right)^{1/2}.$$

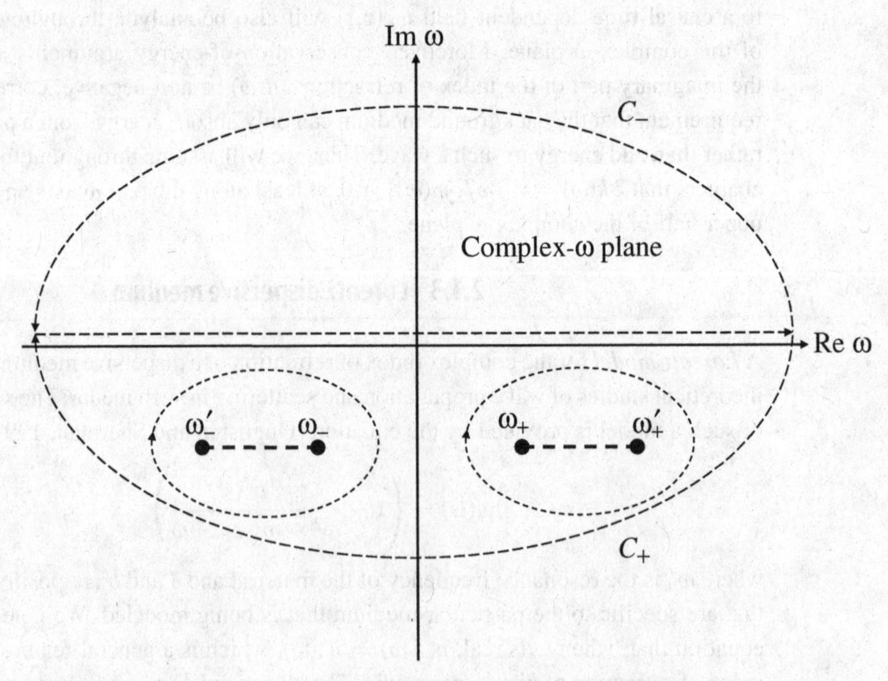

Fig. 2.1 The distribution of singularities of the complex index of refraction for a Lorentz medium having a single resonance. $n(\omega)$ is free of singularities in the u.h.p. but has four branch points in the l.h.p. Integration around the contour C_- yields zero, while integration around C_+ yields contributions from the four branch points.

This quantity defines an analytic function of ω over the entire complex-ω plane. This continuation is seen to be free of singularities in the lower half of the complex-ω plane but has singularities in the form of four branch points at $\omega_\pm'^*$ and $\omega_\pm^*$ in the upper half of the complex-ω plane.

Example 2.1 It is easy to show using the Cauchy–Riemann (CR) equations satisfied by $n(\omega)$ that the continuation of its complex conjugate into the complex-ω plane from its boundary value on the real-ω axis is analytic. On setting $n(\omega) = n_\mathrm{r}(\omega_\mathrm{r}, \omega_\mathrm{i}) + in_\mathrm{i}(\omega_\mathrm{r}, \omega_\mathrm{i})$, with $\omega = \omega_\mathrm{r} + i\omega_\mathrm{i}$ and n_r and n_i the real and imaginary components of $n(\omega)$, the CR equations satisfied by $n(\omega)$ are given by

$$\frac{\partial n_\mathrm{r}(\omega_\mathrm{r}, \omega_\mathrm{i})}{\partial \omega_\mathrm{r}} = \frac{\partial n_\mathrm{i}(\omega_\mathrm{r}, \omega_\mathrm{i})}{\partial \omega_\mathrm{i}}, \qquad \frac{\partial n_\mathrm{r}(\omega_\mathrm{r}, \omega_\mathrm{i})}{\partial \omega_\mathrm{i}} = -\frac{\partial n_\mathrm{i}(\omega_\mathrm{r}, \omega_\mathrm{i})}{\partial \omega_\mathrm{r}}. \tag{2.5}$$

The continuation of $n^*(\omega)$ from its boundary value on the real-ω axis is equal to $n^*(\omega^*) = n_\mathrm{r}(\omega_\mathrm{r}, -\omega_\mathrm{i}) - in_\mathrm{i}(\omega_\mathrm{r}, -\omega_\mathrm{i})$. If we thus write this continued function in the form $n^*(\omega^*) = u(\omega_\mathrm{r}, \omega_\mathrm{i}) + iv(\omega_\mathrm{r}, \omega_\mathrm{i})$ we have that

$$u(\omega_\mathrm{r}, \omega_\mathrm{i}) = n_\mathrm{r}(\omega_\mathrm{r}, -\omega_\mathrm{i}), \qquad v(\omega_\mathrm{r}, \omega_\mathrm{i}) = -n_\mathrm{i}(\omega_\mathrm{r}, -\omega_\mathrm{i}). \tag{2.6}$$

We will now show that the real and imaginary parts u and v of $n^*(\omega^*)$ satisfy the CR equations. By making use of the definitions of u and v given in Eqs. (2.6) and the first CR equation for $n(\omega)$ given in Eqs. (2.5) we find that

$$\frac{\partial u(\omega_\mathrm{r}, \omega_\mathrm{i})}{\partial \omega_\mathrm{r}} = \frac{\partial n_\mathrm{r}(\omega_\mathrm{r}, -\omega_\mathrm{i})}{\partial \omega_\mathrm{r}} = \frac{\partial n_\mathrm{i}(\omega_\mathrm{r}, -\omega_\mathrm{i})}{\partial \omega_\mathrm{i}} = \frac{\partial v(\omega_\mathrm{r}, \omega_\mathrm{i})}{\partial \omega_\mathrm{i}},$$

which is the first CR equation for $n^*(\omega^*)$. Similarly, we find that

$$\frac{\partial u(\omega_\mathrm{r}, \omega_\mathrm{i})}{\partial \omega_\mathrm{i}} = \frac{\partial n_\mathrm{r}(\omega_\mathrm{r}, -\omega_\mathrm{i})}{\partial \omega_\mathrm{i}} = \frac{\partial n_\mathrm{i}(\omega_\mathrm{r}, -\omega_\mathrm{i})}{\partial \omega_\mathrm{r}} = -\frac{\partial v(\omega_\mathrm{r}, \omega_\mathrm{i})}{\partial \omega_\mathrm{r}},$$

which is the second CR equation for $n^*(\omega^*)$.

2.1.4 The Sommerfeld radiation condition in dispersive media

As was the case for the inhomogeneous wave equation Eq. (1.1), the inhomogeneous Helmholtz equation does not possess a unique solution and, in particular, given any solution $U(\mathbf{r}, \omega)$ we can obtain a new solution by adding any field $\delta U(\mathbf{r}, \omega)$ that satisfies the *homogeneous Helmholtz equation*

$$[\nabla^2 + k^2]\delta U(\mathbf{r}, \omega) = 0. \tag{2.7}$$

The different solutions obtained in this way will satisfy the same defining equation Eq. (2.3) but different *boundary conditions*. The choice of boundary conditions is dictated by the physics of the problem at hand, and for the case of the radiation problem in an infinite homogeneous background medium the appropriate boundary condition is the *Sommerfeld radiation condition* (SRC) (Sommerfeld, 1967). The SRC was discussed at some length in Chapter 1 and, as shown in that chapter, is equivalent to the requirement that the associated time-domain field and Green function be causal.

The SRC can be stated in either of the two forms (cf. Eqs. (1.45a) and (1.48))

$$\lim_{r\to\infty} r\left[\frac{\partial U_+(\mathbf{r},\omega)}{\partial r} - ikU_+(\mathbf{r},\omega)\right] \to 0, \tag{2.8a}$$

$$U_+(\mathbf{r},\omega) \sim f(\mathbf{s},\omega)\frac{e^{ikr}}{r}, \tag{2.8b}$$

where $\mathbf{s} = \mathbf{r}/r$ is the unit vector along the $\mathbf{r}$ direction and, as usual, we have used the subscript $+$ to denote the field that satisfies the SRC. The function $f(\mathbf{s},\omega)$, called the *radiation pattern* and previously introduced in Section 1.5.1, plays an important role in a host of inverse problems, as we will see in later chapters. The approximation of U by the first term in Eq. (2.8b) is generally referred to as the *far-field approximation*. We will employ both forms of the SRC in this chapter.

In the traditional form of the SRC the wavenumber $k = \omega/c$ is a real-valued quantity and the SRC defines an *outgoing-wave field*. In particular, if we substitute the SRC in the form of Eq. (2.8b) into Eq. (2.2) we conclude that the time-dependent radiated field behaves asymptotically like

$$u_+(\mathbf{r},t) \sim \frac{1}{2\pi}\int_{-\infty}^{\infty} d\omega f(\mathbf{s},\omega)\frac{e^{-i\omega(t-r/c)}}{r} = \frac{F_+(\mathbf{s},t-r/c)}{r}, \tag{2.9a}$$

where

$$F_+(\mathbf{s},t) = \frac{1}{2\pi}\int_{-\infty}^{\infty} d\omega f(\mathbf{s},\omega)e^{-i\omega t} \tag{2.9b}$$

is the so-called *time-domain radiation pattern* introduced in Section 1.5.1 of Chapter 1. It is clear from Eq. (2.9a) that u_+ behaves asymptotically as an outward expanding spherical wave; i.e., wavefronts defined by $t - r/c =$ constant are spherical and expand outward from the origin as time t increases.

In the case of dispersive media where k is complex with a positive imaginary part the SRC is still equivalent to causality in the time domain and still requires that the radiated field behave as an outgoing spherical wave as the field point $\mathbf{r}$ tends to infinity, but this condition also requires that its amplitude decay *exponentially* with increasing distance r from the origin. This, of course, is a requirement that is imposed by the fact that $\Im k > 0$, corresponding to an absorbing background medium that attenuates any propagating wave.

2.1.5 Incoming- and conjugate-wave radiation conditions

As discussed above, the SRC is sometimes referred to as the "outgoing-wave radiation condition" due to the fact that the associated time-domain wavefield propagates "outward" from the source region into the surrounding space; i.e., it is causally "created" by the source. It is also possible to construct wavefields that propagate "inward" from space toward their "source" region. Such wavefields satisfy a far-field condition that is completely analogous to the SRC. This condition, called the *incoming-wave radiation condition*, can be stated in either of the two forms

$$\lim_{r\to\infty} r\left[\frac{\partial U_-(\mathbf{r},\omega)}{\partial r} + ikU_-(\mathbf{r},\omega)\right] \to 0, \tag{2.10a}$$

$$U_-(\mathbf{r},\omega) \sim f_-(\mathbf{s},\omega)\frac{e^{-ikr}}{r}, \tag{2.10b}$$

where the function $f_-(\mathbf{s},\omega)$ is the incoming-wave radiation-condition counterpart to the usual outgoing-wave radiation pattern $f(\mathbf{s},\omega)$. The incoming-wave radiation condition when employed as a boundary condition to the inhomogeneous Helmholtz equation generates a unique solution, which, in the case of non-dispersive media where $k = \omega/c$, yields an acausal time-domain wavefield $u_-(\mathbf{r},t)$ (see the discussion below).

In a realizable dispersive medium where $\Im k(\omega) > 0$ the incoming wavefield will grow exponentially fast with increasing distance r from the source region. Because of this it is desirable in dispersive media to sometimes employ an incoming wave that decays rather than grows with increasing distance r from the source but reduces to the usual incoming wave when the medium is non-dispersive and k is real-valued. The required wavefield is obtained by noting that if we replace the wavenumber k in the Helmholtz equation by its complex conjugate k^* then the incoming-wave solution of this new Helmholtz equation is perfectly stable and reduces to the incoming-wave solution of the usual Helmholtz equation in the special case in which k is real-valued. We will name this class of wavefields *conjugate waves* and define them to be solutions to the inhomogeneous Helmholtz equation *with wavenumber* k^* that satisfy the *conjugate-wave radiation condition*

$$\lim_{r\to\infty} r\left[\frac{\partial U(\mathbf{r},\omega)}{\partial r} + ik^*U(\mathbf{r},\omega)\right] \to 0, \tag{2.11a}$$

$$U(\mathbf{r},\omega) \sim f_c(\mathbf{s},\omega)\frac{e^{-ik^*r}}{r}, \tag{2.11b}$$

where f_c is the "conjugate-wave" radiation pattern.

Conjugate waves and time reversal

For real-valued functions of time t the process of time reversal whereby $t \to -t$ corresponds, in the frequency domain, to the complex-conjugation operation. It then follows that the time-reversed field $u_+(\mathbf{r},-t)$ has a Fourier transform given by $U_+^*(\mathbf{r},\omega)$. When the wavenumber k is real we have that

$$[\nabla^2 + k^2]U_+^*(\mathbf{r},\omega) = Q^*(\mathbf{r},\omega), \tag{2.12a}$$

and the complex conjugate of the SRC yields

$$\lim_{r\to\infty} r\left[\frac{\partial U_+^*(\mathbf{r},\omega)}{\partial r} + ikU_+^*(\mathbf{r},\omega)\right] = 0. \tag{2.12b}$$

The time-reversed field is thus obtained as the incoming-wave solution to the Helmholtz equation with a time-reversed source term. Since the time-reversed field $u_+(\mathbf{r}, -t)$ is *acausal* it then follows that in the case of real wavenumbers (non-dispersive media) *the incoming-wave radiation condition in the frequency domain implies acausality in the time domain*.

More generally, in dispersive media where k is a complex function of ω the above conclusions are no longer true and the time-reversed field is not obtained from the incoming-wave solution to the Helmholtz equation Eq. (2.12a). In particular, in the case of dispersive media the complex conjugate of the Helmholtz equation yields the result[1]

$$[\nabla^2 + k^{*2}]U_+^*(\mathbf{r},\omega) = Q^*(\mathbf{r},\omega), \tag{2.12c}$$

and the complex conjugate of the SRC becomes the conjugate-wave radiation condition defined in Eqs. (2.11). In this more general case we conclude that the Fourier transform $U_+^*(\mathbf{r},\omega)$ of the acausal, time-reversed field $u_+(\mathbf{r}, -t)$ *is the incoming-wave solution to the Helmholtz equation for a medium having k^* as its wavenumber rather than for the original medium in which the original field was radiated.* In other words, in the frequency domain, the time-reversed field $U_+^*(\mathbf{r},\omega)$ is the conjugate-wave field as we have defined it above.

These conclusions regarding the conjugate wave and time reversal in a dispersive medium make perfect sense when we interpret the transformation $k(\omega) \to k^*(\omega)$. In particular, this operation corresponds to *time reversing the material in which the wave propagates.* Thus, to obtain true time reversal in a dispersive medium it is necessary to time reverse the source $Q(\mathbf{r},\omega) \to Q^*(\mathbf{r},\omega)$, the medium $k(\omega) \to k^*(\omega)$ and the SRC.

2.2 Green functions

We define a *Green function* to the Helmholtz equation to be any solution to the partial differential equation

$$[\nabla^2 + k^2]G(\mathbf{r},\mathbf{r}',\omega) = \delta(\mathbf{r}-\mathbf{r}') \tag{2.13}$$

within some region τ, possibly infinite, bounded by a surface $\partial\tau$. A Green function can be interpreted as being the field generated by an impulsive source located at the space point $\mathbf{r}'$. As is the case for any solution to the inhomogeneous Helmholtz equation, a Green function is non-unique and it is necessary to append appropriate boundary conditions over $\partial\tau$ in order to obtain a unique Green function. In the case of the radiation problem where we

[1] Here, and throughout this chapter, the quantity $k^*(\omega)$ is taken to mean the analytic continuation of this quantity from its values along the real-ω axis (cf. the discussion in Section 2.1.3).

seek a causal solution to the associated time-dependent problem in an infinite medium the physically appropriate boundary condition is that G satisfy the SRC. The SRC is equivalent to the requirement of causality in the time domain and the resulting Green function is the *outgoing-wave Green function*

$$G_+(\mathbf{R}, \omega) = -\frac{1}{4\pi}\frac{e^{ikR}}{R}, \tag{2.14}$$

derived by Fourier transformation of the causal time-domain Green function[2] (retarded Green function) $g_+(\mathbf{r}-\mathbf{r}', t-t')$ of the wave equation in Section 1.2.2 of Chapter 1. The outgoing-wave Green function can also be obtained directly from its Fourier-integral representation

$$G_+(\mathbf{R}, \omega) = \frac{1}{(2\pi)^3}\int d^3K \frac{e^{i\mathbf{K}\cdot\mathbf{R}}}{k^2 - K^2}, \tag{2.15}$$

where the condition $\Im k > 0$ guarantees that G_+ will satisfy the SRC. The incoming-wave Green function G_- is obtained from G_+ by simply replacing k by $-k$, while the conjugate-wave Green function is obtained simply by taking the complex conjugate of G_+. The incoming-wave Green function G_- is a solution to the Helmholtz equation Eq. (2.13) while the conjugate-wave Green function is a solution to the equation

$$[\nabla^2 + k^{*2}]G^*(\mathbf{r}, \mathbf{r}', \omega) = \delta(\mathbf{r}-\mathbf{r}') \tag{2.16}$$

and, hence, is actually a Green function to the Helmholtz equation with wavenumber k^* rather than with wavenumber k. We will show below that the time-domain conjugate-wave Green function is acausal both for dispersive and for non-dispersive media, while the time-domain incoming-wave Green function is acausal only for non-dispersive media.

Example 2.2 Consider the one-dimensional Helmholtz equation

$$\left[\frac{\partial^2}{\partial z^2} + k^2\right] G(z, \omega) = \delta(z), \tag{2.17}$$

with the wavenumber k strictly real-valued. Equation (2.17) is easily converted to Fourier space and we obtain the result

$$G(z, \omega) = \frac{1}{2\pi}\int_{-\infty}^{\infty} dK \frac{e^{iKz}}{-K^2 + k^2}. \tag{2.18}$$

The poles in the integrand of Eq. (2.18) occur at $K = \pm k$ and, thus, lie on the real-K axis when the wavenumber is real. The outgoing- and incoming-wave Green functions are obtained as above by selecting the contour to lie above the pole at $K = -k$ and below the pole at $K = +k$ (outgoing-wave Green function) or below the pole at $K = -k$ and above the pole at $K = +k$ (incoming-wave Green function). However, it is also possible to use contours that pass either above both poles or below both poles. By making use of

[2] Although the derivation of G_+ obtained in Chapter 1 was for the wave equation, the functional form as given in Eq. (2.14) remains valid in dispersive media and can be directly derived from the Fourier-integral representation given in Eq. (2.15).

Cauchy's residue theorem the Green functions corresponding to these various contours are easily computed and we obtain

$$G_+(z,\omega) = -\frac{i}{2k}e^{ik|z|},$$

$$G_-(z,\omega) = +\frac{i}{2k}e^{-ik|z|},$$

$$G_{++}(z,\omega) = \begin{cases} 0 & \text{if } z > 0, \\ -[i/(2k)][e^{-ikz} - e^{ikz}] & \text{if } z < 0, \end{cases}$$

$$G_{--}(z,\omega) = \begin{cases} +[i/(2k)][e^{-ikz} - e^{ikz}] & \text{if } z > 0, \\ 0 & \text{if } z < 0. \end{cases}$$

The Green functions G_+ and G_- are the 1D forms of the 3D incoming- and outgoing-wave Green functions, while the Green functions G_{++} and G_{--} are linear combinations of G_+ and G_- of solutions to the 1D homogeneous Helmholtz equation.

It is instructive to verify by direct differentiation that the above Green functions all satisfy the defining equation Eq. (2.17). For example, we can express G_{--} in the form

$$G_{--} = \frac{\sin(kz)}{k}\Theta(z),$$

where $\Theta(z)$ is the step function. We then find that

$$\frac{\partial}{\partial z}G_{--} = \frac{1}{k}[k\cos(kz)\Theta(z) + \sin(kz)\delta(z)]$$

$$\frac{\partial^2}{\partial z^2}G_{--} = \frac{1}{k}\left[-k^2\sin(kz)\Theta(z) + 2k\cos(kz)\delta(z) + \sin(kz)\frac{\partial}{\partial z}\delta(z)\right],$$

where $\delta(z)$ is the delta function. Using the above results, we conclude that

$$\left[\frac{\partial^2}{\partial z^2} + k^2\right]G_{--}(z,\omega) = 2\cos(kz)\delta(z) + \frac{\sin(kz)}{k}\frac{\partial}{\partial z}\delta(z) = \delta(z),$$

where we have used the results that

$$f(z)\delta(z) = f(0)\delta(z),$$

$$f(z)\frac{\partial}{\partial z}\delta(z) = -\frac{\partial}{\partial z}f(z)\delta(z) = -\frac{\partial}{\partial z}f(z)|_{z=0}\delta(z)$$

for any continuous and differentiable function $f(z)$.

2.2.1 Green functions in two space dimensions

The outgoing-wave Green function in two space dimensions is derived directly from the 2D Fourier-integral representation

$$G_+(\mathbf{R},\omega) = \frac{1}{(2\pi)^2}\int d^2K\,\frac{e^{i\mathbf{K}\cdot\mathbf{R}}}{k^2 - K^2},$$

where $\mathbf{K}$ and $\mathbf{R}$ are now 2D vectors and $\Im k > 0$ corresponding to the requirement of causality in the time domain. The incoming-wave Green function G_- is obtained from G_+ by making the transformation $k \to -k$ or, alternatively, from G_+^* under the transformation $k^* \to k$.

The actual computation of the 2D Green functions is left as a problem at the end of the chapter. One finds that

$$G_+(\mathbf{R}, \omega) = \frac{-i}{4} H_0^+(kR), \tag{2.19a}$$

where $H_0^+(\cdot)$ is the zeroth-order Hankel function of the first kind. The conjugate-wave Green function is found by taking the complex conjugate of Eq. (2.19a). We obtain

$$G_+^*(\mathbf{R}, \omega) = \frac{i}{4} H_0^-(k^* R),$$

where we have used the relationship

$$H_0^{+*}(kR) = H_0^-(k^* R).$$

The incoming-wave Green function is obtained from G_+ by replacing k by $-k$ or from G_+^* by replacing k^* by k. Either method results in

$$G_-(\mathbf{R}, \omega) = \frac{i}{4} H_0^-(kR). \tag{2.19b}$$

The 2D Green functions are important since they reduce the size and complexity of simulation studies in inverse scattering and diffraction tomography, and will be employed extensively later in the book in such studies.

2.3 Time-domain Green functions

The time-domain Green functions are obtained from the frequency-domain quantities by making use of the inverse Fourier transform Eq. (2.2). Corresponding to the outgoing- and incoming-wave Green functions we obtain[3]

$$g_+(\mathbf{R}, \tau) = \frac{1}{2\pi} \int_{-\infty}^{\infty} d\omega \, \frac{e^{ik(\omega)R}}{-4\pi R} e^{-i\omega\tau}, \tag{2.20a}$$

$$g_-(\mathbf{R}, \tau) = \frac{1}{2\pi} \int_{-\infty}^{\infty} d\omega \, \frac{e^{-ik(\omega)R}}{-4\pi R} e^{-i\omega\tau}, \tag{2.20b}$$

where we have denoted the time-domain outgoing- and incoming-wave Green function by g_+ and g_-, respectively. The time-domain conjugate-wave Green function is, of course, simply the time-reversed version of g_+ and is thus given by $g_+(\mathbf{R}, -\tau)$, a result that follows directly from Eq. (2.20a) on making use of the fact that $k(-\omega) = -k^*(\omega)$ when ω is real-valued. In the case of non-dispersive media $k(\omega) = \omega/c$ and Eq. (2.20a) then yields the

[3] As in Chapter 1 we use the Greek symbol τ here to denote a time difference. This symbol is also used to denote regions of space, but its meaning should be clear from the context in which it is employed.

retarded Green function, while Eq. (2.20b) yields the advanced Green function that was derived in Section 1.2.1 of Chapter 1; i.e.,

$$g_{\pm}(\mathbf{R}, \tau) = -\frac{1}{4\pi}\frac{\delta(\tau \mp R/c)}{R},$$

where $\delta(\cdot)$ is the 1D Dirac delta function.

As discussed in Section 1.2.1 and as is apparent from their definition, the retarded and advanced Green functions vanish off the *light-cone* $\tau^2 = R^2/c^2$, with g_+ vanishing unless $\tau = R/c$ and g_- vanishing unless $\tau = -R/c$. Setting $\tau = t - t'$ and $\mathbf{R} = \mathbf{r} - \mathbf{r}'$, we then conclude that these two Green functions correspond to propagating pulses that are emitted at the space-time point $\mathbf{r}', t'$ and are observed at the space point $\mathbf{r}$ at the *retarded and advanced times* $t = t_{\pm}$ defined according to the equation

$$t_{\pm} = t' \pm \frac{|\mathbf{r} - \mathbf{r}'|}{c}.$$

In the case of dispersive media the wavenumber depends on frequency and the integrals in Eqs. (2.20) cannot, in general, be evaluated in closed form. Approximations to the time-domain Green functions that are valid in certain asymptotic regimes can, however, be obtained by making use of the method of steepest descents as first developed by Sommerfeld and Brillouin (Stratton, 1941; Oughstun and Sherman, 1997; Oughstun, 2006). In this book we will not delve into these methods, since most of our work is performed in the frequency domain where the various inverse problems that we will treat are best formulated and solved. However, it is useful to review a few key features of the inversion integrals Eqs. (2.20) that will be of use later in the present chapter.

2.3.1 Key features of the time-domain Green functions

A physically realizable dispersive medium is characterized by a complex wavenumber of the form $k(\omega) = (\omega/c)n(\omega)$, where the complex index of refraction is analytic with no singularities in the upper half of the complex-ω plane and, in addition, satisfies the condition

$$\lim_{|\omega|\to\infty} n(\omega) = 1. \tag{2.21}$$

These two properties allow the ω contour of integration in Eq. (2.20a) to be closed in the upper half of the complex-ω plane if $R/c - \tau > 0$ and that of Eq. (2.20b) to be closed in this half-plane if $-R/c - \tau > 0$. Since the integrands are analytic with no singularities throughout the u.h.p., we conclude from Cauchy's integral theorem that $g_+(\mathbf{r} - \mathbf{r}', t - t')$ will be causal relative to the retarded time t_+ and $g_-(\mathbf{r} - \mathbf{r}', t - t')$ will be causal relative to the advanced time t_-; i.e.,

$$\begin{aligned} g_+(\mathbf{r} - \mathbf{r}', t - t') &= 0, \quad t < t' + \frac{|\mathbf{r} - \mathbf{r}'|}{c}, \\ g_-(\mathbf{r} - \mathbf{r}', t - t') &= 0, \quad t < t' - \frac{|\mathbf{r} - \mathbf{r}'|}{c}. \end{aligned} \tag{2.22}$$

The time-dependent conjugate-wave Green function $g_+(\mathbf{R}, -\tau)$ will vanish if $R/c+\tau > 0$, which corresponds to the time t being greater than the advanced time $t_- = t' - R/c$. This time-domain Green function is thus acausal relative to the advanced time.

If $R/c - \tau < 0 \to t > t_+$ it is not possible to close the ω contour of integration in Eq. (2.20a) in the u.h.p. and this integral will not vanish. Similarly, if $-R/c - \tau < 0 \to t > t_-$ it is not possible to close the ω contour of integration in Eq. (2.20b) in the u.h.p. and this integral also will not vanish. It is possible, however, to close the contours of integration of these two integrals in the l.h.p., where the wavenumber $k(\omega)$ has singularities in the form of branch points. The integral around the branch lines connecting the various branch points will contribute to the time-domain Green functions so that $g_+(\mathbf{r}-\mathbf{r}', t-t')$ will be a pulse that begins oscillating at the retarded time $t = t_+$ and continues, in principle, to infinity, while the incoming-wave Green function $g_-(\mathbf{r}-\mathbf{r}', t-t')$ will be a pulse that begins oscillating at the advanced time $t = t_-$ and also continues ideally to $t = \infty$. The time-reversed Green function will be a pulse that begins oscillating at the advanced time $t = t_-$ and terminates at $t = -\infty$.

Example 2.3 We consider the case of a *Lorentz medium* characterized by the index of refraction given in Eq. (2.4) of Section 2.1.3. It is clear from this model that $n(\omega)$ satisfies Eq. (2.21) and has no singularities in the upper half of the complex-ω plane but has singularities in the form of four branch points in the lower half of this plane as illustrated in Fig. 2.1. It then follows from the discussion presented above that the ω contour of integration in the integral Eq. (2.20a) can be closed in the u.h.p. using the contour labeled C_- in Fig. 2.1 for g_+ if $\tau < R/c$, thus leading to the conclusion reached above that g_+ is causal relative to the retarded time t_+. If, on the other hand, $\tau > R/c$ the ω contour must be closed in the l.h.p. using the contour C_+ and will yield contributions from the integrations around the branch cuts as illustrated in Fig. 2.1. These integrations can only be evaluated approximately using the method of steepest descents and lead to the famous Sommerfeld and Brillouin *precursors*.

2.4 Green-function solution of the radiation problem

The solution to the radiation problem in the frequency domain for non-dispersive media governed by the wave equation was found in Section 1.5 of Chapter 1 to be given by

$$U_+(\mathbf{r}, \omega) = \int_{\tau_0} d^3r' \, G_+(\mathbf{r}-\mathbf{r}', \omega) Q(\mathbf{r}', \omega), \tag{2.23}$$

where $k = \omega/c$. This result applies also to a dispersive medium, where now $k = k(\omega)$ in the outgoing-wave Green function G_+. Since the solutions to the radiation problem both for dispersive and for non-dispersive media are formally identical in the frequency domain all of the results obtained in that section carry over to the case of dispersive media and we find that

$$U_+(\mathbf{r},\omega) \sim f(\mathbf{s},\omega)\frac{e^{ikr}}{r} \quad \text{as} \quad r \to \infty, \tag{2.24a}$$

where

$$f(\mathbf{s},\omega) = -\frac{1}{4\pi}\int_{\tau_0} d^3r'\, Q(\mathbf{r}',\omega)e^{-ik\mathbf{s}\cdot\mathbf{r}'} \tag{2.24b}$$

is the *radiation pattern* of the source. The only difference between the results established earlier in Section 1.5 and here is that now the wavenumber k is complex with a positive imaginary part and will be an analytic function of the temporal frequency ω. Because of the complex nature of k the *far-field* expression given in Eq. (2.24a) will decay exponentially with increasing distance r from the source region. As was pointed out in Section 1.5.1, the expression for the radiation pattern given in Eq. (2.24b) identifies this quantity as being proportional to the spatial Fourier transform of the source,

$$\tilde{Q}(\mathbf{K},\omega) = \int_{\tau_0} d^3r'\, Q(\mathbf{r}',\omega)e^{-i\mathbf{K}\cdot\mathbf{r}'},$$

evaluated on the surface of the (generally complex) sphere $\mathbf{K} = k\mathbf{s}$; i.e.,

$$f(\mathbf{s},\omega) = -\frac{1}{4\pi}\tilde{Q}(\mathbf{K},\omega)|_{\mathbf{K}=k\mathbf{s}}. \tag{2.25}$$

The surface $\mathbf{K} = k\mathbf{s}$ is a special case of the more general *Ewald sphere* that plays a fundamental and dominant role in the inverse scattering theory treated later in the book. Again, as pointed out earlier in our treatment of the radiation problem in non-dispersive media, the source spatial Fourier transform $\tilde{Q}(\mathbf{K},\omega)$ is an entire analytic function of the three Cartesian components of the $\mathbf{K}$ vector so that its boundary value $\tilde{Q}(k\mathbf{s},\omega)$ and, hence, the radiation pattern $f(\mathbf{s},\omega)$ are entire analytic functions of the unit vector $\mathbf{s}$; e.g., of the direction cosines of this quantity. We emphasize that these quantities are analytic functions of $\mathbf{s}$ over the entire *complex* unit sphere; i.e., for all real and complex values of $\mathbf{s}$ such that $\mathbf{s}\cdot\mathbf{s} = 1$, and their analyticity depends on the source having a compact support volume τ_0.

Example 2.4 As an example of Eq. (2.25) consider the source

$$Q(\mathbf{r},\omega) = \begin{cases} g(r,\omega)Y_l^m(\hat{\mathbf{r}}) & \text{if } r \le a_0, \\ 0 & \text{else,} \end{cases}$$

where $g(r,\omega)$ is an arbitrary bounded function of frequency and the radial coordinate r, and $Y_l^m(\hat{\mathbf{r}}) = Y_l^m(\theta,\phi)$ are the spherical harmonics defined in Section 3.3 of Chapter 3, which are functions of the polar and azimuthal angles θ and ϕ of the unit vector $\hat{\mathbf{r}}$. The spatial Fourier transform of this source is given by

$$\tilde{Q}(\mathbf{K},\omega) = \int_0^{a_0} r^2\,dr\,g(r,\omega)\int_{4\pi} d\Omega\,Y_l^m(\hat{\mathbf{r}})e^{-i\mathbf{K}\cdot\mathbf{r}}$$
$$= 4\pi(-i)^l Y_l^m(\hat{\mathbf{K}})\int_0^{a_0} r^2\,dr\,g(r,\omega)j_l(Kr) = G(K,\omega)Y_l^m(\hat{\mathbf{K}}), \qquad (2.26)$$

where

$$G(K,\omega) = 4\pi(-i)^l\int_0^{a_0} r^2\,dr\,g(r,\omega)j_l(Kr)$$

and we have made use of the result (see Example 3.4 of Chapter 3)

$$j_l(Kr)Y_l^m(\hat{\mathbf{K}}) = \frac{i^l}{4\pi}\int_{4\pi} d\Omega\,Y_l^m(\hat{\mathbf{r}})e^{i\mathbf{K}\cdot\mathbf{r}}.$$

The radiation pattern of the source is found from Eq. (2.26) to be given by

$$f(\mathbf{s},\omega) = -\frac{1}{4\pi}G(k,\omega)Y_l^m(\mathbf{s}),$$

which is an entire analytic function of the polar and azimuthal angles of the unit vector $\mathbf{s}$.

Example 2.5 The sphere $\mathbf{K} = k\mathbf{s}$ with $\mathbf{s}\cdot\mathbf{s} = 1$ can be complex either due to dispersion where k is complex or because the components of the unit vector $\mathbf{s}$ are themselves complex. To see this we set

$$\mathbf{s} = \sin\alpha\cos\beta\,\hat{\mathbf{x}} + \sin\alpha\sin\beta\,\hat{\mathbf{y}} + \cos\alpha\,\hat{\mathbf{z}}, \qquad (2.27)$$

from which we conclude that

$$\mathbf{s}\cdot\mathbf{s} = \sin^2\alpha\cos^2\beta + \sin^2\alpha\sin^2\beta + \cos^2\alpha = 1$$

for all real and complex values of the polar α and azimuthal β angles defining the unit vector $\mathbf{s}$. The *complex* unit sphere is thus spanned by all unit vectors of the form given in Eq. (2.27) with arbitrary complex or real polar and azimuthal angles. We will employ plane waves having complex unit propagation vectors of this general type in later chapters in connection with plane-wave expansions of the field.

2.4.1 Solution of the radiation problem in two space dimensions

The above development can be repeated in two space dimensions, where $\mathbf{r}$ denotes position on the plane, τ and $\tau^\perp$ are now regions on this plane interior and exterior to the closed boundary $\partial\tau$ and G_+ is the 2D Green function defined in Eq. (2.19a). In place of Eq. (2.23) we now have

$$U_+(\mathbf{r},\omega) = \int_{\tau_0} d^2r'\,G_+(\mathbf{r}-\mathbf{r}',\omega)Q(\mathbf{r}',\omega), \qquad (2.28)$$

where G_+ is the 2D outgoing-wave Green function given in Eq. (2.19a) and $\tau_0 \subset \tau$ is a closed planar region. The field radiation pattern in the 2D case is obtained using the asymptotic expansion

$$H_0^+(kr) \sim \sqrt{\frac{2}{\pi kr}}e^{i(kr-\frac{\pi}{4})}, \quad kr\to\infty. \qquad (2.29)$$

On making use of Eq. (2.29) we find that

$$G_+(\mathbf{r}-\mathbf{r}',\omega) \sim -\frac{i}{4}\sqrt{\frac{2}{\pi k}}e^{-i\frac{\pi}{4}}e^{-ik\mathbf{s}\cdot\mathbf{r}'}\frac{e^{ikr}}{\sqrt{r}}$$

as $kr \to \infty$ along the direction of the unit vector $\mathbf{s}$. On substituting the above equation into Eq. (2.28) we obtain the result

$$U_+(\mathbf{r},\omega) \sim f(\mathbf{s},\omega)\frac{e^{ikr}}{\sqrt{r}},$$

where

$$f(\mathbf{s},\omega) = -\sqrt{\frac{1}{8\pi k}}e^{i\frac{\pi}{4}}\tilde{Q}(k\mathbf{s},\omega) \tag{2.30}$$

is the 2D radiation pattern and

$$\tilde{Q}(\mathbf{K},\omega) = \int_{\tau_0} d^2r'\, Q(\mathbf{r}',\omega)e^{-i\mathbf{K}\cdot\mathbf{r}'}$$

is the spatial Fourier transform of the 2D source.

All of our remarks made in connection with the solution of the radiation problem in three dimensions extend to the 2D case. In particular, the 2D spatial Fourier transform of a compactly supported source is an entire analytic function of $\mathbf{K}$ so that the radiation pattern is an analytic function of the unit vector $\mathbf{s}$, which now lies on the complex unit circle rather than on the complex unit sphere. We will make use of the 2D solution to the radiation problem in later chapters in computer simulations and examples since it requires much less CPU time and is easier to implement than the 3D solution but retains all of the salient features of the 3D case.

2.5 The Kirchhoff–Helmholtz representation of the radiated field

Fourier transformation of the time-domain Kirchhoff–Helmholtz equations obtained in Section 1.3.2 yields the frequency-domain versions

$$\int_{\partial\tau} dS'\left[G_+\frac{\partial}{\partial n'}U_+ - U_+\frac{\partial}{\partial n'}G_+\right] = U_+(\mathbf{r},\omega), \quad \mathbf{r}\in\tau^{\perp}, \tag{2.31a}$$

$$\int_{\partial\tau} dS'\left[G_+\frac{\partial}{\partial n'}U_+ - U_+\frac{\partial}{\partial n'}G_+\right] = 0, \quad \mathbf{r}\in\tau, \tag{2.31b}$$

which hold both for dispersive media and for non-dispersive media. In these equations $\tau \supset \tau_0$ and the normal derivatives are directed out of τ into the infinite exterior region $\tau^{\perp}$ bounded by $\partial\tau$ and a sphere at infinity.

As discussed in Section 1.3.2, Eqs. (2.31) are, together, a statement of the famed *Kirchhoff–Helmholtz theorem* and are often referred to as the *Helmholtz identities*. The first Helmholtz identity expresses the field throughout the exterior region $\tau^{\perp}$ in terms of the boundary value of the field and its normal derivative on the interior boundary $\partial\tau$ of $\tau^{\perp}$, while the second identity holds over the interior region τ. The second Helmholtz identity is, in fact, an integral equation that relates the boundary values of the field and its normal derivative over $\partial\tau$. As discussed in Chapter 1 the Kirchhoff–Helmholtz-type field representations are formal identities that must be satisfied by the field and boundary values, and are not properly posed solutions to a *boundary-value problem*, which will be covered in a later section.

The Kirchhoff–Helmholtz representation Eq. (2.31a) is not the solution of a *properly posed* boundary-value problem for the Helmholtz equation; it is, however, the solution to an over-specified boundary-value problem for this equation. In particular, it is easily verified that this representation satisfies the homogeneous Helmholtz equation throughout $\tau^{\perp}$ as well as the outgoing-wave radiation condition (the SRC) and possesses the limits

$$\lim_{\mathbf{r}\to\mathbf{r}_0}\int_{\partial\tau} dS'\left[G_+\frac{\partial}{\partial n'}U_+ - U_+\frac{\partial}{\partial n'}\right] = U_+(\mathbf{r}_0,\omega),$$

$$\lim_{\mathbf{r}\to\mathbf{r}_0}\frac{\partial}{\partial n}\int_{\partial\tau} dS'\left[G_+\frac{\partial}{\partial n'}U_+ - U_+\frac{\partial}{\partial n'}G_+\right] = \frac{\partial}{\partial n_0}U_+(\mathbf{r}_0,\omega),$$

where the limits are taken from the exterior region $\tau^{\perp}$ onto the boundary $\partial\tau$. The above limits are readily verified using a procedure almost identical to that employed in Section 1.8 of Chapter 1 to establish the same limits in the case of the Kirchhoff–Helmholtz representation for the wave equation (see also Section 2.12 later in this chapter). It thus follows that the first Helmholtz identity is, in fact, the unique solution to the over-specified boundary-value problem stated above.

2.5.1 The interior field solution and field back propagation

Fourier transformation of the interior field solutions for the wave equation given in Eqs. (1.37) of Section 1.3.3 yields the set of equations

$$\Phi(\mathbf{r},\omega) + \int_{\tau_0} d^3r'\, G_-(\mathbf{r}-\mathbf{r}',\omega)Q(\mathbf{r}',\omega) = U_+(\mathbf{r},\omega), \qquad \mathbf{r}\in\tau, \tag{2.32a}$$

$$\Phi(\mathbf{r},\omega) + \int_{\tau_0} d^3r'\, G_-(\mathbf{r}-\mathbf{r}',\omega)Q(\mathbf{r}',\omega) = 0, \qquad \mathbf{r}\in\tau^{\perp}, \tag{2.32b}$$

where, as before, $\tau^{\perp}$ denotes the complement to τ and

$$\Phi(\mathbf{r},\omega) = \int_{\partial\tau} dS'\left[U_+(\mathbf{r}',\omega)\frac{\partial}{\partial n'}G_-(\mathbf{r}-\mathbf{r}',\omega) - G_-(\mathbf{r}-\mathbf{r}',\omega)\frac{\partial}{\partial n'}U_+(\mathbf{r}',\omega)\right], \tag{2.33a}$$

where the normal derivatives are directed out of the interior τ and into the exterior $\tau^{\perp}$. Equations (2.32) and (2.33a) hold both for dispersive and for non-dispersive media and, hence, are the generalizations of the interior field solutions for the wave equation obtained in Section 1.3.3 for dispersive media. We identified the time-domain version of Φ in Section 1.4.2 of Chapter 1 to be the field *back propagated* from the boundary $\partial\tau$ and thus Eq. (2.33a) is the frequency-domain expression for the back-propagated field both in dispersive and in non-dispersive media.

The back-propagated field as defined in Eq. (2.33a) is not equal to the actual radiated field $U_+(\mathbf{r},\omega)$ but, as shown in Section 1.3.3, its time-domain equivalent $\phi(\mathbf{r},t)$ will, in non-dispersive media, equal the actual radiated field $u_+(\mathbf{r},t)$ for all times t exceeding the turn-off time T_0 of the source. Unfortunately, this will not be true in dispersive media due to the fact that, as shown in Section 2.3, the time-domain incoming-wave Green function $g_-(\mathbf{r}-\mathbf{r}',t-t')$ is not acausal, so there is no guarantee that the source term in the time-domain version of Eq. (2.32a) will vanish if $t > T_0$. However, we will show later that the frequency-domain back-propagated field defined according to Eq. (2.33a) both for dispersive and for non-dispersive media is an excellent approximation to the radiated field U_+ everywhere outside the source region τ_0 so long as the boundary $\partial\tau$ is many wavelengths removed from τ_0. For now we note that if we use the primary field solution on the r.h.s. of Eq. (2.32a) we obtain an integral equation that relates the source to the back-propagated field; i.e.,

$$\Phi(\mathbf{r},\omega) = \int_{\tau_0} d^3r'\, G_f(\mathbf{r}-\mathbf{r}',\omega)Q(\mathbf{r}',\omega), \tag{2.33b}$$

where

$$G_f(\mathbf{R},\omega) = G_+(\mathbf{R},\omega) - G_-(\mathbf{R},\omega),$$

is the frequency-domain free-field propagator whose time-domain counterpart

$$g_f(\mathbf{R},\tau) = g_+(\mathbf{R},\tau) - g_-(\mathbf{R},\tau)$$

was first encountered in Section 1.4 of the last chapter. Equation (2.33b) is known as the *Porter–Bojarski* (PB) integral equation (Bojarski, 1982a; Porter, 1970) and forms the basis for one approach for solving the *inverse source problem* (ISP).

2.6 Radiated power and energy

The power and energy radiated by a compactly supported source $q(\mathbf{r},t)$ in a dispersive medium can be computed using a similar procedure to that employed in Section 1.6 of Chapter 1. The radiated energy and energy spectrum are of most interest in the frequency domain and, by following lines of reasoning identical to those employed in Section 1.6, we find that the total energy radiated out of a surface $\partial\tau$ surrounding the source is given by

$$e_{\partial\tau} = \frac{1}{2\pi}\int_{-\infty}^{\infty} d\omega\, E_{\partial\tau}(\omega), \tag{2.34a}$$

where

$$E_{\partial\tau}(\omega) = 2\kappa\omega\Im \int_{\partial\tau} dS\, U_+^*(\mathbf{r},\omega)\frac{\partial U_+(\mathbf{r},\omega)}{\partial n} \tag{2.34b}$$

is the *energy spectrum* (energy per unit frequency) of the radiated field. In this equation κ is a real and positive constant that depends on the nature of the radiation (e.g., acoustic, optical, etc.).

An important and very useful property of the energy spectrum and radiated energy of the source in the case of the wave equation was that these quantities are independent of the surface $\partial\tau$ over which they are computed as long as the surface completely encloses the source region τ_0. This result was established in Theorem 1.1 in Section 1.6, which also gave a simple expression for the energy spectra in terms of the radiation pattern of the source. Unfortunately, these results and this theorem do not carry over in the general case of dispersive media. To see this, we consider two surfaces $\Sigma_<$ and $\Sigma_>$ that completely surround the source and such that $\Sigma_>$ lies entirely outside (i.e., encloses) $\Sigma_<$. Since these two surfaces both lie outside τ_0 we have that

$$[\nabla^2 + k^2]U_+(\mathbf{r},\omega) = 0, \qquad [\nabla^2 + k^{*2}]U_+^*(\mathbf{r},\omega) = 0$$

throughout the region $\delta\tau$ enclosed between these two surfaces. It then follows that

$$\int_{\delta\tau} d^3r\{U_+^*(\mathbf{r},\omega)\nabla^2 U_+(\mathbf{r},\omega) - U_+(\mathbf{r},\omega)\nabla^2 U_+^*(\mathbf{r},\omega)\}$$
$$- (k^2 - k^{*2})\int_{\delta\tau} d^3r|U_+(\mathbf{r},\omega)|^2 = 0.$$

On using Green's theorem we then obtain the result

$$\int_{\Sigma_>} dS\left[U_+^*(\mathbf{r},\omega)\frac{\partial U_+(\mathbf{r},\omega)}{\partial n} - U_+(\mathbf{r},\omega)\frac{\partial U_+^*(\mathbf{r},\omega)}{\partial n}\right]$$
$$= \int_{\Sigma_<} dS\left[U_+^*(\mathbf{r},\omega)\frac{\partial U_+(\mathbf{r},\omega)}{\partial n} - U_+(\mathbf{r},\omega)\frac{\partial U_+^*(\mathbf{r},\omega)}{\partial n}\right] + (k^2 - k^{*2})\int_{\delta\tau} d^3r|U_+(\mathbf{r},\omega)|^2,$$

from which we conclude that

$$E_{\Sigma_>}(\omega) = E_{\Sigma_<}(\omega) - 2\kappa\omega\Im(k^2)\mathcal{E}_{\delta\tau}(\omega), \tag{2.35}$$

where

$$\mathcal{E}_{\delta\tau}(\omega) = \int_{\delta\tau} d^3r|U_+(\mathbf{r},\omega)|^2$$

is the "energy" of the radiated field within the volume $\delta\tau$ contained between the two surfaces $\Sigma_<$ and $\Sigma_>$.

Equation (2.35) states that the energy density (and total energy) radiated out of the larger surface $\Sigma_>$ is equal to that radiated out of the smaller surface $\Sigma_<$ decreased by an amount proportional to the energy of the radiated field over the volume contained between these two surfaces. This decrease in energy is proportional to the imaginary part of the square of the wavenumber of the medium and clearly represents bulk energy loss due to absorption in the medium. In the limiting case of a loss-free medium where k is real this loss term disappears and we regain the results of Theorem 1.1 of Chapter 1.

The second part of Theorem 1.1 established that for dispersion-free media the energy spectrum and radiation pattern are related via the simple equation

$$E_{\partial\tau}(\omega) = 2\kappa\omega k \int_{4\pi} d\Omega_s |f(\mathbf{s},\omega)|^2, \tag{2.36a}$$

where $\partial\tau$ is *any* closed surface completely surrounding the source volume τ_0. By again following identical lines to those used to establish the above result we find in the general case of dispersive media that

$$\begin{aligned} E_{\Sigma_R}(\omega) &= -i\kappa\omega \int_{\partial\tau} R^2\, d\Omega_s \left[f^*(\mathbf{s},\omega)\frac{e^{-ik^*R}}{R} f(\mathbf{s},\omega)\frac{ike^{ikR}}{R} \right. \\ &\qquad \left. - f(\mathbf{s},\omega)\frac{e^{ikR}}{R} f^*(\mathbf{s},\omega)\frac{-ik^* e^{-ik^*R}}{R} \right] \\ &= 2\kappa\omega\Re k \int_{4\pi} d\Omega_s |f(\mathbf{s},\omega)|^2, \end{aligned} \tag{2.36b}$$

where Σ_R is the surface of an asymptotically large sphere centered on the origin and having radius R. Equation (2.36b) is the generalization of Eq. (2.36a) to the case of dispersive media and reduces to this earlier result when the medium is lossless and $\Im k = 0$. It is important to note that, while the energy spectrum defined in Eq. (2.36b) depends on the radius R of the reference sphere Σ_R over which it is computed, this dependence disappears in the case of a non-dispersive medium, as guaranteed by Theorem 1.1.

Although we have lost Theorem 1.1 in the case of dispersive media, we still have the important and intuitively obvious result that the energy radiated out of any surface cannot increase with propagation distance; i.e., that $E_{\Sigma_>}(\omega) \leq E_{\Sigma_<}(\omega)$ if $\Sigma_>$ lies outside $\Sigma_<$. We also mention that the general results established above are quite useful in certain areas of time-reversal imaging in random media and form the basis for important work relating to Green-function estimation from two-point correlation measurements in such media.

Example 2.6 In the case of non-absorbing media where the wavenumber k is real-valued we can express the energy spectrum of the radiated field given in Eq. (2.36b) in a particularly simple form in terms of the source Q. For this case the energy spectrum is given by

$$E_{\Sigma_R}(\omega) = 2\kappa\omega k \int_{4\pi} d\Omega_s \left| \overbrace{-\frac{1}{4\pi}\int_{\tau_0} d^3r\, Q(\mathbf{r},\omega)e^{-ik\mathbf{s}\cdot\mathbf{r}}}^{f(\mathbf{s},\omega)} \right|^2,$$

where we have made use of Eq. (2.24b). The above equation can be simplified to become

$$E_{\Sigma_R}(\omega) = \frac{\kappa\omega k}{8\pi^2}\int_{\tau_0} d^3r \int_{\tau_0} d^3r'\, Q^*(\mathbf{r},\omega)Q(\mathbf{r}',\omega)\left\{\int_{4\pi} d\Omega_s\, e^{ik\mathbf{s}\cdot(\mathbf{r}-\mathbf{r})}\right\}.$$

The quantity in curly brackets in the above equation can be computed using Eq. (2.26) of Example 2.4 for the special case of $l = 0$, $m = 0$. In particular, we have from that equation that

$$j_0(kR) = \frac{1}{4\pi}\int_{4\pi} d\Omega_s\, e^{ik\mathbf{s}\cdot\mathbf{R}},$$

so that

$$E_{\Sigma_R}(\omega) = \frac{\kappa\omega\Re k e^{-2\Im kR}}{2\pi} \int_{\tau_0} d^3r \int_{\tau_0} d^3r'\, Q^*(\mathbf{r},\omega)Q(\mathbf{r}',\omega) j_0(k|\mathbf{r}-\mathbf{r}'|).$$

2.7 Non-radiating and essentially non-radiating sources in dispersive media

The basic definition and all of the results pertaining to frequency-domain non-radiating (NR) sources embedded in non-dispersive media developed in Section 1.7.1 carry over to dispersive media. In particular, such sources are still characterized by Definition 1.1, which gives as necessary and sufficient conditions for a source to be non-radiating at some given frequency ω that its radiated field $U_+(\mathbf{r},\omega)$ as given by Eq. (2.23) must vanish everywhere outside its spatial support τ_0 at that particular frequency. Piecewise continuous compactly supported NR sources are constructed using the recipe (cf., Eq. (1.57))

$$Q_{\rm nr}(\mathbf{r},\omega) = [\nabla^2 + k^2]\Pi(\mathbf{r},\omega), \tag{2.37}$$

where $\Pi(\mathbf{r},\omega)$ is a function that is compactly supported in the spatial volume τ_0 at any given frequency ω and possesses continuous first partial spatial derivatives throughout this volume but is otherwise arbitrary. The field radiated by the NR source defined above is found using Eq. (2.23):

$$\begin{aligned} U_{\rm nr}(\mathbf{r},\omega) &= \int_{\tau_0} d^3r'\, G_+(\mathbf{r}-\mathbf{r}',\omega)\overbrace{\{[\nabla_{r'}^2 + k^2]\Pi(\mathbf{r}',\omega)\}}^{Q_{\rm nr}(\mathbf{r}',\omega)} \\ &= \int_{\tau_0} d^3r'\, \overbrace{[\nabla_{r'}^2 + k^2]G_+(\mathbf{r}-\mathbf{r}',\omega)}^{\delta(\mathbf{r}-\mathbf{r}')}\,\Pi(\mathbf{r}',\omega) = \Pi(\mathbf{r},\omega), \end{aligned}$$

where we have twice integrated by parts and dropped the surface terms due to the assumption that Π is continuously differentiable throughout τ_0. The field $U_{\rm nr}$ vanishes outside τ_0, which then establishes that the source defined via Eq. (2.37) is an NR source at frequency ω and, moreover, that the field it generates is the compactly supported function $\Pi(\mathbf{r},\omega)$.

2.7.1 Non-radiating sources and the radiation pattern

In our treatment of NR sources for the wave equation in Chapter 1 we established Theorem 1.2, which stated that a compactly supported and piecewise continuous source to the wave equation whose energy spectrum vanished at any given frequency must be NR at that frequency. Since the vanishing of the energy spectrum is equivalent to the vanishing of the radiation pattern of the field it follows that this theorem also establishes that *the radiation pattern generated from a compactly supported and piecewise continuous source will vanish if and only if the source is NR.* This, in turn, implies that the vanishing of the radiation pattern of such sources guarantees that the field will also vanish everywhere outside the source region τ_0. We state this in the form of the following theorem.

Theorem 2.1 *Let $Q(\mathbf{r}, \omega)$ be a piecewise continuous source compactly supported in τ_0 whose radiation pattern $f(\mathbf{s}, \omega)$ vanishes for all directions $\mathbf{s}$ at some given frequency ω. Then $Q(\mathbf{r}, \omega)$ is NR at that frequency and the radiated field will vanish everywhere outside the source volume τ_0.*

As a point of interest we note that, since the radiation pattern is an entire analytic function of the (real or complex) unit vector $\mathbf{s}$, the above theorem applies even if the radiation pattern $f(\mathbf{s}, \omega)$ vanishes over some finite section of the complex unit sphere. In particular, the source will be NR and the field will vanish everywhere outside the source volume τ_0 if the radiation pattern vanishes over any finite but arbitrarily small region of the real unit sphere.

2.7.2 Essentially non-radiating sources

A frequency-domain non-radiating source generates *zero field* outside the source's space volume τ_0 at one or more frequencies ω. We showed in Section 1.7.3 of Chapter 1 that there exist also sources to the wave equation that radiate negligible power and energy over a given set of frequencies. Although these *essentially non-radiating sources* (essentially NR sources) do not generate a zero field outside their support, the field is exponentially damped and is, for all practical purposes, unobservable. As we discussed in Chapter 1, the class of NR sources plays a dominant role in the uniqueness question for the inverse source problem (ISP), while the class of essentially NR sources impacts on the well-posedness and stability of solutions of this problem.

As was the case with NR sources, the treatment of essentially NR sources presented in Section 1.7.3 of Chapter 1 is virtually unchanged when applied to dispersive media. In particular, the radiated energy out of a large sphere of radius R is given by Eq. (2.36b) and will be small, *irrespective of the amount of absorption in the medium*, so long as

$$\int_{4\pi} d\Omega_s |f(\mathbf{s}, \omega)|^2 < \epsilon(k), \tag{2.38}$$

where, as in Section 1.7.3, $\epsilon(k)$ is a small parameter that characterizes the essentially NR source. Following arguments identical to those employed in Section 1.7.3, it is easily verified that the above condition is equivalent to the requirement that the *multipole moments*

$$q_l^m(k) = \int_{\tau_0} d^3r\, Q(\mathbf{r}, \omega) Y_l^{m*}(\mathbf{s}) j_l(kr) \tag{2.39a}$$

satisfy the condition

$$|q_l^m(k)|^2 \leq \mathcal{E}_Q \overbrace{\int_0^{a_0} r^2\, dr |j_l(kr)|^2}^{\mu_l^2(ka_0)} = \mathcal{E}_Q \mu_l^2(ka_0), \tag{2.39b}$$

where

$$\mathcal{E}_Q(\omega) = \int_{\tau_0} d^3r |Q(\mathbf{r}, \omega)|^2 \tag{2.39c}$$

is the L^2-norm square of the source (source energy) and a_0 is the radius of the smallest bounding sphere to τ_0.

The conditions Eq. (2.38) and, equivalently, Eq. (2.39b) guarantee that the energy of the field radiated out of an asymptotically large sphere of radius R will be small irrespective of the amount of absorption in the medium. We will take these inequalities as a working definition of an essentially NR source. We showed in Section 1.7.3 that the parameters $\mu_l^2(ka_0)$ are exponentially decreasing functions of their index l for all $l > l_0 = ka_0$. It then follows that an essentially NR source is any source for which the quantities defined in Eq. (2.39a) are arbitrarily small for all $l \leq l_0$. As mentioned in Section 1.7.3, essentially NR sources play a critical role in the inverse source problem treated in Chapter 5.

We also showed in Section 1.7.3 that the exponential decay of the parameters $\mu_l^2(ka_0)$ for $l > ka_0$ limits the size of the smallest angular periods of the source radiation pattern to being larger than λ/a_0. This, in turn, sets a minimum central lobe size achievable by any finite-norm source, a result that plays an important role in the inverse source problem and antenna design, as we will see in Chapter 5.

2.8 Boundary-value problems for the Helmholtz equation

The Kirchhoff–Helmholtz representation of the radiated field given in Eq. (2.31a) is, in fact, a solution to the *homogeneous* Helmholtz equation throughout the infinite region $\tau^\perp$ lying outside the volume τ bounded by the surface $\partial\tau$. As pointed out in the discussion presented below this equation, the surface integral must vanish at all points within the interior of τ, which establishes the fact that the boundary values of the radiated field and its normal derivative over $\partial\tau$ are not independent but rather are connected via this homogeneous integral equation. Indeed, it is possible to replace the outgoing-wave Green function in the Kirchhoff–Helmholtz representation by a Green function that vanishes on the boundary $\partial\tau$ and, hence, transform this representation into a form that employs only the field and not its normal derivative. In this form the Kirchhoff–Helmholtz representation of the radiated field outside τ becomes the solution of an *exterior boundary-value problem for the homogeneous Helmholtz equation.* Loosely speaking, a solution to a (properly posed) boundary-value problem for the homogeneous Helmholtz equation is a Kirchhoff–Helmholtz type of representation that involves only the boundary value of the field, its normal derivative, or a linear combination of the two. We will review this important class of problems in this section.

We can formally define a boundary-value problem for the homogeneous Helmholtz equation to be that of computing a solution to this equation within some region τ that satisfies prescribed *boundary conditions* on the boundary $\partial\tau$ of the region. The region τ can be simply or multiply connected[4] and can be finite or infinite in extent, but the field must satisfy the *homogeneous* Helmholtz equation throughout the entire region, not just over some subset of this region. We will consider three types of boundary-value problems defined according to the character of the region τ.

- The *interior boundary-value problem*, where τ is a finite region that can be simply or multiply connected and whose boundary $\partial\tau$ entirely encloses the region.
- The *exterior boundary-value problem*, where τ is the *exterior* of a finite region.
- The *Rayleigh–Sommerfeld boundary-value problem*, where τ is a semi-infinite region that is bounded by an infinite plane and a hemisphere at infinity.

Examples of these three cases are illustrated in Fig. 2.2.

In the case of the interior problem the boundary conditions can consist of specification of the field, its normal derivative, or a linear combination of these two quantities over the entire bounding surface $\partial\tau$ to the region. We refer to these three types of boundary

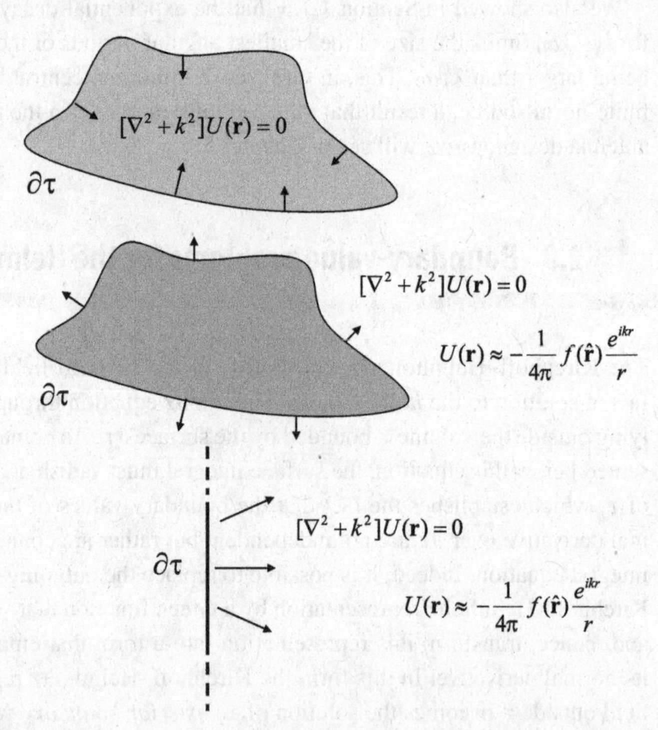

Fig. 2.2 Examples of the *interior* (top), *exterior* (middle) and *Rayleigh–Sommerfeld* (bottom) boundary-value problems. The arrows indicate the direction in which the field propagates.

[4] A 3D region is said to be "simply connected" if the interior of every closed surface within the region contains only points within the region. A region that is not simply connected is said to be "multiply connected." An example of a multiply connected region is the interior of a donut.

conditions as inhomogeneous Dirichlet, Neumann and mixed conditions, respectively. In this book we will restrict our attention to Dirichlet and Neumann conditions, although the results are easily generalized to general inhomogeneous mixed conditions. In the case of infinite (the exterior problem) and semi-infinite regions, Dirichlet, Neumann or mixed conditions are specified over a portion of the boundary and homogeneous mixed conditions, in the form of a radiation condition (Sections 2.1.4 and 2.1.5) are specified over the portion of the boundary located at infinity.

2.8.1 The interior boundary-value problem

In the interior boundary-value problem the region τ throughout which the field satisfies the homogeneous Helmholtz equation is finite and the boundary $\partial\tau$ entirely encloses τ. For a simply connected region, such as that illustrated in Fig. 2.2, $\partial\tau$ is the single exterior surface to τ, while in a multiply connected (donut-shaped) region $\partial\tau$ is the union of its interior and exterior surfaces. We can formally solve the interior boundary-value problem by applying standard Green-function techniques to the pair of equations

$$[\nabla_{r'}^2 + k^2]G(\mathbf{r}, \mathbf{r}', \omega) = \delta(\mathbf{r} - \mathbf{r}'), \tag{2.40a}$$

$$[\nabla_{r'}^2 + k^2]U(\mathbf{r}', \omega) = 0, \quad \mathbf{r}' \in \tau, \tag{2.40b}$$

where the subscript on the Laplacian operator indicates that this operator acts on the primed coordinate vector and G is, for the moment, an unspecified Green function to the Helmholtz equation that satisfies no particular boundary conditions. Following the same general procedure as was employed earlier in Section 2.5, we obtain the result

$$\int_{\partial\tau} dS' \left[U(\mathbf{r}', \omega)\frac{\partial}{\partial n'}G(\mathbf{r}, \mathbf{r}', \omega) - G(\mathbf{r}, \mathbf{r}', \omega)\frac{\partial}{\partial n'}U(\mathbf{r}', \omega) \right] = U(\mathbf{r}, \omega), \tag{2.41a}$$

if $\mathbf{r} \in \tau$, and

$$\int_{\partial\tau} dS' \left[U(\mathbf{r}', \omega)\frac{\partial}{\partial n'}G(\mathbf{r}, \mathbf{r}', \omega) - G(\mathbf{r}, \mathbf{r}', \omega)\frac{\partial}{\partial n'}U(\mathbf{r}', \omega) \right] = 0, \tag{2.41b}$$

otherwise, where $\partial\tau$ is the entire boundary to the region τ and the normal derivatives are directed out of the volume τ. Equations (2.41) are Helmholtz identities satisfied by a field obeying the homogeneous Helmholtz equation throughout the interior region τ.

The first Helmholtz identity Eq. (2.41a) is a formal solution to the boundary-value problem that consists of determining the solution to the homogeneous Helmholtz equation throughout τ which achieves specified boundary conditions in the form of $U(\mathbf{r}', \omega)$ and $\partial U(\mathbf{r}', \omega)/\partial n'$ at all points $\mathbf{r}'$ on the boundary $\partial\tau$. However, this boundary-value problem is not well posed in the sense that the boundary values $U(\mathbf{r}', \omega)$ and $\partial U(\mathbf{r}', \omega)/\partial n'$ are not independent and therefore cannot be assigned arbitrarily. As discussed earlier, these two quantities are coupled by the second Helmholtz identity, which implies that Dirichlet, Neumann or mixed conditions are necessary and sufficient in order to yield a unique solution to a properly posed boundary-value problem.[5] For example, it is clear from the first Helmholtz identity in Eqs. (2.41a) that the field within the region τ can be expressed entirely in terms of the field boundary value (Dirichlet conditions) on $\partial\tau$ if we require that the Green function vanish on $\partial\tau$ (i.e., satisfy homogeneous Dirichlet conditions on $\partial\tau$) and, similarly, can be expressed entirely in terms of Neumann conditions if the Green function satisfies homogeneous Neumann conditions of $\partial\tau$. The solution to a properly posed boundary-value problem thus reduces to determining a Green function that satisfies the homogeneous form of the boundary conditions which are satisfied by the field.

Restricting our attention to the two cases of Dirichlet and Neumann boundary conditions, we find using Eqs. (2.41a) that the solution to the interior boundary-value problem is given by

$$U(\mathbf{r},\omega) = \begin{cases} \int_{\partial\tau} dS'\, U(\mathbf{r}',\omega)(\partial/\partial n')G_{\mathrm{D}}(\mathbf{r},\mathbf{r}',\omega) & \text{Dirichlet conditions,} \\ \\ -\int_{\partial\tau} dS'(\partial/\partial n')U(\mathbf{r}',\omega)G_{\mathrm{N}}(\mathbf{r},\mathbf{r}',\omega) & \text{Neumann conditions,} \end{cases} \tag{2.42}$$

where the subscripts D and N refer, respectively, to Dirichlet and Neumann cases and the derivatives are directed outward from the interior region τ. The Dirichlet and Neumann Green functions are required to satisfy the homogeneous conditions

$$G_{\mathrm{D}}(\mathbf{r},\mathbf{r}',\omega)|_{\mathbf{r}'\in\partial\tau} = 0, \qquad \frac{\partial}{\partial n'}G_{\mathrm{N}}(\mathbf{r},\mathbf{r}',\omega)|_{\mathbf{r}'\in\partial\tau} = 0,$$

with $\mathbf{r} \in \tau$. We will show below that a Green function satisfying the above homogeneous Dirichlet or Neumann conditions with respect to source coordinates $\mathbf{r}'$ will automatically satisfy these conditions with respect to the field coordinate $\mathbf{r}$; i.e., the above conditions imply that

$$G_{\mathrm{D}}(\mathbf{r},\mathbf{r}',\omega)|_{\mathbf{r}\in\partial\tau} = 0, \qquad \frac{\partial}{\partial n}G_{\mathrm{N}}(\mathbf{r},\mathbf{r}',\omega)|_{\mathbf{r}\in\partial\tau} = 0,$$

where now $\mathbf{r}' \in \tau$.

It is important to note that the solutions to the boundary-value problem as given in Eqs. (2.42) are, in fact, algorithms that allow the user to propagate an arbitrarily imposed boundary condition into a wavefield that satisfies the homogeneous Helmholtz equation throughout the interior region τ and, in addition, reduces to the imposed boundary condition on the bounding surface $\partial\tau$ to τ.

2.8.2 The exterior boundary-value problem for closed boundaries

We first consider the case of the exterior problem where the field satisfies the homogeneous Helmholtz equation throughout the exterior $\tau^{\perp}$ (complement) of some finite region

[5] Here, we exclude the possibility of resonances, which can occur in the *interior boundary-value problem*. Resonances occur for certain discrete values of the wavenumber and for special surface geometries for which it is possible for the field or its normal derivative to vanish everywhere over a closed surface.

τ having boundary $\partial\tau$ as well as the Sommerfeld radiation condition (SRC) at infinity. The Helmholtz identities are, in this case, found to be given by

$$U(\mathbf{r},\omega) = \int_{\partial\tau\cup\Sigma_\infty} dS' \left[U(\mathbf{r}',\omega)\frac{\partial}{\partial n'}G(\mathbf{r},\mathbf{r}',\omega) - G(\mathbf{r},\mathbf{r}',\omega)\frac{\partial}{\partial n'}U(\mathbf{r}',\omega)\right], \quad \mathbf{r}\in\tau^\perp,$$

$$\int_{\partial\tau\cup\Sigma_\infty} dS' \left[U(\mathbf{r}',\omega)\frac{\partial}{\partial n'}G(\mathbf{r},\mathbf{r}',\omega) - G(\mathbf{r},\mathbf{r}',\omega)\frac{\partial}{\partial n'}U(\mathbf{r}',\omega)\right] = 0, \quad \mathbf{r}\in\tau,$$

where $\partial\tau \cup \Sigma_\infty$ denotes the union of $\partial\tau$ with the surface Σ_∞ of an infinite sphere, and the normal derivatives are directed *outward* from the exterior region $\tau^\perp$ into the interior region τ and into the surface at infinity. On the infinite sphere Σ_∞ the field and Green function are required to satisfy the outgoing-wave radiation condition (SRC) expressed in Eqs. (2.8), from which it follows that the integrals over Σ_∞ in the above equations vanish and we are left with a surface integral only over $\partial\tau$; i.e.,

$$U(\mathbf{r},\omega) = \int_{\partial\tau} dS' \left[U(\mathbf{r}',\omega)\frac{\partial}{\partial n'}G(\mathbf{r},\mathbf{r}',\omega) - G(\mathbf{r},\mathbf{r}',\omega)\frac{\partial}{\partial n'}U(\mathbf{r}',\omega)\right], \tag{2.43a}$$

if $\mathbf{r} \in \tau^\perp$, and

$$\int_{\partial\tau} dS' \left[U(\mathbf{r}',\omega)\frac{\partial}{\partial n'}G(\mathbf{r},\mathbf{r}',\omega) - G(\mathbf{r},\mathbf{r}',\omega)\frac{\partial}{\partial n'}U(\mathbf{r}',\omega)\right] = 0, \tag{2.43b}$$

if $\mathbf{r} \in \tau$.

All of the discussion given for the interior problem continues to apply in the case of the exterior problem. In particular, the second Helmholtz identity Eq. (2.43b) implies that Green functions can be found that satisfy homogeneous Dirichlet or Neumann conditions on $\partial\tau$ as well as the SRC. The field in the exterior region $\tau^\perp$ can then be represented in the general form according to Eqs. (2.42), where the Green functions must satisfy homogeneous Dirichlet or Neumann conditions on $\partial\tau$ *as well as the outgoing-wave radiation condition at infinity.* The fields so constructed will satisfy the same radiation condition at infinity. As in the case of the interior boundary-value problem, these "solutions" are formal in the sense that they require that the appropriate Green function be computed.

Representation of the radiated field as the solution of an exterior boundary-value problem

The field U_+ radiated by the source Q satisfies the homogeneous Helmholtz equation everywhere outside the source spatial volume τ_0 and the SRC and, hence, can be expressed as the solution to an exterior boundary-value problem outside any surface $\partial\tau$ that completely surrounds the source volume. In contrast to the *Kirchhoff–Helmholtz representation* given in Eq. (2.31a) the field representation as the solution of a *properly posed* boundary-value problem in terms of Dirichlet or Neumann conditions on $\partial\tau$ is not over-determined and is entirely equivalent to the primary field solution given in Eq. (2.23). Incidentally, it should be clear from the equivalence of these two representations (primary and via the solution of a boundary-value problem) that the radiated field has only two degrees of freedom outside the source volume τ_0. This indicates that a source and its radiated field outside τ_0

are not in a 1:1 correspondence and, in particular, that there exist *NR sources* that generate zero field outside τ_0. Non-radiating sources were treated within the time domain for non-dispersive media in Section 1.7 of Chapter 1 and earlier in Section 2.7 of this chapter for the more general case of dispersive media.

2.8.3 The exterior boundary-value problem for open boundaries

Consider now the case in which $\tau^{\perp}$ is a semi-infinite half-space bounded by an open surface $\partial\tau$ and a hemisphere Σ_∞ located at infinity. We will again require that the field and Green function both satisfy the outgoing-wave radiation condition on the hemisphere and that the field satisfy inhomogeneous Dirichlet or Neumann conditions over the open surface $\partial\tau$ (see Fig. 2.2). Using arguments identical to those employed in Section 2.4, it is easy to verify that the integral over Σ_∞ vanishes and we are left with a surface integral over only $\partial\tau$. All of the analysis and results obtained for the exterior problem for infinite regions continue to apply in the case of semi-infinite regions, where now the boundary $\partial\tau$ is the infinite open surface bounding the given half-space. In particular, the solution of the boundary-value problem for a field satisfying the SRC and inhomogeneous Dirichlet or Neumann conditions on the surface $\partial\tau$ is given by Eqs. (2.42), where the Green functions must satisfy a homogeneity condition on $\partial\tau$ as well as the outgoing-wave radiation condition. As in the case of the interior and infinite-region exterior boundary-value problems, these solutions require that the appropriate Green function be computed and, as mentioned above, this is generally non-trivial. One case for which a solution is easily constructed is when $\partial\tau$ is a planar boundary. In this case the boundary-value problem is sometimes referred to as the *Rayleigh–Sommerfeld problem* after the two scientists who first solved it. We will treat this problem in Section 2.9, where we will compute the required Green functions using the method of images first employed by Rayleigh and Sommerfeld in their classic solution of this problem. These Green functions will also be computed using the second Helmholtz identity Eq. (2.41b) in Section 2.10.

As was the case for the exterior boundary-value problem in an infinite region, we can also represent the radiated field U_+ in a semi-infinite region as the solution of a boundary-value problem. In particular, it should be clear that this field representation can be used to represent the radiated field throughout any half-space that lies outside the source volume τ_0 in terms of Dirichlet or Neumann conditions over the bounding plane to the half-space.

Uniqueness

The solutions of the Dirichlet and Neumann boundary-value problems as given in Eq. (2.42) are unique; i.e, they are the *only* solutions to the homogeneous Helmholtz equation throughout the region τ that satisfy the prescribed Dirichlet or Neumann conditions on the bounding surface $\partial\tau$. The same is true for infinite and semi-infinite regions, where the fields must satisfy the outgoing-wave radiation condition at infinity and Dirichlet or Neumann conditions on $\partial\tau$. To show this, assume to the contrary that there exists some other solution $U'(\mathbf{r},\omega)$ that also satisfies the homogeneous Helmholtz equation as well as the prescribed boundary conditions. Then the difference field $\delta U = U - U'$ must satisfy the homogeneous Helmholtz equation, and *homogeneous* boundary conditions on $\partial\tau$ and the SRC. The difference field δU is then also given by Eq. (2.42), where, however, the boundary value of δU or its first derivative is zero and, hence, the δU vanishes identically so that $U = U'$.

2.8.4 Symmetry of the Green functions

We can employ a procedure similar to that used to derive the Helmholtz identities to prove that a Green function to the interior or exterior boundary-value problems must be a symmetric function of its arguments. Here we prove this only for the interior boundary-value problem since the proof for the exterior problem follows entirely similar lines.

To prove the result for the interior problem, we consider a Green function computed for source points located at $\mathbf{r}_1$ and $\mathbf{r}_2$ within some closed bounded region τ with boundary $\partial\tau$; i.e.,

$$[\nabla_{r'}^2 + k^2]G(\mathbf{r}_j, \mathbf{r}', \omega) = \delta(\mathbf{r}_j - \mathbf{r}'), \quad \mathbf{r}_j \in \tau,$$

where $j = 1, 2$. Following the same general procedure as that used in Section 2.1.2, we find that

$$\int_{\partial\tau} dS' \left[G(\mathbf{r}_2, \mathbf{r}', \omega)\frac{\partial}{\partial n'}G(\mathbf{r}_1, \mathbf{r}', \omega) - G(\mathbf{r}_1, \mathbf{r}', \omega)\frac{\partial}{\partial n'}G(\mathbf{r}_2, \mathbf{r}', \omega)\right]$$
$$= G(\mathbf{r}_2, \mathbf{r}_1, \omega) - G(\mathbf{r}_1, \mathbf{r}_2, \omega).$$

Now, since G must satisfy homogeneous Dirichlet or Neumann conditions with respect to $\mathbf{r}'$ on $\partial\tau$, the surface integral must vanish and $G(\mathbf{r}_1, \mathbf{r}_2, \omega) = G(\mathbf{r}_2, \mathbf{r}_1, \omega)$. In fact, it is simple to show that the same result is obtained if the Green function satisfies general homogeneous mixed conditions on the boundary.

We note that the symmetry of a Green function with respect to its arguments implies important symmetry conditions for Green functions satisfying homogeneous Dirichlet or Neumann conditions on the boundary $\partial\tau$. In particular, it follows from the symmetry of such Green functions that

$$G_{\mathrm{D}}(\mathbf{r}_1, \mathbf{r}_2, \omega)|_{\mathbf{r}_2\in\partial\tau} = 0 \Rightarrow G_{\mathrm{D}}(\mathbf{r}_2, \mathbf{r}_1, \omega)|_{\mathbf{r}_2\in\partial\tau} = 0, \tag{2.44a}$$

$$\frac{\partial}{\partial n_2}G_{\mathrm{N}}(\mathbf{r}_1, \mathbf{r}_2, \omega)|_{\mathbf{r}_2\in\partial\tau} = 0 \Rightarrow \frac{\partial}{\partial n_2}G_{\mathrm{N}}(\mathbf{r}_2, \mathbf{r}_1, \omega)|_{\mathbf{r}_2\in\partial\tau} = 0, \tag{2.44b}$$

where G_{D} and G_{N} are Green functions satisfying homogeneous Dirichlet and Neumann conditions over the surface $\partial\tau$. The above equations state that if a Green function satisfies homogeneous Dirichlet or Neumann conditions with respect to the source coordinate $\mathbf{r}'$ then it automatically satisfies the same condition with respect to the field coordinate. We note that this *does not* require that the normal derivative of the Neumann Green function with respect to a given coordinate vector vanish when either $\mathbf{r}$ or $\mathbf{r}'$ lies on the surface $\partial\tau$; e.g.,

$$\frac{\partial}{\partial n'}G_{\mathrm{N}}(\mathbf{r}, \mathbf{r}', \omega)|_{\mathbf{r}'\in\partial\tau} = 0 \not\Rightarrow \frac{\partial}{\partial n'}G_{\mathrm{N}}(\mathbf{r}, \mathbf{r}', \omega)|_{\mathbf{r}\in\partial\tau} = 0,$$

a conclusion that one might be tempted to draw from Eq. (2.44a).

2.9 The Rayleigh–Sommerfeld boundary-value problem

As an example of the Green-function method for solving a boundary-value problem for the Helmholtz equation we consider the classic problem of determining the solution to the homogeneous Helmholtz equation in a half-space that satisfies prescribed Dirichlet or Neumann boundary conditions on the bounding plane of the half-space and the Sommerfeld radiation condition on the infinite hemisphere enclosing the half-space. For the sake of simplicity we will take the half-space to be either the right half-space $z > 0$ or the left half-space $z < 0$, with the understanding that the general case of a half-space bounded by the plane $z = z_0 \neq 0$ is trivially obtained by a simple coordinate-system translation. The formal solution to these problems was shown in Section 2.8.2 to be given by Eqs. (2.42) with $\partial\tau$ equal to the surface of the plane $z = 0$ and where the Green function is required to satisfy homogeneous Dirichlet or Neumann conditions on $\partial\tau$ and the SRC on the infinite hemisphere enclosing the half-space in which the solution is sought.

The boundary-value problems stated above were first solved using the method of images by Sommerfeld, and we will refer to them as the *Rayleigh–Sommerfeld problems* (RS problems). Here, we will outline the method of images for the case of the RS problem under inhomogeneous Dirichlet conditions (the RS Dirichlet problem) on the plane $z = 0$ and leave it as an exercise for the reader to apply the same method to the RS Neumann problem. Before solving the full 3D RS problem it is instructive to examine this problem in the case of the 1D Helmholtz equation, for which the algebra is reduced to a minimum.

Example 2.7 Consider the 1D boundary-value problem of solving the homogeneous Helmholtz equation

$$\left[\frac{\partial^2}{\partial z^2} + k^2\right] U(z,\omega) = 0$$

on the half-line $z > 0$ subject to an outgoing-wave condition on this half-line and inhomogeneous Dirichlet conditions at $z = 0$. The general solution to this problem is given by the 1D version of Eq. (2.42) (written for the exterior problem):

$$U(z,\omega) = -U_0(k)\frac{\partial}{\partial z'}G_D(z,z',\omega)|_{z'=0}, \tag{2.45}$$

where $U_0(k) = U(z = 0,\omega)$ is the Dirichlet boundary condition and the Dirichlet Green function is required to satisfy the 1D SRC on the right-half line $z > 0$ and to vanish when $z' = 0$. Note that we have used the fact that the outward directed unit normal to the half-line $z > 0$ is $-\hat{\mathbf{z}}$ so that $\partial/\partial n' = -\partial/\partial z'$.

In the "method of images" we consider general field and source points z and z' both contained on the right-half line $z, z' > 0$ and a mirror-image source point $-z'$ about the point $z' = 0$. We consider the quantity

$$\Delta(z,z',\omega) = G_+(z - z',\omega) - G_+(z + z',\omega),$$

where

$$G_+(z-z',\omega) = -\frac{i}{2k}e^{ik|z-z'|}$$

is the 1D outgoing-wave Green function derived in Example 2.2. The quantity Δ satisfies the equation

$$\left[\frac{\partial^2}{\partial z^2} + k^2\right]\Delta(z,z',\omega) = \delta(z-z') - \delta(z+z')$$

and, hence, is itself a Green function if we require both z and z' to lie in the same half-line $z, z' > 0$. Moreover, it is easily shown that $\Delta(z,z',\omega)$ satisfies the SRC and vanishes at the point $z' = 0$. It then follows that $\Delta(z,z',\omega) = G_{\rm D}(z,z',\omega)$ is the Dirichlet Green function for the half-line problem.

For $z \geq z'$ we have that

$$\frac{\partial}{\partial z'}G_{\rm D}(z,z',\omega) = -\frac{i}{2k}\frac{\partial}{\partial z'}[e^{ik(z-z')} - e^{ik(z+z')}] = -\frac{1}{2}[e^{ik(z-z')} + e^{ik(z-z')}],$$

so that

$$\frac{\partial}{\partial z'}G_{\rm D}(z,z',\omega)|_{z'=0} = -e^{ikz},$$

which, on using Eq. (2.45), yields the solution

$$U(z,\omega) = U_0(k)e^{ikz},$$

which is valid throughout the right half-space $z \geq 0$. It is easy to verify that the above solution satisfies the homogeneous Helmholtz equation for all $z > 0$ and the Dirichlet condition at $z = 0$ and is outgoing on the half-line $z > 0$.

Returning to the 3D case, we consider general field and source points $\mathbf{r}$ and $\mathbf{r}'$ both contained in either the right half-space $V_+ = \{\mathbf{r} : z > 0\}$ or the left half-space $V_- = \{\mathbf{r} : z < 0\}$ and a mirror-image source point $\tilde{\mathbf{r}}' = (x', y', -z')$ about the plane $z' = 0$ such as illustrated in Fig. 2.3. Thus, if $\mathbf{r}' \in V_+$ then $\tilde{\mathbf{r}}' \in V_-$. We will now show that the Rayleigh–Sommerfeld Dirichlet and Neumann Green functions in either half-space are given by

$$G_{\rm D}(\mathbf{r},\mathbf{r}',\omega) = G_+(\mathbf{r}-\mathbf{r}',\omega) - G_+(\mathbf{r}-\tilde{\mathbf{r}}',\omega), \tag{2.46a}$$

$$G_{\rm N}(\mathbf{r},\mathbf{r}',\omega) = G_+(\mathbf{r}-\mathbf{r}',\omega) + G_+(\mathbf{r}-\tilde{\mathbf{r}}',\omega). \tag{2.46b}$$

We first note that $G_{\rm D}$ is in fact a Green function so long as the source and field points $\mathbf{r}$ and $\mathbf{r}'$ both lie in the same half-space; i.e.,

$$[\nabla^2_{r,r'} + k^2]G_{\rm D}(\mathbf{r},\mathbf{r}',\omega) = \delta(\mathbf{r}-\mathbf{r}') - \delta(\mathbf{r}-\tilde{\mathbf{r}}') = \delta(\mathbf{r}-\mathbf{r}')$$

so long as $\mathbf{r} \neq \tilde{\mathbf{r}}'$, which will be the case as long as both $\mathbf{r}$ and $\mathbf{r}'$ lie in the same half-space (either V_+ or V_-). It is also clear that this Green function satisfies the SRC both in V_+ and in V_-. We need then show only that $G_{\rm D}$ vanishes if $\mathbf{r}'$ lies on the boundary plane $z' = 0$. To establish this, we note that if either $\mathbf{r}$ or $\mathbf{r}'$ lies on the boundary plane then

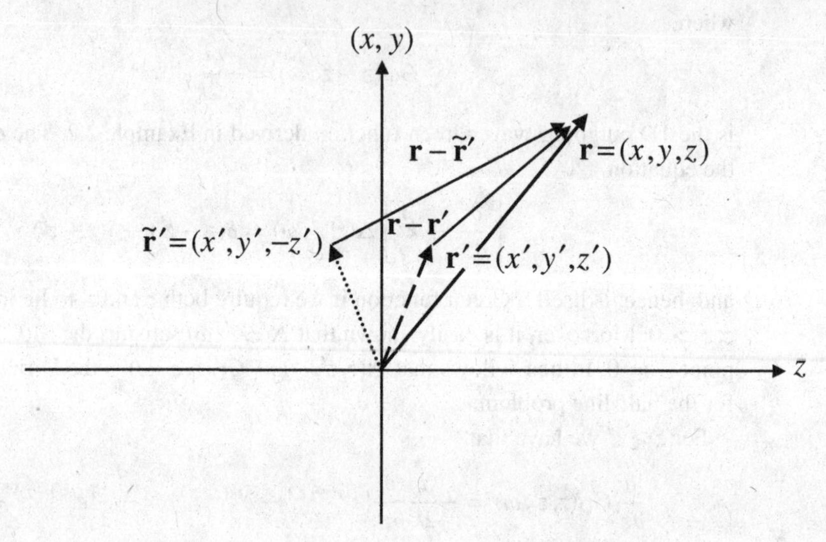

Fig. 2.3 The geometry for computing the Rayleigh–Sommerfeld Green functions. Each source point $\mathbf{r}' = (x', y', z')$ has a corresponding "image point" $\tilde{\mathbf{r}}' = (x', y', -z')$ so that $|\mathbf{r}-\mathbf{r}'| = |\mathbf{r}-\tilde{\mathbf{r}}'|$ when the source point lies on the boundary plane $z' = 0$.

$$|\mathbf{r}-\tilde{\mathbf{r}}'| = |\mathbf{r}-\mathbf{r}'|,$$

so that G_D vanishes in either of these situations. This then establishes that G_D is, indeed, the Dirichlet Green function for the problem.

By entirely analogous reasoning it is clear that G_N as defined in Eq. (2.46b) satisfies the SRC and is a proper Green function so long as the source and field points $\mathbf{r}$ and $\mathbf{r}'$ both lie in the same half-space. Moreover,

$$\frac{\partial}{\partial z'}G_+(\mathbf{r}-\mathbf{r}',\omega) = -\frac{\partial}{\partial z'}G_+(\mathbf{r}-\tilde{\mathbf{r}}',\omega),$$

so that

$$\frac{\partial}{\partial z'}[G_+(\mathbf{r}-\mathbf{r}',\omega) + G_+(\mathbf{r}-\tilde{\mathbf{r}}',\omega)] = 0$$

when either $\mathbf{r}$ or $\mathbf{r}'$ lies on the boundary plane. This then establishes G_N as defined in Eq. (2.46b) as the appropriate Neumann Green function for the RS problem.

We now note that if we let the source point $\mathbf{r}'$ lie on the plane $z' = 0$ we have that

$$\frac{\partial}{\partial n'}G_D(\mathbf{r},\mathbf{r}',\omega)|_{\mathbf{r}'=\boldsymbol{\rho}'} = -\frac{\partial}{\partial z'}G_D(\mathbf{r},\mathbf{r}',\omega)|_{\mathbf{r}'=\boldsymbol{\rho}'} = -2\frac{\partial}{\partial z'}G_+(\mathbf{r}-\boldsymbol{\rho}',\omega),$$

where $\boldsymbol{\rho}' = (x', y')$ denotes an integration point on the (x, y) plane and we have used the shorthand notation $(\partial/\partial z')G_+(\mathbf{r}-\boldsymbol{\rho}',\omega) = (\partial/\partial z')G_+(\mathbf{r}-\mathbf{r}',\omega)|_{z'=0}$. If we now substitute the above expression into Eq. (2.42) we obtain the result

$$U(\mathbf{r},\omega) = \mp 2\int dS'\, U_0(\boldsymbol{\rho}',\omega)\frac{\partial}{\partial z'}G_+(\mathbf{r}-\boldsymbol{\rho}',\omega), \tag{2.47a}$$

where $dS' = d^2\rho'$ is the differential area on the $\boldsymbol{\rho}'$ plane. Here the top sign is used for the solution in the right half-space and the bottom sign for the solution in the left half-space, and $U_0(\boldsymbol{\rho}', \omega)$ is an arbitrarily assigned Dirichlet boundary condition; i.e., an arbitrary function of the surface coordinate $\boldsymbol{\rho}'$. A completely parallel development leads to the following solution to the Neumann boundary-value problem:

$$U(\mathbf{r}, \omega) = \pm 2 \int dS'\, U_0'(\boldsymbol{\rho}', \omega) G_+(\mathbf{r} - \boldsymbol{\rho}', \omega), \tag{2.47b}$$

where $U_0'(\boldsymbol{\rho}', \omega) = (\partial/\partial z')U|_{z'=0}$ is the arbitrarily assigned Neumann boundary value on the plane $z = 0$.

Although for the solutions to the RS Dirichlet and Neumann boundary-value problems given in Eqs. (2.47) it is assumed that the boundary plane is the plane $z = 0$, the general case of a boundary plane located at $z = z_0$ is easily obtained via a coordinate-system translation along the z axis. In particular, if we note that the arguments on the Green functions in Eqs. (2.47) depend only on the difference vector $\mathbf{r} - \boldsymbol{\rho}'$ we can translate the origin of the coordinate system so that the data plane $z = 0$ is translated to $z = z_0$, and it is easy to show that Eqs. (2.47) yield the result

$$U(\mathbf{r}, \omega) = \begin{cases} \mp 2 \int dS'\, U_{z_0}(\boldsymbol{\rho}', \omega)(\partial/\partial z') G_+(\mathbf{r} - \mathbf{r}_0', \omega) & \text{Dirichlet conditions,} \\ \pm 2 \int dS'\, U_{z_0}'(\boldsymbol{\rho}', \omega) G_+(\mathbf{r} - \mathbf{r}_0', \omega) & \text{Neumann conditions,} \end{cases} \tag{2.48}$$

where $\mathbf{r}_0' = \boldsymbol{\rho}' + \hat{\mathbf{z}} z_0$ is the position vector on the z_0 plane and the integrations are performed over this plane, and we have re-labeled the boundary conditions with the subscript z_0 to indicate that these conditions are now applied on the plane $z = z_0$ rather than on $z = 0$. All of the remarks made above continue to hold with the plane $z = 0$ now replaced by the plane $z = z_0$. In particular, the top sign in each equation is used in the half-space $z > z_0$ and the bottom expression in the half-space $z < z_0$, and the quantities U_{z_0} and U_{z_0}' are arbitrary functions of the transverse component $\boldsymbol{\rho}'$ of the surface coordinate vector $\mathbf{r}_0'$.

The classical Rayleigh–Sommerfeld problem corresponds physically to the process of *propagation* whereby a wavefield is propagated *outward* from its boundary value on the plane $z = z_0$ into either the right half-space $z > z_0$ or the left half-space $z < z_0$. In most applications the wavefield is physically generated by a (causal) source or set of sources located in the half-space opposite to that into which the field is being propagated; e.g., the sources are located in the left half-space if the field is being propagated into the right half-space and vice versa. The solution to the RS boundary-value problem allows the wavefield to be determined from its boundary values as opposed to the primary sources that actually generate the field.

2.9.1 The Rayleigh–Sommerfeld solution for two-dimensional wavefields

In two space dimensions a wavefield satisfies the 2D Helmholtz equation

$$\left[\frac{\partial^2}{\partial x^2}+\frac{\partial^2}{\partial z^2}+k^2\right]U(\mathbf{r},\omega)=0,$$

where $\mathbf{r}$ is now the position vector on the (x,z) plane. The outgoing-wave Green function for the 2D case was found in Section 2.2.1 to be given by

$$G_+(\mathbf{R},\omega)=\frac{-i}{4}H_0^+(kR).$$

The 2D RS Dirichlet and Neumann Green functions are obtained directly from Eqs. (2.46) by substituting for G_+ from the above equation, and the solutions to the RS boundary-value problems for data specified on the plane $z=z_0$ are given by Eqs. (2.48) using the above outgoing-wave Green function.

Example 2.8 As an example we consider the RS Dirichlet problem in 2D in which the boundary-value data are specified on the plane $z=0$ and the field is to be computed throughout the r.h.s. $z>0$. The solution to this problem is given by Eq. (2.47a) with G_+ equal to the 2D outgoing-wave Green function and with the integration performed over the line $z'=0$. In order to compute the propagated field we need the normal derivative of this Green function, which is found using the well-known derivative relation

$$\frac{d}{dz}H_0^+(z)=-H_1^+(z).$$

It then follows that

$$\frac{\partial}{\partial z'}G_+(\mathbf{r}-\mathbf{r}',\omega)=\frac{ik}{4}H_1^+(k|\mathbf{r}-\mathbf{r}'|)\frac{\partial}{\partial z'}|\mathbf{r}-\mathbf{r}'|=\frac{-ik}{4}H_1^+(k|\mathbf{r}-\mathbf{r}'|)\frac{z-z'}{|\mathbf{r}-\mathbf{r}'|},$$

which, when substituted into Eq. (2.47a), yields the result

$$U(\mathbf{r},\omega)=\frac{ikz}{2}\int_{-\infty}^{\infty}dx'\,U_0(x',\omega)\frac{H_1^+(k|\mathbf{r}-\mathbf{r}'|)}{|\mathbf{r}-\mathbf{r}'|}. \tag{2.49}$$

2.9.2 Rayleigh–Sommerfeld representation of the radiated field

As mentioned earlier, the field U_+ radiated by the source Q can be expressed in terms of Dirichlet or Neumann conditions over any plane $z=z_0$ that lies outside the source spatial volume τ_0 via the RS field representations given in Eqs. (2.48). In contrast to the Kirchhoff–Helmholtz field representations given in Eq. (2.31a), the RS representation is not over-determined and can be used to represent the radiated field throughout any half-space that lies outside the source volume τ_0. By appropriate selection of this half-space it is thus possible to express the radiated field everywhere outside the smallest convex region that contains τ_0 by such a representation.

Example 2.9 As an example of the RS boundary-value problem consider the problem of computing a *plane wave* $\exp(ik\mathbf{s}\cdot\mathbf{r})$ propagating into the right half-space $z > z_0$ from its Neumann condition

$$\frac{\partial}{\partial z'} e^{ik\mathbf{s}\cdot\mathbf{r}'}|_{z'=z_0} = iks_z e^{ik\mathbf{s}\cdot\mathbf{r}'_0}. \tag{2.50}$$

In the above equation $\mathbf{s}$ is a unit vector directed along the direction of propagation of the plane wave and $\mathbf{r}'_0$ is a general field point on the plane $z' = z_0$. Because the plane wave is propagating *into* the right half-space the z component of the plane wave $s_z > 0$.

On substituting the Neumann condition Eq. (2.50) into the RS formula Eq. (2.48) we obtain the result

$$e^{ik\mathbf{s}\cdot\mathbf{r}} = 2iks_z \int dS'\, e^{ik\mathbf{s}\cdot\mathbf{r}'_0} G_+(\mathbf{r}-\mathbf{r}'_0, \omega), \tag{2.51}$$

which is valid for all $z \geq z_0$. The above equation represents the plane wave as a superposition of *spherical waves* G_+, each centered at a "source point" on the plane $z = z_0$. This seemingly unimportant result will be the basis for the important *slant-stack* operation introduced in Chapter 6 that allows the data collected in a suite of scattering experiments employing spherical incident waves to be converted into the data that would have been obtained in a suite of scattering experiments employing incident plane waves.

2.10 Solution of the RS problem using the Helmholtz identities

A nifty derivation of the RS formulas Eqs. (2.48) is possible using only the Helmholtz identities Eqs. (2.43) with $G = G_+$ and the fact that the outgoing-wave Green function $G_+(\mathbf{R}, \omega)$ is a function only of the magnitude $R = |\mathbf{R}|$ of the vector $\mathbf{R}$. For the sake of simplicity we take our data boundary $\partial\tau$ to be the plane $z = 0$ and look for a solution to the RS problem throughout the half-space $z > 0$. The two Helmholtz identities then become

$$U(\mathbf{r}, \omega) = -\int_{z'=0} dS' \left[U \frac{\partial}{\partial z'} G_+(\mathbf{r}-\mathbf{r}', \omega) - G_+(\mathbf{r}-\mathbf{r}', \omega)\frac{\partial}{\partial z'} U\right], \quad z > 0, \tag{2.52a}$$

and

$$\int_{z'=0} dS' \left[U \frac{\partial}{\partial z'} G_+(\mathbf{r}-\mathbf{r}', \omega) - G_+(\mathbf{r}-\mathbf{r}', \omega)\frac{\partial}{\partial z'} U\right] = 0, \quad z < 0, \tag{2.52b}$$

where the normal derivatives are directed into the positive r.h.s. $z > 0$.

Consider now a general field point $\mathbf{r} = (x, y, z)$ lying in the r.h.s. $z > 0$ and its mirror image $\tilde{\mathbf{r}} = (x, y, -z)$. Since $\tilde{\mathbf{r}}$ lies in the l.h.s. we conclude from the second Helmholtz identity that

$$\int_{z'=0} dS' \left[U(\mathbf{r}', \omega)\frac{\partial}{\partial z'} G_+(\tilde{\mathbf{r}}-\mathbf{r}', \omega) - G_+(\tilde{\mathbf{r}}-\mathbf{r}', \omega)\frac{\partial}{\partial z'} U(\mathbf{r}', \omega)\right] = 0, \quad z > 0.$$

But

$$G_+(\tilde{\mathbf{r}} - \mathbf{r}', \omega)|_{z'=0} = G_+(\mathbf{r} - \mathbf{r}', \omega)|_{z'=0},$$
$$\frac{\partial}{\partial z'} G_+(\tilde{\mathbf{r}} - \mathbf{r}', \omega)|_{z'=0} = -\frac{\partial}{\partial z'} G_+(\mathbf{r} - \mathbf{r}', \omega)|_{z'=0},$$

from which we conclude that

$$\int_{z'=0} dS'\, U(\mathbf{r}', \omega) \frac{\partial}{\partial z'} G_+(\mathbf{r} - \mathbf{r}', \omega) = -\int_{z'=0} dS'\, G_+(\mathbf{r} - \mathbf{r}', \omega) \frac{\partial}{\partial z'} U(\mathbf{r}', \omega), \quad z > 0.$$

Using this result we can then rewrite the first Helmholtz identity Eq. (2.52a) in either of the two forms

$$U(\mathbf{r}, \omega) = -2 \int_{z'=0} dS'\, U(\mathbf{r}', \omega) \frac{\partial}{\partial z'} G_+(\mathbf{r} - \mathbf{r}', \omega) \tag{2.53a}$$

and

$$U(\mathbf{r}, \omega) = 2 \int_{z'=0} dS'\, G_+(\mathbf{r} - \mathbf{r}', \omega) \frac{\partial}{\partial z'} U(\mathbf{r}', \omega). \tag{2.53b}$$

The above two equations are recognized as being identical to the solutions for the RS problem in the half-space $z > 0$ given in Eqs. (2.47). The solutions for the l.h.s. $z < 0$ can be obtained in an entirely analogous manner and the results extended to any arbitrary plane $z = z_0$ using the same procedure as employed in Section 2.9.

2.11 Back propagation and the inverse RS boundary-value problem

The solutions to exterior boundary-value problems such as that to the RS boundary-value problems in Eqs. (2.48) are solutions to "forward" propagation problems. By this we mean that boundary values of an outgoing-wave field U_+ over some surface $\partial\tau$ that surrounds the source of the field are used to compute the field at points that are exterior to this surface and, hence, further removed from the source than is the surface. An interesting and important twist to this forward-propagation problem is the inverse or "back-propagation" problem in which these boundary values are used to compute the field at points that are closer to the source than is the boundary-value surface. The latter problem is not equivalent to the interior boundary-value problem treated in Section 2.8.1 since the field is generated from sources located within the boundary $\partial\tau$ and, hence, does not satisfy the homogeneous Helmholtz equation within the interior τ of the boundary over which the boundary values are specified. In fact, the back-propagation problem is not a classical boundary-value problem and cannot be solved exactly using the tools and techniques we have developed so far but does have an exact, albeit unstable, solution that we will obtain in Chapter 4. Although it cannot be solved exactly as a boundary-value problem, it can be solved approximately as such, and the approximation will be found to be excellent so long as the boundary over which the outgoing-wave boundary values are specified and the field point $\mathbf{r}$ to which the field is back propagated are many wavelengths removed from the source.

One approximate solution to the back-propagation problem was obtained in Sections 1.4.2 of Chapter 1 and in Section 2.5.1 of this chapter using over-specified field data on a boundary $\partial\tau$ that completely surrounds the source. This over-determined solution as provided by Eq. (2.33a) can be converted into one that is not over-determined by replacing the incoming-wave Green function G_- by either the incoming-wave Dirichlet Green function G_{D_-} or the incoming-wave Neumann Green function G_{N_-} appropriate to the boundary $\partial\tau$. For example, on replacing G_- by G_{D_-} in Eq. (2.33a) we obtain the following alternative expression for the back-propagated field

$$\Phi_{\mathrm{D}}(\mathbf{r},\omega) = \int_{\partial\tau} dS'\, U_+(\mathbf{r}',\omega)\frac{\partial}{\partial n'}G_{\mathrm{D}_-}(\mathbf{r},\mathbf{r}',\omega), \tag{2.54a}$$

where the field point $\mathbf{r} \in \tau$ and the normal derivative is directed out of the interior τ and into the exterior $\tau^\perp$. It is important to note that this particular back-propagated field will not be equal to the one generated using over-specified data from Eq. (2.33a)! This conclusion is readily verified if we note that under the replacement $G_- \to G_{\mathrm{D}_-}$ the Porter–Bojarski integral equation Eq. (2.33b) relating the back-propagated field to the source becomes

$$\Phi_{\mathrm{D}}(\mathbf{r},\omega) = \int_{\tau_0} d^3r'[G_+(\mathbf{r}-\mathbf{r}',\omega) - G_{\mathrm{D}_-}(\mathbf{r},\mathbf{r}',\omega)]Q(\mathbf{r}',\omega),$$

which is clearly different from Eq. (2.33b). We could also use the incoming-wave Neumann Green function and generate a back-propagated field according to the formula

$$\Phi_{\mathrm{N}}(\mathbf{r},\omega) = -\int_{\partial\tau} dS'\, \frac{\partial}{\partial n'}U_+(\mathbf{r}',\omega)G_{\mathrm{N}_-}(\mathbf{r},\mathbf{r}',\omega), \tag{2.54b}$$

which is different from both Φ and Φ_{D}.

Although there are various choices for a "back-propagation algorithm" we will find that they all generate almost identical results as long as the boundary $\partial\tau$ and the field point $\mathbf{r}$ are many wavelengths removed from the source. Justifying this claim will have to wait until we develop the angular-spectrum expansion in Chapter 4, but for now we can make this claim plausible by looking at the inverse RS boundary-value problem, which will involve back propagation from planar boundaries.

2.11.1 The inverse RS boundary-value problem

In this section we will apply back propagation to obtain an approximate solution to the *inverse RS boundary-value problem.* As its name suggests, the inverse RS boundary-value problem consists of computing the field at points that lie closer to the source than a boundary plane over which Dirichlet or Neumann boundary conditions are specified. For example, if the source of the field is located in the left half-space $z \le z^+ < z_0$ and its radiated field is observed over the plane z_0, then the inverse RS boundary-value problem consists of determining the field to the left of this plane within the strip $z^+ \le z \le z_0$ from Dirichlet or Neumann boundary conditions specified over the plane $z = z_0$. As mentioned above, this problem is not a standard boundary-value problem of the type considered earlier and cannot be solved using any Green-function technique. We will solve it exactly (but unstably) in

Chapter 4 using the *angular-spectrum expansion* and will here employ field back propagation as described above to obtain an approximate solution.

We first consider the inverse RS boundary-value problem described above in terms of Dirichlet or Neumann data specified over a plane $z = z_0$ located to the right of a source that is supported in the left half-space $z \leq z^+ < z_0$. We can then obtain an approximate solution to this inverse problem by using either of the back-propagation algorithms given above. The incoming-wave Dirichlet and Neumann Green functions are given by Eqs. (2.46) with G_+ replaced by G_-. The fields back propagated into the l.h.s. from Dirichlet or Neumann data on the plane $z = z_0$ according to Eqs. (2.54) are then given by Eqs. (2.48) with G_+ replaced by G_- and where the bottom signs of each equation are used since we are back propagating into the l.h.s.:

$$\Phi_D(\mathbf{r}, \omega) = 2 \int dS' \, U_{z_0}(\mathbf{r}'_0, \omega) \frac{\partial}{\partial z'} G_-(\mathbf{r} - \mathbf{r}'_0, \omega), \tag{2.55a}$$

$$\Phi_N(\mathbf{r}, \omega) = -2 \int dS' \, U'_{z_0}(\mathbf{r}'_0, \omega) G_-(\mathbf{r} - \mathbf{r}'_0, \omega), \tag{2.55b}$$

where $\mathbf{r}'_0 = (x', y', z_0)$ denotes a point on the boundary-value plane, U_{z_0} denotes the Dirichlet data on this plane and U'_{z_0} denotes the Neumann data on this plane. It is easy to verify that the above expressions also apply to the inverse RS boundary-value problem posed in the l.h.s so long as the signs are reversed; i.e., for the case in which the source is confined to the half-space $z > z_-$ and the boundary-value plane is located to the left of the source at $z_0 < z_-$.

The above back-propagated fields can be seen to be the solutions of the standard RS boundary-value problem formulated for a field that satisfies the homogeneous Helmholtz equation in the l.h.s. and the *incoming-wave* boundary condition at infinity in that half-space. This suggests an interesting interpretation of these fields and the back-propagation process. In particular, *we can consider an outgoing wave propagating into the r.h.s. also as an incoming wave propagating from the l.h.s.* Thus, we can view the back-propagated fields defined in Eqs. (2.55) as being solutions to the homogeneous Helmholtz equation throughout the half-space $z < z_0$ that satisfy the incoming-wave radiation condition in this half-space and reduce to specified Dirichlet or Neumann boundary values on the boundary plane $z = z_0$. This is only an approximate solution to the inverse RS boundary-value problem since the l.h.s. $z < z_0$ is source-free in this formulation of the problem but contains an actual source in the exact statement of the problem. We will show in Chapter 4 that the two approximate solutions found above are identical and give excellent approximations to the field radiated by a source located in the l.h.s. $z < z_+ < z_0$ so long as the distances of the field point and boundary-value plane from the source are large compared with the wavelength of the field.

2.11.2 Connection with wavefield time reversal

The approximate solution of the inverse RS boundary-value problem obtained using a back-propagation algorithm is intimately connected with the process of wavefield time reversal. To see this, we consider the case of Dirichlet data over a plane z_0 lying to the right of the source for which the back-propagated field is given by Eq. (2.55a). If we take the complex conjugate of both sides of the above equation, we conclude that

$$\Phi_D^*(\mathbf{r},\omega) = 2\int dS'\, U_{z_0}^*(\mathbf{r}_0',\omega)\frac{\partial}{\partial z'}G_-^*(\mathbf{r}-\mathbf{r}_0',\omega). \tag{2.56}$$

The quantity

$$G_-^*(\mathbf{r}-\mathbf{r}_0',\omega) = -\frac{1}{4\pi}\frac{e^{ik^*R}}{R}$$

is recognized as being the outgoing-wave Green function for the conjugate (time-reversed) medium having wavenumber k^*. It then follows that Eq. (2.56) is the solution to the normal outgoing-wave RS boundary-value problem in the conjugate-medium half-space $z < z_0$ for Dirichlet data equal to the complex conjugate of U_{z_0}. In words, this result indicates that the *the time-reversed back-propagated field $\phi_D(\mathbf{r}, -t)$ is equal to the field radiated by the time-reversed boundary-value data into the time-reversed medium.* In other words, if the Dirichlet data are first time-reversed and allowed to radiate back into the source region but in the time-reversed medium then the resulting field will be equal to $\phi(\mathbf{r}, -t)$. We will return to this point later when we implement the back-propagation algorithms using the angular-spectrum expansion in Section 4.4.

2.12 Surface sources in dispersive media

The treatment of surface sources for the wave equation presented in Section 1.8 of Chapter 1 carries over to dispersive media if we simply reformulate that earlier treatment in the frequency domain. In particular, the most general surface source will consist of a *singlet–doublet* pair $Q_s(\mathbf{r}_0,\omega)$, $Q_d(\mathbf{r}_0,\omega)$ defined over some boundary $\partial\tau_0$ that separates a region τ_0 and its complement $\tau_0^\perp$. This surface source pair then radiates a field according to the equation

$$U_+(\mathbf{r},\omega) = \int_{\partial\tau_0} dS_0\left[Q_s(\mathbf{r}_0,\omega)G_+(\mathbf{r}-\mathbf{r}_0,\omega) - Q_d(\mathbf{r}_0,\omega)\frac{\partial}{\partial n_0}G_+(\mathbf{r}-\mathbf{r}_0,\omega)\right], \tag{2.57}$$

where the normal derivative is, by convention, directed out of the region τ_0 and into its complement $\tau_0^\perp$, although this selection is arbitrary (see the discussion in Section 1.8). The field $U_+(\mathbf{r},\omega)$ radiated by the singlet–doublet source pair clearly satisfies the SRC and the homogeneous Helmholtz equation at all space points $\mathbf{r}$ that lie away from the boundary $\partial\tau_0$. This boundary can be finite with interior τ_0 and infinite exterior $\tau_0^\perp$ or infinite in extent as in the Rayleigh–Sommerfeld boundary-value problem.

As in the case of surface sources for the wave equation, a surface source that is distributed over a separable surface (see Chapter 3) defined by generalized coordinates (ξ_1,ξ_2,ξ_3) with $\xi_3 = \xi_{30} =$ constant can be expressed in the frequency domain as a 3D source distribution in the form

$$Q(\mathbf{r},\omega) = Q_s(\xi_1,\xi_2,\omega)\frac{\delta(\xi_3-\xi_{30})}{h_3(\mathbf{r}_0)} + Q_d(\xi_1,\xi_2,\omega)\frac{h_1(\mathbf{r}_0)h_2(\mathbf{r}_0)}{h_1(\mathbf{r})h_2(\mathbf{r})}\frac{\frac{\partial}{\partial\xi_3}\delta(\xi_3-\xi_{30})}{h_3(\mathbf{r})}, \quad (2.58)$$

where $\delta(\cdot)$ is the Dirac delta function and h_1, h_2, h_3 are the scale factors for the coordinate system. The field radiated by Q is then given by the primary field representation Eq. (2.23) with the source given by Eq. (2.58), which reduces to Eq. (2.57). The justification of the use of the terms "singlet" for Q_s and "doublet" for Q_d is apparent from this source representation in terms of a delta function (singlet) and the derivative of the delta function (doublet).

Again, as was the case of surface sources to the wave equation, the singlet–doublet pair is completely arbitrary and, in particular, its elements are *not, necessarily, equal to the boundary values of the field U_+ and its normal derivative on $\partial\tau_0$*. As was established for the wave equation in Section 1.8, this will occur if and only if U_+ vanishes throughout one of the two regions τ_0 ($\tau_0^\perp$) in which case the limit of U_+ and its normal derivative taken from the region $\tau_0^\perp$ (τ_0) onto the boundary $\partial\tau_0$ will equal Q_d and Q_s, respectively. More generally, the field U_+ radiated by an arbitrary singlet–doublet pair will not vanish within either region τ_0 or $\tau_0^\perp$ and the limiting boundary values of U_+ from either region will be determined from the pair of equations (cf. Eqs. (1.71) of Section 1.8)

$$\lim_{\mathbf{r}\to\mathbf{r}_0} U_+(\mathbf{r},\omega) = U_{+\mathrm{P}}(\mathbf{r}_0,\omega) \pm \frac{1}{2}Q_d(\mathbf{r}_0,\omega), \quad (2.59a)$$

$$\lim_{\mathbf{r}\to\mathbf{r}_0} \frac{\partial}{\partial n}U_+(\mathbf{r},\omega) = \frac{\partial}{\partial n_0}U_{+\mathrm{P}}(\mathbf{r}_0,\omega) \pm \frac{1}{2}Q_s(\mathbf{r}_0,\omega), \quad (2.59b)$$

where the + sign is used when the limit is taken from $\tau_0^\perp$ and the minus sign when the limit is taken from τ_0, and the subscript P stands for the principal value of the quantity being subscripted.

2.12.1 Non-radiating surface sources

The NR surface sources within a dispersive medium are defined to be any singlet–doublet pair $Q_{s\mathrm{nr}}, Q_{d\mathrm{nr}}$ that radiates a field U_+ that vanishes throughout one of the two regions τ_0 or $\tau_0^\perp$ at one or more temporal frequencies ω. If the field vanishes in τ_0 we say that the source pair is NR relative to τ_0 and vice versa if the field vanishes throughout $\tau_0^\perp$. In analogy with the surface NR sources to the wave equation treated in Section 1.8 of Chapter 1, the most general surface NR source results from the boundary values of a field radiated by a 3D source supported within τ_0 or $\tau_0^\perp$. This conclusion is a consequence of the fact that any such field must satisfy the second Helmholtz identity throughout the complement region $\tau_0^\perp$ (τ_0) and, hence, by definition, constitute an NR surface source for that region. The proof that *any* NR surface source can be so constructed is not much deeper and is established in Example 2.10 below.

By making use of the fact that NR surface sources result from the boundary values of fields radiated by 3D sources it is possible to give a simple prescription for specifying any NR surface source. In analogy with Eqs. (1.72) of Section 1.8 we conclude that, if $\Psi_{\hat{\tau}_0}(\mathbf{r},\omega)$ is a field that is radiated by a 3D source $Q_{\hat{\tau}_0}(\mathbf{r},\omega)$ supported within a volume $\hat{\tau}_0$ that is located entirely within τ_0 or its complement $\tau_0^\perp$, then for any point $\mathbf{r}_0$ lying on the boundary $\partial\tau_0$

$$Q_{\mathrm{snr}}(\mathbf{r}_0,\omega) = \frac{\partial}{\partial n_0}\Psi_{\hat{\tau}_0}(\mathbf{r}_0,\omega) = -\frac{1}{4\pi}\int_{\hat{\tau}_0} d^3r'\, Q_{\hat{\tau}_0}(\mathbf{r}',\omega)\frac{\partial}{\partial n_0}\frac{e^{ik|\mathbf{r}_0-\mathbf{r}'|}}{|\mathbf{r}_0-\mathbf{r}'|}, \tag{2.60a}$$

$$Q_{\mathrm{dnr}}(\mathbf{r}_0,\omega) = \Psi_{\hat{\tau}_0}(\mathbf{r}_0,\omega) = -\frac{1}{4\pi}\int_{\hat{\tau}_0} d^3r'\, Q_{\hat{\tau}_0}(\mathbf{r}',\omega)\frac{e^{ik|\mathbf{r}_0-\mathbf{r}'|}}{|\mathbf{r}_0-\mathbf{r}'|}, \tag{2.60b}$$

where $\hat{\tau}_0$ is either within τ_0 or $\tau_0^\perp$ and the singlet–doublet pair is NR into the same region τ_0 or $\tau_0^\perp$ as that in which the 3D primary source is located.

Again, as was the case for the wave equation, no surface sources exist that are NR simultaneously in both τ_0 and $\tau_0^\perp$. This is easily proven using Eqs. (2.59). Also, as was the case for the wave equation, it is possible to have essentially NR surface sources that radiate a field that possesses an energy spectrum that is essentially zero at one or more temporal frequencies. Such sources are generated by the boundary values of essentially NR 3D volume sources supported within either the interior region τ_0 or its exterior complement $\tau_0^\perp$ and produce a field identical to that which is radiated by the primary 3D source.

Finally, we mention the important result that NR surface sources can be employed to reduce the complexity of a surface source by removing the singlet or doublet component without affecting the field radiated by the source into a specific region τ_0 or $\tau_0^\perp$. Consider, for example, a singlet–doublet pair Q_s, Q_d that radiates a field according to Eq. (2.57) into the two regions τ_0 and $\tau_0^\perp$. If we require that the field radiated by this pair remain unchanged in only one of these two regions then we can add to this pair of surface sources the singlet–doublet pair Q_{snr}, Q_{dnr} that is NR within the region where the original field is to remain unchanged. The new field that is radiated by the pair $Q_s + Q_{\mathrm{snr}}$, $Q_d + Q_{\mathrm{dnr}}$ is then identical to the original field within the region where the pair Q_{snr}, Q_{dnr} is NR but is changed within the other region due to the radiation generated by the NR pair within this region (recall that a surface source cannot be NR within both τ_0 and $\tau_0^\perp$ simultaneously). On selecting the NR pair such that $Q_{\mathrm{snr}} = -Q_s$ the field will be radiated by the doublet component $Q_d + Q_{\mathrm{dnr}}$ alone, whereas on taking $Q_{\mathrm{dnr}} = -Q_d$ the field will be radiated by the pure singlet $Q_s + Q_{\mathrm{snr}}$.

Example 2.10 We wish to prove that any surface source pair Q_{snr}, Q_{dnr} that radiates a field $U_{+\mathrm{nr}}$ and that is NR into a given region (say, τ_0) must be equal to the boundary values of a field radiated by a 3D volume source supported within the same region. The proof of this follows from the fact that the vanishing of the field $U_{+\mathrm{nr}}$ throughout τ_0 requires that the singlet–doublet pair be related to the boundary values of $U_{+\mathrm{nr}}$ over $\partial\tau_0$ via the equations

$$Q_{\mathrm{snr}}(\mathbf{r}_0,\omega) = \frac{\partial}{\partial n_0}U_{+\mathrm{nr}}(\mathbf{r}_0,\omega), \qquad Q_{\mathrm{dnr}}(\mathbf{r}_0,\omega) = U_{+\mathrm{nr}}(\mathbf{r}_0,\omega),$$

where the r.h.s. of the above two equations are limits of the field quantities taken from the region $\tau_0^\perp$ onto the boundary $\partial\tau_0$. The above results were derived in the time domain for the wave equation in Section 1.8 of Chapter 1 and can be easily derived in the frequency domain by making use of Eqs. (2.59) in conjunction with the requirement that $U_{+\mathrm{nr}}$ vanish within τ_0.

On substituting the above into Eq. (2.57) we find that the field $U_{+\text{nr}}$ can be expressed in the form

$$U_{+\text{nr}}(\mathbf{r},t) = \int_{\partial\tau_0} dS' \left[G_+(\mathbf{r}-\mathbf{r}',\omega)\frac{\partial}{\partial n'}U_{+\text{nr}}(\mathbf{r}',\omega) - U_{+\text{nr}}(\mathbf{r}',\omega)\frac{\partial}{\partial n'}G_+(\mathbf{r}-\mathbf{r}',\omega)\right].$$

We can convert the surface integral into a volume integral over the interior τ_0 of $\partial\tau_0$ by using Green's theorem. We obtain the result

$$\begin{aligned} U_{+\text{nr}}(\mathbf{r},\omega) &= \int_{\tau_0} d^3r' [G_+ \nabla_{r'}^2 U_{+\text{nr}} - U_{+\text{nr}} \nabla_{r'}^2 G_+] \\ &= \int_{\tau_0} d^3r' [G_+(\nabla_{r'}^2 + k^2)U_{+\text{nr}} - U_{+\text{nr}}(\nabla_{r'}^2 + k^2)G_+]. \end{aligned} \tag{2.61}$$

If we now use the fact that

$$(\nabla_{r'}^2 + k^2)G_+(\mathbf{r}-\mathbf{r}',\omega) = \delta(\mathbf{r}-\mathbf{r}'),$$

define the 3D volume source

$$Q(\mathbf{r},\omega) = (\nabla_{r'}^2 + k^2)U_{+\text{nr}}(\mathbf{r}',\omega)$$

and restrict the field point $\mathbf{r}$ to lie within the complement region $\tau_0^{\perp}$ we find that Eq. (2.61) reduces to

$$U_{+\text{nr}}(\mathbf{r},\omega) = \int_{\tau_0} d^3r'\, Q(\mathbf{r}',\omega)G_+(\mathbf{r}-\mathbf{r}',\omega),$$

which establishes the desired result. A similar development can be employed to establish the result for surface sources that are NR into the complement region $\tau_0^{\perp}$.

Further reading

Most of the books and papers referenced in Chapter 1 also apply to this chapter. In addition to those references the book by Goodman (Goodman, 1968) contains an excellent treatment of the Rayleigh–Sommerfeld boundary-value problem, while the SRC is discussed in Sommerfeld (1967). The Porter–Bojarksi integral equation is reviewed in the paper by Bojarski (Bojarski, 1982a) and is used in the inverse source problem by R. P. Porter (Porter, 1970). The papers by Wolf (Shewell and Wolf, 1968) and Sherman (Sherman, 1967) are among the first to deal with back propagation and inverse diffraction in optics. The excellent book by Stamnes (Stamnes, 1986) contains an account of the focusing and back propagation of water waves. Stratton (1941) has a basic review of dispersive media theory, while complete and thorough treatments are contained in Oughstun (2006) and Oughstun and Sherman (1997). Additional treatments of NR sources and their fields are contained in Marengo and Ziolkowski (2000) and Marengo and Devaney (1998). The problem of reducing over-specified boundary conditions for the Helmholtz equation is a form of the so-called "Dirichlet-to-Neumann" map and has been treated extensively in the literature (Fokas, 2008). It plays a crucial role in a number of inverse problems (Sylvester and Uhlmann, 1990; Kirsch, 1993).

Problems

2.1 Prove that the real and imaginary parts of the analytic signal are connected by Hilbert transforms. Hint: use the fact that $f^{(+)}(t)$ is analytic in the l.h.p. and goes to zero as $|t| \to \infty$ in that half-plane.

2.2 Let

$$\epsilon_k(\tau) = \frac{1}{2\pi} \int_{-\infty}^{\infty} d\omega\, k^2(\omega) e^{-i\omega\tau},$$

where $k(\omega)$ is the wavenumber of a causal medium.

1. Prove that $\epsilon_k(\tau)$ is causal, i.e., vanishes for negative τ.
2. Compute $\epsilon_k(\tau)$ as a generalized function for a non-dispersive medium in which $k^2(\omega) = \omega^2/c^2$.
3. Prove that the real and imaginary parts of $k^2(\omega)-(\omega/c)^2$ are connected via Hilbert transforms and derive these transforms.

2.3 Derive the equation satisfied by the time-dependent field $u_+(\mathbf{r}, t)$ radiated by the source $q(\mathbf{r}, t)$ if $k^2(\omega) \Leftrightarrow \epsilon_k(\tau)$. Show that this equation reduces to the usual wave equation in the special case in which $k^2(\omega) = (\omega/c)^2$.

2.4 Prove that $k(-\omega) = -k^*(\omega)$ if ω is real-valued.

2.5 Compute the incoming- and outgoing-wave Green functions $G_-(\mathbf{R}, \omega)$ and $G_+(\mathbf{R}, \omega)$ from their Fourier-integral representations using contour-integration techniques.

2.6 Using the Fourier-integral representations of the frequency-domain outgoing- and incoming-wave Green functions $G_\pm(\mathbf{R}, \omega)$ and Cauchy's integral theorem show that the free-field propagator $G_f(\mathbf{R}, \omega) = G_+(\mathbf{R}, \omega) - G_-(\mathbf{R}, \omega)$ satisfies the homogeneous Helmholtz equation.

2.7 Use the second Helmholtz identity in the time-domain Eq. (1.36b) to derive Eq. (2.31b). Can Eq. (1.36b) be derived from Eq. (2.31b) for dispersive media? What does this say about the validity of this identity in the time domain for dispersive media?

2.8 Show that if $\mathbf{r} \in \tau$ the back-propagated field $\Phi(\mathbf{r}, \omega)$ defined in Eq. (2.33a) can be expressed in terms of the free-field propagator in the form

$$\Phi(\mathbf{r}, \omega) = -\int_{\partial\tau} dS' \left[U_+(\mathbf{r}', \omega) \frac{\partial}{\partial n'} G_f(\mathbf{r} - \mathbf{r}', \omega) - G_f(\mathbf{r} - \mathbf{r}', \omega) \frac{\partial}{\partial n'} U_+(\mathbf{r}', \omega) \right].$$

2.9 Verify that the contribution from the integral over the surface Σ_∞ in the derivation of Eqs. (2.43) vanishes.

2.10 State and prove the frequency-domain version of the source decomposition Theorem 1.3.

2.11 Show by using a coordinate-system translation along the z axis that the RS solutions given in Eqs. (2.47) yield the solutions in Eqs. (2.48), which are valid for a boundary-value plane $z = z_0$, with z_0 being arbitrary.

2.12 Solve the 1D inverse RS problem in the r.h.s. $z > z^+$ in terms of Dirichlet data on the line $z = z_0 > z^+$. Compare this solution with the exact field radiated by the source into this half-space.

2.13 Compute a 1D NR surface source distributed on the two points $z = z^{\pm}$. Hint: find a pair of coupled algebraic conditions that must be satisfied.

2.14 Use the general procedure employed in Section 2.8 to compute the field radiated by a source located in the l.h.s. in the presence of a Dirichlet plane (a plane over which the field vanishes) at $z = 0$. Express your answer in terms of a Dirichlet Green function.

2.15 Interpret the radiated field found in the previous problem in terms of a mirror-image source relative to the plane $z = 0$.

2.16 Determine a cloaking surface source over a closed surface $\partial\tau_0$ directly in the frequency domain using NR surface sources constructed from the incident wavefield and its normal derivative over the cloaking surface.

2.17 Prove that the cloaking surface source found in the previous problem with the incident wavefield replaced by the total wavefield (incident plus that generated from the cloaked object) radiates an identical wavefield into the interior τ_0 as that of the original surface source.

2.18 Give an argument for why the surface source found in the previous problem also cloaks the region τ_0; i.e., both cancels out the incident wavefield within τ_0 and is NR outside of τ_0.

3 Eigenfunction expansions of solutions to the Helmholtz equation

The "solutions" of boundary-value problems presented in Section 2.8 are merely formal in that the appropriate Green functions must be found, and finding these Green functions is extremely difficult except in certain special cases such as the Rayleigh–Sommerfeld problem. Moreover, even when the appropriate Green functions can be found they are often expressed in the form of superpositions of elementary solutions of the homogeneous Helmholtz equation that are especially suited for dealing with boundaries of a specific shape. One example of a set of such elementary solutions is the *plane waves* which arise from applying the method of separation of variables to the homogeneous Helmholtz equation using a Cartesian coordinate system and, as we will see below, form a complete set of basis functions for fitting boundary-value data specified on plane surfaces; e.g., for the RS problems. However, the plane waves have limited utility in solving boundary-value problems involving non-planar boundaries such as spherical boundaries. In that case the method of separation of variables is applied to the Helmholtz equation using a spherical polar coordinate system and the so-called *multipole fields* arise as a set of elementary solutions that form a basis for fitting boundary-value data specified on spherical boundaries. In this chapter we will briefly review the method of separation of variables for the Helmholtz equation and obtain the resulting eigenfunctions for the important cases of Cartesian, spherical polar and cylindrical coordinate systems. We mention that, although one of the principal uses of eigenfunction representations is in solving boundary-value problems, they are also extremely useful in general analytic studies of wave propagation and scattering and in the implementation of general algorithms associated with wave phenomena. We will use these representations heavily in developing inversion algorithms both for inverse source problems and for inverse scattering problems.

3.1 Separation of variables and the Sturm–Liouville problem

In the method of separation of variables applied to the homogeneous Helmholtz equation one decides on a specific coordinate system (e.g., Cartesian, cylindrical, spherical) and looks for a solution of the homogeneous Helmholtz equation with wavenumber $k = \omega/(cn(\omega))$ of the form

$$U(\mathbf{r}, \omega) = U(\xi_1, \xi_2, \xi_3) = U_1(\xi_1)U_2(\xi_2)U_3(\xi_3), \tag{3.1}$$

where the $U_j(\xi_j)$ are functions only of the jth generalized coordinate in the selected coordinate system and, possibly, the wavenumber k. Equation (3.1) is then substituted into the homogeneous Helmholtz equation with the D'Alembertian operator represented in the ξ_1, ξ_2, ξ_3 system. For certain choices of the coordinate system the resulting equation then *separates* into three ordinary differential equations, one for each component function $U_j(\xi_j)$. It is known that for the scalar Helmholtz equation there exist exactly 11 coordinate systems for which the separation method works. We will be mainly concerned with three of them: (i) Cartesian coordinates, (ii) spherical coordinates and (iii) cylindrical coordinates, although we will also have occasion to deal with others such as the spheroidal coordinates whose corresponding eigenfunctions play a central role in certain inverse problems associated with the Fourier transform (Slepian and Pollak, 1961).

The three ordinary differential equations arising from the separation-of-variables procedure form a two-parameter coupled set of Sturm–Liouville problems, which can be jointly solved to yield a complete set of basis functions for solving the homogeneous scalar wave Helmholtz equation. When solving boundary-value problems for the homogeneous Helmholtz equation one selects the two free parameters to be eigenvalues associated with the generalized coordinates of the surface on which the boundary-value data are specified (say, ξ_3 = constant). In such a way it is then possible to generate a two-parameter set of functions that satisfy the homogeneous Helmholtz equation and, in addition, form a complete set of functions for fitting Dirichlet, Neumann or mixed boundary conditions specified on the surface(s) ξ_3 = constant. We will employ this procedure in the following sections to solve the Dirichlet and Neumann boundary-value problems for Cartesian, spherical and cylindrical boundary surfaces.

3.1.1 The Sturm–Liouville problem

In the separable systems of interest here separation of variables applied to the homogeneous Helmholtz equation

$$[\nabla^2 + k^2]U(\mathbf{r}, k) = 0 \tag{3.2}$$

will lead to a set of coupled linear first-order ordinary differential equations of the general form

$$\left[\frac{d}{d\xi}\left[p(\xi)\frac{d}{d\xi}\right] + q(\xi) + \lambda w(\xi)\right]\psi(\xi, \lambda) = 0, \tag{3.3}$$

where $p(\xi)$ and $q(\xi)$ are real-valued analytic functions of the real variable ξ, $w(\xi) > 0$ is a real-valued weight function and λ is an eigenvalue that depends on the wavenumber k. The differential operator

$$\mathcal{L} = \frac{d}{d\xi}\left[p(\xi)\frac{d}{d\xi}\right] + q(\xi) \tag{3.4}$$

is Hermitian under the standard inner product in the space $L^2(a, b)$ of square-integrable functions on $[a, b]$; i.e.,

$$\int_a^b d\xi [\mathcal{L}\psi(\xi, \lambda)]\chi(\xi, \lambda) = p(\xi)[\psi'\chi - \psi\chi']|_a^b + \int_a^b d\xi\, \psi(\xi, \lambda)[\mathcal{L}\chi(\xi, \lambda)], \tag{3.5}$$

so that

$$\langle \mathcal{L}\psi, \chi \rangle = \langle \psi, \mathcal{L}\chi \rangle \tag{3.6a}$$

so long as the functions ψ and χ satisfy the condition

$$p(\xi)[\psi'(\xi,\lambda)\chi(\xi,\lambda) - \psi(\xi,\lambda)\chi'(\xi,\lambda)]|_a^b = 0. \tag{3.6b}$$

It is important to note that for the operator $\mathcal{L}$ to be Hermitian it is required that it be of the general form given in Eq. (3.4) *and* that the eigenfunctions satisfy the boundary conditions given in Eq. (3.6b).

It is well known and easily proven that the eigenvalues of an Hermitian operator are real-valued and that eigenfunctions corresponding to different eigenvalues are orthogonal. More difficult to prove, but also true, is the fact that the eigenfunctions are complete in $L^2(a, b)$; i.e., under the induced norm

$$||\psi|| = \sqrt{\langle \psi, \psi \rangle}.$$

Finally, it can also be shown that the eigenvalues are discrete if the interval $[a, b]$ is finite and will form a continuum otherwise. Both possibilities will occur in our applications.

3.2 Cartesian coordinates

Returning to the homogeneous Helmholtz equation, we select the three generalized coordinates ξ_j to be the Cartesian coordinates x, y, z and we find that

$$[\nabla^2 + k^2]U_x(x)U_y(y)U_z(z) = 0,$$

which results in the separated system of equations

$$\left[\frac{\partial^2}{\partial x^2} + K_x^2\right] U_x(x) = 0, \qquad \left[\frac{\partial^2}{\partial y^2} + K_y^2\right] U_y(y) = 0, \qquad \left[\frac{\partial^2}{\partial z^2} + K_z^2\right] U_z(z) = 0,$$

where the three *separation constants* K_x^2, K_y^2 and K_z^2 must satisfy the constraint equation (dispersion relation)

$$K_x^2 + K_y^2 + K_z^2 = k^2. \tag{3.7}$$

The general solution of the coupled system is easily found to be

$$U(\mathbf{r}, \omega) = U_x(x)U_y(y)U_z(z) = Ae^{iK_x x}e^{iK_y y}e^{iK_z z} = Ae^{i\mathbf{K}\cdot\mathbf{r}}, \tag{3.8}$$

where A is an arbitrary constant and the wave vector $\mathbf{K} = K_x\hat{\mathbf{x}} + K_y\hat{\mathbf{y}} + K_z\hat{\mathbf{z}}$ has arbitrary direction but a (possibly complex) length $K = \sqrt{\mathbf{K}\cdot\mathbf{K}} = k$ as required by the constraint equation Eq. (3.7).

The most general solution of the homogeneous Helmholtz equation with fixed wavenumber $k = \omega/(cn(\omega))$ can be expressed as an integral of the separated solutions Eq. (3.8) over all wave vectors $\mathbf{K}$ satisfying the constraint equation Eq. (3.7). Because of this constraint there are only two degrees of freedom for the wave vector $\mathbf{K}$. They can be selected in a number of ways. For example, we can take K_x and K_y free and constrain K_z to satisfy the constraint equation. Alternatively, we can represent the wave vector in spherical coordinates, take the radial component equal to the wavenumber k and take the polar and azimuthal angles α and β as the free variables. The latter procedure leads to a plane-wave expansion of the form

$$U(\mathbf{r}, \omega) = \int d\Omega_{\mathbf{s}} \, A(k\mathbf{s}, \omega) e^{ik\mathbf{s}\cdot\mathbf{r}}, \tag{3.9a}$$

where the *unit propagation vector* $\mathbf{s}$ is given in terms of its polar and azimuthal angles α and β by

$$\mathbf{s} = \sin\alpha\cos\beta\,\hat{\mathbf{x}} + \sin\alpha\sin\beta\,\hat{\mathbf{y}} + \cos\alpha\,\hat{\mathbf{z}} \tag{3.9b}$$

and the integration over $\mathbf{s}$ in Eq. (3.9a) is over a set of real or complex angles. We will find below that different choices for the angular integration region correspond to different properties of the wavefield U and, in particular, to different boundary conditions for the field U.

3.2.1 Homogeneous plane-wave expansions

If we demand that the unit propagation vector $\mathbf{s}$ defined in Eq. (3.9b) be strictly real then the polar and azimuthal angles α and β must be strictly real. The most general form of the corresponding plane-wave expansion is then given by Eq. (3.9a), with the angular integration region taken to be 4π steradians. This plane-wave expansion converges everywhere under very weak conditions on the plane-wave amplitude $A(k\mathbf{s}, \omega)$ and, hence, satisfies the homogeneous Helmholtz equation over all of space.

In the important case in which the wavenumber k is real-valued and the unit propagation vector $\mathbf{s}$ is also real, the elemental plane waves entering into the plane-wave expansion Eq. (3.9a) are all *homogeneous* plane waves. By this we mean that they have strictly real wave vectors $\mathbf{K} = k\mathbf{s}$ and unit magnitude over all of space and, in particular, do not decay along any direction in space. In this case we say that Eq. (3.9a) is a *homogeneous plane-wave expansion.* A plane wave that decays (or grows) in amplitude as it propagates is called an *inhomogeneous* plane wave. In the case of a dispersive medium where $\Im k > 0$ the wave vector $k\mathbf{s}$ will be complex even if the unit propagation vector $\mathbf{s}$ is real-valued. Thus, it is not possible to have purely homogeneous plane waves in a dispersive medium, and what are homogeneous plane waves in a non-dispersive medium become *inhomogeneous* plane waves in a dispersive medium.

Plane-wave expansion in the time domain

We can construct solutions to the homogeneous wave equation Eq. (1.3) by using a superposition of elementary plane waves both over frequencies ω and over propagation directions **s**. In this case the wavenumber k is real-valued and we can construct the time-dependent plane-wave expansion

$$u(\mathbf{r},t) = \frac{1}{(2\pi)^3}\int_{-\infty}^{\infty} k^2\,dk \int d\Omega_{\mathbf{s}}\, A(k\mathbf{s},\omega)e^{ik(\mathbf{s}\cdot\mathbf{r}-ct)}, \tag{3.10}$$

where we have used the wavenumber $k = \omega/c$ as the integration variable and included the multiplying factor k^2 in the k integral and the factor $1/(2\pi)^3$ for later notational simplicity. It is easy to verify that the field U constructed in Eq. (3.10) satisfies the homogeneous wave equation throughout its region of (uniform) convergence.

Let's consider the special, but important, case in which the integration angles α and β are real and the angular range in the plane-wave expansion Eq. (3.10) is over a full 4π steradians. In this case the expansion Eq. (3.10) is a *homogeneous plane-wave expansion* since the wave vectors $k\mathbf{s}$ of the plane waves comprising the expansion are all real-valued. It then follows that the expansion will converge and satisfy the homogeneous wave equation over all of space and for all time. Moreover, for this case we can express the wave vector $\mathbf{K} = k\mathbf{s}$ in terms of its Cartesian components

$$K_x = k\sin\alpha\cos\beta, \quad K_y = k\sin\alpha\sin\beta, \quad K_z = k\cos\alpha.$$

If we then make the change of variable from the polar coordinates k, α, β to the Cartesian coordinates K_x, K_y, K_z in Eq. (3.10) we obtain the result

$$\begin{aligned} u(\mathbf{r},t) &= \frac{1}{(2\pi)^3}\int d^3K\, A(\mathbf{K},cK)e^{i(\mathbf{K}\cdot\mathbf{r}-cKt)} \\ &+ \frac{1}{(2\pi)^3}\int d^3K\, A(\mathbf{K},-cK)e^{i(\mathbf{K}\cdot\mathbf{r}+cKt)}. \end{aligned} \tag{3.11}$$

The first term on the r.h.s. of Eq. (3.11) represents the contribution of positive frequencies in Eq. (3.10) while the second represents the contribution of negative frequencies. Equation (3.10) is referred to as the *angle-variable* (or polar) form of the plane-wave expansion and Eq. (3.11) as the *Cartesian-variable* form of this expansion.

Example 3.1 As an example of the use of the plane-wave expansion Eq. (3.11) we consider the initial-value problem addressed in Chapter 1 using Green-function techniques. It follows from this expansion that the plane-wave amplitudes $A(\mathbf{K}, cK)$ and $A(\mathbf{K}, -cK)$ must satisfy the Cauchy conditions

$$\begin{aligned} u_{t_0}(\mathbf{r}) &= \frac{1}{(2\pi)^3}\int d^3K\, A(\mathbf{K},cK)e^{i(\mathbf{K}\cdot\mathbf{r}-cKt_0)} \\ &+ \frac{1}{(2\pi)^3}\int d^3K\, A(\mathbf{K},-cK)e^{i(\mathbf{K}\cdot\mathbf{r}+cKt_0)}, \\ u'_{t_0}(\mathbf{r}) &= \frac{1}{(2\pi)^3}\int d^3K\, -icKA(\mathbf{K},cK)e^{i(\mathbf{K}\cdot\mathbf{r}-cKt_0)} \\ &+ \frac{1}{(2\pi)^3}\int d^3K\, icKA(\mathbf{K},-cK)e^{i(\mathbf{K}\cdot\mathbf{r}+cKt_0)}, \end{aligned}$$

where

$$u_{t_0}(\mathbf{r}) = u(\mathbf{r},t)|_{t=t_0}, \qquad u'_{t_0}(\mathbf{r}) = \frac{\partial}{\partial t}u(\mathbf{r},t)|_{t=t_0}.$$

On spatial Fourier transforming the above set of equations we then find that

$$A(\mathbf{K},cK)e^{-icKt_0} + A(\mathbf{K},-cK)e^{icKt_0} = \tilde{u}_{t_0}(\mathbf{K}), \tag{3.12a}$$

$$A(\mathbf{K},cK)e^{-icKt_0} - A(\mathbf{K},-cK)e^{icKt_0} = \frac{i}{cK}\tilde{u}'_{t_0}(\mathbf{K}), \tag{3.12b}$$

where

$$\tilde{u}_{t_0}(\mathbf{K}) = \int d^3r\, u_{t_0}(\mathbf{r})e^{-i\mathbf{K}\cdot\mathbf{r}},$$

$$\tilde{u}'_{t_0}(\mathbf{K}) = \int d^3r\, u'_{t_0}(\mathbf{r})e^{-i\mathbf{K}\cdot\mathbf{r}},$$

are the spatial Fourier transforms of the Cauchy data at t_0. The coupled set of equations Eqs. (3.12) are easily solved and we obtain the result

$$A(\mathbf{K},cK) = \frac{1}{2}\left[\tilde{u}_{t_0}(\mathbf{K}) + \frac{i}{cK}\tilde{u}'_{t_0}(\mathbf{K})\right]e^{icKt_0}, \tag{3.13a}$$

$$A(\mathbf{K},-cK) = \frac{1}{2}\left[\tilde{u}_{t_0}(\mathbf{K}) - \frac{i}{cK}\tilde{u}'_{t_0}(\mathbf{K})\right]e^{-icKt_0}. \tag{3.13b}$$

The plane-wave expansion Eq. (3.11), with the plane-wave amplitudes $A(\mathbf{K},cK)$ and $A(\mathbf{K},-cK)$ defined in Eqs. (3.13), satisfies the homogeneous wave equation and the specified Cauchy conditions and hence is the unique solution to the initial-value problem under this Cauchy data.

3.2.2 Plane-wave expansions that include inhomogeneous plane waves

Homogeneous plane waves were defined in the previous section as being plane waves whose wave vectors $\mathbf{K}$ are real-valued so that they have unit magnitude over all of space. Superpositions of these plane waves (called "homogeneous plane-wave expansions") can be employed, as we have just seen in the above example, to solve the Cauchy initial-value problem for the homogeneous wave equation and, in certain cases, can also be used to represent solutions to the inhomogeneous Helmholtz equation in a non-dispersive medium or to the wave equation over a restricted region of space such as within a finite volume lying outside the source volume τ_0.

We can obtain a generalization to the homogeneous plane-wave expansion that can be employed in dispersive media by simply allowing the wavenumber in Eq. (3.9a) to be complex but keeping the unit propagation vector $\mathbf{s}$ real-valued. The resulting plane waves comprising the expansion are no longer homogeneous, since the wave vectors $\mathbf{K} = k\mathbf{s}$ are now complex due to the complex nature of k, but they have the special property that they become homogeneous in the limit when $\Im k \to 0$. We will generally refer to such waves as *weakly inhomogeneous plane waves*. Plane-wave expansions comprised of weakly inhomogeneous plane waves satisfy the homogeneous Helmholtz equation in a dispersive medium and can thus be used to represent certain classes of solutions to this equation.

Unfortunately, the generalized form of the plane-wave expansion comprised of weakly inhomogeneous plane waves has limited applicability for solving boundary-value problems for the Helmholtz equation due to the fact that this set of plane waves does not form a complete set of functions on any closed or open boundary-value surface.[1] For example, to solve a boundary-value problem for Dirichlet or Neumann data specified on the (x, y) plane, it is required that K_x and K_y vary over the entire (K_x, K_y) plane in order for the plane waves defined in Eq. (3.8) to be complete on this plane. This, in turn, requires that $K_z = \pm\sqrt{k^2 - K_x^2 - K_y^2}$ assume complex values even in the limit when $\Im k \to 0$. It is thus necessary to admit into the expansion inhomogeneous plane waves that have complex wave vectors even in the limit where the wavenumber is purely real-valued. Such plane waves are called *evanescent* plane waves. It is clear that the basic form Eq. (3.9a) of the plane-wave expansion does not exclude such plane waves and, in particular, this class of inhomogeneous plane waves arises when the polar integration angles α and β are allowed to assume complex values.

Weakly inhomogeneous and evanescent plane waves

To understand better the nature of weakly inhomogeneous and evanescent plane waves, we define the transverse wavenumber $\mathbf{K}_\rho = K_x\hat{\mathbf{x}} + K_y\hat{\mathbf{y}}$ and take K_x and K_y to be the two free variables in the general plane wave defined in Eq. (3.8). The allowable values for K_z are then given by $K_z = \pm\gamma$, where

$$\gamma = \begin{cases} \sqrt{k^2 - K_\rho^2} & \text{if } K_\rho^2 < \Re k^2, \\ i\sqrt{K_\rho^2 - k^2} & \text{if } K_\rho^2 > \Re k^2. \end{cases} \tag{3.14}$$

The quantity γ is the principal root of the multivalued function $\sqrt{k^2 - K_\rho^2}$ and has positive real and imaginary parts for all real and positive values of K_ρ and $\Re k$.[2] We note that in the limit of a non-dispersive medium in which $\Im k \to 0$ γ is pure real (and positive) for $K_\rho^2 < \Re k^2$ and pure positive imaginary for $K_\rho^2 > \Re k^2$. The plane waves having transverse wave vectors $\mathbf{K}_\rho$ for which $K_\rho^2 < \Re k^2$ are, thus, inhomogeneous plane waves due to the fact that the medium in which they propagate is dispersive, and become homogeneous in the limit $\Im k \to 0$. These plane waves have the same character as those having a wave vector $\mathbf{K} = k\mathbf{s}$ with k complex and $\mathbf{s}$ real, and are thus what we have defined to be "weakly inhomogeneous plane waves."

[1] Note, in this connection, that the homogeneous plane waves used in Example 3.1 form a complete set of functions for expanding the Cauchy conditions for the initial-value problem. This is the reason why the plane-wave expansion using homogeneous plane waves is ideally suited for this problem.

[2] Here, and in much of what is to follow, we tacitly assume that $\Re k > 0$. As mentioned at the beginning of Chapter 2, the positive frequencies $\omega > 0$ completely and uniquely define real-valued fields of the type considered in this book, so no generality is lost in making this assumption.

On the other hand, the plane waves having transverse wave vectors $\mathbf{K}_\rho$ for which $K_\rho^2 > \Re k^2$ are inhomogeneous plane waves of a different type altogether. These plane waves are inhomogeneous even if the medium is non-dispersive and k is strictly real-valued. To distinguish them from the weakly inhomogeneous plane waves they are called *evanescent* plane waves.

Example 3.2 To get a better feel for the two types of inhomogeneous plane wave we examine the special case in which the medium absorption is small and set $k^2 = \Re k^2 + i\Im k^2$. We then have that

$$\gamma = \sqrt{k^2 - K_\rho^2} \approx \sqrt{\Re k^2 - K_\rho^2} + i\frac{\Im k^2}{2\sqrt{\Re k^2 - K_\rho^2}}, \qquad K_\rho^2 < \Re k^2, \tag{3.15a}$$

$$\gamma = i\sqrt{K_\rho^2 - k^2} \approx i\sqrt{K_\rho^2 - \Re k^2} + \frac{\Im k^2}{2\sqrt{K_\rho^2 - \Re k^2}}, \qquad K_\rho^2 > \Re k^2. \tag{3.15b}$$

When $\Re k^2 - K_\rho^2 > 0$, γ has a non-zero real part and a (small) imaginary part that vanishes in the limit $\Im k \to 0$. Thus, the plane waves resulting from $K_z = \pm\gamma$ in these cases are homogeneous plane waves in the limit $\Im k \to 0$ and, hence, are what we have called "weakly inhomogeneous plane waves." On the other hand, when $\Re k^2 - K_\rho^2 < 0$ the first term in Eq. (3.15b) is pure imaginary and γ is complex even in the limit $\Im k \to 0$. The associated set of plane waves with the set of wave vectors with $K_z = \pm\gamma$ when $\Re k^2 - K_\rho^2 < 0$ thus corresponds to the class of *evanescent plane waves.*

3.2.3 Plane-wave expansions involving evanescent plane waves

We can construct a plane-wave expansion using the plane waves defined in Eq. (3.8) with $K_z = \pm\gamma$. This expansion will contain both weakly inhomogeneous plane waves having transverse wave numbers $\mathbf{K}_\rho$ whose magnitude squared $K_\rho^2 < \Re k^2$ and evanescent plane waves for which $K_\rho^2 > \Re k^2$. The most general form for such an expansion is given by

$$U(\mathbf{r},\omega) = \overbrace{\frac{i}{2\pi}\int \frac{d^2K_\rho}{\gamma} A^{(+)}(\mathbf{k}^+,\omega)e^{i\mathbf{k}^+\cdot\mathbf{r}}}^{U^{(+)}(\mathbf{r},\omega)} + \overbrace{\frac{i}{2\pi}\int \frac{d^2K_\rho}{\gamma} A^{(-)}(\mathbf{k}^-,\omega)e^{i\mathbf{k}^-\cdot\mathbf{r}}}^{U^{(-)}(\mathbf{r},\omega)}, \tag{3.16a}$$

where

$$\mathbf{k}^\pm = \mathbf{K}_\rho \pm \gamma\hat{\mathbf{z}}, \tag{3.16b}$$

with γ defined in Eq. (3.14) and where we have expressed the above expansion in a form that is most convenient for later applications. Clearly, the wave vectors $\mathbf{k}^\pm$ satisfy the constraint equation Eq. (3.7) due to the definition of γ so that both components satisfy the homogeneous Helmholtz equation in the spatial regions in which they converge.

Both components, $U^{(+)}$ and $U^{(-)}$, of the general plane-wave expansion given in Eq. (3.16a) are seen to consist of superpositions of plane waves, each of which satisfies the homogeneous Helmholtz equation. The plane waves comprising $U^{(+)}$ all propagate in the positive-z direction, while the plane waves comprising $U^{(-)}$ all propagate in the negative-z direction. Both components include both weakly inhomogeneous and evanescent plane waves, with the latter decaying or growing exponentially with increasing $|z|$ from the $z = 0$ plane. The wavefield $U^{(+)}$ is appropriate for representing wavefields propagating into the right half-space $z > 0$, while $U^{(-)}$ is used for representing wavefields that propagate into the left half-space $z < 0$. The plane-wave amplitudes $A^{(\pm)}(\mathbf{k}^{\pm}, \omega)$ can be arbitrarily chosen but are, in principle, uniquely specified by appropriate boundary conditions.

The two fields $U^{(+)}$ and $U^{(-)}$ defined by the plane-wave expansions in Eqs. (3.16a) satisfy the Sommerfeld radiation condition (SRC) in the half-spaces $z > 0$ and $z < 0$, respectively. This conclusion can be arrived at by applying the method of stationary phase to their respective plane-wave expansions, and one finds that (Born and Wolf, 1999; Mandel and Wolf, 1995)

$$U^{(\pm)}(r\mathbf{s}, \omega) \sim A^{(\pm)}(\mathbf{k}^{\pm}, \omega)\frac{e^{ikr}}{r}, \tag{3.17}$$

as $r \to \infty$ along the direction of the unit vector $\mathbf{s} = \mathbf{r}/r$. In this equation $\mathbf{K}_\rho = (ks_x, ks_y) = (k \sin\theta \cos\phi, k \sin\theta \sin\phi)$ and $K_z = ks_z = k\cos\theta$, with (r, θ, ϕ) being the polar coordinates of the field point $\mathbf{r}$. We will not digress here to discuss this result or its ramifications, since we will return to this issue in Section 4.2.2 of the following chapter, where we will derive the result directly from the Green-function representation of the radiated field. Because of the fact that these plane-wave expansions satisfy the SRC they are ideally suited for solving the Rayleigh–Sommerfeld boundary-value problem, as we mentioned above and as the following example illustrates.

Example 3.3 The plane-wave expansion derived above can be employed to solve the Rayleigh–Sommerfeld problem treated in Section 2.9 using the Green-function approach. Here we will solve the RS problem in the right half-space $z > 0$ in terms of Dirichlet or Neumann data on the plane $z = 0$. To solve this boundary-value problem we employ the plane-wave expansion $U^{(+)}(\mathbf{r}, \omega)$ comprising the first term in Eq. (3.16a) and compute the plane-wave amplitude from boundary conditions on the plane $z = 0$. It is clear from Eq. (3.17) that U^{+} satisfies the SRC in the r.h.s. and, hence, is an appropriate representation for the solution to the RS problem in this half-space.

To determine the plane-wave amplitude we note using $U^{(+)}$ as defined in Eq. (3.16a) that on the plane $z = 0$

$$U_0(\boldsymbol{\rho}, \omega) = \frac{i}{2\pi} \int \frac{d^2K_\rho}{\gamma} A^{(+)}(\mathbf{k}^{+}, \omega) e^{i\mathbf{K}_\rho \cdot \boldsymbol{\rho}},$$

$$U_0'(\boldsymbol{\rho}, \omega) = \frac{i}{2\pi} \int \frac{d^2K_\rho}{\gamma} i\gamma A^{(+)}(\mathbf{k}^{+}, \omega) e^{i\mathbf{K}_\rho \cdot \boldsymbol{\rho}},$$

where U_0 is the Dirichlet boundary condition on the plane $z = 0$ and $U_0' = (\partial/\partial z)U$ is the Neumann condition on this plane. In addition, we have defined the vector

$$\boldsymbol{\rho} = x\hat{\mathbf{x}} + y\hat{\mathbf{y}},$$

which is recognized as the projection of the field point $\mathbf{r}$ onto the (x, y) plane. If we Fourier invert both sides of the above equations, we then find that

$$A^{(+)}(\mathbf{k}^+, \omega) = \begin{cases} -[i\gamma/(2\pi)]\tilde{U}_0(\mathbf{K}_\rho, \omega) & \text{Dirichlet conditions,} \\ -[1/(2\pi)]\tilde{U}_0'(\mathbf{K}_\rho, \omega) & \text{Neumann conditions,} \end{cases} \tag{3.18}$$

where the tilde $\tilde{}$ denotes the spatial Fourier transform; e.g.,

$$\tilde{U}_0(\mathbf{K}_\rho, \omega) = \int d^2\rho\, U_0(\boldsymbol{\rho}, \omega) e^{-i\mathbf{K}_\rho \cdot \boldsymbol{\rho}}.$$

As a final step we substitute the expressions for the plane-wave amplitudes directly into the plane-wave expansion for U_+ to obtain

$$U_{\mathrm{D}}^{(+)}(\mathbf{r}, \omega) = \frac{1}{(2\pi)^2} \int d^2K_\rho\, \tilde{U}_0(\mathbf{K}_\rho, \omega) e^{i\mathbf{k}^+ \cdot \mathbf{r}}, \tag{3.19a}$$

$$U_{\mathrm{N}}^{(+)}(\mathbf{r}, \omega) = \frac{-i}{(2\pi)^2} \int d^2K_\rho \frac{\tilde{U}_0'(\mathbf{K}_\rho, \omega)}{\gamma} e^{i\mathbf{k}^+ \cdot \mathbf{r}}, \tag{3.19b}$$

where we have used the subscripts D and N to denote the solutions to the Dirichlet and Neumann RS boundary-value problems in the half-space $z > 0$. It is readily verified that these solutions satisfy the homogeneous Helmholtz equation and reduce to the required boundary conditions on the plane $z = 0$.

The plane-wave expansions of $U^{(+)}$ and $U^{(-)}$ derived in the example presented above employ the plane $z = 0$ as the boundary-value plane and represent the field either in the half-space $z > 0$ or in the half-space $z < 0$. However, since the orientation of our Cartesian system is arbitrary it is clear that these representations can be employed in any half-space within which the field satisfies the homogeneous Helmholtz equation and the SRC. Moreover, since the plane-wave expansion U_+ converges throughout the right-half space $z > 0$ and satisfies the SRC in that half-space this expansion actually solves the Rayleigh–Sommerfeld boundary-value problem for the general case of specified Dirichlet or Neumann conditions on any plane $z = z_0 \geq 0$. For example, it follows from the plane-wave expansion for the Dirichlet problem addressed in Example 3.3 that

$$\tilde{U}_{z_0}(\mathbf{K}_\rho, \omega) = \tilde{U}_0(\mathbf{K}_\rho, \omega) e^{i\gamma z_0},$$

where $\tilde{U}_{z_0}(\mathbf{K}_\rho, \omega)$ is the spatial Fourier transform of the field on the plane $z = z_0 \geq 0$. Solving for $\tilde{U}_0$ in terms of $\tilde{U}_{z_0}$ and substituting into the plane-wave expansion then yields

$$U_{\mathrm{D}}^{(+)}(\mathbf{r}, \omega) = \frac{1}{(2\pi)^2} \int d^2K_\rho\, \tilde{U}_{z_0}(\mathbf{K}_\rho, \omega) e^{-i\gamma z_0} e^{i\mathbf{k}^+ \cdot \mathbf{r}} \tag{3.20}$$

for the solution of the boundary-value problem with Dirichlet conditions specified on the plane $z_0 > 0$. A similar result is obtained for the left half-space using $U^{(-)}$ in place of $U^{(+)}$.

Spatial resolution of homogeneous and evanescent plane waves

We see from the above example that the plane-wave amplitudes of the fields resulting from Dirichlet or Neumann conditions over a boundary plane are related to the boundary conditions via a spatial Fourier transform. Each spatial frequency component of the field or its normal derivative over the plane is mapped into a specific plane wave that then propagates into a half-space as a component of the solution to the boundary-value problem. The division between weakly inhomogeneous and evanescent plane waves occurs at $K_\rho^2 = \Re k^2$, corresponding to a spatial period equal to $L = 2\pi/\sqrt{\Re k^2}$, which for non-dispersive media is $L = \lambda$, the wavelength of the field. It then follows that *weakly inhomogeneous plane waves radiate from spatial periods larger than the spatial period* $L = 2\pi/\sqrt{\Re k^2}$ *and evanescent plane waves from spatial periods smaller than L.* For non-dispersive and weakly dispersive media the conclusion is simpler: *homogeneous plane waves radiate from spatial periods larger than the wavelength and evanescent plane waves from spatial periods smaller than the wavelength.* These observations will be used many times throughout the book to interpret both wave-propagation phenomena and wavefield-inversion algorithms.

The angle-variable form of the plane-wave expansion

It is also possible to represent the plane-wave expansion given in Eq. (3.16a) in the general form of the homogeneous plane-wave expansion given in Eq. (3.9a) using spherical polar coordinates for the integration variables. We first make a change from Cartesian to cylindrical polar integration variables

$$K_x = K_\rho \cos\beta, \qquad K_y = K_\rho \sin\beta,$$
$$\mathbf{k}^\pm = K_\rho \cos\beta\,\hat{\mathbf{x}} + K_\rho \sin\beta\,\hat{\mathbf{y}} \pm \gamma\hat{\mathbf{z}}$$

to find that the component fields $U^\pm$ in Eq. (3.16a) assume the form

$$U^{(\pm)}(\mathbf{r},\omega) = \frac{i}{2\pi}\int_{-\pi}^{\pi} d\beta \int_0^\infty \frac{K_\rho\, dK_\rho}{\gamma} A^{(\pm)}(\mathbf{k}^\pm,\omega) e^{i(K_\rho\cos\beta\, x + K_\rho \sin\beta\, y\, \pm\, \gamma z)}, \tag{3.21}$$

where we have used the result that

$$d^2K_\rho = K_\rho\, dK_\rho\, d\beta.$$

We now make the transformation

$$K_\rho = k\sin\alpha,$$
$$K_\rho\, dK_\rho = k^2 \sin\alpha\cos\alpha\, d\alpha,$$
$$\pm\gamma = k\sqrt{1-\sin^2\alpha} = k\cos\alpha,$$

so that

$$\mathbf{k}^\pm = K_\rho\cos\beta\,\hat{\mathbf{x}} + K_\rho\sin\beta\,\hat{\mathbf{y}} \pm \gamma\hat{\mathbf{z}} \rightarrow k\overbrace{(\sin\alpha\cos\beta\,\hat{\mathbf{x}} + \sin\alpha\sin\beta\,\hat{\mathbf{y}} + \cos\alpha\,\hat{\mathbf{z}})}^{\mathbf{s}}. \tag{3.22}$$

The integration ranges of the angle α must be selected so that K_ρ varies from zero to infinity and $k\cos\alpha = +\gamma$ or $-\gamma$ depending on whether the field is in U^+ or U^-. In the important case in which the medium is non-dispersive and k is real-valued it is easily verified that the required integration ranges are as follows:

- $+\gamma$ **case:** from 0 to $\pi/2$ along the real-α axis and then from $\pi/2$ to $\pi/2 - i\infty$ along the straight line $\Re\alpha = \pi/2$;
- $-\gamma$ **case:** from $\pi/2 + i\infty$ to $\pi/2$ along the straight line $\Re\alpha = \pi/2$ and then from $\pi/2$ to π along the real-α axis.

The integration contours corresponding to the above two cases are illustrated in Fig. 3.1. In the case of dispersive media the α contours are similar but are displaced above the contours shown in the figure. In most of our applications the plane-wave amplitude $A^{(\pm)}(\mathbf{k}^{\pm}, \omega)$ will be an entire analytic function of $\mathbf{K}_\rho$ and, hence, of the integration variables α and β so that the two α contours can be arbitrarily deformed so long as they extend from 0 to $\pi/2 - i\infty$ when $z > 0$ (the $+\gamma$ case) and from $\pi/2 + i\infty$ to π when $z < 0$ (the $-\gamma$ case).

Under the above transformation we find that Eq. (3.21) can be written in the form

$$U^{(\pm)}(\mathbf{r}, \omega) = \frac{ik}{2\pi} \int_{-\pi}^{\pi} d\beta \int_{C_\pm} \sin\alpha \, d\alpha \, A^{(\pm)}(k\mathbf{s}, \omega) e^{ik\mathbf{s}\cdot\mathbf{r}}, \tag{3.23}$$

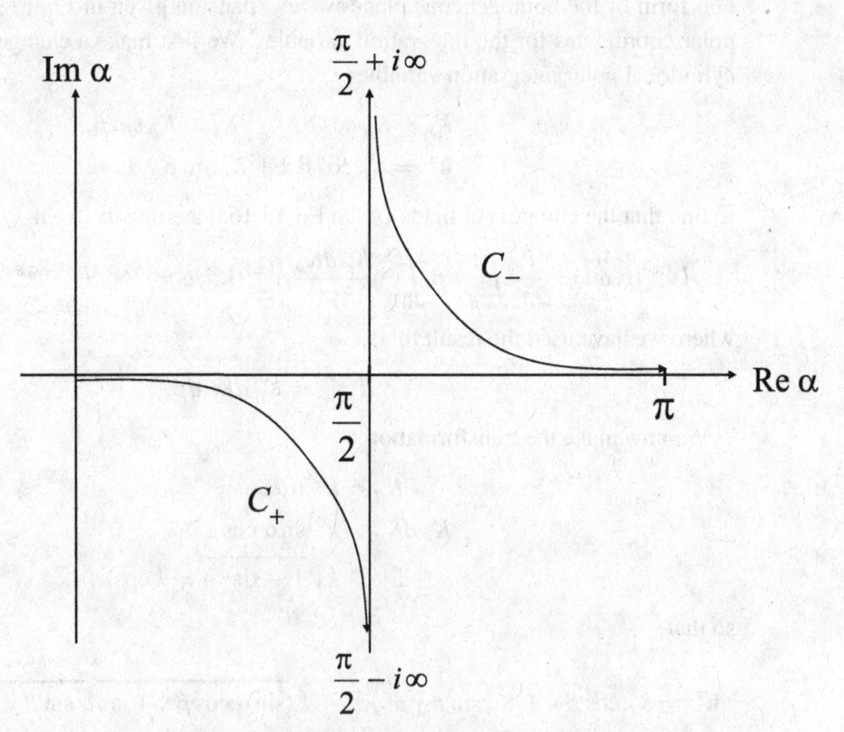

Fig. 3.1 Choices for the α integration contour in the angular-spectrum expansion. The contour labeled C_+ applies for $z > 0$ and the contour C_- for $z < 0$. Because of analyticity the contours can be deformed as illustrated in the figure.

where $C_{\pm}$ are the α contours of integration illustrated in Fig. 3.1 and $\mathbf{s}$ is the complex unit-propagation vector defined in Eq. (3.22). In this equation the plane-wave amplitude is given by

$$A^{(\pm)}(k\mathbf{s},\omega) = \begin{cases} A^{(+)}(\mathbf{k}^{+},\omega) & \text{if } \alpha \in C_{+}, \\ A^{(-)}(\mathbf{k}^{-},\omega) & \text{if } \alpha \in C_{-}. \end{cases} \tag{3.24}$$

Equation (3.23) is the angle-variable form of the plane-wave expansion Eq. (3.16a). We note that, as mentioned above, the integrand is usually an entire analytic function of the polar angles α and β so that the α integration contours can be arbitrarily deformed so long as they extend from 0 to $\pi/2 - i\infty$ when $z > 0$ and from $\pi/2 + i\infty$ to π when $z < 0$. It is easily verified that the field $U^{(+)}$ satisfies the SRC in the half-space $z > 0$ and the field $U^{(-)}$ satisfies the SRC in the half-space $z < 0$. In fact, it follows directly from Eq. (3.17) that

$$U^{(\pm)}(r\mathbf{s},\omega) \sim A^{(\pm)}(k\mathbf{s},\omega)\frac{e^{ikr}}{r}, \tag{3.25}$$

where we have made use of Eq. (3.24). It is now apparent why we have expressed the general plane-wave expansion in the form used in Eq. (3.16a): it results in an angle-variable form where the plane-wave amplitudes $A^{(\pm)}(k\mathbf{s},\omega)$ are equal to the radiation patterns in the two half-spaces $z > 0$ and $z < 0$, respectively.

3.3 Spherical coordinates

Selecting the three generalized coordinates ξ_j to be the spherical coordinates r, θ, ϕ we have that

$$\nabla^2 = \frac{1}{r^2}\frac{\partial}{\partial r}\left(r^2\frac{\partial}{\partial r}\right) - \frac{L^2}{r^2},$$

where

$$L^2 = \mathbf{L}\cdot\mathbf{L} = -\frac{1}{\sin\theta}\frac{\partial}{\partial\theta}\left(\sin\theta\frac{\partial}{\partial\theta}\right) - \frac{1}{\sin^2\theta}\frac{\partial^2}{\partial\phi^2}, \tag{3.26a}$$

is the square of the *angular-momentum operator*

$$\mathbf{L} = -i\mathbf{r}\times\nabla = i\left[\hat{\boldsymbol{\theta}}\frac{1}{\sin\theta}\frac{\partial}{\partial\phi} - \hat{\boldsymbol{\phi}}\frac{\partial}{\partial\theta}\right]. \tag{3.26b}$$

The homogeneous Helmholtz equation then becomes

$$\left[\frac{1}{r^2}\frac{\partial}{\partial r}\left(r^2\frac{\partial}{\partial r}\right) - \frac{L^2}{r^2} + k^2\right]U(r,\theta,\phi) = 0.$$

Assuming a separable solution in the form of Eq. (3.1), we then conclude that

$$\frac{1}{U_r(r)}\frac{\partial}{\partial r}\left(r^2\frac{\partial}{\partial r}\right)U_r(r) - \frac{L^2 U_\theta(\theta)U_\phi(\phi)}{U_\theta(\theta)U_\phi(\phi)} + k^2r^2 = 0,$$

which requires that

$$\frac{\partial}{\partial r}\left(r^2\frac{\partial}{\partial r}\right)U_r(r) + [k^2r^2 - l(l+1)]U_r(r) = 0, \tag{3.27a}$$

$$L^2 U_\theta(\theta)U_\phi(\phi) = l(l+1)U_\theta(\theta)U_\phi(\phi), \tag{3.27b}$$

where $l(l+1)$ is a separation constant.

Equation (3.27b) can itself be separated. In particular, we have that

$$L^2 U_\theta(\theta)U_\phi(\phi) = -\frac{U_\phi(\phi)}{\sin\theta}\frac{\partial}{\partial\theta}\left(\sin\theta\,\frac{\partial}{\partial\theta}\right)U_\theta(\theta) - \frac{U_\theta(\theta)}{\sin^2\theta}\frac{\partial^2}{\partial\phi^2}U_\phi(\phi),$$

which when used in Eq. (3.27b) yields the equation

$$-\frac{\sin\theta}{U_\theta(\theta)}\frac{\partial}{\partial\theta}\left(\sin\theta\frac{\partial}{\partial\theta}\right)U_\theta(\theta) - \frac{1}{U_\phi(\phi)}\frac{\partial^2}{\partial\phi^2}U_\phi(\phi) = l(l+1)\sin^2\theta,$$

from which we conclude that

$$\frac{1}{U_\phi(\phi)}\frac{\partial^2}{\partial\phi^2}U_\phi(\phi) = -m^2, \tag{3.28a}$$

$$\frac{1}{\sin\theta}\frac{\partial}{\partial\theta}\left(\sin\theta\,\frac{\partial}{\partial\theta}\right)U_\theta(\theta) - \frac{m^2}{\sin^2\theta} + l(l+1)U_\theta(\theta) = 0. \tag{3.28b}$$

Equation (3.27a) for integer l is the equation satisfied by the spherical Bessel and Neumann functions

$$j_l(kr) = \sqrt{\frac{\pi}{2kr}}J_{l+\frac{1}{2}}(kr), \tag{3.29a}$$

$$n_l(kr) = \sqrt{\frac{\pi}{2kr}}N_{l+\frac{1}{2}}(kr), \tag{3.29b}$$

where $J_{l+\frac{1}{2}}$ and $N_{l+\frac{1}{2}}$ are the ordinary Bessel and Neumann functions of half-integer order. Related to the spherical Bessel and Neumann functions are the spherical Hankel functions

$$h_l^{\pm}(z) = j_l(z) \pm i n_l(z), \tag{3.29c}$$

so $h_l^{\pm}(kr)$ are also solutions to Eq. (3.27a). The solutions to Eq. (3.28a) are the complex exponentials

$$U_\phi(\phi) = e^{\pm im\phi},$$

where m must be an integer in order for the solution to be single-valued in the angle ϕ. Finally, the solutions to Eq. (3.28b) are the associated Legendre polynomials with argument $\cos\theta$:

$$U_\theta(\theta) = P_l^m(\cos\theta).$$

We note for future reference that the associated Legendre polynomials satisfy the parity relationship

$$P_l^m(-x) = (-1)^{l+m} P_l^m(x),$$

and that the spherical Bessel and Neumann functions are both real-valued if their arguments are real, so that it follows from the Schwartz reflection principle that

$$j_l^*(kr) = j_l(k^*r), \qquad n_l^*(kr) = n_l(k^*r), \qquad h_l^{\pm *}(kr) = h_l^{\mp}(k^*r) \tag{3.30}$$

in a dispersive medium having a complex wavenumber k (with, of course, r real-valued). The above conditions satisfied by the spherical Bessel, Neumann and Hankel functions will be important in our treatment of the inverse source problem presented in Chapter 5, while we will make use of the parity condition satisfied by the associated Legendre polynomials in an example presented below.

It is customary to combine the two angular functions $U_\theta(\theta)$ and $U_\phi(\phi)$ into a single function of θ, ϕ that satisfies Eq. (3.27b) and, hence, is an eigenfunction of the square of the angular-momentum operator with eigenvalue $l(l+1)$. Thus, we introduce the *spherical harmonics*

$$Y_l^m(\theta,\phi) = (-1)^m \sqrt{\frac{2l+1}{4\pi}\frac{(l-m)!}{(l+m)!}} P_l^m(\cos\theta) e^{im\phi}, \tag{3.31a}$$

where m is an integer in the range $-l \leq m \leq l$ and

$$L^2 Y_l^m(\theta,\phi) = l(l+1) Y_l^m(\theta,\phi). \tag{3.31b}$$

The spherical harmonics are orthonormal on the unit sphere:

$$\int d\Omega\, Y_l^m(\theta,\phi) Y_{l'}^{m'*}(\theta,\phi) = \delta_{l,l'}\delta_{m,m'}, \tag{3.31c}$$

where $d\Omega = \sin\theta\, d\theta\, d\phi$ is the differential solid angle, $\delta_{l,l'}$ and $\delta_{m,m'}$ are the Kronecker delta functions and the superscript asterisk denotes the complex conjugate.

Separable solutions to the Helmholtz equation in spherical coordinates are then given by

$$U_l^m(\mathbf{r},\omega) = a_l^m(\omega) f_l(kr) Y_l^m(\theta,\phi),$$

where $a_l^m(\omega)$ is an arbitrary constant and $f_l(kr)$ is a linear combination of spherical Bessel and Neumann functions. The most general solution to the homogeneous Helmholtz equation is then given by a sum of the above elementary solutions having the general form

$$U(\mathbf{r},\omega) = \sum_{l=0}^{\infty}\sum_{m=-l}^{l} a_l^m(\omega) f_l(kr) Y_l^m(\theta,\phi), \tag{3.32}$$

where the specific combination of spherical Bessel and Neumann functions $f_l(kr)$ and the expansion constants $a_l^m(\omega)$ are determined from the boundary conditions. We will refer to the expansion Eq. (3.32) as a *multipole expansion* and the expansion coefficients as *multipole moments*.

The addition theorem for spherical harmonics

An important result involving the spherical harmonics is the *addition theorem*. Let θ_1, ϕ_1 and θ_2, ϕ_2 be two sets of polar and azimuthal angles associated with two unit vectors $\mathbf{s}_1$ and $\mathbf{s}_2$; i.e., $\mathbf{s}_1 = \sin\theta_1 \cos\phi_1 \,\hat{\mathbf{x}} + \sin\theta_1 \sin\phi_1 \,\hat{\mathbf{y}} + \cos\theta_1 \,\hat{\mathbf{z}}$ and similarly for $\mathbf{s}_2$. Also let χ denote the angle formed between these two unit vectors so that $\mathbf{s}_1 \cdot \mathbf{s}_2 = \cos\chi$. Then the addition theorem states that

$$P_l(\cos\chi) = \frac{4\pi}{2l+1} \sum_{m=-l}^{l} Y_l^m(\theta_1, \phi_1) Y_l^{m*}(\theta_2, \phi_2)$$

$$= \frac{4\pi}{2l+1} \sum_{m=-l}^{l} Y_l^{m*}(\theta_1, \phi_1) Y_l^m(\theta_2, \phi_2).$$

Note the important result that the complex-conjugation operation can be applied to either spherical harmonic. The addition theorem can be used to simplify expressions involving spherical-harmonic expansions as well as to establish various results involving these functions that are otherwise difficult to prove.

Vector spherical harmonics

In dealing with vector-valued fields such as the electromagnetic field we will need to employ the so-called vector spherical harmonics, which are generated from the spherical harmonics Y_l^m by application of the angular-momentum operator:

$$\mathbf{Y}_l^m(\theta, \phi) = \mathbf{L} Y_l^m(\theta, \phi). \tag{3.33}$$

By integrating by parts and using the fact that the spherical harmonics are orthonormal eigenfunctions of L^2 with eigenvalue $l(l+1)$ it is easily established that the vector spherical harmonics are orthogonal over the unit sphere with square norm equal to $l(l+1)$:

$$\begin{aligned}\int d\Omega\, \mathbf{Y}_{l'}^{m'*}(\theta, \phi) \cdot \mathbf{Y}_l^m(\theta, \phi) &= \int d\Omega\, \mathbf{L} Y_{l'}^{m'*}(\theta, \phi) \cdot \mathbf{L} Y_l^m(\theta, \phi) \\ &= \int d\Omega\, Y_l^{m*}(\theta, \phi) L^2 Y_l^m(\theta, \phi) \\ &= l(l+1)\delta_{l,l'}\delta_{m,m'}.\end{aligned} \tag{3.34}$$

We showed above that the spherical harmonics are the angular functions that arise when the (scalar) Helmholtz equation is separated in spherical coordinates. In a similar fashion we will find that the vector spherical harmonics result from separation of variables of the vector Helmholtz equation

$$\nabla \times \nabla \times \mathbf{E}(\mathbf{r}) - k^2 \mathbf{E}(\mathbf{r}) = 0$$

in spherical coordinates. We will not investigate this further at this point, but will develop the theory in some detail in Chapter 11, where we address direct and inverse problems involving the electromagnetic field.

Example 3.4 A plane wave $\exp(ik\mathbf{s}\cdot\mathbf{r})$ satisfies the homogeneous Helmholtz equation over all of space and can, thus, be represented in a multipole expansion of the form Eq. (3.32) with the radial functions equal to the spherical Bessel functions $j_l(kr)$. This expansion, which will be important in later applications, can be derived as a solution to the interior boundary-value problem with the bounding surface $\partial V = \Sigma_\infty$ being the surface of an infinite sphere centered at the origin. The computation of the multipole moments $a_l^m(\omega)$ for this expansion is straightforward, albeit tedious, and leads to the result

$$e^{ik\mathbf{s}\cdot\mathbf{r}} = 4\pi \sum_{l=0}^{\infty} \sum_{m=-l}^{l} i^l j_l(kr) Y_l^m(\hat{\mathbf{r}}) Y_l^{m*}(\mathbf{s}), \tag{3.35}$$

where we have denoted the azimuthal and polar angles of the unit propagation vector and field point $\mathbf{r}$ in the arguments of the spherical harmonics by $\mathbf{s}$ and $\hat{\mathbf{r}}$, respectively. This shorthand notation for the arguments of the spherical harmonics is very useful and will be used almost exclusively from this point on in the book.

Equation (3.35) represents a plane wave in a multipole expansion. It is sometimes useful to have the reverse situation, namely the plane-wave expansion of a multipole field $j_l(kr)Y_l^m(\hat{\mathbf{r}})$. This expansion can be obtained directly from Eq. (3.35) by making use of the orthogonality relation Eq. (3.31c) for the spherical harmonics. In particular, by multiplying both sides of Eq. (3.35) by a spherical harmonic $Y_l^m(\mathbf{s})$ and integrating the result over 4π steradians we obtain

$$j_l(kr)Y_l^m(\hat{\mathbf{r}}) = \frac{(-i)^l}{4\pi} \int d\Omega_s\, Y_l^m(\mathbf{s}) e^{ik\mathbf{s}\cdot\mathbf{r}}. \tag{3.36}$$

Both Eqs. (3.35) and (3.36) will be used extensively in later chapters.

Finally, we note that it is sometimes useful to have a multipole expansion of the plane wave $\exp(-ik\mathbf{s}\cdot\mathbf{r})$, where the wavenumber k may be complex. This expansion can be obtained directly from Eq. (3.35) by noting that

$$e^{ik^*\mathbf{s}\cdot\mathbf{r}} = 4\pi \sum_{l=0}^{\infty} \sum_{m=-l}^{l} i^l j_l^*(kr) Y_l^m(\hat{\mathbf{r}}) Y_l^{m*}(\mathbf{s}),$$

where we have made use of Eq. (3.30). On taking the complex conjugate of the above equation we then obtain the desired expansion

$$e^{-ik\mathbf{s}\cdot\mathbf{r}} = 4\pi \sum_{l=0}^{\infty} \sum_{m=-l}^{l} (-i)^l j_l(kr) Y_l^{m*}(\hat{\mathbf{r}}) Y_l^m(\mathbf{s}). \tag{3.37}$$

3.4 Multipole expansions

The outgoing-wave Green function can be represented in a multipole expansion given by

$$G_+(\mathbf{r}-\mathbf{r}',\omega) = -ik\sum_{l=0}^{\infty}\sum_{m=-l}^{l} j_l(kr_<)h_l^+(kr_>)Y_l^m(\hat{\mathbf{r}})Y_l^{m*}(\hat{\mathbf{r}}'), \tag{3.38a}$$

where $r_< = \min r, r'$ and $r_> = \max r, r'$, and where we have used the unit vectors $\hat{\mathbf{r}} = \mathbf{r}/r$ and $\hat{\mathbf{r}}' = \mathbf{r}'/r'$ to denote the arguments θ, ϕ and θ', ϕ' of the spherical harmonics $Y_l^m(\theta,\phi)$ and $Y_l^m(\theta',\phi')$. By making use of the addition theorem the above multipole expansion can also be expressed in the form

$$G_+(\mathbf{r}-\mathbf{r}',\omega) = -ik\sum_{l=0}^{\infty}\frac{2l+1}{4\pi} j_l(kr_<)h_l^+(kr_>)P_l(\cos\chi), \tag{3.38b}$$

where χ is the angle formed between the unit vectors $\hat{\mathbf{r}}$ and $\hat{\mathbf{r}}'$. It also follows from the addition theorem that the complex-conjugate operations in Eq. (3.38a) can be interchanged.

We can derive the multipole expansion Eq. (3.38a) starting from the Fourier-integral representation of the outgoing-wave Green function and making use of the multipole expansions of the plane waves given in Example 3.4. In particular, on substituting the expansions from Eqs. (3.35) and (3.37) into the Fourier-integral representation of G_+ given in Eq. (2.14) from Chapter 2 we find that

$$G_+(\mathbf{r}-\mathbf{r},\omega) = \frac{2}{\pi}\int d^3K \sum_{l,m}\sum_{l',m'} i^l(-i)^{l'}\frac{j_l(Kr)j_{l'}(Kr')}{k^2-K^2}Y_l^m(\hat{\mathbf{r}})Y_l^{m*}(\hat{\mathbf{K}})Y_{l'}^{m'}(\hat{\mathbf{K}})Y_{l'}^{m'*}(\hat{\mathbf{r}}')$$

$$= \frac{2}{\pi}\sum_{l=0}^{\infty}\sum_{m=-l}^{l} Y_l^m(\hat{\mathbf{r}})Y_l^{m*}(\hat{\mathbf{r}}')\int_0^{\infty} K^2\,dK\,\frac{j_l(Kr)j_l(Kr')}{k^2-K^2},$$

where we have made use of the orthogonality of the spherical harmonics according to Eq. (3.31c).

Consider now the integral

$$I_l(r,r') = \int_0^{\infty} K^2\,dK\,\frac{j_l(Kr)j_l(Kr')}{k^2-K^2} = \frac{1}{2}\int_{-\infty}^{\infty} K^2\,dK\,\frac{j_l(Kr)j_l(Kr')}{k^2-K^2},$$

in terms of which

$$G(\mathbf{r},\mathbf{r}') = \frac{2}{\pi}\sum_{l=0}^{\infty}\sum_{m=-l}^{l} I_l(r,r')Y_l^m(\hat{\mathbf{r}})Y_l^{m*}(\hat{\mathbf{r}}'). \tag{3.39}$$

We now assume that $r > r'$ and express the spherical Bessel function $j_l(Kr)$ in terms of the spherical Hankel functions using Eq. (3.29c). We then find that

$$
\begin{aligned}
I(r, r') &= \frac{1}{2}\int_{-\infty}^{\infty} K^2\, dK \frac{j_l(Kr) j_l(Kr')}{k^2 - K^2} \\
&= \frac{1}{4}\int_{-\infty}^{\infty} K^2\, dK \frac{h_l^+(Kr) j_l(Kr')}{k^2 - K^2} + \frac{1}{4}\int_{-\infty}^{\infty} K^2\, dK \frac{h_l^-(Kr) j_l(Kr')}{k^2 - K^2}.
\end{aligned}
$$

It follows from well-known properties of the spherical Bessel and Hankel functions that $h_l^+(Kr)j_l(Kr')$ goes to zero exponentially fast in the upper half of the complex K plane and $h_l^-(Kr)j_l(Kr')$ goes to zero exponentially fast in the lower half of the complex K plane when $r > r'$. We can then close the contours in the above equation in the upper and lower half-planes, respectively. Since $\Im k > 0$ the first closed contour integral contains an interior pole at $K = k$ and the second has one at $K = -k$ so that we obtain

$$
I(r, r') = -\frac{\pi i}{4} k h_l^+(kr) j_l(kr') - \frac{\pi i}{4} k h_l^-(-kr) j_l(-kr') = -\frac{\pi i}{2} k h_l^+(kr) j_l(kr'), \tag{3.40}
$$

with the final equality following from the parity conditions

$$
j_l(-x) = (-1)^l j_l(x), \qquad h_l^-(-x) = (-1)^l h_l^+(x).
$$

The multipole expansion Eq. (3.38b) when $r > r'$ then follows directly upon substituting Eq. (3.40) into Eq. (3.39). The derivation of the expansion for $r' > r$ follows identical lines.

The multipole expansion of the conjugate wave Green function is readily obtained from the above expansions by simply taking its complex conjugate. We find that

$$
G^*(\mathbf{r} - \mathbf{r}', \omega) = ik^* \sum_{l=0}^{\infty} \sum_{m=-l}^{l} j_l(k^* r_<) h_l^-(k^* r_>) Y_l^{m*}(\hat{\mathbf{r}}) Y_l^m(\hat{\mathbf{r}}'),
$$

which, on account of the addition theorem, can also be written in the form

$$
G^*(\mathbf{r} - \mathbf{r}', \omega) = ik^* \sum_{l=0}^{\infty} \sum_{m=-l}^{l} j_l(k^* r_<) h_l^-(k^* r_>) Y_l^m(\hat{\mathbf{r}}) Y_l^{m*}(\hat{\mathbf{r}}'). \tag{3.41a}
$$

The incoming-wave Green function is obtained from the conjugate-wave Green function by letting $k \to k^*$. Upon making this transformation we then obtain

$$
G_-(\mathbf{r} - \mathbf{r}', \omega) = ik \sum_{l=0}^{\infty} \sum_{m=-l}^{l} j_l(kr_<) h_l^-(kr_>) Y_l^m(\hat{\mathbf{r}}) Y_l^{m*}(\hat{\mathbf{r}}'). \tag{3.41b}
$$

Example 3.5 In this example we turn our attention to the *exterior boundary-value problem* in which Dirichlet or Neumann (or mixed) conditions are specified on a closed surface ∂V bounding some region V and the solution satisfying the outgoing radiation condition is desired in the region of space *outside* or *exterior* to V. Here, we address the exterior boundary-value problem for the case in which the interior region V is a sphere centered at the origin and having radius $a > 0$ and Dirichlet or Neumann conditions are specified on the surface ∂V of the sphere. Although Eqs. (2.42) formally give the solutions to these boundary-value problems these "solutions" are incomplete since the Green functions G_D and/or G_N need to be computed. An alternative solution method is to use an eigenfunction expansion of the field in the exterior region $r > a$ that can fit arbitrary boundary-value data on the bounding surface ∂V. We present such a solution here.

We can represent the solution throughout the exterior of the region V in the general multipole expansion Eq. (3.32) where the radial functions $f_n(kr)$ must be chosen so as to satisfy the SRC. It is easily determined that the appropriate combination of spherical Bessel and Neumann functions is given by the *spherical Hankel function of the first kind*:

$$h_l^+(kr) = j_l(kr) + in_l(kr). \tag{3.42}$$

The general solution satisfying the SRC is then given by

$$U_+(\mathbf{r}, \omega) = \sum_{l=0}^{\infty} \sum_{m=-l}^{l} a_l^m(\omega) h_l^+(kr) Y_l^m(\theta, \phi), \tag{3.43}$$

where the expansion coefficients (multipole moments) must be determined from the Dirichlet or Neumann conditions on the surface $r = a$.

If we set $r = a$ in the multipole expansion Eq. (3.43) and use the fact that the spherical harmonics are orthonormal over the unit sphere (cf. Eq. (3.31c)) we find that

$$a_{\mathrm{D}}{}_l^m(\omega) = \frac{1}{h_l^+(ka)} \int d\Omega\, Y_l^{m*}(\theta, \phi) U_+(r = a, \theta, \phi, \omega), \tag{3.44a}$$

where $U_+(r = a, \theta, \phi, \omega)$ is the Dirichlet boundary condition on the sphere $r = a$ and the subscript D denotes the solution to the Dirichlet boundary-value problem. In a similar fashion we find the multipole moments for the Neumann problem to be given by

$$a_{\mathrm{N}}{}_l^m = \frac{1}{k h_l^{+\prime}(ka)} \int d\Omega\, Y_l^{m*}(\theta, \phi) U_+'(r = a, \theta, \phi, \omega), \tag{3.44b}$$

where $U_+'(r = a, \theta, \phi, \omega)$ is the Neumann condition on the sphere $r = a$ and $h_l^{+\prime}$ denotes the derivative of the spherical Hankel function.

The multipole expansion Eq. (3.32) with the radial functions taken to be the spherical Hankel functions of the first kind leads directly to a solution of the exterior boundary-value problem for Dirichlet or Neumann (or mixed) conditions specified on the surface of a sphere of radius $a > 0$. The multipole expansion with a different choice for the radial functions can also be used for solving the interior boundary-value problem for the sphere, which consists of determining the solution to the homogeneous Helmholtz equation within the interior of a sphere of radius $a > 0$ from Dirichlet or Neumann conditions on the surface of the sphere. In this case we require that the field be well behaved throughout the interior of the sphere, which requires the radial functions to be the spherical Bessel functions $j_n(kr)$.[3] The solution to the interior problem is then given by an expansion of the form of Eq. (3.43) of Example 3.5 but with the spherical Hankel functions replaced by the spherical Bessel functions. The multipole moments are given by Eqs. (3.44) of this example where, again, the spherical Hankel functions are replaced by spherical Bessel functions.

[3] The spherical Neumann functions have a logarithmic singularity at the origin $r = 0$ and, hence, must be excluded from the radial functions in order to have a well-behaved field at the origin.

3.4.1 Multipole expansions of the Dirichlet and Neumann Green functions

The solutions to the exterior and interior boundary-value problems for spherical boundaries presented in the above two examples can also be obtained using the Dirichlet and Neumann Green functions appropriate to spherical boundaries. These Green functions are obtained in the form of the sum of multipole expansions of the outgoing-wave Green function G_+ and a free field that satisfies the homogeneous Helmholtz equation over the region τ in which the boundary-value problem is to be solved. For the exterior boundary-value problem the free field must also satisfy a radiation condition over the sphere at infinity. The multipole moments (expansion coefficients) of the free field component are then found by requiring the sum or its normal derivative to vanish over the sphere on which the boundary conditions are to be imposed.

For example, we can express the Dirichlet Green function for the interior boundary-value problem via the multipole expansion

$$G_{\rm D}(\mathbf{r},\mathbf{r}',\omega) = \overbrace{-ik\sum_{l=0}^{\infty}\sum_{m=-l}^{l} j_l(kr)h_l^+(kr')Y_l^m(\hat{\mathbf{r}})Y_l^{m*}(\hat{\mathbf{r}}')}^{G_+(\mathbf{r}-\mathbf{r}',\omega)}$$
$$+\overbrace{\sum_{l=0}^{\infty}\sum_{m=-l}^{l} a_l^m j_l(kr)j_l(kr')Y_l^m(\hat{\mathbf{r}})Y_l^{m*}(\hat{\mathbf{r}}')}^{\text{free field}},$$

where it is assumed that $r < r'$ appropriate to an interior boundary-value problem. We select the multipole moments a_l^m by requiring that $G_{\rm D} = 0$ over the data sphere $r' = a$. We then find that

$$a_l^m = ik\frac{h_l^+(ka)}{j_l(ka)},$$

so that the required multipole expansion of the Dirichlet Green function for the interior boundary-value problem is found to be

$$G_{\rm D}(\mathbf{r},\mathbf{r}',\omega) = -ik\sum_{l=0}^{\infty}\sum_{m=-l}^{l} j_l(kr)h_l^+(kr')Y_l^m(\hat{\mathbf{r}})Y_l^{m*}(\hat{\mathbf{r}}')$$
$$+\overbrace{ik\sum_{l=0}^{\infty}\sum_{m=-l}^{l}\frac{h_l^+(ka)}{j_l(ka)}j_l(kr)j_l(kr')Y_l^m(\hat{\mathbf{r}})Y_l^{m*}(\hat{\mathbf{r}}')}^{\text{free field}}, \tag{3.45a}$$

where $r < r'$. A completely parallel development yields the following multipole expansion of the Neumann Green function for the interior boundary-value problem:

$$G_{\rm N}(\mathbf{r},\mathbf{r}',\omega) = -ik\sum_{l=0}^{\infty}\sum_{m=-l}^{l} j_l(kr)h_l^+(kr')Y_l^m(\hat{\mathbf{r}})Y_l^{m*}(\hat{\mathbf{r}}')$$
$$+ik\sum_{l=0}^{\infty}\sum_{m=-l}^{l}\frac{h_l^{+\prime}(ka)}{j_l'(ka)}j_l(kr)j_l(kr')Y_l^m(\hat{\mathbf{r}})Y_l^{m*}(\hat{\mathbf{r}}'). \tag{3.45b}$$

It is easy to verify that G_D and $\partial G_N/\partial r' = 0$ both vanish when $r' = a$.

The Dirichlet and Neumann Green functions for the exterior problem for the sphere not only must vanish on the boundary $r' = a$ but also must satisfy a radiation condition. For the case of the outgoing-wave radiation condition (SRC) the Green functions are expressed in the general form

$$G(\mathbf{r},\mathbf{r}',\omega) = \overbrace{-ik\sum_{l=0}^{\infty}\sum_{m=-l}^{l} j_l(kr')h_l^+(kr)Y_l^m(\hat{\mathbf{r}})Y_l^{m*}(\hat{\mathbf{r}}')}^{G_+(\mathbf{r}-\mathbf{r}',\omega)}$$

$$+\overbrace{\sum_{l=0}^{\infty}\sum_{m=-l}^{l} a_l^m h_l^+(kr)h_l^+(kr')Y_l^m(\hat{\mathbf{r}})Y_l^{m*}(\hat{\mathbf{r}}')}^{\text{free field}},$$

where now $r > r'$. Again the multipole moments a_l^m are selected to make the Green function G satisfy homogeneous Dirichlet or Neumann conditions on the data sphere $r' = a$. We find that

$$G_D(\mathbf{r},\mathbf{r}',\omega) = -ik\sum_{l=0}^{\infty}\sum_{m=-l}^{l}\left\{j_l(kr')h_l^+(kr) - \frac{j_l(ka)}{h_l^+(ka)}h_l^+(kr)h_l^+(kr')\right\}Y_l^m(\hat{\mathbf{r}})Y_l^{m*}(\hat{\mathbf{r}}'), \tag{3.46a}$$

$$G_N(\mathbf{r},\mathbf{r}',\omega) = -ik\sum_{l=0}^{\infty}\sum_{m=-l}^{l}\left\{j_l(kr')h_l^+(kr) - \frac{j_l'(ka)}{h_l^{+'}(ka)}h_l^+(kr)h_l^+(kr')\right\}Y_l^m(\hat{\mathbf{r}})Y_l^{m*}(\hat{\mathbf{r}}'). \tag{3.46b}$$

The conjugate-wave Green functions both for the interior and for the exterior boundary-value problem are obtained by simply taking the complex conjugate of those found above, and the incoming-wave Green functions are, of course, equal to the conjugate-wave Green functions with the wavenumber k replaced by its complex conjugate k^*.

Example 3.6 We solved the exterior Dirichlet and Neumann problems with an outgoing-wave radiation condition in Example 3.5. Here we solve those problems using the outgoing-wave Dirichlet and Neumann Green functions given in Eqs. (3.46). The Green-function solutions to the Dirichlet and Neumann exterior boundary-value problems are given in Eqs. (2.42) of Section 2.8.2 of Chapter 2. For the special case of a spherical data boundary these equations become[4]

$$U(\mathbf{r},\omega) = -a^2\int d\Omega'\, U(\mathbf{r}',\omega)|_{r'=a}\frac{\partial}{\partial r'}G_D(\mathbf{r},\mathbf{r}',\omega)|_{r'=a}$$

for inhomogeneous Dirichlet conditions and

[4] Note that in the exterior boundary-value problem the normal derivatives in Eqs. (2.42) are directed outward from the exterior into the interior region so that $\partial/\partial n' = -\partial/\partial r'$ when the data boundary is a sphere.

$$U(\mathbf{r},\omega) = a^2 \int d\Omega' \frac{\partial}{\partial r'} U(\mathbf{r}',\omega)|_{r'=a} G_{\mathrm{N}}(\mathbf{r},\mathbf{r}',\omega)|_{r'=a}$$

for inhomogeneous Neumann conditions. The solution to the boundary-value problem for inhomogeneous Neumann conditions is immediately obtained by substituting the multipole expansion of G_{N} given in Eq. (3.46b) into the above equation. We first note that

$$G_{\mathrm{N}}(\mathbf{r},\mathbf{r}',\omega)|_{r'=a}$$

$$= -ik \sum_{l=0}^{\infty} \sum_{m=-l}^{l} \left\{ j_l(ka) h_l^+(kr) - \frac{j_l'(ka)}{h_l^{+'}(ka)} h_l^+(kr) h_l^+(ka) \right\} Y_l^m(\hat{\mathbf{r}}) Y_l^{m*}(\hat{\mathbf{r}}')$$

$$= -ik \sum_{l=0}^{\infty} \sum_{m=-l}^{l} \frac{\overbrace{h_l^{+'}(ka) j_l(ka) - j_l'(ka) h_l^+(ka)}^{i/(ka)^2}}{h_l^{+'}(ka)} h_l^+(kr) Y_l^m(\hat{\mathbf{r}}) Y_l^{m*}(\hat{\mathbf{r}}')$$

$$= \frac{1}{ka^2} \sum_{l=0}^{\infty} \sum_{m=-l}^{l} \frac{h_l^+(kr)}{h_l^{+'}(ka)} Y_l^m(\hat{\mathbf{r}}) Y_l^{m*}(\hat{\mathbf{r}}'),$$

where we have used the Wronskian relationship

$$h_l^{+'}(ka) j_l(ka) - j_l'(ka) h_l^+(ka) = \frac{i}{(ka)^2}.$$

On substituting this expansion into the Green-function solution of the Neumann problem given above we obtain the result

$$U(\mathbf{r},\omega) = a^2 \int d\Omega' \frac{\partial}{\partial r'} U(\mathbf{r}',\omega)|_{r'=a} \frac{1}{ka^2} \sum_{l=0}^{\infty} \sum_{m=-l}^{l} \frac{h_l^+(kr)}{h_l^{+'}(ka)} Y_l^m(\hat{\mathbf{r}}) Y_l^{m*}(\hat{\mathbf{r}}')$$

$$= \sum_{l=0}^{\infty} \sum_{m=-l}^{l} a_{\mathrm{N}}{}_l^m h_l^+(kr) Y_l^m(\hat{\mathbf{r}}),$$

where the multipole moments are given in Eq. (3.44b) of Example 3.5. We thus arrive at precisely the same solution as that which we obtained previously by matching boundary conditions using the multipole expansion of the field. A completely parallel development yields the solution for the Dirichlet problem in Example 3.5.

Example 3.7 As a second example we solve the interior Dirichlet boundary-value problem using the interior Dirichlet Green function given in Eq. (3.45a). The Green-function solution to this problem is given in Eqs. (2.42) of Section 2.8.1 of Chapter 2. For the special case of a spherical data boundary this solution assumes the form

$$U(\mathbf{r},\omega) = a^2 \int d\Omega' \, U(\mathbf{r}',\omega)|_{r'=a} \frac{\partial}{\partial r'} G_{\mathrm{D}}(\mathbf{r},\mathbf{r}',\omega)|_{r'=a}, \tag{3.47}$$

where we have used the fact that the normal derivatives in the interior boundary-value problem are directed outward from the interior of the sphere so that $\partial/\partial n' = \partial/\partial r'$. A straightforward calculation similar to that employed in the above problem for the Neumann Green function yields the result

$$\frac{\partial}{\partial r'}G_D(\mathbf{r},\mathbf{r}',\omega)|_{r'=a} = \frac{1}{a^2}\sum_{l=0}^{\infty}\sum_{m=-l}^{l}\frac{j_l(kr)}{j_l(ka)}Y_l^m(\hat{\mathbf{r}})Y_l^{m*}(\hat{\mathbf{r}}'),$$

which, when substituted into Eq. (3.47), yields the solution

$$U(\mathbf{r},\omega) = \sum_{l=0}^{\infty}\sum_{m=-l}^{l}\frac{u_l^m}{j_l(ka)}j_l(kr)Y_l^m(\hat{\mathbf{r}}), \tag{3.48a}$$

where

$$u_l^m = \int d\Omega'\, Y_l^{m*}(\hat{\mathbf{r}}')U(\mathbf{r}',\omega)|_{r'=a}. \tag{3.48b}$$

It is readily verified that Eq. (3.48a) satisfies the prescribed boundary condition when $r = a$. An entirely parallel development can be used to compute the solution of the interior boundary-value problem for the case of Neumann data.

3.4.2 Plane-wave expansions of the multipole fields

We have already derived the plane-wave expansion of the free multipole fields $j_l(kr)Y_l^m(\hat{\mathbf{r}})$ in Example 3.4. It is also possible to represent the outgoing-wave multipole fields $h_l^+(kr)Y_l^m(\hat{\mathbf{r}})$ employed in Example 3.5 in such an expansion, where, however, we have to include both evanescent plane waves and the weakly inhomogeneous plane waves that are employed exclusively in the plane-wave expansion of the free multipole fields. Although the rigorous derivation of the plane-wave expansion for the outgoing-wave fields is somewhat complicated (Devaney and Wolf, 1974), it is possible to obtain the correct result by making use of the far-field result given in Eq. (3.25) of Section 3.2.3. This equation states that the plane-wave amplitude in the angle-variable form of the plane-wave expansion of an outgoing-wave field is simply the radiation pattern of this field! For the outgoing-wave multipole fields we have that

$$h_l^+(kr)Y_l^m(\hat{\mathbf{r}}) \sim \overbrace{\frac{(-i)^{(l+1)}}{k}Y_l^m(\hat{\mathbf{r}})}^{f(\hat{\mathbf{r}},\omega)}\frac{e^{ikr}}{r},$$

from which we conclude that the plane-wave expansion for this field is given by

$$h_l^+(kr)Y_l^m(\hat{\mathbf{r}}) = \frac{ik}{2\pi}\int_{-\pi}^{\pi}d\beta\int_{C_\pm}\sin\alpha\, d\alpha\, \overbrace{A^{(\pm)}(k\mathbf{s},\omega)}^{f(\mathbf{s},\omega)}\, e^{ik\mathbf{s}\cdot\mathbf{r}},$$

which becomes

$$h_l^+(kr)Y_l^m(\hat{\mathbf{r}}) = \frac{(-i)^l}{2\pi}\int_{-\pi}^{\pi}d\beta\int_{C_\pm}\sin\alpha\, d\alpha\, Y_l^m(\mathbf{s})e^{ik\mathbf{s}\cdot\mathbf{r}}, \tag{3.49}$$

where the contours of integration $C_\pm$ are illustrated in Fig. 3.1 and C_+ is used in the r.h.s. $z > 0$ and C_- in the l.h.s. $z < 0$.

On comparing the result Eq. (3.49) with the plane-wave expansion of the free multipole fields given in Eq. (3.36) of Example 3.4 we see that they differ by a trivial multiplicative factor of two and by the replacement of the integration over the entire unit sphere for the case of the free multipole fields by the integration over the complex contours shown in Fig. 3.1 for the outgoing-wave fields. This again is an indication that homogeneous and weakly inhomogeneous plane waves are associated with free fields; i.e., fields that satisfy the homogeneous Helmholtz or wave equations while evanescent plane waves are associated with fields satisfying the inhomogeneous Helmholtz or wave equations and a radiation condition. We will return to these issues in the next chapter, where we investigate in more detail plane-wave and multipole expansions of radiated fields.

The evanescent-wave component of the multipole fields

We showed in Section 3.2.3 that there is a one-to-one correspondence between the spatial variations of the boundary value of a field or its normal derivative over a plane surface and the decomposition of the field radiated from that surface into a plane-wave expansion of weakly inhomogeneous and evanescent plane waves. We can use this observation to deduce a general property of the multipole fields that will be of importance when we study the inverse source problem (ISP) in Chapter 5. This general property follows from the fact that the spherical harmonics $Y_l^m(\hat{\mathbf{r}})$ possess angular periods in the polar and azimuthal angles θ and ϕ varying from a largest equal to 2π at $l = 1$, $m = \pm 1$ to a smallest of $2\pi/l$ for general $l > 1$. The corresponding spatial periods associated with the multipole fields at some radial distance r vary then from a maximum of $R = 2\pi r$ to a minimum of $R = 2\pi r/l$ for general l. As shown in Section 3.2.3, the division between weakly inhomogeneous and evanescent plane waves in a non-dispersive medium occurs at a field spatial period equal to the wavelength λ. If we then set the smallest spatial period of the multipole field $R = 2\pi r/l$ equal to the wavelength, we conclude that *the multipole fields in a non-dispersive medium consist of mostly homogeneous plane waves if $l < kr$ and will then include increasingly more evanescent plane waves when $l > kr$.* A similar conclusion is reached in dispersive media if we replace the wavelength λ by $2\pi/\sqrt{\Re k^2}$.

The above observations can be interpreted to mean that *sub-wavelength field information at any radial distance r is carried by the multipole fields for which $l > kr$.* As mentioned above, this observation will be of use in our treatment of the ISP later in the book.

3.5 Circular cylindrical coordinates

Selecting the three generalized coordinates ξ_j to be the circular cylindrical coordinates ρ, ϕ, z, we have that

$$\nabla^2 = \frac{1}{\rho}\frac{\partial}{\partial\rho}\left(\rho\frac{\partial}{\partial\rho}\right) + \frac{1}{\rho^2}\frac{\partial^2}{\partial\phi^2} + \frac{\partial^2}{\partial z^2}. \tag{3.50}$$

The homogeneous Helmholtz equation then becomes

$$\left[\frac{1}{\rho}\frac{\partial}{\partial\rho}\left(\rho\frac{\partial}{\partial\rho}\right)+\frac{1}{\rho^2}\frac{\partial^2}{\partial\phi^2}+\frac{\partial^2}{\partial z^2}+k^2\right]U(\rho,\phi,z)=0.$$

Assuming a separable solution in the form of Eq. (3.1), we conclude that

$$\frac{U_\phi(\phi)U_z(z)}{\rho}\frac{\partial}{\partial\rho}\left(\rho\frac{\partial U_\rho(\rho)}{\partial\rho}\right)+\frac{U_\rho(\rho)U_z(z)}{\rho^2}\frac{\partial^2 U_\phi(\phi)}{\partial\phi^2}$$
$$+U_\rho(\rho)U_\phi(\phi)\frac{\partial^2 U_z(z)}{\partial z^2}+k^2 U_\rho(\rho)U_\phi(\phi)U_z(z)=0,$$

which requires that

$$\frac{\partial^2 U_z(z)}{\partial z^2}=-h^2 U_z(z), \tag{3.51a}$$

$$\frac{1}{\rho U_\rho(\rho)}\frac{\partial}{\partial\rho}\left(\rho\frac{\partial U_\rho(\rho)}{\partial\rho}\right)+\frac{1}{\rho^2 U_\phi(\phi)}\frac{\partial^2 U_\phi(\phi)}{\partial\phi^2}+k^2=h^2, \tag{3.51b}$$

where h is a separation constant.

Equation (3.51b) can be further separated. In particular, we find that

$$\frac{\partial^2 U_\phi(\phi)}{\partial\phi^2}=-n^2 U_\phi(\phi), \tag{3.52a}$$

$$\rho\frac{\partial}{\partial\rho}\left(\rho\frac{\partial U_\rho(\rho)}{\partial\rho}\right)+[(k^2-h^2)\rho^2-n^2]U_\rho(\rho)=0, \tag{3.52b}$$

where n is another separation constant. Equations (3.51a) and (3.52a) have the complex exponentials $\exp(\pm ihz)$ and $\exp(\pm in\phi)$ as solutions, while Eq. (3.52b) is Bessel's equation, for which the solutions are linear combinations of the Bessel and Neumann functions J_n and N_n of order n and having argument $\sqrt{k^2-h^2}\rho$. Associated with these two elementary solutions are the Hankel functions

$$H_n^{\pm}(z)=J_n(z)\pm N_n(z). \tag{3.53}$$

Separable solutions to the homogeneous Helmholtz equation in circular cylindrical coordinates are then given by

$$U_{n,h}(\rho,\phi,z,\omega)=Z_n(\sqrt{k^2-h^2}\rho)e^{in\phi}e^{\pm ihz}, \tag{3.54a}$$

where $Z_n(\cdot)$ is a linear combination of Bessel and Hankel functions of order n.

Of particular interest is the special case when $h=0$, which corresponds to *two-dimensional wave propagation in the* (x,y) *plane*. In this case the separable solutions given in Eq. (3.54a) reduce to

$$U_n(\rho,\phi,\omega)=Z_n(k\rho)e^{in\phi}, \tag{3.54b}$$

and, thus, depend on the single separation constant n, which can assume any positive or negative integer values. The most general solution to the homogeneous Helmholtz equation in the two-dimensional case is then given by a sum of the elementary solutions Eq. (3.54b) having the general form

$$U(\mathbf{r},\omega) = \sum_{n=-\infty}^{\infty} a_n(\omega) Z_n(k\rho) e^{in\phi}, \tag{3.55}$$

where the specific combination of Bessel and Hankel functions $Z_n(k\rho)$ and the expansion constants $a_n(\omega)$ are determined from the boundary conditions.

3.6 Two-dimensional wavefields

It is often useful to specialize our analysis to 2D wavefields that satisfy the 2D Helmholtz equation

$$\left[\frac{\partial^2}{\partial x^2} + \frac{\partial^2}{\partial y^2} + k^2\right] U(\mathbf{r},\omega) = Q(\mathbf{r},\omega), \tag{3.56}$$

where now $\mathbf{r} = (x, y)$ denotes a point on the (x, y) plane. The 2D case is of importance for testing and evaluating wavefield propagation and inversion algorithms in computer simulation studies since it reduces the computational burden to a minimum while still retaining all of the flavor of the 3D case. In addition, certain 3D problems reduce to a 2D formulation such as scattering of plane waves off cylindrical structures. We presented the 2D incoming- and outgoing-wave Green functions in Section 2.2.1 of Chapter 2 and the solution of the 2D radiation problem in Section 2.4.1 of that chapter. We now turn to computation of the 2D eigenfunctions and eigenfunction expansions for 2D wavefields.

3.6.1 Polar coordinates

Polar coordinates in the plane correspond to circular cylindrical coordinates with the axial coordinate z set equal to zero. We then conclude from Eq. (3.50) that

$$\nabla^2 = \frac{1}{r}\frac{\partial}{\partial r}\left(r\frac{\partial}{\partial r}\right) + \frac{1}{r^2}\frac{\partial^2}{\partial \phi^2},$$

where $r = \sqrt{x^2 + y^2}$ and ϕ denotes the polar angle made between the coordinate vector $\mathbf{r}$ and the positive-y axis. The homogeneous Helmholtz equation then becomes

$$\left[\frac{1}{r}\frac{\partial}{\partial r}\left(r\frac{\partial}{\partial r}\right) + \frac{1}{r^2}\frac{\partial^2}{\partial \phi^2} + k^2\right] U(r,\phi) = 0. \tag{3.57}$$

Assuming a separable solution in the form of Eq. (3.1) we then find that

$$\frac{U_\phi(\phi)}{r}\frac{\partial}{\partial r}\left(r\frac{\partial U_r(r)}{\partial r}\right) + \frac{U_r(r)}{r^2}\frac{\partial^2 U_\phi(\phi)}{\partial \phi^2} + k^2 U_r(r) U_\phi(\phi) = 0,$$

which requires that

$$\frac{\partial^2 U_\phi(\phi)}{\partial \phi^2} = -n^2 U_\phi(\phi), \tag{3.58a}$$

$$\left[r\frac{\partial}{\partial r}\left(r\frac{\partial U_r(r)}{\partial r}\right) + k^2r^2 - n^2 \right] U_r(r) = 0, \tag{3.58b}$$

where n is a separation constant. Equation (3.58a) has the complex exponentials $\exp(in\phi)$ as solutions while Eq. (3.58b) is Bessel's equation for which the solutions are the Bessel and Hankel functions of order n and having argument kr. Separable solutions to the 2D homogeneous Helmholtz equation in polar coordinates are then given by

$$U_n(r, \phi, \omega) = Z_n(kr)e^{in\phi}, \tag{3.59}$$

where $Z_n(kr)$ is a linear combination of Bessel and Hankel functions of order n, where n is any positive or negative integer or zero.

The separable solutions given in Eq. (3.59) are precisely those obtained by simply letting the axial coordinate $z \to 0$ in the 3D separable solutions corresponding to circular cylindrical coordinates given in Eq. (3.54a). The most general solution to the 2D homogeneous Helmholtz equation in polar coordinates is then given by

$$U(\mathbf{r}, \omega) = \sum_{n=-\infty}^{\infty} a_n(\omega) Z_n(kr)e^{in\phi}, \tag{3.60}$$

where the combination of Bessel and Hankel functions $Z_n(kr)$ are selected to satisfy specific boundary conditions. We will sometimes refer to expansions of the form Eq. (3.60) as *2D multipole expansions.*

Multipole expansion of the 2D Green function

In analogy to the multipole expansions given in Eqs. (3.38) and (3.41) of the 3D outgoing-, conjugate- and incoming-wave Green functions we can expand the 2D Green functions into 2D multipole expansions using the 2D multipole fields $Z_n(kr)e^{in\phi}$. In particular, one finds that

$$G_\pm(\mathbf{r} - \mathbf{r}') = \mp\frac{i}{4}H_0^\pm(kR) = \mp\frac{i}{4}\sum_{n=-\infty}^{\infty} J_n(kr_<)H_n^\pm(kr_>)e^{in(\phi-\phi')}, \tag{3.61}$$

where $r_< = \min r, r'$ and $r_> = \max r, r'$. Here ϕ and ϕ' are the polar angles of $\mathbf{r}$ and $\mathbf{r}'$, respectively. The conjugate Green function is simply obtained by taking the complex conjugate of G_+.

Example 3.8 Consider the exterior boundary-value problem of determining a solution to the *2D Helmholtz equation*

$$\left[\frac{\partial^2}{\partial x^2} + \frac{\partial^2}{\partial y^2} + k^2\right] U(\mathbf{r}, \omega) = 0, \tag{3.62}$$

which satisfies the SRC and inhomogeneous Dirichlet conditions on the circle $r = \sqrt{x^2+y^2} = a$. The general solution can be expressed in the form Eq. (3.55), where the particular combination of Bessel functions and the expansion coefficients are determined by the boundary conditions.

The most general combination of Bessel functions Z_n can be expressed as a sum of the Bessel function J_n of the first kind of order n and the Hankel function H_n^+ of the first kind of order n. These two quantities behave asymptotically as

$$J_n(kr) \sim \sqrt{\frac{2}{\pi kr}} \cos\left(kr - \frac{2n+1}{4}\pi\right),$$

$$H_n(kr) \sim \sqrt{\frac{2}{\pi kr}} \exp\left[i\left(kr - \frac{2n+1}{4}\pi\right)\right].$$

The SRC in two space dimensions requires the field to have the asymptotic dependence

$$U(\mathbf{r},\omega) \sim f(\hat{\mathbf{r}},\omega)\frac{e^{ikr}}{\sqrt{r}}, \quad kr \to \infty,$$

from which we conclude that only the Hankel function satisfies the SRC. Thus, the solution to the exterior boundary-value problem is given by an expansion of the form Eq. (3.55) with the functions $Z_n(kr)$ equal to the Hankel functions $H_n(kr)$. The expansion coefficients are readily determined from the Dirichlet boundary conditions via the formula

$$a_n(\omega) = \frac{1}{2\pi H_n(ka)} \int_0^{2\pi} d\phi\, U(\mathbf{r},\omega)|_{r=a} e^{-in\phi}. \tag{3.63}$$

Example 3.9 The 3D plane wave $\exp(ik\mathbf{s}\cdot\mathbf{r})$ was expanded into a multipole expansion in Example 3.4. Here we derive the multipole expansion of the 2D plane wave, where now $\mathbf{s} = (s_x, s_y)$ lies on the unit circle and $\mathbf{r} = (x, y)$ lies on the plane. It is tempting to simply put $z = 0$ in the 3D expansion, but this is not a good approach to this problem since it would yield an expansion involving the spherical Bessel functions, which are not appropriate to a 2D geometry. The proper approach is to employ the general 2D multipole expansion Eq. (3.60) with the coefficients $a_n(\omega)$ and radial functions $Z_n(kr)$ treated as unknowns to be determined.

We write the expansion Eq. (3.60) for the special case of a plane wave propagating along the positive-x axis in the form

$$e^{ik\mathbf{s}\cdot\mathbf{r}} = e^{ikr\cos\phi} \sum_{n=-\infty}^{\infty} X_n(r) e^{in\phi},$$

where $X_n(r) = a_n(\omega)Z_n(kr)$ is an unknown to be determined. We conclude from the above equation that

$$X_n(r) = \frac{1}{2\pi}\int_0^{2\pi} d\phi\, e^{ikr\cos\phi} e^{-in\phi} = i^n J_n(kr), \tag{3.64}$$

leading to the result

$$e^{ikr\cos\phi} = \sum_{n=-\infty}^{\infty} i^n J_n(kr) e^{in\phi}, \tag{3.65}$$

which is generally known as the "Jacobi–Anger expansion." The general case of a plane wave propagating at an angle ϕ_0 relative to the positive-x axis is obtained from Eq. (3.65) by simply replacing ϕ by $\phi - \phi_0$:

$$e^{i\mathbf{ks}\cdot\mathbf{r}} = e^{ikr\cos(\phi-\phi_0)} = \sum_{n=-\infty}^{\infty} i^n e^{-in\phi_0} J_n(kr) e^{in\phi}. \tag{3.66}$$

Further reading

The books by Arfken (Arfken and Weber, 2001) and Vaughn (Vaughn, 2007) contain readable treatments of the Sturm–Liouville problem and separation of variables for the Helmholtz equation. Advanced material in these areas is contained in Morse and Feshbach (1953), Stratton (1941) and Jackson (1998). An excellent, but advanced, treatise on these and other related topics is the book by Claus Muller (Muller, 1969). The book by Mandel and Wolf (Mandel and Wolf, 1995) contains an excellent exposition on the angular-spectrum plane-wave representation and its asymptotic expansion. The time-domain theory of plane-wave expansions of both free and radiating scalar wavefields is presented in Devaney and Sherman (1973) and an excellent account is presented in the book on antenna theory by Hansen and Yaghjian (Hansen and Yaghjian, 1999).

Problems

3.1 Determine the Cauchy initial conditions such that the solution to the initial-value problem has only positive frequency components. Discuss the consequences of this. Hint: Example 3.1.

3.2 Compute the plane-wave expansion of the free-field propagator $g_f(\mathbf{R}, \tau) = g_+(\mathbf{R}, \tau) - g_-(\mathbf{R}, \tau)$ by employing the general procedure described in Example 3.1. Hint: See Problem 1.12.

3.3 Compute the plane-wave expansion Eq. (3.11) of the field radiated by a source $q(\mathbf{r}, t)$ for times t exceeding the turn-off time $t = T_0$ of the source. Hint: express the field for $t = T_0$ in terms of the free-field propagator.

3.4 Determine a source $q(\mathbf{r}, t)$ supported on the space-time boundary $t = 0$ that radiates a field for $t > 0$ that has prescribed Cauchy conditions at $t = t_0 > 0$.

3.5 Find the plane-wave expansion in the form of Eq. (3.16a) of a wavefield that satisfies the homogeneous Helmholtz equation over all of space and whose Dirichlet and Neumann conditions on the plane $z = 0$ are $U_0(x, y, \omega)$ and $U_0'(x, y, \omega)$, respectively. What must be true of these boundary conditions if the field is to be finite over all of space?

3.6 Compute the plane-wave expansion found in Problem 3.5 for a single plane wave propagating along the positive-z axis; i.e., $U(\mathbf{r}, \omega) = \exp(ikz)$. Verify that the resulting expansion reduces to the plane wave.

3.7 Compute the plane-wave amplitudes and plane-wave expansion for a monochromatic wavefield that propagates into the r.h.s. $z > 0$ and whose Dirichlet condition over the plane $z = 0$ is the Rect function

$$\text{Rect}(x) = \begin{cases} 1 & -X_0 \leq x \leq +X_0, \\ 0 & \text{else.} \end{cases}$$

3.8 Use the method of stationary phase to derive Eq. (3.17) from the plane-wave expansions in Eqs. (3.16a).

3.9 Use the multipole expansion of the plane wave given in Eq. (3.35) of Example 3.4 in the plane-wave expansion Eq. (3.9a) to obtain a multipole expansion for the field represented by this plane-wave expansion. Determine the multipole moments in terms of the plane-wave amplitude $A(k\mathbf{s}, \omega)$.

3.10 Use the plane-wave expansion of the multipole field $j_l(kr)Y_l^m(\hat{\mathbf{r}})$ given in Eq. (3.36) of Example 3.4 in the multipole expansion of the solution to the interior boundary-value problem for Dirichlet conditions over a sphere given in Example 3.7 to obtain a plane-wave expansion of the solution to this problem.

3.11 Use the multipole expansion of the Dirichlet Green function in Section 3.4 and the solution of the exterior boundary-value problem in Section 2.8.2 to show that the radiation pattern of a field radiated by a source confined to a sphere of radius a_0 centered at the origin admits the expansion

$$f(\mathbf{s}) = \sum_{l,m} f_l^m(\omega) Y_l^m(\mathbf{s}), \tag{3.67a}$$

where the expansion coefficients are given in terms of Dirichlet conditions over the sphere by

$$f_l^m(\omega) = \frac{(-i)^{(l+1)}}{k h_l^+(ka_0)} \int_{4\pi} d\Omega_r\, U(a_0\hat{\mathbf{r}}) Y_l^{m*}(\hat{\mathbf{r}}), \tag{3.67b}$$

where Ω_r is the solid angle on the unit sphere.

3.12 Compute the radiation pattern of a field radiated by a source confined to a sphere of radius a_0 centered at the origin from the solution of the exterior Dirichlet problem for a sphere presented in Example 3.5. Verify that the solution you obtain is identical to that given in the previous problem.

4 Angular-spectrum and multipole expansions

The Green-function solution to the radiation problem given in Eq. (2.23) of Chapter 2 represents this solution in terms of a superposition of outgoing spherical waves with each spherical wave being weighted by the source amplitude at that point. This solution was derived starting from the fact that the Helmholtz equation is linear and, hence, can be represented as a superposition of elementary solutions to the equation when excited by delta functions; i.e., as a convolution of the source term with a Green function that satisfies the same outgoing-wave condition, namely the Sommerfeld radiation condition (SRC), as is satisfied by the radiated field. Alternative representations of the field can also be obtained by making use of the linearity of the Helmholtz equation and the fact that the radiated field satisfies the homogeneous Helmholtz equation everywhere outside the source region τ_0. In particular, as we have seen in the last chapter, it is possible to represent the field in such regions in terms of an expansion of *eigenfunctions* of the homogeneous Helmholtz equation such as the plane waves or multipole fields. Indeed, in Examples 3.3 and 3.5 of Chapter 3 we expanded outgoing-wave fields such as the radiated field in a plane-wave expansion and a multipole expansion, respectively, with the expansion coefficients (plane-wave amplitudes and multipole moments) determined directly from boundary values of the field. We continue with this task in this chapter, where we develop plane-wave and multipole expansions for the radiated field directly in terms of the source Q rather than in terms of the boundary value of the radiated field. We first derive the so-called *angular-spectrum expansion* of the field, which is a superposition of weakly inhomogeneous and evanescent plane waves of the type introduced in Section 3.2.2 of the previous chapter. We then turn our attention to the *multipole expansion* of the radiated field, which is a superposition of elementary multipole fields of the type considered in Section 3.3 of Chapter 3. In both cases we derive these expansions directly from the primary (Green-function) representation of the radiated field given in Eq. (2.23) and, hence, are able to compute the plane-wave amplitude and multipole moments directly in terms of the source term $Q(\mathbf{r}, \omega)$. Both of these eigenfunction expansions are extremely useful when solving certain classes of inverse problems associated with the wave and Helmholtz equations and will be used extensively in later chapters.

4.1 The Weyl expansion

Although there are several ways of deriving the angular-spectrum expansion of the field U_+, the most direct procedure is to expand the *outgoing-wave Green function* G_+ in an angular-spectrum expansion and then substitute this expansion into the Green-function solution for U_+ given in Eq. (2.23). The angular-spectrum expansion of the outgoing-wave Green function, originally due to Weyl (Weyl, 1919) and called the *Weyl expansion*, is derived directly from the Fourier-integral representation of the outgoing-wave Green function given in Eq. (2.15). The outgoing-wave Green function is derived directly from this Fourier-integral representation by transforming to spherical polar coordinates and performing the resulting integrations. Our goal here, however, is not to obtain a closed-form expression for G_+ (which we already have) but, rather, to express G_+ as a superposition of plane waves all of which satisfy the homogeneous Helmholtz equation.[1]

To derive the Weyl expansion we introduce a specific Cartesian coordinate system x, y, z and represent both $\mathbf{R}$ and $\mathbf{K}$ in this system according to the equations

$$\mathbf{R} = \mathbf{R}_\rho + Z\hat{\mathbf{z}}, \tag{4.1a}$$

$$\mathbf{K} = \mathbf{K}_\rho + K_z\hat{\mathbf{z}}, \tag{4.1b}$$

where $\hat{\mathbf{z}}$ is the unit vector along the z axis of the selected coordinate system, $\mathbf{R}_\rho$ and $\mathbf{K}_\rho$ denote the *transverse* coordinates; i.e., are the projections of the vectors $\mathbf{R}$ and $\mathbf{K}$ onto the (x, y) plane of this system, and Z and K_z are the z coordinates of these two vectors. Using this coordinate system we can express Eq. (2.15) in the form

$$G_+(\mathbf{R}, \omega) = \frac{-1}{(2\pi)^3} \int_{-\infty}^{\infty} d^2K_\rho \int_{-\infty}^{\infty} dK_z \, \frac{e^{i(\mathbf{K}_\rho \cdot \mathbf{R}_\rho + K_z Z)}}{K_z^2 - \gamma^2}, \tag{4.2}$$

where γ is that root of $\sqrt{k^2 - K_\rho^2}$ that has positive real and imaginary parts for $K_\rho = \sqrt{K_x^2 + K_y^2}$ real and positive and was defined previously in Eq. (3.14), which we repeat here for convenience[2]

$$\gamma = \begin{cases} \sqrt{k^2 - K_\rho^2} & \text{if } K_\rho^2 < \Re k^2, \\ i\sqrt{K_\rho^2 - k^2} & \text{if } K_\rho^2 > \Re k^2. \end{cases} \tag{4.3}$$

The integrand in Eq. (4.2) has poles at $K_z = \pm\gamma$. Since both the real part and the imaginary part of γ are positive for all values of K_x and K_y in the integral Eq. (4.2), it follows that the pole located at $K_z = +\gamma$ lies in the upper half of the complex-K_z plane while the pole at $K_z = -\gamma$ lies in the lower half of this plane. Since the integrand of Eq. (4.2) tends to zero in the upper half of the complex-K_z plane if $Z > 0$ and in the l.h.p. if $Z < 0$, we can close the K_z contour of integration in the u.h.p. if $Z > 0$ and in the l.h.p. if $Z < 0$, and we find using Cauchy's integral formula that

[1] It might be argued that the Fourier-integral representation Eq. (2.15) is, itself, a "plane-wave expansion." However, the plane waves in this expansion have wavenumbers K that can vary from $K = 0$ to $K = \infty$ so that the expansion does not satisfy the homogeneous Helmholtz equation with a given fixed wavenumber and, hence, is not a true plane-wave expansion of the type considered in this chapter and the previous one.

[2] Here, and throughout the remainder of this chapter, we will tacitly assume that $\Re k > 0$. As discussed in Section 2.1.1 of Chapter 2, there is no loss in generality in making this assumption since the (time-dependent) wavefields considered in this book are assumed to be real-valued and, hence, are completely and uniquely defined by their positive-frequency components.

$$G_+(\mathbf{R},\omega) = \frac{-i}{8\pi^2}\int_{-\infty}^{\infty} d^2K_\rho \frac{e^{i(\mathbf{K}_\rho\cdot\mathbf{R}_\rho \pm \gamma Z)}}{\gamma},$$

which we can write in the form

$$G_+(\mathbf{R},\omega) = \frac{-i}{8\pi^2}\int_{-\infty}^{\infty} d^2K_\rho \frac{e^{i\mathbf{k}^\pm\cdot\mathbf{R}}}{\gamma}, \tag{4.4a}$$

where the plus sign is used if $Z > 0$ and the minus sign if $Z < 0$, and where we have defined

$$\mathbf{k}^\pm = \mathbf{K}_\rho \pm \gamma\hat{\mathbf{z}}. \tag{4.4b}$$

The Weyl expansion of the outgoing-wave Green function given in Eq. (4.4a) is a plane-wave expansion of the type introduced in Section 3.2.2 of the last chapter involving both weakly inhomogeneous and evanescent plane waves, both types satisfying the homogeneous Helmholtz equation with (generally complex) wavenumber k. As was discussed in that section, the weakly inhomogeneous plane waves are those for which $K_\rho^2 < \Re k^2$ and have a complex wave vector due to the dispersive nature of the medium in which they propagate. If the loss in this medium as characterized by $\Im k$ were to vanish, these particular plane waves would become homogeneous plane waves and have unit magnitude over all of space. The plane waves for which $K_\rho^2 > \Re k^2$ are *evanescent* plane waves and have a complex wave vector that does not become real in the limit $\Im k \to 0$. These plane waves derive their inhomogeneous character from the fact that we allow the (K_x, K_y) components of the wave vector $\mathbf{K} = K_x\hat{\mathbf{x}} + K_y\hat{\mathbf{y}} + K_z\hat{\mathbf{z}}$ to vary over the entire (K_x, K_y) plane, thus requiring the z component $K_z = \pm\gamma$ to be inherently complex when $K_x^2 + K_y^2 > \Re k^2$.

Many of the important applications of the Weyl expansion are in non-dispersive or weakly dispersive media where $\Im k$ can be taken to be zero. In this non-dispersive limit the wave vectors $\mathbf{k}^\pm$ are purely real if $K_\rho^2 < k^2$ and are complex with z components that are purely imaginary when $K_\rho^2 > k^2$. In this case, the Weyl expansion Eq. (4.4a) thus decomposes the outgoing-wave Green function into a superposition of homogeneous and evanescent plane waves. We note that, while each plane wave satisfies the homogeneous Helmholtz equation, the superposition of plane waves comprising the Weyl expansion satisfies the homogeneous Helmholtz equation only if the integral Eq. (4.4a) converges uniformly, which occurs only if $|Z| > 0$.

4.1.1 The angular-spectrum expansion for the conjugate-wave Green function

The angular-spectrum expansion of the conjugate-wave Green function is found by taking the complex conjugate of the Weyl expansion Eq. (4.4a):

$$G_+^*(\mathbf{R},\omega) = \frac{i}{8\pi^2}\int_{-\infty}^{\infty} d^2K_\rho \frac{e^{-i\mathbf{k}^{\pm *}\cdot\mathbf{R}}}{\gamma^*}.$$

By making the transformation $\mathbf{K}_\rho \to -\mathbf{K}_\rho$ and making use of the definition Eq. (4.4b) of $\mathbf{k}^\pm$ we can express this expansion in the simplified form

$$G_+^*(\mathbf{R},\omega) = \frac{i}{8\pi^2}\int_{-\infty}^{\infty}\frac{d^2K_\rho}{\gamma^*}\,e^{i\mathbf{K}_\rho\cdot\mathbf{R}_\rho}e^{\mp i\gamma^* Z}, \tag{4.5a}$$

where

$$\gamma^* = \begin{cases}\gamma(k^*), & K_\rho < \sqrt{\Re k^2},\\ -\gamma(k^*), & K_\rho > \sqrt{\Re k^2},\end{cases} \tag{4.5b}$$

and where the top sign is used if $Z > 0$ and the bottom sign if $Z < 0$. On decomposing the above expansion into weakly inhomogeneous and evanescent plane-wave components we obtain the result

$$G_+^*(\mathbf{R},\omega) = \frac{i}{8\pi^2}\int_{K_\rho<\sqrt{\Re k^2}} d^2K_\rho\,\frac{e^{i\mathbf{k}^\mp(k^*)\cdot\mathbf{R}}}{\gamma(k^*)} - \frac{i}{8\pi^2}\int_{K_\rho>\sqrt{\Re k^2}} d^2K_\rho\,\frac{e^{i\mathbf{k}^\pm(k^*)\cdot\mathbf{R}}}{\gamma(k^*)}, \tag{4.6}$$

where again the top sign is used if $Z > 0$ and the bottom sign if $Z < 0$, and where $\mathbf{k}^\pm(k^*)$ is simply $\mathbf{k}^\pm$ as defined in Eq. (4.4b) with $\gamma(k)$ replaced by $\gamma(k^*)$.

4.1.2 The angular-spectrum expansion of the incoming-wave Green function

The incoming-wave Green function is related to the conjugate-wave Green function under the replacement of k by k^*; i.e.,

$$G_-(\mathbf{R},k) = G_+^*(\mathbf{R},k^*).$$

It then follows that the angular-spectrum expansion of this Green function is obtained from that of the conjugate-wave Green function by simply replacing the wavenumber k by its complex conjugate k^*. We then find using Eq. (4.5a) that

$$G_-(\mathbf{R},\omega) = \frac{i}{8\pi^2}\int_{-\infty}^{\infty}\frac{d^2K_\rho}{\gamma^*(k^*)}\,e^{i\mathbf{K}_\rho\cdot\mathbf{R}_\rho}e^{\mp i\gamma^*(k^*)Z}, \tag{4.7a}$$

while from Eq. (4.6) we obtain the result

$$G_-(\mathbf{R},\omega) = \frac{i}{8\pi^2}\int_{K_\rho<\sqrt{\Re k^2}} d^2K_\rho\,\frac{e^{i\mathbf{k}^\mp\cdot\mathbf{R}}}{\gamma} - \frac{i}{8\pi^2}\int_{K_\rho>\sqrt{\Re k^2}} d^2K_\rho\,\frac{e^{i\mathbf{k}^\pm\cdot\mathbf{R}}}{\gamma}, \tag{4.7b}$$

with $\mathbf{k}^\pm$ defined in Eq. (4.4b) and where, again, the top sign is used if $Z > 0$ and the bottom sign if $Z < 0$.

On comparing the angular-spectrum expansion of G_- in the form Eq. (4.7b) with the Weyl expansion Eq. (4.4a) we see that the two expansions are identical over the evanescent region $K_\rho > k$ and differ by overall sign and by a shift from outgoing to incoming plane waves over the weakly homogeneous region $K_\rho < \sqrt{\Re k^2}$. In contrast with the expansion for the outgoing-wave Green function, the plane waves over the homogeneous region thus propagate *inward* from the half-space containing the field point $\mathbf{R}$ toward the origin while the evanescent plane waves still propagate on the plane $Z = 0$ and decay exponentially with distance $|Z|$ from this plane. These observations are consistent with the fact that G_- is an *incoming-wave* Green function, so it is to be expected on intuitive grounds that it be composed of a superposition of incoming plane waves.

Example 4.1 In problems involving non-dispersive media where the wavenumber k is strictly real, the *free-field propagator*

$$G_{\mathrm{f}}(\mathbf{R},\omega) = G_{+}(\mathbf{R},\omega) - G_{-}(\mathbf{R},\omega) = -\frac{i}{2\pi}\frac{\sin(kR)}{R} \tag{4.8}$$

plays an important role in a host of inverse problems related to the wave and Helmholtz equations. This quantity can be represented in a plane-wave expansion by making use of the angular-spectrum expansions for the outgoing-wave and incoming-wave Green functions given in Eqs. (4.4a) and (4.7b). In particular, on substituting these expansions into the definition of G_{f} given above and noting that the two expansions are identical over the evanescent region we find that

$$G_{\mathrm{f}}(\mathbf{R},\omega) = -\frac{i}{8\pi^2}\int_{K_\rho<k}\frac{d^2K_\rho}{\gamma}\,[e^{i\mathbf{k}^{\pm}\cdot\mathbf{R}} + e^{i\mathbf{k}^{\mp}\cdot\mathbf{R}}],$$

which can be written in the form

$$G_{\mathrm{f}}(\mathbf{R},\omega) = -\frac{i}{8\pi^2}\int_{K_\rho<k}\frac{d^2K_\rho}{\gamma}\,e^{i\mathbf{K}_\rho\cdot\mathbf{R}_\rho}[e^{i\gamma Z} + e^{-i\gamma Z}], \tag{4.9}$$

where $\mathbf{R}_\rho$ and Z are defined in Eq. (4.1a) and which holds for all $\mathbf{R}$. Equation (4.9) and its angle-variable form (see Example 4.2 below) will reappear in a number of inverse problems that are treated in later chapters.

4.1.3 Angle-variable forms of the Green-function expansions

We can also express the angular-spectrum expansions of the outgoing- and conjugate-wave Green functions in the angle-variable form given in Eq. (3.23) of Section 3.2.3. The angle-variable forms of the angular-spectrum expansions are particularly elegant and important in theoretical studies and will be employed throughout the book. Restricting our attention, for the moment, to the outgoing-wave Green function G_+, we follow the same general procedure as was employed in Section 3.2.3 and make a change of integration variable in the Weyl expansion Eq. (4.4a) from the transverse spatial frequency vector $\mathbf{K}_\rho$ to the polar α and azimuthal β angles relative to the fixed Cartesian (x, y, z) system used in the derivation of Eq. (4.4a). Upon making this transformation, the propagation vectors $\mathbf{k}^{\pm}$ become

$$\mathbf{k}^{\pm} = k\mathbf{s} = k\overbrace{(\sin\alpha\cos\beta\,\hat{\mathbf{x}} + \sin\alpha\sin\beta\,\hat{\mathbf{y}} + \cos\alpha\,\hat{\mathbf{z}})}^{\mathbf{s}},$$

where the azimuthal angle β varies from 0 to 2π and the polar angle α varies along the contour C_+ in Fig. 4.1 for the wave vector $\mathbf{k}^+$ and over the contour C_- in this figure for the wave vector k^-. For the outgoing-wave Green function we find on using the results of that section that

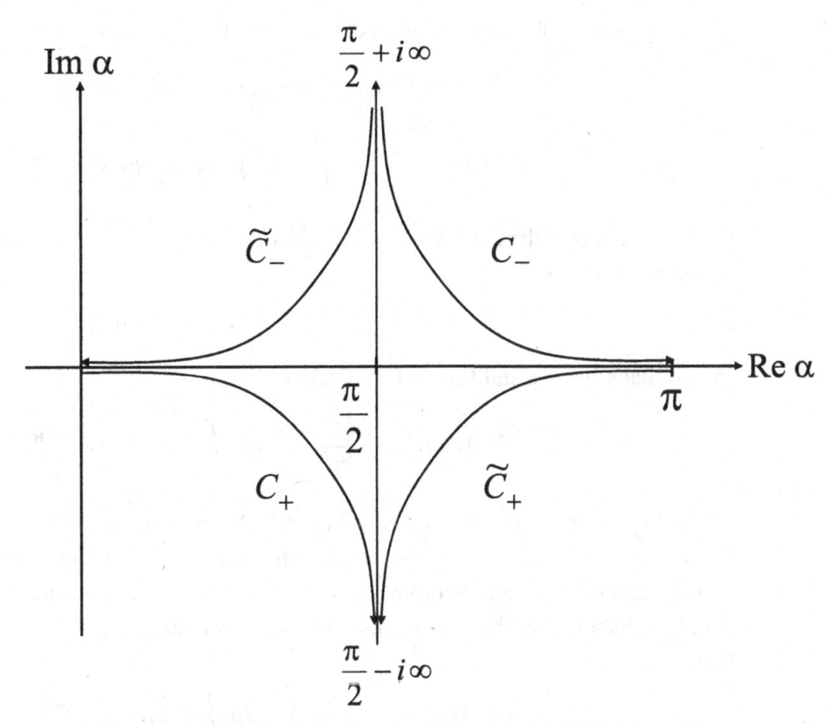

Fig. 4.1 The α integration contours for the Weyl expansions of $G_{\pm}$. The contours labeled $C_{\pm}$ are used for the outgoing-wave Green function G_+ while the contours $\tilde{C}_{\pm}$ are used for G_-. In both cases the contours labeled with the + sign are employed if $Z > 0$ and the ones labeled with the minus sign are used if $Z < 0$. It can be seen that $\tilde{C}_{\pm}$ are simply the mirror images of $C_{\pm}$ about the line $\Re\alpha = \pi/2$.

$$G_+(\mathbf{R}, \omega) = -\frac{ik}{8\pi^2}\int_{-\pi}^{\pi} d\beta \int_{C_{\pm}} \sin\alpha \, d\alpha \, e^{ik\mathbf{s}\cdot\mathbf{R}}, \tag{4.10}$$

where the contour C_+ is used if $Z > 0$ and C_- if $Z < 0$. It should be noted that the integrand in the above angular-spectrum expansion is an entire analytic function of the angles α and β, so the precise shape of the integration contours $C_{\pm}$ is unimportant. As discussed in Section 3.2.3, the decomposition of the α contour to lie along the real axis and along the line $\Re\alpha = \pi/2$ corresponds to a separation of the plane waves in the expansions into weakly inhomogeneous plane waves in the first case or evanescent plane waves in the second case.

Expansion of the conjugate-wave and incoming-wave Green functions

On taking the complex conjugate of Eq. (4.10) we obtain

$$G_+^*(\mathbf{R}, \omega) = \frac{ik^*}{8\pi^2}\int_{-\pi}^{\pi} d\beta \int_{C_{\pm}} \sin\alpha^* \, d\alpha^* \, e^{-ik^*\mathbf{s}^*\cdot\mathbf{R}}. \tag{4.11a}$$

If we now make the transformation $\alpha' = \alpha^*$ we find that Eq. (4.11a) becomes

$$\mathbf{s}^* = \mathbf{s}' = [\sin\alpha' \cos\beta, \sin\alpha' \sin\beta, \cos\alpha'],$$

$$G_+^*(\mathbf{R}, \omega) = \frac{ik^*}{8\pi^2} \int_{-\pi}^{\pi} d\beta \int_{C_\pm^*} \sin\alpha' \, d\alpha' \, e^{-ik^* \mathbf{s}' \cdot \mathbf{R}}, \tag{4.11b}$$

where $C_\pm^*$ denotes the complex conjugates of the contours $C_\pm$. As a final step we make the transformations

$$\beta \to \beta + \pi, \qquad \alpha' \to \pi - \alpha$$

to find that $\mathbf{s}' \to -\mathbf{s}$ and Eq. (4.11b) becomes

$$G_+^*(\mathbf{R}, \omega) = -\frac{ik^*}{8\pi^2} \int_{-\pi}^{\pi} d\beta \int_{\tilde{C}_\pm} \sin\alpha \, d\alpha \, e^{ik^* \mathbf{s} \cdot \mathbf{R}}, \tag{4.11c}$$

where $\tilde{C}_\pm$ are the contours shown in Fig. 4.1. As in the case of the outgoing-wave Green function, the precise shape of the integration contours $\tilde{C}_\pm$ is unimportant.

The expansion of the incoming-wave Green function is obtained from that of G_+^* by simply replacing the wavenumber k by its complex conjugate k^*. We find using Eq. (4.11c) that

$$G_-(\mathbf{R}, \omega) = -\frac{ik}{8\pi^2} \int_{-\pi}^{\pi} d\beta \int_{\tilde{C}_\pm} \sin\alpha \, d\alpha \, e^{ik \mathbf{s} \cdot \mathbf{R}}. \tag{4.12}$$

If we take the α contours to lie along the real axis and along the line $\Re\alpha = \pi/2$ it is easily verified that the expansions for G_+ and G_- are identical over the evanescent part of the spectra and differ by a sign and a shift from outgoing to incoming plane waves over the homogeneous part of the spectra.

Example 4.2 The angle-variable form of the plane-wave expansion of the *free-field propagator* defined in Eq. (4.8) of Example 4.1 is obtained by substituting the above angle-variable forms of the angular-spectrum expansions of G_+ and G_- with k real-valued into Eq. (4.8) of that example. We obtain the result

$$G_\mathrm{f}(\mathbf{R}, \omega) = -\frac{ik}{8\pi^2} \int_{-\pi}^{\pi} d\beta \int_{C_\pm} \sin\alpha \, d\alpha \, e^{ik \mathbf{s} \cdot \mathbf{R}} + \frac{ik}{8\pi^2} \int_{-\pi}^{\pi} d\beta \int_{\tilde{C}_\pm} \sin\alpha \, d\alpha \, e^{ik \mathbf{s} \cdot \mathbf{R}}.$$

First consider using the contours C_+ and $\tilde{C}_+$ corresponding to $Z > 0$. If we deform these two contours so that C_+ lies along the real axis from $\alpha = 0$ to $\alpha = \pi/2$ and $\tilde{C}_+$ lies along the real axis from $\alpha = \pi$ to $\alpha = \pi/2$, we conclude that the integrals extending from $\pi/2$ to $\pi/2 - i\infty$ cancel out while the integrals along the real-α axis add and we obtain the result

$$G_\mathrm{f}(\mathbf{R}, \omega) = -\frac{ik}{8\pi^2} \int_{-\pi}^{\pi} d\beta \int_0^{\pi} \sin\alpha \, d\alpha \, e^{ik \mathbf{s} \cdot \mathbf{R}}. \tag{4.13}$$

It is not difficult to show that the same result is obtained for $Z < 0$ using the contours C_- and $\tilde{C}_-$, so Eq. (4.13) holds over all of space.

The above result can also be obtained directly from the plane-wave expansion of the free multipole fields $j_l(kr)Y_l^m(\hat{\mathbf{r}})$ given in Example 3.5 of Chapter 3. In particular, we showed in that example that

$$j_l(kr)Y_l^m(\hat{\mathbf{r}}) = \frac{(-i)^l}{4\pi}\int d\Omega_s\, Y_l^m(\mathbf{s})e^{ik\mathbf{s}\cdot\mathbf{r}}.$$

On setting $l = 0$ we then find that

$$j_0(kR) = \frac{\sin(kR)}{kR} = \frac{1}{4\pi}\int d\Omega_s\, e^{ik\mathbf{s}\cdot\mathbf{R}},$$

from which we conclude that

$$G_f(\mathbf{R},\omega) = -\frac{ik}{2\pi}j_0(kR) = -\frac{ik}{8\pi^2}\int_{-\pi}^{\pi} d\beta \int_0^{\pi} \sin\alpha\, d\alpha\, e^{ik\mathbf{s}\cdot\mathbf{R}},$$

which is identical to the result given in Eq. (4.13).

4.2 The angular-spectrum expansion of the radiated field

If we substitute the Weyl expansion given in Eq. (4.4a) into the expression for the radiated field given in Eq. (2.23) of Chapter 2 we obtain

$$U_+(\mathbf{r},\omega) = \frac{-i}{2(2\pi)^2}\int_{\tau_0} d^3r'\, Q(\mathbf{r}',\omega)\int_{-\infty}^{\infty} d^2K_\rho\, \frac{e^{i\mathbf{k}^\pm\cdot(\mathbf{r}-\mathbf{r}')}}{\gamma}, \tag{4.14}$$

where the plus sign is used if $z > z'$ and the minus sign if $z < z'$. We now assume that the source spatial volume τ_0 is entirely contained within a strip $z^- \le z \le z^+$ and restrict our attention to field points $\mathbf{r}$ whose z coordinates lie outside this strip. For such field points $\mathbf{k}^\pm$ will be either $\mathbf{k}^+$ (if $z > z^+$) or $\mathbf{k}^-$ (if $z < z^-$), and we can interchange the orders of integration in Eq. (4.14) to obtain

$$U_+(\mathbf{r},\omega) = \frac{i}{2\pi}\int_{-\infty}^{\infty}\frac{d^2K_\rho}{\gamma}A(\mathbf{k}^\pm,\omega)e^{i\mathbf{k}^\pm\cdot\mathbf{r}}, \tag{4.15a}$$

where the plus sign is used if $z > z^+$ and the minus sign if $z < z^-$, and the spectral amplitude $A(\mathbf{k}^\pm,\omega)$ is known as the *angular spectrum* and is given in terms of the source by

$$A(\mathbf{k}^\pm,\omega) = \frac{-1}{4\pi}\tilde{Q}(\mathbf{K},\omega)|_{\mathbf{K}=\mathbf{k}^\pm}, \tag{4.15b}$$

where

$$\tilde{Q}(\mathbf{K},\omega) = \int_{\tau_0} d^3r'\, Q(\mathbf{r}',\omega)e^{-i\mathbf{K}\cdot\mathbf{r}'}$$

is the space-time Fourier transform of the source.

Like the Weyl expansion, the angular-spectrum expansion of the field is in the form of a superposition of plane waves that individually satisfy the homogeneous Helmholtz equation and divide into the two classes of weakly inhomogeneous ($K_\rho^2 < \Re k^2$) and evanescent ($K_\rho^2 > \Re k^2$) plane waves. Since the angular-spectrum expansion converges uniformly so long as the z coordinate of the observation point $\mathbf{r}$ lies outside the source volume τ_0, the expansion also satisfies the homogeneous Helmholtz equation and, hence, is a *mode expansion* of the field throughout this region. Clearly, the orientation of our coordinate system is arbitrary, so that it is possible to obtain such an expansion outside any strip whose parallel planes completely contain the source spatial volume τ_0. See Fig. 4.2.

The angular-spectrum expansion of the radiated field given above is seen to be of the same general form as that given in Eq. (3.16a) of the last chapter where, however, the two plane-wave amplitudes (angular spectra) $A^{(+)}(\mathbf{k}^+,\omega)$ and $A^{(-)}(\mathbf{k}^-,\omega)$ have the same functional form as defined in terms of the source transform via Eq. (4.15b). It is not difficult to show from Eq. (4.15b) that in the evanescent region

$$|A(\mathbf{k}^\pm,\omega)| \le Ce^{|\gamma||z^\pm|}, \tag{4.16}$$

where C is a constant. Since the evanescent plane waves decay exponentially with $|z|$, it then follows that a good approximation to the field, valid for field points $\mathbf{r}$ whose z coordinates are more than a few wavelengths from the source volume τ_0, is given by Eq. (4.15a) with the $\mathbf{K}_\rho$ integration limited to the weakly inhomogeneous region; i.e.,

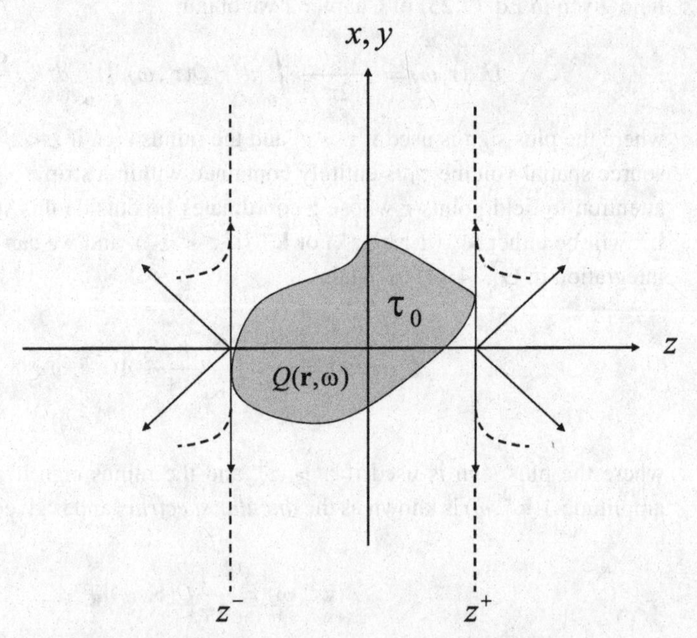

Fig. 4.2 The geometry for expanding a radiated field into an angular-spectrum expansion. The expansion is valid outside the strip $z^- < z < z^+$, with A^+ and $\mathbf{k}^+$ employed in the r.h.s. $z > z^+$ and A^- and $\mathbf{k}^-$ employed in the l.h.s. $z < z^-$. The weakly inhomogeneous plane waves propagate into either the right half-space or the left half-space and decay weakly with propagation distance while the evanescent plane waves propagate in the (x, y) plane and decay exponentially fast with increasing $|z|$.

$$U_+(\mathbf{r},\omega) \approx \frac{i}{2\pi}\int_{K_\rho<\sqrt{\Re k^2}} \frac{d^2K_\rho}{\gamma} A(\mathbf{k}^\pm,\omega)e^{i\mathbf{k}^\pm\cdot\mathbf{r}}. \tag{4.17}$$

The approximate plane-wave expansions of U_+ given in Eq. (4.17) converge uniformly over all of space and, hence, satisfy the homogeneous Helmholtz equation everywhere. It then follows that the expansion using $\mathbf{k}^+$ is a *free field*[3] that will closely approximate the radiated field if $z \gg z^+$ while the expansion using $\mathbf{k}^-$ is a free field that will closely approximate this field if $z \ll z^-$.

4.2.1 The angle-variable form of the radiated field expansion

The angle-variable form of the angular-spectrum expansion for the radiated field can be obtained by making the same change of integration variables in the Cartesian form given in Eq. (4.15a) as was used in deriving the angle-variable form of the general plane-wave expansion in Section 3.2.3 of the last chapter. It can also be obtained by simply substituting the angle-variable form of the Weyl expansion given in Eq. (4.10) into Eq. (2.23). By employing this second scheme and following steps almost identical to those used in deriving Eq. (4.15a) we obtain the result

$$U_+(\mathbf{r},\omega) = \frac{ik}{2\pi}\int_{-\pi}^{\pi} d\beta \int_{C_\pm} \sin\alpha\, d\alpha\, A(k\mathbf{s},\omega)e^{ik\mathbf{s}\cdot\mathbf{r}}, \tag{4.18a}$$

where the angular spectrum $A(k\mathbf{s},\omega)$ is given in terms of the source via the equation

$$A(k\mathbf{s},\omega) = -\frac{1}{4\pi}\tilde{Q}(k\mathbf{s},\omega), \tag{4.18b}$$

and where the α integration contour C_+ is used if $z > z^+$ and C_- if $z < z^-$. The source transform $\tilde{Q}(k\mathbf{s},\omega)$ is an entire analytic function of the unit vector $\mathbf{s}$ continued onto the complex unit sphere; i.e., is an entire function of its polar and azimuthal angles within any specific Cartesian reference system. Thus, the exact shape of the α contour of integration in Eq. (4.18a) is not important as long as it begins and terminates at the proper points in the complex-α plane.

4.2.2 The angular spectrum and radiation pattern

The *radiation pattern* of the field, was introduced in Section 1.5.1 of Chapter 1 and further treated in Section 2.4 of Chapter 2, where it was shown to be related to the space-time Fourier transform of the source via the equation

$$f(\mathbf{s},\omega) = -\frac{1}{4\pi}\tilde{Q}(\mathbf{K},\omega)|_{\mathbf{K}=k\mathbf{s}}. \tag{4.19}$$

[3] A "free field" is loosely defined to be a field that satisfies the homogeneous Helmholtz equation over all of space.

On comparing Eqs. (4.18b) and (4.19) we see that the angular spectrum in angle-variable form is related to the radiation pattern through the equation

$$A(k\mathbf{s}, \omega) = f(\mathbf{s}, \omega), \tag{4.20a}$$

a result that was previously derived in Section 3.2.3 of the previous chapter. A less elegant statement of this same result can be expressed in Cartesian variable form from Eq. (4.15b):

$$A(\mathbf{k}^{\pm}, \omega) = f(\mathbf{s}^{\pm}, \omega), \tag{4.20b}$$

where $\mathbf{s}^{\pm} = \mathbf{k}^{\pm}/k$ is the unit vector along the direction of the wave vectors $\mathbf{k}^{\pm}$.

The radiation pattern $f(\mathbf{s}, \omega)$ is the complex amplitude of the field $U_{+}(r\mathbf{s}, \omega)$ in the limit $r \to \infty$. Its argument $\mathbf{s}$ is thus restricted to the real unit sphere corresponding to real observable field points $\mathbf{r} = r\mathbf{s}$, so Eqs. (4.20b) have meaning only for transverse wave vectors $\mathbf{K}_{\rho}$ corresponding to the weakly inhomogeneous part of the spectrum $K_{\rho}^{2} < \Re k^{2}$. In a similar way Eq. (4.20a) also has meaning only over the weakly inhomogeneous part of the spectrum which corresponds to $\mathbf{s}$ lying on the real unit sphere. However, because of analyticity these two quantities are, in fact, uniquely determined for all values of their arguments via the process of *analytic continuation* from their values over the observable part of the spectrum. Indeed, they are both uniquely specified by the radiation pattern over any arbitrary section of the real unit sphere.[4] For example, if the angular spectrum $A^{+}(\mathbf{K}_{\rho}, \omega)$ is specified over any area on the $\mathbf{K}_{\rho}$ plane it is, in principle, uniquely determined over the entire $\mathbf{K}_{\rho}$ plane. In a similar fashion $A(k\mathbf{s}, \omega)$ is uniquely determined from the radiation pattern $f(\mathbf{s}, \omega)$ specified over any arbitrary section of the real unit sphere. This fact also indicates that the angular spectra in the two regions (physical and evanescent) cannot be defined independently: any change in the physically observable radiation pattern corresponding to the weakly inhomogeneous part of the angular spectrum is accompanied by an associated change in the evanescent region and vice versa. This observation has important implications in the inverse source and antenna synthesis problems (Hansen, 1981), as we will see in Chapter 5.

The fact that the radiation pattern of compactly supported sources uniquely determines the angular spectrum establishes the important result that it also uniquely determines the field everywhere outside the smallest convex region[5] which surrounds the source. In particular, since the choice of the orientation of our Cartesian coordinate system is arbitrary, the radiated field can be represented in an angular-spectrum expansion that will be valid outside *any* plane surface that bounds the source region and the angular spectrum for this particular expansion can then be determined from the radiation pattern via Eqs. (4.20). By this means the field can, in principle, be uniquely and completely determined everywhere

[4] An entire analytic function of N variables is uniquely determined from its values over an N-dimensional hyper-volume of arbitrarily small size. Thus, for example, in the common case of a function of a single variable it is uniquely specified by its value over any line segment of arbitrary length. In our case of a function of two (generally) complex variables we require specification over an area.

[5] A convex region has the defining property that any two points within the region can be connected by a straight line that does not intersect the surface of the region. The smallest convex region that encloses the source volume τ_0 is called the "convex hull" of this region.

outside the convex hull of the source region.[6] Of course this conclusion is merely formal in the sense that it would involve an *analytic continuation* of the radiation pattern from the physical to the evanescent region, and this process is computationally unstable and cannot be performed in practice.

4.2.3 The radiation pattern of a non-radiating source

In our treatment of non-radiating (NR) sources for the Helmholtz equation in Chapter 2 we established Theorem 2.1, which stated that a necessary and sufficient condition for a compactly supported and piecewise continuous source to be NR at frequency ω is that its radiation pattern vanish over the entire real unit sphere. The necessity of the condition is an immediate consequence of the fact that an NR source generates a zero field outside its support and, hence, generates a zero radiation pattern. The sufficiency condition is easily established using the angular-spectrum expansion. In particular, we consider any particular bounding plane to the source, which we are free to take as the (x, y) plane located at $z = z_0$ with the source located in the half-space $z < z_0$. The field throughout the half-space $z \geq z_0$ can be represented via the angle-variable form of the angular-spectrum expansion Eq. (4.18a), where the angular spectrum is proportional to the analytic continuation of the radiation pattern onto the complex unit sphere and, hence, must vanish if the radiation pattern vanishes over the entirety, or, indeed, any finite region, of the real unit sphere. This then requires that the field itself vanish throughout the half-space $z > z_0$. Since the orientation of the coordinate system is arbitrary, it follows that the field must vanish everywhere outside the convex hull of τ_0, which then establishes the vanishing of the radiation pattern as a sufficient condition for a source to be NR.

4.3 Forward and back propagation using the angular spectrum

We have already seen in Example 3.3 of Chapter 3 that the angular-spectrum expansion can be used to solve the Rayleigh–Sommerfeld (RS) boundary-value problems. Indeed, this conclusion follows immediately upon taking the spatial Fourier transforms of both sides of Eq. (4.15a) over any plane $z = z_0$ lying outside the source strip $z^- \leq z \leq z^+$:

$$A(\mathbf{k}^{\pm}, \omega) = \frac{\gamma}{2\pi i} \tilde{U}_{+}(\mathbf{K}_{\rho}, z_0, \omega) e^{\mp i\gamma z_0}, \tag{4.21a}$$

where $\tilde{U}_{+}(\mathbf{K}_{\rho}, z_0, \omega)$ is the spatial Fourier transform of the field on the plane $z = z_0$ and the top sign is used if $z_0 > z^+$ and the bottom sign if $z_0 < z^-$. An entirely analogous procedure allows the angular spectra to be determined in terms of Neumann conditions from the equation

$$A(\mathbf{k}^{\pm}, \omega) = \mp \frac{1}{2\pi} \tilde{U}'_{+}(\mathbf{K}_{\rho}, z_0, \omega) e^{\mp i\gamma z_0}, \tag{4.21b}$$

[6] Actually, we will show later using the multipole expansion (see Section 4.8) that the radiation pattern uniquely determines the field *everywhere* outside the source support τ_0, not just outside the convex hull of this support.

where $\tilde{U}'_+$ is the spatial Fourier transform of the Neumann boundary value on the plane $z = z_0$ and again the top sign is used if $z_0 > z^+$ and the bottom sign if $z_0 < z^-$. The angular spectra as computed above when used in Eq. (4.15a) then yield the solution to the RS problem in either of the two half-spaces $z \lessgtr z_0$ depending on whether z_0 lies to the right or left of the source support strip.

The two equations Eqs. (4.21) are the generalization of Eqs. (3.18) of Example 3.3 of Chapter 3 to boundary-value planes other than the plane $z = 0$ and, in addition, apply to propagation into both the left half-space $z < z_0 \leq z^-$ and the right half-space $z > z_0 \geq z^+$. It is important to note that the data plane $z = z_0$ is arbitrary in these equations so long as it lies outside the source strip $z^- < z < z^+$. Because of this we can employ these equations to implement the process of *field back propagation* whereby the angular spectrum is computed from a plane $z = z_0$ and the result used in Eq. (4.15a) to compute the field over other planes *that lie closer to the source than the data plane* $z = z_0$. This process is distinct from the normal process of field forward propagation whereby boundary values of the field over surfaces near the source are used to compute the field at points further removed from the source than the data surface. Thus *forward propagation* implemented via the angular spectrum corresponds to computing the angular spectrum from data on some plane $z = z_0$ and using this quantity in the angular-spectrum expansion to compute the field for field points whose z coordinates are further removed from the source region than z_0. *Back propagation*, on the other hand, uses the angular spectrum that is computed from data on the plane $z = z_0$ to compute the field at field points whose z coordinates are closer to the source than z_0.

The processes of forward and back propagation are easy to implement directly from Eqs. (4.21) when we note that in any given half-space the angular spectrum $A(\mathbf{k}^{\pm}, \omega)$ is independent of z_0. It then follows from Eq. (4.21a) that

$$\tilde{U}_+(\mathbf{K}_\rho, z, \omega) = \tilde{U}_+(\mathbf{K}_\rho, z_0, \omega) e^{\pm i\gamma(z - z_0)}, \tag{4.22a}$$

with the result holding for any values of z_0, z lying outside the source strip $z^- < z < z^+$ and in the same half-space and where the upper (+) sign applies if $z, z_0 > z^+$ and the lower (−) sign if $z, z_0 < z^-$. When z is further removed from the source than z_0 then Eq. (4.22a) performs the process of normal field propagation (is equivalent to the RS boundary-value problem), but when z is closer to the source than z_0 this equation performs the process of field *back propagation*. The actual field is then computed via an inverse spatial Fourier transform:

$$U_+(\boldsymbol{\rho}, z, \omega) = \frac{1}{(2\pi)^2} \int d^2K_\rho\, \tilde{U}_+(\mathbf{K}_\rho, z, \omega) e^{i\mathbf{K}_\rho \cdot \boldsymbol{\rho}},$$

thus leading to

$$U_+(\boldsymbol{\rho}, z, \omega) = \frac{1}{(2\pi)^2} \int d^2K_\rho\, \tilde{U}_+(\mathbf{K}_\rho, z_0, \omega) e^{\pm i\gamma(z - z_0)} e^{i\mathbf{K}_\rho \cdot \boldsymbol{\rho}}, \tag{4.22b}$$

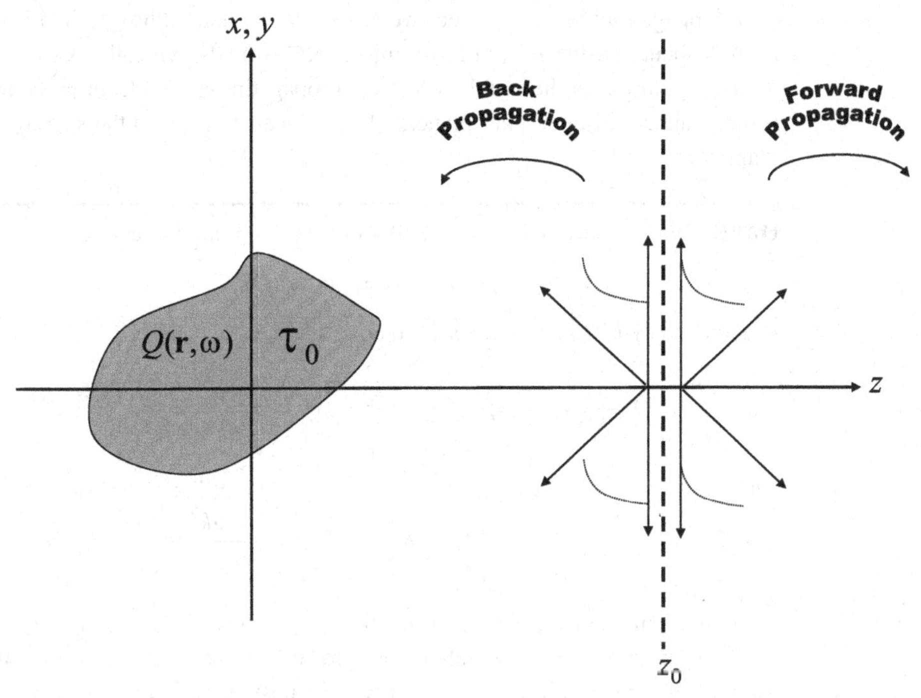

Fig. 4.3 The processes of forward and back propagation from a boundary-value plane $z = z_0$ implemented in the r.h.s. of some radiating source using the angular-spectrum expansion. Forward propagation is stable since the evanescent plane waves decay exponentially with increasing distance from the boundary-value plane if $z > z_0$, whereas back propagation is unstable since the evanescent plane waves grow exponentially with increasing distance from this boundary-value plane if $z < z_0$.

with an analogous relationship connecting the field to Neumann conditions over any given plane (see the problems at the end of this chapter). We illustrate the processes of forward and back propagation in Fig. 4.3.

We emphasize that the processes of forward and back propagation as encoded in Eqs. (4.22) apply both to dispersive and to non-dispersive media and are valid for all frequencies ω, hence allowing the time-dependent field $u_+(\mathbf{r}, t)$ to be computed for all times throughout any half-space not containing the source from Dirichlet or Neumann data specified over any plane in this half-space. The quantity γ has a positive imaginary part both in the weakly inhomogeneous part and in the evanescent part of the spectra,[7] so the process of forward propagation is well posed and equivalent to the solutions of the RS boundary-value problems in terms of Green functions presented in Section 2.8. On the other hand, for the same reason, the process of back propagation is unstable and requires some modification in order for it to be applicable to real data. We will return to this issue later in this section, where we will compare two stabilized forms of field back propagation with exact field

[7] As discussed in Section 3.2.2 of Chapter 3, $\Im\gamma = 0$ over the homogeneous part of the spectrum in non-dispersive media, but will have $\Im\gamma > 0$ over this part of the spectrum in normal dispersive media where $\Im k > 0$.

back propagation. For the present we re-emphasize that, although field back propagation as implemented using Eqs. (4.22) is unstable, it is mathematically exact and does provide a formal solution to the problem of back propagating a field from its boundary value on some plane z_0 to some other plane z_1 that is located closer to the source than is the data plane z_0.

Example 4.3 We showed in Example 3.2 of Chapter 3 that if we set

$$k^2 = \Re k^2 + i\Im k^2$$

and assume that $\Im k^2 \ll |K_\rho^2 - \Re k^2|$ then

$$\gamma = \sqrt{k^2 - K_\rho^2} \approx \sqrt{\Re k^2 - K_\rho^2} + i\frac{\Im k^2}{2\sqrt{\Re k^2 - K_\rho^2}}, \qquad K_\rho^2 < \Re k^2,$$

and

$$\gamma = i\sqrt{K_\rho^2 - k^2} \approx i\sqrt{K_\rho^2 - \Re k^2} + \frac{\Im k^2}{2\sqrt{K_\rho^2 - \Re k^2}}, \qquad K_\rho^2 > \Re k^2.$$

It can be seen from these expressions that $\Im\gamma > 0$ so long as both the real part and the imaginary part of the wavenumber k are greater than zero. It then follows that back propagation implemented via Eq. (4.22b) will be unstable due to exponential growth of the plane waves *both over the weakly inhomogeneous part and over the evanescent part of the spectra of the field data.* However, the exponential growth of the weakly inhomogeneous plane waves will be small since it is clear from the above expressions that the imaginary part of γ over this region of the spectra will be small. On the other hand, the evanescent plane waves grow exponentially fast in proportion to $\sqrt{K_\rho^2 - \Re k^2}$ and, hence, can become unbounded.

Example 4.4 On comparing Eqs. (4.21a) and (4.21b) we conclude that

$$\tilde{U}'_+(\mathbf{K}_\rho, z_0, \omega) = \pm i\gamma\tilde{U}_+(\mathbf{K}_\rho, z_0, \omega), \tag{4.23}$$

where the top sign (+) is used in the r.h.s. $z_0 > z^+$ and the bottom sign in the l.h.s. $z_0 < z^-$ where the source to the field U_+ is assumed to lie in the strip $[z^-, z^+]$. The above result, which is a special case of the so-called "Dirichlet-to-Neumann map" (Sylvester and Uhlmann, 1990), is a consequence of the fact that the radiated field and its normal derivative over any infinite plane boundary that lies outside this strip are not independent and, in particular, must satisfy the second Helmholtz identity given in Eq. (2.31b) of Section 2.5. In fact we can derive Eq. (4.23) directly from this identity by substituting the Weyl expansion Eq. (4.4a) directly into Eq. (2.31b) with the volume τ throughout which this equation must be satisfied taken to be the region $z < z_0$ if $z_0 > z^+$ and equal to $z > z_0$ if $z_0 < z^+$. Here we will derive the result only for $z_0 > z^+$, in which case the second Helmholtz identity becomes

$$\int_{z_0} dS_0 \left[G_+(\mathbf{r} - \mathbf{r}_0, \omega)\frac{\partial}{\partial z_0}U_+(\mathbf{r}_0, \omega) - U_+(\mathbf{r}_0, \omega)\frac{\partial}{\partial z_0}G_+(\mathbf{r} - \mathbf{r}_0, \omega)\right] = 0, \quad z < z_0.$$

If we substitute the Weyl expansion Eq. (4.4a) into the above equation and perform some algebra (see the problems at the end of this chapter) we obtain the result

$$\frac{-i}{8\pi^2}\int_{-\infty}^{\infty}\frac{d^2K_\rho}{\gamma}e^{i\mathbf{K}_\rho\cdot\boldsymbol{\rho}}e^{-i\gamma(z-z_0)}[\tilde{U}'_+(\mathbf{K}_\rho,z_0,\omega)-i\gamma\tilde{U}_+(\mathbf{K}_\rho,z_0,\omega)]=0,\quad z<z_0.$$

If we now perform an inverse spatial Fourier transform over any plane $z < z_0$ we then arrive at the relationship Eq. (4.23) for the case in which $z_0 > z^+$ as required. A completely parallel development gives the result for $z_0 < z^-$.

4.3.1 Back propagation from the radiation pattern

If we express the radiated field via the angle-variable form of the angular-spectrum expansion given in Eq. (4.18a) and make use of Eq. (4.20a) we obtain

$$U_+(\mathbf{r},\omega)=\frac{ik}{2\pi}\int_{-\pi}^{\pi}d\beta\int_{C_\pm}\sin\alpha\,d\alpha\,f(\mathbf{s},\omega)e^{ik\mathbf{s}\cdot\mathbf{r}},\tag{4.24}$$

where $f(\mathbf{s},\omega)$ is the radiation pattern of the field and the α integration contour C_+ is used if $z > z^+$ and C_- if $z < z^-$. Equation (4.24) expresses the field everywhere outside the source strip $z^- < z < z^+$ in terms of the radiation pattern $f(\mathbf{s},\omega)$. Thus it performs the operation of back propagation from far-field data. As was the case for back propagation from boundary-value data, the above expansion is unstable both due to inherent absorption in a dispersive medium and due to the presence of evanescent plane waves in the expansion. The reason for the instability of the expansion is not as clear as in the case of boundary-value data, where the factor $\exp[\pm i\gamma(z-z_0)]$ in Eq. (4.22b) clearly grows exponentially fast if $z < z_0$ in the r.h.s. and if $z > z_0$ in the l.h.s. both over the evanescent part and over the weakly inhomogeneous part of the spectrum (cf. Example 4.3). In the case of the expansion Eq. (4.24) the instability over the evanescent part of the spectrum arises due to the fact that the actual far-field data specify only the radiation pattern over the real unit sphere so that $f(\mathbf{s},\omega)$ is directly known from the data only for weakly inhomogeneous components of the field. In order to determine the evanescent components it is necessary to perform an analytic continuation: this is a process that is unstable and that would generate exponentially large errors in the angular-spectrum expansion from arbitrarily small errors in the field data.

4.4 Stabilized field back propagation and the inverse boundary-value problem

We consider the process of field back propagation from Dirichlet data specified over a plane $z = z_0$ that lies outside the source strip $[z^-, z^+]$ to field points $\mathbf{r} = (\boldsymbol{\rho}, z)$ that lie within the interior strips $z^+ < z \le z_0$ or $z_0 \le z < z^-$. The boundary-value plane lies to the right of the source in the first case and to the left of the source in the second. In our previous discussion we concluded that this process is exact but unstable due to the exponential growth of the evanescent plane waves in these interior strips. The process can be stabilized if we limit the integral in Eq. (4.22b) to the weakly inhomogeneous region of the spectrum. We obtain the approximation

$$U_+(\boldsymbol{\rho}, z, \omega) \approx \frac{1}{(2\pi)^2} \int_{K_\rho^2 < \Re k^2} d^2K_\rho\, \tilde{U}_+(\mathbf{K}_\rho, z_0, \omega) e^{\pm i\gamma(z-z_0)} e^{i\mathbf{K}_\rho \cdot \boldsymbol{\rho}}, \tag{4.25}$$

where the upper (plus) sign is used to the right of the source strip (for back propagation from a plane lying to the right of the source) and the lower (minus) sign is used to the left of this strip (for back propagation from a plane lying to the left of the source). An analogous relationship connecting the back-propagated field to Neumann conditions over any given plane lying outside the source strip is easily obtained using the results in Section 4.3. The accuracy of the approximation depends on the distances of the field point $\mathbf{r} = (\boldsymbol{\rho}, z)$ and boundary-value plane z_0 from the source strip. If these distances are much larger than the wavelength λ then the accuracy is excellent since the evanescent components of the radiated field will be highly damped and give a negligible contribution to the value of the field. On the other hand, the approximation can be quite poor in the very near field, where the evanescent field components can be large.[8]

An approximate form of field back propagation was obtained earlier, in Section 2.11 of Chapter 2, where we addressed the problem of approximately back propagating the field into the interior strips $z^+ < z \leq z_0$ or $z_0 \leq z < z^-$ from Dirichlet or Neumann conditions specified over the z_0 plane. The approximate form of field back propagation arrived at in that treatment was based on the solution to an RS boundary-value problem using the incoming-wave radiation condition and was expressed in terms of the incoming-wave Dirichlet and Neumann Green functions. In this section we will represent this earlier solution in angular-spectrum form and compare the result with the approximate form of field back propagation given in Eq. (4.25). We will consider two cases: (i) back propagation using the incoming-wave Green function and (ii) back propagation using the conjugate-wave Green function.

4.4.1 Back propagation using the incoming-wave Green function

Using the notation employed in this chapter we showed in Section 2.11 that for the case of Dirichlet data an approximate solution to the problem of back propagation from a boundary-value plane z_0 lying outside the source support region is given by (cf. Eq. (2.55a))

$$U_+(\boldsymbol{\rho}, z, \omega) \approx \pm 2 \int d^2\rho'\, U_+(\boldsymbol{\rho}', z_0, \omega) \frac{\partial}{\partial z'} G_-(\mathbf{r} - \mathbf{r}_0', \omega)|_{z'=z_0}, \tag{4.26}$$

where G_- is the incoming-wave Green function to the Helmholtz equation and the upper sign is used for field points lying in the interior strip $z^+ < z \leq z_0$ that lie to the right of

[8] As discussed in Example 4.3 and in the preceding section, there will also be a small amount of instability introduced in dispersive media over the weakly inhomogeneous part of the spectrum. However, this instability will be small and can usually be ignored.

the source and to the left of a boundary-value plane located at $z = z_0$ and the lower sign is used for field points lying in the interior strip $z_0 \leq z < z^-$ that lie to the left of the source and to the right of a boundary-value plane located at $z = z_0$. A similar result was obtained for the case of Neumann data. If we make use of the angular-spectrum expansion of the incoming-wave Green function given in Eq. (4.7a) we can express Eq. (4.26) in the form

$$U_+(\boldsymbol{\rho}, z, \omega) \approx \frac{1}{(2\pi)^2} \int d^2K_\rho \, \tilde{U}_+(\mathbf{K}_\rho, z_0, \omega) e^{\pm i\gamma^*(k^*)(z-z_0)} e^{i\mathbf{K}_\rho \cdot \boldsymbol{\rho}}, \tag{4.27}$$

where again, the top sign is used in the strip $z^+ < z \leq z_0$ and the bottom sign in the strip $z_0 \leq z < z^-$.

The approximate solution to the inverse RS boundary-value problem as given in Eq. (4.27) is seen to be identical to the exact back-propagated field as given in Eq. (4.22b) except for the replacement of γ by $\gamma^*(k^*)$. Over the homogeneous region of the spectrum $\gamma^*(k^*) = \sqrt{k^2 - K_\rho^2} = \gamma$ so that this component of the expansion is identical to the stabilized back-propagated field as given in Eq. (4.25). Over the evanescent region of the spectrum $\Im\gamma^*(k^*) = -\sqrt{K_\rho^2 - k^2} = -\Im\gamma < 0$, so the plane waves comprising this component of the expansion Eq. (4.27) will decay exponentially with $|z - z_0|$ in the two strips $z^+ < z \leq z_0$ and $z_0 \leq z < z^-$. The overall conclusion is that the approximate solution to the back-propagation problem as given in Eq. (4.27) will be a good approximation under precisely the same conditions as those for which the approximation given in Eq. (4.25) is accurate; i.e., if the boundary-value plane z_0 and field point $\mathbf{r} = (\boldsymbol{\rho}, z)$ are more than a few wavelengths from the source strip $[z^-, z^+]$ so that evanescent-wave components can be neglected.

Example 4.5 The approximation Eq. (4.25) for the back-propagated field from Dirichlet data on a planar boundary turns out to be an *exact* expression for the field back propagated from *over-specified data* on this boundary using the incoming-wave Green function. As an example we consider the field back propagated using G_- from over-specified data on the plane $z = z_0$ that lies to the right of the source. The field back propagated using G_- into the half-space $z < z_0$ is given by Eq. (2.33a) of Section 2.5.1 of Chapter 2

$$\Phi_+(\mathbf{r}, \omega) = \int_{z'=z_0} d^2\rho' \left[U_+ \frac{\partial}{\partial z'} G_- - G_- \frac{\partial}{\partial z'} U_+ \right].$$

If we now substitute the angular-spectrum expansions of the incoming-wave Green function and its normal derivative into the above equation and simplify the resulting expression we obtain

$$\Phi_+(\mathbf{r}, \omega) = \frac{1}{8\pi^2} \int d^2K_\rho \, A_+(\mathbf{K}_\rho, \omega) e^{i\gamma^*(k^*)(z-z_0)} e^{i\mathbf{K}_\rho \cdot \boldsymbol{\rho}}, \tag{4.28a}$$

where

$$A_+(\mathbf{K}_\rho, \omega) = \left[\tilde{U}_+(\mathbf{K}_\rho, z_0, \omega) - \frac{i}{\gamma^*(k^*)} \widetilde{\frac{\partial}{\partial z_0} U_+(\mathbf{K}_\rho, z_0, \omega)} \right] \tag{4.28b}$$

and $\tilde{U}_+(\mathbf{K}_\rho, z_0, \omega)$ and $\widetilde{(\partial/\partial z_0) U_+(\mathbf{K}_\rho, z_0, \omega)}$ are the spatial Fourier transforms of the field $U_+(\boldsymbol{\rho}, z, \omega)$ and its normal derivative over the plane $z = z_0$.

We now use Eq. (4.23) from Example 4.4, which relates the two transforms $\widetilde{U_+}(\mathbf{K}_\rho, z_0, \omega)$ and $(\partial/\partial z_0)U_+(\mathbf{K}_\rho, z_0, \omega)$, to find that Eq. (4.28b) reduces to

$$A_+(\mathbf{K}_\rho, \omega) = \begin{cases} 2\widetilde{U_+}(\mathbf{K}_\rho, z_0, \omega) & \text{if } K_\rho^2 < \Re k^2 k, \\ 0 & \text{if } K_\rho^2 > \Re k^2, \end{cases}$$

which, when used in Eq. (4.28a), yields the result

$$\Phi_+(\mathbf{r}, \omega) = \frac{1}{(2\pi)^2} \int_{K_\rho^2 < \Re k^2} d^2 K_\rho \, \widetilde{U_+}(\mathbf{K}_\rho, z_0, \omega) e^{i\gamma(z - z_0)} e^{i\mathbf{K}_\rho \cdot \boldsymbol{\rho}},$$

which is identical to Eq. (4.25) for the case of a plane lying to the right of the source. A parallel development can be employed for the field back propagated using G_- from over-specified data on a plane that lies to the left of the source.

4.4.2 Back propagation using the conjugate-wave Green function and field time reversal

Back propagation implemented with the incoming-wave Green function can suffer instability problems in high-loss dispersive media due the positive imaginary part of γ over the weakly inhomogeneous region of the spectrum (cf. Example 4.3). A fully stabilized approximate form of back propagation over both the weakly inhomogeneous region and the evanescent region of the spectra is obtained by replacing the incoming-wave Green function by the conjugate-wave Green function. Thus in place of Eq. (4.26) we use

$$U_+(\boldsymbol{\rho}, z, \omega) \approx \pm 2 \int d^2\rho' \, U_+(\boldsymbol{\rho}', z_0, \omega) \frac{\partial}{\partial z'} G_+^*(\mathbf{r} - \mathbf{r}_0', \omega)|_{z'=z_0}, \tag{4.29}$$

which, of course, reduces to Eq. (4.26) when k is real-valued. The angular-spectrum expansion of this latter approximation is identical in form to the angular-spectrum expansion given in Eq. (4.27), where, however, $\gamma^*(k^*)$ is now replaced by $\gamma^*(k)$.

Back propagation and time reversal

We showed in Section 2.11 that the time-reversed back-propagated field obtained using the incoming-wave Green function is equal to the field radiated by the time-reversed boundary-value data into the time-reversed medium. In other words, if the Dirichlet data are first time-reversed and allowed to radiate back into the source region but in the time-reversed medium then the resulting field will be equal to the time-domain back-propagated field defined in Eq. (4.26). The approximate solution to the back-propagation problem implemented using the conjugate-wave Green function is also intimately connected with the process of wave-field time reversal. In particular, on taking the complex conjugate of both sides of Eq. (4.29) we conclude that the time-domain field $u_+(\boldsymbol{\rho}, z, -t)$ (obtained by inverse Fourier transformation of $U_+^*(\boldsymbol{\rho}, z, \omega)$) results from forward propagation of the time-reversed boundary-value field $u_+(\boldsymbol{\rho}, z_0, -t)$ into the original medium having wavenumber k. Thus, both versions of field back propagation are also versions of field time reversal and differ only in terms of the medium in which the time reversal is performed. When k is real-valued the conjugate-wave Green function is identical to the incoming-wave Green function so that the two forms of time reversal coincide.

4.5 The angular-spectrum expansion of the scalar wavelet field

A popular canonical radiation pattern is that of a so-called *scalar wavelet field*, which is given by (Kaiser, 2003, 2004, 2005)

$$f(\mathbf{s},\omega) = f(\alpha,\omega) = \frac{e^{ka\cos\alpha}}{e^{ka}}, \tag{4.30a}$$

where $a > 0$ is a positive real parameter and α is the polar angle at which the radiation pattern is measured. We will show in Chapter 5 that the wavelet source is compactly supported within a sphere of radius a that is centered at the origin, which gives physical meaning to the parameter a in its radiation pattern. For the sake of simplicity we will assume throughout this section that the wavenumber k is strictly real, corresponding to a non-dispersive background medium. The results are easily generalized to dispersive media.

The wavelet radiation pattern is seen to be rotationally symmetric about the polar (z) axis, which peaks in the forward direction ($\alpha = 0$) where it is unity. We show in Fig. 4.4 the radiation pattern for two values of ka together, for comparison, with the radiation patterns of a planar circular source supported on the $z = 0$ plane having the same radius a as that of the 3D wavelet source. This radiation pattern, normalized to unity in the forward direction, is found to be given by

$$f(\alpha,\omega) = 2\frac{J_1(ka\sin\alpha)}{ka\sin\alpha}. \tag{4.30b}$$

On examining Fig. 4.4 it can be seen that, although the disk radiation patterns have smaller central lobes than the wavelet patterns for a given radius, they have the unfortunate properties of possessing large side lobes and radiating equally in the right and left half-spaces.[9] The wavelet radiation pattern, by comparison, has no side lobes and very little back radiation.

The scalar wavelet field can be computed outside the source region (for $r > a$) using the angular-spectrum expansion in either of the two forms given in Eqs. (4.15a) and (4.18a) with the angular spectrum computed from the scalar wavelet radiation pattern using Eqs. (4.20). We will use the Cartesian form of the expansion here and leave the computation of the field via the angle-variable form as a problem at the end of the chapter.

On making use of Eqs. (4.4b) and (4.20b) and using cylindrical coordinates for $\mathbf{r}$ we can express the angular-spectrum expansion of the wavelet field in the form

[9] The back radiation can be significantly reduced or even removed entirely by employing both a singlet and a doublet component to the surface disk source (see Sections 1.8, 2.12 and Section 5.2 in the following chapter).

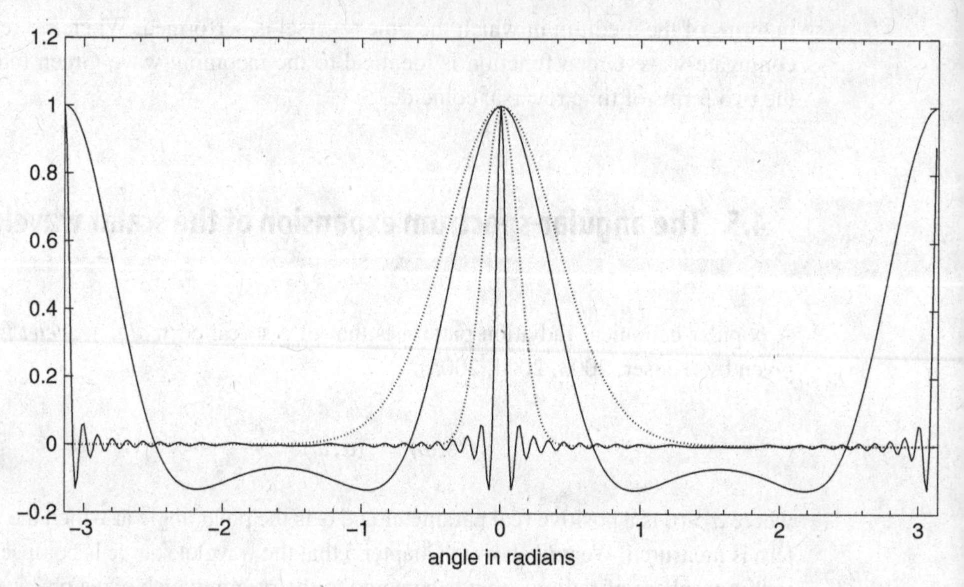

Fig. 4.4 The radiation patterns of a wavelet field for a wavelet parameter $a = 1$ and for wavelengths of $\lambda = a/10 = 0.1$ and $\lambda = a = 1$ (dotted). Shown for comparison (drawn solid) are the radiation patterns of a circular disk source supported on the plane $z = 0$ and having the same radius as that of the wavelet source. The wavelet and disk radiation patterns with the smaller central lobes correspond to the smaller of the two wavelengths.

$$U_+(\mathbf{r},\omega) = \frac{1}{(2\pi)^2}\int_{-\infty}^{\infty} d^2K_\rho \overbrace{\frac{2\pi i\, e^{\pm a\gamma}}{\gamma\ e^{ka}}}^{A(\mathbf{k}^\pm,\omega)} e^{i(\mathbf{K}_\rho\cdot\boldsymbol{\rho}\pm\gamma z)}.$$

If we now transform to cylindrical coordinates for the spatial frequency variable $\mathbf{K}_\rho$ the above expression simplifies to become

$$\begin{aligned} U_+(\mathbf{r},\omega) &= i\int_0^\infty \frac{K_\rho\, dK_\rho}{\gamma}\frac{e^{\pm a\gamma}}{e^{ka}} e^{\pm i\gamma z} \overbrace{\frac{1}{2\pi}\int_{-\pi}^{\pi} d\beta\, e^{iK_\rho\cos(\beta-\phi)}}^{J_0(K_\rho\rho)} \\ &= i\int_0^\infty \frac{K_\rho\, dK_\rho}{\gamma}\frac{e^{\pm a\gamma}}{e^{ka}} J_0(K_\rho\rho)e^{\pm i\gamma z}. \end{aligned} \tag{4.31}$$

We show in Fig. 4.5 mesh plots of the real (top) and imaginary (bottom) parts of the homogeneous part of the wavelet field computed using Eq. (4.31) band-limited to $K_\rho < k$. We employed a wavelet parameter $a = 1$ (equal to the radius of the wavelet source), propagation distances $z = 0$, $z = a$ and $z = 2a$ and a wavelength of $\lambda = a = 1$. All three plots can be considered to be stabilized back propagations from the far field (as given by the radiation pattern) resulting from removing all evanescent components of the field (see the discussion in Section 4.3). The plane $z = 0$ is internal to the wavelet source and, hence, the field computed over this internal plane has no direct physical meaning, while the field computed over the bounding plane to the source (at $z = a$) is in error due to the non-inclusion of the evanescent-wave components of the field. On the other hand, the field

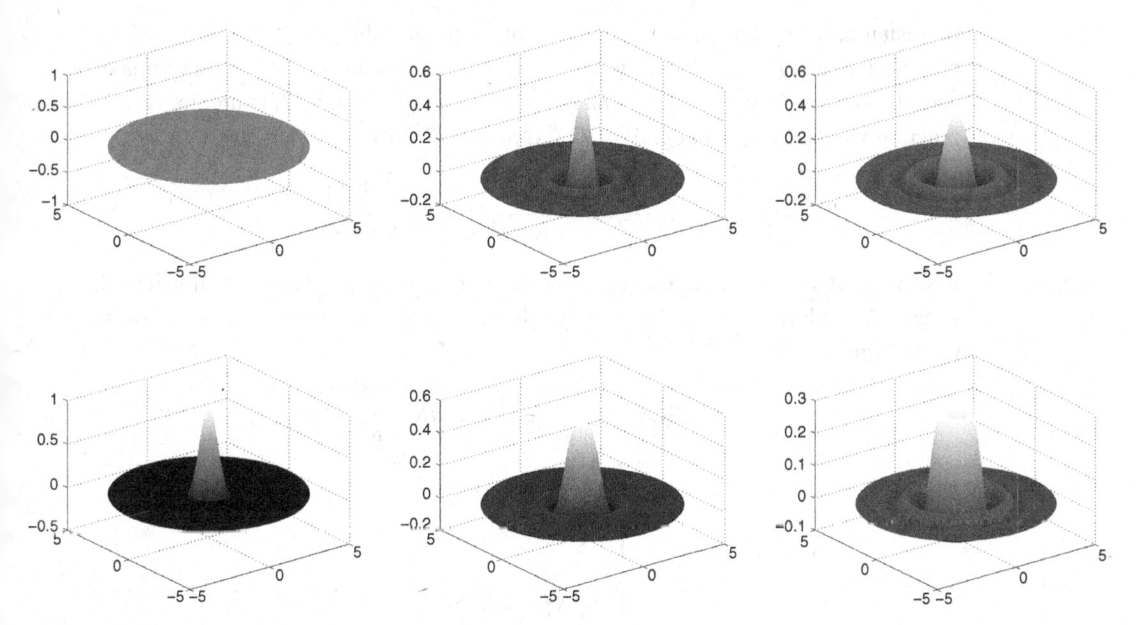

Fig. 4.5 Mesh plots of the real (top) and imaginary (bottom) parts of the wavelet field having the parameter $a = \lambda$ band-limited to homogeneous plane waves over the planes $z = 0$, $z = a$ and $z = 2a$ for a wavelength $\lambda = a = 1$.

computed over the plane $z = 2a$ will be quite accurate since it is a wavelength from the source for the wavelength of $\lambda = 1$ so that the evanescent-wave components of the field would be considerably damped, with the result that the homogeneous component of the field would be a good approximation to the total field.

4.6 Angular-spectrum expansions in two space dimensions

In this section we derive the plane-wave (angular-spectrum) expansions for the 2D Green function and the solutions to the 2D radiation problem as well as the 2D Dirichlet and Neumann boundary-value problems. These expansions can be employed for both forward and back propagation of 2D wavefields in a manner completely parallel to their 3D counterparts. The advantage of these 2D expansions over their 3D counterparts is computational speed in computer simulations both of forward and of inverse problems in wave propagation and scattering.

We can obtain the angular-spectrum expansion of the 2D Green function following steps identical to those used in obtaining the Weyl expansion of the 3D Green function in Section 4.1. The Weyl expansion is not a coordinate-free representation of the Green function and depends, in particular, on the selection of a specific coordinate axis relative to which the decomposition into homogeneous and evanescent plane waves is performed. In the 3D case this axis is, by convention, taken to be the z axis of the 3D right-handed

Cartesian x, y, z system and it is convenient to employ the same coordinate axis for this purpose in the 2D case. We thus will employ a 2D Cartesian system with coordinate axes x and z (corresponding to the $y = 0$ plane in 3D) and represent the 2D outgoing-wave Green function via the Fourier integral in 2D (cf. Section 2.2.1 of Chapter 2):

$$G_+(\mathbf{R}, \omega) = \frac{-1}{(2\pi)^2} \int d^2K \, \frac{e^{i\mathbf{K}\cdot\mathbf{R}}}{K^2 - k^2},$$

where now $\mathbf{R} = (X, Z)$ and $\mathbf{K} = (K_x, K_z)$ are defined on the (x, z) plane. Following the same general procedure as was used to derive the Weyl expansion, we write the above integral in the form

$$G_+(\mathbf{R}, \omega) = \frac{-1}{(2\pi)^2} \int dK_x \, dK_z \, \frac{e^{i(K_x X + K_z Z)}}{K_z^2 - \gamma^2}, \tag{4.32a}$$

where

$$\gamma = \begin{cases} \sqrt{k^2 - K_x^2} & \text{if } K_x^2 < \Re k^2, \\ i\sqrt{K_x^2 - k^2} & \text{if } K_x^2 > \Re k^2. \end{cases} \tag{4.32b}$$

Following steps identical to those used in deriving the Weyl expansion, we perform the K_z integration in Eq. (4.32a) to obtain the angular-spectrum expansion of the 2D Green function (which we will refer to as the 2D Weyl expansion):

$$G_+(\mathbf{R}, \omega) = \frac{-i}{4\pi} \int_{-\infty}^{\infty} dK_x \, \frac{e^{i\mathbf{k}^{\pm}\cdot\mathbf{R}}}{\gamma}, \tag{4.33}$$

where the plus sign is used if $Z > 0$ and the minus sign if $Z < 0$, and we have defined

$$\mathbf{k}^{\pm} = \hat{\mathbf{x}} K_x \pm \gamma \hat{\mathbf{z}}.$$

Incoming- and conjugate-wave Green functions

The angular-spectrum expansion of the conjugate-wave Green function in 2D is obtained by taking the complex conjugate of Eq. (4.33). We find that

$$G_+^*(\mathbf{R}, \omega) = \frac{i}{4\pi} \int_{-\infty}^{\infty} dK_x \, \frac{e^{i\mathbf{k}^{\mp *}\cdot\mathbf{R}}}{\gamma^*},$$

where the top sign is used if $Z > 0$ and the bottom sign if $Z < 0$, and

$$\gamma^* = \begin{cases} \sqrt{k^{*2} - K_x^2} & \text{if } K_x^2 < \Re k^2, \\ -i\sqrt{K_x^2 - k^{*2}} & \text{if } K_x^2 > \Re k^2, \end{cases}$$

and

$$\mathbf{k}^{\mp *} = \hat{\mathbf{x}}K_x \mp \gamma^* \hat{\mathbf{z}}$$
$$= \begin{cases} \hat{\mathbf{x}}K_x \mp \sqrt{k^{*2} - K_x^2}\,\hat{\mathbf{z}} & \text{if } K_x^2 < \Re k^2, \\ \hat{\mathbf{x}}K_x \pm i\sqrt{K_x^2 - k^{*2}}\,\hat{\mathbf{z}} & \text{if } K_x^2 > \Re k^2. \end{cases}$$

The angular-spectrum expansion of the incoming-wave Green function G_- is obtained from that of G_+^* by simply replacing k by k^*.

All of the discussion relating to the angular-spectrum expansion of the 3D conjugate- and incoming-wave Green functions presented in earlier sections applies equally well to the 2D expansions of these quantities. In particular, the expansion for the incoming-wave Green function is identical to that for the outgoing-wave Green function over the evanescent region $K_\rho > k$ and differs by an overall sign and shift from outgoing to incoming plane waves over the homogeneous region $K_\rho < k$. In contrast with the expansion for the outgoing-wave Green function, the plane waves over the homogeneous region thus propagate *inward* from the half-space containing the field point $\mathbf{R}$ toward the origin, while the evanescent plane waves still propagate in or near to the plane $Z = 0$ and decay exponentially with distance $|Z|$ from this plane. As mentioned earlier, these observations are consistent with the fact that G_- is an *incoming-wave* Green function so that it is to be expected on intuitive grounds that it be composed of a superposition of incoming plane waves.

4.6.1 The angular-spectrum expansion of the solution to the 2D radiation problem

Now consider a 2D source $Q(\mathbf{r}, \omega)$ compactly supported within the 2D region τ_0 that is bounded by two infinite lines parallel to the x axis and located at $z = z^\pm$. The radiated field is given by Eq. (2.28) of Chapter 2,

$$U_+(\mathbf{r}, \omega) = \int_{\tau_0} d^2r'\, G_+(\mathbf{r} - \mathbf{r}', \omega) Q(\mathbf{r}', \omega),$$

where $\mathbf{r} = (x, z)$ and G_+ is the 2D outgoing-wave Green function to the Helmholtz equation. On substituting the 2D Weyl expansion into the above equation and restricting our attention to field points lying outside the strip $z^- \leq z \leq z^+$ and performing some simplifying algebra we obtain the result

$$U_+(\mathbf{r}, \omega) = \sqrt{\frac{k}{2\pi}} e^{i\frac{\pi}{4}} \int_{-\infty}^{\infty} \frac{dK_x}{\gamma} A(\mathbf{k}^\pm, \omega) e^{i\mathbf{k}^\pm \cdot \mathbf{r}}, \qquad (4.34a)$$

where the plus sign is used if $z > z^+$ and the minus sign if $z < z^-$ and the "angular spectrum" $A(\mathbf{k}^\pm, \omega)$ is given in terms of the source by

$$A(\mathbf{k}^{\pm}, \omega) = -i\sqrt{\frac{1}{8\pi k}}e^{-i\frac{\pi}{4}}\tilde{Q}(\mathbf{K}, \omega)|_{\mathbf{K}=\mathbf{k}^{\pm}}, \tag{4.34b}$$

with

$$\tilde{Q}(\mathbf{K}, \omega) = \int_{\tau_0} d^2r' \, Q(\mathbf{r}', \omega)e^{-i\mathbf{K}\cdot\mathbf{r}'}$$

being the 2D spatial Fourier transform of the source $Q(\mathbf{r}, \omega)$.

We have used the specific form of the angular-spectrum expansion given in Eq. (4.34a) in order that the angular spectra $A(\mathbf{k}^{\pm}, \omega)$ reduce to the radiation pattern of the 2D radiated field expressed in terms of the propagation vectors $\mathbf{k}^{\pm}$. In particular, if we express that radiation pattern given in Eq. (2.30) of Chapter 2 in terms of the propagation vectors $\mathbf{k}^{\pm}$ we obtain

$$f(\mathbf{s}^{\pm}, \omega) = -\sqrt{\frac{1}{8\pi k}}e^{i\frac{\pi}{4}}\tilde{Q}(\mathbf{k}^{\pm}, \omega) = A(\mathbf{k}^{\pm}, \omega), \tag{4.35}$$

where $\mathbf{s}^{\pm} = \mathbf{k}^{\pm}/k$ is the unit vector along the direction of the propagation vector $\mathbf{k}^{\pm}$.

4.6.2 Two-dimensional forward and back propagation

The above results are seen to completely parallel those obtained in Sections 4.1 and 4.2. Indeed, it can be easily verified that all of those results apply to the 2D case with minor modification. For example, forward and back propagation are directly implemented using Eq. (4.34a) by noting that the 1D spatial Fourier transform of the field or its normal derivative over any line z lying outside the source strip $[z^-, z^+]$ is related to the angular spectrum via the 2D version of Eqs. (4.21a) and (4.21b); i.e.,

$$\tilde{U}_+(K_x, z, \omega) = \sqrt{2\pi k}\frac{e^{i\frac{\pi}{4}}}{\gamma}A(\mathbf{k}^{\pm}, \omega)e^{\pm i\gamma z}, \tag{4.36a}$$

where

$$\tilde{U}_+(K_x, z, \omega) = \int dx \, U_+(x, z, \omega)e^{-iK_x x}$$

is the 1D spatial Fourier transform of the field on the line z. An entirely analogous argument yields the result

$$\tilde{U}'_+(K_x, z, \omega) = \pm i\sqrt{2\pi k}e^{i\frac{\pi}{4}}A(\mathbf{k}^{\pm}, \omega)e^{\pm i\gamma z}, \tag{4.36b}$$

where $\tilde{U}'_+$ is the 1D spatial Fourier transform of the Neumann boundary value on the line z, and the $+$ sign is used if $z > z^+$ and the minus sign if $z < z^-$.

The location of the line z is arbitrary so long as it lies outside the source strip $[z^-, z^+]$ so that the above equations can be used to relate the spatial Fourier transforms of the fields over any two lines lying on the same side of the source strip. In particular, in analogy to Eq. (4.22a) we have that

$$\tilde{U}_+(K_x, z, \omega) = \tilde{U}_+(K_x, z_0, \omega)e^{\pm i\gamma(z-z_0)}, \tag{4.37}$$

where the plus sign is used if z and z_0 lie to the right of the source strip and the minus sign applies if they both lie to the left of the source strip. Again, as in the 3D case, forward propagation of the boundary value field $\tilde{U}_+(K_x, z_0, \omega)$ corresponds to the case in which $\exp[\pm i\gamma(z - z_0)]$ decays with increasing $|z - z_0|$ and back propagation of this boundary-value field to the case in which $\exp[\pm i\gamma(z - z_0)]$ increases (grows) with increasing $|z - z_0|$. The actual fields are, of course, computed via an inverse 1D Fourier transform of $\tilde{U}_+(K_x, z, \omega)$. The case of Neumann field data is treated in an entirely parallel manner.

Example 4.6 As an example we implemented the 2D angular spectra and 2D Rayleigh–Sommerfeld (RS) solution given by Eq. (2.48) of Chapter 2. The 2D RS solution was shown in Example 2.8 of Chapter 2 to be given by

$$U_+(x, z) = \pm \frac{ik}{2} \int_{-\infty}^{\infty} dx' \, U_+(x', z_0) H_1^+(k|\mathbf{r} - \mathbf{r}'|) \frac{z - z'}{|\mathbf{r} - \mathbf{r}'|} \bigg|_{z'=z_0}, \tag{4.38}$$

where H_1^+ is the first-order Hankel function of the first kind and the plus sign is used for propagation into the r.h.s. $z > z_0 > z^+$ and the minus sign for propagation into the l.h.s. $z < z_0 < z^-$, where $[z^-, z^+]$ is the source strip.

Unlike the RS solution, the angular-spectrum expansion can be used for both forward and back propagation and is given by Eq. (4.34a), with the angular spectrum determined from Dirichlet or Neumann data over any line lying outside the source strip via Eqs. (4.36). As in the 3D case, the process of forward propagation is stable and, in principle, identical to the RS solution. On the other hand, the process of back propagation is unstable due to the growth of the evanescent plane waves when $|z| < |z_0|$, where z_0 is the boundary-value line and z the line onto which the propagated or back-propagated field is to be evaluated. To stabilize the back-propagation operation, we can either limit the angular-spectrum expansion to only homogeneous plane waves as discussed in Section 4.4 or use the approximate form of back propagation given in Eq. (4.26) or Eq. (4.27) of that section. It is computationally easier to employ Eq. (4.27), which, for the non-dispersive media assumed in these simulations, simply requires us to replace γ by γ^* in the angular-spectrum expansion of the fields.

We show in Fig. 4.6 plots of the forward-propagated field from a rectangular-shaped boundary-value field having a total extent of 20λ on the line $z_0 = 0$, for propagation distances ranging from $z = 50\lambda$ to $z = 200\lambda$ with 50λ separation. The boundary-value field can be interpreted as being that generated by an infinite slit when illuminated by a plane wave normally incident to the plane of the slit. The solid curves were computed using the RS formula given in Eq. (4.38), while the dotted curves were computed using the 2D angular-spectrum expansion Eq. (4.34a) with the angular spectrum computed from the Fourier transform of the boundary value according to Eq. (4.36a). It is apparent from Fig. 4.6 that the difference between the two computations increases as the propagation distance z increases. We have magnified the plots over an interval about the central peak of the propagated wave fields in Fig. 4.7, which clearly shows the increasing error with increasing propagation distance. This increase in error with increasing propagation distance is due to the fact that the angular-spectrum expansion is implemented using an FFT with a fixed sampling interval δx on the x axis and fixed overall interval size $L_x = N\,\delta x$, where N is the

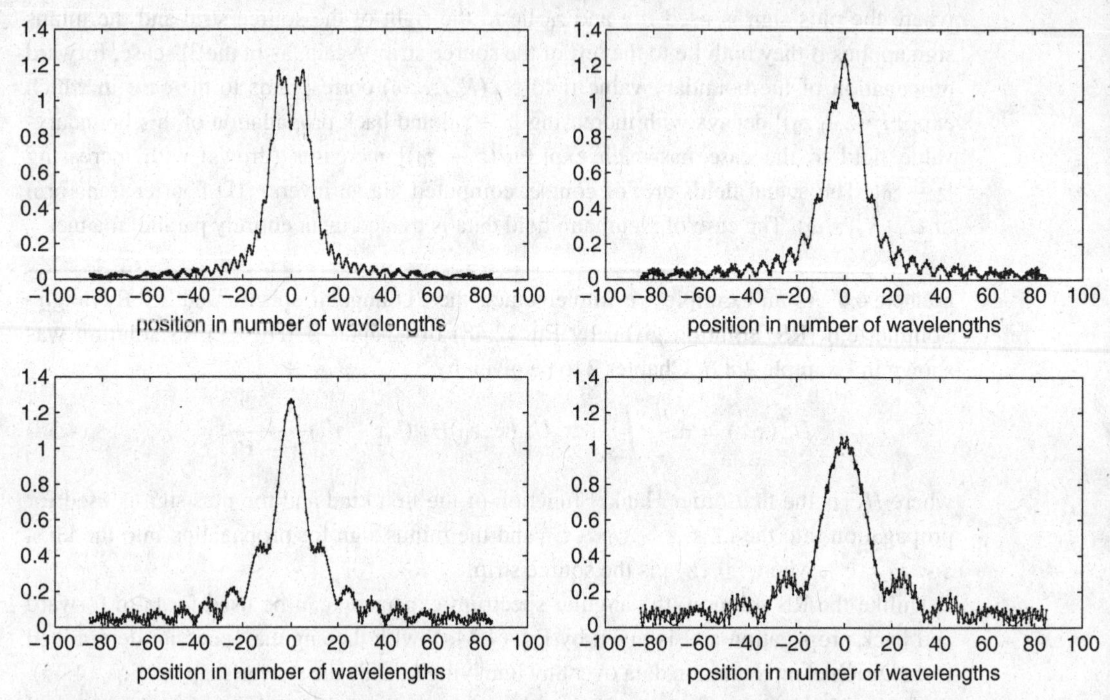

Fig. 4.6 The forward-propagated fields from a rectangular-shaped boundary-value field of width equal to 20λ at $z_0 = 0$ computed using the 2D RS formula (drawn solid) and the angular spectrum truncated to only homogeneous plane waves (dotted). The curves correspond to propagation distances of $z = 50\lambda$ (top left), $z = 100\lambda$ (top right), $z = 150\lambda$ (lower left) and $z = 200\lambda$ (lower right).

number of points used in the FFT. As the propagation distance increases, the field expands and eventually spills outside the (fixed) interval, with the result that the angular-spectrum computation will eventually suffer from aliasing in the space domain. This aliasing effect can be reduced by increasing the interval size L on the boundary-value line (by increasing the total number of points N used in the FFT). However, as we will discuss later in connection with the Fresnel transform, the required interval size increases linearly with z, so such a buffering scheme will require extremely large array sizes, which limits the use of the angular-spectrum expansion for field-propagation computations at large propagation distances.

Example 4.7 As a second example we implemented 2D back propagation using the approximate form of field back propagation given by the 2D version of Eq. (4.27) of Section 4.4. The field data were computed over a line $z > z_0 \geq 0$ using the RS formula given in the previous example and the back propagation was performed by taking the inverse Fourier transform of these data and using Eq. (4.27). We show in Fig. 4.8 the stabilized back-propagated fields computed from two of the four RS forward-propagated fields shown in

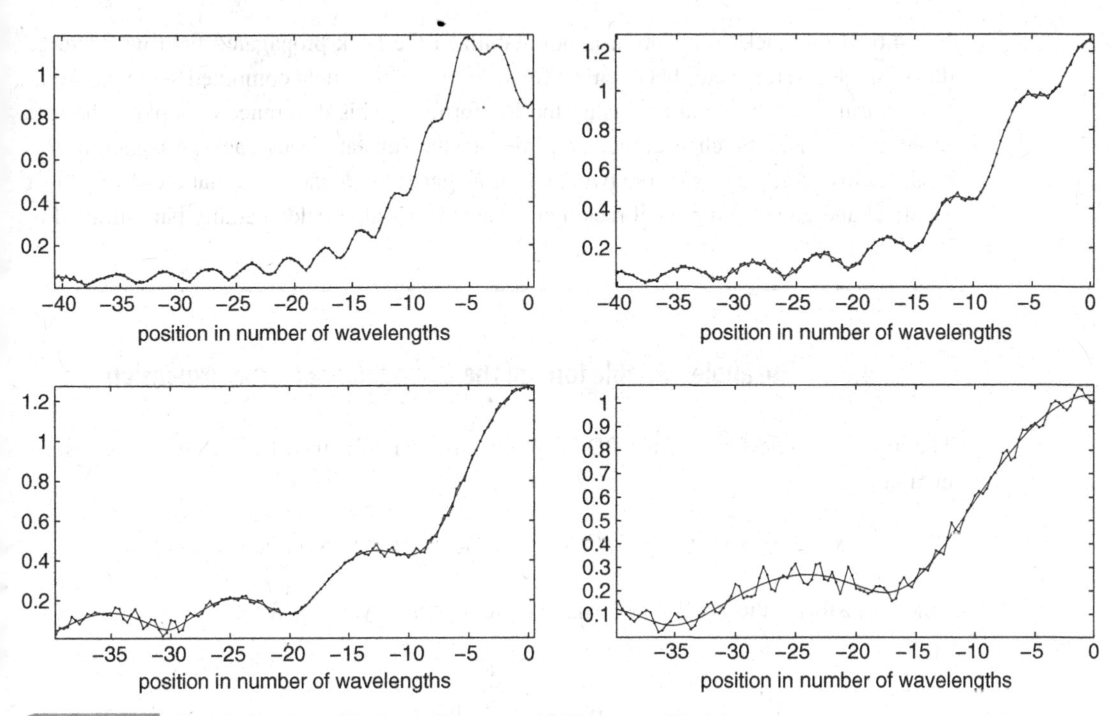

Fig. 4.7 Magnified versions of the forward-propagated fields shown in Fig. 4.6.

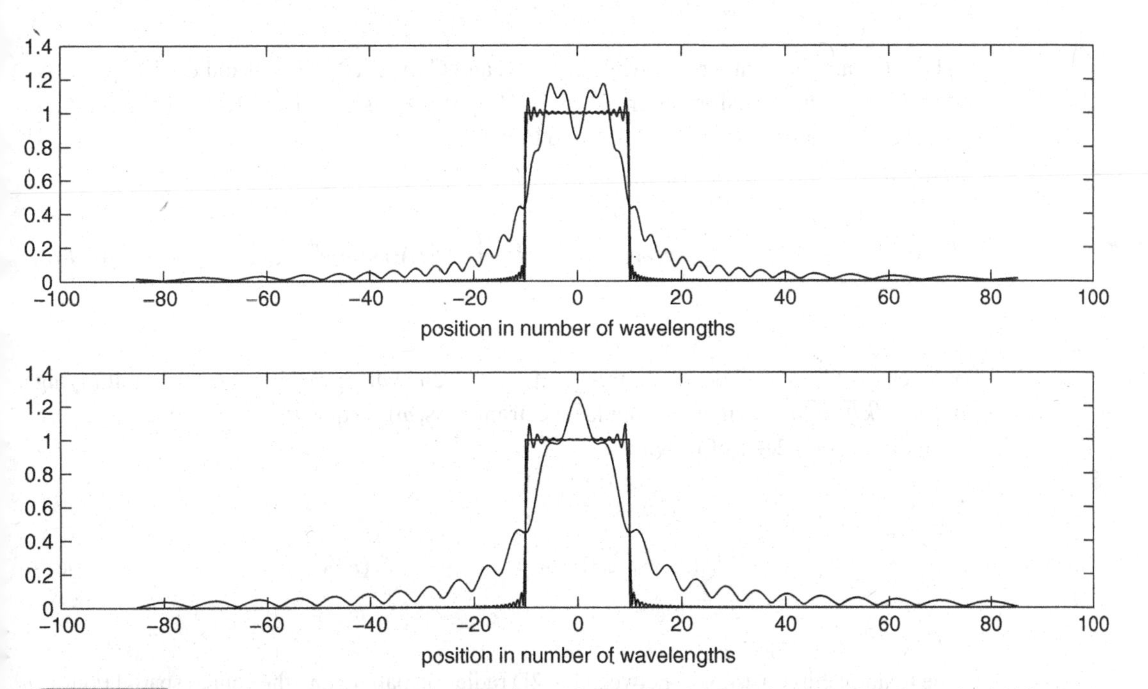

Fig. 4.8 The forward-propagated fields, back-propagated fields and ideal rectangular-shaped boundary-value field for propagation and back-propagation distances of $z = 50\lambda$ (top) and $z = 100\lambda$ (bottom).

Fig. 4.6. If the back-propagation is not stabilized the back-propagated field will diverge due to small discrepancies between the forward-propagated field computed using the angular spectrum and that obtained using the RS formula. This difference thus plays the role of additive noise, which becomes amplified in the (unstabilized) back-propagation process. A close comparison of the two back-propagated fields indicates that they are almost identical and equal to the ideal rectangular boundary-value field spatially band-limited to $|K_x| < k$.

4.6.3 The angle-variable form of the 2D angular-spectrum expansion

The angle-variable form of the 2D Weyl expansion is obtained by making the transformation

$$K_x = k \sin\alpha, \qquad \gamma = k\cos\alpha, \qquad \mathbf{k}^{\pm} \to k\mathbf{s} = k\sin\alpha\,\hat{\mathbf{x}} + k\cos\alpha\,\hat{\mathbf{z}},$$

which transforms the 2D Weyl expansion into the angle-variable form

$$G_+(\mathbf{R}, \omega) = \frac{-i}{4\pi}\int_{C_\pm} d\alpha\, e^{ik\mathbf{s}\cdot\mathbf{R}}, \tag{4.39a}$$

where $C_\pm$ are the contours shown in Fig. 4.9, and C_+ is used if $Z > 0$ and C_- if $Z < 0$. By making the same transformation as in Eq. (4.34a) the angle-variable form of the angular-spectrum expansion of the field is found to be

$$U_+(\mathbf{r}, \omega) = \sqrt{\frac{k}{2\pi}} e^{i\frac{\pi}{4}} \int_{C_\pm} d\alpha\, A(k\mathbf{s}, \omega) e^{ik\mathbf{s}\cdot\mathbf{r}}, \tag{4.39b}$$

where now C_+ is used if $z > z^+$ and C_- if $z < z^-$, and where we have used the multiplying factor $\sqrt{k/(8\pi)}e^{i\frac{\pi}{4}}$ so that the angular spectrum $A(k\mathbf{s}, \omega)$ is equal to the 2D radiation pattern given in Section 2.4.1 of Chapter 2:

$$A(k\mathbf{s}, \omega) = f(\mathbf{s}, \omega) = -\sqrt{\frac{1}{8\pi k}} e^{i\frac{\pi}{4}} \tilde{Q}(k\mathbf{s}, \omega). \tag{4.39c}$$

The relationship Eq. (4.39c) between the 2D radiation pattern and the source spatial Fourier transform forms the basis for one formulation of the inverse source problem (ISP) in two space dimensions that will be treated in Chapter 5.

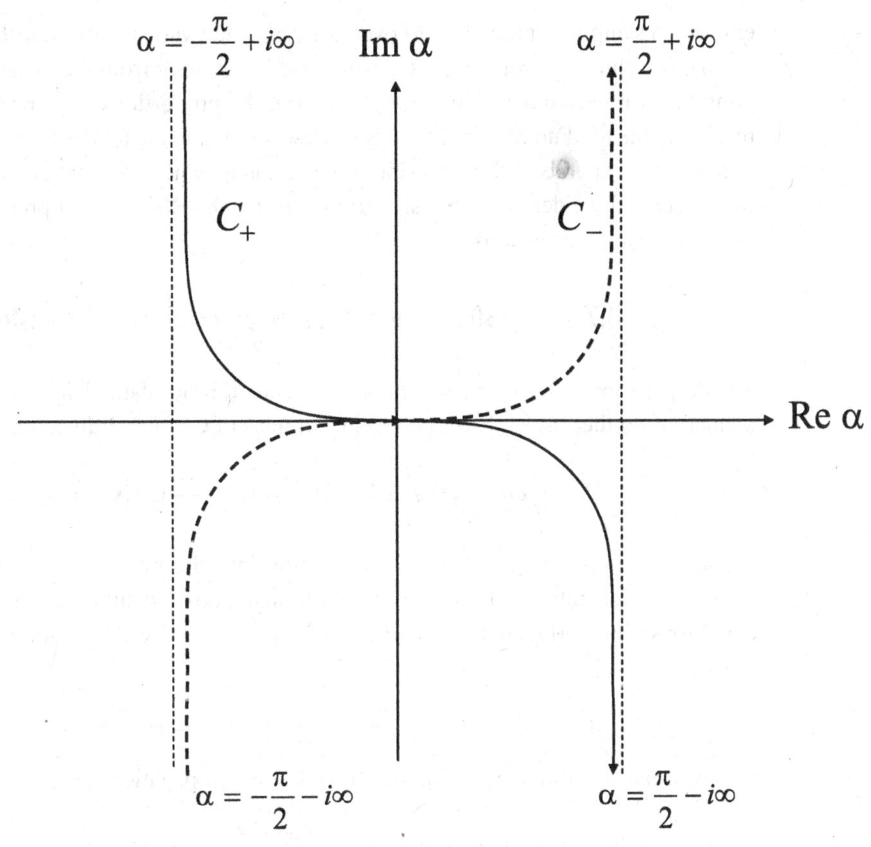

Fig. 4.9 The α integration contours of integration for the 2D Weyl expansions of G_+. The solid-line contour labeled C_+ is used if $Z > 0$ and the dashed contour labeled C_- if $Z < 0$.

4.7 The Fresnel approximation and Fresnel transform

The RS boundary-value problems can be solved using either the Rayleigh–Sommerfeld (RS) formulas or the angular-spectrum expansion. The angular-spectrum expansion has the apparent advantage that it can be implemented digitally using the fast Fourier transform (FFT). However, as we found in Example 4.6, this apparent advantage disappears if the propagation distance is much larger than the wavelength. As we discussed in that example, the reason for this is that a freely propagating outgoing wave will eventually expand as it propagates so that the number of spatial samples required in the FFT increases with propagation distance. A way around this difficulty is possible by use of the Fresnel approximation (Goodman, 1968), which, when implemented digitally, is known as the *Fresnel transform* (Hamam and de Bougrenet de la Tocnaye, 1995). This transform can be used to implement free-space propagation and back propagation between two planes z_0 and z if the propagation distance $h = |z - z_0|$ between the two planes is much larger than the wavelength. The advantage of the Fresnel transform is that it is *self-scaling* in that it employs a sampling interval $\delta x(h)$ over any plane that varies linearly with the propagation distance h. Thus, for example, if the wavefield is outgoing from the plane z_0 then the sampling interval required on the plane $z > z_0$ will be proportional to $h = |z - z_0|$, with the result that the total number of samples is preserved while the total area over which the field is computed expands with increasing propagation distance and spatial aliasing is avoided. In this section we derive the Fresnel transform for the RS Dirichlet problem for both two and three space dimensions.

4.7.1 The 3D Fresnel approximation and Fresnel transform

The propagation of outgoing wavefields from an arbitrary data plane $z = z_0$ can be implemented using the Rayleigh–Sommerfeld formulas of Chapter 2. In particular, we have that

$$U_+(\mathbf{r},\omega) = \mp 2 \int_{z_0} dS'\, U_+(\mathbf{r}',\omega)|_{z'=z_0} \frac{\partial}{\partial z'} G_+(\mathbf{r}-\mathbf{r}')|_{z'=z_0}, \tag{4.40}$$

where the minus sign is used for propagation into the r.h.s. $z > z_0$ and the plus sign for propagation into the l.h.s. $z < z_0$, with analogous formulas for Neumann boundary conditions. The normal derivative of the outgoing-wave Green function is found to be

$$\frac{\partial}{\partial z'} G_+(\mathbf{r}-\mathbf{r}') = -\frac{1}{4\pi} \frac{e^{ik|\mathbf{r}-\mathbf{r}'|}}{|\mathbf{r}-\mathbf{r}'|} \left(-ik \frac{z-z'}{|\mathbf{r}-\mathbf{r}'|} + \frac{z-z'}{|\mathbf{r}-\mathbf{r}'|^2} \right),$$

which, when substituted into Eq. (4.40), yields the propagation formula

$$U_+(\mathbf{r},\omega) = \pm \frac{1}{2\pi} \int_{z_0} dS'\, U_+(\mathbf{r}',\omega)|_{z'=z_0} \frac{e^{ik|\mathbf{r}-\mathbf{r}'|}}{|\mathbf{r}-\mathbf{r}'|} \left(-ik \frac{z-z_0}{|\mathbf{r}-\mathbf{r}'|} + \frac{z-z_0}{|\mathbf{r}-\mathbf{r}'|^2} \right)\Bigg|_{z'=z_0}.$$

The 3D Fresnel approximation

In three space dimensions we have that

$$|\mathbf{r}-\mathbf{r}'| = |z-z'| \sqrt{1 + \frac{(x-x')^2 + (y-y')^2}{(z-z')^2}} \sim |z-z'| + \frac{(x-x')^2 + (y-y')^2}{2|z-z'|},$$

which is valid if $|z-z'| \gg \sqrt{(x-x')^2 + (y-y')^2}$. We conclude that in this case

$$\begin{aligned}
\frac{\partial}{\partial z'} G_+(\mathbf{r}-\mathbf{r}') &= -\frac{1}{4\pi} \frac{e^{ik|\mathbf{r}-\mathbf{r}'|}}{|\mathbf{r}-\mathbf{r}'|} \left(-ik \frac{z-z'}{|\mathbf{r}-\mathbf{r}'|} + \frac{z-z'}{|\mathbf{r}-\mathbf{r}'|^2} \right) \\
&\sim -\frac{1}{4\pi} \frac{e^{ik|z-z'| + ik\frac{(x-x')^2+(y-y')^2}{2|z-z'|}}}{|z-z'|} \left(-ik \frac{z-z'}{|z-z'|} + \frac{z-z'}{|z-z'|^2} \right) \\
&= \pm \frac{ik}{4\pi} \frac{e^{ik|z-z'| + ik\frac{(x-x')^2+(y-y')^2}{2|z-z'|}}}{|z-z'|},
\end{aligned}$$

where the plus sign is used if $z > z_0$ and the minus sign if $z < z_0$. The Fresnel approximation in 3D is then given by

$$U_+(\mathbf{r},\omega) = -\frac{ik}{2\pi}\int_{z_0} dS'\, U_+(\mathbf{r}')|_{z'=z_0} \frac{e^{ik|z-z_0|+ik\frac{(x-x')^2+(y-y')^2}{2|z-z_0|}}}{|z-z_0|}. \tag{4.41}$$

A sufficient condition for the validity of the 3D Fresnel approximation is that

$$|z-z_0|^3 \gg \frac{\pi}{4\lambda}\,\delta\rho^4|_{\max},$$

where $\delta\rho = \sqrt{(x-x')^2+(y-y')^2}$ and the maximum is to be taken relative to all source points in the boundary-value plane and all field points on the plane z for which the field is to be computed.

The 3D Fresnel transform

If we expand out the exponentials in Eq. (4.41) we can write the Fresnel approximation in the form

$$U_+(\mathbf{r},\omega) = -\frac{ik}{2\pi}\frac{e^{ik|z-z_0|}}{|z-z_0|} e^{i\frac{k}{2}\frac{x^2+y^2}{|z-z_0|}} \int_{z_0} dS'\, U_+(\mathbf{r}',\omega)|_{z'=z_0} \times e^{i\frac{k}{2}\frac{x'^2+y'^2}{|z-z_0|}} e^{-i\frac{k}{|z-z_0|}(xx'+yy')}. \tag{4.42}$$

Equation (4.42) can be implemented in three steps.

1. Multiplication of the boundary-value field $U_+(\mathbf{r}',\omega)|_{z'=z_0}$ by $\exp(i\frac{k}{2}\frac{x'^2+y'^2}{|z-z_0|})$, yielding

$$F(x',y') = U_+(\mathbf{r}',\omega)|_{z'=z_0} e^{i\frac{k}{2}\frac{x'^2+y'^2}{|z-z_0|}}.$$

2. Two-dimensional spatial Fourier transformation of F. After the 2D Fourier transform has been computed the integral in Eq. (4.42) is given by

$$\int_{z_0} dS'\, F(x',y')e^{-i\frac{k}{|z-z_0|}(xx'+yy')} = \tilde{F}(K_x,K_y)|_{K_x=\frac{k}{|z-z_0|}x,K_y=\frac{k}{|z-z_0|}y}, \tag{4.43}$$

where $\tilde{F}(K_x,K_y)$ is the 2D spatial Fourier transform of F.

3. Multiplication of the result obtained in Eq. (4.43) by

$$C(x,y) = -\frac{ik}{2\pi}\frac{e^{ik|z-z_0|}}{|z-z_0|} e^{i\frac{k}{2}\frac{x^2+y^2}{|z-z_0|}}.$$

It is important to note that in a computer implementation of the Fresnel transform the sampling intervals on the boundary-value plane δx_0 and δy_0 are *different* from those on the observation plane located at z. In particular, due to the FFT we have that

$$\delta K_x\,\delta x_0 = \frac{2\pi}{N}, \qquad \delta K_y\,\delta y_0 = \frac{2\pi}{N},$$

where N is the total number of sample points for x and y on the boundary-value plane $z = z_0$. But, from Eq. (4.43),

$$K_x = \frac{k}{|z - z_0|}x, \qquad K_y = \frac{k}{|z - z_0|}y,$$

so that

$$\delta x = \frac{|z - z_0|}{k}\delta K_x = \frac{|z - z_0|}{k}\frac{2\pi}{N\,\delta x_0} = \lambda\frac{|z - z_0|}{N\,\delta x_0}, \tag{4.44a}$$

$$\delta y = \frac{|z - z_0|}{k}\delta K_y = \frac{|z - z_0|}{k}\frac{2\pi}{N\,\delta y_0} = \lambda\frac{|z - z_0|}{N\,\delta y_0}, \tag{4.44b}$$

where δx and δy are the sample spacings on the plane z. Thus, the sampling interval on the observation plane increases linearly with propagation distance. For this reason, the Fresnel transform is sometimes referred to as a *self-scaling* transform.

Back propagation: the inverse Fresnel transform

We can invert the Fresnel transform to approximately recover the boundary-value field on the plane $z = z_0$ from the propagated field on the plane z. To do this we simply invert the three steps that were used above to compute the field at z from the boundary value at z_0.

1. Multiplication of the field $U_+(x, y, z, \omega)$ on the plane z by

$$C^{-1}(x, y) = \frac{2\pi i}{k}|z - z_0|e^{-ik|z-z_0|}e^{-i\frac{k}{2}\frac{x^2+y^2}{|z-z_0|}}.$$

2. Inverse Fourier transformation of $\tilde{F} = C^{-1}(x, y)U_+(x, y, z, \omega)$. Once the inverse transform of $\tilde{F}$ has been computed the field on the boundary-value plane is given by

$$U_+(x, y, z_0, \omega) = \exp\left(-i\frac{k}{2}\frac{x^2 + y^2}{|z - z_0|}\right)F(x, y, z_0, \omega),$$

where $F(x, y, z_0, \omega)$ is the inverse Fourier transform of $\tilde{F}$.

Note again that, in a computer implementation using the FFT, care has to be taken that the correct sample spacing is employed at z and at z_0. On the observation plane the sample spacing is defined by Eq. (4.44), while on the boundary-value plane it is δx_0 and δy_0.

Finally, we note that the forward and inverse Fresnel transforms form a perfect pair just as the forward and inverse discrete Fourier transforms (DFTs) do. Thus, for example, if an arbitrary boundary-value field U_0 over the plane $z = 0$ is forward propagated some distance z using the forward Fresnel transform and then the resulting field over the z plane (or line) is back propagated to $z = 0$ using the inverse Fresnel transform, the original boundary-value field U_0 will be *exactly reconstructed*. However, if the forward propagation is computed using the (in principle exact) RS formula then application of the inverse Fresnel transform to this forward-propagated field will not exactly yield the original boundary-value field U_0. The reason for this is that the exactly forward-propagated field will not include any evanescent-wave components, so, at best, a low-pass-filtered version of U_0 will result (see Example 4.7). However, the reconstruction will also be in error due to the approximate nature of the Fresnel approximation which forms the basis of the forward and inverse Fresnel transforms.

4.7.2 The 2D Fresnel approximation

In two space dimensions the RS formulas are given by (cf. Eqs. (2.49) of Chapter 2)

$$U_+(x,z) = \pm\frac{ik}{2}\int_{-\infty}^{\infty} dx'\, U_+(x',z_0)H_1^+(k|\mathbf{r}-\mathbf{r}'|)\frac{z-z'}{|\mathbf{r}-\mathbf{r}'|}\bigg|_{z'=z_0}, \tag{4.45}$$

where the plus sign is used for propagation into the r.h.s. $z > z_0$ and the minus sign for propagation into the l.h.s. $z < z_0$.

If the propagation distance $|z - z'|$ is sufficiently large relative to the maximum lateral extent of the boundary-value field we can make the approximation

$$|\mathbf{r}-\mathbf{r}'| = |z-z'|\sqrt{1+\frac{(x-x')^2}{(z-z')^2}} \approx |z-z'| + \frac{1}{2}\frac{(x-x')^2}{|z-z'|}. \tag{4.46}$$

Moreover, for large values of its argument we have that

$$H_1^+(k|\mathbf{r}-\mathbf{r}'|) \sim \sqrt{\frac{2}{k\pi|\mathbf{r}-\mathbf{r}'|}}e^{-i3/(2\pi)}e^{ik|\mathbf{r}-\mathbf{r}'|},$$

which, when coupled with Eq. (4.46), yields the result

$$H_1^+(k|\mathbf{r}-\mathbf{r}'|) \sim \sqrt{\frac{2}{k\pi|z-z'|}}e^{-i3/(2\pi)}e^{ik\left(|z-z'|+\frac{1}{2}\frac{(x-x')^2}{|z-z'|}\right)}. \tag{4.47}$$

Substituting Eq. (4.47) into Eq. (4.45) and using the fact that

$$\frac{z-z'}{|\mathbf{r}-\mathbf{r}'|} \sim \frac{z-z'}{|z-z'|} = \pm 1, \quad |z-z'| \gg 0,$$

then yields the 2D Fresnel approximation

$$U_+(x,z) = \mp\sqrt{\frac{k}{2\pi}}\frac{e^{ik|z-z_0|}e^{i\frac{k}{2}\frac{x^2}{|z-z_0|}}}{\sqrt{|z-z_0|}}\int_{-\infty}^{\infty} dx'\, U_+(x',z_0)e^{i\frac{k}{2}\frac{x'^2}{|z-z_0|}}e^{-ik\frac{x'x}{|z-z_0|}}, \tag{4.48}$$

where the plus sign is used if $z > z_0$ and the minus sign if $z < z_0$ and the approximation requires that $|z - z_0| \gg \lambda$.

The 2D Fresnel transform

The 2D Fresnel approximation defined in Eq. (4.48) can be implemented in three steps.

1. Multiplication of the boundary-value field $U_+(x', z_0)$ by $\exp(i\frac{k}{2}\frac{x'^2}{|z-z_0|})$, yielding

$$F(x') = U_+(x', z_0)e^{i\frac{k}{2}\frac{x'^2}{|z-z_0|}}.$$

2. Spatial Fourier transformation of F. After the Fourier transform has been computed the integral in Eq. (4.48) is given by

$$\int_{-\infty}^{\infty} dx'\, F(x')e^{-i\frac{k}{|z-z_0|}xx'} = \tilde{F}(K_x)|_{K_x=\frac{k}{|z-z_0|}x}, \tag{4.49}$$

where $\tilde{F}(K_x)$ is the spatial Fourier transform of F.

3. Multiplication of the result obtained in Eq. (4.49) by

$$C(x) = \mp\sqrt{\frac{k}{2\pi}}\frac{e^{ik|z-z_0|}e^{i\frac{k}{2}\frac{x^2}{|z-z_0|}}}{\sqrt{|z-z_0|}}.$$

As mentioned earlier in connection with the 3D Fresnel transform, in a computer implementation of the 2D Fresnel transform the spatial Fourier transform would be implemented using the FFT. In this implementation the sampling interval on the boundary-value plane z_0 is *different* from the sampling interval on the observation plane located at $z > z_0$. In particular, the sampling interval on the observation plane increases linearly with propagation distance and is connected with the sampling interval on the boundary-value plane at z_0 via the equation

$$\delta x = \frac{|z-z_0|}{k}\delta K_x = \frac{|z-z_0|}{k}\frac{2\pi}{N\,\delta x_0} = \lambda\frac{|z-z_0|}{N\,\delta x_0}. \tag{4.50}$$

Example 4.8 As an example we implemented the 2D Fresnel transform and 2D Raleigh–Sommerfeld (RS) solution given by Eq. (4.38) in Example 4.6. We show in Fig. 4.10 plots of the real and imaginary parts of the forward-propagated fields from a rectangular-shaped boundary-value field having a total extent of 20λ on the line $z_0 = 0$ and for propagation distances ranging from $z = 10\lambda$ to $z = 40\lambda$ in 10λ steps. The boundary-value field can be interpreted as being that generated by an infinite slit when illuminated by a plane wave normally incident to the plane of the slit. The solid curves were computed using the RS formula given in Eq. (4.38) of Example 4.6, while the dotted curves were computed using the 2D Fresnel transform as outlined above. It is apparent from Fig. 4.10 that the difference between the results from the two computations *decreases* as the propagation distance z increases. Thus, while the angular-spectrum expansion implemented in Example 4.6 using the FFT degrades with increasing propagation distance due to spatial aliasing, the Fresnel transform becomes more accurate as the propagation distance increases. The two computational schemes thus complement each other.

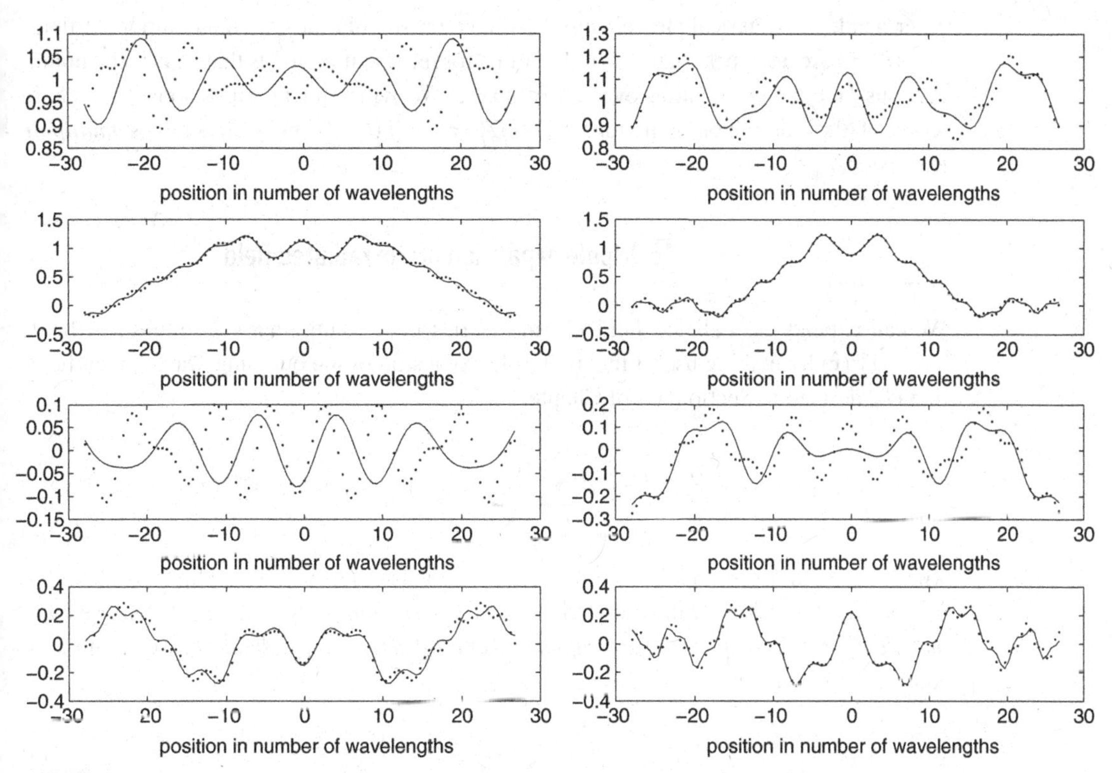

Fig. 4.10 The real (top four) and imaginary (bottom four) forward-propagated fields from a rectangular-shaped boundary-value field of width equal to 20λ at $z_0 = 0$ computed using the 2D RS formula (drawn solid) and the 2D Fresnel transform (dotted). The curves correspond to propagation distances of $z = 10\lambda$ (top left) to $z = 40\lambda$ (bottom right) in steps of 10λ.

4.8 Multipole expansions

We showed in Section 3.3 that the component fields $f_l(kr)Y_l^m(\hat{\mathbf{r}})$ with $l = 0, 1, \ldots,$ $m = -l, -l+1, \ldots, l$ form a complete set of functions for expanding solutions to the Helmholtz equation. The radial functions $f_l(kr)$ are linear combinations of spherical Bessel and Neumann functions, and we have denoted the azimuthal and polar angles of the field point $\mathbf{r}$ in the argument of the spherical harmonics by the unit vector $\hat{\mathbf{r}}$. The particular selection of the radial functions depends on the far-field behavior of the field and, as shown in Examples 3.5 and 3.4 of Section 3.3, the radial functions to be selected are the spherical Hankel functions of the first kind $h_l^+(kr)$ of order l if the field is required to satisfy the outgoing-wave radiation condition (SRC) and are the spherical Bessel functions $j_l(kr)$ of order l if the field is the difference between an outgoing- and an incoming-wave field at infinity. The spherical Hankel functions have singularities at the origin $r = 0$ and satisfy the SRC, so the component fields $h_l^+(kr)Y_l^m(\hat{\mathbf{r}})$ are an appropriate set of functions for representing radiated fields. On the other hand, the spherical Bessel functions have no singularities, so the component fields $j_l(kr)Y_l^m(\hat{\mathbf{r}})$ are an appropriate set of functions for representing fields that satisfy the homogeneous Helmholtz equation over all of space. We will refer to expansions of fields in terms of the component functions $h_l^+(kr)Y_l^m(\hat{\mathbf{r}})$ or $j_l(kr)Y_l^m(\hat{\mathbf{r}})$ in general as *multipole expansions.*

4.8.1 Multipole expansion of the radiated field

We can expand the radiated field U_+ in a series of the outgoing-wave multipole fields $h_l^+(kr)Y_l^m(\hat{\mathbf{r}})$ by making use of the multipole expansion of the outgoing-wave Green function G_+ derived in Section 3.4 of Chapter 3:

$$G_+(\mathbf{r}-\mathbf{r}',\omega) = -ik\sum_{l=0}^{\infty}\sum_{m=-l}^{l} j_l(kr_<)h_l^+(kr_>)Y_l^m(\hat{\mathbf{r}})Y_l^{m*}(\hat{\mathbf{r}}'), \tag{4.51}$$

where $r_< = \min r, r'$ and $r_> = \max r, r'$. If we substitute the above expansion into the primary field solution for the radiated field and *restrict our attention to field points* $\mathbf{r}$ *lying outside the smallest sphere that completely contains the source volume* τ_0, we obtain after some algebraic manipulation

$$U_+(\mathbf{r},\omega) = -ik\sum_{l=0}^{\infty}\sum_{m=-l}^{l} q_l^m(\omega)h_l^+(kr)Y_l^m(\hat{\mathbf{r}}), \tag{4.52a}$$

where it is assumed that $r > a_0$, with a_0 being the radius of the smallest sphere that is centered at the origin and completely encloses the source volume τ_0, and the *multipole moments* $q_l^m(\omega)$ are given by

$$q_l^m(\omega) = \int_{\tau_0} d^3r'\, Q(\mathbf{r}',\omega)j_l(kr')Y_l^{m*}(\hat{\mathbf{r}}'). \tag{4.52b}$$

The multipole moments $q_l^m(\omega)$ were first introduced in connection with *essentially nonradiating sources* in Section 1.7.3 of Chapter 1 and Section 2.7.2 of Chapter 2. In those discussions it was shown that the multipole moments $q_l^m(\omega)$ decay exponentially fast with index $l > \Re ka_0$. It then follows that the multipole expansion of the field radiated by a source with radius a_0 is effectively limited to l values for which $l < \Re ka_0$. This is in analogy with the angular-spectrum expansion of the field, which is effectively limited to weakly inhomogeneous plane-wave components. The multipole expansion Eq. (4.52a) plays an important role in various inverse problems associated with the wave and Helmholtz equations and will be re-visited many times in later chapters.

Example 4.9 The conjugate-wave Green function $G_+^*(\mathbf{R}, k)$ admits the multipole expansion

$$G_+^*(\mathbf{r}-\mathbf{r}',\omega) = ik^* \sum_{l=0}^{\infty} \sum_{m=-l}^{l} j_l(k^* r_<) h_l^-(k^* r_>) Y_l^m(\hat{\mathbf{r}}) Y_l^{m*}(\hat{\mathbf{r}}'), \tag{4.53}$$

where we have used the easily proven relationships

$$j_l^*(kr_<) = j_l(k^* r_<), \qquad h_l^{+*}(kr_>) = h_l^-(k^* r_>), \tag{4.54}$$

where h_l^- is the spherical Hankel function of the second kind. In the special case of non-dispersive media in which $\Im k = 0$, we then find that

$$\begin{aligned}\frac{\sin(k|\mathbf{r}-\mathbf{r}'|)}{|\mathbf{r}-\mathbf{r}'|} &= i2\pi[G_+(\mathbf{r}-\mathbf{r}',\omega) - G_+^*(\mathbf{r}-\mathbf{r}',\omega)] \\ &= 4\pi k \sum_{l=0}^{\infty} \sum_{m=-l}^{l} j_l(kr_>) j_l(kr_<) Y_l^m(\hat{\mathbf{r}}) Y_l^{m*}(\hat{\mathbf{r}}') \\ &= 4\pi k \sum_{l=0}^{\infty} \sum_{m=-l}^{l} j_l(kr) j_l(kr') Y_l^m(\hat{\mathbf{r}}) Y_l^{m*}(\hat{\mathbf{r}}'),\end{aligned} \tag{4.55}$$

where we have used the result that $h_l^+(kr_>) + h_l^-(kr_>) = 2j_l(kr_>)$ and obtained the last line by noting that the product $j_l(kr_>)j_l(kr_<)$ is a symmetric function of $r_>$ and $r_<$ and hence could be replaced by $j_l(kr)j_l(kr')$.

4.8.2 Forward and back propagation using the multipole expansion

The multipole expansion can also be the basis for field forward- and back-propagation algorithms. In particular, if we evaluate the field over some sphere having a radius a larger than the source radius a_0, we find using Eq. (4.52a) that

$$U_+(\mathbf{r},\omega)|_{r=a} = -ik \sum_{l=0}^{\infty} \sum_{m=-l}^{l} q_l^m(\omega) h_l^+(ka) Y_l^m(\hat{\mathbf{r}}), \tag{4.56a}$$

from which we find that

$$q_l^m(\omega) = \frac{i}{k} \frac{u_l^m(\omega)}{h_l^+(ka)}, \tag{4.56b}$$

where

$$u_l^m(\omega) = \int d\Omega\, U_+(\mathbf{r},\omega)|_{r=a} Y_l^{m*}(\hat{\mathbf{r}}) \tag{4.56c}$$

are the projections of the boundary field onto the spherical harmonics. The multipole moments computed using Eq. (4.56b) determine the field everywhere outside the source sphere $r > a_0$ via the multipole expansion Eq. (4.52a). If $r > a$ then the resulting expansion is identical to the solution to the exterior boundary-value problem found in Example 3.5 of Chapter 3 and the expansion implements the process of field forward propagation.

However, if $r < a$ this expansion performs the process of field back propagation whereby the multipole moments are computed from boundary-value data over the sphere $r = a$ and the results used in Eq. (4.52a) to compute the field over points *that lie closer to the source than the data sphere* $r = a$.

The process of computing the field closer to the source than the data sphere is the process of field back propagation developed in Section 4.3 using the angular-spectrum expansion. Indeed, as was the case for back propagation implemented using the angular-spectrum expansion, back propagation implemented using the multipole expansion is also exact but unstable and corresponds to the solution of an ill-posed problem rather than being the solution of a true interior boundary-value problem. The reason for the instability can be seen if we substitute the expression Eq. (4.56b) into Eq. (4.52a) to express the field directly in terms of the $u_l^m(\omega)$. We find that

$$U_+(\mathbf{r}, \omega) = \sum_{l=0}^{\infty} \sum_{m=-l}^{l} \frac{h_l^+(kr)}{h_l^+(ka)} u_l^m(\omega) Y_l^m(\hat{\mathbf{r}}). \tag{4.57}$$

Now, although Eq. (4.57) is, in principle, exact for $r > a_0$, the factor

$$A_l = \frac{h_l^+(kr)}{h_l^+(ka)}$$

will grow exponentially fast with index l if $l > \Re kr$. The reason for this is that the spherical Hankel functions $h_l^+(x)$ grow exponentially fast with index l for $l > x$. It then follows that, since $r < a$ in field back propagation, the factor A_l will grow exponentially fast with l for $l > \Re kr$. Thus, any small error in the boundary-value coefficients $u_l^m(\omega)$ will be amplified exponentially for such index values, leading to large errors in the back-propagated field. In order to stabilize the back propagation it is necessary to limit the l values to $l < \Re kr$. However, as discussed earlier, the radiated field multipole moments and, hence, the boundary-value field expansion coefficients $u_l^m(\omega)$ will be effectively zero for $l > \Re ka_0$, so it is prudent to limit the expansion Eq. (4.57) to this l range since the expansion coefficients in the intermediate range $\Re ka_0 < l < \Re kr$ will be effectively zero anyway.

Example 4.10 The multipole expansion can be used to relate the generalized Fourier coefficients of the Dirichlet and Neumann boundary conditions over any sphere lying outside the source volume as was done for plane boundaries via the angular-spectrum expansion in Example 4.4. In particular, it follows from Eq. (4.57) that

$$v_l^m(\omega) = \int d\Omega \, \frac{\partial}{\partial r} U_+(\mathbf{r}, \omega)|_{r=a} Y_l^{m*}(\hat{\mathbf{r}}) = \frac{k h_l^{+\prime}(ka)}{h_l^+(ka)} u_l^m(\omega),$$

which states that the projections of U_+ and its normal derivative onto the spherical harmonics are related via the equation

$$v_l^m(\omega) = \frac{k h_l^{+\prime}(ka)}{h_l^+(ka)} u_l^m(\omega). \tag{4.58}$$

Equation (4.58), which can also be derived from the second Helmholtz identity, is the spherical-wave equivalent to Eq. (4.23) of Example 4.4 and is also a special case of a Dirichlet-to-Neumann map; it allows the Neumann boundary condition over any sphere lying outside the source region to be computed from the Dirichlet condition and vice versa.

4.8.3 Back propagation in the interior boundary-value problem

Up to this point we have limited our discussion of the process of field back propagation to outgoing-wave fields such as those radiated by compactly supported sources and to cases in which the goal was to compute the field interior to some closed surface $\partial\tau$ that completely encloses the source from boundary-value data over $\partial\tau$. However, it is also possible to back propagate wavefields that satisfy the homogeneous Helmholtz equation and are solutions of an interior boundary-value problem. In this case the "source" of the field is boundary-value data over a closed surface $\partial\tau$ that completely encloses a *source-free* region τ and back propagation consists of computing the field throughout the annular region lying between $\partial\tau$ and some closed surface $\partial\tau_<$ that is interior to $\partial\tau$ from boundary-value data specified over $\partial\tau_<$.

As an example of field back propagation such as described above we consider a source-free region bounded by two concentric spheres centered at the origin and having radii a and a_0, with $a_0 < a$. The interior boundary-value problem for Dirichlet or Neumann data specified over the exterior sphere is given by multipole expansions in Example 3.7 of Chapter 3. For example, for the case of Dirichlet data we found in that example that

$$U_+(\mathbf{r},\omega) = \sum_{l=0}^{\infty}\sum_{m=-l}^{l} \frac{u_l^m(a)}{j_l(ka)} j_l(kr) Y_l^m(\hat{\mathbf{r}}), \tag{4.59a}$$

where

$$u_l^m(a) = \int d\Omega'\, Y_l^{m*}(\hat{\mathbf{r}}') U_+(\mathbf{r}',\omega)|_{r'=a} \tag{4.59b}$$

are the generalized Fourier coefficients of the boundary-value data over the bounding sphere having radius a. The multipole expansion Eq. (4.59a) converges throughout the interior of the bounding sphere and, in particular, is valid everywhere over the interior concentric sphere having radius $a_0 < a$. We thus conclude that we can compute the multipole moments $u_l^m(a)/j_l(ka)$ from Dirichlet data over this sphere according to the formula

$$\frac{u_l^m(a)}{j_l(ka)} = \frac{u_l^m(a_0)}{j_l(ka_0)}, \tag{4.60a}$$

where

$$u_l^m(a_0) = \int d\Omega'\, Y_l^{m*}(\hat{\mathbf{r}}') U_+(\mathbf{r}',\omega)|_{r'=a_0}. \tag{4.60b}$$

The multipole expansion Eq. (4.59a) can then be written in terms of the Fourier coefficients $u_l^m(a_0)$ in the form

$$\bar{U}_+(\mathbf{r}, \omega) = \sum_{l=0}^{\infty} \sum_{m=-l}^{l} \frac{u_l^m(a_0)}{j_l(ka_0)} j_l(kr) Y_l^m(\hat{\mathbf{r}}). \tag{4.61}$$

Equation (4.61) converges everywhere within the exterior bounding sphere having radius $a > a_0$ and thus performs the process of field back propagation from the surface of the interior sphere to all points within the annular region bounded by the two concentric spheres. However, just as the field back propagation was unstable for outgoing-wave fields, the same is true for back propagation implemented according to Eq. (4.61). In particular, the spherical Bessel functions $j_l(x)$ decay exponentially fast with $l > x$, so the ratio

$$\eta(r, a_0) = \frac{j_l(kr)}{j_l(ka_0)}$$

will be well behaved for all l values if $r < a_0$ but will grow exponentially fast for $r > a_0$ when $l > ka_0$. Thus, in practice only a stabilized version of Eq. (4.61) is useful where the expansion is limited to $l < ka_0$. This, of course, is completely analogous to truncating the angular-spectrum expansion to homogeneous plane waves. In fact, as we showed in Section 3.4.2 of Chapter 3, the evanescent component of outgoing-wave fields corresponds to multipole components of such fields for which $l > ka_0$, where a_0 is the source radius.

Although back propagation for solutions to the homogeneous Helmholtz equation as developed above is unstable, it is, nevertheless, mathematically exact and can be used to establish fundamental results and theorems related to this equation. In particular, it can be used to continue a solution to the homogeneous Helmholtz equation specified over a given region $\tau_<$ into all points lying in a larger region $\tau_> \supset \tau_<$ in much the same way as an analytic function is extended into larger regions of the complex plane via the process of analytic continuation. For example, we showed in Section 4.2.2 that the radiation pattern uniquely determines the radiated field everywhere outside the convex hull of the source support τ_0. If we then construct a sphere tangent to the convex hull and having radius a_0 we can represent the field via Eq. (4.61), where all the expansion coefficients are computed from the known radiated field over the surface of this sphere. This expansion will then converge within a possibly larger sphere of radius $a > a_0$ and in this way we can then extend the region where the radiation pattern has uniquely determined the radiated field. By continuing in this manner we can then determine the field everywhere outside the support volume of the source.

4.8.4 Back propagation from the radiation pattern

By making use of the well-known result

$$h_l^+(kr) \sim (-i)^{l+1} \frac{e^{ikr}}{kr} \quad \text{as} \quad kr \to \infty$$

we find using the multipole expansion Eq. (4.52a) that

$$U_+(\mathbf{r},\omega) \sim \left\{ -i \sum_{l=0}^{\infty} \sum_{m=-l}^{l} (-i)^{l+1} q_l^m(\omega) Y_l^m(\mathbf{s}) \right\} \frac{e^{ikr}}{r},$$

where $\mathbf{s} = \mathbf{r}/r$ is the unit vector along the direction of $\mathbf{r}$. We conclude that the radiation pattern $f(\mathbf{s},\omega)$ is given by

$$f(\mathbf{s},\omega) = \sum_{l=0}^{\infty} \sum_{m=-l}^{l} f_l^m(\omega) Y_l^m(\mathbf{s}), \tag{4.62a}$$

where the expansion coefficients

$$f_l^m(\omega) = \int d\Omega\, f(\mathbf{s},\omega) Y_l^{m*}(\mathbf{s})$$

are related to the multipole moments via the equation

$$f_l^m(\omega) = -(-i)^l q_l^m(\omega). \tag{4.62b}$$

The expansion of the radiation pattern in terms of spherical harmonics in Eq. (4.62a) will be used in the following chapter in one formulation of the inverse source problem. We note that this expansion also establishes the result that the radiation pattern uniquely determines the field everywhere outside the source region and in particular allows us to back propagate the field from the radiation pattern using the multipole expansion. In this case the result is slightly less general than the corresponding result established using the angular-spectrum expansion in the last section in that the multipole expansion is valid only outside the smallest sphere that completely contains the source. It then follows that the multipole moments as determined from the radiation pattern via Eq. (4.62b) will determine the field only outside this sphere and not everywhere outside the smallest convex region that encloses τ_0. However, the field can be further continued inside this sphere by using the solution of the interior boundary-value problem developed in the previous section.

Finally, we note that back propagation from the radiation pattern, like that from data prescribed on a sphere surrounding the source considered in the last section, will be ill-posed (unstable) due to the requirement that the source multipole moments $q_l^m(\omega)$ are required to decay exponentially fast with index $l > ka_0$. Thus, unless the generalized Fourier coefficients of the radiation pattern satisfy this requirement, the back-propagated field will grow exponentially fast as the field point recedes from the sphere at infinity. This also indicates that an ideal radiation pattern (one generated from a compactly supported source) must have Fourier coefficients $f_l^m(\omega)$ that effectively vanish when $l > ka_0$, thus limiting the resolution (central lobe size) of any such radiation pattern. We will return to this issue in the following chapter, where we treat the antenna synthesis problem formulated as an inverse source problem.

4.9 Multipole expansions of two-dimensional wavefields

We introduced the eigenfunctions appropriate to 2D wavefields in polar coordinates in Section 3.6 of the last chapter and turn now to using these eigenfunctions to develop the multipole expansion of the field radiated by a 2D source. We will employ a Cartesian coordinate system with horizontal axis x, vertical axis y and polar coordinates r and ϕ so that $x = r\cos\phi$ and $y = r\sin\phi$. The radiated field in two space dimensions is given by Eq. (2.28) of Chapter 2,

$$U_+(\mathbf{r},\omega) = \int_{\tau_0} d^2r'\, G_+(\mathbf{r}-\mathbf{r}',\omega)Q(\mathbf{r}',\omega),$$

where

$$G_+(\mathbf{r}-\mathbf{r}',\omega) = -\frac{i}{4}H_0^+(k|\mathbf{r}-\mathbf{r}'|)$$

is the outgoing-wave Green function of the 2D Helmholtz equation. This Green function was expanded in a 2D multipole expansion in Section 3.6 of Chapter 3:

$$G_+(\mathbf{r}-\mathbf{r}',\omega) = -\frac{i}{4}\sum_{n=-\infty}^{\infty} J_n(kr_<)H_n^+(kr_>)e^{in(\phi-\phi')},$$

where $r_< = \min r, r'$ and $r_> = \max r, r'$. Here ϕ and ϕ' are the polar angles of $\mathbf{r}$ and $\mathbf{r}'$, respectively. On making use of this expansion and assuming that the field point $\mathbf{r}$ lies outside the smallest circle that completely encloses the source region τ_0 we obtain after some minor algebra the result

$$U_+(\mathbf{r},\omega) = -\frac{i}{4}\sum_{n=-\infty}^{\infty} q_n(\omega)H_n^+(kr)e^{in\phi}, \tag{4.63a}$$

where

$$q_n(\omega) = \int_{\tau_0} d^2r'\, Q(\mathbf{r}',\omega)J_n(kr')e^{-in\phi'} \tag{4.63b}$$

are the 2D *multipole moments*.

By making use of the asymptotic expression

$$H_n^+(kr) \sim \sqrt{\frac{2}{\pi k}}e^{-i(n+\frac{1}{2})\frac{\pi}{2}}\frac{e^{ikr}}{\sqrt{r}}, \quad kr \to \infty,$$

we find that the 2D multipole expansion Eq. (4.63a) becomes

$$U_+(\mathbf{r},\omega) \sim \overbrace{\left\{-\frac{i}{4}\sqrt{\frac{2}{\pi k}}e^{-i\frac{\pi}{4}}\sum_{n=-\infty}^{\infty}(-i)^n q_n(\omega)e^{in\phi}\right\}}^{f(\mathbf{s},\omega)}\frac{e^{ikr}}{\sqrt{r}}.$$

It then follows that the 2D radiation pattern admits the orthonormal expansion

$$f(\mathbf{s},\omega) = \frac{1}{2\pi}\sum_{n=0}^{\infty} f_n(\omega)e^{in\phi}, \tag{4.64a}$$

where the expansion coefficients $f_n(\omega)$ are related to the source multipole moments $q_n(\omega)$ via the equation

$$f_n(\omega) = -\frac{i}{4}\sqrt{\frac{8\pi}{k}}e^{-i\frac{\pi}{4}}(-i)^n q_n(\omega). \tag{4.64b}$$

As was the case for the 2D angular-spectrum expansions treated in Section 4.6, all of the results for 3D multipole fields have their 2D versions. In particular, the multipole expansion Eq. (4.63a) can be expressed in terms of Dirichlet or Neumann boundary conditions over any circle completely enclosing the source region τ_0. For example, over any such circle of radius $a > a_0$, where a_0 is the radius of the smallest circle enclosing the source, we find that

$$U_+(\mathbf{r},\omega)|_{r=a} = -\frac{i}{4}\sum_{n=-\infty}^{\infty} q_n(\omega)H_n^+(ka)e^{in\phi},$$

from which it follows that

$$u_n(\omega) = \frac{1}{2\pi}\int_0^{2\pi} d\phi\, U_+(\mathbf{r},\omega)|_{r=a}e^{-in\phi} = -\frac{i}{4}q_n(\omega)H_n^+(ka),$$

so that

$$U_+(\mathbf{r},\omega) = \sum_{n=-\infty}^{\infty}\frac{H_n^+(kr)}{H_n^+(ka)}u_n(\omega)e^{in\phi}. \tag{4.65}$$

4.10 Connection between the angular-spectrum and multipole expansions

The outgoing-wave multipole field $h_l^+(kr)Y_l^m(\hat{\mathbf{r}})$ was expanded into an angular-spectrum expansion in Eq. (3.49) of Section 3.4.2 of Chapter 3:

$$h_l^+(kr)Y_l^m(\hat{\mathbf{r}}) = \frac{(-i)^l}{2\pi}\int_{-\pi}^{\pi} d\beta \int_{C_\pm} \sin\alpha\, d\alpha\, Y_l^m(\mathbf{s})e^{ik\mathbf{s}\cdot\mathbf{r}}, \tag{4.66a}$$

where, as usual, the contour C_+ is used if $z > 0$ and the contour C_- if $z < 0$. Conversely, the plane waves can be expanded into the free multipole expansions (see Example 3.4 of Chapter 3)

$$e^{ik\mathbf{s}\cdot\mathbf{r}} = 4\pi\sum_{l=0}^{\infty}\sum_{m=-l}^{l} i^l j_l(kr)Y_l^m(\hat{\mathbf{r}})Y_l^{m*}(\mathbf{s}). \tag{4.66b}$$

Equations (4.66) provide a bridge between the angular-spectrum and multipole expansions that allows one to pass from one to the other with relative ease.

As an example of the connection between the two types of expansions we now derive the multipole expansion Eq. (4.52a) directly from the angle-variable form of the angular-spectrum expansion given in Eq. (4.18a). In order to do this we first note that, as shown in Section 4.2.2, the angular spectrum $A(k\mathbf{s},\omega)$ over the weakly inhomogeneous part of the spectra is equal to the radiation pattern $f(\mathbf{s},\omega)$ of the field and is the analytic continuation of this quantity over the evanescent part of the spectrum. If we then use the spherical-harmonic expansion of the radiation pattern derived above, we conclude that

$$A(k\mathbf{s},\omega) = -\sum_{l=0}^{\infty}\sum_{m=-l}^{l}(-i)^l q_l^m(\omega)Y_l^m(\mathbf{s}),$$

where we have made use of Eq. (4.62b). If we now substitute the above expansion into the angle-variable form of the angular-spectrum expansion we obtain the result

$$\begin{aligned}U_+(\mathbf{r},\omega) &= \frac{ik}{2\pi}\int_{-\pi}^{\pi}d\beta\int_{C_\pm}\sin\alpha\,d\alpha\,\overbrace{\left\{-\sum_{l=0}^{\infty}\sum_{m=-l}^{l}(-i)^l q_l^m(\omega)Y_l^m(\mathbf{s})\right\}}^{A(k\mathbf{s},\omega)}e^{ik\mathbf{s}\cdot\mathbf{r}}\\ &= -ik\sum_{l=0}^{\infty}\sum_{m=-l}^{l}q_l^m(\omega)\left\{\frac{(-i)^l}{2\pi}\int_{-\pi}^{\pi}d\beta\int_{C_\pm}\sin\alpha\,d\alpha\,Y_l^m(\mathbf{s})e^{ik\mathbf{s}\cdot\mathbf{r}}\right\}\\ &= -ik\sum_{l=0}^{\infty}\sum_{m=-l}^{l}q_l^m(\omega)h_l^+(kr)Y_l^m(\hat{\mathbf{r}}),\end{aligned}$$

where we have made use of Eq. (4.66a).

We have thus been able to derive the multipole expansion directly from the angular-spectrum expansion. We note in this connection that in obtaining the above result we have interchanged an infinite sum with an indefinite integral. A more careful analysis of this interchange indicates that it is allowable only if the z coordinate of the field point $\mathbf{r}$ lies outside the source volume τ_0 that gave rise to the field U_+. Thus, the multipole expansion will converge only outside the smallest sphere surrounding the source, as we have already determined in Section 4.8.

Example 4.11 The multipole expansion of the outgoing-wave multipole fields derived in Section 3.4.2 of Chapter 3 was based on the relationship Eq. (4.20a) between the angular spectra of an outgoing-wave field and its radiation pattern. In particular, for the outgoing-wave multipole fields we have that

$$h_l^+(kr)Y_l^m(\hat{\mathbf{r}}) \sim \overbrace{\frac{(-i)^{(l+1)}}{k}Y_l^m(\hat{\mathbf{r}})}^{f(\hat{\mathbf{r}},\omega)}\frac{e^{ikr}}{r},$$

which, on making use of Eq. (4.20a), yields the result

$$A(\mathbf{s}) = \frac{(-i)^{(l+1)}}{k}Y_l^m(\mathbf{s}),$$

from which we obtained the expansion given in Eq. (4.66a).

This expansion can also be derived from the multipole and angular-spectrum expansions of the outgoing-wave Green function. In particular, if we assume an angular-spectrum expansion for the multipole fields in the standard form

$$h_l^+(kr)Y_l^m(\hat{\mathbf{r}}) = \frac{ik}{2\pi}\int_{-\pi}^{\pi} d\beta \int_{C_\pm} \sin\alpha \, d\alpha \, A_l^m(\mathbf{s})e^{ik\mathbf{s}\cdot\mathbf{r}}$$

and substitute this into the r.h.s. of the multipole expansion Eq. (4.51), we obtain the result

$$G_+(\mathbf{r}-\mathbf{r}',\omega) = -ik\sum_{l=0}^{\infty}\sum_{m=-l}^{l} \overbrace{\frac{ik}{2\pi}\int_{-\pi}^{\pi} d\beta \int_{C_\pm} \sin\alpha \, d\alpha \, A_l^m(\mathbf{s})e^{ik\mathbf{s}\cdot\mathbf{r}}}^{h_l^+(kr)Y_l^m(\hat{\mathbf{r}})} j_l(kr')Y_l^{m*}(\hat{\mathbf{r}}')$$

$$= \int_{-\pi}^{\pi} d\beta \int_{C_\pm} \sin\alpha \, d\alpha \left\{ \frac{k^2}{2\pi}\sum_{l=0}^{\infty}\sum_{m=-l}^{l} A_l^m(\mathbf{s}) j_l(kr')Y_l^{m*}(\hat{\mathbf{r}}') \right\} e^{ik\mathbf{s}\cdot\mathbf{r}},$$

where we have assumed that $r > r'$. If we now make use of the Weyl expansion Eq. (4.10), we conclude that

$$\frac{k^2}{2\pi}\sum_{l=0}^{\infty}\sum_{m=-l}^{l} A_l^m(\mathbf{s}) j_l(kr')Y_l^{m*}(\hat{\mathbf{r}}') = -\frac{ik}{8\pi^2}e^{-ik\mathbf{s}\cdot\mathbf{r}'}.$$

As a final step we make use of the multipole expansion of the plane wave $\exp(-ik\mathbf{s}\cdot\mathbf{r}')$ presented in Example 3.4 of Chapter 3:

$$e^{-ik\mathbf{s}\cdot\mathbf{r}'} = 4\pi\sum_{l=0}^{\infty}\sum_{m=-l}^{l} (-i)^l j_l(kr')Y_l^{m*}(\hat{\mathbf{r}}')Y_l^m(\mathbf{s}),$$

from which it follows that

$$\sum_{l=0}^{\infty}\sum_{m=-l}^{l} A_l^m(\mathbf{s}) j_l(kr')Y_l^{m*}(\hat{\mathbf{r}}') = -\frac{i}{k}\sum_{l=0}^{\infty}\sum_{m=-l}^{l} (-i)^l j_l(kr')Y_l^{m*}(\hat{\mathbf{r}}')Y_l^m(\mathbf{s}),$$

which yields

$$A(\mathbf{s}) = \frac{(-i)^{(l+1)}}{k}Y_l^m(\mathbf{s}).$$

4.11 Radiated energy out of plane and spherical boundaries

The energy and energy spectrum (energy per unit frequency) radiated by a source embedded in a non-dispersive medium were treated in Section 1.6 of Chapter 1 and were generalized to dispersive media in Section 2.6 of Chapter 2. In those sections we showed that the total energy radiated out of a surface $\partial\tau$ surrounding the source is given by

$$e_{\partial\tau} = \frac{1}{2\pi}\int_{-\infty}^{\infty} d\omega \, E_{\partial\tau}(\omega),$$

where

$$E_{\partial\tau}(\omega) = 2\kappa\omega\Im\int_{\partial\tau} dS \, U_+^*(\mathbf{r},\omega)\frac{\partial U_+(\mathbf{r},\omega)}{\partial n} \tag{4.67}$$

is the energy spectrum of the radiated field over the surface $\partial\tau$, where κ is an unessential real and positive constant.

In the case of non-dispersive media we showed in Section 1.6 that the energy spectrum is independent of the surface $\partial\tau$ and could be expressed in terms of the field radiation pattern via the equation

$$E_{\partial\tau}(\omega) = 2\kappa\omega k \int_{4\pi} d\Omega_s |f(\mathbf{s},\omega)|^2. \tag{4.68a}$$

In the more general case of dispersive media treated in Section 2.6 the energy spectrum depends on the particular surface $\partial\tau$ over which it is computed and an expression in terms of the radiation pattern is possible only when $\partial\tau = \Sigma_R$ is the surface of an asymptotically large sphere centered on the origin and having radius $R \gg 0$. In this case we showed in Section 2.6 that

$$E_{\Sigma_R}(\omega) = 2\kappa\omega\Re k e^{-2\Im kR} \int_{4\pi} d\Omega_s |f(\mathbf{s},\omega)|^2, \tag{4.68b}$$

which is a generalization of Eq. (4.68a) to the case of dispersive media.

Equation (4.68b) allows the energy spectrum to be computed over asymptotically large spheres. However, it is useful to have expressions for this quantity over general surfaces that lie at arbitrary distances from the source volume τ_0. Of particular interest are expressions for this quantity over spheres of arbitrary radius $a > a_0$ and over infinite plane surfaces bounding the source volume τ_0. In this section we employ the angular-spectrum and multipole expansions of the radiated field to obtain such expressions in terms of the angular-spectrum and multipole moments of the field.

4.11.1 Radiated energy into an infinite half-space

Let's assume that a source is located in the half-space $z < z_0$. Then the radiated field can be expressed throughout the half-space $z \geq z_0$ via the angular-spectrum expansion

$$U_+(\mathbf{r},\omega) = \frac{1}{(2\pi)^2} \int_{-\infty}^{\infty} d^2K_\rho A^+(\mathbf{K}_\rho,\omega) e^{i\mathbf{k}^+\cdot\mathbf{r}},$$

where $\mathbf{k}^+ = \mathbf{K}_\rho + \gamma\hat{\mathbf{z}}$ and the angular spectrum A^+ is determined from the boundary value of the field on the plane z_0 via its inverse spatial Fourier transform

$$A^+(\mathbf{K}_\rho,\omega) = e^{-i\gamma z_0} \int d^2\rho\, U_+(\mathbf{r},\omega)|_{z=z_0} e^{-i\mathbf{K}_\rho\cdot\boldsymbol{\rho}} = e^{-i\gamma z_0} \widetilde{U}_+(\mathbf{K}_\rho, z_0, \omega), \tag{4.69}$$

where $\widetilde{U}_+(\mathbf{K}_\rho, z_0, \omega)$ denotes the spatial Fourier transform of the field over the plane $z = z_0$. If we now substitute the angular-spectrum expansion into the expression for the energy spectrum over any plane $z \geq z_0$, we find on using Eq. (4.67) that

$$E_z(\omega) = 2\kappa\omega\Im \int_z d^2\rho \left\{ \frac{1}{(2\pi)^4} \int_{-\infty}^{\infty} d^2K_\rho \int_{-\infty}^{\infty} d^2K'_\rho \, i\gamma(\mathbf{K}'_\rho) A^{+*}(\mathbf{K}_\rho, \omega) \right.$$
$$\left. \times A^+(\mathbf{K}'_\rho, \omega) e^{i(\mathbf{k}^{+\prime} - \mathbf{k}^{+*})\cdot\mathbf{r}} \right\}$$
$$\Downarrow$$
$$E_z(\omega) = \frac{\kappa\omega}{2\pi^2} \int_{-\infty}^{\infty} d^2K_\rho (\Re\gamma) |A^+(\mathbf{K}_\rho, \omega)|^2 e^{-2\Im\gamma z}, \tag{4.70a}$$

where $E_z(\omega)$ is the energy spectra computed over the plane $z \geq z_0$. In deriving the above we have used the result that

$$\frac{1}{(2\pi)^2} \int_z d^2\rho \, e^{i(\mathbf{k}^{+\prime} - \mathbf{k}^{+*})\cdot\mathbf{r}} = e^{-2\Im\gamma z} \delta(\mathbf{K}'_\rho - \mathbf{K}_\rho).$$

We can also express the result Eq. (4.70a) directly in terms of the field spatial Fourier transform over the plane $z > z_0$. In particular, if we substitute the expression for the angular spectrum A^+ in Eq. (4.69) into Eq. (4.70a), we obtain the result

$$E_z(\omega) = \frac{\kappa\omega}{2\pi^2} \int_{-\infty}^{\infty} d^2K_\rho (\Re\gamma) |\tilde{U}_+(\mathbf{K}_\rho, z_0, \omega)|^2 e^{-2\Im\gamma(z - z_0)}. \tag{4.70b}$$

When the medium is non-dispersive and the wavenumber is real-valued γ is pure real over the homogeneous part of the spectrum and pure positive imaginary over the evanescent part of the spectrum. In this case the contribution to the energy spectrum from the evanescent plane waves in Eqs. (4.70) vanishes and we obtain

$$E_z(\omega) = \frac{\kappa\omega}{2\pi^2} \int_{-k}^{k} d^2K_\rho \, \gamma |A^+(\mathbf{K}_\rho, \omega)|^2 = \frac{\kappa\omega}{2\pi^2} \int_{-k}^{k} d^2K_\rho \, \gamma |\tilde{U}_+(\mathbf{K}_\rho, z_0, \omega)|^2, \tag{4.71}$$

which is independent of the location z of the plane over which the energy spectrum is computed. This result is in accord with Theorem 1.1, which requires that the energy spectrum in a non-dispersive medium be invariant with respect to the location of the surface $\partial\tau$ over which it is computed. On the other hand, when the medium is dispersive then γ is generally complex both in the weakly inhomogeneous part and in the evanescent part of the spectrum, and the energy spectrum will be z-dependent and will decay with propagation distance z as is required by the general results established in Eqs. (4.70).

The radiated energy spectrum of an NR source

We recall that an NR source has to generate zero radiated energy so that $E_z(\omega)$ must vanish if z lies outside the source support. However, it is possible for a source to generate zero radiated energy out of its support region τ_0 and yet not be an NR source. For example, consider a source in a non-dispersive medium that is confined to a region bounded by two infinite parallel planes. It follows from Eq. (4.71) that the energy density radiated out of this region will be zero so long as the angular spectrum $A^+(\mathbf{K}_\rho, \omega)$ vanishes over the weakly inhomogeneous region of the spectrum (for $|\mathbf{K}_\rho| \leq k$). However, as pointed out earlier in Section 4.2.3, this does not require the angular spectrum to vanish over the evanescent region $|\mathbf{K}_\rho| > k$. Only in the case of compactly supported sources whose angular spectra are entire functions of $\mathbf{K}_\rho$ will the vanishing of the angular spectrum over the homogeneous part of the spectrum guarantee that it also vanishes over the evanescent region. Indeed, it is easily verified that the source defined by

$$Q(\mathbf{r},\omega) = \begin{cases} \sin(\kappa x) & |z| < a, \\ 0 & \text{otherwise}, \end{cases}$$

where $\kappa > k$, generates zero angular spectrum over the homogeneous part of the spectrum but radiates a pure evanescent plane wave outside the strip $|z| < a$.

4.11.2 Radiated energy from a spherical region

We now consider a source located within a sphere that is centered at the origin and has a radius a_0. In this case we represent the radiated field in the multipole expansion Eq. (4.52a) and we find upon using Eq. (4.67) that

$$E_a(\omega) = 2\kappa\omega a^2 \Im \int_{4\pi} d\Omega \left[U_+^*(\mathbf{r},\omega) \frac{\partial U_+(\mathbf{r},\omega)}{\partial r} \right]\Bigg|_{r=a}$$

$$= 2\kappa\omega |k|^{2a^2} \Im \int_{4\pi} d\Omega \left\{ \sum_{l=0}^{\infty} \sum_{m=-l}^{l} \sum_{l'=0}^{\infty} \sum_{m'=-l'}^{l'} q_{l'}^{m'}(\omega) q_l^{m*}(\omega) k h_{l'}^{+'}(ka) \times Y_{l'}^{m'}(\hat{\mathbf{r}}) h_l^{+*}(ka) Y_l^{m*}(\hat{\mathbf{r}}) \right\},$$

where $E_a(\omega)$ denotes the energy spectrum of the radiated field computed over any sphere $r = a \geq a_0$ and $h_{l'}^{+'}$ denotes the derivative of the spherical Hankel function. On making use of the orthonormality of the spherical harmonics the above expression simplifies to become

$$E_a(\omega) = 2\kappa\omega |k|^2 a^2 \Im \left\{ k \sum_{l=0}^{\infty} \sum_{m=-l}^{l} |q_l^m(\omega)|^2 h_l^{+'}(ka) h_l^{+*}(ka) \right\}. \tag{4.72a}$$

In the case of non-dispersive media where k is real-valued we have that

$$\Im[h_l^{+'}(ka) h_l^{+*}(ka)] = \frac{1}{2i}[h_l^{+'}(ka) h_l^{+*}(ka) - h_l^{+'*}(ka) h_l^{+}(ka)]$$

$$= \frac{1}{2i}[h_l^{+'}(ka) h_l^{-}(ka) - h_l^{-'}(ka) h_l^{+}(ka)] = \frac{1}{k^2 a^2},$$

where we have used the fact that $h_l^+(x)^* = h_l^-(x)$ when x is real-valued and also the Wronskian relationship satisfied by the two spherical Hankel functions. On substituting the above into Eq. (4.72a) we find that

$$E_a(\omega) = 2\kappa\omega k \sum_{l=0}^{\infty} \sum_{m=-l}^{l} |q_l^m(\omega)|^2, \tag{4.72b}$$

which is independent of the radius $a > a_0$ of the sphere over which the energy spectrum is computed. Again this is in agreement with Theorem 1.1, as was our result presented above for plane boundaries. If the medium is dispersive then k is complex and the general expression Eq. (4.72a) cannot be reduced to simple form. In this case the energy spectra will be a monotonically decreasing function of the radius a of the sphere over which it is computed, in accordance with the results presented in Section 2.6 of Chapter 2.

Further reading

Excellent treatments of the angular-spectrum expansion can be found in Born and Wolf (1999), Stratton (1941), Goodman (1968) and Stamnes (1986), as well as in Mandel and Wolf (1995) and Oughstun (2006). The multipole expansion for both scalar and vector (electromagnetic) fields is covered in Born and Wolf (1999), Stratton (1941), Jackson (1998), Papas (1988) and Chew (1990) and was developed within the time domain in Heyman and Devaney (1996) and Marengo and Devaney (1998). Field back propagation implemented with the angular-spectrum expansion is treated in Sherman (1967) and Shewell and Wolf (1968), while Devaney and Sherman (1973) contains an in-depth treatment of plane-wave expansions of solutions to the radiation problem. Interesting treatments of the Fresnel transform are contained in Gori (1981) and James and Agarwal (1996).

Problems

4.1 Prove that the Weyl expansion Eq. (4.4a) satisfies the homogeneous Helmholtz equation if $|Z| > 0$.

4.2 Use the angular-spectrum expansion to prove that the solution to the RS boundary-value problem is unique; i.e., that Dirichlet or Neumann conditions over any infinite plane bounding a half-space over which a field satisfies the Sommerfeld radiation condition uniquely determine the field throughout this half-space.

4.3 Derive the plane-wave expansion of the radiated field given in Eq. (4.15a) by performing the K_z integration using contour integration and the calculus of residues in the Fourier-integral representation of the radiated field

$$U_+(\mathbf{r},\omega) = \frac{1}{(2\pi)^3}\int d^3K\,\frac{\tilde{Q}(\mathbf{K},\omega)}{k^2 - K^2}e^{i\mathbf{K}\cdot\mathbf{r}}.$$

4.4 Derive a plane-wave expansion in the time-domain involving only homogeneous plane waves that is valid for a field radiated by a causal source in a non-dispersive medium for all times after the source has ceased to radiate.

4.5 Comment on the relationship between the evanescent plane waves in the *time-domain* angular-spectrum expansion of the field radiated by the source in the previous problem and the homogeneous waves in the plane-wave expansion of this field for times exceeding the turn-off time of the source.

4.6 Derive the radiation pattern of a uniform circular disk surface source confined to the plane $z = 0$ employed in Section 4.5 for comparison with the wavelet radiation pattern.

4.7 Express the back-propagated field in Eq. (4.25) in terms of Neumann conditions.

4.8 Derive the inequality given in Eq. (4.16).

4.9 Use the Weyl expansion in the expression for the field radiated by a source located in the left half-space within the strip $z^- < z^+ < 0$ in the presence of a Dirichlet plane (a plane over which the field vanishes) located at $z = 0$ to develop an angular-spectrum expansion of the radiated field in the half-space $z < z^-$; i.e., to the left of the source (cf. Problems 2.14 and 2.15 from Chapter 2).

4.10 Represent the radiated field in an angular-spectrum expansion for the previous problem in the interior strip $z^+ < z < 0$ lying between the source and the Dirichlet plane.

4.11 Use the angular-spectrum expansion of the outgoing-wave multipole fields given in Eq. (4.66a) to derive the angular-spectrum expansion of the radiated field from the multipole expansion of this field given in Eq. (4.52a).

4.12 Derive the angle-variable form of the expression for the wavelet field in Eq. (4.31) of Section 4.5 using the angle-variable form of the angular-spectrum expansion. Check your result by transforming your expression into Cartesian-variable form.

4.13 Use the scheme given in Example 4.11 to derive the 2D angular-spectrum expansion of the 2D outgoing-wave multipole fields.

4.14 Use the 2D angular-spectrum expansion of the 2D outgoing-wave multipole fields found in the previous problem to derive the 2D multipole expansion of a radiated field from its angular-spectrum expansion.

4.15 Use the 2D angular spectrum found in Problem 4.13 to derive the angular-spectrum expansion of a 2D outgoing-wave field from its 2D multipole expansion.

4.16 Derive the expression for the radiation-pattern expansion coefficients $f_l^m(\omega)$ given in Problem 3.11 of the last chapter directly from the multipole expansion of the radiated field.

5 The inverse source problem

The formulas Eq. (1.33) of Chapter 1 represent the solution to the *radiation problem* in a non-dispersive medium governed by the wave equation; i.e., they give the radiated field $u_+(\mathbf{r}, t)$ in terms of a known source $q(\mathbf{r}, t)$. These formulas were generalized to dispersive media in Chapter 2, where the radiation problem was solved directly in the frequency domain for a known source embedded in a uniform dispersive background medium. The *inverse source problem* (ISP), as its name indicates, is the inverse to the radiation problem, and in this problem one seeks the source $q(\mathbf{r}, t)$ from knowledge of its radiated field $u_+(\mathbf{r}, t)$. The question of what applications require a solution to an inverse source problem naturally arises. There are basically two such applications that consist of (i) imaging (reconstructing) the interior of a volume source from observations of the field radiated by the source and (ii) designing a volume source to act as a multi-dimensional antenna to radiate a prescribed field. In the first application actual field measurements are employed, thereby generating data that are then used to "solve" the ISP and thus "reconstruct" the interior of the source, whereas in the second application desired field data are used to "design" a source that will generate those data. Regarding the ISP, the two applications are essentially identical, differing only in emphasis; in application (i) we have to contend with measurement error and noisy data, whereas in application (ii) we have to contend with inconsistencies between the desired data and the constraints required of the source (antenna). However, both of these issues, as well as several others, ultimately reduce to considerations of the impact of non-radiating and essentially non-radiating sources on the ISP and can be treated in a systematic manner using tools already developed in previous chapters.

5.1 The ISP for the wave equation

We will begin our treatment of the ISP for the case of non-dispersive media governed by the wave equation

$$\left[\nabla^2 - \frac{1}{c^2}\frac{\partial^2}{\partial t^2}\right] u(\mathbf{r}, t) = q(\mathbf{r}, t),$$

where the source q is assumed to be supported in the space-time region $\{\mathcal{S}_0 | \mathbf{r} \in \tau_0, t \in [0, T_0]\}$. The ISP then consists of determining the source from measurements of the radiated field over some space-time domain. Of course, if we know the field u_+ over the source space-time support region $\mathcal{S}_0$ then the solution to the inverse problem is trivial. In particular, we simply apply the D'Alembertian operator

$$\mathcal{W} = \nabla^2 - \frac{1}{c^2}\frac{\partial^2}{\partial t^2}$$

to the field to recover the source according to the wave equation. The catch is that in *practical* versions of the ISP we can measure the radiated field only in restricted regions of space-time that lie *outside* the source's space-time support $\mathcal{S}_0$. For example, in one common form of the ISP the source is surrounded by an external closed surface $\partial\tau$ and the field u_+ or its normal derivative $\partial u_+/\partial n$ is measured at each point on the surface for all time $t > 0$. The ISP then consists of deducing the source from the measured field data. In a second version of this problem Cauchy data (field and first time derivative) are measured over all of space at some time t_0 greater then the turn-off time T_0 of the source and the source is to be determined from these data. An ideal data set would be measurements of the radiated field specified everywhere outside the source space-time region $\mathcal{S}_0$.

Fortunately, although it is unrealistic to have complete knowledge of the field everywhere outside $\mathcal{S}_0$, it is also not necessary. In particular, according to Theorem 1.4 in Chapter 1 *the radiated field is uniquely determined over all space-time points lying outside its space-time support $\mathcal{S}_0$ by its boundary values (field and normal derivative) over any closed surface $\partial\tau$ that completely surrounds τ_0 or by Cauchy data specified at any time t_0 exceeding the turn-off time T_0 of the source.* We emphasize that this theorem is a *uniqueness theorem* and neither indicates *how* to compute the field from the boundary value or Cauchy data nor considers the effect of noise or measurement error on the field determination. It simply states that under ideal conditions (perfect field data) either of these data sets contains, in principle, all the information required to compute the field everywhere outside the (known) space-time support $\mathcal{S}_0$ of the source and, hence, contains complete information as regards the ISP for the wave equation. In fact, as we found in Chapter 4, there is no stable algorithm for computing the radiated field everywhere outside the spatial support volume τ_0 from boundary values on a boundary $\partial\tau$ that is more than a few wavelengths removed from τ_0. Rather, it is possible merely to compute this field exactly and stably for space points $\mathbf{r}$ that lie outside the measurement boundary $\partial\tau$ or approximately and stably for space points $\mathbf{r}$ that lie inside this boundary. We also note that the theorem requires that both the field and the normal derivative be specified on $\partial\tau$ to insure uniqueness. However, only one of the two quantities is actually required, since the two data sets are connected via the second Helmholtz identity Eq. (1.36b) of Chapter 1 and either one can, in principle, be computed from the other.

Finally, we note that the ISP cannot have a unique solution even under zero-noise, ideal measurement conditions due to the possible presence of NR sources within the source region $\mathcal{S}_0$. In particular, if q_1 is any "solution" to the ISP (i.e., generates a specified field everywhere outside $\mathcal{S}_0$) then $q_2 = q_1 + q_{\text{nr}}$ will also be a solution, where q_{nr} is *any* NR source supported in $\mathcal{S}_0$. A good deal of the "mathematics" associated with the ISP is related to this inherent non-uniqueness of the problem and finding inversion methods that select out that particular source which is most physically meaningful. An equally important issue is the stability problem created by the possible presence of *essentially NR* source components within the source volume τ_0. As we discussed in Section 1.7.3 of Chapter 1, such sources radiate negligible energy and, hence, are difficult to detect, let alone determine, from field measurements performed outside their region of localization.

5.1.1 The ISP integral equation

We first address the ISP for the case of "ideal data" as defined by the field uniqueness theorem. We will discuss the problem within the context of limited and noisy field data later in this chapter. According to the field uniqueness theorem boundary-value data acquired over any closed surface $\partial\tau$ that completely surrounds the source spatial volume τ_0 or Cauchy data specified at any time t_0 exceeding the turn-off time T_0 generate the maximal amount of information about the source that is available from unlimited field measurements performed exterior to its space-time support volume $\{S_0|\mathbf{r} \in \tau_0,\ t \in [0, T_0]\}$. We can thus pose the ISP in terms of ideal data as represented by these two data sets without any loss of generality.

The ISP in terms of boundary-value data

We have already obtained a relationship between over-specified data on any bounding surface $\partial\tau \supseteq \partial\tau_0$ to the source volume τ_0 and the source term $q(\mathbf{r}, t)$ in the form of the *interior field solution* to the wave equation in Section 1.3.3 of Chapter 1. By making use of the two Helmholtz identities derived in Section 1.3.2 we can convert the interior field solution to the following integral equation which holds over all of space-time; e.g., for $\mathbf{r}$ located either within or outside of τ:

$$\int_0^{T_0} dt' \int_{\tau_0} d^3r'\, g_f(\mathbf{r} - \mathbf{r}', t - t')q(\mathbf{r}', t') = \phi(\mathbf{r}, t), \tag{5.1a}$$

where

$$g_f(\mathbf{R}, \tau) = g_+(\mathbf{R}, \tau) - g_-(\mathbf{R}, \tau)$$

is the "free-field propagator" first introduced in Chapter 1, with g_+ being the retarded Green function and g_- the advanced Green function, and

$$\phi(\mathbf{r}, t) = -\int_{-\infty}^{\infty} dt' \int_{\partial\tau} dS' \left[u_+ \frac{\partial}{\partial n'} g_f - g_f \frac{\partial}{\partial n'} u_+\right], \tag{5.1b}$$

with the normal derivatives being directed out of the interior τ. If we impose the restriction that the observation point $\mathbf{r}$ lies inside the region τ bounded by $\partial\tau$ then the contribution to ϕ in Eq. (5.1b) from the retarded Green function g_+ will vanish due to the second Helmholtz identity, so an alternative expression for ϕ in terms of boundary-value data valid within τ is given by

$$\phi(\mathbf{r}, t) = \int_{-\infty}^{\infty} dt' \int_{\partial\tau} dS' \left[u_+ \frac{\partial}{\partial n'} g_- - g_- \frac{\partial}{\partial n'} u_+ \right], \quad \mathbf{r} \in \tau. \tag{5.1c}$$

Note that, although the integration over t' in Eqs. (5.1b) and (5.1c) formally extends from $-\infty$ to $+\infty$, the radiated field over the boundary $\partial\tau$ will be non-zero only over some finite time interval so that these integrals will, in fact, extend only over that finite interval of time.

The field $\phi(\mathbf{r}, t)$ is equal to the actual radiated field $u_+(\mathbf{r}, t)$ for all values of t exceeding the source turn-off time T_0. This follows from Eq. (5.1a) on noting that the advanced-Green-function component of g_f is acausal, so $\phi(\mathbf{r}, t) = u_+(\mathbf{r}, t)$ if $t > T_0$. Although the field $\phi(\mathbf{r}, t)$ is not equal to the radiated field for times $t < T_0$, this quantity is still everywhere finite and well defined for such times and, thus, can be used as an *estimate* of the radiated field over the source's space-time region. Moreover, Eq. (5.1a) provides a relationship between this field and the source q that is valid over all of space and for all time, even for times $t < T_0$. In particular, Eq. (5.1a) can be regarded as an integral equation relating the source to the field ϕ computed from boundary-value data on $\partial\tau$ via Eq. (5.1b) or Eq. (5.1c). We will refer to this equation as the *ISP integral equation*.

The field $\phi(\mathbf{r}, t)$ satisfies the *homogeneous* wave equation everywhere within τ but is equal to the actual radiated field if $t > T_0$ and is a continuation of this field for times $t < T_0$. In a sense, then, it is the radiated field without the radiated field's singularities, just like the free-field propagator g_f is the retarded Green function to the wave equation without the Green function's singularities. In fact, for field points $\mathbf{r} \in \tau$, ϕ is the *back-propagated field* defined in Section 1.4.2 of Chapter 1 and the ISP integral equation Eq. (5.1a) is nothing more than the interior field solution to the radiation problem given in Eq. (1.37a) of Section 1.3.3 of that chapter (see also the treatment of back propagation in the frequency domain in Section 2.5.1 and stabilized field back propagation and the inverse RS boundary-value problem in Section 2.11 of Chapter 2). For this reason we will refer to ϕ as the *back-propagated field* even though it is only a stabilized version of the exactly back-propagated field (cf. the discussion in Section 4.4 of Chapter 4).

Finally, we mention that the back-propagated field ϕ computed from boundary-value data via Eq. (5.1b) is *over-specified* in that the radiated field and its normal derivative over $\partial\tau$ are not independent and, in particular, must together satisfy the second Helmholtz identity Eq. (1.36b) of Chapter 1 throughout the interior region τ. We show in an example presented below that this redundancy in the data can be removed and that ϕ can be computed in terms of the field (Dirichlet data) or its normal derivative (Neumann data), or, indeed, any linear combination of these two quantities over $\partial\tau$. In this connection we also note that the back-propagated field computed using Eq. (5.1b) is not the solution to a properly or improperly posed interior boundary-value problem. The reason for this is that the boundary values u_+ and $\partial u_+/\partial n$ are of the radiated field and its normal derivative and, hence, of a field that does not satisfy the homogeneous wave equation within the interior region τ of the data boundary $\partial\tau$. This means, in particular, that the limits of ϕ and its normal derivative $\partial\phi/\partial n$ as the field point $\mathbf{r}$ tends to $\partial\tau$ from τ will not be equal to the boundary values u_+ and $\partial u_+/\partial n$. We will show in Section 5.1.3 that a proper interpretation of ϕ is in terms of a field generated from singlet and doublet sources distributed over $\partial\tau$ according to the theory developed in Section 1.8 in Chapter 1.

The ISP in terms of Cauchy data

The back-propagated field expressed in terms of Cauchy data is, in fact, nothing more than the representation of the radiated field in terms of the solution to the initial-value problem presented in Section 1.4 of Chapter 1. In particular, the radiated field $u_+(\mathbf{r}, t)$ satisfies the homogeneous wave equation for all times $t > T_0$ and, hence, can be expressed over all of space and for all times $t > t_0 > T_0$ in terms of Cauchy data acquired at any time $t_0 > T_0$ via Eq. (1.41) of that section. It then follows that

$$\phi(\mathbf{r}, t) = \frac{1}{c^2} \int d^3r' [u_+(\mathbf{r}', t_0) g_f'(\mathbf{r} - \mathbf{r}', t - t_0) - g_f(\mathbf{r} - \mathbf{r}', t - t_0) u_+'(\mathbf{r}', t_0)], \quad (5.2)$$

where the primes denote derivatives w.r.t. t_0, will be equal to the radiated field if $t > T_0$ and is the continuation of this field for $t < T_0$. The fact that the back-propagated field ϕ computed either from boundary-value data via Eq. (5.1b) or from Cauchy data according to Eq. (5.2) is related to the source via the same integral equation means, of course, that these two quantities are identical. Thus, in particular, ϕ computed from either data set will equal the actual radiated field over all of space and for all times $t > T_0$ and contains, according to Theorem 1.4, the maximum amount of information concerning the source that is available from full knowledge of the field everywhere outside S_0.

We note also that the representation of the back-propagated field ϕ in terms of Cauchy data continues the field radiated backwards in time from the Cauchy data; i.e., in a certain sense, performs the process of field time reversal. This should not be surprising, considering the close connection between the processes of time reversal and back propagation (cf. Section 2.11 of Chapter 2 and Section 5.1.3 below). Of course, the back-propagated field so computed will satisfy the homogeneous wave equation over all of space and time and, hence, will not be equal to the actual radiated field for times t smaller than the source turn-off time T_0.

Finally, we note that, unlike the computation of ϕ from boundary-value data using Eq. (5.1b) or Eq. (5.1c), its computation using Cauchy data according to Eq. (5.2) is not over-specified. In particular, the radiated field $u_+(\mathbf{r}, t)$ at any time $t > T_0$ satisfies the homogeneous wave equation and is thus uniquely determined by Cauchy data as the solution of the initial-value problem covered in Section 1.4 of Chapter 1.

5.1.2 The Porter–Bojarski integral equation

The Porter–Bojarski integral equation is the ISP integral equation in the frequency domain and was previously derived in Section 2.5.1 for the case of boundary-value data, where it was found to be given by

$$\Phi(\mathbf{r},\omega) = \int_{\tau_0} d^3r'\, G_{\rm f}(\mathbf{r}-\mathbf{r}',\omega)Q(\mathbf{r}',\omega), \tag{5.3a}$$

where

$$G_{\rm f}(\mathbf{R},\omega) = G_+(\mathbf{R},\omega) - G_-(\mathbf{R},\omega) = -\frac{i}{2\pi}\frac{\sin(kR)}{R},$$

is the frequency-domain free-field propagator and

$$\Phi(\mathbf{r},\omega) = -\int_{\partial\tau} dS'\left[U_+(\mathbf{r}',\omega)\frac{\partial}{\partial n'}G_{\rm f}(\mathbf{r}-\mathbf{r}',\omega) - G_{\rm f}(\mathbf{r}-\mathbf{r}',\omega)\frac{\partial}{\partial n'}U_+(\mathbf{r}',\omega)\right], \tag{5.3b}$$

which holds over all of space, and

$$\Phi(\mathbf{r},\omega) = \int_{\partial\tau} dS'\left[U_+(\mathbf{r}',\omega)\frac{\partial}{\partial n'}G_-(\mathbf{r}-\mathbf{r}',\omega) - G_-(\mathbf{r}-\mathbf{r}',\omega)\frac{\partial}{\partial n'}U_+(\mathbf{r}',\omega)\right], \tag{5.3c}$$

which holds within the region τ bounded by $\partial\tau$. The back-propagated field can also be expressed in terms of Cauchy data for the special case of the wave equation[1] by Fourier transformation of Eq. (5.2):

$$\Phi(\mathbf{r},\omega) = \frac{e^{i\omega t_0}}{c^2}\int d^3r'[i\omega u_+(\mathbf{r}',t_0) - u'_+(\mathbf{r}',t_0)]G_{\rm f}(\mathbf{r}-\mathbf{r}',\omega). \tag{5.3d}$$

Example 5.1 The inhomogeneous terms in the time-domain ISP integral equation Eq. (5.1a) and its frequency-domain equivalent Eq. (5.3a) are computed from over-specified boundary-value data. In this example we will remove this over-specification for the special case in which the data boundary $\partial\tau$ is a sphere and express the back-propagated field in terms of Dirichlet data alone. We will work in the frequency domain and take the volume τ to be a sphere of radius $r_0 > a_0$, where a_0 is the radius of the smallest bounding sphere to τ_0, and will compute the back-propagated field using Eq. (5.3b). An entirely parallel development based on Eq. (5.3c) leads to the same result at all interior points to the data boundary at $r = r_0$. Since the geometry is spherical we will use the multipole expansion of the free-field propagator, which is found from Example 4.9 to be given by

[1] The initial-value problem and, hence, the representation of the back-propagated field in terms of Cauchy data is applicable only to the wave equation, not to general dispersive media.

$$G_f(\mathbf{R},\omega) = -\frac{i}{2\pi}\frac{\sin(kR)}{R} = -2ik\sum_{l=0}^{\infty}\sum_{m=-l}^{l} j_l(kr)j_l(kr')Y_l^m(\hat{\mathbf{r}})Y_l^{m*}(\hat{\mathbf{r}}'), \tag{5.4}$$

where j_l, $l = 0, 1, \ldots$, are the spherical Bessel functions of order l and Y_l^m, $m = -l, -l+1, \ldots, l$, the spherical harmonics of degree l and order m; $\hat{\mathbf{r}}$ and $\hat{\mathbf{r}}'$ denote the polar and azimuthal angles θ, ϕ and θ', ϕ' of the general field points $\mathbf{r}$ and $\mathbf{r}'$. On substituting the above expansion into Eq. (5.3b), with $\partial\tau$ taken to be the surface of the sphere having radius r_0, and performing some minor algebra we obtain the result

$$\Phi(\mathbf{r},\omega) = 2ikr_0^2\sum_{l=0}^{\infty}\sum_{m=-l}^{l}[kj_l'(kr_0)u_l^m(\omega) - j_l(kr_0)v_l^m(\omega)]j_l(kr)Y_l^m(\hat{\mathbf{r}}),$$

where

$$u_l^m(\omega) = \int d\Omega\, U_+(\mathbf{r},\omega)|_{r=r_0}Y_l^{m*}(\hat{\mathbf{r}}), \qquad v_l^m(\omega) = \int d\Omega\,\frac{\partial}{\partial r}U_+(\mathbf{r},\omega)|_{r=r_0}Y_l^{m*}(\hat{\mathbf{r}})$$

are the generalized Fourier coefficients of the field and its normal derivative over the surface of the sphere. These coefficients are not independent but rather are related via the equation

$$v_l^m(\omega) = \frac{kh_l^{+\prime}(kr_0)}{h_l^+(kr_0)}u_l^m(\omega), \tag{5.5}$$

which was derived in Example 4.10. We can use Eq. (5.5) to remove the data redundancy in Eq. (5.1) and express Φ entirely in terms of Dirichlet or Neumann data. For example, for the case of Dirichlet data we obtain the result

$$\Phi(\mathbf{r},\omega) = 2i(kr_0)^2\sum_{l=0}^{\infty}\sum_{m=-l}^{l}\frac{j_l'(kr_0)h_l^+(kr_0) - h_l^{+\prime}(kr_0)j_l'(kr_0)}{h_l^+(kr_0)}u_l^m(\omega)j_l(kr)Y_l^m(\hat{\mathbf{r}})$$

$$\Downarrow$$

$$\Phi(\mathbf{r},\omega) = 2\sum_{l=0}^{\infty}\sum_{m=-l}^{l}\frac{u_l^m(\omega)}{h_l^+(kr_0)}j_l(kr)Y_l^m(\hat{\mathbf{r}}), \tag{5.6}$$

where we have used the Wronskian relationship

$$j_l'(kr_0)h_l^+(kr_0) - h_l^{+\prime}(kr_0)j_l'(kr_0) = -\frac{i}{(kr_0)^2}.$$

A similar development can be employed to express Φ and, hence, ϕ in terms of Neumann data over the sphere.

Example 5.2 We now consider the case in which the field measurements are performed over a closed sphere Σ_∞ having an arbitrarily large radius. The back-propagated field is then given by Eq. (5.3b) with $\partial\tau = \Sigma_\infty$. Over this sphere we have that

$$U_+(\mathbf{r}',\omega) \sim f(\mathbf{s},\omega)\frac{e^{ikr'}}{r'},$$

$$\frac{\partial}{\partial n'}U_+(\mathbf{r}',\omega) \sim ikf(\mathbf{s},\omega)\frac{e^{ikr'}}{r'},$$

$$G_{\mathrm{f}}(\mathbf{r}-\mathbf{r}',\omega) \sim -\frac{1}{4\pi r'}\{e^{-ik\mathbf{s}\cdot\mathbf{r}}e^{ikr'} - e^{ik\mathbf{s}\cdot\mathbf{r}}e^{-ikr'}\},$$

$$\frac{\partial}{\partial n'}G_{\mathrm{f}}(\mathbf{r}-\mathbf{r}',\omega) \sim -\frac{ik}{4\pi r'}\{e^{-ik\mathbf{s}\cdot\mathbf{r}}e^{ikr'} + e^{ik\mathbf{s}\cdot\mathbf{r}}e^{-ikr'}\},$$

where $\mathbf{s} = \mathbf{r}'/r'$ and $kr' \to \infty$ on Σ_∞ and $f(\mathbf{s},\omega)$ is the radiation pattern of the field. By substituting the above into Eq. (5.3b) with $\partial\tau = \Sigma_\infty$ we then obtain

$$\Phi(\mathbf{r},\omega) = \int_{\Sigma_\infty} dS' \left[f(\mathbf{s},\omega)\frac{e^{ikr'}}{r'}\frac{ik}{4\pi r'}\{e^{-ik\mathbf{s}\cdot\mathbf{r}}e^{ikr'} + e^{ik\mathbf{s}\cdot\mathbf{r}}e^{-ikr'}\} \right.$$
$$\left. - \frac{1}{4\pi r'}\{e^{-ik\mathbf{s}\cdot\mathbf{r}}e^{ikr'} - e^{ik\mathbf{s}\cdot\mathbf{r}}e^{-ikr'}\}ikf(\mathbf{s},\omega)\frac{e^{ikr'}}{r'}\right]$$

$$\Downarrow$$

$$\Phi(\mathbf{r},\omega) = \frac{ik}{2\pi}\int_{4\pi} d\Omega_{\mathbf{s}}\, f(\mathbf{s},\omega)e^{ik\mathbf{s}\cdot\mathbf{r}}. \tag{5.7a}$$

If we expand the radiation pattern in a series of spherical harmonics with expansion coefficients (cf. Section 4.8.4)

$$f_l^m(\omega) = \int_{4\pi} d\Omega\, f(\mathbf{s},\omega)Y_l^{m*}(\mathbf{s})$$

we obtain using Eq. (5.7a)

$$\Phi(\mathbf{r},\omega) = \sum_{l=0}^{\infty}\sum_{m=-l}^{l} 2i^{l+1}kf_l^m(\omega)\overbrace{\frac{(-i)^l}{4\pi}\int_{4\pi} d\Omega_{\mathbf{s}}\, Y_l^m(\mathbf{s})e^{ik\mathbf{s}\cdot\mathbf{r}}}^{j_l(kr)Y_l^m(\hat{\mathbf{r}})}, \tag{5.7b}$$

where we have made use of the plane-wave expansion of the free multipole fields derived in Example 3.4 of Chapter 3. The plane-wave expansion Eq. (5.7a) or the multipole expansion Eq. (5.7b) allows the back-propagated field to be computed directly from the radiation pattern of the field.

5.1.3 Time reversal and the back-propagated field

We have seen that the back-propagated field $\phi(\mathbf{r},t)$ when generated by Cauchy data acquired at a time $t_0 > T_0$ can be interpreted as being a version of the actual radiated field $u_+(\mathbf{r},t)$ continued backward into times $t < t_0$. A similar interpretation can be applied to this field when generated by boundary-value data over $\partial\tau$. Moreover, when these boundary data are over-specified (the field and its normal derivative) then the field $\phi(\mathbf{r},-t)$ can be interpreted as being generated by singlet and doublet sources distributed over $\partial\tau$ that are time-reversed versions of the radiated field and its (inward-directed) normal derivative measured over $\partial\tau$.

To see this, we restrict our attention to field points $\mathbf{r} \in \tau$ for which the back-propagated field can be expressed in terms of the advanced Green function via Eq. (5.1c). If in this equation we make the transformations $t \to -t$ and $t' \to -t'$ and replace the outward-directed normal derivatives by *inward-directed* normal derivatives $\partial/\partial\eta'$ (from $\tau^{\perp}$ to τ) then the equation becomes

$$\phi(\mathbf{r},-t) = \int_{-\infty}^{\infty} dt' \int_{\partial\tau} dS' \left[g_{+}(\mathbf{r}-\mathbf{r}',t-t')\frac{\partial}{\partial\eta'}u_{+}(\mathbf{r}',-t') - u_{+}(\mathbf{r}',-t')\frac{\partial}{\partial\eta'}g_{+}(\mathbf{r}-\mathbf{r}',t-t')\right], \quad \mathbf{r}\in\tau. \tag{5.8}$$

On comparing Eq. (5.8) with Eq. (1.69) of Section 1.8 of Chapter 1 we conclude that $\phi(\mathbf{r},-t)$ can be interpreted as being the field radiated into the interior region τ by the *singlet* and *doublet* surface sources

$$q_{\mathrm{s}}(\mathbf{r},t) = \frac{\partial}{\partial\eta}u_{+}(\mathbf{r},-t), \qquad q_{\mathrm{d}}(\mathbf{r},t) = u_{+}(\mathbf{r},-t), \quad \mathbf{r}\in\partial\tau. \tag{5.9}$$

The above development shows that the time-reversed back-propagated field $\phi(\mathbf{r},-t)$ is radiated into the interior region τ containing the source by the time-reversed boundary values $u_{+}(\mathbf{r},-t)$ and its normal derivative $\partial/\partial\eta u_{+}(\mathbf{r},-t)$ directed from the exterior $\tau^{\perp}$ into τ. Since $\phi(\mathbf{r},-t) = u_{+}(\mathbf{r},-t)$ if $t < -T_0$ we conclude that the time-reversed boundary-value fields radiate the time-reversed back-propagated field within τ $\forall t < -T_0$, illustrating the close connection between field time reversal and field back propagation and the ISP. A similar connection was obtained in Section 2.11, where we showed that the time-reversed approximate solution to the inverse RS boundary-value problem generated from field back propagation was equal to the field radiated by the time-reversed Dirichlet or Neumann boundary conditions.

5.1.4 The ISP in terms of Dirichlet or Neumann boundary-value data

Generating a singularity-free version of the radiated field by back propagating *both* the field and its normal derivative over $\partial\tau$ is only one of a number of ways of constructing such a field. For example, we used stable back propagation of only the field or its normal derivative over infinite plane boundaries in Section 2.11 of Chapter 2. A similar approach can be used in the ISP if we replace the free-field propagator by the difference between the retarded Green function and the time-reversed Dirichlet Green function appropriate to the boundary $\partial\tau$; i.e., we replace g_{f} by

$$g(\mathbf{r},\mathbf{r}',t-t') = g_{+}(\mathbf{r}-\mathbf{r}',t-t') - g_{\mathrm{d}}(\mathbf{r},\mathbf{r}',t'-t), \tag{5.10}$$

where the time-reversed Dirichlet Green function $g_{\mathrm{d}}(\mathbf{r},\mathbf{r}',t'-t) = 0$, with $\mathbf{r},\mathbf{r}' \in \partial\tau$. The quantity $g(\mathbf{r},\mathbf{r}',t-t')$ is the difference between two Green functions to the wave equation and, like the free-field propagator, satisfies the homogeneous wave equation. All of the development presented using the free-field propagator can thus be repeated and we obtain, in place of Eqs. (5.1), the set of equations

$$\int_0^{T_0} dt' \int_{\tau_0} d^3r'\, g(\mathbf{r}, \mathbf{r}', t-t') q(\mathbf{r}', t') = \chi(\mathbf{r}, t), \tag{5.11a}$$

where

$$\chi(\mathbf{r}, t) = -\int_{-\infty}^{\infty} dt' \int_{\partial\tau} dS' \left[u_+ \frac{\partial}{\partial n'} g - g \frac{\partial}{\partial n'} u_+ \right] \tag{5.11b}$$

$$= \int_{-\infty}^{\infty} dt' \int_{\partial\tau} dS'\, u_+(\mathbf{r}', t') \frac{\partial}{\partial n'} g_{\mathrm{d}}(\mathbf{r}, \mathbf{r}', t'-t), \quad \mathbf{r} \in \tau, \tag{5.11c}$$

where in deriving Eq. (5.11c) we have made use of the fact that the Dirichlet Green function vanishes on $\partial\tau$. Equation (5.11a) is a new ISP integral equation that is just as valid as our original one defined in Eq. (5.1a), while Eqs. (5.11b) and (5.11c) are the generalizations of Eqs. (5.1b) and (5.1c).

If we restrict our attention to field points $\mathbf{r} \in \tau$, we see from Eq. (5.11c) that only the value of the field and not its normal derivative is required in order to generate the inhomogeneous term $\chi(\mathbf{r}, t)$ in the ISP integral equation Eq. (5.11a). Moreover,

$$\chi(\mathbf{r}, -t) = \int_{-\infty}^{\infty} dt' \int_{\partial\tau} dS'\, u_+(\mathbf{r}', -t') \frac{\partial}{\partial n'} g_{\mathrm{d}}(\mathbf{r}, \mathbf{r}', t-t'), \quad \mathbf{r} \in \tau,$$

which shows that the field $\chi(\mathbf{r}, -t)$ is radiated into the interior region τ from time-reversed Dirichlet data over $\partial\tau$. The field $\chi(\mathbf{r}, -t)$ is thus generated as a properly posed interior boundary-value problem from the time-reversed radiated field in a manner completely parallel to that employed in our discussion of the inverse RS boundary-value problem and stabilized field back propagation in Section 2.11 of Chapter 2!

Other "ISP integral equations" can be constructed using other replacements for the free-field propagator so long as the replacement is the difference between the retarded Green function and a time-reversed second Green function to the wave equation. All of them will have a kernel $g(\mathbf{r}, \mathbf{r}', t-t')$ that satisfies the homogeneous wave equation over τ which forces them to be mathematically equivalent. It is important to note, however, that these various versions of the ISP integral equation will have different kernels and different inhomogeneous terms. For example, the inhomogeneous terms ϕ and χ in the two versions of the ISP developed above are different and will, of course, remain different if the redundancy of the boundary-value data in the first version of this integral equation given in Eq. (5.1a) is removed so that ϕ is computed, say, from Dirichlet data alone. In this case ϕ is still the same as when computed using over-specified boundary-value data and, in particular, will *not* be equal to χ in our second version of the ISP computed from the Dirichlet data via Eq. (5.11c). The two integral equations will, however, possess the same set of solutions to the ISP.

5.2 The ISP for surface sources

The simplest type of source to the wave equation is, of course, the delta function $\delta(\mathbf{r}-\mathbf{r}')\delta(t-t')$ which radiates the retarded Green function $g_+(\mathbf{r}-\mathbf{r}',t-t')$. The next simplest types would be those sources that are distributed on a line or surface in 3D space. This latter class of source was treated in the time-domain in Section 1.8 of Chapter 1 and in the frequency domain in Section 2.12, where we considered sources distributed over an open or closed surface $\partial\tau_0$ that separates two disjoint regions τ_0 and $\tau_0^\perp$. The surface source radiates a field into both regions according to the formula

$$u_+(\mathbf{r},t) = \int_0^\infty dt' \int_{\partial\tau_0} dS_0 \left[q_s(\mathbf{r}_0,t')g_+(\mathbf{r}-\mathbf{r}_0,t-t') - q_d(\mathbf{r}_0,t')\frac{\partial}{\partial n_0}g_+(\mathbf{r}-\mathbf{r}_0,t-t')\right],$$

where $\mathbf{r}_0 \in \partial\tau_0$ denotes a point on the surface $\partial\tau_0$ which can be closed with finite interior τ_0 and infinite exterior $\tau_0^\perp$ or the two regions can be infinite with common boundary $\partial\tau_0$. The normal derivative in the above equation can be selected to be directed out of τ_0 and into the $\tau_0^\perp$ or vice versa. The surface source is seen to be characterized by the two components q_s and q_d, which are referred to as the "singlet" and "doublet" components of the source due to the fact that they radiate a monopole field and a dipole field, respectively. The ISP for surface sources consists of determining these two source components in terms of the back-propagated field $\phi(\mathbf{r},t)$ computed from Cauchy or boundary-value data. We mention in this connection that a surface source requires both a singlet component and a doublet component since a single component (singlet or doublet) would radiate an identical field into the two regions separated by the surface τ. We will elaborate on this point later in connection with our solution of the ISP presented below.

Surface sources are idealizations of 3D (volume) sources that are distributed over a thin layer or shell that is much smaller than the shortest wavelength $\lambda_{\min}$ radiated by the source. As such they provide convenient and simple limiting forms of a large class of realizable volume sources that can be employed as a benchmark against which the performance of the actual volume source can be compared. Their simplicity also makes them an ideal first source with which to study and apply the formulation of the ISP developed in the preceding sections. In this section we will solve the ISP for a surface source distributed over an infinite plane with boundary-value data specified over two bounding planes on each side of the source plane. A similar development (see the problems at the end of the chapter) can be employed to solve the ISP for a source distributed over the surface of a sphere with boundary-value data specified over the surface of a concentric larger sphere.

5.2.1 The ISP for a planar surface source

We will work in the frequency domain and consider a source distributed over the plane $z = 0$ with boundary-value data acquired over two parallel planes located at $z = \pm z_0$. We showed in Section 1.8 of Chapter 1 and Section 2.12 of Chapter 2 that a source that is distributed over a separable surface for the scalar wave and Helmholtz equations defined by generalized coordinates (ξ_1, ξ_2, ξ_3) with $\xi_3 = \xi_{30} =$ constant can be expressed in the frequency domain as a 3D source distribution in the form

$$Q(\mathbf{r},\omega) = Q_s(\xi_1,\xi_2,\omega)\frac{\delta(\xi_3-\xi_{30})}{h_3(\mathbf{r}_0)} + Q_d(\xi_1,\xi_2,\omega)\frac{h_1(\mathbf{r}_0)h_2(\mathbf{r}_0)}{h_1(\mathbf{r})h_2(\mathbf{r})}\frac{\frac{\partial}{\partial\xi_3}\delta(\xi_3-\xi_{30})}{h_3(\mathbf{r})}, \quad (5.12)$$

where $\delta(\cdot)$ is the Dirac delta function and h_1, h_2, h_3 are the scale factors for the coordinate system. On selecting the separable system to be the Cartesian coordinate system with $\xi_1 = x$, $\xi_2 = y$ and $\xi_3 = z$ with $\xi_{30} = 0$ the above source becomes

$$Q(\mathbf{r},\omega) = Q_s(\boldsymbol{\rho},\omega)\delta(z) + Q_d(\boldsymbol{\rho},\omega)\frac{\partial}{\partial z}\delta(z),$$

where $\boldsymbol{\rho}_0 = (x_0, y_0)$ denotes a point on the source plane.

On substituting the expression for the surface source given above into the r.h.s. of the frequency-domain ISP integral equation (the Porter–Bojarski integral equation) Eq. (5.3a) we obtain the integral equation

$$\Phi(\mathbf{r},\omega) = \int_{z=0} d^2\rho' \left[G_f(\mathbf{r}-\boldsymbol{\rho}',\omega)Q_s(\boldsymbol{\rho}',\omega) + Q_d(\boldsymbol{\rho}',\omega)\frac{\partial}{\partial z}G_f(\mathbf{r}-\boldsymbol{\rho}',\omega)\right], \quad (5.13)$$

where $\mathbf{r} = \boldsymbol{\rho} + z\hat{\mathbf{z}}$ is a general field point within τ. The back-propagated field generated from the boundary-value data over the two planes at $z = \pm z_0$ can be expressed as the sum of the two field components

$$\Phi(\mathbf{r},\omega) = \Phi_+(\mathbf{r},\omega) + \Phi_-(\mathbf{r},\omega),$$

where

$$\Phi_\pm(\mathbf{r},\omega) = \pm\int d^2\rho' \left[U_+(\boldsymbol{\rho}',z',\omega)\frac{\partial}{\partial z'}G_-(\boldsymbol{\rho}-\boldsymbol{\rho}',z-z',\omega) - G_-(\boldsymbol{\rho}-\boldsymbol{\rho}',z-z',\omega)\frac{\partial}{\partial z'}U_+(\boldsymbol{\rho}',z',\omega)\right]\Bigg|_{z'=\pm z_0},$$

where the top (plus) sign is used for the plane $z' = +z_0$ and the bottom (minus) sign for the plane at $z' = -z_0$. In deriving the above expression we have made use of the fact that $\partial/\partial n' = \pm\partial/\partial z'$, with the plus sign holding over the surface at z_0 and the minus sign for the surface at $-z_0$. The ISP integral equation for the planar surface source is thus given by Eq. (5.13) with the back-propagated field computed from the boundary-value data via the above set of equations.

5.2.2 Solving the ISP integral equation

To proceed we represent the r.h.s. of the ISP integral equation Eq. (5.13) in terms of a plane-wave (angular-spectrum) expansion by using the plane-wave expansion of the free-field propagator developed in Example 4.1. This choice of expansion is dictated by the fact that the source and data are defined over planar boundaries, thus immediately suggesting the use of plane waves (see the discussion at the beginning of Chapter 3). For the case of non-dispersive media with real wavenumber k we find from Example 4.1

$$G_{\mathrm{f}}(\boldsymbol{\rho}-\boldsymbol{\rho}',z-z',\omega)=-\frac{i}{8\pi^2}\int_{K_\rho\le k}\frac{d^2K_\rho}{\gamma}\,e^{i\mathbf{K}_\rho\cdot(\boldsymbol{\rho}-\boldsymbol{\rho}')}[e^{i\gamma(z-z')}+e^{-i\gamma(z-z')}], \tag{5.14}$$

where $\gamma=\sqrt{k^2-K_\rho^2}$. On substituting the plane-wave expansion into the r.h.s. of the integral equation Eq. (5.13) and making some algebraic simplifications we obtain the result

$$\Phi(\mathbf{r},\omega)=\frac{i}{8\pi^2}\int_{K_\rho\le k}\frac{d^2K_\rho}{\gamma}\,e^{i\mathbf{K}_\rho\cdot\boldsymbol{\rho}}\Big[(\tilde{Q}_{\mathrm{s}}(\mathbf{K}_\rho,\omega)-i\gamma\tilde{Q}_{\mathrm{d}}(\mathbf{K}_\rho,\omega))e^{i\gamma z}$$
$$-(\tilde{Q}_{\mathrm{s}}(\mathbf{K}_\rho,\omega)+i\gamma\tilde{Q}_{\mathrm{d}}(\mathbf{K}_\rho,\omega))e^{-i\gamma z}\Big], \tag{5.15}$$

where $\tilde{Q}_{\mathrm{s}}(\mathbf{K}_\rho,\omega)$ and $\tilde{Q}_{\mathrm{d}}(\mathbf{K}_\rho,\omega)$ are the spatial Fourier transforms of the singlet and doublet components of the source over the source plane. It should be noted that Eqs. (5.15) involve only the spatial Fourier transforms of the source components over the homogeneous region of the spectra $K_\rho<k$. This means that these transforms can be determined from the solution of the ISP integral equation only over the homogeneous region, which, in turn, means that only a low-pass-filtered version of the source is, in general, possible (however, see the discussion below).

We can also express the back-propagated fields $\Phi_\pm$ from the two data planes in angular-spectrum expansions using the angular-spectrum expansion of the incoming-wave Green function given in Eq. (4.5a) of Section 4.1 of Chapter 4. In Example 4.5 of that chapter we computed these plane-wave expansions and obtained the result

$$\Phi_\pm(\mathbf{r},\omega)=\frac{1}{(2\pi)^2}\int_{K_\rho^2<k^2}d^2K_\rho\,\widetilde{U_+}(\mathbf{K}_\rho,\pm z_0)e^{\pm i\gamma(z-z_0)}e^{i\mathbf{K}_\rho\cdot\boldsymbol{\rho}}, \tag{5.16}$$

where $\widetilde{U_+}(\mathbf{K}_\rho,\pm z_0)$ are the 2D spatial Fourier transforms of $U_+(\mathbf{r},\omega)$ over the two data planes $z=\pm z_0$. It should be noted that the two back-propagated fields, even though they are computed from over-specified boundary-value data (field plus normal derivative) have plane-wave expansions that depend only on Dirichlet conditions on the two planes and also include only homogeneous plane-wave components; i.e., there are no evanescent plane waves in the expansions. The reason for this is that the transforms $\widetilde{U_+}(\mathbf{K}_\rho,\pm z_0)$ and $\widetilde{(\partial/\partial z_0)U_+}(\mathbf{K}_\rho,z_0)$ of the Dirichlet and Neumann boundary values are linearly related (see below), which allows the plane-wave amplitudes in the plane-wave expansions of $\Phi_\pm$ to be expressed entirely in terms of Dirichlet or Neumann boundary values and also causes these plane-wave amplitudes to vanish over the evanescent region of the spectra (cf. Example 4.5).

On adding the two component fields we find that

$$\Phi(\mathbf{r},\omega)=\frac{1}{(2\pi)^2}\int_{K_\rho^2<k^2}d^2K_\rho\Big[\widetilde{U_+}(\mathbf{K}_\rho,+z_0)e^{i\gamma(z-z_0)}$$
$$+\widetilde{U_+}(\mathbf{K}_\rho,-z_0)e^{-i\gamma(z-z_0)}\Big]e^{i\mathbf{K}_\rho\cdot\boldsymbol{\rho}}. \tag{5.17}$$

As a final step we equate Eqs. (5.15) and (5.17) and solve for the Fourier transforms of the two source components to obtain

$$\tilde{Q}_s(\mathbf{K}_\rho,\omega) = i\gamma[\widetilde{U}_+(\mathbf{K}_\rho,-z_0)e^{i\gamma z_0} + \widetilde{U}_+(\mathbf{K}_\rho,+z_0)e^{-i\gamma z_0}], \tag{5.18a}$$

$$\tilde{Q}_d(\mathbf{K}_\rho,\omega) = \widetilde{U}_+(\mathbf{K}_\rho,-z_0)e^{i\gamma z_0} - \widetilde{U}_+(\mathbf{K}_\rho,+z_0)e^{-i\gamma z_0}, \tag{5.18b}$$

where K_ρ is limited to the homogeneous region $K_\rho < k$. The solution of the ISP then yields the following low-pass-filtered (spatially band-limited) versions of the source components

$$\hat{Q}_s(\boldsymbol{\rho},\omega) = \frac{1}{(2\pi)^2}\int_{K_\rho<k} d^2K_\rho\, \tilde{Q}_s(\mathbf{K}_\rho,\omega)e^{i\mathbf{K}_\rho\cdot\boldsymbol{\rho}}, \tag{5.19a}$$

$$\hat{Q}_d(\boldsymbol{\rho},\omega) = \frac{1}{(2\pi)^2}\int_{K_\rho<k} d^2K_\rho\, \tilde{Q}_d(\mathbf{K}_\rho,\omega)e^{i\mathbf{K}_\rho\cdot\boldsymbol{\rho}}, \tag{5.19b}$$

with $\tilde{Q}_s$ and $\tilde{Q}_d$ defined in Eqs. (5.18).

5.2.3 Interpretation of the solution

The expressions for the low-pass-filtered singlet and doublet source components given in Eqs. (5.19) have simple interpretations in terms of the back propagation of Dirichlet and Neumann boundary conditions on the two boundary-value planes $z = \pm z_0$. Consider first the doublet component, which, on substituting for $\tilde{Q}_d$ from Eq. (5.18b), becomes

$$\hat{Q}_d(\boldsymbol{\rho},\omega) = \frac{1}{(2\pi)^2}\int_{K_\rho<k} d^2K_\rho\, \widetilde{U}_+(\mathbf{K}_\rho,-z_0)e^{i\gamma z_0}e^{i\mathbf{K}_\rho\cdot\boldsymbol{\rho}}$$
$$-\frac{1}{(2\pi)^2}\int_{K_\rho<k} d^2K_\rho\, \widetilde{U}_+(\mathbf{K}_\rho,+z_0)e^{-i\gamma z_0}e^{i\mathbf{K}_\rho\cdot\boldsymbol{\rho}}.$$

It follows from the angular-spectrum expansions of the two back-propagated fields given in Eq. (5.16) that the first term on the r.h.s. of the above equation is the field back propagated from Dirichlet data on the boundary plane at $z = -z_0$ lying to the left of the source plane, while the second term is the field back propagated from the Dirichlet data on the boundary plane at $z = +z_0$ lying to the right of the source plane. The doublet source component is thus simply the difference of these two back-propagated fields.

Consider now the singlet source component. It is shown in Example 4.4 that the spatial Fourier transforms of the Dirichlet and Neumann boundary values over any plane $z = z_0$ are related via the equation

$$\widetilde{U}'_+(\mathbf{K}_\rho,\pm z_0) = \pm i\gamma\,\widetilde{U}_+(\mathbf{K}_\rho,\pm z_0),$$

which, when used in Eq. (5.18a), yields the result

$$\tilde{Q}_s(\mathbf{K}_\rho,\omega) = -\widetilde{U}'_+(\mathbf{K}_\rho,-z_0)e^{i\gamma z_0} + \widetilde{U}'_+(\mathbf{K}_\rho,+z_0)e^{-i\gamma z_0}.$$

On substituting this into Eq. (5.19a) we then obtain

$$\hat{Q}_s(\boldsymbol{\rho}, \omega) = -\frac{1}{(2\pi)^2} \int_{K_\rho < k} d^2K_\rho \, \widetilde{U'_+}(\mathbf{K}_\rho, -z_0) e^{i\gamma z_0} e^{i\mathbf{K}_\rho \cdot \boldsymbol{\rho}} + \frac{1}{(2\pi)^2} \int_{K_\rho < k} d^2K_\rho \, \widetilde{U'_+}(\mathbf{K}_\rho, +z_0) e^{-i\gamma z_0} e^{i\mathbf{K}_\rho \cdot \boldsymbol{\rho}}.$$

Again, we conclude from the plane-wave expansions given in Eq. (5.16) that we can interpret the first term on the r.h.s. of the above equation as being the field back propagated using Neumann data on the boundary plane at $z = -z_0$ lying to the left of the source plane,[2] while the second term is the field back propagated from Neumann data over the boundary plane at $z = +z_0$ lying to the right of the source plane. The singlet source component is thus simply the sum of these two back-propagated fields.

Finally we comment on the fact that the solution of the ISP yielded only low-pass-filtered versions of the surface source rather than the exact, or complete, source. This would appear to contradict Theorem 1.4, which states that the boundary-value data over the two bounding planes to the source should uniquely determine the source up to NR components, and we have shown in Section 1.8 of Chapter 1 that NR surface sources do not exist. The apparent contradiction is resolved when we recall that Theorem 1.4 applies only to causal sources supported in a finite spatial region and our solution to the ISP as formulated above includes no such constraint. Thus, if we demand that the source be supported in a finite region on the (x, y) plane then we can, in principle, compute the exact surface source. The basic reason for this is that sources that are supported in finite spatial regions have spatial Fourier transforms $\tilde{Q}(\mathbf{K}, \omega)$ that are entire analytic functions of the spatial frequency vector $\mathbf{K}$. For surface sources this means that the transforms $\tilde{Q}_s(\mathbf{K}_\rho, \omega)$ and $\tilde{Q}_d(\mathbf{K}_\rho, \omega)$ are entire functions of the 2D spatial frequency vector $\mathbf{K}_\rho$ and can, in principle, be determined for all $\mathbf{K}_\rho$ from specification of the transforms over the homogeneous region (the interior of a circle of radius k).

5.3 The ISP for 3D sources supported in plane-parallel slabs

The surface source computed in the preceding section has a number of drawbacks, the most serious being that it consists of the sum of a delta-function component (singlet) and a derivative of a delta-function component (doublet). These components are, of course, not realizable in any real system and can only be approximated by thin 3D source distributions. Thus, in terms of *source synthesis*, where the goal is to design a source to radiate a specified field, the surface source has limited applicability. Another goal of the ISP is to compute an existing source distribution from measured field data. Again, the ISP formulated in terms of surface sources has limited applicability to real-world problems since any physically

[2] Note that the normal derivatives of the boundary-value fields in this equation are taken with respect to positive z so that $-\widetilde{U'_+}(\mathbf{K}_\rho, -z_0, \omega)$ is actually the spatial Fourier transform of the Neumann boundary value on the plane $z = -z_0$ with the derivative taken with respect to the outward-directed normal (in the $-z$ direction) over this plane.

realizable source will be 3D. We are thus led to consider the general formulation of the ISP for 3D sources in terms of the ISP integral equation Eq. (5.1a) or its frequency-domain equivalent, the Porter–Bojarski integral equation, Eq. (5.3a).

We first consider the generalization of the surface source to that of a 3D source supported within a slab bounded by two infinite parallel planes located at $z = \pm a_0$. As in the preceding sections, we will consider only the case of a non-dispersive medium where k is real-valued and will also limit our attention to boundary-value data acquired over the two parallel planes at $z = \pm z_0$ with $z_0 \geq a_0$ as was done for the surface source in the last section. After solving the ISP we will find that our solution reduces to that obtained for the surface source in the limit where the source depth $2a_0 \to 0$.

The Porter–Bojarski integral equation for the case under consideration assumes the form

$$\Phi(\mathbf{r}, \omega) = \int d^2\rho' \int_{-a_0}^{a_0} dz'\, G_{\mathrm{f}}(\mathbf{r} - \mathbf{r}', \omega) Q(\mathbf{r}', \omega),$$

where $\mathbf{r} = \boldsymbol{\rho} + z\hat{\mathbf{z}}$ and $\mathbf{r}' = \boldsymbol{\rho}' + z'\hat{\mathbf{z}}$ are field points within $\tau = [-z_0 \leq z \leq +z_0]$ and $\tau_0 = [-a_0 \leq z \leq +a_0]$, respectively. If we then make use of the plane-wave expansion of the free-field propagator in Eq. (5.14) we obtain after some simplification

$$\Phi(\mathbf{r}, \omega) = -\frac{i}{8\pi^2} \int_{K_\rho \leq k} \frac{d^2 K_\rho}{\gamma} e^{i\mathbf{K}_\rho \cdot \boldsymbol{\rho}} [\tilde{Q}(\mathbf{K}_\rho, +\gamma, \omega) e^{i\gamma z} + \tilde{Q}(\mathbf{K}_\rho, -\gamma, \omega) e^{-i\gamma z}],$$

where $\tilde{Q}(\mathbf{K}_\rho, \pm\gamma, \omega)$ denotes the spatial Fourier transform of $\tilde{Q}(\mathbf{K}, \omega)$ over the surfaces $\mathbf{K} = \mathbf{K}_\rho \pm \gamma\hat{\mathbf{z}}$. The back-propagated field from the two planes at $z = \pm z_0$ was computed in the last section and is given in Eq. (5.17). If we then substitute this expression into the l.h.s. of the above equation and inverse Fourier transform both sides of the resulting equation over the $\boldsymbol{\rho}$ plane we obtain

$$\begin{aligned}
&\tilde{Q}(\mathbf{K}_\rho, +\gamma, \omega) e^{i\gamma z} + \tilde{Q}(\mathbf{K}_\rho, -\gamma, \omega) e^{-i\gamma z} \\
&\quad = 2i\gamma [\widetilde{U}_+(\mathbf{K}_\rho, +z_0) e^{i\gamma(z-z_0)} + \widetilde{U}_+(\mathbf{K}_\rho, -z_0) e^{-i\gamma(z-z_0)}],
\end{aligned} \tag{5.20}$$

which must hold over the homogeneous region $K_\rho \leq k$. As a final step we note that since the above equation has to hold $\forall z \in [-z_0, +z_0]$ it requires that

$$\tilde{Q}(\mathbf{K}_\rho, \pm\gamma, \omega) = 2i\gamma \widetilde{U}_+(\mathbf{K}_\rho, \pm z_0) e^{\mp i\gamma z_0}, \qquad K_\rho \leq k. \tag{5.21}$$

5.3.1 Solving for the source

We showed in Theorem 1.3 in Chapter 1 that any source to the wave equation can be decomposed into the sum of an NR component $q_{\mathrm{nr}}(\mathbf{r}, t)$ and a second component $\hat{q}(\mathbf{r}, t)$ such that these two components are orthogonal over $\{\mathcal{S}_0 | \mathbf{r} \in \tau_0,\ t \in [0, T_0]\}$ and $\hat{q}(\mathbf{r}, t)$ satisfies the homogeneous wave equation in $\mathcal{S}_0$. In the frequency domain these two conditions become

$$[\nabla^2 + k^2] \hat{Q}(\mathbf{r}, \omega) = 0, \qquad \langle Q_{\mathrm{nr}}, \hat{Q} \rangle_{L^2(\tau_0)} = 0, \tag{5.22}$$

where, for the problem under consideration, τ_0 is the infinite region lying between the two planes $z = \pm a_0$ and

$$\langle Q_{\rm nr}, \hat{Q}\rangle_{L^2(\tau_0)} = \int_{\tau_0} d^3r\, Q^*_{\rm nr}(\mathbf{r},\omega)\hat{Q}(\mathbf{r},\omega),$$

stands for the standard inner product in the Hilbert space $L^2(\tau_0)$ of square integral functions supported within τ_0. We have already shown that the NR component of a source cannot be determined from field data acquired outside the source's space-time support, so the best that we can expect to determine from such field data is the component $\hat{Q}(\mathbf{r},\omega)$. This solution to the ISP satisfies the homogeneous Helmholtz equation within τ_0 and also possesses a minimum-L^2 norm of all possible solutions $Q = \hat{Q} + Q_{\rm nr}$; i.e.,

$$||\hat{Q} + Q_{\rm nr}||^2 = ||\hat{Q}||^2 + ||Q_{\rm nr}||^2 \geq ||\hat{Q}||^2,$$

which establishes $\hat{Q}$ as the *minimum-norm* solution to the ISP. It is the solution to the ISP that we will seek throughout this chapter.

Our goal is thus to determine that particular 3D source supported in τ_0 that satisfies Eqs. (5.21) as well as the homogeneous Helmholtz equation within τ_0. If we perform a spatial Fourier transform of the homogeneous Helmholtz equation satisfied by $\hat{Q}$ over the (x, y) plane we obtain

$$\left[\frac{\partial^2}{\partial z^2} + \gamma^2\right]\overline{\hat{Q}}(\mathbf{K}_\rho, z, \omega) = 0, \quad -a_0 \leq z \leq +a_0, \tag{5.23}$$

where

$$\overline{\hat{Q}}(\mathbf{K}_\rho, z, \omega) = \int d^2\rho\, \hat{Q}(\mathbf{r},\omega)e^{-i\mathbf{K}_\rho\cdot\boldsymbol{\rho}}$$

is the 2D spatial Fourier transform of the minimum-norm source over the (x, y) plane. The most general solution to Eq. (5.23) is given by

$$\overline{\hat{Q}}(\mathbf{K}_\rho, z, \omega) = A_+(\mathbf{K}_\rho,\omega)e^{i\gamma z} + A_-(\mathbf{K}_\rho,\omega)e^{-i\gamma z}, \quad -a_0 \leq z \leq +a_0, \tag{5.24}$$

where $A_\pm(\mathbf{K}_\rho,\omega)$ are arbitrary functions of the transverse wavenumber $\mathbf{K}_\rho$ and frequency ω that vanish over the evanescent region $K_\rho > k$.

On taking the spatial Fourier transform of $\overline{\hat{Q}}(\mathbf{K}_\rho, z, \omega)$ defined in Eq. (5.24) w.r.t. z and substituting the result into Eq. (5.21) we obtain after some minor algebra the matrix equation

$$\begin{bmatrix} 1 & j_0(2\gamma a_0) \\ j_0(2\gamma a_0) & 1 \end{bmatrix}\begin{bmatrix} A_+(\mathbf{K}_\rho,\omega) \\ A_-(\mathbf{K}_\rho,\omega)\end{bmatrix} = \frac{i\gamma}{a_0}\begin{bmatrix} \widetilde{U}_+(\mathbf{K}_\rho, +z_0)e^{-i\gamma z_0} \\ \widetilde{U}_+(\mathbf{K}_\rho, -z_0)e^{+i\gamma z_0}\end{bmatrix},$$

whose solution is readily found to be

$$A_\pm(\mathbf{K}_\rho,\omega) = \frac{i\gamma/a_0}{1 - j_0^2(2\gamma a_0)}[\widetilde{U}_+(\mathbf{K}_\rho, \pm z_0)e^{\mp i\gamma z_0} - j_0(2\gamma a_0)\widetilde{U}_+(\mathbf{K}_\rho, \mp z_0)e^{\pm i\gamma z_0}],$$

where $j_0(\cdot)$ is the zeroth-order spherical Bessel function and, again, $\mathbf{K}_\rho$ lies in the homogeneous region $K_\rho \leq k$. On substituting this result into Eq. (5.24) we then find that

$$\bar{\hat{Q}}(\mathbf{K}_\rho, z, \omega) = \frac{i\gamma/a_0}{1 - j_0^2(2\gamma a_0)} \times \{[\widetilde{U}_+(\mathbf{K}_\rho, +z_0)e^{-i\gamma z_0} - j_0(2\gamma a_0)\widetilde{U}_+(\mathbf{K}_\rho, -z_0)e^{+i\gamma z_0}]e^{i\gamma z} + [\widetilde{U}_+(\mathbf{K}_\rho, -z_0)e^{+i\gamma z_0} - j_0(2\gamma a_0)\widetilde{U}_+(\mathbf{K}_\rho, +z_0)e^{-i\gamma z_0}]e^{-i\gamma z}\}, \quad (5.25a)$$

if $z \in [-a_0, +a_0]$ and zero outside this interval. After a bit of algebra the above expression can be expressed in the following simplified form that will lead to a direct comparison with the results obtained in the preceding section for the planar source:

$$\bar{\hat{Q}}(\mathbf{K}_\rho, z, \omega) = \frac{i\gamma/a_0}{1 + j_0(2\gamma a_0)}[\widetilde{U}_+(\mathbf{K}_\rho, +z_0)e^{-i\gamma z_0} + \widetilde{U}_+(\mathbf{K}_\rho, -z_0)e^{+i\gamma z_0}]\cos(\gamma z) - \frac{\gamma/a_0}{1 - j_0(2\gamma a_0)}[\widetilde{U}_+(\mathbf{K}_\rho, +z_0)e^{-i\gamma z_0} - \widetilde{U}_+(\mathbf{K}_\rho, -z_0)e^{+i\gamma z_0}]\sin(\gamma z), \quad (5.25b)$$

where, again, $-a_0 \leq z \leq +a_0$. As a final step we take an inverse spatial Fourier transform of $\bar{\hat{Q}}(\mathbf{K}_\rho, z, \omega)$ to obtain

$$\hat{Q}(\mathbf{r}, \omega) = \frac{1}{(2\pi)^2} \int_{K_\rho < k} d^2K_\rho \, e^{i\mathbf{K}_\rho \cdot \boldsymbol{\rho}} \times \left\{ \frac{i\gamma/a_0 \cos(\gamma z)}{1 + j_0(2\gamma a_0)}[\widetilde{U}_+(\mathbf{K}_\rho, +z_0)e^{-i\gamma z_0} + \widetilde{U}_+(\mathbf{K}_\rho, -z_0)e^{i\gamma z_0}] - \frac{\gamma/a_0 \sin(\gamma z)}{1 - j_0(2\gamma a_0)}[\widetilde{U}_+(\mathbf{K}_\rho, +z_0)e^{-i\gamma z_0} - \widetilde{U}_+(\mathbf{K}_\rho, -z_0)e^{i\gamma z_0}] \right\}. \quad (5.26)$$

5.3.2 Limiting form as a surface source

If we take the limit $a_0 \to 0$ in the source Eq. (5.26) we should recover the surface source derived in Section 5.2. To this end we note that for $-a_0 \leq z \leq +a_0$ we have that

$$\lim_{a_0 \to 0} \frac{\cos(\gamma z)}{a_0(1 + j_0(2\gamma a_0))} \to \delta(z), \qquad \lim_{a_0 \to 0} \frac{\gamma \sin(\gamma z)}{a_0(1 - j_0(2\gamma a_0))} \to -\delta'(z). \quad (5.27)$$

The above results are easily verified by using the definitions of the delta function and doublet; i.e.,

$$\int_{-\infty}^{\infty} g(z)\delta(z)dz = g(0), \qquad \int_{-\infty}^{\infty} g(z)\delta'(z)dz = -g'(0),$$

for any function $g(z)$ that is analytic in the neighborhood of the origin. On making use of Eqs. (5.27) we find that Eq. (5.26) yields

$$\lim_{a_0\to 0}\hat{Q}(\mathbf{r},\omega) =$$

$$\overbrace{\frac{1}{(2\pi)^2}\int_{K_\rho<k} d^2K_\rho\, e^{i\mathbf{K}_\rho\cdot\boldsymbol{\rho}} i\gamma[\widetilde{U_+}(\mathbf{K}_\rho,+z_0)e^{-i\gamma z_0} + \widetilde{U_+}(\mathbf{K}_\rho,-z_0)e^{i\gamma z_0}]\,\delta(z)}^{Q_s(\boldsymbol{\rho},\omega)}$$

$$+\overbrace{\frac{1}{(2\pi)^2}\int_{K_\rho<k} d^2K_\rho\, e^{i\mathbf{K}_\rho\cdot\boldsymbol{\rho}}[\widetilde{U_+}(\mathbf{K}_\rho,-z_0)e^{i\gamma z_0} - \widetilde{U_+}(\mathbf{K}_\rho,+z_0)e^{-i\gamma z_0}]\,\delta'(z)}^{Q_d(\boldsymbol{\rho},\omega)},$$

which is the solution we obtained for the surface source in Section 5.2.

5.3.3 Time-reversal imaging for slab geometry

In the limit where the wavelength is small relative to the width a_0 of the source and the field boundary-value data $U_+(\boldsymbol{\rho},\pm z_0)$ are effectively band-limited to spatial frequencies $\mathbf{K}_\rho$ much smaller than the wavenumber k we have that

$$\gamma = \sqrt{k^2 - K_\rho^2} \to k, \qquad j_0(2\gamma a_0) \to j_0(2ka_0) \to 0,$$

so that Eq. (5.25a) becomes

$$\lim_{ka_0\to\infty} \bar{\hat{Q}}(\mathbf{K}_\rho,z,\omega) = i\frac{k}{a_0}[\widetilde{U_+}(\mathbf{K}_\rho,+z_0)e^{-i\gamma z_0} + \widetilde{U_+}(\mathbf{K}_\rho,-z_0)e^{+i\gamma z_0}].$$

The source $\hat{Q}(\mathbf{r},\omega)$ then becomes

$$\lim_{ka_0\to\infty} \hat{Q}(\mathbf{r},\omega) \to \frac{1}{(2\pi)^2}\int_{K_\rho<k} d^2K_\rho\, e^{i\mathbf{K}_\rho\cdot\boldsymbol{\rho}}\bar{\hat{Q}}(\mathbf{K}_\rho,z,\omega) = i\frac{k}{a_0}\Phi(\mathbf{r},\omega). \tag{5.28}$$

The above development shows that in non-dispersive media $i(k/a_0)\Phi(\mathbf{r},\omega)$ is a good approximation to the minimum-norm solution to the ISP if the source is much larger than a wavelength in size and radiates a field having limited angular resolution. The Porter–Bojarski (PB) integral equation need not be solved and it is necessary only to compute the back-propagated field directly from over-determined boundary-value data via Eq. (5.3b) or Dirichlet or Neumann data using Eq. (5.17). The back-propagated field can also be computed in terms of Cauchy data in the case of non-dispersive media using Eq. (5.3d). The underlying reason for this is that under these conditions the source $Q(\mathbf{r},\omega)$ is a slowly varying function of position within the source volume relative to the wavelength so that the additional smoothing introduced by the kernel of the PB integral equation is of little consequence and, hence, a deconvolution is not required and the back-propagated field itself is, to within a multiplicative constant, a good estimate of the source.

5.4 The Hilbert-space formulation of the ISP

The "classical" treatment of the ISP based on the PB integral equation has several limitations that include it being generally limited to sources embedded in non-dispersive media (e.g., the wave equation) and the requirement of having a full data set in the form of over-determined data over a closed surface surrounding the source. In this section we will formulate the ISP directly within the frequency domain in a form that is valid for dispersive backgrounds as well as limited data sets. The reformulated problem also has the advantage that it is in a "standard form" that will transition easily and naturally into the *inverse scattering problem* (ISCP) that will be treated in later chapters and is much more important from an application viewpoint than the ISP. This "standard form" consists of a linear mapping $\hat{T} : \mathcal{H}_Q \to \mathcal{H}_f$ from a Hilbert space of source functions $\mathcal{H}_Q$ to a Hilbert space of data $\mathcal{H}_f$ that is very general and includes the ISP and ISCP as well as other inverse problems such as inverse diffraction, which will be considered in a later chapter. Because the two Hilbert spaces $\mathcal{H}_Q$ and $\mathcal{H}_f$ are generally different and the linear operator $\hat{T}$ is not generally Hermitian, it will be necessary to delve rather deeply into Hilbert-space theory and employ more powerful mathematical techniques than are used when basing the ISP on the PB integral equation. Most of the required Hilbert-space theory will be developed within this section, although it is recommended that the reader consult one or more of the excellent references that are listed at the end of the chapter.

The ISP is governed by an equation of the general form

$$\hat{T}Q = f, \tag{5.29}$$

where $f \in \mathcal{H}_f$ is the "data" contained in some Hilbert space $\mathcal{H}_f$ and $\hat{T} : \mathcal{H}_Q \to \mathcal{H}_f$ is a linear mapping that relates the source $Q \in \mathcal{H}_Q$ to the data $f \in \mathcal{H}_f$. An "ideal" data set would be Dirichlet or Neumann boundary values over any closed surface $\partial\tau$ surrounding the source support volume τ_0, which, of course, includes as a special case the radiation pattern $f(\mathbf{s}, \omega)$ specified over the surface of the real unit sphere. The general formulation of the ISP as defined via Eq. (5.29) applies also to cases of incomplete field data as well as to dispersive backgrounds where the wavenumber k is complex. What is required, of course, is a systematic method of inverting such mappings for the source in terms of the data.

One important form of the ISP is the "antenna-synthesis problem" in which the goal is to synthesize a 3D source that will radiate a specified radiation pattern. In this case $\mathcal{H}_Q = L^2(\tau_0)$ is the space of square-integrable functions supported in τ_0 and the data f are the radiation pattern of the source. Equation (5.29) reduces to

$$f(\mathbf{s}, \omega) = -\frac{1}{4\pi}\int_{\tau_0} d^3r\, Q(\mathbf{r}, \omega)e^{-ik\mathbf{s}\cdot\mathbf{r}}, \tag{5.30}$$

where $\mathbf{s}$ is a real or complex unit vector ($\mathbf{s} \cdot \mathbf{s} = 1$) and k is the wavenumber, which can be either real- or complex-valued. As mentioned in previous chapters, the radiation pattern for compactly supported sources is an analytic function of the unit vector $\mathbf{s}$ over the entire complex unit sphere ($\mathbf{s} \cdot \mathbf{s} = 1$) and, hence, can be uniquely determined via the process of analytic continuation from its specification over the real unit sphere. There is no loss in generality on then restricting the range of the mapping Eq. (5.30) to functions defined over the real unit sphere. In this case, then, $\mathcal{H}_f$ is the Hilbert space of square-integrable functions defined on the real unit sphere, which we denote by $L^2(\Omega)$. This Hilbert space has an inner product and induced norm defined by

$$\langle f_1, f_2 \rangle_{\mathcal{H}_f} = \int_{4\pi} d\Omega_s f_1^*(\mathbf{s}, \omega) f_2(\mathbf{s}, \omega), \qquad ||f||_{\mathcal{H}_f} = \sqrt{\langle f, f \rangle_{\mathcal{H}_f}}. \tag{5.31}$$

We will use the antenna-synthesis problem governed by the mapping defined in Eq. (5.30) with $\mathcal{H}_f = L^2(\Omega)$ as a prototype ISP throughout this section, with the understanding that most of our results and discussion can be generalized to apply to any form of the ISP governed by any linear mapping $\hat{T} : \mathcal{H}_Q \to \mathcal{H}_f$ between the Hilbert space of source functions $\mathcal{H}_Q$ and the Hilbert space of data $\mathcal{H}_f$.

We can write Eq. (5.30) in the form of Eq. (5.29), where $\mathcal{H}_f = L^2(\Omega)$ and $f \in L^2(\Omega)$ is a radiation pattern and $\hat{T} : \mathcal{H}_Q \to \mathcal{H}_f$ is the linear mapping (operator)[3] that is defined via the equation

$$\hat{T}Q = -\frac{1}{4\pi} \int_{\tau_0} d^3r\, e^{-ik\mathbf{s}\cdot\mathbf{r}} Q(\mathbf{r}, \omega) \tag{5.32}$$

for all functions $Q \in \mathcal{H}_Q$. We will sometimes express the operator $\hat{T}$ in the short-hand form

$$\hat{T} = -\frac{1}{4\pi} \int_{\tau_0} d^3r\, e^{-ik\mathbf{s}\cdot\mathbf{r}},$$

where it is understood that the domain of the operator is source functions in the Hilbert space $\mathcal{H}_Q = L^2(\tau_0)$.

Example 5.3 The operator $\hat{T}$ defined above is a *bounded* operator, by which we mean that

$$||\hat{T}Q||^2_{\mathcal{H}_f} < \infty, \quad \forall Q \in \mathcal{H}_Q. \tag{5.33}$$

On substituting the definition of $\hat{T}Q$ from Eq. (5.32) we find that

$$\begin{aligned} ||\hat{T}Q||^2_{\mathcal{H}_f} &= \left(\frac{1}{4\pi}\right)^2 \int d\Omega_s \left\{ \int_{\tau_0} d^3r\, e^{ik^*\mathbf{s}\cdot\mathbf{r}} Q^*(\mathbf{r}, \omega) \right\} \left\{ \int_{\tau_0} d^3r'\, e^{-ik\mathbf{s}\cdot\mathbf{r}'} Q(\mathbf{r}', \omega) \right\} \\ &= \left(\frac{1}{4\pi}\right)^2 \int_{\tau_0} d^3r \int_{\tau_0} d^3r'\, Q^*(\mathbf{r}, \omega) Q(\mathbf{r}', \omega) \int d\Omega_s\, e^{ik^*\mathbf{s}\cdot\mathbf{r}} e^{-ik\mathbf{s}\cdot\mathbf{r}'}. \end{aligned} \tag{5.34}$$

[3] Strictly speaking, an operator maps a Hilbert space into itself, so the linear mapping $\hat{T}$ is not an operator within this strict definition. However, it is common practice to stretch this definition to include mappings, such as $\hat{T}$, that map one Hilbert space into a different Hilbert space, and we will sometimes use the term "operator" according to this more general definition.

On setting $k = k_r + ik_i$ with $k_i \geq 0$ we find that

$$\left|\int d\Omega_s\, e^{ik^*\mathbf{s}\cdot\mathbf{r}} e^{-ik\mathbf{s}\cdot\mathbf{r}'}\right| = \left|\int d\Omega_s\, e^{k_i\mathbf{s}\cdot(\mathbf{r}+\mathbf{r}')} e^{ik_r\mathbf{s}\cdot(\mathbf{r}-\mathbf{r}')}\right| \leq e^{2k_i a_0}\left|\int d\Omega_s\, e^{ik_r\mathbf{s}\cdot(\mathbf{r}-\mathbf{r}')}\right|$$
$$= 4\pi e^{2k_i a_0}|j_0(k_r|\mathbf{r}-\mathbf{r}'|)|,$$

where a_0 is the radius of the smallest support sphere to τ_0 and we have used the plane-wave expansion of the spherical Bessel function j_0 given in Example 4.2 of Chapter 4. On substituting the above inequality into Eq. (5.34) we obtain

$$||\hat{T}Q||^2_{\mathcal{H}_f} \leq \frac{e^{2k_i a_0}}{4\pi}\int_{\tau_0} d^3r \int_{\tau_0} d^3r' |Q^*(\mathbf{r},\omega)Q(\mathbf{r}',\omega)||j_0(k_r|\mathbf{r}-\mathbf{r}'|)| < \frac{\tau_0 e^{2k_i a_0}}{4\pi}||Q||^2_{\mathcal{H}_Q}$$

$\forall Q \in \mathcal{H}_Q$, which establishes Eq. (5.33).

The fact that $\hat{T}$ maps source functions in $Q \in \mathcal{H}_Q$ into the Hilbert space $\mathcal{H}_f$ follows from the fact that, as we have just shown, $\hat{T}$ is a bounded operator so that $||\hat{T}Q||_{\mathcal{H}_f} < \infty$ for all functions $Q \in \mathcal{H}_Q$. However, this operator has much better properties than being bounded. It is also a *Hilbert–Schmidt* operator, from which it follows that it is also a *compact operator*.[4] Compact operators are the closest thing in infinite-dimensional Hilbert spaces to the finite-dimensional operators and matrices of standard linear algebra. In particular, they are defined by the property that they result from convergent Cauchy sequences of finite-dimensional operators. Almost all of the operators that we will deal with in this book are compact.

Example 5.4 It is not difficult to prove that the operator $\hat{T}$ defined in Eq. (5.32) is Hilbert–Schmidt and, hence, compact. In particular, the operator $\hat{T} : \mathcal{H}_Q \to \mathcal{H}_f$ is Hilbert–Schmidt if there exists a complete orthonormal sequence $e_n \in \mathcal{H}_Q$ such that

$$\sum_n ||\hat{T}e_n||^2_{\mathcal{H}_f} < \infty.$$

To prove that $\hat{T}$ is Hilbert–Schmidt we select any orthonormal sequence $e_n(\mathbf{r}) \in \mathcal{H}_Q$ so that

$$\hat{T}e_n = -\frac{1}{4\pi}\int_{\tau_0} d^3r\, e^{-ik\mathbf{s}\cdot\mathbf{r}} e_n(\mathbf{r}) = -\frac{1}{4\pi}\langle e^{ik^*\mathbf{s}\cdot\mathbf{r}}, e_n\rangle_{\mathcal{H}_Q}.$$

It then follows that

$$||\hat{T}e_n||^2_{\mathcal{H}_f} = \left(\frac{1}{4\pi}\right)^2 \int_{4\pi} d\Omega_s |\langle e^{ik^*\mathbf{s}\cdot\mathbf{r}}, e_n\rangle_{\mathcal{H}_Q}|^2,$$

[4] The term "completely continuous" is sometimes used in place of "compact" for such operators.

from which we find that

$$\sum_n ||\hat{T}e_n||^2_{\mathcal{H}_f} = \left(\frac{1}{4\pi}\right)^2 \sum_n \int_{4\pi} d\Omega_s |\langle e^{ik^*\mathbf{s}\cdot\mathbf{r}}, e_n\rangle_{\mathcal{H}_Q}|^2$$

$$= \left(\frac{1}{4\pi}\right)^2 \int_{4\pi} d\Omega_s \sum_n |\langle e^{ik^*\mathbf{s}\cdot\mathbf{r}}, e_n\rangle_{\mathcal{H}_Q}|^2 \leq \left(\frac{1}{4\pi}\right)^2 \int_{4\pi} d\Omega_s ||e^{-ik\mathbf{s}\cdot\mathbf{r}}||^2_{\mathcal{H}_Q}, \tag{5.35}$$

where we have used Bessel's inequality. As a final step we set $k = k_r + ik_i$ to find that

$$||e^{-ik\mathbf{s}\cdot\mathbf{r}}||^2_{\mathcal{H}_Q} = \left|\int_{\tau_0} d^3r\, e^{2k_i\mathbf{s}\cdot\mathbf{r}}\right| \leq e^{2k_i a_0}\tau_0,$$

where a_0 is the radius of the smallest sphere that entirely encloses the source volume τ_0. Using this result in Eq. (5.35), we then conclude that

$$\sum_n ||\hat{T}e_n||^2_{\mathcal{H}_f} \leq \frac{e^{2k_i a_0}\tau_0}{4\pi},$$

which establishes the desired result.

5.4.1 The adjoint operator

The *adjoint* mapping $\hat{T}^\dagger : \mathcal{H}_f \to \mathcal{H}_Q$ carries elements $f \in \mathcal{H}_f$ into the source Hilbert space $\mathcal{H}_Q$. The adjoint mapping is defined via the equation

$$\langle f, \hat{T}Q\rangle_{\mathcal{H}_f} = \langle \hat{T}^\dagger f, Q\rangle_{\mathcal{H}_Q},$$

which must hold for all $Q \in \mathcal{H}_Q$ and all $f \in \mathcal{H}_f$. By making use of the definitions of the standard inner products in the two spaces and the definition of $\hat{T}$ given in Eq. (5.32) we find that

$$\hat{T}^\dagger f = -\frac{1}{4\pi}\mathcal{M}_{\tau_0} \int_{4\pi} d\Omega\, e^{ik^*\mathbf{s}\cdot\mathbf{r}} f(\mathbf{s}, \omega), \tag{5.36}$$

where

$$\mathcal{M}_{\tau_0} = \begin{cases} 1 & \text{if } \mathbf{r} \in \tau_0 \\ 0 & \text{else} \end{cases}$$

is a so-called masking operator. The masking operator is required in order to guarantee that the adjoint maps arbitrary elements in $\mathcal{H}_f$ to elements in $\mathcal{H}_Q$, which must be compactly supported in the volume τ_0. As was the case with the operator $\hat{T}$, we will sometimes express the adjoint operator using the shorthand notation

$$\hat{T}^\dagger = -\frac{1}{4\pi}\mathcal{M}_{\tau_0} \int_{4\pi} d\Omega\, e^{ik^*\mathbf{s}\cdot\mathbf{r}},$$

where the domain of the operator is functions $f(\mathbf{s},\omega) \in \mathcal{H}_f$. It should be noted that in constructing the adjoint operator we have allowed for the possibility that the wavenumber k is complex. The ISP formulated using this operator is thus valid for sources embedded in dispersive as well as in non-dispersive media.

Example 5.5 We consider the 1D radiation problem

$$\left(\frac{\partial^2}{\partial z^2} + k^2\right) U_+(z,\omega) = Q(z,\omega),$$

where the source is assumed to be supported in some finite interval $L_a = [-a, a]$ about the origin of the z axis. The outgoing-wave solution to the above equation is given by

$$U_+(z,\omega) = -\frac{i}{2k}\int_{-a}^{a} dz'\, Q(z',\omega)e^{ik|z-z'|},$$

where we have used the 1D outgoing-wave Green function derived in Example 2.3 of Chapter 2. The far field is easily found to be

$$U_+(z,\omega) \sim -\frac{i}{2k}\int_{-a}^{a} dz'\, Q(z',\omega)e^{\mp ikz'}e^{\pm ikz}, \quad |z| \to \infty,$$

where the top sign applies if $z > 0$ and the bottom sign if $z < 0$. The radiation pattern is the coefficient to the term $\exp(\pm ikz)$ in the above equation and can be written in the form

$$f(s,\omega) = -\frac{i}{2k}\int_{-a}^{a} dz'\, Q(z',\omega)e^{-iksz'}, \tag{5.37}$$

where $f(s,\omega)$ with $s = -1$ is the radiation pattern along the left half-line $z < 0$ and $f(s,\omega)$ with $s = +1$ is the radiation pattern along the right half-line $z > 0$. The far-field ISP consists of inverting Eq. (5.37) for the source from the radiation pattern specified for both $z > 0$ and $z < 0$; i.e., from both $f(1,\omega)$ and $f(-1,\omega)$.

We can formulate the 1D ISP described above in a Hilbert-space framework. Thus, we introduce the Hilbert space $L^2(L_a)$ of square-integrable sources $Q(z,\omega)$ compactly supported on the line interval $L_a = [-a, a]$ with $l_a = 2a$ being the length of the interval. We also introduce the Hilbert space C^2 consisting of all pairs of complex numbers $f(s,\omega)$, $s = \pm 1$; i.e., $C^2 = \{f(-1,\omega), f(+1,\omega) | f(\pm 1,\omega) \in C\}$. The mapping from the space $L^2(L_a)$ to the space C^2 is governed by Eq. (5.37), so the operator $\hat{T} : L^2(L_a) \to C^2$ is

$$\hat{T} = -\frac{i}{2k}\int_{L_0} dz'\, e^{-iksz'},$$

which is the 1D version of the $\hat{T}$ operator defined in Eq. (5.32).

The inner products in the two spaces are given by

$$\langle Q_1, Q_2\rangle_{L^2(L_a)} = \int_{L_a} dz\, Q_1^*(z,\omega)Q_2(z,\omega),$$

$$\langle f_1, f_2\rangle_{C^2} = \sum_{s=\pm 1} f_1^*(s,\omega)f_2(s,\omega),$$

where the inner products defined above are the standard inner products in $L^2(L_a)$ and C^2, respectively.

The adjoint operator $\hat{T}^\dagger$ maps the space C^2 into the space $L^2(L_a)$ and is obtained using its definition

$$\langle f, \hat{T}Q\rangle_{C^2} = \langle \hat{T}^\dagger f, Q\rangle_{L^2(L_a)}.$$

On substituting the definition of $\hat{T}$ we find that

$$\langle f, \hat{T}Q\rangle_{C^2} = \sum_{s=\pm 1} f^*(s,\omega)\overbrace{\left\{-\frac{i}{2k}\int_{L_0} dz'\, e^{-iksz'}Q(z',\omega)\right\}}^{\hat{T}Q}$$

$$= \int_{L_0} dz' \left\{\frac{i}{2k^*}\sum_{s=\pm 1} f(s,\omega)e^{ik^*sz'}\right\}^* Q(z',\omega),$$

from which we conclude that

$$\hat{T}^\dagger = \frac{i}{2k^*}\mathcal{M}\sum_{s=\pm 1} e^{ik^*sz'},$$

where $\mathcal{M}$ is the masking operator defined by

$$\mathcal{M} = \begin{cases} 1 & \text{if } |z| < a, \\ 0 & \text{else.} \end{cases}$$

Back propagation and the adjoint operator

The adjoint operator $\hat{T}^\dagger$ maps the data space $\mathcal{H}_f$ back into the Hilbert space $\mathcal{H}_Q$ of source functions just as the back-propagation operation maps data into the back-propagated field $\Phi(\mathbf{r},\omega)$, which is related to the source via the PB integral equation. Indeed, for the special case of a non-dispersive medium and far-field data the back propagated field was found in Example 5.2 to be given by

$$\Phi(\mathbf{r},\omega) = \frac{ik}{2\pi}\int_{4\pi} d\Omega_{\mathbf{s}}\, f(\mathbf{s},\omega)e^{ik\mathbf{s}\cdot\mathbf{r}} = -2ik\hat{T}^\dagger f(\mathbf{r},\omega), \quad \mathbf{r}\in\tau_0.$$

We thus see that back propagation is simply an implementation of the adjoint operator and, as such, plays the important role of mapping the data space into back-propagated fields that are related to the source via an integral equation that then must be solved to recover the source. The great thing about our Hilbert-space formulation is that the adjoint operator and, hence, the back-propagation operation resulted directly from the formal mathematics of this formulation without any recourse to the physics of the underlying problem. This approach to the problem is basically a "turn the crank" approach that is guaranteed to yield the "best" solution to the problem in the fastest and most economical way. We will find later that the back-propagation process implemented in terms of the adjoint operator yields an easily computed approximate solution to a host of inverse problems related to the wave and Helmholtz equations that includes the ISP as well as the ISCP.

5.4.2 Singular value decomposition

In the case in which the source volume τ_0 is finite, the operator $\hat{T}$ and its adjoint $\hat{T}^\dagger$ are *compact*,[5] from which it follows that there exist a countable set of orthonormal basis functions $v_p \in \mathcal{H}_Q$ and a second, companion set, of orthonormal basis elements $u_p \in \mathcal{H}_f$ that are coupled via the set of equations

$$\hat{T}v_p = \sigma_p u_p, \qquad \hat{T}^\dagger u_p = \sigma_p v_p, \tag{5.38a}$$

where the σ_p are a discrete set of non-negative real numbers and $p = 0, 1, \ldots$ is an integer index that labels the singular set. The functions v_p and u_p are referred to as *singular functions*, the constants σ_p as *singular values* and the triplet $\{v_p, u_p, \sigma_p\}$ as the singular system associated with the operator $\hat{T}$. Since the two sets $\{v_p, u_p\}$ are complete in their respective Hilbert spaces we can represent the operators $\hat{T}$ and $\hat{T}^\dagger$ in the expansions

$$\hat{T} = \int_{\tau_0} d^3r \sum_p \sigma_p u_p(\mathbf{s}) v_p^*(\mathbf{r}), \qquad \hat{T}^\dagger = \int_{4\pi} d\Omega \sum_p \sigma_p v_p(\mathbf{r}) u_p^*(\mathbf{s}), \tag{5.38b}$$

which are called the "singular value decompositions" (SVDs) of the operators $\hat{T}$ and $\hat{T}^\dagger$. Note that the presence of the singular values σ_p in these expansions means that they involve only the singular vectors associated with non-zero singular values. We will not prove the existence of the SVD here but refer the interested reader to the excellent literature on this subject.

The singular vectors v_p and u_p each satisfy a separate set of equations known as the *normal equations*, which are readily derived from the defining equations Eqs. (5.38) of the SVD. In particular, we find that

$$\hat{T}^\dagger \hat{T} v_p = \sigma_p^2 v_p, \qquad \hat{T}\hat{T}^\dagger u_p = \sigma_p^2 u_p. \tag{5.39}$$

Both of the composite operators $\hat{T}^\dagger \hat{T}$ and $\hat{T}\hat{T}^\dagger$ are clearly Hermitian and compact (since $\hat{T}$ and $\hat{T}^\dagger$ are compact) and, because of this, it is well known that each operator possesses a complete set of orthonormal eigenvectors with a discrete spectrum. Thus, the singular set $\{v_p, u_p, \sigma_p\}$ can be computed by solving the two eigenvector equations Eqs. (5.39). In practice it is necessary only to solve for those singular functions associated with non-zero singular values $\sigma_p > 0$ and to solve one of the normal equations for either v_p or u_p and then use the defining equations Eqs. (5.38a) to compute its partner.

[5] That $\hat{T}$ is compact has been established in Example 5.4. That its adjoint $\hat{T}^\dagger$ is also compact is easily established using an almost identical treatment.

Example 5.6 For our prototype operator $\hat{T}$ defined in Eq. (5.32) we find that the composite operator $\hat{T}^\dagger\hat{T}$ is given by

$$\hat{T}^\dagger\hat{T} = \left(\frac{1}{4\pi}\right)^2 \mathcal{M}_{\tau_0} \int_{4\pi} d\Omega\, e^{ik^*\mathbf{s}\cdot\mathbf{r}} \int_{\tau_0} d^3r'\, e^{-ik\mathbf{s}\cdot\mathbf{r}'},$$

so

$$\hat{T}^\dagger\hat{T}v_p = \left(\frac{1}{4\pi}\right)^2 \mathcal{M}_{\tau_0} \int_{4\pi} d\Omega\, e^{ik^*\mathbf{s}\cdot\mathbf{r}} \langle e^{ik^*\mathbf{s}\cdot\mathbf{r}}, v_p\rangle_{\mathcal{H}_Q} = \sigma_p^2 v_p.$$

It then follows immediately that if $\sigma_p > 0$ then

$$[\nabla^2 + k^{*2}]v_p = 0, \quad \mathbf{r} \in \tau_0;$$

i.e., the singular vectors v_p of the ISP operator $\hat{T}$ that have non-zero singular values satisfy the homogeneous Helmholtz equation with wavenumber k^* within the source volume τ_0.

Example 5.7 A simple yet important linear transformation is the 1D Fourier transform of a time- or space-limited function. In terms of space-dependent functions this transform takes the form

$$\hat{T}g(K) = \frac{1}{\sqrt{2\pi}} \int_{-a_0}^{a_0} dx\, g(x)e^{-iKx}, \quad -\infty < K < +\infty. \tag{5.40a}$$

Using the standard inner product in the spaces $L^2([-a_0, +a_0])$ and $L^2(-\infty, \infty)$, we find that

$$\hat{T}^\dagger = \mathcal{M}_{a_0} \frac{1}{\sqrt{2\pi}} \int_{-\infty}^{\infty} dK\, e^{iKx}, \tag{5.40b}$$

where

$$\mathcal{M}_{a_0} = \begin{cases} 1 & -a_0 \le x \le a_0, \\ 0 & \text{else.} \end{cases}$$

The normal equations for the singular functions $\{v_p(x), u_p(K)\}$ are found to be

$$\hat{T}^\dagger\hat{T}v_p = \mathcal{M}_{a_0} \frac{1}{2\pi} \int_{-\infty}^{\infty} dK\, e^{iKx} \int_{-a_0}^{a_0} dx'\, e^{-iKx'} v_p(x') = \sigma_p^2 v_p(x),$$

$$\hat{T}\hat{T}^\dagger u_p = \frac{1}{2\pi} \int_{-a_0}^{a_0} dx\, e^{-iKx} \int_{-\infty}^{\infty} dK'\, e^{iK'x} u_p(K') = \sigma_p^2 u_p(K).$$

It is not difficult to verify (see the problems at the end of the chapter) that the singular system satisfying the above normal equations is given by

$$v_p(x) = \mathcal{M}_{a_0} \frac{e^{i\frac{\pi}{a_0}px}}{\sqrt{2a_0}}, \qquad u_p(K) = \frac{\text{sinc}[(a_0/\pi)(K - (\pi/a_0)p)]}{\sqrt{\pi/a_0}}, \qquad \sigma_p = 1,$$

where

$$\text{sinc}\, x = \frac{\sin(\pi x)}{\pi x}$$

is the "sinc function." The set of orthonormal singular functions v_p, $p = -\infty, \ldots, +\infty$, comprises the classical basis for Fourier-series expansions of square-integrable functions within the interval $[-a_0, a_0]$, while the orthonormal singular functions u_p, $p = -\infty, \ldots, +\infty$, form the basis for the classical Whittaker–Shannon sinc-function expansion of band-limited functions.

Example 5.8 The linear mapping $\hat{T}$ considered in the above example maps the Hilbert space of square-integrable functions over the finite interval $-a_0 \leq x \leq +a_0$ to square-integrable functions defined over the entire line $-\infty < K < +\infty$. A related mapping is the Fourier transform of time- or space-limited functions into frequency-limited functions. Using the same notation as in the previous example, this mapping is defined as

$$\hat{T}g(K) = \mathcal{M}_{K_0} \frac{1}{\sqrt{2\pi}} \int_{-a_0}^{a_0} dx\, g(x) e^{-iKx}, \tag{5.41}$$

where

$$\mathcal{M}_{K_0} = \begin{cases} 1 & -K_0 \leq K \leq K_0, \\ 0 & \text{else.} \end{cases}$$

It is thus similar to the mapping defined in Eq. (5.40a) of the previous example, *with the important difference* that the Fourier transform is now frequency-limited to the finite band pass $-K_0 \leq K \leq +K_0$. This mapping is extremely important in a host of applications ranging from communication theory to inverse problems in wave propagation and scattering. It was treated in the famous paper by Slepian and Pollak (Slepian and Pollak, 1961) who showed that the SVD of this mapping is given by the *angular prolate spheroidal wavefunctions* which satisfy the equation

$$\kappa_n \overbrace{S_{0,n}(c, \omega)}^{\widetilde{S_{0,n}(c, c\omega)}} = \int_{-1}^{1} d\xi\, S_{0,n}(c, \xi) e^{ic\omega\xi}, \quad -1 < \omega < +1, \tag{5.42}$$

where c is a constant parameter and

$$\kappa_n = \frac{2i^n}{\sqrt{2\pi}} R_{on}^{(1)}(c, 1),$$

with $R_{on}^{(1)}(c, 1)$ being the radial prolate spheroidal wavefunctions. Moreover, and most importantly, these functions are orthogonal over the intervals $-1 < \xi < +1$ and $-1 < \omega < +1$ with norm

$$||S_{0,n}||^2 = \int_{-1}^{+1} d\xi\, |S_{0,n}(c, \xi)|^2.$$

Starting from Eq. (5.42) and setting $c = K_0 a_0$ it is not difficult (see the problems at the end of the chapter) to show that the SVD of the mapping Eq. (5.41) is given by

$$v_p(x) = \mathcal{M}_{a_0} \frac{i^{-p/2} S_{0p}(c, x/a_0)}{\sqrt{a_0}||S_{0p}||}, \qquad u_p(K) = \mathcal{M}_{K_0} \frac{i^{p/2} S_{0p}(c, K/K_0)}{\sqrt{K_0}||S_{0p}||},$$
$$\sigma_p = \sqrt{K_0 a_0} |\kappa_p|.$$

The singular functions $\{v_p(x)\}$ form an orthonormal basis in $L^2(-a_0, +a_0)$, while the singular functions $\{u_p(K)\}$ form an orthonormal basis in $L^2(-K_0, +K_0)$. These two sets then can be used to (exactly) invert the truncated 1D Fourier transform of a space-limited function.

5.4.3 The range and null space of $\hat{T}$

We will require a few definitions for our subsequent development of the solution to the ISP. The first of these is of the *range* $\mathcal{R}(\hat{T})$ of a linear mapping $\hat{T}$, which is simply the set of all data f that result from applying the operator $\hat{T}$ to a $Q \in \mathcal{H}_Q$. If we make use of the SVD of $\hat{T}$ given in Eq. (5.38b) we find that

$$\hat{T}Q = \sum_p \sigma_p \langle v_p, Q \rangle_{\mathcal{H}_Q} u_p(\mathbf{s}), \tag{5.43}$$

from which we conclude that $\mathcal{R}(\hat{T})$ is spanned by the set of singular functions $u_p(\mathbf{s})$ having non-zero singular values $\sigma_p > 0$. The range, however, is not actually itself a Hilbert space in that not every Cauchy convergent series formed from this basis is an image of a source $Q \in \mathcal{H}_Q$, which is another way of saying that not all Cauchy sequences formed from elements $f \in \mathcal{R}(\hat{T})$ are themselves in $\mathcal{R}(\hat{T})$. The closure of the range, denoted by $\overline{\mathcal{R}(\hat{T})}$, is obtained by adding the limits of all Cauchy sequences to $\mathcal{R}(\hat{T})$ and is, thus, a proper Hilbert space.

Example 5.9 The closure of the range $\overline{\mathcal{R}(\hat{T})}$ of an operator $\hat{T}$ has the orthonormal basis $\{u_p(\mathbf{s}), \sigma_p > 0\}$ so that any element in this subspace can be expanded into the series

$$f = \sum_{\sigma_p > 0} \langle u_p, f \rangle_{\mathcal{H}_f} u_p(\mathbf{s}),$$

with

$$||f||^2_{\mathcal{H}_f} = \sum_{\sigma_p > 0} |\langle u_p, f \rangle_{\mathcal{H}_f}|^2 < \infty. \tag{5.44}$$

If f is also in the range $\mathcal{R}(\hat{T})$ of $\hat{T}$ it can be expanded into the series Eq. (5.43) so that

$$\langle u_p, f \rangle_{\mathcal{H}_f} = \sigma_p \langle v_p, Q \rangle_{\mathcal{H}_Q} \to \langle v_p, Q \rangle_{\mathcal{H}_Q} = \frac{\langle u_p, f \rangle_{\mathcal{H}_f}}{\sigma_p}, \quad \sigma_p > 0,$$

for some $Q \in \mathcal{H}_Q$. It then follows that

$$||Q||^2_{\mathcal{H}_Q} = \sum_p |\langle v_p, Q\rangle_{\mathcal{H}_Q}|^2 = \sum_{\sigma_p>0} \left|\frac{\langle u_p, f\rangle_{\mathcal{H}_f}}{\sigma_p}\right|^2 < \infty, \tag{5.45}$$

which is the *Picard condition* required of the expansion coefficients $\langle u_p, f\rangle_{\mathcal{H}_f}$ for a data function $f \in \mathcal{R}(\hat{T})$. It should be clear that a function f can be in the closure of $\mathcal{R}(\hat{T})$ and satisfy Eq. (5.44) but not satisfy the Picard condition Eq. (5.45) and, hence, not be in the range $\mathcal{R}(\hat{T})$.

A second definition that we will require is that of the null space $\eta(\hat{T})$ of a linear mapping $\hat{T}$. The null space is the set of all $Q \in \mathcal{H}_Q$ such that $\hat{T}Q = 0$. It follows from Eq. (5.43) that the null space is spanned by all singular vectors v_p having zero singular values $\sigma_p = 0$. For the ISP this space is the space of NR sources $Q_{\rm nr}$ that generate zero field outside their support volume. The null space is a proper Hilbert subspace of $\mathcal{H}_Q$ in that the limits of all Cauchy sequences within this space remain within this space.[6]

It is not difficult to show using the results of Example 5.6 that the NR sources lie in the null space of our prototype operator $\hat{T}$. To show this, we make use of the fact that the orthogonal complement $\eta(\hat{T})^\perp$ of the null space is spanned by the singular vectors v_p having non-zero singular values and, as shown in Example 5.6, these singular vectors satisfy the homogeneous Helmholtz equation with wavenumber k^*. It then follows that if v_p has non-zero singular value $\sigma_p > 0$ and hence lies in $\eta(\hat{T})^\perp$ then

$$\langle Q_{\rm nr}, v_p\rangle_{\mathcal{H}_Q} = \langle \overbrace{[\nabla^2 + k^2]\Pi}^{Q_{nr}}, v_p\rangle_{\mathcal{H}_Q} = \langle \Pi, [\nabla^2 + k^{*2}]v_p\rangle_{\mathcal{H}_Q} = 0.$$

This then establishes that $Q_{\rm nr} \in \eta(\hat{T})$ since the singular vectors v_p, $\sigma_p \neq 0$ form a basis in $\eta(\hat{T})^\perp$. In fact the NR sources span the null space since any solution of $\hat{T}v_p = 0$ must be a source that radiates a zero field outside of τ_0 and, hence, must be NR.

5.4.4 The least-squares pseudo-inverse

The ISP operator $\hat{T}$ possesses a null space $\eta(\hat{T})$ formed of all NR sources, so the ISP does not possess a unique solution. Moreover, generally the data will be noisy or the field measurements non-ideal so that the data need not be in the range $\mathcal{R}(\hat{T})$ of $\hat{T}$ and, hence, Eq. (5.29) might not even possess any solution! In this section we will address both of these issues by computing the so-called *least-squares pseudo-inverse* $\hat{Q} \in \mathcal{H}_Q$ of the mapping Eq. (5.29), and we will employ that as our fundamental solution to the ISP. The least-squares pseudo-inverse has the following properties.

[6] This is reminiscent of the often-heard phrase "Everything that takes place in Vegas stays in Vegas."

1. $\hat{Q}$ minimizes the squared error

$$\mathcal{E}_f = ||\hat{T}Q - f||^2_{\mathcal{H}_f} \tag{5.46a}$$

among all sources $Q \in \mathcal{H}_Q$.
2. $\hat{Q}$ has minimum norm among all sources that minimize $\mathcal{E}_f$; i.e.,

$$||\hat{Q}||^2_{\mathcal{H}_Q} = \int_{\tau_0} d^3r |\hat{Q}(\mathbf{r},\omega)|^2 \tag{5.46b}$$

is minimum among all Q that minimize the squared error defined in Eq. (5.46a).

Least-squares solution

We first address the possibility that the data $f \in \mathcal{H}_f$ might not be in the range of the ISP operator $\hat{T}$. This is addressed by noting that

$$\hat{T}^\dagger f = \sum_p \sigma_p \langle u_p, f \rangle_{\mathcal{H}_f} v_p(\mathbf{r})$$

and thus involves only that part of f which lies in the closure $\overline{\mathcal{R}(\hat{T})}$ of $\mathcal{R}(\hat{T})$. If we then apply the adjoint operator $\hat{T}^\dagger$ to both sides of the ISP mapping defined in Eq. (5.29) we annihilate that component of f that lies outside $\overline{\mathcal{R}(\hat{T})}$ and obtain the *normal equations*

$$\hat{T}^\dagger \hat{T} Q = \hat{T}^\dagger f. \tag{5.47}$$

If the data are in the range of the operator $\hat{T}$ ($f \in \mathcal{R}(\hat{T})$) then Eqs. (5.29) and (5.47) are equivalent and will possess the same solution set. If, on the other hand, $f \notin \mathcal{R}(\hat{T})$ then the original equations Eq. (5.29) will not possess a solution but the normal equations Eq. (5.47) *may* possess a solution set that is the least-squares solution set of Eq. (5.29); i.e., will be such that any solution in this set will minimize the squared error $\mathcal{E}_f$ defined in Eq. (5.46a). Whether or not a least-squares solution exists depends on whether or not the projection of the data into $\overline{\mathcal{R}(\hat{T})}$ satisfies the Picard condition and, hence, is in the range of $\hat{T}$. The normal equations Eq. (5.47) will have a solution set that is the least-squares solution set of Eq. (5.29) if and only if the projection of the data into $\overline{\mathcal{R}(\hat{T})}$ satisfies the Picard condition or, equivalently, if the data themselves satisfy this condition.

The pseudo-inverse

A second problem arises in connection with the normal equations Eq. (5.47) even if the data satisfy the Picard condition in that the null space $\eta(\hat{T})$ of $\hat{T}$ is not empty and includes, in particular, all NR sources in $\mathcal{H}_Q$. This is apparent from Eq. (5.43), which indicates that the projection of the source Q onto the singular vectors v_p having zero singular value $\sigma_p = 0$ will vanish so that such source components are NR and cannot be determined from the data. The composite operator $\hat{T}^\dagger\hat{T}$ possesses the same null space, so the normal equations Eq. (5.47) will also not possess a unique solution. Although they do not possess a *unique* solution, they can, nevertheless, be solved (if the data satisfy the Picard condition), and the so-called *pseudo-inverse* is that *particular* solution that results from projecting the source onto the subspace spanned by all singular vectors v_p having non-zero singular values $\sigma_p \neq 0$. This subspace is the orthogonal complement of the null space and is denoted by $\eta(\hat{T})^\perp$, so this solution, which we will denote by $\hat{Q}$, is then given by

$$\hat{Q}(\mathbf{r},\omega) = \sum_{\sigma_p>0} \langle v_p, Q\rangle_{\mathcal{H}_Q} v_p(\mathbf{r}).$$

If the data $f \in \mathcal{R}(\hat{T})$ then the pseudo-inverse is the projection of the source into $\eta(\hat{T})^\perp$, whereas if the data $f \notin \mathcal{R}(\hat{T})$ but satisfy the Picard condition then $\hat{Q}$ is the projection of the least-squares solution to the ISP onto this subspace. In either case $\hat{Q}$ has no components in the null space $\eta(\hat{T})$ of the operator $\hat{T}$ and, for this reason, we will often refer to $\hat{Q}$ as the *minimum-norm* solution to the ISP.

If we make use of Eqs. (5.38b) we find that

$$\hat{T}^\dagger\hat{T} = \int_{\tau_0} d^3r' \sum_p \sigma_p^2 v_p(\mathbf{r}) v_p^*(\mathbf{r}').$$

The pseudo-inverse $[\hat{T}^\dagger\hat{T}]^+$ of this operator must generate the projection operator onto $\eta(\hat{T})^\perp$, from which we conclude that

$$[\hat{T}^\dagger\hat{T}]^+ = \int_{\tau_0} d^3r' \sum_{\sigma_p>0} \frac{1}{\sigma_p^2} v_p(\mathbf{r}) v_p^*(\mathbf{r}').$$

Note that $[\hat{T}^\dagger\hat{T}]^+$ (with a plus sign $+$, not a dagger $\dagger$) denotes the pseudo-inverse of the composite operator $[\hat{T}^\dagger\hat{T}]$, *not* its adjoint. Applying the pseudo-inverse to both sides of Eq. (5.47) then yields the result

$$\hat{Q}(\mathbf{r},\omega) = [\hat{T}^\dagger\hat{T}]^+\hat{T}^\dagger f. \tag{5.48a}$$

As a final step we can represent the pseudo-inverse in Eq. (5.48a) in terms of the singular system $\{v_p, u_p, \sigma_p\}$ by using the expansions of $[\hat{T}^\dagger\hat{T}]^+$ and $\hat{T}^\dagger$ into that system. We find that

$$\hat{Q}(\mathbf{r},\omega) = \overbrace{\int_{\tau_0} d^3r' \sum_{\sigma_p>0} \frac{1}{\sigma_p^2} v_p(\mathbf{r}) v_p^*(\mathbf{r}')}^{[\hat{T}^\dagger\hat{T}]^+} \overbrace{\int_{4\pi} d\Omega \sum_{p'} \sigma_{p'} v_{p'}(\mathbf{r}') u_{p'}^*(\mathbf{s}) f(\mathbf{s},\omega)}^{\hat{T}^\dagger},$$

which reduces to

$$\hat{Q}(\mathbf{r},\omega) = \sum_{\sigma_p>0} \frac{\langle u_p, f\rangle_{\mathcal{H}_f}}{\sigma_p} v_p. \tag{5.48b}$$

5.4.5 Filtered back propagation and back-propagation imaging

The operator expression for the pseudo-inverse solution of the ISP given in Eq. (5.48a) can be expressed in an alternative form that is useful in a number of applications and, indeed, forms the basis for so-called *filtered* back projection and back-propagation algorithms in computed tomography (CT) and diffraction tomography (DT). In particular, by making use of the definitions of $\hat{T}$ and $\hat{T}^\dagger$ and following a similar set of steps to those used above, we find that

$$\hat{T}\hat{T}^\dagger = \int_{4\pi} d\Omega_{s'} \sum_p \sigma_p^2 u_p(\mathbf{s}) u_p^*(\mathbf{s}'),$$

from which it follows that the pseudo-inverse is given by

$$[\hat{T}\hat{T}^\dagger]^+ = \int_{4\pi} d\Omega_{s'} \sum_{\sigma_p>0} \frac{1}{\sigma_p^2} u_p(\mathbf{s}) u_p^*(\mathbf{s}').$$

We then conclude that

$$\hat{T}^\dagger[\hat{T}\hat{T}^\dagger]^+ = \int_{4\pi} d\Omega_{s'} \sum_{\sigma_p>0} \frac{1}{\sigma_p} v_p(\mathbf{r}) u_p^*(\mathbf{s}'),$$

which, upon comparison with Eqs. (5.48b), yields the following alternative expression for the pseudo-inverse solution to the ISP:

$$\hat{Q}(\mathbf{r},\omega) = \hat{T}^\dagger[\hat{T}\hat{T}^\dagger]^+ f. \tag{5.49}$$

The expression for the pseudo-inverse given in Eq. (5.49) can be interpreted as a two-step procedure whereby in the first step the data f are *filtered* in the Hilbert space $\mathcal{H}_f$ using the filter $[\hat{T}\hat{T}^\dagger]^+ : \mathcal{H}_f \to \mathcal{H}_f$ and then *back projected (propagated)* from $\mathcal{H}_f$ onto $\mathcal{H}_Q$ using the adjoint operator $\hat{T}^\dagger : \mathcal{H}_f \to \mathcal{H}_Q$. In contrast, the expression for the pseudo-inverse given in Eq. (5.48a) performs back projection from $\mathcal{H}_f$ onto $\mathcal{H}_Q$ followed by filtering in the space $\mathcal{H}_Q$ by the filter $[\hat{T}^\dagger\hat{T}]^+ : \mathcal{H}_Q \to \mathcal{H}_Q$. The importance of this difference is that filtering in the space $\mathcal{H}_f$ is much more efficient than filtering in the space $\mathcal{H}_Q$ and this difference can be important in a number of applications (cf. the discussion in Section 8.1.1 of Chapter 8).

It is interesting to note that the final expression Eq. (5.48b) for the pseudo-inverse expressed in terms of the SVD is automatically in the form of the filtering of the data, expressed by

$$[\hat{T}\hat{T}^\dagger]^+ f = \frac{\langle u_p, f\rangle_{\mathcal{H}_f}}{\sigma_p} \tag{5.50}$$

followed by back propagation expressed by the summation over singular functions $v_p(\mathbf{r})$ with expansion coefficients equal to the filtered data. As mentioned earlier, this is another advantage of the SVD-based solution to the ISP and other inverse problems: it automatically yields the most economical form of the solution to the problem. We will find this to be especially important in the (linearized) inverse scattering problem where the SVD generates the filtered back-propagation algorithm of diffraction tomography.

Back-propagation imaging

In many cases of interest the singular values σ_p are approximately constant up to some cutoff value P after which they decay exponentially fast to zero. In such cases we can basically ignore the inverse filtering step as given in Eq. (5.50) and approximate the filtered back-propagation algorithm via

$$\hat{Q}(\mathbf{r}, \omega) \approx \kappa \hat{T}^\dagger f, \tag{5.51}$$

where κ is a constant. Using the expression for the adjoint operator given in Eq. (5.38b), we find that this approximate solution can be expressed in the form

$$\hat{Q}(\mathbf{r}, \omega) \approx \sum_p \sigma_p \langle u_p, f\rangle_{\mathcal{H}_f} v_p,$$

which differs from the exact solution given in Eq. (5.48b) by having the singular values in the numerator rather than in the denominator. The above approximation will thus be accurate within a constant multiplier so long as the singular values behave as described above. This procedure of forming an approximate solution to the ISP by basically ignoring the inverse filtering of the data required in the exact inversion and simply back propagating the data via $\hat{T}^\dagger$ will be referred to as *back-propagation imaging*[7] and, as one might expect, plays a large role in certain inverse scattering applications.

5.5 The antenna-synthesis problem

In this section we will employ the Hilbert-space framework and the SVD to solve the ISP for the case of far-field data in the form of the radiation pattern $f(\mathbf{s}, \omega)$ specified for all observation directions $\mathbf{s}$ and a single frequency ω. This problem is one form of the

[7] Back-propagation imaging is also sometimes referred to as *adjoint imaging*.

antenna-synthesis problem, where the goal is to determine a 3D source ("antenna") that radiates a specified radiation pattern. We will solve the problem for the special case of sources confined to spherical regions where $\tau_0 = \{\mathbf{r}|r \leq a_0\}$, with a_0 being the radius of the source sphere assumed centered at the origin. The problem can be solved for other source geometries and data sets using the same general Hilbert-space formulation, but here we will restrict our attention to the classic problem of determining a source confined to a sphere from specification of its radiation pattern.

The ISP in terms of far-field data is governed by our prototype mapping $\hat{T}Q = f$ with $\hat{T}$ defined in Eqs. (5.32). We obtained the minimum-norm inverse of this mapping in Section 5.4.4, where it was found to be

$$\hat{Q} = [\hat{T}^\dagger \hat{T}]^+ \hat{T}^\dagger f = \sum_{\sigma_p > 0} \frac{\langle u_p, f \rangle_{\mathcal{H}_f}}{\sigma_p} v_p(\mathbf{r}),$$

where $[\hat{T}^\dagger \hat{T}]^+$ is the so-called pseudo-inverse of the operator $\hat{T}^\dagger \hat{T}$ and is given in terms of the singular system $\{v_p, u_p, \sigma_p\}$ in Eq. (5.48b). As discussed in that section, $\hat{Q}$ is the minimum-norm solution to the ISP if the data f are in the range $\mathcal{R}(\hat{T})$ of $\hat{T}$ and will be the minimum-norm least-squares solution to the ISP if $f \notin \mathcal{R}(\hat{T})$ but satisfies the Picard condition

$$\sum_{\sigma_p > 0} \left| \frac{\langle u_p, f \rangle_{\mathcal{H}_f}}{\sigma_p} \right|^2 < \infty.$$

5.5.1 Implementation of the SVD

In order to compute $\hat{Q}$ it is necessary to compute the singular system of the operator $\hat{T}$ for the specific case of a spherical source support volume. Because τ_0 is the interior of a sphere and the data are specified over the surface of the unit sphere it is natural to employ a spherical coordinate system in implementing the solution given in general form in Eq. (5.48b). This can be accomplished by using the multipole expansion of the plane wave $\exp(ik\mathbf{s} \cdot \mathbf{r})$ given in Example 3.4 of Chapter 3 to express $\hat{T}$ and its adjoint $\hat{T}^\dagger$ via the equations

$$\begin{aligned}\hat{T} &= -\frac{1}{4\pi} \int_{\tau_0} d^3r \, e^{-ik\mathbf{s}\cdot\mathbf{r}} \\ &= -\sum_{l=0}^{\infty} \sum_{m=-l}^{l} (-i)^l Y_l^m(\mathbf{s}) \int_{r' \leq a_0} d^3r \, j_l(kr) Y_l^{m*}(\hat{\mathbf{r}})\end{aligned} \tag{5.52a}$$

and

$$\begin{aligned}\hat{T}^\dagger &= -\frac{1}{4\pi} \mathcal{M}_{\tau_0} \int_{4\pi} d\Omega \, e^{ik^*\mathbf{s}\cdot\mathbf{r}} \\ &= -\mathcal{M}_{\tau_0} \sum_{l=0}^{\infty} \sum_{m=-l}^{l} i^l j_l(k^* r) Y_l^m(\hat{\mathbf{r}}) \int_{4\pi} d\Omega_s \, Y_l^{m*}(\mathbf{s}),\end{aligned} \tag{5.52b}$$

where $\hat{\mathbf{r}}$ is the unit vector along the $\mathbf{r}$ direction having polar angle θ and azimuthal angle ϕ, we have taken the source volume τ_0 to be a sphere of radius a_0 centered at the origin and

$$\mathcal{M}_{\tau_0} = \begin{cases} 1 & \text{if } r \leq a_0, \\ 0 & \text{else.} \end{cases} \tag{5.52c}$$

We have that $j_l(k^* r) = j_l^*(kr)$, so we can also express the adjoint operator in the alternative form

$$\hat{T}^\dagger = -\mathcal{M}_{\tau_0} \sum_{l=0}^{\infty} \sum_{m=-l}^{l} i^l j_l^*(kr) Y_l^m(\hat{\mathbf{r}}) \int_{4\pi} d\Omega_s \, Y_l^{m*}(\mathbf{s}), \tag{5.52d}$$

which we will employ in the following development.

The singular system $\{v_p, u_p, \sigma_p\}$ is computed using the normal equations Eqs. (5.39) and, hence, we have need of the composite operators $\hat{T}^\dagger \hat{T} : \mathcal{H}_Q \to \mathcal{H}_Q$ and $\hat{T}\hat{T}^\dagger : \mathcal{H}_f \to \mathcal{H}_f$. These composite operators can be constructed directly from Eqs. (5.52) and we find that

$$\hat{T}^\dagger \hat{T} = \mathcal{M}_{\tau_0} \sum_{l=0}^{\infty} \sum_{m=-l}^{l} j_l^*(kr) Y_l^m(\hat{\mathbf{r}}) \int_{r' \leq a_0} d^3 r' \, j_l(kr') Y_l^{m*}(\hat{\mathbf{r}}'), \tag{5.53a}$$

$$\hat{T}\hat{T}^\dagger = \sum_{l=0}^{\infty} \sum_{m=-l}^{l} \mu_l^2(ka_0) Y_l^m(\mathbf{s}) \int d\Omega_{s'} \, Y_l^{m*}(\mathbf{s}'), \tag{5.53b}$$

where

$$\mu_l^2(ka_0) = \int_0^{a_0} r^2 \, dr |j_l(kr)|^2 \tag{5.54}$$

was first encountered in connection with essentially NR sources in Section 1.7.3 of Chapter 1.

The singular functions v_p and u_p satisfy the normal equations Eqs. (5.39), which, upon making use of Eqs. (5.53), can be expressed in the form

$$\mathcal{M}_{\tau_0} \sum_{l=0}^{\infty} \sum_{m=-l}^{l} \langle j_l^* Y_l^m, v_p \rangle_{\mathcal{H}_Q} j_l^*(kr) Y_l^m(\hat{\mathbf{r}}) = \sigma_p^2 v_p(\mathbf{r}, \omega), \tag{5.55a}$$

$$\sum_{l=0}^{\infty} \sum_{m=-l}^{l} \mu_l^2 \langle Y_l^m, u_p \rangle_{\mathcal{H}_f} Y_l^m(\mathbf{s}) = \sigma_p^2 u_p(\mathbf{s}, \omega), \tag{5.55b}$$

where

$$\langle j_l^* Y_l^m, v_p \rangle_{\mathcal{H}_Q} = \int_{r' \leq a_0} d^3 r' \, j_l(kr') Y_l^{m*}(\hat{\mathbf{r}}') v_p(\mathbf{r}', \omega),$$

$$\langle Y_l^m, u_p \rangle_{\mathcal{H}_f} = \int d\Omega_s' \, Y_l^{m*}(\mathbf{s}) u_p(\mathbf{s}, \omega).$$

Equations (5.55) constitute the normal equations for the SVD of the far-field ISP. These equations are standard eigenfunction equations satisfied by the two singular functions v_p and u_p.

The singular functions v_p

We first consider Eq. (5.55a) satisfied by the singular functions v_p. It follows from this equation that the singular functions $v_p(\mathbf{r}, \omega)$ having non-zero singular values are linear combinations of the free multipole fields $j_l^*(kr)Y_l^m(\hat{\mathbf{r}})$. It then follows that since this set of functions is orthogonal over the source volume the singular functions $v_p(\mathbf{r}, \omega)$ having non-zero singular value can be selected to be normalized versions of the free multipole fields having wavenumber k^*. In particular, we have that

$$v_{l,m}(\mathbf{r}, \omega) = \frac{-i^l}{\mu_l(ka_0)} \mathcal{M}_{\tau_0} j_l^*(kr)Y_l^m(\hat{\mathbf{r}}), \tag{5.56a}$$

where $\mu_l(ka_0)$ is the positive square root of $\mu_l^2(ka_0)$ defined in Eq. (5.54), the index p is replaced by the pair l, m and we have included the factor $-i^l$ for reasons soon to become apparent.

Non-zero singular values

To determine the non-zero singular values $\sigma_p = \sigma_{l,m} > 0$ we take the inner product of both sides of Eq. (5.55a) with respect to the functions $j_l^*(kr)Y_l^m(\hat{\mathbf{r}})$ to find that

$$\langle j_l^* Y_l^m, v_p \rangle_{\mathcal{H}_Q} \sigma_{l,m}^2 = \langle j_l^* Y_l^m, v_p \rangle_{\mathcal{H}_Q} \mu_l^2(ka_0),$$

from which we conclude that $\sigma_{l,m} > 0$ are equal to the positive root of the $\mu_l^2(ka_0)$ defined in Eq. (5.54).

The singular functions u_p

To compute the singular functions $u_p = u_{l,m}(\mathbf{s}, \omega) \in \mathcal{H}_f$ we can solve the normal equations Eqs. (5.55b). Alternatively, these functions corresponding to non-zero singular values $\mu_l(ka_0) > 0$ can be computed directly from the set $v_{l,m}$ found above using the first of the two defining equations for the SVD given in Eqs. (5.38a). In particular, if we make use of Eqs. (5.52a) and (5.56a) we find that

$$\begin{aligned}
\hat{T} v_{l,m} &= \mu_l(ka_0) u_{l,m}(\mathbf{s}) \\
&= \overbrace{-\sum_{l'=0}^{\infty} \sum_{m'=-l'}^{l'} (-i)^{l'} Y_{l'}^{m'}(\mathbf{s}) \int_{r' \le a_0} d^3r' \, j_{l'}(kr') Y_{l'}^{m'*}(\hat{\mathbf{r}}')}^{\hat{T}} \overbrace{\frac{-i^l}{\mu_l(ka_0)} j_l^*(kr) Y_l^m(\hat{\mathbf{r}})}^{v_{l,m}} \\
&= \frac{1}{\mu_l(ka_0)} \int_0^{a_0} r^2 \, dr |j_l^2(kr)|^2 Y_l^m(\mathbf{s}) = \mu_l(ka_0) Y_l^m(\mathbf{s}),
\end{aligned}$$

from which we conclude that

$$u_{l,m}(\mathbf{s}) = Y_l^m(\mathbf{s}), \qquad \mu_l(ka_0) > 0. \tag{5.56b}$$

The singular functions $u_{l,m}$ having non-zero singular values $\mu_l(ka_0) > 0$ are thus the spherical harmonics and are clearly complete in the space $\mathcal{H}_f$ of square-integrable functions on the unit sphere; i.e., in this particular problem the closure of the range of the mapping $\hat{T}Q = f$ is the whole space $\mathcal{H}_f$! Our motivation for including the factor $-i^l$ in the definition of the singular functions $v_{l,m}$ is now apparent in that it resulted in the singular functions $u_{l,m}$ being the spherical harmonics with no multiplying phase factors.

5.5.2 The solution to the far-field ISP

The minimum-norm least-squares solution to the ISP is given in general form in Eq. (5.48b). On using the singular system computed in the previous section for the far field ISP and a spherical source volume this solution reduces to

$$\hat{Q}(\mathbf{r},\omega) = \sum_{l=0}^{\infty}\sum_{m=-l}^{l} \frac{\langle Y_l^m, f\rangle_{\mathcal{H}_f}}{\mu_l(ka_0)} v_{l,m}(\mathbf{r},\omega) = \sum_{l=0}^{\infty}\sum_{m=-l}^{l} \frac{f_l^m(\omega)}{\mu_l} v_{l,m}(\mathbf{r},\omega)$$

$$= -\sum_{l=0}^{\infty}\sum_{m=-l}^{l} i^l \frac{f_l^m(\omega)}{\mu_l^2(ka_0)} \mathcal{M}_{\tau_0} j_l^*(kr) Y_l^m(\hat{\mathbf{r}}), \tag{5.57a}$$

where

$$f_l^m(\omega) = \langle Y_l^m, f\rangle_{\mathcal{H}_f} = \int d\Omega_s f(\mathbf{s},\omega) Y_l^{m*}(\mathbf{s})$$

are the radiation-pattern Fourier coefficients first introduced in Section 4.8.4 of Chapter 4. We will sometimes refer to the minimum-norm solution $\hat{Q}$ as the *minimum-norm source*.

We obtained a multipole expansion and plane-wave expansion of the field back propagated from the radiation pattern in Example 5.2. On comparison of the minimum-norm solution given above with the multipole expansion in that example we conclude that $\hat{Q}(\mathbf{r},\omega)$ can be considered to be *the back propagation of a filtered version of the radiation pattern in a medium with wavenumber* k^*. In particular, we can write the minimum-norm solution in the form

$$\hat{Q}(\mathbf{r},\omega) = \mathcal{M}_{\tau_0} \sum_{l=0}^{\infty}\sum_{m=-l}^{l} 2i^{l+1} k \overline{f_l^m}(\omega) j_l^*(kr) Y_l^m(\hat{\mathbf{r}}),$$

where

$$\overline{f_l^m}(\omega) = \frac{i}{2k}\frac{f_l^m(\omega)}{\mu_l^2(ka_0)}$$

are the filtered versions of the radiation-pattern generalized Fourier coefficients $f_l^m(\omega)$.

If we make use of the plane-wave expansion of the free multipole fields given in Example 3.4 of Chapter 3 we can also express $\hat{Q}(\mathbf{r}, \omega)$ in a plane-wave expansion with the plane-wave amplitudes equal to a filtered version of the radiation pattern. In particular, we have from that example

$$j_l^*(kr)Y_l^m(\hat{\mathbf{r}}) = \frac{(-i)^l}{4\pi} \int_{4\pi} d\Omega_s\, Y_l^m(\mathbf{s})e^{ik^*\mathbf{s}\cdot\mathbf{r}},$$

which, when substituted into Eq. (5.57a), yields after some minor algebra

$$\hat{Q}(\mathbf{r}, \omega) = \frac{ik^*}{2\pi}\mathcal{M}_{\tau_0} \int_{4\pi} d\Omega_s \bar{f}(\mathbf{s}, \omega)e^{ik^*\mathbf{s}\cdot\mathbf{r}}, \tag{5.57b}$$

with

$$\bar{f}(\mathbf{s}, \omega) = -\frac{1}{2ik^*}\sum_{l=0}^{\infty}\sum_{m=-l}^{l} \frac{f_l^m(\omega)}{\mu_l^2(ka_0)} Y_l^m(\mathbf{s}).$$

Again, on comparison of Eq. (5.57b) with the plane-wave expansion of the field back propagated from the radiation pattern found in Example 5.2 we see that the minimum-norm solution is the back propagation of a filtered version of the radiation pattern.

Field radiated by the minimum-norm source

According to Theorem 1.4 the radiation pattern uniquely determines the field everywhere outside the source spatial volume τ_0. It then follows that the field radiated by the minimum-norm source has to be equal to the field radiated by the actual source everywhere outside the source sphere. We can prove this result directly by making use of the multipole expansion of the field developed in Section 4.8 of Chapter 4. In particular, we showed in that section that the field radiated everywhere outside the source sphere can be expressed in terms of the radiation-pattern Fourier coefficients via the multipole expansion

$$U_+(\mathbf{r}, \omega) = ik\sum_{l=0}^{\infty}\sum_{m=-l}^{l} i^l f_l^m(\omega)h_l^+(kr)Y_l^m(\hat{\mathbf{r}}).$$

We can thus establish that the field radiated by the minimum-norm source $\hat{Q}$ is equal to the actual radiated field U_+ everywhere outside the source sphere if we can prove that this field possesses the above multipole expansion.

To establish this we first represent the field radiated by $\hat{Q}$ via the basic multipole expansion

$$U_{\hat{Q}}(\mathbf{r}, \omega) = -ik\sum_{l=0}^{\infty}\sum_{m=-l}^{l} q_l^m(k)h_l^+(kr)Y_l^m(\hat{\mathbf{r}}), \tag{5.58}$$

where we have denoted the field radiated by $\hat{Q}$ by $U_{\hat{Q}}$. In this expansion the multipole moments $q_l^m(k)$ are given in terms of the source by Eq. (4.52b) of Chapter 4:

$$\begin{aligned} q_l^m(k) &= \int_{r'\leq a_0} d^3r'\, \hat{Q}(\mathbf{r}',\omega) j_l(kr') Y_l^{m*}(\hat{\mathbf{r}}') \\ &= -\sum_{l'=0}^{\infty}\sum_{m'=-l'}^{l'} i^{l'} \frac{f_{l'}^{m'}(k)}{\mu_{l'}^2(ka_0)} \int_{r'\leq a_0} d^3r'\, j_{l'}^*(kr') Y_{l'}^{m'}(\hat{\mathbf{r}}') j_l(kr') Y_l^{m*}(\hat{\mathbf{r}}') \\ &= -i^l f_l^m(k). \end{aligned}$$

On substituting this result into Eq. (5.58) we then find that

$$U_{\hat{Q}}(\mathbf{r},\omega) = ik\sum_{l=0}^{\infty}\sum_{m=-l}^{l} i^l f_l^m(\omega) h_l^+(kr) Y_l^m(\hat{\mathbf{r}}), \tag{5.59}$$

which establishes that $U_{\hat{Q}} = U_+$ everywhere outside the source sphere.

Source efficiency and essentially NR source components

On comparing the multipole expansion Eq. (5.59) with the expansion of the minimum-norm source Eq. (5.57a) we see that there is a one-to-one correspondence between the terms in the two series. In particular, each component of the source radiates a single multipole field according to the correspondence

$$\overbrace{v_{l,m}(\mathbf{r},\omega)}^{\text{Source}} \Rightarrow \overbrace{i^{l+1} k\mu_l(ka_0) h_l^+(kr) Y_l^m(\hat{\mathbf{r}})}^{\text{Radiated Field}}. \tag{5.60}$$

We can loosely define the "energy" of a source as the square of its norm, which, for each component $v_{l,m}$ of $\hat{Q}$, is unity. On the other hand, the real field energy radiated by the multipole field out of a large sphere of radius $R \gg a_0$ is given by Eq. (2.36b) of Section 2.6 of Chapter 2 and found to be

$$E(\omega) = 2\kappa\omega\Re k e^{-2\Im kR}\int_{4\pi} d\Omega_s |f(\mathbf{s},\omega)|^2 = 8\pi\kappa\omega\frac{\Re k}{|k|^2} e^{-2\Im kR}\mu_l^2(ka_0),$$

where we have used the fact that the radiation pattern of a single multipole field is equal to $[(-i)^{l+1}/k]Y_l^m(\mathbf{s})$. The field energy can be seen to be proportional to $\mu_l^2(ka_0)$, which decays exponentially fast when $l > ka_0$, while the source "energy" is constant and equal to unity. It then follows that the source components having $l > |k|a_0$ are very inefficient radiators. In fact, they are precisely the essentially NR sources that are discussed in some depth in Chapters 1 and 2 as well as earlier in this and the previous chapter. We can thus decompose the minimum-norm source into the two components

$$\hat{Q}(\mathbf{r},\omega) = -\overbrace{\sum_{l=0}^{ka_0}\sum_{m=-l}^{l} i^l \frac{f_l^m(\omega)}{\mu_l^2(ka_0)} j_l^*(kr) Y_l^m(\hat{\mathbf{r}})}^{\text{Radiating}} \overbrace{- \sum_{l=ka_0+1}^{\infty}\sum_{m=-l}^{l} i^l \frac{f_l^m(\omega)}{\mu_l^2(ka_0)} j_l^*(kr) Y_l^m(\hat{\mathbf{r}})}^{\text{Essentially Non-radiating}}, \tag{5.61}$$

a result that has obvious ramifications in the stability of the solution of the ISP.

5.5.3 The algorithm point-spread function

One way of assessing the quality of an inversion of a linear mapping is to compute the *point-spread function* (PSF) of the algorithm used to generate the inverse. This quantity plays a central role in linearized inverse theory and will be used extensively in the inverse scattering theory treated in later chapters of the book. The PSF for the ISP is the inversion generated from data resulting from a delta-function source: $Q(\mathbf{r},\omega) = \delta(\mathbf{r} - \mathbf{r}')$. We note that such a source is not square-integrable. However, if we formally expand this source into a generalized Fourier series using the SVD basis $\{v_p\}$ we find that

$$\delta(\mathbf{r} - \mathbf{r}') = \sum_p \langle v_p, \delta(\mathbf{r} - \mathbf{r}')\rangle_Q v_p(\mathbf{r}) = \sum_p v_p(\mathbf{r}) v_p^*(\mathbf{r}'),$$

which is the so-called "completeness relationship" for the basis v_p. Now the minimum-norm solution $\hat{Q}$ of the ISP is a projection of the actual source Q onto the singular functions v_p having non-zero singular values $\sigma_p > 0$ so that the PSF for the ISP is given by the completeness relation truncated to singular functions having non-zero singular values; i.e.,

$$H(\mathbf{r},\mathbf{r}',k) = \sum_{\sigma_p>0} v_p(\mathbf{r}) v_p^*(\mathbf{r}'). \tag{5.62}$$

The PSF for the antenna-synthesis problem is then given by this general expression with the singular functions and singular values given by Eqs. (5.56a) and (5.54), respectively. We then find that

$$H(\mathbf{r},\mathbf{r}',\omega) = \mathcal{M}(r,r')\sum_{l=0}^{\infty}\sum_{m=-l}^{l} \frac{j_l^*(kr) j_l(kr')}{|\mu_l(ka_0)|^2} Y_l^m(\hat{\mathbf{r}}) Y_l^{m*}(\hat{\mathbf{r}}'), \tag{5.63}$$

where

$$\mathcal{M}(r,r') = \begin{cases} 1 & \text{if } r, r' \le a_0, \\ 0 & \text{otherwise.} \end{cases} \tag{5.64}$$

We will return to the PSF in a later section where we solve the antenna-synthesis problem in two space dimensions, where we can obtain a detailed and complete picture of the PSF and the quality of the minimum-norm solution to the ISP.

5.5.4 Time reversal and back-propagation imaging

We found in Section 5.3.3 that for sources confined to plane-parallel slabs in a non-dispersive background, in the limit that the wavelength is small relative to the width of the slab, the minimum-norm solution to the ISP became proportional to the back-propagated field with proportionality constant ik/a_0, where $2a_0$ is the total width of the slab. The same result holds here in the special case in which the background is non-dispersive (k is real-valued). In particular, in the limit that the wavelength is much smaller than the source radius a_0 and the wavenumber is real-valued we have that

$$\begin{aligned}\mu_l^2(ka_0) &= \frac{a_0^3}{2}[j_l^2(ka_0) - j_{l-1}(ka_0)j_{l+1}(ka_0)] \\ &\sim \frac{a_0^3}{2(ka_0)^2}\left\{\cos^2\left(ka_0 - \frac{l+1}{2}\pi\right) - \cos\left(ka_0 - \frac{l}{2}\pi\right)\cos\left(ka_0 - \frac{l+2}{2}\pi\right)\right\} \\ &= \frac{a_0^3}{4(ka_0)^2}\{1 + \cos(2ka_0 - (l+1)\pi) - \cos(2ka_0 - (l+1)\pi) - \cos(\pi)\}\end{aligned}$$

$$\Downarrow$$

$$\mu_l^2(ka_0) \to \frac{a_0}{2k^2}, \tag{5.65}$$

where we have used the result that

$$j_l(ka_0) \sim \frac{\cos\left(ka_0 - \frac{l+1}{2}\pi\right)}{ka_0}, \quad ka_0 \gg l.$$

If we now assume that the minimum-norm solution given by the series in Eq. (5.57a) terminates at some finite $l = l_0$, we find on substituting the above limiting value for $\mu_l^2(ka_0)$ into this series and using the fact that k is real

$$\hat{Q}(\mathbf{r},\omega) \to -\frac{2k^2}{a_0}\sum_{l=0}^{\infty}\sum_{m=-l}^{l} i^l f_l^m(\omega)\mathcal{M}_{\tau_0} j_l(kr)Y_l^m(\hat{\mathbf{r}}) = \frac{ik}{a_0}\Phi(\mathbf{r},\omega), \tag{5.66}$$

where we have made use of the multipole expansion of the back-propagated field obtained in Example 5.2.

The approximate solution to the ISP given in Eq. (5.66) can be interpreted as being a form of the back-propagation imaging described earlier in Section 5.4.5. In particular, if we approximate the inverse filter in Eq. (5.49) by $[\hat{T}\hat{T}^\dagger]^+ \approx 2k^2/a_0$ we find that

$$\hat{Q}(\mathbf{r},\omega) = \hat{T}^\dagger[\hat{T}\hat{T}^\dagger]^+ f \approx \frac{2k^2}{a_0}\overbrace{\left[-\mathcal{M}_{\tau_0}\sum_{l=0}^{\infty}\sum_{m=-l}^{l} i^l f_l^m(\omega) j_l^*(kr)Y_l^m(\hat{\mathbf{r}})\right]}^{\hat{T}^\dagger f}, \tag{5.67}$$

where we have made use of the expression for $\hat{T}^\dagger$ given in Eq. (5.52d). Equation (5.67) is the generalization of the approximate solution given in Eq. (5.66) to lossy media with complex k and allows us to interpret the image of the source formed by application of the adjoint operator to the data as being the back-propagated image as described in Section 5.4.5. We thus conclude that back-propagation imaging implemented via the simple application of the adjoint operator to the data will generate a good approximation to the solution to the ISP so long as the wavelength is small relative to the overall size of the source region τ_0.

5.6 Picard's condition and minimum-sized sources

Picard's condition that must be satisfied by the expansion coefficients of a minimum-norm solution to the ISP is, in fact, a condition that places a lower bound on the size required of a source in order for it to radiate a given field. We have the following theorem.

Theorem 5.1 (Minimum-sized-source theorem) *Let a finite-norm source $Q \in L^2(\tau_0)$ be supported within the source volume $\tau_0 = \{\mathbf{r}|r \leq a_0\}$ and radiate a field $U_+(\mathbf{r}, \omega)$. Then there exists a minimum radius $a_{\min} \leq a_0$ below which no finite-norm source supported within $\tau_0' = \{\mathbf{r}|r \leq a_0' \leq a_0\}$ can radiate that same field at all points outside the source region τ_0. Moreover, this minimum radius is given by*

$$a_{\min} = \frac{2}{|k|} \operatorname{Max} \lim_{l \to \infty} l \frac{|f_{l+1}^m(\omega)|}{|f_l^m(\omega)|}, \tag{5.68}$$

where $f_l^m(\omega)$ are the expansion coefficients of the radiation pattern into the spherical harmonics and the maximum is taken with respect to the index m.

Before proving this theorem we mention that, while the original source Q is supported within the sphere $\tau_0 = \{\mathbf{r}|r \leq a_0\}$, it is certainly possible that this source has NR components within this region. An equivalent minimum-norm source that will radiate the same field everywhere outside of τ_0 may, thus, be supported within a *smaller* sphere having radius $a_0' < a_0$. The above theorem then states that, although this is certainly possible, there is, in fact, a *minimum* value for a_0' that is required in order for the minimum-norm source to radiate the field everywhere outside of the original source volume τ_0.

To prove the theorem let's assume that a source is confined to a sphere τ_0 having radius a_0 and radiates a field having a specified radiation pattern $f(\mathbf{s}, \omega)$. The minimum-norm source that radiates this field is given in Eq. (5.57a), where the expansion coefficients are required to satisfy Picard's condition, which takes the form

$$||\hat{Q}||^2 = \sum_{l=0}^{\infty} \sum_{m=-l}^{l} \frac{|f_l^m(\omega)|^2}{\mu_l^2(ka_0')} < \infty. \tag{5.69}$$

By application of the ratio test we then conclude that

$$\lim_{l \to \infty} \frac{|f_{l+1}^m(\omega)|^2}{|f_l^m(\omega)|^2} \frac{\mu_l^2(ka_0')}{\mu_{l+1}^2(ka_0')} < 1,$$

which can be rewritten in the form

$$\lim_{l\to\infty} \frac{\mu_{l+1}^2(ka_0')}{\mu_l^2(ka_0')} > \lim_{l\to\infty} \frac{|f_{l+1}^m(\omega)|^2}{|f_l^m(\omega)|^2}. \tag{5.70}$$

For the index $l \gg |x|$ we have that

$$j_l(x) \sim \frac{l!}{(2l+1)!}(2x)^l, \tag{5.71}$$

so that

$$\mu_l^2(ka_0') = \int_0^{a_0'} r^2\, dr |j_l|^2(kr) \sim \frac{2^{2l} l!^2\, (|k|a_0')^{2l} a_0'^3}{(2l+1)!^2\,(2l+3)}. \tag{5.72}$$

Using the above result we conclude that

$$\lim_{l\to\infty} \frac{\mu_{l+1}^2(ka_0')}{\mu_l^2(ka_0')} \to \left(\frac{|k|a_0'}{2l}\right)^2,$$

from which we conclude from Eq. (5.70) that

$$|k|a_0' > 2 \lim_{l\to\infty} l \frac{|f_{l+1}^m(\omega)|}{|f_l^m(\omega)|}, \tag{5.73}$$

which establishes the theorem.

It is interesting to note that a square-integrable radiation pattern requires that

$$\lim_{l\to\infty} \frac{|f_{l+1}^m(\omega)|}{|f_l^m(\omega)|} < 1,$$

which is not sufficient for the limit in Eq. (5.68) to be finite and, thus, for there to exist a finite-norm source that will generate that radiation pattern. In other words, one cannot simply prescribe a (square-integrable) radiation pattern and expect that a finite-norm and finite-sized source will radiate this pattern.

Example 5.10 Consider the scalar wavelet field (Kaiser, 2003) in a non-dispersive medium from Section 4.5 of the previous chapter. The generalized Fourier coefficients of the radiation pattern of this field are easily computed and found to be given by

$$f_l^m(\omega) = \sqrt{4\pi(2l+1)}(-i)^l j_l(ika).$$

The condition Eq. (5.68) for the minimum source radius for this radiation pattern is found to be

$$ka_0' > 2 \lim_{l\to\infty} l \frac{|j_{l+1}(ika)|}{|j_l(ika)|}.$$

For index $l \gg |x|$, the spherical Bessel functions $j_l(x)$ obey the asymptotic relationship Eq. (5.71), which, when substituted into the above inequality, yields the result

$$ka_0' > 2 \lim_{l\to\infty} l \frac{[(l+1)!/(2l+3)!](2ka)^{l+1}}{[l!/(2l+1)!](2ka)^l} = ka.$$

We thus conclude that the minimum source radius that will radiate the scalar wavelet field is $a_0 = a$, where a is the wavelet field parameter defining its radiation pattern.

Finally, we note that the condition Eq. (5.68) implies that any radiation pattern that has a finite number of terms in its spherical-harmonic expansion can be radiated by a finite-norm source having an arbitrarily small radius a_0. However, this condition is not the whole story, since it is also necessary to consider the source norm, which, although finite, can become exponentially large if a_0 is smaller than a threshold value determined by the highest-order term in the spherical-harmonic expansion of the radiation pattern. In particular, the condition Eq. (5.68) guarantees that the Picard condition Eq. (5.69) will be satisfied but does not place an upper bound on the source norm $||\hat{Q}||$. This issue is addressed in the following example.

Example 5.11 We consider a radiation pattern of the general form

$$f(\mathbf{s},\omega) = \sum_{l=0}^{L}\sum_{m=-l}^{l} f_l^m(\omega) Y_l^m(\mathbf{s}),$$

where $L > 1$ is a constant and $\forall l, m$, $C_1 \leq |f_l^m(\omega)| \leq C_2$, where C_1 and C_2 are two constants. The minimum-norm source that is supported within a sphere of radius a_0 and that will radiate this pattern is given by

$$\hat{Q}(\mathbf{r},\omega) = -\sum_{l=0}^{L_0}\sum_{m=-l}^{l} i^l \frac{f_l^m(\omega)}{\mu_l^2(ka_0)} j_l(kr) Y_l^m(\hat{\mathbf{r}}), \quad r \leq a_0.$$

The norm square of this radiation pattern is given by

$$||f||^2 = \int d\Omega_s |f(\mathbf{s},\omega)|^2 = \sum_{l=0}^{L}\sum_{m=-l}^{l} |f_l^m(\omega)|^2 \leq (L+1)C_2^2.$$

The squared norm of the source $\hat{Q}$ is given by

$$||\hat{Q}||^2 = \int d^3r |\hat{Q}(\mathbf{r},\omega)|^2 = \sum_{l=0}^{L_0}\sum_{m=-l}^{l} \frac{|f_l^m(\omega)|^2}{\mu_l^2(ka_0)} \geq C_1^2 \sum_{l=0}^{L_0}\sum_{m=-l}^{l} \frac{1}{\mu_l^2(ka_0)}.$$

The problem that arises is that the $\mu_l^2(ka_0)$ go to zero exponentially fast for values of $l > ka_0$ (see Section 1.7.3 Of Chapter 1). Thus, although the norm of the radiation pattern will be finite for any value of L, the norm of the source will become exponentially large if $ka_0 < L$, thus placing a practical limit on the size of a source that will radiate a radiation pattern having a finite number of terms.

5.7 Antenna synthesis and the far-field ISP in 2D space

The ISP in 2D space in terms of far-field data is governed by Eq. (4.39c) of Chapter 4, which relates the radiation pattern to the 2D spatial Fourier transform of the 2D source. To simplify the algebra it is convenient to rewrite this basic relationship in the form

$$\hat{T}Q = d, \tag{5.74a}$$

where the data $d \in L^2([-\pi, \pi])$ are related to the 2D radiation pattern $f(\alpha, \omega)$ via the equation

$$d(\alpha, \omega) = 2\sqrt{k}e^{i\frac{3\pi}{4}} f(\alpha, \omega)$$

and $\hat{T} : \mathcal{H}_Q \to L^2([-\pi, \pi])$ is the operator

$$\hat{T} = \frac{1}{\sqrt{2\pi}} \int_{\tau_0} d^2r\, e^{-ikr\cos(\phi-\alpha)}. \tag{5.74b}$$

In these equations r is the radial coordinate in a 2D Cartesian system having $x = r\cos\phi$, $y = r\sin\phi$ and α is the polar angle at which the radiation pattern $f(\alpha, \omega)$ is observed.

The adjoint operator $\hat{T}^\dagger$ is obtained in the usual way

$$\langle d, \hat{T}Q\rangle_{L^2([-\pi,\pi])} = \langle \hat{T}^\dagger d, Q\rangle_{\mathcal{H}_Q},$$

with the inner products defined by

$$\langle d_1, d_2\rangle_{L^2([-\pi,\pi])} = \int_{-\pi}^{\pi} d\alpha\, d_1^*(\alpha) d_2(\alpha), \qquad \langle Q_1, Q_2\rangle_{\mathcal{H}_Q} = \int_{\tau_0} d^2r\, Q_1^*(\mathbf{r}) Q_2(\mathbf{r}).$$

Using the definition of the adjoint, we find that

$$\hat{T}^\dagger = \frac{1}{\sqrt{2\pi}} \int_{-\pi}^{\pi} d\alpha\, e^{ik^* r\cos(\phi-\alpha)}. \tag{5.75}$$

We can expand the plane wave $\exp(-ikr\cos(\phi - \alpha))$ in the plane-wave expansion (cf. Example 3.9 of Chapter 3)

$$e^{-ikr\cos(\phi-\alpha)} = \sum_{n=-\infty}^{\infty} (-i)^n e^{in\alpha} J_n(kr) e^{-in\phi}, \tag{5.76}$$

from which we conclude that

$$e^{ik^* r\cos(\phi-\alpha)} = \sum_{n=-\infty}^{\infty} i^n e^{-in\alpha} J_n^*(kr) e^{in\phi}.$$

On substituting these expansions into Eqs. (5.74b) and (5.75) we obtain the 2D versions of Eqs. (5.52):

$$\hat{T} = \frac{1}{\sqrt{2\pi}} \sum_{n=-\infty}^{\infty} (-i)^n e^{in\alpha} \int_0^{a_0} r\,dr \int_{-\pi}^{\pi} d\phi\, J_n(kr) e^{-in\phi}, \tag{5.77a}$$

$$\hat{T}^\dagger = \mathcal{M}_{\tau_0} \frac{1}{\sqrt{2\pi}} \sum_{n=-\infty}^{\infty} i^n e^{in\phi} J_n^*(kr) \int_{-\pi}^{\pi} d\alpha\, e^{-in\alpha}, \tag{5.77b}$$

where we have assumed that the source region τ_0 is a circle of radius a_0 centered at the origin and $\mathcal{M}_{\tau_0}$ is the masking operator defined in Eq. (5.52c) but with r now the radial coordinate on the plane.

5.7.1 Implementation of the SVD

The composite operators $\hat{T}^\dagger \hat{T}$ and $\hat{T}\hat{T}^\dagger$ are easily found to be given by

$$\hat{T}^\dagger \hat{T} = \mathcal{M}_{\tau_0} \sum_{n=-\infty}^{\infty} e^{in\phi} J_n^*(kr) \int_0^{a_0} r'\,dr' \int_{-\pi}^{\pi} d\phi'\, J_n(kr') e^{-in\phi'}, \tag{5.78a}$$

$$\hat{T}\hat{T}^\dagger = \sum_{n=-\infty}^{\infty} \nu_n^2(ka_0) e^{in\alpha} \int_{-\pi}^{\pi} d\alpha'\, e^{-in\alpha'}, \tag{5.78b}$$

where

$$\nu_n^2(ka_0) = \int_0^{a_0} r\,dr |J_n(kr)|^2 \tag{5.79}$$

is the so-called "second Lommel integral," which can be expressed in the closed form

$$\nu_n^2(ka_0) = \frac{a_0^2}{2}[J_n^2(ka_0)^2 - J_{n-1}(ka_0) J_{n+1}(ka_0)].$$

The parameters $\nu_n^2(ka_0)$ are functionally identical to the parameters $\mu_l^2(ka_0)$ that we encountered in the 3D version of the ISP in the previous section and are proportional to the latter quantities if $n = l + 1/2$. Like their 3D counterparts, the parameters $\nu_n^2(ka_0)$ possess a cutoff index $n = ka_0$ beyond which they go to zero exponentially fast with increasing index n. This is illustrated in Fig. 5.1, which should be compared with the corresponding plot for the $\mu_l^2(ka_0)$ given in Fig. 1.2 of Chapter 1.

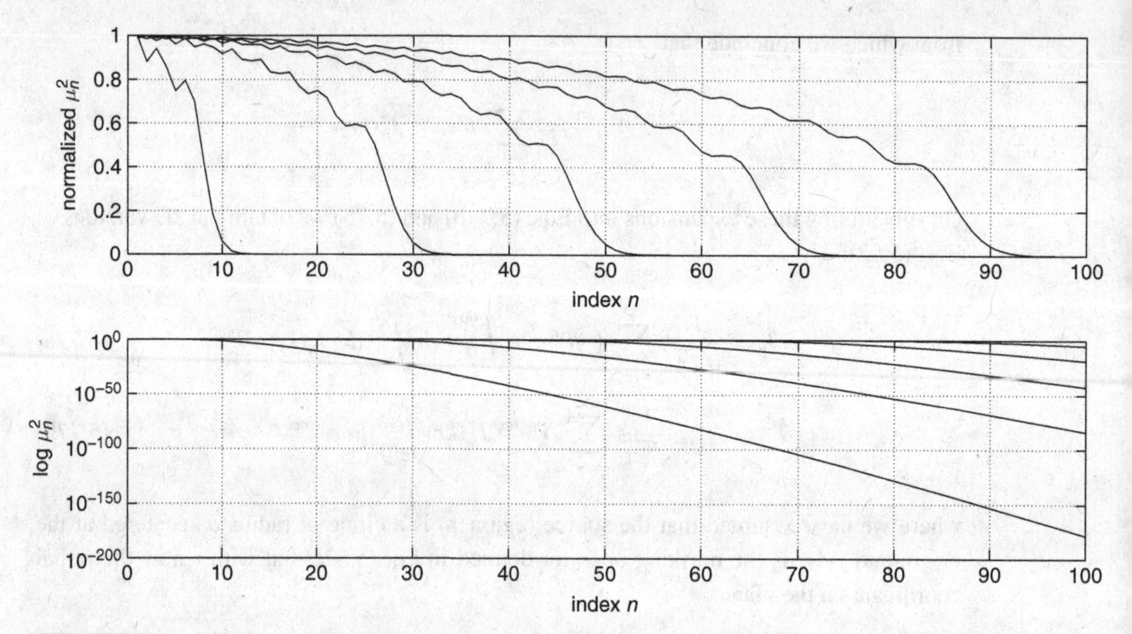

Fig. 5.1 (Top) Plots of $\nu_n^2(ka_0)$ normalized by their peak value for each value of ka_0 with k taken to be real. The parameters are plotted as a function of the index n for various values of ka_0. (Bottom) Plots of the log to base 10 of $\nu_n^2(ka_0)$ plotted as a function of the index n for various values of ka_0.

The singular functions v_p and singular values σ_p

The singular functions v_p satisfy the normal equations Eqs. (5.39), which, using Eq. (5.78a), can be written in the form

$$\mathcal{M}_{\tau_0} \sum_{n=-\infty}^{\infty} \langle J_n^*(kr')e^{in\phi'}, v_p \rangle_{\mathcal{H}_Q} J_n^*(kr)e^{in\phi} = \sigma_p^2 v_p(\mathbf{r}, \omega),$$

from which it follows, using an argument completely parallel to that employed in Section 5.5.1, that the index $p = n$ and the v_p can be taken to be proportional to the 2D free multipole fields $J_p^*(kr)\exp(ip\phi)$ of the homogeneous Helmholtz equation with wavenumber k^*:

$$v_p(\mathbf{r}, \omega) = \frac{i^p}{\sqrt{2\pi}\,\nu_p(ka_0)} \mathcal{M}_{\tau_0} J_p^*(kr)e^{ip\phi}, \tag{5.80}$$

where we have included the phase factor i^p for later notational convenience and p is any positive or negative integer or zero. The non-zero singular values are found to be proportional to the $\nu_p(ka_0)$:

$$\sigma_p = \sqrt{2\pi}\,\nu_p(ka_0) = \sqrt{2\pi \int_0^{a_0} r\,dr |J_p(kr)|^2},$$

which are guaranteed to be real and positive for any $a_0 > 0$ and for all integer p.

The singular functions u_p

The singular functions $u_p(\alpha, \omega)$ can be found by substituting the singular functions $v_p(\mathbf{r}, \omega)$ found above into the defining equation $\hat{T}v_p = \sigma_p u_p$. We find that

$$\frac{1}{\sqrt{2\pi}} \sum_{n=-\infty}^{\infty} (-i)^n e^{in\alpha} \int_0^{a_0} r\,dr \int_{-\pi}^{\pi} d\phi\, J_n(kr) e^{-in\phi} \overbrace{\left\{ \frac{i^p}{\sqrt{2\pi}\, v_p(ka_0)} J_l^*(kr) e^{ip\phi} \right\}}^{v_p(\mathbf{r},\omega)}$$
$$= v_p(ka_0) e^{ip\alpha} = \sigma_p u_p(\alpha, \omega),$$

from which we conclude that

$$u_p(\alpha, \omega) = \frac{1}{\sqrt{2\pi}} e^{ip\alpha}.$$

The inclusion of the phase factor i^p in the definition of the singular functions v_p in Eq. (5.80) should now be evident, since it allowed the singular functions u_p to be the set of orthonormal complex exponentials on the unit circle.

5.7.2 The solution to the 2D far-field ISP and algorithm PSF

The minimum-norm solution to the 2D far-field ISP is given in general form in Eq. (5.48b), which reduces to

$$\hat{Q}(\mathbf{r}, \omega) = \sum_{p=-\infty}^{\infty} \frac{\langle u_p, d \rangle_{L^2([-\pi,\pi])}}{\sigma_p} v_p(\mathbf{r}, \omega)$$
$$= \frac{\sqrt{k}}{\pi} e^{i\frac{3\pi}{4}} \mathcal{M}_{\tau_0} \sum_{p=-\infty}^{\infty} \frac{f_p(\omega)}{v_p^2(ka_0)} i^p J_p^*(kr) e^{ip\phi}, \tag{5.81a}$$

where

$$f_p(\omega) = \langle u_p, f \rangle_{L^2([-\pi,\pi])} = \frac{1}{\sqrt{2\pi}} \int_{-\pi}^{\pi} d\alpha\, f(\mathbf{s}, \omega) e^{-ip\alpha} \tag{5.81b}$$

are the Fourier coefficients of the radiation pattern over the unit circle.

Comments on the minimum-norm solution

As should be expected, the minimum-norm solution to the 2D ISP given in Eq. (5.81a) shares a number of properties with the 3D solution given in Eqs. (5.57a). Of particular importance is the fact, illustrated in Fig. 5.2 for real wavenumbers, that the Bessel functions $J_p(kr)$ are exponentially damped for $kr < p$ so that each term in the expansion Eq. (5.81a) corresponds to a source component that is essentially supported in a circle having an inner radius equal to $a_i = p/k$ and an outer radius equal to a_0. If $p > ka_0$ then $a_i > a_0$, so the effective support of these terms lies outside the actual source radius a_0. It then follows that the contribution of these terms to the minimum-norm source $\hat{Q}$ is negligible, so the expansion Eq. (5.81a) effectively terminates at a maximum p value of ka_0. That these source terms are not important can also be traced to the fact that they radiate mostly

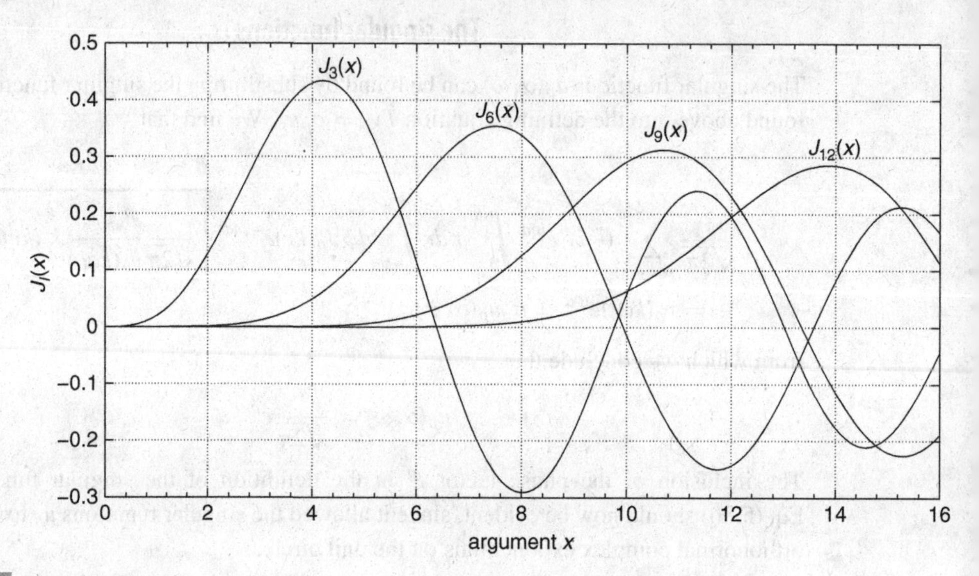

Fig. 5.2 Plots of the Bessel functions $J_l(x)$ for $l = 3, 6, 9$ and 12 and for strictly real wavenumber k.

evanescent waves. This result follows from the fact that these component sources have an angular period of oscillation equal to $2\pi/p$ and thus possess a spatial period at a radial distance r of $2\pi r/p$. We plot a set of four source components with real wavenumber k and $p = 10, 20, 30$ and 40 in Fig. 5.3. When $p \geq ka_0$ the spatial periods of these source components at a distance r will then be smaller than $2\pi(r/ka_0) = (r/a_0)\lambda \leq \lambda$ and they will radiate mostly evanescent plane waves that will be undetectable outside the source except in certain special experimental situations. These components of the source are, in fact, the "essentially non-radiating" sources that we discussed earlier in connection with the solution of the 3D ISP.

The point-spread function

The 2D point-spread function (PSF) is defined in Eq. (5.62) as the projection of the delta function onto singular functions v_p having non-zero singular values σ_p. For the 2D case these singular functions are given in Eq. (5.80) so that

$$H(\mathbf{r}, \mathbf{r}', \omega) = \frac{1}{2\pi}\mathcal{M}(r, r') \sum_{p=-\infty}^{\infty} \frac{J_p(kr)J_p^*(kr')}{\nu_p^2(ka_0)} e^{ip(\phi-\phi')}$$
$$= \frac{1}{\pi}\mathcal{M}(r, r') \sum_{p=0}^{\infty} \epsilon_p \frac{J_p(kr)J_p^*(kr')}{\nu_p^2(ka_0)} \cos[p(\phi - \phi')], \tag{5.82}$$

where

$$\mathcal{M}(r, r') = \begin{cases} 1 & \text{if } r, r' \leq a_0 \\ 0 & \text{otherwise} \end{cases}$$

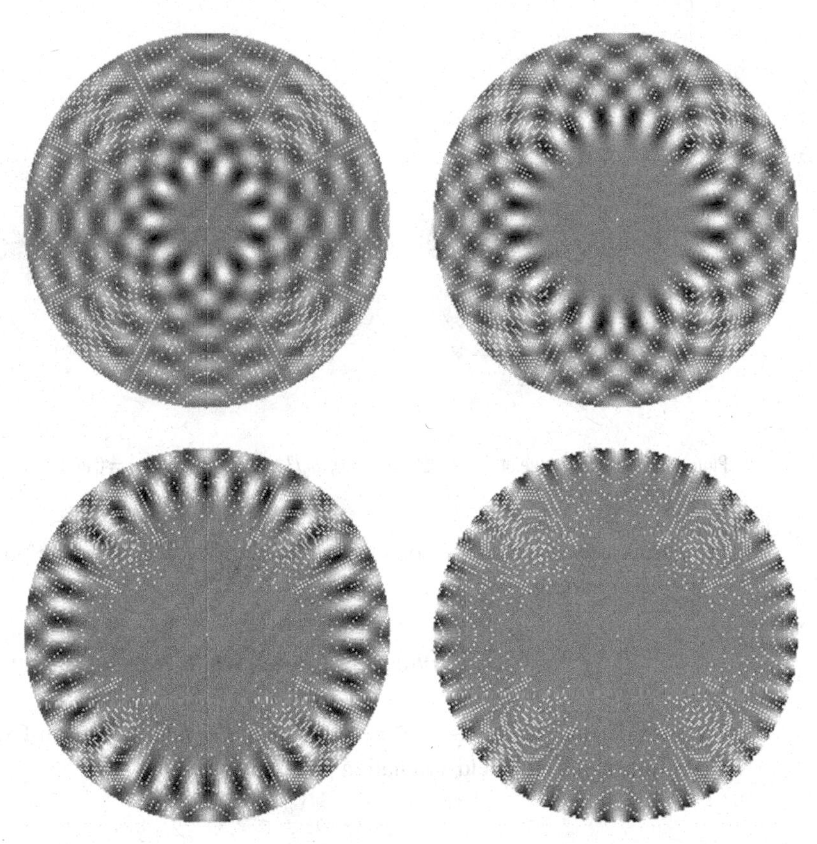

Fig. 5.3 Plots of the real parts of the component sources $J_p(kr)\exp(ip\phi)$ for $p = 10, 20, 30$ and 40 and for real k with $ka_0 = 40$. These component sources are effectively supported within the interval $p/k < r < r_0$ and have spatial periods at any radial distance of $2\pi(r/p)$.

and $\epsilon_p = 1$ if $p \neq 0$ and $1/2$ if $p = 0$. The PSF was previously defined for the 3D case in Eq. (5.64). The PSF is the "image" of a point source (delta-function source) located at the space coordinate $\mathbf{r}'$ and is related to the image (reconstruction) of a general 2D source via the equation

$$\hat{Q}(\mathbf{r}, \omega) = \int d^2r' \, Q(\mathbf{r}', \omega) H(\mathbf{r}, \mathbf{r}', \omega). \tag{5.83}$$

We consider first the computation of the PSF defined in Eq. (5.82) as a function of the image point $\mathbf{r}$ for various source points $\mathbf{r}'$. As mentioned above, the PSF is the image of a delta-function source centered at the space point $\mathbf{r}'$, so the PSF is the image generated by the solution to the ISP for this source. We will assume that the various source points $\mathbf{r}'$ are contained within an outer circle centered at the origin and having a radius a_0 and that the medium is non-dispersive, with real wavenumber k. We first computed the PSF for a maximum p value equal to ka_0 corresponding to the stable component of the minimum-norm source. Shown in Fig. 5.4 is the PSF for three different source points $\mathbf{r}'$ located at the radial distances $r' = 0$, $a_0/3$ and $2a_0/3$. Since the PSF is a function only for $\phi - \phi'$ we

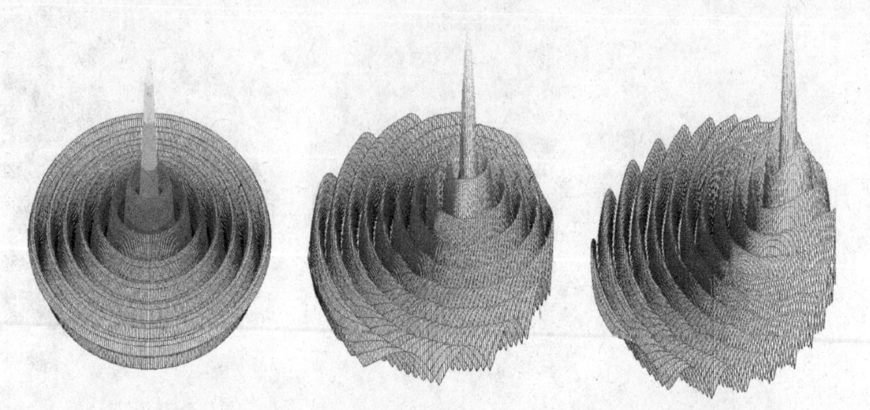

Fig. 5.4 Plots of the absolute value of the 2D PSF for $r' = 0$, $a_0/3$ and $2a_0/3$, with $\phi' = 0$ and max $p = ka_0$.

have arbitrarily selected the source point locations to be along the positive-x (horizontal) axis corresponding to $\phi' = 0$.

5.7.3 Two-dimensional scalar wavelet source

We consider the 2D version of the *scalar wavelet field* treated in Example 5.10. In two space dimensions the field's radiation pattern is given by

$$f(\mathbf{s}, \omega) = e^{ka(\cos\alpha - 1)},$$

where α is the polar angle of the unit vector $\mathbf{s}$, a is a positive constant parameter of the wavelet field and it is assumed that the wavenumber k is real-valued, corresponding to radiation in a non-dispersive medium. The generalized Fourier coefficients of this radiation pattern are found from Eq. (5.81b) to be given by

$$f_p(\omega) = \frac{e^{-ka}}{\sqrt{2\pi}} \int_{-\pi}^{\pi} d\alpha\, e^{ka\cos\alpha} e^{-ip\alpha}.$$

The above integration can be performed if we make use of the expansion Eq. (5.76) with $-ikr = ka$ and $\phi = 0$. We then find that

$$f_p(\omega) = \frac{e^{-ka}}{\sqrt{2\pi}} \int_{-\pi}^{\pi} d\alpha \left\{ \sum_{n=-\infty}^{\infty} (-i)^n e^{in\alpha} J_n(ika) \right\} e^{-ip\alpha} = \sqrt{2\pi}(-i)^p e^{-ka} J_p(ika),$$

which yields the following representation for the 2D wavelet radiation pattern:

$$\begin{aligned} f(\alpha, \omega) &= e^{-ka} \sum_{p=-\infty}^{\infty} (-i)^p J_p(ika) e^{ip\alpha}, \\ &= 2e^{-ka} \sum_{p=0}^{\infty} \epsilon_p (-i)^p J_p(ika) \cos(p\alpha). \end{aligned} \tag{5.84a}$$

The minimum-norm source for the 2D scalar wavelet field is found using Eq. (5.81a) to be

$$\hat{Q}(\mathbf{r},\omega) = \sqrt{\frac{2k}{\pi}} e^{i\frac{3\pi}{4}} e^{-ka} \sum_{p=-\infty}^{\infty} \frac{J_p(ika)}{v_p^2(ka_0)} J_p^*(kr) e^{ip\phi}$$

$$= \sqrt{\frac{8k}{\pi}} e^{i\frac{3\pi}{4}} e^{-ka} \sum_{p=0}^{\infty} \epsilon_p \frac{J_p(ika) J_p^*(kr)}{v_p^2(ka_0)} \cos(p\phi), \quad r \le a_0, \tag{5.84b}$$

where $\epsilon_p = 1$ if $p \neq 0$ and $\epsilon_0 = 1/2$. It is interesting to note the similarity of the form of the minimum-norm wavelet source to the 2D PSF given in Eq. (5.82).

The above expression for the minimum-norm source is merely formal in that the wavelet parameter a and the source radius a_0 must be selected such that the source actually has finite norm. We found in Example 5.10 that the minimum-sized 3D source that will radiate the 3D scalar wavelet field has a radius $a_0 = a$; i.e., the parameter a defining the scalar wavelet radiation pattern is the minimum source radius that will radiate the field. The same is true in the 2D case since it follows from Eq. (5.84b) that

$$||\hat{Q}||^2 = \frac{2k}{\pi} \sum_{p=-\infty}^{\infty} \frac{|J_p(ika)|^2}{v_p^2(ka_0)}.$$

After applying the ratio test, we then conclude that the source norm will be finite only if

$$\lim_{p\to\infty} \frac{|J_{p+1}(ika)|^2}{v_{p+1}^2(ka_0)} \frac{v_p^2(ka_0)}{|J_p(ika)|^2} < 1. \tag{5.85}$$

If we now use the asymptotic expression

$$J_p(x) \sim \frac{x^p}{2^p p!}, \quad l \gg |x|,$$

we find from the definition of the $v_p^2(ka_0)$ that

$$v_p^2(ka_0) = \frac{k^{2p}}{(2^p p!)^2} \int_0^{a_0} dr\, r^{2p+1} \sim \frac{k^{2p}}{(2^p p!)^2} \frac{a_0^{2p+2}}{2p+2},$$

$$|J_p(ika)|^2 \sim \frac{(ka)^{2p}}{(2^p p!)^2}.$$

We then conclude that

$$\frac{|J_{p+1}(ika)|^2}{v_{p+1}^2(ka_0)} \sim \frac{(ka)^{2p+2}(2p+4)}{(ka_0)^{2p+2} a_0^2}, \qquad \frac{v_p^2(ka_0)}{|J_p(ika)|^2} \sim \frac{(ka_0)^{2p} a_0^2}{(ka)^{2p}(2p+2)},$$

which, when substituted into Eq. (5.85), yields the condition

$$\lim_{p\to\infty} \frac{(ka)^{2p+2}(2p+4)}{(ka_0)^{2p+2} a_0^2} \frac{(ka_0)^{2p} a_0^2}{(ka)^{2p}(2p+2)} = \left(\frac{a}{a_0}\right)^2 < 1,$$

which establishes that $a_0 > a$.

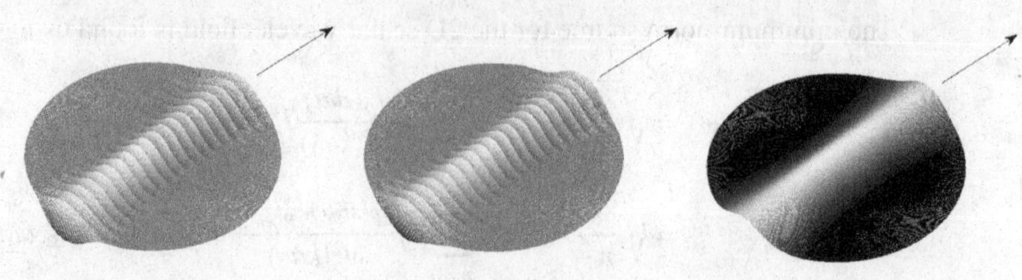

Fig. 5.5 Images of the real (left) and imaginary (center) components and magnitude (right) of the 2D wavelet source computed using a maximum p index value of $3ka_0$. The arrows indicate the main beam direction.

The problem in computing the wavelet source is that, although the series Eq. (5.84b) is guaranteed to converge so long as the source radius a_0 is larger than the wavelet parameter a, the separate terms $J_p^*(kr)J_p(ika)$ and $v_p^2(ka_0)$ each become exponentially small when $p \gg ka_0$ even though in the limit $p \to \infty$ their ratio tends to zero and the series Eq. (5.84b) converges. Because of this it is possible only to perform simple simulations of the stabilized approximation of the wavelet source and radiation pattern obtained by limiting the two series in Eqs. (5.84) to a maximum p value in the vicinity of ka_0. We computed the stabilized wavelet radiation pattern obtained using the series Eq. (5.84a) for the case in which $a_0 = a = 10\lambda$ and the maximum index p is equal to $3ka_0$ and found that it was virtually identical to the exact radiation pattern, indicating that the stabilized source will radiate a far field virtually identical to the exact source. The stabilized source is shown in Fig. 5.5, where the arrows indicate the direction of the main radiation lobe ($\phi = 0$). It should be noted that, although the stabilized source will radiate a field whose radiation pattern is virtually identical to that of the exact wavelet field, this field may differ markedly in the very near field of the source (within a wavelength) due to the higher-order terms that have been neglected in the stabilized version of the source.

5.8 The limited-view problem

Up to this point in our treatment of the ISP we have addressed only the ideal case in which the scattering amplitude $f(\mathbf{s}, \omega)$ is specified over a continuum of view angles over the real unit sphere Ω. This is a reasonable assumption in the antenna-synthesis problem where the antenna designer has in mind a specific radiation pattern that can be specified either functionally or numerically over a continuous or dense set of view directions $\mathbf{s}$. As mentioned at the beginning of the chapter, a second application of the ISP is in "imaging" where a source to a radiated field is to be computed from actual data, which will, of course, be known only over a limited set of observation directions or over a discrete set of points on a bounding surface to the source. For example, in cases of limited data the radiation pattern will be known only over a discrete set of directions $\mathbf{s}_j$, $j = 1, 2, \ldots, N$, and $\mathcal{H}_f$ will now be the space of square-summable complex N-tuples $l^2(C^N)$ with inner product given by

$$\langle f_1, f_2 \rangle_{\mathcal{H}_f} = \sum_{j=1}^{N} f_1^*(\mathbf{s}_j, \omega) f_2(\mathbf{s}_j, \omega), \tag{5.86}$$

while the Hilbert space of source functions $\mathcal{H}_Q$ remains $L^2(\tau_0)$. The basic structure of the Hilbert-space formulation of the limited-view ISP remains the same as for the case of complete data but the details of the solution to the problem change considerably. Here we will just develop the solution of the limited-view problem in two space dimensions for a 2D source confined to a cylinder of radius a_0 and for field data consisting of the 2D radiation pattern specified over a set of N different view directions $\mathbf{s}_j$, $j = 1, 2, \ldots, N$. Other geometries and data sets can be treated in an entirely analogous manner to that presented below.

5.8.1 The 2D limited-view problem

The far-field 2D limited-view ISP problem is defined via Eqs. (5.74), where, however, the data now consist of

$$d_j(\omega) = 2\sqrt{k} e^{i\frac{3\pi}{4}} f(\alpha_j, \omega),$$

where α_j, $j = 1, 2, \ldots, N$ are a discrete set of view angles over $[-\pi, +\pi]$. The data space is now $l^2(C^N)$ and the operators $\hat{T}$ and $\hat{T}^\dagger$ are given by

$$\hat{T} = \frac{1}{\sqrt{2\pi}} \int_{\tau_0} d^2r' \, e^{-ik\mathbf{s}_j\cdot\mathbf{r}}, \qquad \hat{T}^\dagger = \frac{1}{\sqrt{2\pi}} \mathcal{M}_{\tau_0} \sum_{j=1}^{N} e^{ik^*\mathbf{s}_j\cdot\mathbf{r}}. \tag{5.87}$$

In the full-view 2D problem treated in the previous section we represented the operators $\hat{T}$ and $\hat{T}^\dagger$ in cylindrical coordinates appropriate to the cylindrical source geometry. This was appropriate for that problem since this representation leads directly to orthogonal basis functions into which we expanded the singular functions $v_p(\mathbf{r})$ and $u_p(\mathbf{s})$ for $\sigma_p > 0$. However, this advantage disappears in the limited-view case due to the new inner product Eq. (5.86) appropriate to the limited-view problem. It is thus better to stay with the basic definitions of these two operators in terms of plane waves as given in Eqs. (5.87).

It is advantageous to define the set of functions

$$\chi_j(\mathbf{r}) = \frac{1}{\sqrt{2\pi}} \mathcal{M}_{\tau_0} e^{ik^*\mathbf{s}_j\cdot\mathbf{r}},$$

where it is easily shown that

$$\langle \chi_j, \chi_{j'} \rangle_{\mathcal{H}_Q} = \frac{1}{2\pi} \sum_{n=-\infty}^{\infty} \nu_n^2(ka_0) e^{in(\alpha_j - \alpha_{j'})}, \tag{5.88}$$

with $\nu_n^2(ka_0)$ defined in Eq. (5.79). Note that since $\nu_{-n}^2(ka_0) = \nu_n^2(ka_0)$ the matrix $\langle \chi_j, \chi_{j'} \rangle_{\mathcal{H}_Q}$ is real and symmetric and, hence, possesses an orthonormal set of eigenvectors with real eigenvalues. In terms of the functions χ_j we find that

$$\hat{T} = \int d^2r' \, \chi_j^*(\mathbf{r}), \qquad \hat{T}^\dagger = \sum_{j=1}^{N} \chi_j(\mathbf{r}),$$

and the normal equations for the singular functions become

$$\sum_{j=1}^{N} \chi_j(\mathbf{r})\langle \chi_j, v_p\rangle_{\mathcal{H}_Q} = \sigma_p^2 v_p, \tag{5.89a}$$

$$\int d^2r\, \chi_j^*(\mathbf{r})\langle \chi_{j'}^*(\mathbf{r}), u_p\rangle_{\mathcal{H}_f} = \sigma_p^2 u_p. \tag{5.89b}$$

5.8.2 Computing the singular system

It follows from Eq. (5.89a) that the singular functions v_p corresponding to non-zero singular values $\sigma_p > 0$ are linear combinations of the functions χ_j; i.e.,

$$v_p(\mathbf{r}) = \sum_{j=1}^{N} C_j(p)\chi_j(\mathbf{r}), \quad \sigma_p > 0.$$

On substituting the above expansion into the normal equation Eq. (5.89a) we find that

$$\sum_{j=1}^{N}\sum_{j'=1}^{N} C_{j'}(p)\chi_j(\mathbf{r})\langle \chi_j, \chi_{j'}\rangle_{\mathcal{H}_Q} = \sigma_p^2 \sum_{j=1}^{N} C_j(p)\chi_j(\mathbf{r}),$$

from which it follows from the linear independence of the set χ_j that

$$\sum_{j'=1}^{N} C_{j'}(p)\langle \chi_j, \chi_{j'}\rangle_{\mathcal{H}_Q} = \sigma_p^2 C_j(p). \tag{5.90}$$

The expansion coefficients are thus the eigenvectors of the real symmetric matrix defined in Eq. (5.88) with real and non-negative eigenvalues $\sigma_p^2 \geq 0$. Since the matrix $\langle \chi_j, \chi_{j'}\rangle_{\mathcal{H}_Q}$ is $N \times N$ there will be exactly N real non-negative eigenvalues, yielding N real and non-negative singular values and N orthonormal singular functions $v_p(\mathbf{r})$, $p = 1, 2, \ldots, N$.

The singular vectors u_p are N-tuples $u_p(j)$, $j = 1, 2, \ldots, N$, so

$$\langle \chi_j^*(\mathbf{r}), u_p\rangle_{\mathcal{H}_f} = \sum_{j'=1}^{N} \chi_{j'}(\mathbf{r})u_p(j')$$

and, hence, we can write the normal equation Eq. (5.89b) in the form

$$\sum_{j'=1}^{N} u_p(j')\langle \chi_j, \chi_{j'}\rangle_{\mathcal{H}_Q} = \sigma_p^2 u_p(j). \tag{5.91}$$

On comparing this equation with Eq. (5.90) we conclude that *the singular vectors $u_p(j)$ are equal to the eigenvectors $C_j(p)$ of the matrix $\langle \chi_j, \chi_{j'}\rangle_{\mathcal{H}_Q}$*. The solution to the limited-view 2D ISP is then given by the usual expression

$$\hat{Q}(\mathbf{r}, \omega) = \sum_{\sigma_p > 0} \frac{\langle u_p, d\rangle_{\mathcal{H}_f}}{\sigma_p} v_p(\mathbf{r}),$$

where, for $\sigma_p > 0$,

$$\langle u_p, d\rangle_{\mathcal{H}_f} = \frac{1}{\sqrt{2\pi}} \sum_{j=1}^{N} C_j^*(p) d_j(\omega) = \frac{\sqrt{2k} e^{i\frac{3\pi}{4}}}{2\pi} \sum_{j=1}^{N} C_j^*(p) f(\alpha_j, \omega), \tag{5.92a}$$

$$v_p(\mathbf{r}) = \sum_{j=1}^{N} C_j(p) \chi_j(\mathbf{r}) = \frac{1}{\sqrt{2\pi}} \mathcal{M}_{\tau_0} \sum_{j=1}^{N} C_j(p) e^{ik^* \mathbf{s}_j \cdot \mathbf{r}}. \tag{5.92b}$$

Further reading

There are a number of excellent texts on Hilbert space, among which I highly recommend Young (1988), Naylor and Sell (1982) and the treatment in Vaughn (2007). Strong proponents of the SVD and its use in imaging and wavefield inversion have been Bertero, Pike and co-workers (Bertero *et al.*, 1985; 1988). Good overall treatments of the mathematics of imaging and wavefield inversion include Colton and Kress (1992) and Bertero and Poccacci (1998). Stability issues with SVD inversion and regularization schemes are discussed in Bertero (1986, 1989) and Hansen (1988). One of the first treatments of the inverse source problem was due to Bleistein and Cohen (Bleistein and Cohen, 1977), while the so-called Porter–Bojarski integral equation first appeared in Porter (1970) and Bojarski (1982a). Fiddy and co-workers have long been involved with the inverse source problem as well as with the inverse scattering problem (Byrne and Fiddy, 1987; Fiddy and Testorf, 2006; Ross *et al.*, 1979). Their work is especially important in optical applications where the phase problem (Fienup, 1982; Gerchberg and Saxton, 1972; Gonsalves, 1976; Guizar-Sicairos and Fienup, 2006; Taylor, 1981) is important. Moses (Moses, 1984) has investigated the inverse source problem in the time domain both for scalar and for electromagnetic wavefields. The inverse problem for random sources was formulated and solved for spatially incoherent sources by the author (Devaney, 1979). An outstanding treatment of random wavefields in general is given in Ishimaru's book (Ishimaru, 1999). The formulation of the ISP using a reactive power constraint rather than minimum energy was treated in Marengo *et al.* (2004), and Marengo and Ziolkowski (2000) provide a slightly different approach to the ISP than has been used here. The solution of the ISP for the 2D scalar wavelet presented in Example 5.7.3 was generalized to the 3D case both for scalar and for electromagnetic wavelet fields in Devaney *et al.* (2008).

Problems

5.1 Derive the ISP integral equation Eq. (5.1a) in terms of boundary-value data as given by Eq. (5.1b) by applying standard Green-function techniques to the two wave equations satisfied by the radiated field $u_+(\mathbf{r}', t')$ and the free-field propagator $g_f(\mathbf{r} - \mathbf{r}', t - t')$.

5.2 Derive the ISP integral equation Eq. (5.1a) in terms of Cauchy data as given by Eq. (5.2) by applying standard Green-function techniques to the two wave equations satisfied by the radiated field $u_+(\mathbf{r}', t')$ and the free-field propagator $g_f(\mathbf{r} - \mathbf{r}', t - t')$.

5.3 Prove that the ISP integral equation holds under the replacement of $g_f(\mathbf{r} - \mathbf{r}', t - t')$ by any function $\hat{g}_f(\mathbf{r} - \mathbf{r}', t - t')$ that satisfies the homogeneous wave equation over all of space-time.

5.4 Derive the frequency-domain back-propagated field given in terms of Cauchy data in Eq. (5.3d) by Fourier transformation of Eq. (5.2).

5.5 Use the multipole expansion of the free-field propagator given in Eq. (5.4) of Example 5.1 in Eq. (5.3d) to derive the expansion

$$\Phi(\mathbf{r}, \omega) = \sum_{l,m} \Phi_l^m(\omega) j_l(kr) Y_l^m(\hat{\mathbf{r}}),$$

where

$$\Phi_l^m(\omega) = -2ki \frac{e^{i\omega t_0}}{c^2} \int d^3r' \left[i\omega u_+(\mathbf{r}', t_0) - \frac{\partial}{\partial t_0} u_+(\mathbf{r}', t_0) \right] j_l(kr') Y_l^{m*}(\hat{\mathbf{r}}').$$

5.6 Use the expansions of the back-propagated field obtained in Problem 5.5 and in Example 5.1 to establish a relationship between Dirichlet data on a sphere surrounding the source and Cauchy conditions acquired after a source has ceased to radiate.

5.7 Show that the PB integral equation for a source distributed over the surface of a sphere and for which the data consist of boundary-value data of any kind over the surfaces of two concentric spheres, one interior and one exterior to the source sphere, is incomplete and involves only the data on the exterior sphere. This is an example for which the formulation of the ISP in terms of the PB integral equation fails and the more powerful SVD-based approach is required (cf. Problem 5.18).

5.8 Derive the most general form of a surface source that is distributed over an infinite plane and that is NR throughout one of the two half-spaces bounded by the plane.

5.9 Derive the most general form of a surface source distributed over the surface of a sphere and that is NR throughout the interior (exterior) of the sphere.

5.10 Prove the following theorem, which is the frequency-domain version of the "source decomposition theorem" (Theorem 1.3) proven in Section 1.7 of Chapter 1. Let $Q(\mathbf{r}, \omega)$ be a square-integrable source compactly supported within τ_0. Then this source can be uniquely decomposed into an NR component $Q_{\mathrm{nr}}(\mathbf{r}, \omega)$ and a minimum-norm component $\hat{Q}(\mathbf{r}, \omega)$ such that

$$\int_{\tau_0} d^3r\, Q_{\mathrm{nr}}(\mathbf{r}, \omega) \hat{Q}(\mathbf{r}, \omega) = 0,$$

$$[\nabla_r^2 + k^2] \hat{Q}(\mathbf{r}, \omega) = 0,$$

$$Q_{\mathrm{nr}}(\mathbf{r}, \omega) = [\nabla_r^2 + k^2] \Pi(\mathbf{r}, \omega),$$

where $\Pi(\mathbf{r},\omega)$ is a square-integrable function supported in τ_0 that has continuous first partial derivatives.

5.11 Verify Eqs. (5.27).

5.12 Show that the adjoint of a compact operator is also compact. Hint: Show that it is Hilbert–Schmidt.

5.13 Prove that $\hat{T}^\dagger\hat{T}$ and $\hat{T}\hat{T}^\dagger$ are compact if $\hat{T}$ is compact.

5.14 Let H be an $N \times M$ matrix with complex elements $h_{n,m}$ and possessing the singular set v_p, u_p, σ_p. Then show that it admits the SVD

$$H = U\Sigma V^\dagger,$$

where U is the $N \times N$ matrix with column vectors $u_1, u_2, \ldots, u_N$, V is the $M \times M$ matrix with column vectors $v_1, v_2, \ldots, v_M$ and Σ is the $N \times M$ diagonal matrix with elements $\sigma_1, \sigma_2, \ldots, \sigma_P$, where $P = \min(N, M)$.

5.15 Compute the singular system for the antenna-synthesis problem addressed in Section 5.5 by first solving for the singular functions u_p and then computing the rest of the system from these functions.

5.16 1. Show that the singular values σ_p and singular vectors v_p for the 1D far-field version of the ISP defined in Example 5.5 satisfy the normal equation

$$\overbrace{\frac{1}{4|k|^2}\mathcal{M}\sum_{s=\pm 1} e^{ik^* sz}\int_{L_0} dz'\, e^{-iksz'} v_p(z',\omega)}^{T^\dagger T v_p} = \sigma_p^2 v_p(z,\omega).$$

2. Give an argument for why the singular functions v_p for $\sigma_p > 0$ satisfy the homogeneous Helmholtz equation with wavenumber k^* everywhere *inside* the interval L_0 and are, in fact, a linear combination of the two functions $\exp(\pm ik^* z)$.

3. Using the above result, show that the singular functions can be expressed in the form

$$v_p(z,\omega) = \mathcal{M}\sum_{s'=\pm 1} A_{s'}(p)e^{ik^* s' z}, \quad \sigma_p > 0,$$

where the Fourier coefficients A_s, $s = \pm 1$, satisfy the matrix equation

$$\begin{bmatrix} \mathrm{sinc}[a_0(k-k^*)] & \mathrm{sinc}[a_0(k+k^*)] \\ \mathrm{sinc}[a_0(k+k^*)] & \mathrm{sinc}[a_0(k-k^*)] \end{bmatrix}\begin{bmatrix} A_{-1}(p) \\ A_{+1}(p) \end{bmatrix} = \begin{bmatrix} A_{-1}(p) \\ A_{+1}(p) \end{bmatrix}.$$

5.17 Set up and solve the ISP for a source compactly supported between two parallel planes and Dirichlet data over two bounding parallel planes using the SVD. Compare and contrast your solution with that found in Section 5.3.

5.18 Set up and solve the ISP for a source distributed over the surface of a sphere, where the data consist of Dirichlet data over the surfaces of two concentric spheres, one interior and one exterior to the source sphere.

5.19 Compute the singular system given in Example 5.7.

5.20 Derive the singular system of the Slepian–Pollak problem given in Example 5.8.

5.21 By following identical steps to those used in solving the full-view 2D ISP problem for a cylindrical source region show that in the limited-view problem $\hat{T}$ and $\hat{T}^\dagger$ can be expressed in the form

$$\hat{T} = \frac{1}{\sqrt{2\pi}} \sum_{n=-\infty}^{\infty} (-i)^n e^{in\alpha_j} \int_0^{a_0} r\, dr \int_{-\pi}^{\pi} d\phi\, J_n(kr) e^{-in\phi},$$

$$\hat{T}^\dagger = \mathcal{M}_{\tau_0} \frac{1}{\sqrt{2\pi}} \sum_{n=-\infty}^{\infty} i^n e^{in\phi} J_n^*(kr) \sum_{j=1}^{N} e^{-in\alpha_j}.$$

5.22 Using the expressions for $\hat{T}$ and $\hat{T}^\dagger$ found in the previous problem show that the 2D composite operators $\hat{T}^\dagger \hat{T}$ and $\hat{T}\hat{T}^\dagger$ are given by

$$\hat{T}^\dagger \hat{T} = \mathcal{M}_{\tau_0} \frac{1}{2\pi} \sum_{n,n'} r(n,n') e^{in\phi} J_n^*(kr) \int_0^{a_0} r'\, dr' \int_{-\pi}^{\pi} d\phi'\, J_{n'}(kr') e^{-in'\phi'},$$

$$\hat{T}\hat{T}^\dagger = \frac{1}{2\pi} \sum_{n=-\infty}^{\infty} v_n^2(ka_0) e^{in\alpha_j} \sum_{j=1}^{N} e^{-in\alpha_{j'}},$$

where

$$r(n,n') = i^{(n-n')} \sum_{j=1}^{N} e^{-i(n-n')\alpha_j}.$$

5.23 Derive Eq. (5.88).

6 Scattering theory

In the radiation problem treated in Chapters 1 and 2 a "source" $q(\mathbf{r}, t)$ in the time domain or $Q(\mathbf{r}, \omega)$ in the frequency domain radiated a wavefield that satisfied either the inhomogeneous wave equation in the time domain or the inhomogeneous Helmholtz equation in the frequency domain. In either case the solution to the radiation problem was easily obtained in the form of a convolution of the given source function with the causal Green function of the wave or Helmholtz equation. A key point concerning the radiation problem is that *the source to the radiated field is assumed to be known (specified) and is assumed to be independent of the field that it radiates.* Such sources are sometimes referred to as "primary" sources since the mechanism or process that created them is unknown or, at least, unimportant as regards the field that they radiate.

In this chapter we will also encounter the radiation problem, but with sources that are created by the interaction of a propagating wave incident on a physical obstacle or inhomogeneous region of space. These new types of sources are referred to as "induced" or "secondary" sources and the problem of computing the field that they radiate given the incident wave and a model for the field–obstacle interaction is called the *scattering problem.* We deal with two classes of scattering problem in this book: (i) scattering from so-called "penetrable" scatterers, where the incident wave penetrates into the interior of the obstacle so that the resulting induced source radiates as a conventional volume source of the type treated in earlier chapters; and (ii) scattering from non-penetrable scatterers, where the interaction of the incident wave with the obstacle occurs only over the object's surface. We will review the first class of scattering problems in this chapter, and treat non-penetrable scatterers in the following chapter. We will also make the simplifying assumption that the scattering object is embedded in a uniform, possibly dispersive, background medium. We will treat the more general case of non-uniform backgrounds in Chapter 9.

For the class of penetrable scatterers the induced source to the inhomogeneous Helmholtz equation is related to the field U through a quantity $V(\mathbf{r})$ called the *scattering potential*:[1]

$$Q(\mathbf{r}) = V(\mathbf{r})U(\mathbf{r}). \tag{6.1}$$

The scattering potential is a complex-valued quantity that we will take to be compactly supported in a finite scattering volume τ_0. The scattering problem then consists of computing

[1] We will not carry the frequency ω in the arguments of the various field quantities in most of the presentation in this and the following chapter. Although these quantities will, in general, depend on frequency, this dependence is parametric and does not play a significant role except in certain time-domain problems, where it will be re-introduced as necessary.

the field U in terms of the incident wavefield and the scattering potential V. In this chapter we will present a general formulation of scattering theory for penetrable scatterers defined via a scattering potential. The *Lippmann–Schwinger integral equation* that governs the scattering process will be derived and shown to define a linear mapping between the incident wave and the scattered field but a *non-linear mapping* between the scattering potential and the scattered field. This non-linear character of the scattering problem makes the *inverse scattering problem*, which is a central topic and concern in this book, extremely difficult and one goal of the current chapter is to derive approximations to the scattered field that are linear functionals of the scattering potential. Fortunately, in many applications the Lippmann–Schwinger equation can be approximately linearized, thus yielding an approximate linear formulation of the scattering problem that effectively reduces it to a simple radiation problem of the type treated in Chapter 2.

In the scattering problem for non-penetrable scatterers the field–obstacle interaction occurs only over the surface of the obstacle and cannot be described in terms of a volume source in the form of Eq. (6.1). In the following chapter we will treat two types of such scattering problems: (i) scattering from "Dirichlet" objects, where the total field over the object's surface must vanish; and (ii) scattering from "Neumann" objects, where the normal derivative of the total field over the object's surface must vanish. In the case of electromagnetic fields Dirichlet objects are perfect conductors and Neumann objects are perfect absorbers. As is the case for penetrable scatterers, the solution of the scattering problem for non-penetrable objects is in the form of a linear mapping from incident field to scattered field but a non-linear mapping from the object to the scattered field and we will also obtain linearized approximate models of this non-linear mapping that can be easily and quickly implemented in computer code and that yield simple workable models for the associated inverse scattering problem.

6.1 Potential scattering theory

The discipline of scattering theory goes far beyond the simple case of potential scattering theory, for which the interaction of the wave with physical material is described by Eq. (6.1). However, potential scattering is an extremely important class of scattering theory that has special relevance to practical direct and inverse problems that are encountered in applications in electromagnetics, optics, acoustics and elastic-wave phenomenon. For example, in all of these applications the scattering problem can be cast in terms of a scattering potential of the general form[2]

$$V(\mathbf{r}) = k_0^2[1 - n_r^2(\mathbf{r}, \omega)], \tag{6.2}$$

[2] The scattering problems associated with vector-valued wavefields such as the electromagnetic field treated in the final chapter can vary in detail but have the same general mathematical structure of the potential scattering problem.

where

$$n_r(\mathbf{r},\omega)=\frac{n(\mathbf{r},\omega)}{n_0(\omega)},$$

which is called the "relative index of refraction of the scatterer," is the ratio of the complex index of refraction of the scattering object $n(\mathbf{r},\omega)$ to that of the background medium $n_0(\omega)$ and $k_0=(\omega/c)n_0(\omega)$ is the constant wavenumber of the background. The scattering object is completely defined (from a wave point of view) by its complex index of refraction and, hence, the inverse scattering problem that will be treated in later chapters reduces to the determination of V and, hence, $n(\mathbf{r},\omega)$ from available scattering data.

The basic formulation of potential scattering theory is directly obtained from the formulation of the radiation problem presented in Chapter 2. In particular, the total field *generated in any given scattering experiment* satisfies the inhomogeneous Helmholtz equation Eq. (2.3) with the source given by Eq. (6.1). We have emphasized the phrase "in any given scattering experiment" since it is extremely important for the inverse scattering problem that a suite of scattering experiments be performed rather than a single experiment and that the entire suite of scattered field data be available to aid in the determination (reconstruction) of the scattering potential. If just a single scattering experiment is performed then the inverse scattering problem reduces to the inverse source problem and is solved using the theory and algorithms presented in the previous chapter. However, such solutions are highly non-unique and, in fact, provide very-low-grade reconstructions of the scattering potential and it is thus important to employ as many scattering experiments as possible to generate adequate data to use in solving the inverse scattering problem. In order to keep track of different scattering experiments (with the same scattering potential V) we will write Eq. (2.3) in the form

$$[\nabla^2+k_0^2]U(\mathbf{r},\nu)=\overbrace{V(\mathbf{r})U(\mathbf{r},\nu)}^{Q(\mathbf{r},\nu)}$$

$$\Downarrow$$

$$[\nabla^2+k_0^2-V(\mathbf{r})]U(\mathbf{r},\nu)=0, \tag{6.3}$$

where ν is a parameter that labels the particular scattering experiment and $Q(\mathbf{r},\nu)$ is the induced source generated in the νth experiment. The boundary condition satisfied by the field U is that it reduces to the sum of an *incident* wavefield $U^{(\text{in})}$ plus a scattered field $U^{(\text{s})}$ that is required to satisfy the Sommerfeld radiation condition (SRC); i.e.,

$$U(\mathbf{r},\nu)=U^{(\text{in})}(\mathbf{r},\nu)+U_+^{(\text{s})}(\mathbf{r},\nu)\sim U^{(\text{in})}(\mathbf{r},\nu)+f(\mathbf{s},\nu)\frac{e^{ik_0r}}{r}, \tag{6.4}$$

where the incident wavefield $U^{(\text{in})}$ propagates in the uniform background medium and hence satisfies the homogeneous Helmholtz equation

$$[\nabla^2 + k_0^2]U^{(\text{in})}(\mathbf{r}, \nu) = 0,$$

and $f(\mathbf{s}, \nu)$ is the induced source radiation pattern in the direction of the unit vector $\mathbf{s} = \hat{\mathbf{r}} = \mathbf{r}/r$ for the νth scattering experiment. We have also added the subscript $+$ to the scattered field to denote that this quantity satisfies the outgoing-wave radiation condition.

We should note that the incident field will, in fact, be produced by some source radiating in the background medium and, hence, will actually satisfy the inhomogeneous Helmholtz equation. However, we assume that this source is well separated from the scatterer so that the field $U^{(\text{in})}$ will satisfy the homogeneous Helmholtz equation at least within the scatterer volume τ_0. Moreover, as we discussed in our treatment of the angular-spectrum expansion in Section 4.2 of Chapter 4, the field radiated by a compactly supported source can be accurately approximated as a free field at distances that are more than a few wavelengths from the source support volume. Thus, insofar as the potential scattering problem is concerned the incident field can be modeled as a free field that satisfies the homogeneous Helmholtz equation over all of space.

It is clear from the above that any single scattering experiment is formally equivalent to a radiation problem of the type considered in Chapters 1 and 2. In particular, we can view scattering as being a two-step process whereby in the first step the incident wavefield $U^{(\text{in})}(\mathbf{r}, \nu)$ interacts with the scattering potential generating the induced source $Q(\mathbf{r}, \nu)$, which then radiates the scattered field $U_+^{(s)}(\mathbf{r}, \nu)$ according to Eq. (6.3) in the second step. Indeed, since the incident wave satisfies the homogeneous Helmholtz equation, Eq. (6.3) can be expressed in the form

$$[\nabla^2 + k_0^2]U_+^{(s)}(\mathbf{r}, \nu) = Q(\mathbf{r}, \nu), \tag{6.5}$$

which is identical to the governing Helmholtz equation for the radiation problem treated in Chapters 1 and 2. It then follows that the inverse scattering problem differs from the inverse source problem (ISP) that was treated in Chapter 5 only if one performs a *suite* of scattering experiments using two or more different incident waves. In such cases the induced sources generated by the incident wave in each experiment are different, so the mathematical structure of the inverse scattering problem differs markedly from that of the ISP.

6.2 The Lippmann–Schwinger equation

The scattered field satisfies the inhomogeneous Helmholtz equation Eq. (6.5) and the Sommerfeld radiation condition (SRC) and, hence, is obtained by setting the source Q equal to the induced source defined in Eq. (6.1) in the solution of the radiation problem given, for example, in Eq. (2.23) of Chapter 2. We obtain

$$U_+^{(s)}(\mathbf{r}, \nu) = \int d^3r'\, G_{0+}(\mathbf{r} - \mathbf{r}')V(\mathbf{r}')U(\mathbf{r}', \nu), \tag{6.6a}$$

where

$$G_{0_+}(\mathbf{R}) = -\frac{1}{4\pi}\frac{e^{ik_0R}}{R} \tag{6.6b}$$

is the outgoing-wave Green function of the Helmholtz equation in the background medium having wavenumber k_0 (the so-called "free-space Green function"). This Green function satisfies the inhomogeneous Helmholtz equation

$$[\nabla^2 + k_0^2]G_{0_+}(\mathbf{r} - \mathbf{r}') = \delta(\mathbf{r} - \mathbf{r}')$$

and the SRC. It follows from Eqs. (6.4) and (6.6a) that the total field (incident plus scattered) is given by

$$U(\mathbf{r}, \nu) = U^{(\mathrm{in})}(\mathbf{r}, \nu) + \int d^3r'\, G_{0_+}(\mathbf{r} - \mathbf{r}')V(\mathbf{r}')U(\mathbf{r}', \nu), \tag{6.7a}$$

which is known as the *Lippmann–Schwinger* (LS) integral equation. We can write the LS integral equation in the symbolic form

$$U_\nu = U_\nu^{(\mathrm{in})} + G_{0_+}VU_\nu, \tag{6.7b}$$

where $G_{0_+}V$ stands for the integral operator

$$G_{0_+}V = \int d^3r'\, G_{0_+}(\mathbf{r} - \mathbf{r}')V(\mathbf{r}') \tag{6.7c}$$

and we have used a subscript to denote the dependence of the incident and total waves on the parameter ν.

6.2.1 The Lippmann–Schwinger equation for the full Green function

The free-space Green function $G_{0+}(\mathbf{r} - \mathbf{r}_0)$ can be interpreted as being the field radiated by a point (delta-function) source located at $\mathbf{r}_0$ in the infinite homogeneous background medium. In analogy, we define the "full Green function" $G_+(\mathbf{r}, \mathbf{r}_0)$ as the wavefield radiated by a point source located at $\mathbf{r}_0$ in the background medium containing a scatterer characterized by the scattering potential $V(\mathbf{r})$. This quantity satisfies the equation

$$[\nabla_{r'}^2 + k_0^2 - V(\mathbf{r}')]G_+(\mathbf{r}', \mathbf{r}_0) = \delta(\mathbf{r}' - \mathbf{r}_0) \tag{6.8}$$

and the SRC. Clearly, $G_+ \rightarrow G_{0_+}$ when $V \rightarrow 0$.

The full Green function satisfies an LS equation that is readily derived by rearranging Eq. (6.8) in the form

$$[\nabla_{r'}^2 + k_0^2]G_+(\mathbf{r}', \mathbf{r}_0) = \overbrace{\delta(\mathbf{r}' - \mathbf{r}_0) + V(\mathbf{r}')G_+(\mathbf{r}', \mathbf{r}_0)}^{Q(\mathbf{r}', \mathbf{r}_0)},$$

where $Q(\mathbf{r}', \mathbf{r}_0)$ is the sum of a "primary" source $\delta(\mathbf{r}' - \mathbf{r}_0)$ and an "induced" source equal to $V(\mathbf{r}')G_+(\mathbf{r}', \mathbf{r}_0)$. The formal solution to the above equation is found to be

$$G_+(\mathbf{r}, \mathbf{r}_0) = \int d^3r' \, \overbrace{\{\delta(\mathbf{r}' - \mathbf{r}_0) + V(\mathbf{r}')G_+(\mathbf{r}', \mathbf{r}_0)\}}^{Q(\mathbf{r}', \mathbf{r}_0)} G_{0+}(\mathbf{r} - \mathbf{r}')$$
$$= G_{0+}(\mathbf{r} - \mathbf{r}_0) + \int d^3r' \, G_{0+}(\mathbf{r} - \mathbf{r}')V(\mathbf{r}')G_+(\mathbf{r}', \mathbf{r}_0)$$
$$\Downarrow$$

$$G_+(\mathbf{r}, \mathbf{r}_0) = G_{0+}(\mathbf{r} - \mathbf{r}_0) + \int d^3r' \, G_{0+}(\mathbf{r} - \mathbf{r}')V(\mathbf{r}')G_+(\mathbf{r}', \mathbf{r}_0), \qquad (6.9a)$$

which is the Lippmann–Schwinger equation satisfied by the total Green function. It is clear from this equation that the full Green function can also be interpreted as being the field generated in a scattering experiment whose incident wave is the free-space Green function G_{0+}; i.e., when $U^{(\text{in})}(\mathbf{r}, \nu) = G_{0+}(\mathbf{r} - \mathbf{r}_0)$.

The full Green function satisfies a reciprocity condition that is derived using an argument identical to that employed in Section 2.8.4 to establish the reciprocity condition satisfied by the free-space Dirichlet and Neumann Green functions. In particular, by replacing $\mathbf{r}_0$ with $\mathbf{r}_1$ in Eq. (6.8) and applying standard Green-function techniques to these two equations and using the fact that G_+ must satisfy the SRC we find that

$$G_+(\mathbf{r}_1, \mathbf{r}_0) = G_+(\mathbf{r}_0, \mathbf{r}_1).$$

It is obvious that the free-space Green function G_{0+} satisfies the same condition. If we interchange $\mathbf{r}$ and $\mathbf{r}_0$ in Eq. (6.9a) and make use of the reciprocity condition we find that the LS equation for the full Green function can also be written in the form

$$G_+(\mathbf{r}, \mathbf{r}_0) = G_{0+}(\mathbf{r} - \mathbf{r}_0) + \int d^3r' \, G_+(\mathbf{r}, \mathbf{r}')V(\mathbf{r}')G_{0+}(\mathbf{r}' - \mathbf{r}_0). \qquad (6.9b)$$

6.2.2 The formal solution to the LS equation

We can formally "solve" the LS Eq. (6.7b) for the field U in terms of the (assumed known) scattering potential V and incident wave $U^{(\text{in})}$ by making use of the "full Green function" G_+. To obtain this solution we set $U = U^{(\text{in})} + U^{(\text{s})}$ in Eq. (6.3) and cast this equation in the form

$$[\nabla^2_{r'} + k_0^2 - V(\mathbf{r}')]U^{(\text{s})}_+(\mathbf{r}', \nu) = V(\mathbf{r}')U^{(\text{in})}(\mathbf{r}', \nu).$$

On applying "standard Green function techniques" to this equation and Eq. (6.8) we then find that

$$G_+(\nabla_{r'}^2 + k_0^2 - V)U_+^{(s)} - U_+^{(s)}(\nabla_{r'}^2 + k_0^2 - V)G_+ = G_+VU^{(in)} - U_+^{(s)}\delta,$$

from which we conclude that

$$U_+^{(s)}(\mathbf{r}, \nu) = \int d^3r'\, G_+(\mathbf{r}, \mathbf{r}')V(\mathbf{r}')U^{(in)}(\mathbf{r}', \nu), \tag{6.10}$$

so that the total field (incident plus scattered) can be expressed in the form

$$U_\nu = U_\nu^{(in)} + \int d^3r'\, G_+VU_\nu^{(in)} = [I + G_+V]U_\nu^{(in)}, \tag{6.11a}$$

where G_+V stands for the integral operator

$$G_+V = \int d^3r'\, G_+(\mathbf{r}, \mathbf{r}')V(\mathbf{r}').$$

Equation (6.11a) is a formal solution to the LS equation since this solution requires that the full Green function G_+ be known and its computation involves solving a scattering problem. Using an entirely parallel development, we find that the LS equation for the total Green function Eq. (6.9a) admits the formal solution

$$G_+ = G_{0+} + \int d^3r'\, G_+VG_{0+} = [I + G_+V]G_{0+}, \tag{6.11b}$$

which is, in fact, simply another version of the LS equation Eq. (6.9a) for the total Green function.

We note that the solution as given in Eqs. (6.11a) is in the form of a *linear* mapping from the incident wave to the scattered and total wavefields but is a *non-linear* mapping from the scattering potential V to the scattered and total wavefields. The non-linearity of the $V \rightarrow U$ mapping follows from the fact that the total Green function G_+ depends on the scattering potential V so that the composite operator G_+V is non-linear in V. It then follows that the inverse scattering problem requires the inversion of the set of coupled, non-linear integral Eqs. (6.11a) for the scattering potential V in terms of knowledge of the total or scattered field amplitudes over some set of field points and some set of scattering experiments.

6.3 Scattering from homogeneous penetrable objects

The LS integral equation is general in that it governs scattering from arbitrary penetrable objects described by scattering potentials. However, solving it for any given specified scattering potential (the so-called *forward*-scattering problem) can require a great deal of computational effort and time. An alternative approach for the forward-scattering problem that can be employed for homogeneous scattering objects or even piecewise-homogeneous objects involves constructing the field in pieces and tying the pieces together by matching at the boundaries of the various homogeneous regions of the scattering object. In particular, it follows from Eq. (6.3) that, if $V(\mathbf{r})$ represents a piecewise homogeneous object, then it will be piecewise constant and the total field $U(\mathbf{r}, \nu)$ must then be continuous with continuous first partial derivatives everywhere and, in particular, across the discontinuities of the scattering potential. This conclusion follows from the simple fact that if either U or any of its partials were discontinuous then the ∇^2 operation would generate delta functions supported on the discontinuities, such that Eq. (6.3) would not be satisfied.

A general approach for solving Eq. (6.3) is then to represent the solutions within each homogeneous sub-region of the scattering object using the well-known solutions for waves in homogeneous backgrounds developed in Chapter 3 and then to match these various component parts at the interfaces between the various homogeneous sub-regions using the continuity conditions. The wave component lying outside the support of the scattering potential is also required to reduce to the sum of the incident-wave (assumed known in the forward problem) and outgoing-wave scattered field components according to Eq. (6.4).

6.3.1 Scattering from homogeneous spheres and cylinders

Perhaps the simplest homogeneous scattering objects are homogeneous spheres and cylinders. When these scatterers are centered at the origin they are represented by a scattering potential of the general form

$$V(\mathbf{r}) = \begin{cases} k_0^2[1 - n_{\mathrm{r}}^2] & r \leq R_0, \\ 0 & r > R_0, \end{cases}$$

where R_0 is the radius of the sphere or cylinder, n_{r} is its (constant) relative index of refraction, and $\mathbf{r} = (r, \theta, \phi)$ for the case of the sphere and $\mathbf{r} = (r, \phi)$ for the cylinder. Here, we have represented the scattering potential in terms of its relative complex index of refraction n_{r} and the wavenumber k_0 of the background using Eq. (6.2). The sphere and cylinder are especially simple since their boundaries coincide with one of the coordinates in a separable coordinate system for the scalar wave Helmholtz equation, so the associated eigenfunction expansions for these systems presented in Chapter 3 can be used to obtain the solution to the scattering problem. Matching at the surface of the sphere or cylinder is then simple, since the eigenfunctions are functions only of the surface coordinates of the sphere or cylinder.

6.3.2 Scattering from a homogeneous sphere

We first consider a homogeneous sphere, for which case the appropriate eigenfunctions are the multipole fields developed in Section 3.3 of Chapter 3. For a sphere centered at the origin the total fields within and outside the support τ_0 of the scattering potential can be represented in 3D multipole expansions of the general form given in Eq. (3.32) of Section 3.3, where the radial functions are linear combinations of the spherical Bessel functions j_l and the spherical Hankel functions h_l^+. Within the scatterer support volume the field must be finite at the origin so the appropriate radial functions are the spherical Bessel functions and the multipole expansion is given by

$$U(\mathbf{r},\nu)=\sum_{l=0}^{\infty}\sum_{m=-l}^{l} a_l^m(\nu)j_l(k_0 n_r r)Y_l^m(\hat{\mathbf{r}}), \quad r\le R_0, \tag{6.12a}$$

where n_r is the relative index of refraction of the sphere, with $k_0 n_r = \omega/cn_r$ being the constant wavenumber within the sphere. Here, Y_l^m are the spherical harmonics of degree l and order m, and we have used our customary practice of representing the argument θ, ϕ of these quantities by the unit vector $\hat{\mathbf{r}}=\mathbf{r}/r$ of the field point. Outside of the scattering volume the field must reduce to the sum of the incident wave and an outgoing scattered wave, so the multipole expansion becomes

$$U(\mathbf{r},\nu)=\overbrace{\sum_{l=0}^{\infty}\sum_{m=-l}^{l} {a_0}_l^m(\nu)j_l(k_0 r)Y_l^m(\hat{\mathbf{r}})}^{U^{(\mathrm{in})}(\mathbf{r},\nu)}+\overbrace{\sum_{l=0}^{\infty}\sum_{m=-l}^{l} b_l^m(\nu)h_l^+(k_0 r)Y_l^m(\hat{\mathbf{r}})}^{U_+^{(\mathrm{s})}(\mathbf{r},\nu)}. \tag{6.12b}$$

In the forward-scattering problem the incident wave and, hence, the multipole moments ${a_0}_l^m(\nu)$ are assumed to be known (specified). The unknowns are the multipole moments $a_l^m(\nu)$ of the interior field to the scatterer and the multipole moments $b_l^m(\nu)$ of the scattered field. These quantities are determined by applying the continuity conditions at the scatterer surface. On applying these conditions, we then require that

$$\begin{aligned}{a_0}_l^m(\nu)j_l(k_0R_0)+b_l^m(\nu)h_l^+(k_0R_0)&=a_l^m(\nu)j_l(k_0n_rR_0),\\ {a_0}_l^m(\nu)\frac{\partial}{\partial R_0}j_l(k_0R_0)+b_l^m(\nu)\frac{\partial}{\partial R_0}h_l^+(k_0R_0)&=a_l^m(\nu)\frac{\partial}{\partial R_0}j_l(k_0n_rR_0),\end{aligned} \tag{6.13}$$

where we have made use of the fact that the spherical harmonics are orthonormal over the unit sphere.

The above coupled set of equations involving the two unknowns $a_l^m(\nu)$, $b_l^m(\nu)$ can be readily solved in terms of the known multipole moments ${a_0}_l^m(\nu)$ of the incident wavefield. We find that

$$a_l^m(\nu)=T_l{a_0}_l^m(\nu), \qquad b_l^m(\nu)=R_l{a_0}_l^m(\nu), \tag{6.14a}$$

where T_l and R_l are *generalized transmission and reflection coefficients* given by

$$T_l=\frac{j_l(k_0R_0)h_l^{+\prime}(k_0R_0)-j_l^{\prime}(k_0R_0)h_l^+(k_0R_0)}{j_l(k_0n_rR_0)h_l^{+\prime}(k_0R_0)-n_rj_l^{\prime}(k_0nR_0)h_l^+(k_0R_0)}, \tag{6.14b}$$

$$R_l=\frac{n_rj_l(k_0R_0)j_l^{\prime}(k_0n_rR_0)-j_l^{\prime}(k_0R_0)j_l(k_0n_rR_0)}{j_l(k_0n_rR_0)h_l^{+\prime}(k_0R_0)-n_rj_l^{\prime}(k_0n_rR_0)h_l^+(k_0R_0)}. \tag{6.14c}$$

By making use of the Wronskian relationship

$$j_l(k_0R_0)h_l^{+\prime}(k_0R_0) - j_l'(k_0R_0)h_l^+(k_0R_0) = \frac{i}{(k_0R_0)^2}$$

we can also express the transmission coefficient in the alternative form

$$T_l = \frac{i/(k_0R_0)^2}{j_l(k_0n_rR_0)h_l^{+\prime}(k_0R_0) - n_rj_l'(k_0n_rR_0)h_l^+(k_0R_0)}. \tag{6.14d}$$

It is important to note that the transmission and reflection coefficients are independent of the incident wave and depend only on the radius and relative index of refraction of the sphere.

6.3.3 Scattering from a homogeneous cylinder

A homogeneous cylinder with relative index of refraction $n_r(\omega)$ illuminated by an incident wavefield that propagates in the (x, y) plane perpendicular to the cylinder axis is treated in an entirely parallel way.[3] In such cases the fields are 2D and we can expand the interior and exterior fields in 2D multipole expansions of the form given in Eq. (3.60) of Section 3.6 of Chapter 3 and match them at the boundary of the cylinder. The field expansions take the form

$$U(\mathbf{r}, \nu) = \sum_{l=-\infty}^{\infty} a_l(\nu)J_l(k_0n_rr)e^{il\phi}, \quad r \le R_0, \tag{6.15a}$$

and

$$U(\mathbf{r}, \nu) = \overbrace{\sum_{l=-\infty}^{\infty} a_{0l}(\nu)J_l(k_0r)e^{il\phi}}^{U^{(\mathrm{in})}(\mathbf{r},\nu)} + \overbrace{\sum_{l=-\infty}^{\infty} b_l(\nu)H_l^+(k_0r)e^{il\phi}}^{U_+^{(\mathrm{s})}(\mathbf{r},\nu)}, \quad r \ge R_0, \tag{6.15b}$$

where r, ϕ are the polar coordinates in the (x, y) plane. Matching at $r = R_0$ then yields the set of equations

$$\begin{aligned} a_{0l}(\nu)J_l(k_0R_0) + b_l(\nu)H_l^+(k_0R_0) &= a_l(\nu)J_l(k_0n_rR_0), \\ a_{0l}(\nu)\frac{\partial}{\partial R_0}J_l(k_0R_0) + b_l(\nu)\frac{\partial}{\partial R_0}H_l^+(k_0R_0) &= a_l(\nu)\frac{\partial}{\partial R_0}J_l(k_0n_rR_0), \end{aligned} \tag{6.16}$$

which are then solved for in terms of the multipole moments a_{0l} of the incident wave. We obtain

$$a_l(\nu) = T_la_{0l}(\nu), \qquad b_l(\nu) = R_la_{0l}(\nu), \tag{6.17a}$$

[3] By this we mean a wavefield that varies only as a function of the x, y coordinates of the plane perpendicular to the cylinder axis. An example of such a wavefield is a plane wave with its unit propagation vector $\mathbf{s}_0$ lying in the (x, y) plane.

with the generalized transmission and reflection coefficients given by

$$T_l = \frac{2i/(\pi k_0 R_0)}{J_l(k_0 n_r R_0) H_l^{+\prime}(k_0 R_0) - n_r J_l'(k_0 n_r R_0) H_l^+(k_0 R_0)}, \tag{6.17b}$$

$$R_l = \frac{n_r J_l(k_0 R_0) J_l'(k_0 n_r R_0) - J_l'(k_0 R_0) J_l(k_0 n_r R_0)}{J_l(k_0 n_r R_0) H_l^{+\prime}(k_0 R_0) - n_r J_l'(k_0 n_r R_0) H_l^+(k_0 R_0)}, \tag{6.17c}$$

where we have used the Wronskian

$$J_l(k_0 R_0) H_l^{+\prime}(k_0 R_0) - J_l'(k_0 R_0) H_l^+(k_0 R_0) = \frac{2i}{\pi k_0 R_0}.$$

6.3.4 Scattering from concentric cylinders

It is also easy to compute eigenfunction expansions of the fields scattered by composite objects consisting of sets of concentric homogeneous spheres or cylinders. Here we will work out the simplest case of an object that consists of two concentric cylinders having differing indices of refraction. In particular, we assume that the object has a scattering potential given by

$$V(\mathbf{r}) = \begin{cases} k_0^2[1 - n_{r_1}^2] & r \leq R_1, \\ k_0^2[1 - n_{r_2}^2] & R_1 < r \leq R_2, \\ 0 & r > R_2, \end{cases}$$

where n_{r_1} and n_{r_2} are the relative indices of the two component parts of the object. This object is seen to consist of two concentric cylinders having radii R_1 and $R_2 > R_1$ and such that the relative index of refraction is n_{r_1} within the inner cylinder and n_{r_2} in the region between the two cylinder surfaces; i.e., for $R_1 < r \leq R_2$.

For the sake of simplicity we will again assume that the incident wave to the two concentric cylinders propagates in the (x, y) plane so that the appropriate eigenfunctions are again the 2D multipole fields. The field expansions within and outside the composite scatterer then take the form

$$U(\mathbf{r}, \nu) = \sum_{l=-\infty}^{\infty} a_l^{(1)}(\nu) J_l(k_0 n_{r_1} r) e^{il\phi}, \quad r \leq R_1, \tag{6.18a}$$

$$U(\mathbf{r}, \nu) = \sum_{l=-\infty}^{\infty} a_l^{(2)}(\nu) J_l(k_0 n_{r_2} r) e^{il\phi} + \sum_{l=-\infty}^{\infty} b_l^{(2)}(\nu) H_l^+(k_0 n_{r_2} r) e^{il\phi}, \quad R_1 \leq r \leq R_2, \tag{6.18b}$$

$$U(\mathbf{r}, \nu) = \overbrace{\sum_{l=-\infty}^{\infty} a_{0l}(\nu) J_l(k_0 r) e^{il\phi}}^{U^{(\mathrm{in})}(\mathbf{r},\nu)} + \overbrace{\sum_{l=-\infty}^{\infty} b_l(\nu) H_l^+(k_0 r) e^{il\phi}}^{U_+^{(s)}(\mathbf{r},\nu)}, \quad r \geq R_2, \tag{6.18c}$$

where the various coefficients (multipole moments) are to be determined by matching conditions on the two interfaces of the composite cylinders. Note that the wavefield within the inner cylinder consists only of standing waves, which are finite at the origin, while the wavefield in the region between the two cylinders consists of both standing waves and outgoing cylindrical waves. Matching the wavefields at the interfaces yields the set of equations

$$
\begin{aligned}
-a_l^{(1)}(\nu)J_l(k_0 n_{r_1} R_1) + a_l^{(2)}(\nu)J_l(k_0 n_{r_2} R_1) + b_l^{(2)}(\nu)H_l^+(k_0 n_{r_2} R_1) &= 0, \\
-a_l^{(1)}(\nu)n_{r_1} J_l'(k_0 n_{r_1} R_1) + a_l^{(2)}(\nu)n_{r_2} J_l'(k_0 n_{r_2} R_1) + b_l^{(2)}(\nu)n_{r_2} H_l^{+\prime}(k_0 n_{r_2} R_1) &= 0, \\
a_l^{(2)}(\nu)J_l(k_0 n_{r_2} R_2) - b_l(\nu)H_l^+(k_0 R_2) + b_l^{(2)}(\nu)H_l^+(k_0 n_{r_2} R_2) &= a_{0l}(\nu)J_l(k_0 R_2), \\
a_l^{(2)}(\nu)n_{r_2} J_l'(k_0 n_{r_2} R_2) - b_l(\nu)H_l^{+\prime}(k_0 R_2) + b_l^{(2)}(\nu)n_{r_2} H_l^{+\prime}(k_0 n_{r_2} R_2) &= a_{0l}(\nu)J_l'(k_0 R_2),
\end{aligned}
$$

where the prime denotes the derivative with respect to the total argument of the function.

The above coupled set of equations can, of course, be solved for the interior and scattered field multipole moments in terms of the multipole moments of the incident wave similar to those obtained for the simple single cylinder. However, it is preferable to simply represent them in matrix form, which can then be easily incorporated into Matlab or some other programming language. In matrix form these equations can be expressed in the form

$$Ha = b, \tag{6.19}$$

where

$$
H = \begin{pmatrix}
-J_l(k_0 n_{r_1} R_1) & J_l(k_0 n_{r_2} R_1) & 0 & H_l^+(k_0 n_{r_2} R_1) \\
-n_{r_1} J_l'(k_0 n_{r_1} R_1) & n_{r_2} J_l'(k_0 n_{r_2} R_1) & 0 & n_{r_2} H_l^{+\prime}(k_0 n_{r_2} R_1) \\
0 & \dfrac{J_l(k_0 n_{r_2} R_2)}{J_l(k_0 R_2)} & -\dfrac{H_l^+(k_0 R_2)}{J_l(k_0 R_2)} & \dfrac{H_l^+(k_0 n_{r_2} R_2)}{J_l(k_0 R_2)} \\
0 & n_{r_2}\dfrac{J_l'(k_0 n_{r_2} R_2)}{J_l'(k_0 R_2)} & -\dfrac{H_l^{+\prime}(k_0 R_2)}{J_l'(k_0 R_2)} & n_{r_2}\dfrac{H_l^{+\prime}(k_0 n_{r_2} R_2)}{J_l'(k_0 R_2)}
\end{pmatrix}
$$

and

$$
a = \begin{pmatrix} a_l^{(1)}(\nu) \\ a_l^{(2)}(\nu) \\ b_l(\nu) \\ b_l^{(2)}(\nu) \end{pmatrix}, \qquad
b = \begin{pmatrix} 0 \\ 0 \\ a_{0l}(\nu) \\ a_{0l}(\nu) \end{pmatrix}.
$$

6.4 The scattering amplitude

An important class of incident wavefields consists of the plane waves

$$U^{(\mathrm{in})}(\mathbf{r}, \mathbf{s}_0) = e^{ik_0 \mathbf{s}_0 \cdot \mathbf{r}}, \tag{6.20}$$

where $\mathbf{s}_0$ is a unit vector along the direction of propagation of the plane wave and the parameter ν in the argument of the incident wave is taken to be this unit wave vector. The LS equation for plane-wave incidence assumes the form

$$U(\mathbf{r};\mathbf{s}_0) = e^{ik_0\mathbf{s}_0\cdot\mathbf{r}} + \overbrace{\int d^3r'\, G_{0_+}(\mathbf{r}-\mathbf{r}')V(\mathbf{r}')U(\mathbf{r}';\mathbf{s}_0)}^{U_+^{(s)}(\mathbf{r},\mathbf{s}_0)}. \tag{6.21}$$

The *scattering amplitude* $f(\mathbf{s},\mathbf{s}_0)$ is defined to be *the radiation pattern associated with the induced source in a single scattering experiment employing an incident plane wave.* This quantity is obtained directly from the expression for the scattered field given in Eq. (6.21) by letting $r \to \infty$ along the direction of the unit vector $\mathbf{s}$. On making use of the definition of the outgoing-wave Green function in Eq. (6.6b) we find that

$$G_{0_+}(r\mathbf{s}-\mathbf{r}') \sim -\frac{1}{4\pi}e^{-ik_0\mathbf{s}\cdot\mathbf{r}'}\frac{e^{ik_0r}}{r}, \qquad r\to\infty,$$

from which we obtain the result

$$U^{(s)}(k_0\mathbf{s};\mathbf{s}_0) \sim f(\mathbf{s},\mathbf{s}_0)\frac{e^{ik_0r}}{r},$$

where the scattering amplitude is related to the scattering potential and total field via the equation

$$f(\mathbf{s};\mathbf{s}_0) = \frac{-1}{4\pi}\int d^3r\, V(\mathbf{r})U(\mathbf{r};\mathbf{s}_0)e^{-ik_0\mathbf{s}\cdot\mathbf{r}}. \tag{6.22}$$

The above expression for the scattering amplitude is also directly obtained from the radiation pattern given in Eq. (2.24b) of Chapter 2 upon setting the source Q equal to the induced source VU.

Example 6.1 We consider a homogeneous sphere with constant relative index of refraction n_r centered at the origin and excited by an incident plane wave $U^{(\mathrm{in})}(\mathbf{r},\mathbf{s}_0) = \exp(ik_0\mathbf{s}_0\cdot\mathbf{r})$. The incident plane wave admits the multipole expansion (cf. Example 3.4 in Section 3.3 of Chapter 3)

$$e^{ik_0\mathbf{s}_0\cdot\mathbf{r}} = 4\pi\sum_{l=0}^{\infty}\sum_{m=-l}^{l} i^l j_l(k_0r)Y_l^m(\mathbf{r})Y_l^{m*}(\mathbf{s}_0), \tag{6.23}$$

from which we find the multipole moments $a_0{}_l^m(\mathbf{s}_0)$ of the incident wave in Eq. (6.12b) to be given by

$$a_0{}_l^m(\mathbf{s}_0) = 4\pi i^l Y_l^{m*}(\mathbf{s}_0).$$

The multipole moments of the scattered wave are then found from Eqs. (6.14) to be

$$b_l^m(\mathbf{s}_0) = R_l(\mathbf{s}_0)a_0{}_l^m(\mathbf{s}_0) = 4\pi R_l i^l Y_l^{m*}(\mathbf{s}_0),$$

with R_l given by Eq. (6.14c).

The scattering amplitude is obtained from the multipole expansion of the scattered wave-field given in Eq. (6.12b) with $\nu = \mathbf{s}_0$ by making use of the asymptotic expression (cf. Eq. (4.8.4) of Chapter 4)

$$h_l^+(k_0 r) \sim (-i)^{l+1} \frac{e^{ik_0 r}}{k_0 r}, \quad r \to \infty.$$

We then find that

$$U_+^{(s)}(\mathbf{r}, \mathbf{s}_0) = \sum_{l=0}^{\infty} \sum_{m=-l}^{l} \overbrace{4\pi R_l i^l Y_l^{m*}(\mathbf{s}_0)}^{b_l^m(\mathbf{s}_0)} h_l^+(k_0 r) Y_l^m(\hat{\mathbf{r}})$$
$$\sim -\frac{4\pi i}{k_0} \sum_{l=0}^{\infty} \sum_{m=-l}^{l} R_l Y_l^m(\hat{\mathbf{r}}) Y_l^{m*}(\mathbf{s}_0) \frac{e^{ik_0 r}}{r},$$

from which we conclude that the scattering amplitude is given by

$$f(\mathbf{s}, \mathbf{s}_0) = -\frac{4\pi i}{k_0} \sum_{l=0}^{\infty} \sum_{m=-l}^{l} R_l Y_l^m(\mathbf{s}) Y_l^{m*}(\mathbf{s}_0).$$

6.4.1 The scattering amplitude in 2D space

The scattered field in two space dimensions is given by the obvious generalization of Eq. (6.6a) to 2D:

$$U_+^{(s)}(\mathbf{r}, \nu) = \int d^2r' \, G_{0_+}(\mathbf{r} - \mathbf{r}') V(\mathbf{r}') U(\mathbf{r}', \nu), \tag{6.24}$$

where $G_{0_+}(\mathbf{r} - \mathbf{r}')$ is now the outgoing-wave Green function in two space dimensions. This quantity was derived in Chapter 2 and is given in Eq. (2.19a) of Section 2.2.1:

$$G_+(\mathbf{R}) = \frac{-i}{4} H_0^+(k_0 R), \tag{6.25}$$

where H_0^+ is the zeroth-order Hankel function of the first kind. The 2D Green function admits the asymptotic expansion

$$G_+(\mathbf{r} - \mathbf{r}') \sim -\sqrt{\frac{1}{8\pi k_0}} e^{i\frac{\pi}{4}} e^{-ik_0 \mathbf{s} \cdot \mathbf{r}'} \frac{e^{ik_0 r}}{\sqrt{r}}, \quad r \to \infty, \tag{6.26}$$

where $\mathbf{s} = \mathbf{r}/r$ is the unit vector along the direction of $\mathbf{r}$. On substituting Eq. (6.26) into Eq. (6.24) and setting $\nu = \mathbf{s}_0$ corresponding to plane-wave incidence we obtain

$$U_+^{(s)}(\mathbf{r}, \mathbf{s}_0) \sim f(\mathbf{s}, \mathbf{s}_0) \frac{e^{ik_0 r}}{\sqrt{r}}, \quad r \to \infty,$$

where

$$f(\mathbf{s}, \mathbf{s}_0) = -\sqrt{\frac{1}{8\pi k_0}} e^{i\frac{\pi}{4}} \int d^2r\, V(\mathbf{r}) U(\mathbf{r}; \mathbf{s}_0) e^{-ik_0 \mathbf{s}\cdot\mathbf{r}} \tag{6.27}$$

is the 2D scattering amplitude. Except for the constant multiplying factors and the dimensions of the functions, the expressions for the 3D and 2D scattering amplitudes are seen to be formally identical.

Example 6.2 We consider a homogeneous cylinder with constant relative index of refraction n_r centered at the origin and excited by an incident plane wave propagating in the plane that is perpendicular to the axis of the cylinder. The incident plane wave admits the multipole expansion (cf. Example 3.9 in Section 3.6 of Chapter 3)

$$e^{ik_0\mathbf{s}_0\cdot\mathbf{r}} = \sum_{l=-\infty}^{\infty} i^l e^{-il\phi_0} J_l(k_0 r) e^{il\phi},$$

where (r, ϕ) are polar coordinates in the (x, y) plane of a Cartesian coordinate system whose z axis is aligned along the cylinder axis and ϕ_0 is the polar angle of the incident plane wave as measured from the positive x axis so that $\mathbf{s}_0 = (\cos\phi_0, \sin\phi_0)$. The multipole moments of the incident plane wave are thus given by $a_{0l}(\mathbf{s}_0) = i^l \exp(-il\phi_0)$, from which we conclude that

$$b_l(\mathbf{s}_0) = R_l a_{0l} = i^l \exp(-il\phi_0) R_l,$$

with R_l being the generalized reflection coefficients for the cylinder defined in Eq. (6.17c). The scattering amplitude is found directly from the multipole expansion of the scattered wavefield Eq. (6.15b) with $\nu = \mathbf{s}_0$ on making use of the asymptotic expression (cf. Eq. (4.9) of Chapter 4)

$$H_l^+(k_0 r) \sim \sqrt{\frac{2}{\pi k_0}} e^{-i(l+\frac{1}{2})\frac{\pi}{2}} \frac{e^{ik_0 r}}{\sqrt{r}}, \qquad r \to \infty. \tag{6.28}$$

We find that

$$U_+^{(s)}(\mathbf{r}, \mathbf{s}_0) = \sum_{l=-\infty}^{\infty} \overbrace{i^l R_l e^{-il\phi_0}}^{b_l(\mathbf{s}_0)} H_l^+(k_0 r) e^{il\phi}$$
$$\sim \left\{ \sqrt{\frac{2}{\pi k_0}} e^{-i\frac{\pi}{4}} \sum_{l=-\infty}^{\infty} R_l e^{il(\phi-\phi_0)} \right\} \frac{e^{ik_0 r}}{\sqrt{r}}, \qquad r \to \infty,$$

from which we conclude that

$$f(\mathbf{s}, \mathbf{s}_0) = \sqrt{\frac{2}{\pi k_0}} e^{-i\frac{\pi}{4}} \sum_{l=-\infty}^{\infty} R_l e^{il(\phi-\phi_0)}. \tag{6.29}$$

Example 6.3 The results obtained in the previous example are easily extended to scattering from a pair of concentric cylinders. We employ the same expansion of the incident plane wave as that given in Example 6.2 but now use the multipole moment $b_l(\mathbf{s}_0)$ computed from the matrix equation Eq. (6.19) in place of those computed for the cylinder according to Eq. (6.17c). The scattering amplitude is obtained from the multipole expansion of the scattered wave given in Eq. (6.18c) on making use of the asymptotic expression for the Hankel function given in Eq. (6.28) of the previous example. We then again obtain the expansion Eq. (6.29) of that example with the generalized reflection coefficients R_l given by $R_l = b_l(\mathbf{s}_0)/a_{0l}(\mathbf{s}_0)$, where the $b_l(\mathbf{s}_0)$ are obtained as the solutions to the matrix equation Eq. (6.19), with the $a_{0l}(\mathbf{s}_0)$ again given by $i^l \exp(-il\phi_0)$.

6.4.2 Reciprocity and translation theorems for the scattering amplitude

Like the Green function $G_+(\mathbf{r}, \mathbf{r}')$, the scattering amplitude satisfies a reciprocity condition that can easily be derived from the LS equation Eq. (6.9a) for the total Green function. This theorem, which is established in Appendix A, is useful for checking whether a given function $f(\mathbf{s}, \mathbf{s}_0)$ is a realizable scattering amplitude and is also useful for checking for redundant scattering data in inverse-scattering applications. The theorem can be stated in the following form.

Theorem 6.1 (the scattering-amplitude reciprocity theorem) *The scattering amplitude $f(\mathbf{s}, \mathbf{s}_0)$ for a compactly supported scattering potential embedded in a uniform background medium must satisfy the reciprocity condition*

$$f(\mathbf{s}, \mathbf{s}_0) = f(-\mathbf{s}_0, -\mathbf{s}). \tag{6.30}$$

We illustrate the theorem in Fig. 6.1, where a scattering potential is interrogated by a plane wave propagating in the $\mathbf{s}_0$ direction and the scattering amplitude is measured in the direction defined by the unit vector $\mathbf{s}$. In a second experiment the potential is probed by a plane wave propagating in the $-\mathbf{s}$ direction and the scattering amplitude is measured along the $-\mathbf{s}_0$ direction. The reciprocity theorem then states that the scattering amplitudes acquired in the two experiments are identical.

A second theorem, which is useful in certain inverse-scattering applications, relates the scattering amplitude for a scattering potential centered at a point $\mathbf{X}$ in a homogeneous background medium to the scattering amplitude for the same potential centered at the origin; i.e., at $\mathbf{X} = 0$. This theorem, which is also established in Appendix A, takes the following form.

Theorem 6.2 (the scattering-amplitude translation theorem) *The scattering amplitude $f_{\mathbf{X}}(\mathbf{s}, \mathbf{s}_0)$ for a compactly supported scattering potential centered at $\mathbf{X}$ in a uniform background medium is related to the scattering amplitude $f(\mathbf{s}, \mathbf{s}_0)$ for the same potential when it is centered at the origin $\mathbf{X} = \mathbf{0}$ via the equation*

$$f_{\mathbf{X}}(\mathbf{s}, \mathbf{s}_0) = e^{-ik_0(\mathbf{s}-\mathbf{s}_0)\cdot\mathbf{X}} f(\mathbf{s}, \mathbf{s}_0). \tag{6.31}$$

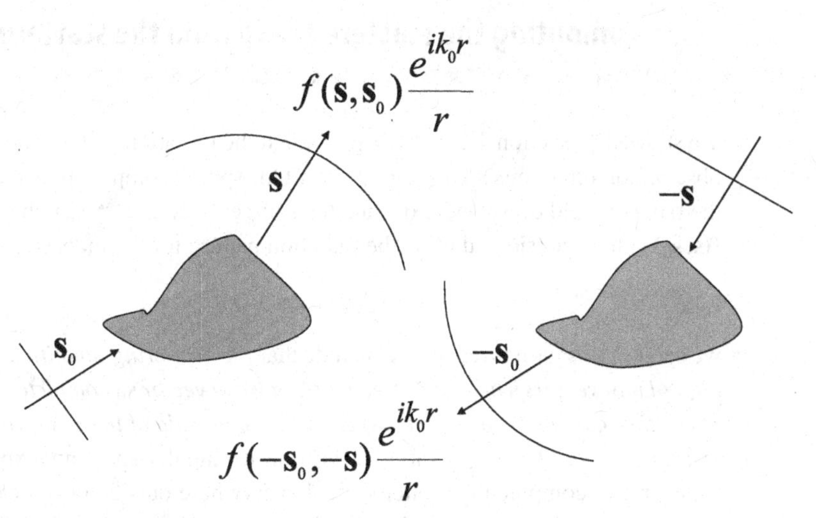

Fig. 6.1 An illustration of the scattering-amplitude reciprocity theorem.

6.4.3 Scattered field energy and the optical theorem

We showed in Section 1.6 of Chapter 1 that the energy radiated from a source having a radiation pattern $f(\mathbf{s},\omega)$ in a non-dispersive medium with wavenumber $k_0 = \omega/c$ is given by

$$E(\omega) = 2\kappa\omega k_0 \int_{4\pi} d\Omega_s |f(\mathbf{s},\omega)|^2,$$

where κ is a real-valued constant that depends on the nature of the scalar field (acoustic, optical, etc.). Since the scattered amplitude for any given scattering experiment (for a given incident wave) is formally equivalent to the radiation pattern for the induced source generated in that experiment, it immediately follows that the scattered field energy is given by the above equation with the radiation pattern replaced by the scattering amplitude; i.e.,

$$E_{\mathbf{s}_0}(\omega) = 2\kappa\omega k_0 \int_{4\pi} d\Omega_s |f(\mathbf{s},\mathbf{s}_0)|^2. \tag{6.32}$$

The scattered field energy for plane-wave scattering is related to the forward-scattering amplitude $f(\mathbf{s}_0,\mathbf{s}_0)$ via a very simple relationship, which is known as the *optical theorem* and is established in Appendix A.

Theorem 6.3 (the optical theorem) *The scattered field energy for plane-wave scattering in a non-dispersive background medium is related to the forward-scattering amplitude via the equation*

$$E_{\mathbf{s}_0}(\omega) = 2\kappa\omega k_0 \int_{4\pi} d\Omega_s |f(\mathbf{s},\mathbf{s}_0)|^2 = 8\pi\kappa\omega\Im f(\mathbf{s}_0,\mathbf{s}_0).$$

6.5 Computing the scattered field from the scattering amplitude

We showed in Section 4.2.2 of Chapter 4 that the radiation pattern $f(\mathbf{s})$ specified for all unit observation directions $\mathbf{s}$ lying on the real unit sphere completely and uniquely determines the radiated field everywhere outside the source volume τ_0. Since the scattering amplitude $f(\mathbf{s}, \mathbf{s}_0)$ can be considered to be the radiation pattern for the induced source

$$Q(\mathbf{r}, \mathbf{s}_0) = V(\mathbf{r})U(\mathbf{r}; \mathbf{s}_0),$$

we can use this earlier result to conclude that *the scattering amplitude* $f(\mathbf{s}, \mathbf{s}_0)$ *specified for all real unit vectors* $\mathbf{s}$ *and any given incident wave vector* $\mathbf{s}_0$ *completely and uniquely determines the scattered field* $U_+^{(s)}(\mathbf{r}, \mathbf{s}_0)$ *everywhere outside of the scattering volume* τ_0. Moreover, we can use the angle-variable form of the angular-spectrum expansion developed in Chapter 4 to compute the scattered field everywhere outside the so-called *convex hull*[4] of the scattering volume τ_0 directly from the (analytically continued) scattering amplitude. In particular, it follows from the analysis presented in Section 4.3 that if the scattering volume is supported within the strip $z^- < z < z^+$ then at all field points lying outside this strip

$$U_+^{(s)}(\mathbf{r}, \mathbf{s}_0) = \frac{ik_0}{2\pi} \int_{-\pi}^{\pi} d\beta \int_{C_\pm} d\alpha \sin\alpha f(\mathbf{s}, \mathbf{s}_0) e^{ik_0 \mathbf{s}\cdot\mathbf{r}}, \tag{6.33}$$

where $\mathbf{s} = (\sin\alpha\cos\beta, \sin\alpha\sin\beta, \cos\alpha)$ and such that the α-integration contour $C_+ = [0 : \pi/2 - i\infty]$ is used in the r.h.s. $z > z^+$ and $C_- = [\pi/2 + i\infty : \pi]$ is used in the l.h.s. $z < z^-$. We note that the two contours $C_\pm$ can be arbitrarily deformed due to the analyticity of the scattering amplitude as a function of the unit vector $\mathbf{s}$ (see the discussion in Section 4.2.2).

Since the orientation of the fixed x, y, z Cartesian coordinate system is arbitrary we can employ the above expansion to compute the scattered field throughout any two half-spaces bounded by parallel tangent planes to the scattering volume and, hence, everywhere outside of the convex hull of the scattering volume. The scattered field so computed can, in principle, be continued into the interior of the convex hull up to the actual boundary of the scattering volume τ_0 using the multipole solution of the interior boundary-value problem developed in Section 4.8 of Chapter 4. However, such a continuation and, indeed, even the exact calculation of the scattered field by means of Eq. (6.33) in the near field of the scattering potential would be computationally unstable. Such calculations serve mainly as mathematical tools by means of which to establish ideal properties of the field. In particular, Eq. (6.33) requires an analytic continuation of the scattering amplitude from scattering directions $\mathbf{s}$ lying on the (observable) real unit sphere onto the complex contour required in this integral. As discussed in Chapter 4, such a continuation is unstable and cannot be used in any practical application.

[4] The convex hull of a simply connected volume τ_0 is the smallest convex region that completely contains τ_0.

Although the scattered field cannot be exactly computed outside the convex hull of the scattering volume τ_0 using Eq. (6.33), it can be approximately computed using a stabilized version of this expansion as described in Section 4.3 of Chapter 4 by simply eliminating the evanescent plane waves from the expansion. We will use this stabilized version of scattered field back propagation from the scattering amplitude in a number of applications in later chapters.

6.5.1 Field scattered by an arbitrary incident wave and the generalized scattering amplitude

The angular-spectrum expansion allows us to construct the wavefield scattered by an incident plane wave everywhere outside the convex hull of the scattering potential from the scattering amplitude. However, since the plane waves form a complete set into which any incident wave can be expanded and the scattering process is represented by a linear transform from incident to scattered wave, we can use the angular-spectrum expansion to represent the wavefield scattered by any arbitrary incident wavefield. To obtain this field representation, we first expand the incident wave in a homogeneous plane-wave expansion of the form (cf. Section 3.2.1 of Chapter 3)

$$U^{(\text{in})}(\mathbf{r},\nu) = \int_{4\pi} d\Omega_{s_0} A(\mathbf{s}_0,\nu) e^{ik_0\mathbf{s}_0\cdot\mathbf{r}}, \tag{6.34}$$

where the integral is over the entire unit sphere of incident unit propagation vectors $\mathbf{s}_0$. If we now substitute this expansion into the expression for the scattered field in terms of the full Green function G_+ given by Eq. (6.11a) we obtain the result

$$\begin{aligned}
U_+^{(\text{s})}(\mathbf{r},\nu) &= \int d^3r'\, G_+(\mathbf{r},\mathbf{r}')V(\mathbf{r}') \overbrace{\left\{\int_{4\pi} d\Omega_{s_0} A(\mathbf{s}_0,\nu) e^{ik_0\mathbf{s}_0\cdot\mathbf{r}'}\right\}}^{U^{(\text{in})}(\mathbf{r}',\nu)} \\
&= \int_{4\pi} d\Omega_{s_0} A(\mathbf{s}_0,\nu) \overbrace{\left\{\int d^3r'\, G_+(\mathbf{r},\mathbf{r}')V(\mathbf{r}') e^{ik_0\mathbf{s}_0\cdot\mathbf{r}'}\right\}}^{U_+^{(\text{s})}(\mathbf{r};\mathbf{s}_0)} \\
&= \int_{4\pi} d\Omega_{s_0} A(\mathbf{s}_0,\nu) \overbrace{\left\{\frac{ik_0}{2\pi}\int_{-\pi}^{\pi} d\beta \int_{C_\pm} d\alpha \sin\alpha\, f(\mathbf{s},\mathbf{s}_0) e^{ik_0\mathbf{s}\cdot\mathbf{r}}\right\}}^{U_+^{(\text{s})}(\mathbf{r};\mathbf{s}_0)} \\
&= \frac{ik_0}{2\pi}\int_{-\pi}^{\pi} d\beta \int_{C_\pm} d\alpha \sin\alpha \left\{\int_{4\pi} d\Omega_{s_0} A(\mathbf{s}_0,\nu) f(\mathbf{s},\mathbf{s}_0)\right\} e^{ik_0\mathbf{s}\cdot\mathbf{r}},
\end{aligned}$$

which will converge throughout the two half-spaces lying outside the support strip $z^- < z < z^+$. We can write this in the compact form

$$U_+^{(\text{s})}(\mathbf{r},\nu) = \frac{ik_0}{2\pi}\int_{-\pi}^{\pi} d\beta \int_{C_\pm} d\alpha \sin\alpha\, f(\mathbf{s},\nu) e^{ik_0\mathbf{s}\cdot\mathbf{r}}, \tag{6.35a}$$

where $f(\mathbf{s}, \nu)$, known as the *generalized scattering amplitude*, is given by

$$f(\mathbf{s}, \nu) = \int_{4\pi} d\Omega_{s_0} A(\mathbf{s}_0, \nu) f(\mathbf{s}, \mathbf{s}_0). \tag{6.35b}$$

The generalized scattering amplitude is simply the induced source radiation pattern for the scattering experiment employing an incident wave parameterized by ν. The genesis of its name follows at once from Eq. (6.4), which states that the scattered wave behaves as an outgoing spherical wave with amplitude equal to $f(\mathbf{s}, \nu)$ as $r \to \infty$ along the direction $\mathbf{s} = \hat{\mathbf{r}} = \mathbf{r}/r$. We see from Eq. (6.35b) that *the generalized scattering amplitude for an incident wave with plane-wave amplitude $A(\mathbf{s}_0, \nu)$ is equal to the convolution of A with the plane-wave scattering amplitude.*

6.5.2 Computing the scattering amplitude from scattered field data over a plane

We showed in Section 4.2 that the angle-variable form of the angular-spectrum expansion can be converted to Cartesian-variable form by means of a simple transformation of integration variables. The Cartesian-variable form of the expansions employs the two Cartesian components of the transverse wavenumber $\mathbf{K}_\rho = (K_x, K_y)$ as integration variables and is especially suited for propagating and back-propagating scattered fields specified over plane surfaces that lie outside the scattering volume τ_0. On making this transformation we have that

$$k_0 \sin\alpha \, d\beta \, d\alpha \Rightarrow \frac{d^2 K_\rho}{\gamma}, \qquad k_0 \mathbf{s} \Rightarrow \mathbf{k}^\pm = \mathbf{K}_\rho \pm \gamma \hat{\mathbf{z}}, \tag{6.36}$$

where $\mathbf{k}^+$ corresponds to $\hat{\mathbf{z}} \cdot \mathbf{s} = s_z > 0$ and $\mathbf{k}^-$ to $s_z < 0$, and

$$\gamma = \begin{cases} \sqrt{k_0^2 - K_\rho^2} & \text{if } K_\rho < k_0, \\ i\sqrt{K_\rho^2 - k_0^2} & \text{if } K_\rho > k_0. \end{cases}$$

On making this transformation in Eq. (6.35a) we obtain

$$U_+^{(s)}(\mathbf{r}, \nu) = \frac{i}{2\pi} \int_{-\infty}^{\infty} \frac{d^2 K_\rho}{\gamma} A(\mathbf{k}^\pm, \nu) e^{i\mathbf{k}^\pm \cdot \mathbf{r}}, \tag{6.37a}$$

where

$$A(\mathbf{k}^\pm = k_0 \mathbf{s}, \nu) = f(\mathbf{s}, \nu), \quad K_\rho < k_0, \tag{6.37b}$$

with $\mathbf{k}^+$ corresponding to the scattering amplitude over the hemisphere $\hat{\mathbf{z}} \cdot \mathbf{s} = s_z > 0$ and $\mathbf{k}^-$ corresponding to the hemisphere $s_z < 0$. In words this result states that over the homogeneous region of the spectra the angular spectra $A(\mathbf{k}^+, \nu)$ for expanding the field in the r.h.s. $z > z^+$ is equal to the scattering amplitude in that half-space and similarly for the l.h.s. $z < z^-$. The angular spectra in the evanescent region $K_\rho > k_0$ is, in principle, obtained from its specification over the homogeneous region via the process of analytic continuation (see the discussion in Section 4.2.2).

By Fourier inverting both sides of Eq. (6.37a) over any plane z_0 that lies outside the support strip $z^- < z < z^+$ we find that

$$\tilde{U}_+^{(s)}(\mathbf{K}_\rho, z_0, \nu) = \frac{2\pi i}{\gamma} A(\mathbf{k}^\pm, \nu) e^{\pm i\gamma z_0}, \qquad (6.38)$$

where

$$\tilde{U}_+^{(s)}(\mathbf{K}_\rho, z_0, \nu) = \int d^2\rho_0\, U_+^{(s)}(\boldsymbol{\rho}, z_0, \nu) e^{-i\mathbf{K}_\rho \cdot \boldsymbol{\rho}_0}$$

is the spatial Fourier transform of the scattered field on the plane $z = z_0$. The above equations allow the angular spectra and, hence, the scattering amplitude to be computed directly from the scattered field lying over plane surfaces lying outside the scattering volume. For example, using these equations and Eqs. (6.37b) we find that

$$f(\mathbf{s}, \nu) = \frac{\gamma}{2\pi i} \tilde{U}_+^{(s)}(\mathbf{K}_\rho, z_0, \nu) e^{\mp i\gamma z_0}, \quad K_\rho < k_0, \qquad (6.39)$$

with

$$\mathbf{K}_\rho = k_0 s_x \hat{\mathbf{x}} + k_0 s_y \hat{\mathbf{y}}, \qquad \gamma = k_0 \sqrt{1 - s_x^2 - s_y^2},$$

where $s_z = \sqrt{1 - s_x^2 - s_y^2} > 0$ if $z_0 > z^+$ and $s_z = -\sqrt{1 - s_x^2 - s_y^2} < 0$ if $z_0 < z^-$. We note that in this equation $\gamma = \sqrt{k_0^2 - K_\rho^2}$ is purely real since the (observable) scattering amplitude is defined only over the homogeneous region of the spectra corresponding to $K_\rho < k_0$. These relationships, which form the basis for diffraction tomography (DT), will be derived using a different approach in our treatment of inverse scattering and diffraction tomography in Section 8.5 of Chapter 8.

6.5.3 Multipole expansion of the scattered field

We employed multipole expansions in our treatment of scattering by homogeneous spheres and cylinders in Section 6.3 and generalize that treatment here to scattering from general potentials. The general treatment is based on the multipole expansion of radiated fields presented in Section 4.8 of Chapter 4, where, as was the case with the angular-spectrum expansion, the multipole expansion of a scattered field is readily obtained by simply replacing the source $Q(\mathbf{r})$ by the induced source $V(\mathbf{r})U(\mathbf{r}, \nu)$ and the radiation pattern $f(\mathbf{s})$ by the (generalized) scattering amplitude $f(\mathbf{s}, \nu)$. We then obtain

$$U_+^{(s)}(\mathbf{r}, \nu) = -ik_0 \sum_{l=0}^{\infty} \sum_{m=-l}^{l} q_l^m(\nu) h_l^+(k_0 r) Y_l^m(\hat{\mathbf{r}}), \qquad (6.40a)$$

which will converge for all field points $\mathbf{r}$ lying outside the smallest sphere that completely encloses the scattering potential. In this expansion $h_l^+(\cdot)$ are the spherical Hankel functions of the first kind and Y_l^m the spherical harmonics of degree l and order m, where we have used the standard practice introduced in Chapter 3 of denoting the argument of the spherical harmonics with a unit vector rather than with the polar and azimuthal angles defining this unit vector. The spherical-harmonic expansion of the radiation pattern derived in Section 4.8.4 also leads to the following analogous expansion of the generalized scattering amplitude:

$$f(\mathbf{s}, \nu) = \sum_{l=0}^{\infty} \sum_{m=-l}^{l} f_l^m(\nu) Y_l^m(\mathbf{s}). \tag{6.40b}$$

The multipole moments $q_l^m(\nu)$ are expressed in terms of the induced source via

$$q_l^m(\nu) = \int_{\tau_0} d^3 r\, V(\mathbf{r}) U(\mathbf{r}, \nu) j_l(k_0 r) Y_l^{m*}(\hat{\mathbf{r}}), \tag{6.41a}$$

and in terms of the scattering amplitude via

$$q_l^m(\nu) = -i^l f_l^m(\nu) = -i^l \int_{4\pi} d\Omega_s\, f(\mathbf{s}, \nu) Y_l^{m*}(\mathbf{s}). \tag{6.41b}$$

It is also possible to express the multipole moments in terms of the scattered field over any sphere surrounding the scattering volume by obvious generalizations of the formulas presented in Section 4.8.

As discussed in Section 4.8, the multipole expansion will converge everywhere outside the smallest sphere that encloses the scattering volume τ_0. Moreover, the multipole moments $q_l^m(\nu)$ as well as the generalized Fourier coefficients $f_l^m(\nu)$ will decay exponentially fast with index l for $l > [k_0 a_0]$, where $[k_0 a_0]$ is the next larger integer to $k_0 a_0$, with a_0 being the radius of the support volume τ_0. In particular, as shown in Section 1.7.3, the multipole moments of any finite-norm source $Q(\mathbf{r}, \nu) = V(\mathbf{r}) U(\mathbf{r}, \nu)$ satisfy the inequality

$$|q_l^m(\nu)|^2 \leq \mathcal{E}_Q \int_0^{a_0} r^2\, dr |j_l(k_0 r)|^2 = \mathcal{E}_Q \overbrace{\frac{a_0^3}{2} [j_l^2(k_0 a_0) - j_{l-1}(k_0 a_0) j_{l+1}(k_0 a_0)]}^{\mu_l^2(k_0 a_0)}, \tag{6.42}$$

where $\mathcal{E}_Q$ is the L^2 squared norm of the induced source. The quantities $\mu_l^2(k_0 a_0)$ are plotted in Fig. 1.2 of Chapter 1, where they are shown to decay exponentially fast with index $l > [k_0 a_0]$. It then follows that the expansion Eq. (6.40a) can be terminated at $l_0 = [k_0 a_0]$ with small error and thus provides a stable scheme for field back propagation from the scattering amplitude.

Finally, we note that the bound for the scattered field multipole moments given in Eq. (6.42) is an upper bound and that these coefficients can decay faster than the $\mu_l^2(k_0a_0)$ due to the fact that the interior field $U(\mathbf{r}, \nu)$ can also decay with index l. For example, for the scattering from a homogeneous sphere considered in Section 6.3.2, the interior field admits the multipole expansion given in Eq. (6.12a), so the multipole moments of the scattered field given in Eq. (6.41a) reduce to

$$\begin{aligned} q_l^m(\nu) &= k_0^2(1-n_r^2)\int_{r<a_0} d^3r \left[\sum_{l'=0}^{\infty}\sum_{m=-l'}^{l'} a_{l'}^{m'}(\nu) j_{l'}(k_0 n_r r) Y_{l'}^{m'}(\hat{\mathbf{r}})\right] j_l(k_0 r) Y_l^{m*}(\hat{\mathbf{r}}) \\ &= k_0^2(1-n_r^2) a_l^m(\nu)\int_0^{a_0} r^2\, dr\, j_l(k_0 n_r r) j_l(k_0 r) \\ &\approx k_0^2(1-n_r^2) a_l^m(\nu)\mu_l(k_0a_0)\mu_l(k_0 n_r a_0), \end{aligned}$$

where n_r is the relative index of refraction of the sphere and we have used the Schwarz inequality. We then conclude that

$$|q_l^m(\nu)|^2 \leq |k_0^2(1-n_r^2)|^2 |a_l^m(\nu)|^2 \mu_l^2(k_0a_0)\mu_l^2(k_0n_ra_0) \tag{6.43}$$

and thus will decay as the product of $\mu_l^2(k_0a_0)$ with $\mu_l^2(k_0n_ra_0)$ rather than simply as $\mu_l^2(k_0a_0)$.

Example 6.4 We computed the multipole expansion of the field scattered from a homogeneous sphere in Section 6.3.2. We found that the scattered field admitted the multipole expansion Eq. (6.40a) with the multipole moments $q_l^m(\nu)$ given by

$$q_l^m(\nu) = \frac{i}{k_0} b_l^m(\nu) = \frac{i}{k_0} R_l a_{0_l}^m(\nu), \tag{6.44}$$

with the R_l being the generalized reflection coefficients defined in Eq. (6.14c) and the $a_{0_l}^m(\nu)$ the multipole moments of the incident wave to the sphere. So long as the incident-wave multipole moments are all finite the scattered field multipole moments should, according to the discussion given above, decay exponentially fast with $l > [k_0a_0]$ *irrespective of the index of refraction of the sphere!* We computed the scattered field multipole moments with the incident-wave multipole moments all set to unity and plot their magnitudes squared, normalized to a peak value of unity, as a function of l for four different index values in Fig. 6.2. We also plot the bounds provided by Eqs. (6.42) and (6.43), also normalized to a peak value of unity. In computing the bound in Eq. (6.43) we used the expression for the multipole moments $a_l^m(\nu) = T_l a_{0_l}^m(\nu)$ of the interior field within the sphere given in Section 6.3.2 with the $a_{0_l}^m(\nu)$ all set to unity. It is clear from Fig. 6.2 that the multipole moments computed from Eqs. (6.44) obey the bound provided by Eq. (6.43).

6.5.4 Multipole expansions of 2D scattered fields

For scattering from a 2D object we employ the multipole expansion of a 2D radiated field developed in Section 4.9, where, again, we replace the source by the induced source and

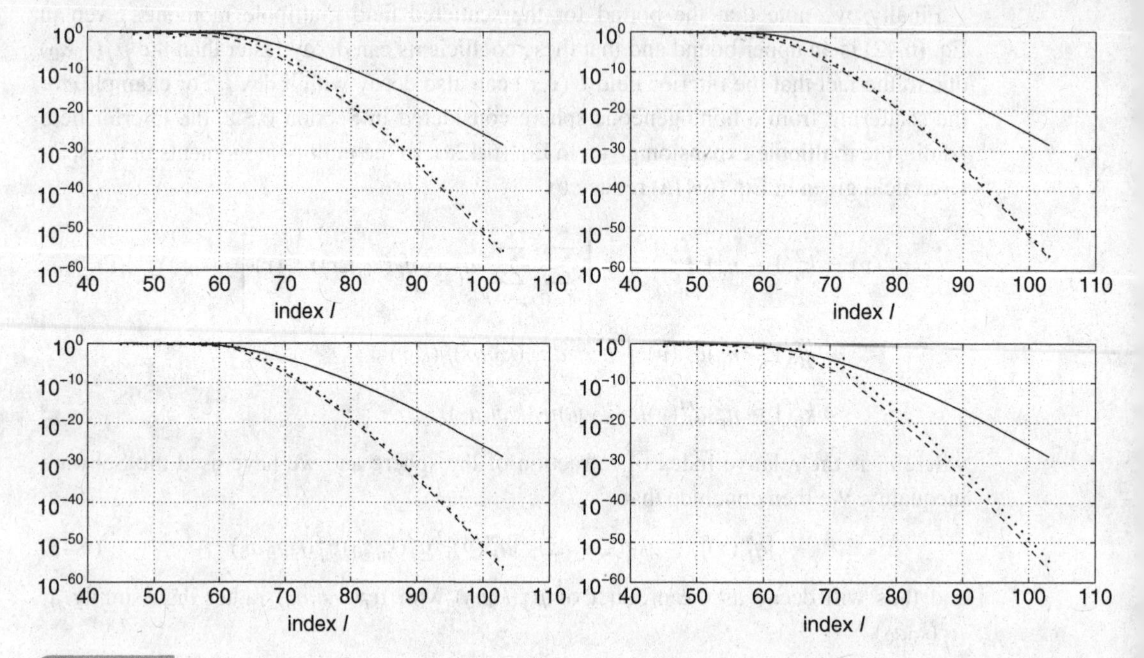

Fig. 6.2 Semi-log plots of $\mu_l^2(k_0a_0)$ (solid), $|a_l^m(\nu)|^2\mu_l^2(k_0a_0)\mu_l^2(k_0n_ra_0)$ (dotted) and $|b_l^m(\nu)|^2$ (dashed), all normalized to a peak value of unity, for $\lambda = 1$, with $a_0 = 10\lambda$, for spheres having relative indices of refraction $n_r = 0.75$ (top left), 0.95 (top right), 1.05 (bottom left) and 1.25 (bottom right).

the radiation pattern by the scattering amplitude. The multipole expansion of a scattered field then assumes the form

$$U_+^{(s)}(\mathbf{r},\nu) = -\frac{i}{4}\sum_{l=-\infty}^{\infty} q_l(\nu)H_l^+(kr)e^{il\phi}, \tag{6.45a}$$

where the multipole moments $q_l(\nu)$ are given in terms of the induced source via

$$q_l(\nu) = \int_{\tau_0} d^2r\, V(\mathbf{r})U(\mathbf{r},\nu)J_l(kr)e^{-il\phi}. \tag{6.45b}$$

The radiation pattern is found to be

$$f(\mathbf{s},\nu) = \frac{1}{2\pi}\sum_{l=0}^{\infty} f_l(\nu)e^{il\phi}, \tag{6.46}$$

where the generalized Fourier coefficients $f_l(\nu)$ are given by

$$f_l(\nu) = \int_0^{2\pi} d\phi\, f(\mathbf{s},\nu)e^{-il\phi}$$

and are related to the source multipole moments via the equation

$$f_l(\nu) = -\frac{i}{4}\sqrt{\frac{8\pi}{k}}e^{-i\frac{\pi}{4}}(-i)^l q_l(\nu).$$

6.6 The transition operator

The *transition operator* is an abstract operator that plays a key role in quantum-mechanical scattering theory and in inverse scattering. This operator maps an incident wave into the product of the scattering potential with the total (incident plus scattered) wave generated in a given scattering experiment. We can define this operator by its action on any complete set of incident wavefields via the equation

$$\hat{T}U_\nu^{(\text{in})} = VU_\nu, \tag{6.47}$$

where $U_\nu^{(\text{in})}$ is an arbitrary incident wave parameterized by ν, U_ν the corresponding total wave generated in the scattering process with the scattering potential V and $\hat{T}$ the transition operator associated with that particular scattering potential. Since the plane waves form a complete set into which any incident wavefield can be expanded (cf. Eq. (6.34)) the transition operator is completely defined by its action on the plane waves, which we can write in the form

$$\hat{T}e^{ik_0\mathbf{s}_0\cdot\mathbf{r}} = V(\mathbf{r})U(\mathbf{r},\mathbf{s}_0), \tag{6.48}$$

where $\mathbf{s}_0$ is any vector on the unit sphere.

The transition operator as defined in Eq. (6.48) maps the plane waves parameterized by propagation vectors $k_0\mathbf{s}_0$ into square-integrable functions supported in the scattering volume τ_0 (the product of the scattering potential with the field generated in the scattering experiment). By taking the spatial Fourier transform of this equation we can then obtain the matrix elements $T(\mathbf{K}, k_0\mathbf{s}_0)$ of the transition operator between the plane waves having wavenumbers k_0 and the plane waves $\exp(i\mathbf{K}\cdot\mathbf{r})$ whose wavenumber is $K = |\mathbf{K}|$; i.e.,

$$\begin{aligned} T(\mathbf{K}; k_0\mathbf{s}_0) &= \langle e^{i\mathbf{K}\cdot\mathbf{r}}, \hat{T}e^{ik_0\mathbf{s}_0\cdot\mathbf{r}}\rangle \\ &= \int d^3r\, e^{-i\mathbf{K}\cdot\mathbf{r}}\{\hat{T}e^{ik_0\mathbf{s}_0\cdot\mathbf{r}}\} = \int d^3r\, V(\mathbf{r})U(\mathbf{r};\mathbf{s}_0)e^{-i\mathbf{K}\cdot\mathbf{r}}, \end{aligned} \tag{6.49}$$

where the bracket $\langle\cdot,\cdot\rangle$ stands for the standard inner product in the Hilbert space of square-integrable functions in R^3. We refer to the matrix $T(\mathbf{K}; k_0\mathbf{s}_0)$ as the *T matrix*. It completely defines the transition operator and, indeed, is the *momentum-space representation* of this operator in the bra–ket language of quantum-mechanical scattering theory.

On comparing Eq. (6.22) with Eq. (6.49) we conclude that the *scattering amplitude* $f(\mathbf{s};\mathbf{s}_0)$ *is proportional to the boundary value of the T matrix over the sphere* $|\mathbf{K}| = K = k_0$. Because of this, the scattering amplitude is easily computed directly from the T matrix for any given set of incident and scattered wave vectors. It is important to note that, while the scattering amplitude can be computed from the T matrix, *it is not evident that the T matrix is computable from the scattering amplitude.* The reason for this is that the T matrix is a function of five variables, namely the three components of the wave vector $\mathbf{K}$ and the two free components of the incident unit wave vector $\mathbf{s}_0$, while the scattering amplitude is a function only of four variables (the two free components each of $\mathbf{s}$ and $\mathbf{s}_0$). Thus, although the scattering amplitude is an analytic function of the four components of the unit vectors $\mathbf{s}$ and $\mathbf{s}_0$ continued onto the complex unit spheres, the T matrix is an analytic function of five complex variables and, hence, cannot be *directly* analytically continued from its boundary value on a four-dimensional (4D) surface.

While the scattering amplitude $f(\mathbf{s}, \mathbf{s}_0)$ uniquely determines the scattered field everywhere outside the scattering volume τ_0, the T matrix allows this field to be determined everywhere, including within the support τ_0 of the scattering potential. In particular, on Fourier inverting Eq. (6.49) and substituting the result into the expression Eq. (6.10) for the scattered field we find that

$$U_+^{(s)}(\mathbf{r}; \mathbf{s}_0) = \int d^3r' \, G_{0+}(\mathbf{r} - \mathbf{r}') \overbrace{\left\{ \frac{1}{(2\pi)^3} \int d^3K \, T(\mathbf{K}; k_0\mathbf{s}_0) e^{i\mathbf{K}\cdot\mathbf{r}'} \right\}}^{V(\mathbf{r}')U(\mathbf{r}';\mathbf{s}_0)}$$
$$= \frac{1}{(2\pi)^3} \int d^3K \, \frac{T(\mathbf{K}; k_0\mathbf{s}_0)}{k_0^2 - K^2} e^{i\mathbf{K}\cdot\mathbf{r}},$$

where we have used the result (cf. Section 2.2 of Chapter 2)

$$\tilde{G}_{0+}(\mathbf{K}) = \frac{1}{k_0^2 - K^2}.$$

Because the T matrix allows the scattered field to be determined within the support of the scattering potential it follows that the scattering potential is uniquely determined from the T matrix; e.g., via the algorithm

$$V(\mathbf{r}) = \frac{[\nabla^2 + k_0^2] U_+^{(s)}(\mathbf{r}; \mathbf{s}_0)}{e^{ik_0\mathbf{s}_0\cdot\mathbf{r}} + U_+^{(s)}(\mathbf{r}; \mathbf{s}_0)}.$$

The inverse scattering problem can thus be viewed as being equivalent to computing the T matrix from the scattering amplitude. However, as mentioned above, this would require some sort of analytic continuation and, hence, would not be a stable procedure.

6.6.1 The Lippmann–Schwinger equation for the transition operator

If we multiply both sides of the abstract form of the Lippmann–Schwinger equation Eq. (6.7b) by the scattering potential we obtain the result

$$VU_\nu = VU_\nu^{(\text{in})} + VG_{0+} VU_\nu.$$

If we now make use of the definition of the transition operator in Eq. (6.47) we obtain the result

$$\hat{T}U_\nu^{(\text{in})} = VU_\nu^{(\text{in})} + VG_{0+} \hat{T}U_\nu^{(\text{in})}.$$

On specializing this equation to the case of plane-wave incidence and using the fact that the plane waves are complete in the space of incident waves we then conclude that the transition operator satisfies the operator Lippmann–Schwinger equation:

$$\hat{T} = V + VG_{0_+}\hat{T}. \tag{6.50}$$

6.7 The Born series

The Lippmann–Schwinger (LS) integral equation Eq. (6.7) can be formally solved using a Liouville–Neumann perturbation expansion commonly known as the *Born series* after Max Born who first employed it in quantum scattering theory. To develop this expansion it is convenient to employ the symbolic form of the LS equation given in Eq. (6.7b). If we associate a perturbation parameter ϵ with the scattering potential (let $V \to \epsilon V$) and expand the field $U(\mathbf{r}, \nu)$ in a power series in this parameter we obtain

$$U = [1 + \epsilon G_{0_+}V + \epsilon^2 G_{0_+}VG_{0_+}V + \cdots]U^{(\text{in})},$$

which, if convergent, tends to a solution of the (forward) scattering problem for the potential ϵV. It is clear from the Born series that the mapping from the incident field to the scattered field is *linear*, while the mapping from the scattering potential to the scattered field is *non-linear*. As discussed earlier, this non-linear character of the scattering-potential-to-field map is the underlying reason for the great difficulty of inverse scattering theory and is the major driving force behind the use of linearizing approximations in this discipline.

If we re-absorb the perturbation parameter ϵ into the scattering potential (i.e., let $\epsilon V \to V$) the Born series assumes the form

$$U = [1 + G_{0_+}V + G_{0_+}VG_{0_+}V + \cdots]U^{(\text{in})}, \tag{6.51}$$

a result that can also be obtained directly from Eq. (6.11a); i.e.,

$$U = (I - G_{0_+}V)^{-1}U^{(\text{in})} = \sum_{n=0}^{\infty}(G_{0_+}V)^n U^{(\text{in})}.$$

For this reason the series solution Eq. (6.51) is sometimes referred to as the inversion of the operator consisting of the identity plus a small perturbation.

Born series for the transition operator

Using a completely parallel development, we can expand the transition operator defined in Eq. (6.50) in the Born series

$$\hat{T} = V + VG_{0_+}V + VG_{0_+}VG_{0_+}V + \cdots,$$

a result that also follows directly from the Born series Eq. (6.51) for the field and the definition of the transition operator Eq. (6.47).

6.7.1 The Born approximation

The celebrated *Born approximation* results from dropping all terms in the Born series Eq. (6.51) except for the first two terms. Thus we have that

$$\begin{aligned} U_{\rm B}(\mathbf{r},\nu) &= [1+G_{0_+}V]U^{(\rm in)} \\ &= U^{(\rm in)}(\mathbf{r},\nu) + \int d^3r'\, G_{0_+}(\mathbf{r}-\mathbf{r}')V(\mathbf{r}')U^{(\rm in)}(\mathbf{r}',\nu), \end{aligned} \tag{6.52}$$

where we have used the subscript B to denote the Born approximation. The Born approximation Eq. (6.52) is seen to be a *linear mapping* from the scattering potential V to the scattered field:

$$U_{\rm B}^{(\rm s)}(\mathbf{r},\nu) = \int d^3r'\, G_{0_+}(\mathbf{r}-\mathbf{r}')V(\mathbf{r}')U^{(\rm in)}(\mathbf{r}',\nu). \tag{6.53}$$

The inverse scattering problem within the Born approximation consists of inverting the *set* of equations Eqs. (6.53) for the scattering potential from measurements of the scattered field obtained in a suite of scattering experiments using a set of incident waves $U^{(\rm in)}(\mathbf{r},\nu)$.

The philosophy employed in the Born approximation to the inverse scattering problem is that *we seek an* **exact** *inversion to an* **approximate** *formulation of the (forward) scattering problem.* Thus we regard the Born scattering model as given by Eqs. (6.52) and (6.53) as *exact* formulations of the forward scattering problem and seek *exact* inversions of these equations; i.e., we assume that these equations accurately model the actual measured scattered field data and seek to recover a scattering potential V that satisfies this set of equations. This philosophy is not the only possible approach to the inverse scattering problem and, for example, we could start with the exact (non-linear) scattering model as described by the LS integral equations Eqs. (6.7) and seek *approximate* solutions to this *exact* forward model. The problem with the latter approach is that it can be addressed only by ad-hoc and, generally, numerically based inversion schemes and does not lend itself to any systematic treatment of the inverse problem. The advantage of the linearized Born model is that we can employ a systematic procedure to develop *analytic* inversion schemes that can later be generalized to include non-linear effects in the actual scattering experiments. For example, we can extend all of the analysis contained in this and the following chapter to non-constant backgrounds characterized by a (known) wavenumber $k_0 = k_0(\mathbf{r})$ that varies with position. Such a generalization is based on the so-called *distorted-wave Born approximation* (DWBA), which will be developed in Chapter 9.

6.7.2 Incident plane waves

For the important class of incident plane waves the scattering amplitude within the Born approximation is obtained from Eq. (6.22) upon setting

$$U(\mathbf{r};\mathbf{s}_0) \approx U^{(\rm in)}(\mathbf{r};\mathbf{s}_0) = e^{ik_0\mathbf{s}_0\cdot\mathbf{r}}.$$

We then obtain the result

$$f_B(\mathbf{s};\mathbf{s}_0) = \frac{-1}{4\pi}\int d^3r'\, V(\mathbf{r}')e^{-ik_0(\mathbf{s}-\mathbf{s}_0)\cdot\mathbf{r}'}.$$

If we define the spatial Fourier transform of the scattering potential via the equation

$$\tilde{V}(\mathbf{K}) = \int d^3r\, V(\mathbf{r})e^{-i\mathbf{K}\cdot\mathbf{r}},$$

we see that the Born approximation to the scattering amplitude is related to $\tilde{V}$ via the equation

$$f_B(\mathbf{s},\mathbf{s}_0) = \frac{-1}{4\pi}\tilde{V}[k_0(\mathbf{s}-\mathbf{s}_0)]. \tag{6.54}$$

Equation (6.54) relates the scattering amplitude to the spatial Fourier transform of the scattering potential evaluated over the set of spatial frequencies defined by the equation

$$\mathbf{K} = k_0(\mathbf{s}-\mathbf{s}_0). \tag{6.55}$$

Let us now assume that the background medium is non-absorbing, so that the wavenumber k_0 is real-valued. In this case the spatial frequencies defined in Eq. (6.55) are real-valued and map out the surface of a sphere of radius k_0 that is centered at the point $\mathbf{K} = -k_0\mathbf{s}_0$ in $\mathbf{K}$ space. This sphere is known as an *Ewald sphere*, so that for non-absorbing background media and within the Born approximation the scattered field generated in any single scattering experiment depends only on the spatial Fourier transform of the scattering potential over the surface of a single Ewald sphere. Conversely, the scattered field data acquired in any single experiment determine this transform over the surface of a single Ewald sphere. If a complete (countably infinite) set of experiments with $\mathbf{s}_0$ varying over the entire unit sphere is performed, the transform $\tilde{V}(\mathbf{K})$ would be determined from this entire suite of data over the union of the surfaces of an infinite number of Ewald spheres, which is easily seen to be the interior of a sphere centered at the origin of $\mathbf{K}$ space that has an outer radius of $2k_0$ (the so-called *Ewald limiting sphere*). The situation is illustrated in Fig. 6.3.

The scattering potential $V(\mathbf{r})$ is compactly supported within the volume τ_0 and everywhere finite. It then follows that its Fourier transform $\tilde{V}(\mathbf{K})$ is an *entire analytic* function of the spatial frequency vector $\mathbf{K} = (K_x, K_y, K_z)$ extended into complex $\mathbf{K}$ space (or, equivalently, into the three complex planes associated with the Cartesian components K_x, K_y, K_z). An entire analytic function of three complex variables is completely determined by its specification over any finite volume in the space of the three variables. Examples of such a volume would be the interior of a cube or sphere in real $\mathbf{K}$ space. Although a single Ewald sphere will not provide such a specification, a *full set* of Ewald spheres will. By a "full set" we mean the Ewald spheres generated from an ideal complete set of scattering experiments in which the scattering amplitude $f(\mathbf{s},\mathbf{s}_0)$ is determined for all pairs of unit vectors $\mathbf{s}$ and $\mathbf{s}_0$. Referring to Fig. 6.3, we see that such a complete data set results in specification of the scattering potential Fourier transform throughout the Ewald limiting sphere, which is a

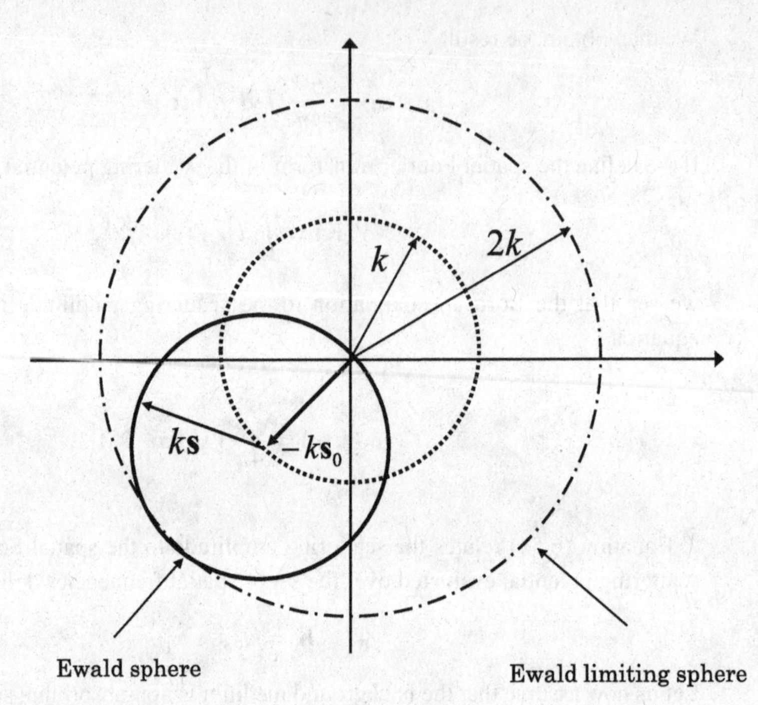

Fig. 6.3 A graphical illustration of an Ewald sphere and an Ewald limiting sphere for a penetrable scatterer embedded in a non-absorbing medium.

volume in 3D **K** *space* and, hence, allows $\tilde{V}(\mathbf{K})$ to be analytically continued into the entire **K** space.

The process of analytic continuation is, of course, an unstable process and, although the above comments indicate that knowledge of $\tilde{V}(\mathbf{K})$ throughout the entire Ewald limiting sphere is, in principle, sufficient to yield an exact solution to the inverse scattering problem for non-absorbing background media within the Born approximation, the situation is quite different in practice. However, it is clear that specification of the transform $\tilde{V}$ throughout the interior of the Ewald limiting sphere leads directly to a stable *low-pass-filtered approximation* of V given by

$$V_{\mathrm{LP}}(\mathbf{r}) = \frac{1}{(2\pi)^3} \int_{|\mathbf{K}| \leq 2k_0} d^3K \, \tilde{V}(\mathbf{K}) e^{i\mathbf{K}\cdot\mathbf{r}}. \tag{6.56}$$

Without prior information regarding the scattering potential, such as support knowledge, the low-pass-filtered approximation V_{LP} defined above is the best that can be hoped for within the Born approximation; it will be our principal goal in Chapter 8.[5]

[5] A better inversion can be obtained if additional information concerning the scattering potential is known, beyond the results from the scattering experiments. For example, if the support volume τ_0 of the scattering potential is known this information can be incorporated into the inversion, as will be discussed in Chapters 8 and 9.

The above discussion assumed that the background medium is non-absorbing so that k_0 is real-valued. This allowed us to use standard Fourier analysis to state and obtain a formal solution to the inverse scattering problem within the Born approximation. If k_0 is complex the above discussion is no longer valid, since the scattering amplitude would specify $\tilde{V}(\mathbf{K})$ over complex spheres so that the standard inverse Fourier transform leading to the solution in Eq. (6.56) would not be applicable. Although there have been proposals regarding how to extend the validity of the Fourier-integral-based solution to absorbing background media, a superior approach that is applicable to general inhomogeneous and dispersive background media and that reduces to the Fourier-based solution given above when the background is uniform and non-absorbing is presented in Chapter 9.

6.8 The Born approximation for spherically and cylindrically symmetric scattering potentials

Especially simple results are obtained for scattering from 3D spherically symmetric and 2D cylindrically symmetric potentials within the Born approximation. Such scattering potentials are very useful in computer-simulation studies of inverse scattering since their (exact) scattering amplitudes are easily and efficiently computed using the algorithms presented in Section 6.3, with the result that the accuracy of the Born and other linearized inversion algorithms can be easily evaluated.

Considering first the 3D case, we expand the scattered field in the multipole expansion developed in Section 6.5.3, where the multipole moments within the Born approximation are found from Eq. (6.41a) to be given by

$$q_l^m(\nu) = \int_{\tau_0} d^3r\, V(r) U^{(\mathrm{in})}(\mathbf{r}, \nu) j_l(k_0 r) Y_l^{m*}(\hat{\mathbf{r}})$$
$$= \int_0^{a_0} r^2\, dr\, V(r) j_l(k_0 r) \int_{4\pi} d\Omega_r\, U^{(\mathrm{in})}(\mathbf{r}, \nu) Y_l^{m*}(\hat{\mathbf{r}}).$$

The incident wave propagates in the homogeneous background and admits a multipole expansion of the general form given in Eq. (6.12b) of Section 6.3 so that

$$\int_{4\pi} d\Omega_r\, U^{(\mathrm{in})}(\mathbf{r}, \nu) Y_l^{m*}(\hat{\mathbf{r}}) = a_{0l}^m(\nu) j_l(k_0 r),$$

with the $a_{0l}^m(\nu)$ being the multipole moments of the incident wave. On making use of the above result we then find that

$$q_l^m(\nu) = a_{0l}^m(\nu) \int_0^{a_0} r^2\, dr\, j_l^2(k_0 r) V(r) = \frac{i}{k_0} R_l^{(\mathrm{B})} a_{0l}^m(\nu),$$

where

$$R_l^{(\mathrm{B})} = -ik_0 \int_0^{a_0} r^2\, dr\, j_l^2(k_0 r) V(r)$$

play the role of generalized reflection coefficients for the symmetric scatterer within the Born approximation (cf. Section 6.3). The multipole expansion of the scattered field within the Born approximation is then found to be

$$U_{\mathrm{B}}^{(s)}(\mathbf{r}) = \sum_{l=0}^{\infty}\sum_{m=-l}^{l} R_l^{(\mathrm{B})} a_{0l}^{\;m}(\nu) h_l^{+}(k_0 r) Y_l^m(\hat{\mathbf{r}}).$$

The Born scattering amplitude is found to be given by Eq. (6.40b) with the expansion coefficients

$$f_l^m(\nu) = -(-i)^l b_l^m(\nu) = \frac{(-i)^{l+1}}{k_0} R_l^{(\mathrm{B})} a_{0l}^{\;m}(\nu).$$

The multipole expansion of a plane wave was derived in Example 3.4 of Chapter 3 where the multipole moments $a_{0l}^{\;m}(\mathbf{s}_0)$ were found to be given by

$$a_{0l}^{\;m}(\mathbf{s}_0) = 4\pi i^l Y_l^{m*}(\mathbf{s}_0) \Rightarrow f_l^m(\mathbf{s}_0) = -\frac{4\pi i}{k_0} R_l^{(\mathrm{B})} Y_l^{m*}(\mathbf{s}_0).$$

The plane-wave scattering amplitude within the Born approximation thus admits the expansion

$$f_{\mathrm{B}}(\mathbf{s},\mathbf{s}_0) = -\frac{4\pi i}{k_0}\sum_{l=0}^{\infty}\sum_{m=-l}^{l} R_l^{(\mathrm{B})} Y_l^m(\mathbf{s}) Y_l^{m*}(\mathbf{s}_0).$$

In two space dimensions the multipole expansion of the scattered field is given by Eq. (6.45a) with the multipole moments computed according to Eq. (6.45b). On making the Born approximation and assuming a cylindrically symmetric scattering potential we then find that

$$\begin{aligned} q_l(\nu) &= \int_{\tau_0} d^2r\, V(r) U^{(\mathrm{in})}(\mathbf{r},\nu) J_l(k_0 r) e^{-il\phi} \\ &= \int_0^{a_0} r\,dr\, V(r) J_l(k_0 r) \int_0^{2\pi} d\phi\, U^{(\mathrm{in})}(\mathbf{r},\nu) e^{-il\phi}. \end{aligned} \tag{6.57}$$

On expanding the incident wave in a multipole expansion as given in Eq. (6.15b) we find that

$$\int_0^{2\pi} d\phi \overbrace{\sum_{l'=-\infty}^{\infty} a_{0l'}(\nu) J_{l'}(k_0 r) e^{il'\phi}}^{U^{(\mathrm{in})}(\mathbf{r},\nu)} e^{-il\phi} = 2\pi J_l(k_0 r) a_{l0}(\nu),$$

where $a_{l0}(\nu)$ are the multipole moments of the incident wave. If we now use this result in Eq. (6.57) we obtain

$$q_l(\nu) = 2\pi \int_0^{a_0} r\,dr\, V(r) J_l^2(k_0 r) a_{l0}(\nu),$$

which then yields the multipole expansion

$$U_{\mathrm{B}}^{(s)}(\mathbf{r}) = \sum_{l=-\infty}^{\infty} \overbrace{R_l^{(\mathrm{B})} a_{0l}(\nu)}^{-\frac{i}{4} q_l(\nu)} H_l^{+}(k_0 r) e^{il\phi}, \tag{6.58a}$$

where

$$R_l^{(B)} = -i\frac{\pi}{2}\int_0^{a_0} r\,dr\,J_l^2(k_0 r)V(r) \tag{6.58b}$$

are the Born approximations to the generalized reflection coefficients R_l for the scatterer.

The Born approximation to the plane-wave scattering amplitude is obtained using the same procedure as was employed in Example 6.2 for a homogeneous cylinder and is given by Eq. (6.29) of that example with the generalized reflection coefficients R_l of the homogeneous cylinder replaced by the $R_l^{(B)}$ defined above. Note that, while the scattering amplitude computed in that example applied only to homogeneous cylinders, the Born approximation computed with $R_l^{(B)}$ employed in place of R_l will be valid for any cylindrically symmetric scattering potential.

6.8.1 Born scattering from homogeneous cylinders

We computed the interior and scattered fields from a homogeneous cylinder and from two concentric homogeneous cylinders in Section 6.3 and used these results in Examples 6.2 and 6.3 to compute the scattering amplitudes for these two scatterers. Within the Born approximation to the scattered fields we obtained the multipole expansions given in Eqs. (6.15b) with the coefficients $b_l(\nu)$ given by

$$b_l(\nu) = R_l^{(B)} a_{0l}(\nu),$$

while the scattering amplitudes are obtained using Eq. (6.29) of Example 6.2 with the generalized reflection coefficients computed using Eq. (6.58b). We then find for the homogeneous cylinder with radius a_0 and index of refraction n_r that

$$R_l^{(B)} = -i\frac{\pi}{2}k_0^2(1-n_r^2)\int_0^{a_0} r\,dr\,J_l^2(k_0 r) = -i\frac{\pi}{2}k_0^2(1-n_r^2)\mu_l^2(k_0 a_0), \tag{6.59a}$$

with

$$\mu_l^2(k_0 a_0) = \int_0^{a_0} r\,dr\,J_l^2(k_0 r) = \frac{a_0^2}{2}[J_l^2(k_0 a_0) - J_{l-1}(k_0 a_0)J_{l+1}(k_0 a_0)].$$

For two concentric cylinders having indices n_1 and n_2 and radii a_1 and $a_2 > a_1$ we obtain

$$\begin{aligned} R_l^{(B)} &= -i\frac{\pi}{2}\left\{k_0^2(1-n_1^2)\int_0^{a_1} r\,dr\,J_l^2(k_0 r) + k_0^2(1-n_2^2)\int_{a_1}^{a_2} r\,dr\,J_l^2(k_0 r)\right\} \\ &= -i\frac{\pi}{2}[k_0^2(n_2^2-n_1^2)\mu_l^2(k_0 a_1) + k_0^2(1-n_2^2)\mu_l^2(k_0 a_2)]. \end{aligned} \tag{6.59b}$$

The Born scattering amplitudes for the two objects are then found to be

$$f_B(\mathbf{s},\mathbf{s}_0) = -\sqrt{\frac{\pi}{2k_0}}e^{i\frac{\pi}{4}}\sum_{l=-\infty}^{\infty} k_0^2(1-n_r^2)\mu_l^2(k_0 a_0)e^{il(\phi-\phi_0)} \tag{6.60a}$$

and

$$f_B(\mathbf{s},\mathbf{s}_0) = -\sqrt{\frac{\pi}{2k_0}}e^{i\frac{\pi}{4}}\sum_{l=-\infty}^{\infty}[k_0^2(n_2^2-n_1^2)\mu_l^2(k_0 a_1)+k_0^2(1-n_2^2)\mu_l^2(k_0 a_2)]e^{il(\phi-\phi_0)}. \tag{6.60b}$$

We computed the exact and Born approximations to the scattering amplitude of a single cylinder parameterized by its relative index of refraction n_r and of two concentric cylinders parameterized by their two radii a_1 and a_2 and relative indices n_1 and n_2. We show in Figs. 6.4 and 6.5 the magnitude and phase of the Born and exact scattering amplitudes computed for the single cylinder having radius $a_0 = 4\lambda$ and indices of refraction varying from 1.01 to 1.07 in steps of 0.02 and in Figs. 6.6 and 6.7 the magnitude and phase of the Born and exact scattering amplitudes computed for the pair of concentric cylinders having radii $a_1 = 2\lambda$ and $a_2 = 4\lambda$ and each having relative indices of refraction of 1.03 and 1.07. Note that, as required, the two concentric cylinders each having the same relative index of refraction (top left and bottom right in Figs. 6.6 and 6.7) yield identical results to the case of the single cylinder having radius 4λ and indices of refraction of 1.03 and 1.07 (top left and bottom right in Figs. 6.4 and 6.5).

6.8.2 The error between the Born and exact scattering amplitudes

The results presented in Figs. 6.4–6.7 indicate that there is relatively good agreement between the Born and exact scattering amplitudes for small index variations and small cylinder radii. To obtain a better idea of the accuracy of the Born approximation we

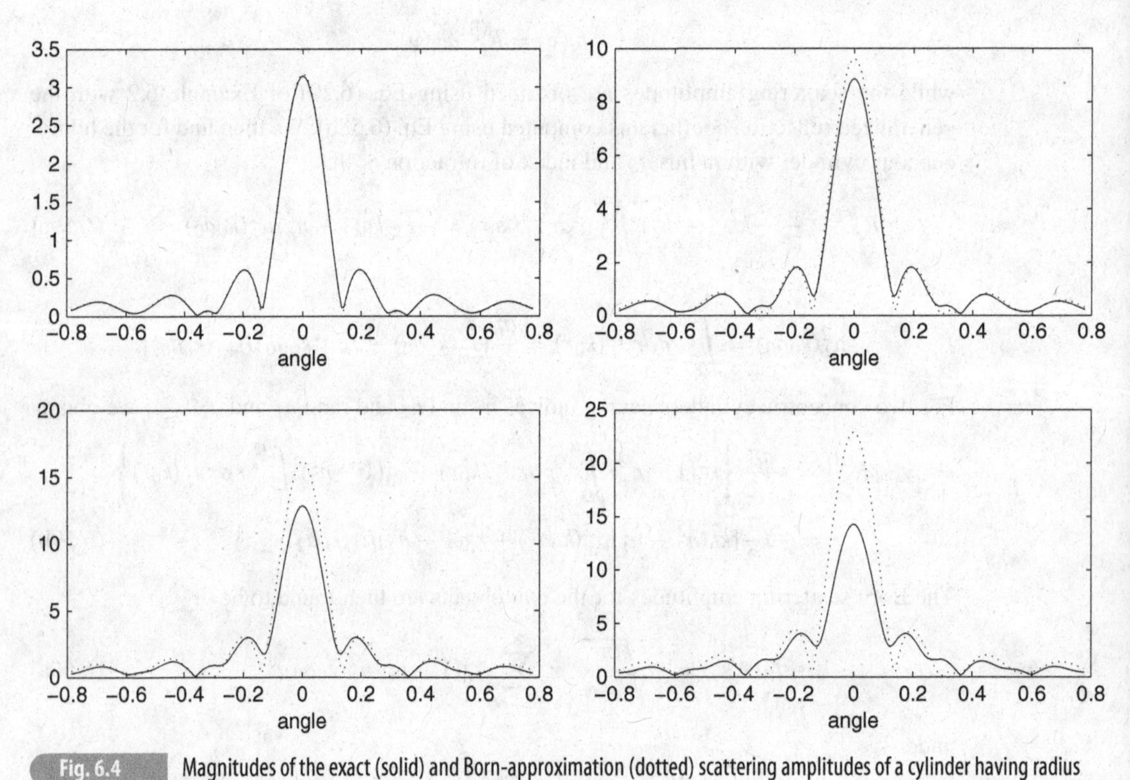

Fig. 6.4 Magnitudes of the exact (solid) and Born-approximation (dotted) scattering amplitudes of a cylinder having radius $a_0 = 4\lambda$ and relative indices of refraction varying from $n_r = 1.01$ (top left) to $n_r = 1.07$ (bottom right) in steps of $\delta n_r = 0.02$.

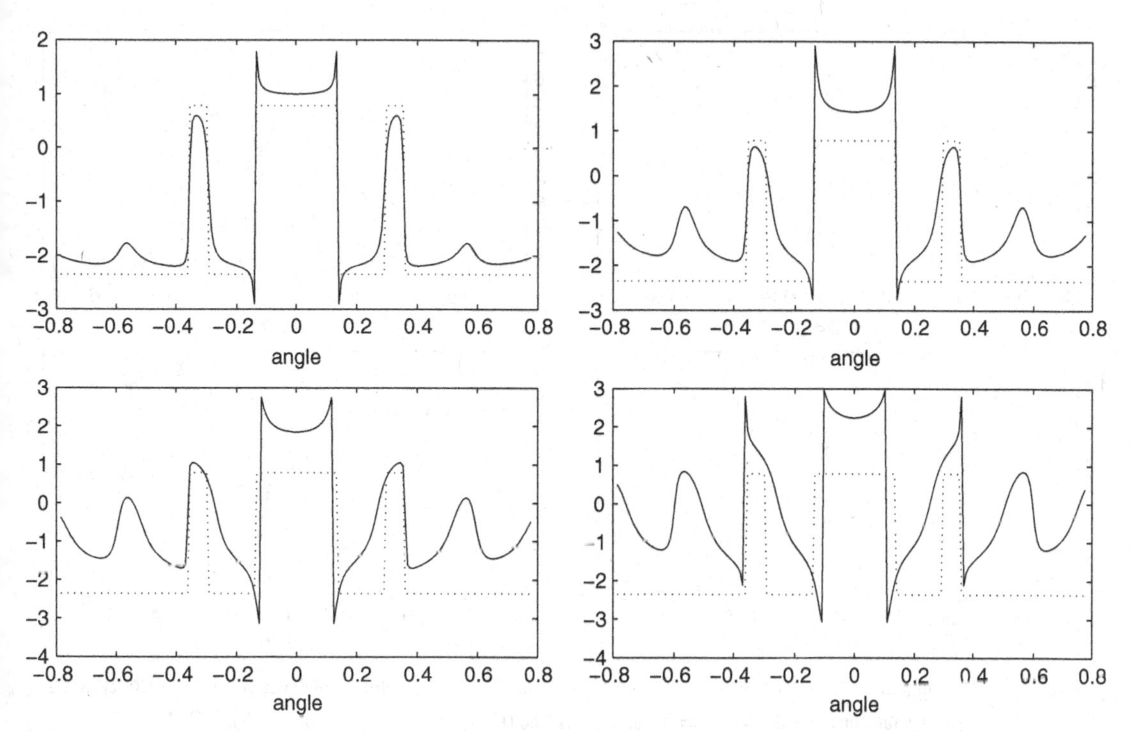

Fig. 6.5 Phases of the exact (solid) and Born-approximation (dotted) scattering amplitudes of a cylinder having radius $a_0 = 4\lambda$ and relative indices of refraction varying from $n_r = 1.01$ (top left) to $n_r = 1.07$ (bottom right) in steps of $\delta n_r = 0.02$.

computed the normalized integrated squared error between the exact and Born scattering amplitudes for a homogeneous cylinder as a function of its relative index of refraction and for four values of its radius. The normalized error is defined according to the equation

$$E(n_r, a_0) = \frac{||f - f_B||}{||f||} = \sqrt{\frac{\int_0^{2\pi} d\phi |f(\mathbf{s}, \mathbf{s}_0) - f_B(\mathbf{s}, \mathbf{s}_0)|^2}{\int_0^{2\pi} d\phi |f(\mathbf{s}, \mathbf{s}_0)|^2}} = \sqrt{\frac{\sum_l |R_l - R_l^{(B)}|^2}{\sum_l |R_l|^2}},$$

where the R_l are given in Eq. (6.14c) and the $R_l^{(B)}$ by Eq. (6.59a).

We show in the top part of Fig. 6.8 the normalized errors in percent plotted as a function of the relative index of the cylinder also in percent for cylinder radii of λ, 2λ, 3λ and 4λ. It is apparent from Fig. 6.8 that for any given relative index the errors appear to increase linearly with the radius. To verify this conclusion we plotted the errors versus the product of the radius with the relative index in the bottom part of Fig. 6.8. It can be seen that the errors in this figure are identical, thus verifying the often-quoted statement that the "validity of the Born approximation requires that the product of the maximum value of the scattering potential with the radius of support of the scatterer must be small." Of course, this simulation applies only to scattering from a cylinder, but certainly supports this conclusion.

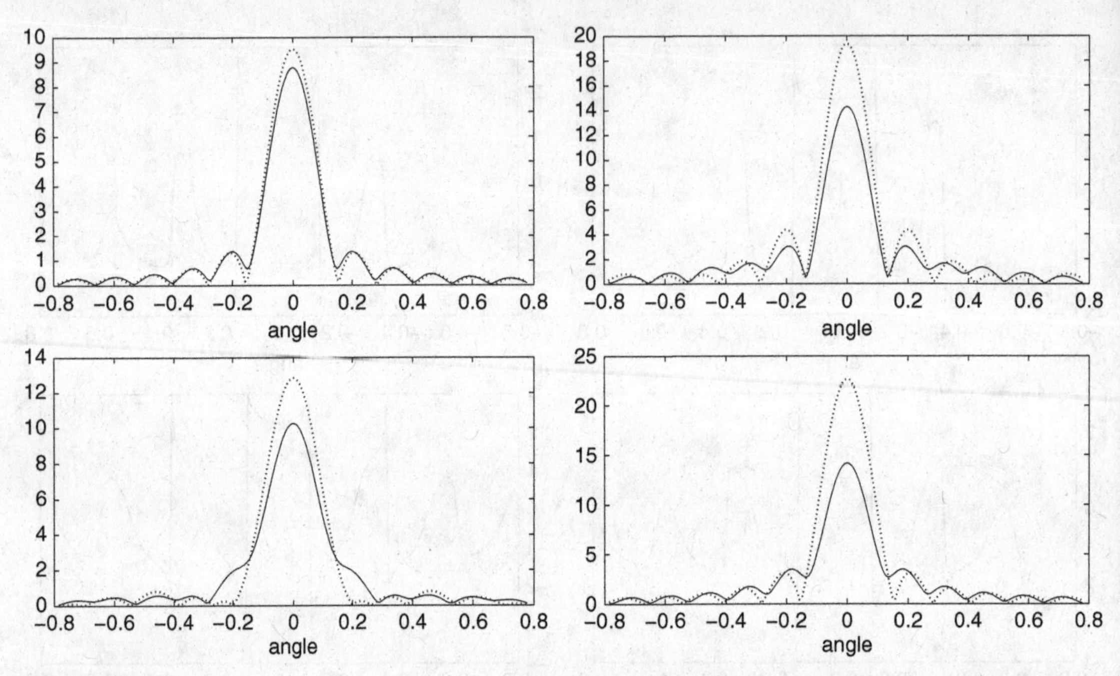

Fig. 6.6 Magnitudes of the exact (solid) and Born-approximation (dotted) scattering amplitudes of two concentric cylinders having radii of $a_1 = 2\lambda$ and $a_2 = 4\lambda$ and relative indices of refraction of $(n_1, n_2) = (1.03, 1.03)$ (top left), (1.03, 1.07) (top right), (1.07, 1.03) (bottom left) and (1.07, 1.07) (bottom right).

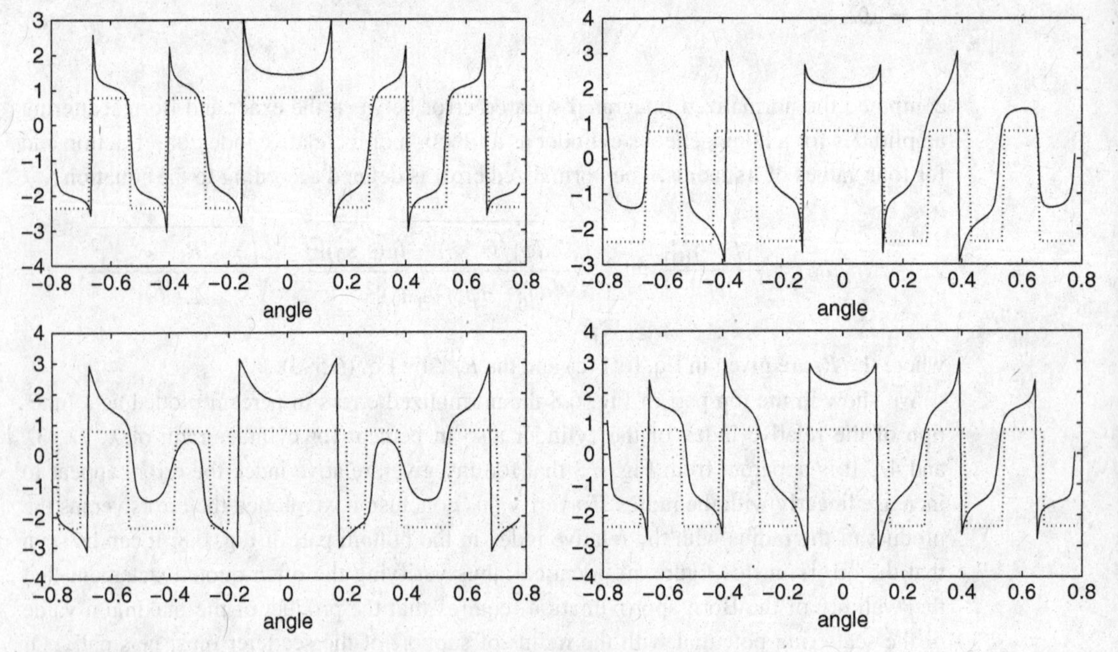

Fig. 6.7 Phases of the exact (solid) and Born-approximation (dotted) scattering amplitudes of two concentric cylinders having radii of $a_1 = 2\lambda$ and $a_2 = 4\lambda$ and relative indices of refraction of $(n_1, n_2) = (1.03, 1.03)$ (top left), (1.03, 1.07) (top right), (1.07, 1.03) (bottom left) and (1.07, 1.07) (bottom right).

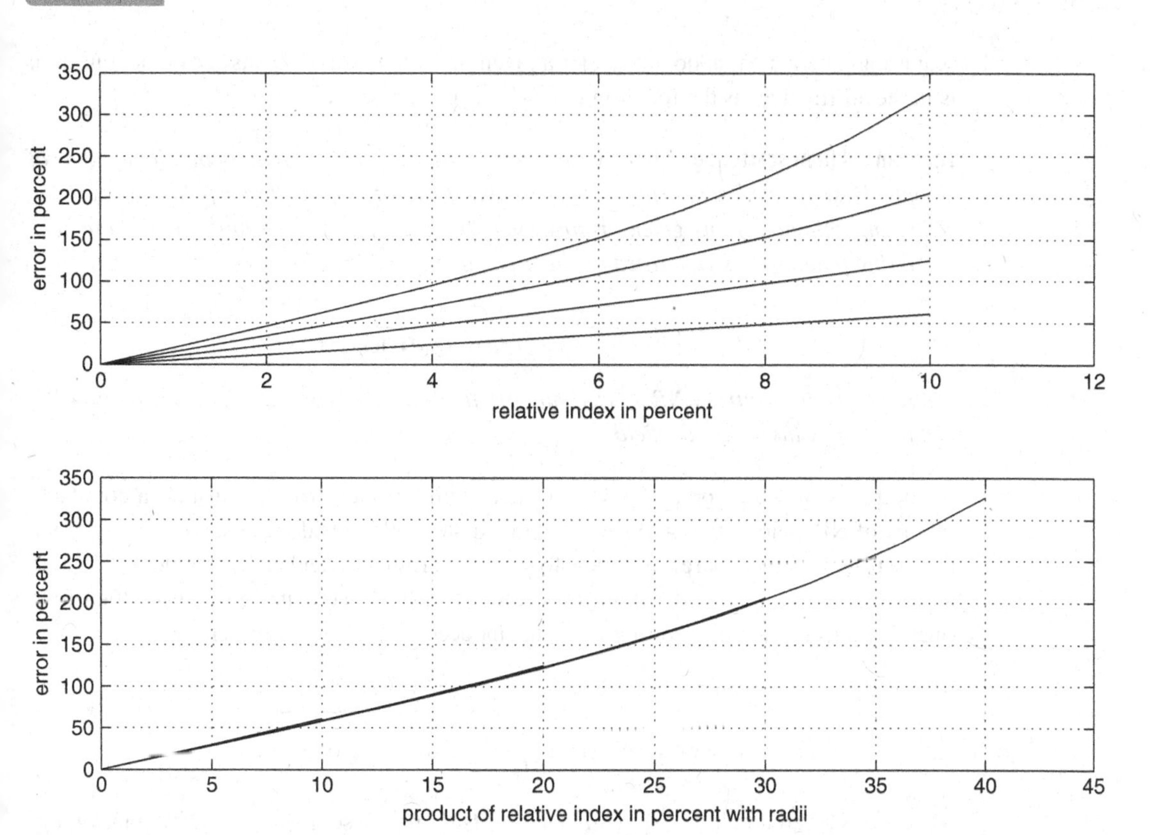

Fig. 6.8 The top plot shows percentage normalized errors for cylinder radii of λ, 2λ, 3λ and 4λ plotted as a function of the percentage relative index of the cylinders. The errors for any given relative index increase with increasing radius. The bottom plot shows percentage errors superimposed for all four cylinders plotted as a function of the product of their radii with their percentage relative index.

6.9 Non-scattering potentials

We showed in Section 6.4 that the scattering amplitude $f(\mathbf{s}, \mathbf{s}_0)$ specified for all unit observation vectors $\mathbf{s}$ and fixed incident unit wave vector $\mathbf{s}_0$ completely and uniquely specifies the scattered field $U_+^{(s)}(\mathbf{r}, \mathbf{s}_0)$ everywhere outside the scattering volume τ_0. It follows that the vanishing of the scattering amplitude over all scattering directions $\mathbf{s}$ for any given incident-wave direction $\mathbf{s}_0$ guarantees that the scattered field will vanish everywhere outside the support of the scatterer for that particular incident plane wave. We will refer to scattering potentials that generate scattered waves that vanish outside the scattering volume as *non-scattering potentials* in analogy with the term *non-radiating sources* that we applied to sources that radiate fields that vanish outside the source volume (cf. Section 1.7 of Chapter 1 and Section 2.7 of Chapter 2).

The first question that needs to be answered is whether non-trivial non-scattering potentials actually exist, at least in a mathematical sense; i.e., do there exist compactly supported scattering potentials that do not scatter a given incident wave? The answer to that question is in the affirmative, as the following theorem establishes.

Theorem 6.4 (non-scattering-potentials theorem) *There exists a countably infinite number of compactly supported scattering potentials that generate no scattered field outside their scattering volume for any given arbitrary incident wave. Moreover, any one of these non-scattering potentials is generated by the algorithm*

$$V_{\rm ns}(\mathbf{r}) = \frac{Q_{\rm nr}(\mathbf{r})}{U^{(\rm in)}(\mathbf{r}) + \int_\tau d^3r'\, Q_{\rm nr}(\mathbf{r}')G_{0+}(\mathbf{r}-\mathbf{r}')}, \tag{6.61}$$

where $Q_{\rm nr}$ is an arbitrary NR source supported within the scattering volume τ_0 and $U^{(\rm in)}$ is an arbitrary incident wavefield.

The proof of the theorem follows at once from the fact that we can find a countable infinity of NR sources that will generate no radiated field outside the source volume. If we then set the induced source resulting from some arbitrary incident wavefield $U^{(\rm in)}$ equal to an NR source supported within the scattering volume we conclude that the total field (incident plus scattered) resulting from this induced (NR) source is given by

$$U(\mathbf{r}) = U^{(\rm in)}(\mathbf{r}) + \overbrace{\int_\tau d^3r'\, Q_{\rm nr}(\mathbf{r}')G_{0+}(\mathbf{r}-\mathbf{r}')}^{U_+^{(\rm s)}(\mathbf{r})},$$

where the scattered field $U_+^{(\rm s)}(\mathbf{r})$ will vanish outside τ_0. On setting $Q_{\rm nr} = UV_{\rm ns}$ and solving for the non-scattering potential $V_{\rm ns}$ we then find that

$$V_{\rm ns}(\mathbf{r}) = \frac{Q_{\rm nr}(\mathbf{r})}{U(\mathbf{r})} = \frac{Q_{\rm nr}(\mathbf{r})}{U^{(\rm in)}(\mathbf{r}) + \int_\tau d^3r'\, Q_{\rm nr}(\mathbf{r}')G_{0+}(\mathbf{r}-\mathbf{r}')},$$

which is an algorithm for constructing a non-scattering potential from an arbitrary NR source and incident field.

Theorem 6.4 allows us to construct a scattering potential that will not generate a scattered wave for any single (specified) incident wave. The question of whether we can construct a potential that will not scatter two or more incident waves naturally arises. Although there does not appear to be any definitive answer to this question at the present time (see, however, the paper by Isakov and Nachman (Isakov and Nachman, 1995)), we can provide an answer as well as a simple algorithm for generating such non-scattering potentials within the Born scattering model.

6.9.1 Non-scattering potentials within the Born approximation

The scattered field within the Born scattering model is given by Eq. (6.53), from which it follows that within this weakly scattering model a non-scattering potential is given by the ratio of an NR source with the incident field:

$$V_{\rm ns}^{\rm B}(\mathbf{r}) = \frac{Q_{\rm nr}(\mathbf{r})}{U^{(\rm in)}(\mathbf{r})}. \tag{6.62}$$

The above result is also obtainable from the expression for a non-scattering potential within the framework of exact scattering theory given by Eq. (6.61) if we perform a perturbation expansion of this expression and drop the terms corresponding to higher-order terms in the Born series. To see this, we first write Eq. (6.61) in the compact form

$$V_{\text{ns}}(\mathbf{r}) = \frac{Q_{\text{nr}}(\mathbf{r})}{U^{(\text{in})}(\mathbf{r}) + \pi(\mathbf{r})}, \tag{6.63a}$$

where

$$\pi(\mathbf{r}) = \int_\tau d^3r'\, Q_{\text{nr}}(\mathbf{r}')G_{0+}(\mathbf{r}-\mathbf{r}')$$

is the field radiated by the NR source Q_{nr}. If we now perform a binomial expansion of the denominator in Eq. (6.63a) we obtain the result

$$V_{\text{ns}}(\mathbf{r}) = \frac{Q_{\text{nr}}(\mathbf{r})}{U^{(\text{in})}(\mathbf{r})}\left\{1 - \frac{\pi(\mathbf{r})}{U^{(\text{in})}(\mathbf{r})} + \left[\frac{\pi(\mathbf{r})}{U^{(\text{in})}(\mathbf{r})}\right]^2 - \cdots\right\}. \tag{6.63b}$$

The lowest-order term in the above expansion corresponds to the Born weak-scattering model and is precisely the result we obtained directly from the Born scattering model in Eq. (6.62).

Another class of scattering potentials within the Born scattering model that generate a zero scattered field are those potentials whose spatial Fourier transforms vanish everywhere within the Ewald limiting sphere. Such scattering potentials would then have $\tilde{V}[k_0(\mathbf{s}-\mathbf{s}_0)] = 0$ for *any* set of incident and scattered wave vectors and, hence, would be truly "invisible" in any scattering experiment.[6] However, such scattering potentials are not compactly supported within a finite scattering volume τ_0 and, hence, are not "non-scattering potentials" within the definition used here.

6.9.2 Incident plane waves

The Born non-scattering potential constructed according to Eq. (6.62) is not the most general non-scattering potential that can be constructed within the Born model for the case in which the incident wave is a plane wave. A more general model for this case can be obtained by working with the Born scattering amplitude $f_B(\mathbf{s}, \mathbf{s}_0)$. This quantity as defined in Eq. (6.54) is proportional to the spatial Fourier transform of the scattering potential $\tilde{V}(\mathbf{K})$ evaluated over the set of spatial frequencies $\mathbf{K} = k_0(\mathbf{s}-\mathbf{s}_0)$. Thus, if for any given incident plane-wave direction $\mathbf{s}_0$ the transform $\tilde{V}[k_0(\mathbf{s}-\mathbf{s}_0)]$ vanishes for all unit vectors $\mathbf{s}$, the scattering amplitude and, hence, the scattered field will vanish everywhere outside the scattering volume τ_0 for that particular incident plane wave. The condition $\mathbf{K} = k_0(\mathbf{s}-\mathbf{s}_0)$ is equivalent to the condition

$$\mathbf{s} = \frac{\mathbf{K}}{k_0} + \mathbf{s}_0,$$

[6] This conclusion follows from the fact that the plane waves form a complete set of wavefields into which any incident wave can be expanded.

which will be satisfied for all unit vectors $\mathbf{s}$ if and only if

$$\mathbf{s}\cdot\mathbf{s} = 1 = \left[\frac{\mathbf{K}}{k_0} + \mathbf{s}_0\right]\cdot\left[\frac{\mathbf{K}}{k_0} + \mathbf{s}_0\right],$$

which reduces to the requirement that $\tilde{V}(\mathbf{K})$ vanish over those spatial frequencies $\mathbf{K}$ that lie on the surface

$$K^2 + 2k_0\mathbf{s}_0\cdot\mathbf{K} = 0.$$

It follows from the above condition that any scattering potential whose spatial Fourier transform is of the general form

$$\tilde{V}^{\mathrm{B}}_{\mathrm{ns}}(\mathbf{K};\mathbf{s}_0) = -(K^2 + 2k_0\mathbf{s}_0\cdot\mathbf{K})\tilde{\chi}(\mathbf{K}), \tag{6.64a}$$

where

$$\tilde{\chi}(\mathbf{K}) = \int_{\tau_0} d^3r\,\chi(\mathbf{r})e^{-i\mathbf{K}\cdot\mathbf{r}}$$

is the spatial Fourier transform of any function $\chi(\mathbf{r})$ that is compactly supported within the scattering volume τ_0, will be non-scattering within the Born approximation for an incident plane wave with unit propagation vector $\mathbf{s}_0$. If we also require that $\chi(\mathbf{r})$ possess continuous first partial derivatives we find that

$$\begin{aligned} V^{\mathrm{B}}_{\mathrm{ns}}(\mathbf{r};\mathbf{s}_0) &= \frac{1}{(2\pi)^2}\int d^3K\,\tilde{V}_{\mathrm{ns}}(\mathbf{K};\mathbf{s}_0)e^{i\mathbf{K}\cdot\mathbf{r}} \\ &= -\frac{1}{(2\pi)^2}\int d^3K(K^2 + 2k_0\mathbf{s}_0\cdot\mathbf{K})\tilde{\chi}(\mathbf{K})e^{i\mathbf{K}\cdot\mathbf{r}} \end{aligned}$$

$$\Downarrow$$

$$V_{\mathrm{ns}}(\mathbf{r};\mathbf{s}_0) = (\nabla^2 + 2ik_0\mathbf{s}_0\cdot\nabla)\chi(\mathbf{r}). \tag{6.64b}$$

The non-scattering potential $V^{\mathrm{B}}_{\mathrm{ns}}(\mathbf{r};\mathbf{s}_0)$ generates a scattering amplitude that vanishes over all directions $\mathbf{s}$ and a scattered field that vanishes everywhere outside the convex hull of the scattering volume τ_0 for an incident plane wave with unit propagation vector $\mathbf{s}_0$. Moreover, the scattering-amplitude reciprocity theorem (Theorem 6.1) states that $f(\mathbf{s},\mathbf{s}_0) = f(-\mathbf{s}_0,-\mathbf{s})$ so that this non-scattering potential will also generate a zero scattering amplitude along the single scattering direction $-\mathbf{s}_0$ for incident plane waves having arbitrary unit propagation vectors. By letting $\mathbf{s}_0 \to -\mathbf{s}$ we can state this in the form that

$$V_{\mathrm{ns}}(\mathbf{r};\mathbf{s}) = (\nabla^2 - 2ik_0\mathbf{s}\cdot\nabla)\chi(\mathbf{r}) \tag{6.65}$$

will possess a Born scattering amplitude $f_{\mathrm{B}}(\mathbf{s},\mathbf{s}_0)$ that vanishes for all $\mathbf{s}_0 \in 4\pi$ steradians. Note that in this case only the scattering amplitude in the single direction $\mathbf{s}$, not the scattered field outside τ_0, will vanish.

6.9.3 The relationship between the two Born non-scattering potentials

To see the relationship between the two models Eq. (6.62) and Eqs. (6.64) for a Born non-scattering potential, we set $U^{(\mathrm{in})}(\mathbf{r}) = \exp(ik_0\mathbf{s}_0\cdot\mathbf{r})$ in Eq. (6.62) and take the spatial Fourier transform of the resulting potential. We obtain the result

$$\begin{aligned}
\tilde{V}^{\mathrm{B}}_{\mathrm{ns}}(\mathbf{K}) &= \int_{\tau_0} d^3r\, Q_{\mathrm{nr}}(\mathbf{r})e^{-i(\mathbf{K}+k_0\mathbf{s}_0)\cdot\mathbf{r}} = \int_{\tau_0} d^3r\, \overbrace{[\nabla^2 + k_0^2]\pi(\mathbf{r})}^{Q_{\mathrm{nr}}(\mathbf{r})}\, e^{-i(\mathbf{K}+k_0\mathbf{s}_0)\cdot\mathbf{r}} \\
&= [-(\mathbf{K}+k_0\mathbf{s}_0)\cdot(\mathbf{K}+k_0\mathbf{s}_0) + k_0^2]\int_{\tau_0} d^3r\, \pi(\mathbf{r})e^{-i(\mathbf{K}+k_0\mathbf{s}_0)\cdot\mathbf{r}} \\
&= -(K^2 + 2k_0\mathbf{s}_0\cdot\mathbf{K})\tilde{\pi}(\mathbf{K}+k_0\mathbf{s}_0),
\end{aligned}$$

where in the first line we have expressed the NR source in terms of its radiated field $\pi(\mathbf{r})$ and integrated by parts twice to obtain the second line. On comparing the above result with Eq. (6.64a) we see that the Born non-scattering model defined in Eq. (6.62) is a special case of the more general model given in Eq. (6.64a), which was obtained by taking

$$\tilde{\chi}(\mathbf{K}) = \tilde{\pi}(\mathbf{K}+k_0\mathbf{s}_0).$$

Although the two models are consistent, we see that in the more general model the function χ is arbitrary and independent of the incident-wave direction, whereas in the more restrictive model defined in Eq. (6.62) χ depends on the incident-wave direction. Because of this it is not possible to extend the restrictive model to cases of multiple experiments using different incident plane waves, whereas the more general model defined in Eqs. (6.64) is easily extended, as we will now show.

6.9.4 Almost-invisible weak scatterers

A truly invisible object would produce no scattering for any arbitrary incident wavefield. However, as we noted earlier, the scattering potential for such an object within the Born scattering model would have to have a spatial Fourier transform that vanished everywhere within the Ewald limiting sphere and such a potential would, of necessity, be infinite in extent. Although no truly invisible compactly supported weakly scattering potential exists, it is simple to construct such potentials that will be invisible in any finite set of scattering experiments employing incident plane waves or incident waves that can be expressed as a superposition of a finite set of plane waves. We will refer to such potentials as "almost-invisible scattering potentials."

An almost-invisible scattering potential within the Born approximation can be generated by cascading the operator $(\nabla^2 + 2ik_0\mathbf{s}_0\cdot\nabla)$ with different values of the unit wave vector $\mathbf{s}_0$ to generate the non-scattering potential

$$V^{\mathrm{B}}_{\mathrm{ns}}(\mathbf{r}) = \prod_{m=1}^{M}(\nabla^2 + 2ik_0\mathbf{s}_{0m}\cdot\nabla)\chi(\mathbf{r}), \tag{6.66a}$$

where $\chi(\mathbf{r})$ is assumed to be compactly supported within a scattering volume τ_0. It is easy to verify that the scattering potential $V^{\rm B}_{\rm ns}(\mathbf{r})$ has a spatial Fourier transform that vanishes over the M Ewald spheres associated with the M unit vectors $\mathbf{s}_{0m}$, and, hence, will not scatter in any of these M scattering experiments.

If we again make use of the scattering-amplitude reciprocity theorem we conclude from Eq. (6.65) that, within the Born approximation, the potential

$$V^{\rm B}_{\rm ns}(\mathbf{r}) = \prod_{m=1}^{M}(\nabla^2 - 2ik_0\mathbf{s}_m \cdot \nabla)\chi(\mathbf{r}) \tag{6.66b}$$

will generate a zero scattering amplitude over the set of M scattering directions $\mathbf{s}_m$, $m = 1, 2, \ldots, M$ for any incident plane wave having an arbitrary unit propagation vector $\mathbf{s}_0$. Moreover, we showed in Section 6.4 that the generalized scattering amplitude for an incident wave having arbitrary plane-wave amplitude $A(\mathbf{s}_0, \nu)$ is given by

$$f(\mathbf{s}, \nu) = \int_{4\pi} d\Omega_{s_0} f(\mathbf{s}, \mathbf{s}_0)A(\mathbf{s}_0, \nu).$$

It then follows that within the Born scattering model *the generalized scattering amplitude of the non-scattering potential defined in Eq. (6.66b) will vanish for any incident wave over the set of M scattering directions* $\mathbf{s}_m$, $m = 1, 2, \ldots, M$. Again we emphasize that this does not guarantee that the scattered field will vanish but only that the generalized scattering amplitude will vanish over that specific set of scattering directions. In order for the scattered field to vanish outside the convex hull of the scattering potential it is necessary that the scattering amplitude vanish for *all* scattering directions $\mathbf{s}$, not just for a discrete set of scattering directions.

6.9.5 Essentially non-scattering potentials

The non-scattering (NS) potentials are the scattering-problem equivalent to the NR sources of the radiation problem. We also encountered "essentially" NR sources in the radiation problem and these sources too have their counterpart in the scattering problem. In analogy with the definition of the essentially NR sources employed in Section 1.7.3 of Chapter 1 we define an essentially non-scattering potential as one that generates "negligible" scattered field energy. On making use of the expression for the scattered field energy from Eq. (6.32) we then require an essentially NS potential to possess a scattering amplitude that satisfies the inequality

$$E_\nu(\omega) = 2\kappa\omega k_0 \int_{4\pi} d\Omega_s |f(\mathbf{s}, \nu)|^2 < \epsilon, \tag{6.67}$$

where ϵ is a small positive parameter used to characterize the essentially NS potential.

Following steps identical to those employed in our treatment of essentially NR sources in Section 1.7.3, we express the generalized scattering amplitude in the spherical harmonic expansion given in Eq. (6.40b) of Section 6.5.3, where the generalized Fourier coefficients

$f_l^m(\nu)$ are related to the induced source and scattering amplitude via Eqs. (6.41) of that section. On substituting this expansion into Eq. (6.67) we obtain the result

$$E_\nu(\omega) = 2\kappa\omega k_0 \sum_{l=0}^{\infty} \sum_{m=-l}^{l} |f_l^m(\nu)|^2. \tag{6.68}$$

We showed in Section 6.5.3 that the quantities $|f_l^m(\nu)|^2$ decay exponentially fast for $l > [k_0 a_0]$, where the bracket $[\cdot]$ stands for the closest next positive integer. It then follows that if the scattering amplitude expansion coefficients $f_l^m(\nu)$ are all negligible for $l < [k_0 a_0]$ then the scattered field energy will be negligible and the scattering potential will be essentially NS. The condition for essential NS can then be expressed in either of the two forms

$$|f_l^m(\nu)| = \left| \int_{4\pi} d\Omega_s f(\mathbf{s}, \nu) Y_l^{m*}(\mathbf{s}) \right| < \epsilon, \quad l < [k_0 a_0], \tag{6.69a}$$

and

$$|f_l^m(\nu)| = \left| \int_{\tau_0} d^3 r\, V(\mathbf{r}) U(\mathbf{r}, \nu) j_l(k_0 r) Y_l^{m*}(\hat{\mathbf{r}}) \right| < \epsilon, \quad l < [k_0 a_0], \tag{6.69b}$$

where ϵ is a small positive parameter. These conditions are precisely those required for the induced source ($Q = VU$) to be essentially NR.

Limits on the angular resolution of a scattering amplitude

We found in our treatments of essentially NR sources in Chapters 1 and 2 that the exponential decay of the parameters $\mu_l^2(k_0 a_0)$ with index $l > [k_0 a_0]$ also limited the achievable angular resolution of the radiation pattern of any finite-norm source. The same result is found for the scattering amplitude of a finite-norm induced source for which the generalized Fourier coefficients of the scattering amplitude $f_l^m(\nu)$ must decay exponentially fast for $l > [k_0 a_0]$. As discussed in our treatments of essentially NR sources, the spherical harmonics $Y_l^m(\mathbf{s})$ are periodic functions of the polar and azimuthal angles α and β, with smallest angular periods of $2\pi/l \geq 2\pi/(k_0 a_0) = \lambda/a_0$ radians. It then follows that the generalized scattering amplitude, like the radiation pattern of a primary source, can oscillate as a function of the polar and azimuthal angles with angular periods no smaller than λ/a_0 radians. This, of course, translates into a minimum primary-lobe width (angular resolution) of the same magnitude.

6.10 The Rytov approximation

In this section we again seek a linearized approximate solution of the scattering problem but will base the theory on the non-linear Ricatti differential equation satisfied by the (complex) space-dependent *phase* of the total field U rather than on the Helmholtz equation satisfied by the field itself. The two formulations of the forward scattering problem (the Helmholtz equation for U and the Ricatti equation for the complex phase of U) are mathematically equivalent and, hence, yield the same (exact) forward solution for the field. However, the linearized versions of the two formulations are different and the one based on the Ricatti equation presented in this section (the Rytov approximation) is generally more accurate and has a greater domain of applicability than the Born approximation developed earlier in the chapter for many of the inverse scattering problems that we will deal with in later chapters. Although the two linearized formulations of the forward scattering problem have different domains of applicability, we will find that the underlying mathematical structure of the solutions to the forward scattering problem within the two formulations is identical so that many of the results obtained earlier for the Born scattering model can again be employed within the Rytov weak-scattering model presented here.

When developing and using the Rytov scattering model it is customary to employ the *complex index of refraction* $n(\mathbf{r})$ in place of the scattering potential $V(\mathbf{r})$. These two quantities are related according to Eq. (6.2). If this relationship is used in the Helmholtz equation Eq. (6.3) we obtain

$$[\nabla^2 + k_0^2 n^2(\mathbf{r})]U(\mathbf{r}, \nu) = 0, \tag{6.70}$$

which will be employed in place of Eq. (6.3) from this point on in this section.

6.10.1 The Ricatti equation for the complex phase of the field

We can express the field U in terms of a complex-valued phase W via the equation

$$U(\mathbf{r}, \nu) = e^{ik_0 W(\mathbf{r},\nu)}, \tag{6.71}$$

where the complex-valued phase function $W(\mathbf{r}, \nu)$ is required to satisfy a differential equation that results from substitution of Eq. (6.71) into Eq. (6.70). On making this substitution we obtain

$$ik_0 \nabla^2 W(\mathbf{r}, \nu) - k_0^2[\nabla W(\mathbf{r}, \nu)]^2 + k_0^2 n^2(\mathbf{r}) = 0, \tag{6.72a}$$

where we have used the shorthand notation $(\nabla W)^2 = \nabla W \cdot \nabla W$. Equation (6.72a) is a non-linear Ricatti equation satisfied by the complex phase function W and is completely equivalent to the Helmholtz equation Eq. (6.70) satisfied by the field U. The phase of the incident field W_0 satisfies the same equation with the index of refraction n set equal to unity:

$$ik_0 \nabla^2 W_0(\mathbf{r}, \nu) - k_0^2[\nabla W_0(\mathbf{r}, \nu)]^2 + k_0^2 = 0. \tag{6.72b}$$

Besides satisfying the Ricatti equation Eq. (6.72a), the phase also has to satisfy a boundary condition corresponding to the requirement that the field satisfy the asymptotic condition Eq. (6.4). If we express the phase W in the form

$$W(\mathbf{r}, \nu) = W_0(\mathbf{r}, \nu) + \delta W(\mathbf{r}, \nu) \tag{6.73}$$

we find that this asymptotic condition can be expressed in the form

$$U(\mathbf{r}, \nu) = e^{ik_0[W_0(\mathbf{r},\nu)+\delta W(\mathbf{r},\nu)]} \sim \overbrace{e^{ik_0 W_0(\mathbf{r},\nu)}}^{U^{(\text{in})}(\mathbf{r},\nu)} + f(\mathbf{s}, \nu)\frac{e^{ik_0 r}}{r}, \quad r \to \infty, \tag{6.74}$$

with $\mathbf{s} = \mathbf{r}/r$. We will employ Eq. (6.74) below where we formally solve the Ricatti equation using a perturbation expansion.

6.10.2 The Liouville–Neumann expansion for the phase

If we make the substitution Eq. (6.73) into the left-hand side of the Ricatti equation Eq. (6.72a) and simplify the result, we find that the phase perturbation δW satisfies the equation

$$ik_0 \nabla^2 \delta W - k_0^2 (\nabla \delta W)^2 - 2k_0^2 \nabla W_0 \cdot \nabla \delta W + k_0^2 (n^2 - 1) = 0. \tag{6.75}$$

The quantity $n^2 - 1$ is a perturbation of the index of refraction from its background value of unity introduced by the presence of the scattering object. This perturbation results in a perturbation δW of the phase of the incident field W_0 governed by Eq. (6.75). If we associate a perturbation parameter ϵ with the index perturbation so that

$$n^2 - 1 \to \epsilon(n^2 - 1)$$

we can expect that the phase perturbation can be expressed in a perturbation expansion (Liouville–Neumann expansion) of the general form

$$\delta W = \sum_{n=1}^{\infty} \delta W^{(n)} \epsilon^n, \tag{6.76}$$

which, when substituted into Eq. (6.75), yields the result

$$ik_0 \sum_{n=1}^{\infty} \nabla^2 \delta W^{(n)} \epsilon^n - k_0^2 \sum_{n'=1}^{\infty} \sum_{n=1}^{\infty} \nabla \delta W^{(n')} \cdot \nabla \delta W^{(n)} \epsilon^{n'+n}$$
$$- 2k_0^2 \sum_{n=1}^{\infty} \nabla W_0 \cdot \nabla \delta W^{(n)} \epsilon^n + k_0^2 (n^2 - 1)\epsilon = 0. \tag{6.77}$$

If we equate terms of equal powers of ϵ we then arrive at the set of equations

$$ik_0\,\nabla^2\delta W^{(1)} - 2k_0^2\,\nabla W_0\cdot\nabla\delta W^{(1)} + k_0^2(n^2-1) = 0,$$
$$ik_0\,\nabla^2\delta W^{(2)} - k_0^2\,\nabla\delta W^{(1)}\cdot\nabla\delta W^{(1)} - 2k_0^2\,\nabla W_0\cdot\nabla\delta W^{(2)} = 0,$$
$$\vdots$$
$$ik_0\,\nabla^2\delta W^{(n)} - k_0^2\sum_{n'+n''=n}\nabla\delta W^{(n')}\cdot\nabla\delta W^{(n'')} - 2k_0^2\,\nabla W_0\cdot\nabla\delta W^{(n)} = 0.$$

The above set of coupled *linear* partial differential equations can be sequentially solved for the various terms $\delta W^{(n)}$ in the perturbation expansion Eq. (6.76). As each equation is solved the resulting phase perturbation $\delta W^{(n)}$ is used in the next equation as an inhomogeneous (driving) term in terms of which the next higher-order perturbation is computed. Note that only the first member of this set explicitly involves the index of refraction $n(\mathbf{r})$ and that this quantity enters the higher-order equations only through their dependence on the first-order phase perturbation $\delta W^{(1)}$.

6.10.3 The Rytov approximation

The Rytov approximation to the phase perturbation δW consists of the first term $\delta W^{(1)}$ in the Liouville–Neumann expansion Eq. (6.76). Denoting the Rytov phase perturbation by δW_{R}, we thus conclude that δW_{R} satisfies the first equation in the coupled set given above, which we write in the form

$$ik_0\,\nabla^2\delta W_{\mathrm{R}}(\mathbf{r},\nu) - 2k_0^2\,\nabla W_0\cdot\nabla\delta W_{\mathrm{R}}(\mathbf{r},\nu) = -k_0^2(n^2-1). \tag{6.78a}$$

In the important case of plane-wave incidence, the phase W_0 of the incident plane wave is given by $W_0 = \mathbf{s}_0\cdot\mathbf{r}$, where $\mathbf{s}_0$ is the unit propagation vector of the incident plane wave. In this case Eq. (6.78a) reduces to

$$ik_0\,\nabla^2\delta W_{\mathrm{R}}(\mathbf{r};\mathbf{s}_0) - 2k_0^2\mathbf{s}_0\cdot\nabla\delta W_{\mathrm{R}}(\mathbf{r};\mathbf{s}_0) = -k_0^2(n^2-1). \tag{6.78b}$$

6.10.4 The short-wavelength limit

We first examine the Rytov approximation within the short-wavelength limit where the wavelength $\lambda\to 0$ ($k_0\to\infty$). In this limit the first term in Eq. (6.78b) is negligible in comparison with the other two terms and we obtain the result

$$\mathbf{s}_0\cdot\nabla\delta w(\mathbf{r};\mathbf{s}_0) = \frac{n^2-1}{2}\approx\delta n, \tag{6.79}$$

where we have denoted the short-wavelength limit of δW_{R} by the lower-case symbol δw and

$$\delta n(\mathbf{r}) = n(\mathbf{r}) - 1.$$

In making the above approximation in Eq. (6.79) we have used the fact that the Rytov approximation requires that $n^2 - 1$ be small.

Equation (6.79) is the ray model for the propagation of a plane wave through a weakly inhomogeneous object. We note that this equation is equivalent to the simple differential equation

$$\frac{\partial}{\partial \eta} \delta w(\mathbf{r}; \mathbf{s}_0) = \delta n,$$

where $\eta = \mathbf{s}_0 \cdot \mathbf{r}$ is the position variable along the Cartesian coordinate axis defined by the direction of propagation of the incident plane wave. The above equation can be immediately integrated to yield

$$\delta w(\mathbf{r}; \mathbf{s}_0)(\mathbf{r})|_{\eta=a}^{\eta=b} = \int_a^b d\eta\, \delta n(\mathbf{r}), \tag{6.80}$$

where the integration has been performed along the *straight-line ray path* that extends from $\eta = a$ to $\eta = b$. If we take the points $\eta = a$ and $\eta = b = l_0$ to lie outside the object then $\delta w|_{\eta=a} = 0$ and the integration on the r.h.s. of Eq. (6.80) can be extended from $-\infty$ to ∞, and we obtain the result

$$\delta w(\mathbf{r})|_{\eta=l_0} = \int_{-\infty}^{\infty} d\eta\, \delta n(\mathbf{r}). \tag{6.81}$$

Equation (6.81) is the underlying mathematical model for computed tomography (CT) and appears here as the limiting model for wave propagation through weakly inhomogeneous scatterers modeled using the Rytov approximation in the short-wavelength limit.

6.10.5 The Rytov transformation

It is possible to convert the linearized Ricatti equations Eqs. (6.78) into Helmholtz equations that can then be easily solved using the methods developed earlier in this chapter. The conversion is based on the use of a transformation originally due to Rytov (Tatarski, 1961; Chernov, 1967) and, for the case of plane-wave incidence, is given by

$$ik_0\, \delta W_R(\mathbf{r}; \mathbf{s}_0) = e^{-ik_0 \mathbf{s}_0 \cdot \mathbf{r}} F(\mathbf{r}), \tag{6.82}$$

where $F(\mathbf{r})$ is a function that satisfies a differential equation that is obtained upon substituting Eq. (6.82) into the linearized Ricatti equation Eq. (6.78b). On making this substitution and simplifying the resulting equation we obtain

$$[\nabla^2 + k_0^2]F(\mathbf{r}) = k_0^2[1 - n^2(\mathbf{r})]e^{ik_0 \mathbf{s}_0 \cdot \mathbf{r}}. \tag{6.83}$$

Equation (6.83) is precisely the equation satisfied by the Born approximation of the scattered field obtained earlier in the chapter. Indeed, the outgoing-wave solution to this equation is given by

$$F(\mathbf{r}) = \int d^3r'\, k_0^2[1 - n^2(\mathbf{r})]e^{ik_0\mathbf{s}_0\cdot\mathbf{r}} G_{0_+}(\mathbf{r} - \mathbf{r}'),$$

which is identical to Eq. (6.53) defining the Born approximation to the scattered field for the case of plane-wave incidence. We thus conclude that the Rytov approximation to the phase perturbation is *formally* related to the Born approximation to the scattered field via the equation

$$\delta W_{\rm R}(\mathbf{r};\mathbf{s}_0) = -\frac{i}{k_0} e^{-ik_0\mathbf{s}_0\cdot\mathbf{r}} U_{\rm B}^{(\rm s)}(\mathbf{r};\mathbf{s}_0). \tag{6.84}$$

Although the two approximations (Born and Rytov) are mathematically related, *they have different domains of validity*. Indeed, as we will find later, the Rytov approximation is generally more accurate than the Born approximation if the scattering object is large compared with the wavelength. On the other hand, the Rytov approximation degrades as the distance between the scattering volume τ and the field observation point $\mathbf{r}$ increases. Indeed, in the far field one obtains the result

$$U_{\rm R}(\mathbf{r};\mathbf{s}_0) = e^{ik_0[\mathbf{s}_0\cdot\mathbf{r} + \delta W_{\rm R}(\mathbf{r};\mathbf{s}_0)]} \sim e^{ik_0\mathbf{s}_0\cdot\mathbf{r}} + U_{\rm B}^{(\rm s)}(\mathbf{r};\mathbf{s}_0), \tag{6.85}$$

which shows that the two approximations become identical in the far field!

6.10.6 A comparison of the Born and Rytov approximations

We compared the Born and Rytov approximations with the exact computation of the fields scattered by a homogeneous cylinder and a pair of concentric homogeneous cylinders that were obtained in Section 6.3. We performed these comparisons for cylinders of differing indices of refraction and radii for which we computed the scattered fields evaluated over lines located at various distances from the cylinder axis. The exact and Born fields were computed using the multipole expansions as described in Sections 6.3, with $r = \sqrt{x^2 + z_0^2}$ and $\phi = \arctan x/z_0$, where x is the position on the line and z_0 the distance of the line from the center of the cylinders. The Rytov *scattered* field was computed according to the equation

$$U_{\rm R}^{(\rm s)}(x,z_0) = \overbrace{e^{ik_0(z_0 + \delta W_{\rm R}(x,z_0))}}^{U_{\rm R}(x,z_0)} - e^{ik_0z_0} = e^{ik_0\left[z_0 - \frac{i}{k_0}e^{-ik_0z_0}U_{\rm B}^{(\rm s)}(x,z_0)\right]} - e^{ik_0z_0},$$

where $\exp(ik_0z_0)$ is the incident plane wave on the line $z = z_0$.

We first computed the Born, Rytov and exact scattered fields for a single cylinder having a radius of $a_0 = 4\lambda$ and parameterized by its index of refraction $n_{\rm r}$, which assumed the values of 1.01 to 1.07 in steps of $\delta n_{\rm r} = 0.02$. We took the measurement line to be located at one wavelength from the cylinder's surface along the z axis; i.e., $z_0 = a_0 + \lambda$. We show in Figs. 6.9 and 6.10 plots of the magnitude and phase of the scattered fields as a function of x on the measurement line. It is clear from these figures that the Rytov approximation far outperforms the Born approximation at larger values of the index of refraction of the cylinder.

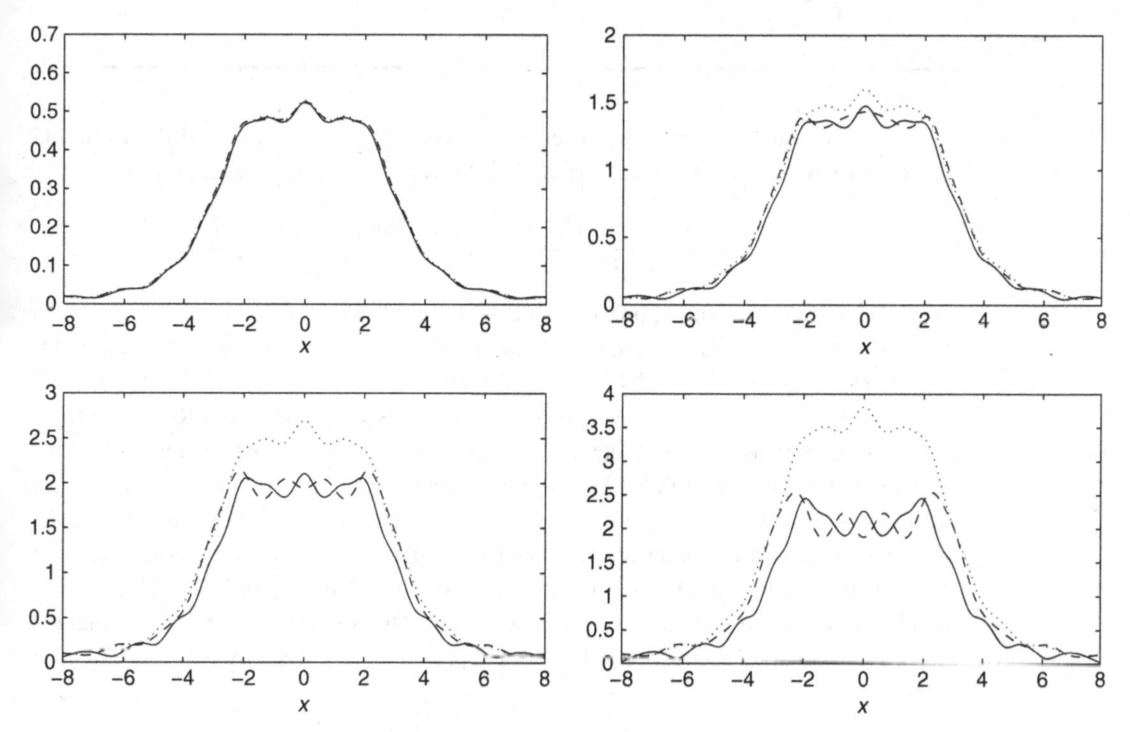

Fig. 6.9 Magnitudes of the exact values (solid line) and of the Born (dotted) and Rytov (dashed) approximations to the field scattered from a cylinder over a line located one wavelength from the cylinder boundary. The cylinder has a radius $a_0 = 4\lambda$ and relative indices of refraction varying from $n_r = 1.01$ (top left) to $n_r = 1.07$ (bottom right).

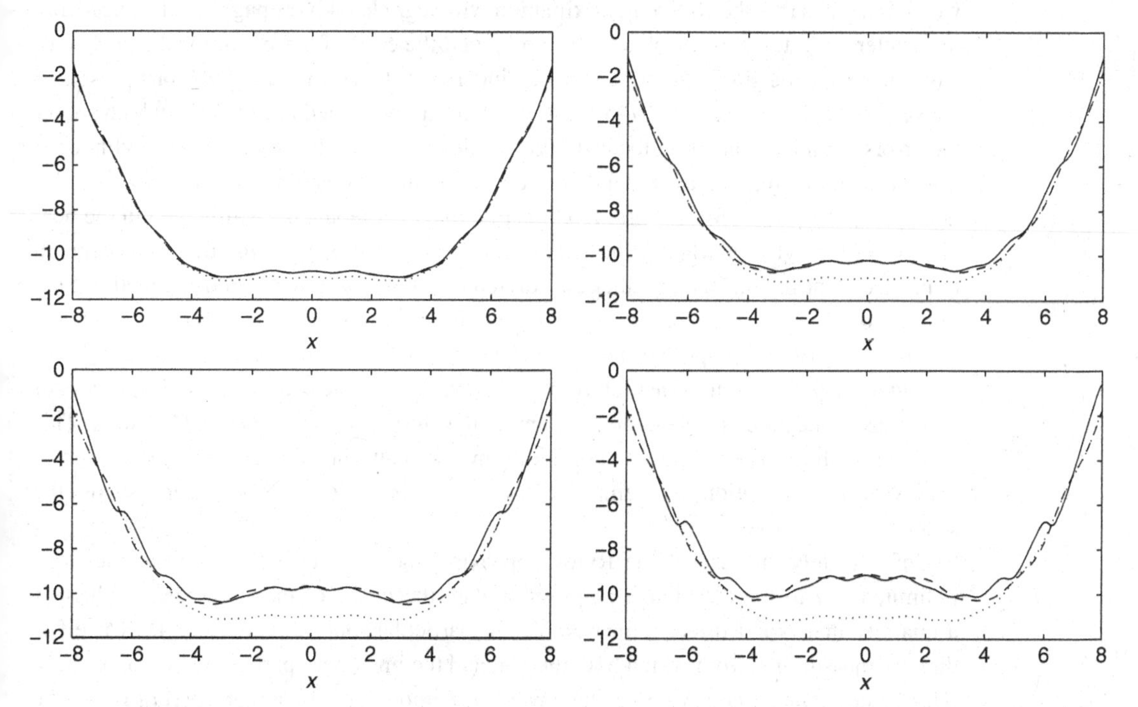

Fig. 6.10 Real parts of the phases of the exact values (solid) and of the Born (dotted) and Rytov (dashed) approximations to the field scattered from a cylinder over a line located one wavelength from the cylinder boundary. The cylinder has a radius $a_0 = 4\lambda$ and relative indices of refraction varying from $n_r = 1.01$ (top left) to $n_r = 1.07$ (bottom right).

We also computed the normalized errors between the exact values and the Born and Rytov approximations to the scattered fields defined according to the equation

$$E(n_r, a_0) = \frac{||U^{(s)} - \hat{U}^{(s)}||}{||U^{(s)}||} = \sqrt{\frac{\int dx |U(x, z_0) - \hat{U}(x, z_0)|^2}{\int dx |U^{(s)}(x, z_0)|^2}},$$

where the hat denotes the approximate scattered field (Born or Rytov). We present in Fig. 6.11 the percentage normalized errors computed for the Born and Rytov approximations plotted as functions of the relative refractive index of the cylinder also in percent for cylinder radii of λ, 2λ, 3λ and 4λ. The top plot is the error in the Rytov scattered field and the bottom, which is identical to that shown in the top part of Fig. 6.8, is the error in the Born scattered field. As discussed in connection with Fig. 6.8, the Born error increases linearly with the cylinder radius at any given cylinder refractive-index value. The Rytov error, on the other hand, is seen to be independent of the cylinder radius, at least for low-to-moderate refractive-index values of the cylinder, and is much less than the Born error for any given refractive-index value. This supports the contention that the Rytov approximation is superior to the Born approximation, especially for large extended scatterers.

6.10.7 The hybrid approximation

We showed above that the Rytov and Born approximations are identical in the far field, which indicates that the Rytov approximation will degrade with propagation distance from the scatterer. To test this conclusion we computed the errors for the Born and Rytov scattered fields for the single homogeneous cylinders employed in the simulations presented above and displayed in Figs. 6.9–6.11. The results are presented in Fig. 6.12, which shows the errors computed on a line located 100 wavelengths from the surface of the cylinder. It is clear from this figure that at that distance the two approximations yield essentially identical errors. To correct for this problem it is possible to compute the Rytov field in the near vicinity of the cylinder where it has small error and then propagate the field so computed to more distant regions using either the Rayleigh–Sommerfeld formula developed in Section 2.9 or the angular-spectrum expansion or Fresnel transform developed in Chapter 4. The resulting two-step approximation has been called the *hybrid approximation* and can, of course, also be used to transition from distant regions to the near field via the process of field back propagation implemented via any of the algorithms developed in Chapter 4. This second possibility is extremely important in inverse scattering applications since it allows the Rytov approximation to be employed for data gathered at arbitrary distances from the scatterer.

Unfortunately, the use of the Rytov approximation in inverse scattering applications is limited due to the fact that it employs as data the *phase* of the (total) field, which is a quantity that is not directly measurable. In particular, only the complex field itself is directly measurable, from which both the real and the imaginary part of the complex phase must be computed, and herein lies the problem. Although the imaginary part of the phase is easily determined as the log amplitude of the magnitude of the field, the real part of the phase can be determined only up to jumps of integral multiples of 2π radians, and the

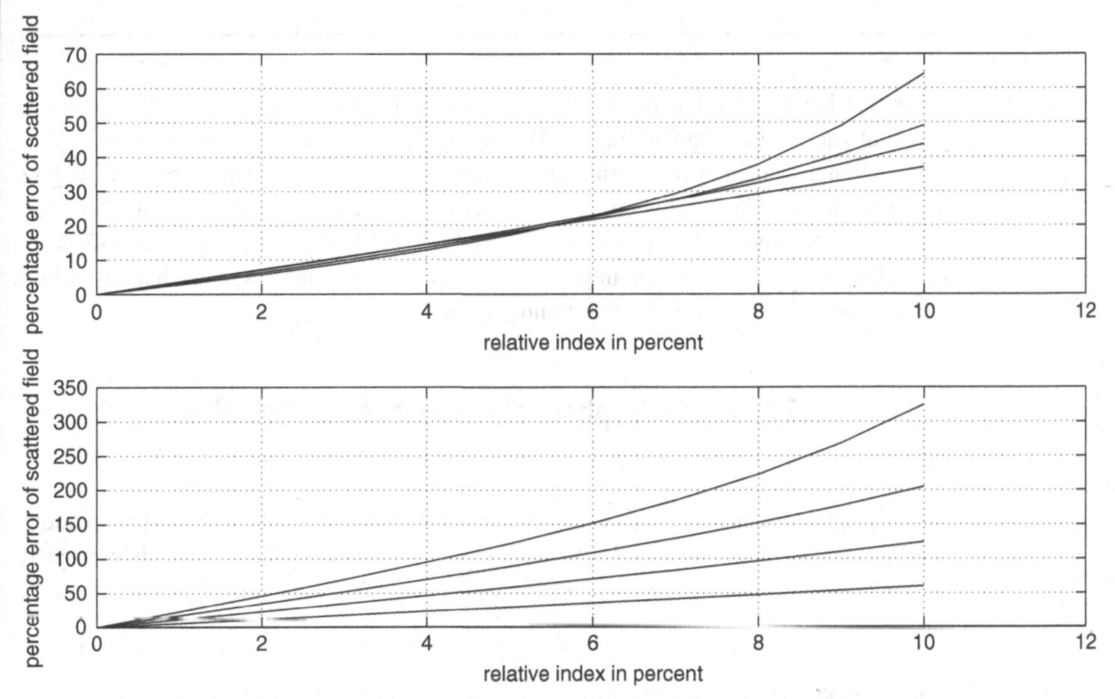

Fig. 6.11 Percentage normalized errors for cylinder radii of λ, 2λ, 3λ and 4λ plotted as a function of the percentage relative index of the cylinders for the scattered fields evaluated on the line $z_0 = a_0 + \lambda$. The error for the Rytov approximation is shown in the top plot and the error for the Born approximation in the bottom plot.

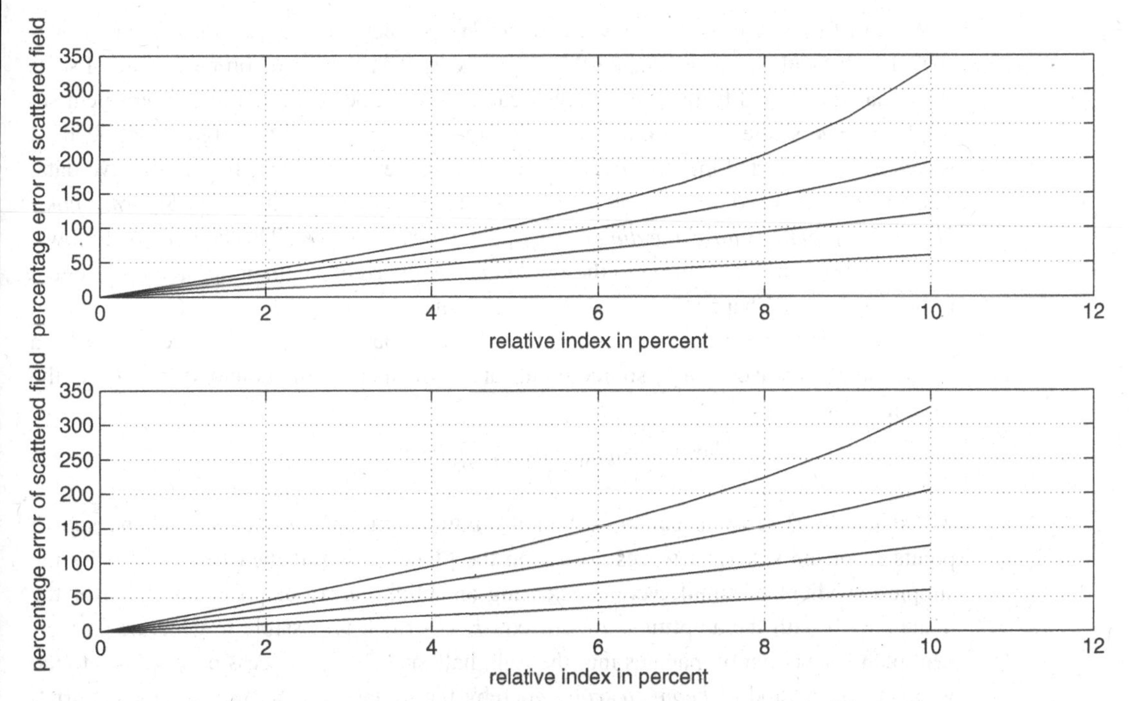

Fig. 6.12 Percentage normalized errors for Rytov (top) and Born (bottom) fields evaluated on the line $z_0 = 1_0 + 100\lambda$ for cylinder radii of λ, 2λ, 3λ and 4λ plotted as a function of the percentage relative index of the cylinders.

phase that is related to the Born approximation via Eq. (6.84) and thus to the scattering potential via the linear model Eq. (6.53) is the *unwrapped phase*; i.e., the real phase that is a continuous function of position and does not possess any artificial jumps of integral multiples of 2π radians. This problem of determining the unwrapped phase of a function on the line or plane is a basic and largely unsolved problem in mathematics and severely limits the Rytov approximation in practice. We will return to this issue in Chapter 8 when we develop the theory of diffraction tomography.

6.11 Incident spherical waves and slant stacking

In many applications it is not possible to generate an incident plane wave and an incident spherical wave is used instead. In the ideal case the incident spherical wave is generated by a point source located at $\mathbf{r} = \mathbf{r}_0$, which results in the outgoing-wave Green function

$$U^{(\text{in})}(\mathbf{r}; \mathbf{r}_0) = G_{0_+}(\mathbf{r} - \mathbf{r}_0), \tag{6.86}$$

where the parameter ν is now equal to the source location $\mathbf{r}_0$. The associated LS equation is then given by

$$U(\mathbf{r}; \mathbf{r}_0) = G_{0_+}(\mathbf{r} - \mathbf{r}_0) + \int d^3r'\, G_{0_+}(\mathbf{r} - \mathbf{r}')V(\mathbf{r}')U(\mathbf{r}'; \mathbf{r}_0). \tag{6.87}$$

We will treat the direct and inverse scattering problems using general incident waves such as spherical waves in Chapter 9 but for now will show how a complete suite of scattering data generated from incident spherical waves can be converted to an equivalent set of plane-wave scattering data, thus allowing the inverse scattering problem for spherical waves to be converted into an equivalent inverse scattering problem for plane-wave data. The key to this procedure is the observation that the spherical waves Eq. (6.86) *form a complete set of functions for expanding other types of incident waves*. This conclusion follows, for example, from the solution to the Rayleigh–Sommerfeld Neumann problem presented in Section 2.9 of Chapter 2. Indeed, it was shown in Example 2.9 of that chapter that an incident plane wave propagating into the right half-space $z > 0$ can be generated by a set of spherical waves whose source points are all located on the boundary $z = 0$ via the formula

$$e^{ik_0\mathbf{s}_0\cdot\mathbf{r}} = 2ik_0 s_{0z} \int_{z_0=0} dS_0\, e^{ik_0\mathbf{s}_0\cdot\mathbf{r}_0} G_{0_+}(\mathbf{r} - \mathbf{r}_0), \tag{6.88}$$

where $s_{0z} > 0$ is the z component of the unit propagation vector $\mathbf{s}_0$ and the source points $\mathbf{r}_0$ of the spherical waves are all on the plane $z = 0$. It then follows that by the simple expedient of simultaneously exciting an infinite array of point sources over the plane $z = 0$ with the amplitude $2ik_0 s_{0z} \exp(ik_0\mathbf{s}_0 \cdot \mathbf{r}_0)$ it is possible to generate an incident plane wave that propagates into the right half-space. This process corresponds to the well-known method of *beam steering* whereby the radiation pattern of an antenna array is *electronically steered* by properly phasing the excitation inputs to the antenna array elements.

The scheme outlined in the above paragraph for generating incident plane waves from point sources has a number of disadvantages in inverse scattering applications. An obvious problem with the procedure is that it appears to require a continuum distribution of sources over the plane $z = 0$. However, the boundary value of a homogeneous plane wave is spatially band-limited to transverse wavenumbers $K_\rho < k_0$ so that a discrete array of sources with sample spacing $\delta \leq \pi/k_0 = \lambda/2$ suffices. Moreover, this array need only be finite in size so as to produce a plane-wave *beam* whose cross-sectional area is much larger than the extent of the scatterer volume τ_0. A more serious problem with the procedure is that a *real array* of sources is usually not available in practical applications and a *synthetic array* formed by sequentially measuring the scattered wave corresponding to different locations of a single point source or, at best, a small group of point sources is the best that can be expected. Fortunately, the data set that results from such a suite of experiments can be converted to an equivalent set that *would have* been obtained using a real (as opposed to synthetic) source array. The reason for this is that, as mentioned earlier in our discussion of the formal solution to the LS equation, the scattering process is a *linear mapping* from the incident wave $U^{(\text{in})}$ to the total wave U. This conclusion can also be arrived at by formally inverting the LS equation Eq. (6.7b) to obtain

$$(I - G_{0_+}V)U = U^{(\text{in})} \rightarrow U = (I - G_{0_+}V)^{-1}U^{(\text{in})}, \tag{6.89}$$

where I is the identity operator and $(I - G_{0_+}V)^{-1}$ is the inverse of $(I - G_{0_+}V)$. It follows from the linearity of the scattering process as embodied in Eq. (6.89) that *a full suite of scattering data obtained using incident spherical waves can be converted to a corresponding suite of scattering data generated by incident plane waves.*

To derive an algorithm for converting a suite of spherical-wave scattered field data to a suite of plane-wave scattered field data we use the LS equation Eq. (6.87) for the case of incident spherical waves $G_{0_+}(\mathbf{r}' - \mathbf{r}_0)$:

$$U(\mathbf{r};\mathbf{r}_0) = G_{0_+}(\mathbf{r} - \mathbf{r}_0) + \int d^3r'\, G_{0_+}(\mathbf{r} - \mathbf{r}')V(\mathbf{r}')U(\mathbf{r}';\mathbf{r}_0).$$

Let us now apply the *slant-stack operator*

$$\mathcal{S}_{\mathbf{s}_0} = 2ik_0 s_{0z} \int_{z_0=0} dS_0\, e^{ik_0\mathbf{s}_0\cdot\mathbf{r}_0} \tag{6.90}$$

to both sides of the above LS equation, where the subscript $\mathbf{s}_0$ indicates that the slant-stack operator depends parametrically on the propagation vector $\mathbf{s}_0$. We obtain the result

$$\begin{aligned}\mathcal{S}_{\mathbf{s}_0}U(\mathbf{r};\mathbf{r}_0) &= \mathcal{S}_{\mathbf{s}_0}G_{0_+}(\mathbf{r} - \mathbf{r}_0) + \int d^3r'\, G_{0_+}(\mathbf{r} - \mathbf{r}')V(\mathbf{r}')\mathcal{S}_{\mathbf{s}_0}U(\mathbf{r}';\mathbf{r}_0)\\ &= e^{ik_0\mathbf{s}_0\cdot\mathbf{r}} + \int d^3r'\, G_{0_+}(\mathbf{r} - \mathbf{r}')V(\mathbf{r}')\mathcal{S}_{\mathbf{s}_0}U(\mathbf{r}';\mathbf{r}_0),\end{aligned}$$

where we have used the fact that the slant-stack operator operates on the $\mathbf{r}_0$ coordinates and have also made use of Eq. (6.88). On comparing the above result with the LS equation Eq. (6.21) corresponding to an incident plane wave we conclude that

$$U(\mathbf{r};\mathbf{s}_0) = \mathcal{S}_{\mathbf{s}_0}U(\mathbf{r};\mathbf{r}_0). \tag{6.91}$$

Thus, by the simple process of slant stacking a set of scattered field data generated by incident spherical waves we can obtain a set of scattered field data that would have resulted from a suite of incident plane waves.

6.11.1 Slant stacking from arbitrary surfaces

Equation (6.91) allows the wavefield generated by an incident plane wave $\exp(ik_0\mathbf{s}_0 \cdot \mathbf{r})$ propagating into the half-space $z > 0$ to be computed from scattered field data generated from a suite of spherical waves all having their source points $\mathbf{r}_0$ lying on the plane $z = 0$. A simple modification of the slant-stack operator defined in Eq. (6.90) would yield analogous results for suites of spherical waves having source points located on any infinite plane surface. More generally, we can construct slant-stack operators that generate plane waves from a suite of spherical waves having source points $\mathbf{r}_0$ located over any closed or infinite surface Σ_0. These operators are obtained using the appropriate Neumann Green function for the particular surface Σ_0 and have the defining property

$$e^{ik_0\mathbf{s}_0\cdot\mathbf{r}} = \mathcal{S}_{\mathbf{s}_0} G(\mathbf{r}, \mathbf{r}_0), \tag{6.92}$$

where G is the Neumann Green function which is, in fact, a spherical wave with source point $\mathbf{r}_0 \in \Sigma_0$. From this point on we will interpret the slant-stack operator in the general form as defined by Eq. (6.92) as an operator that converts spherical waves located on an arbitrary closed or infinite surface Σ_0 to plane wavefields having unit wave vectors $\mathbf{s}_0$.

6.11.2 Slant-stack computation of the scattering amplitude

On making use of the definition Eq. (6.22) of the scattering amplitude and Eq. (6.91) relating the field generated by incident plane waves to that generated by incident spherical waves we conclude that

$$\begin{aligned} f(\mathbf{s};\mathbf{s}_0) &= \frac{-1}{4\pi}\int d^3r\, V(\mathbf{r})U(\mathbf{r};\mathbf{s}_0)e^{-ik_0\mathbf{s}\cdot\mathbf{r}} \\ &= \frac{-1}{4\pi}\int d^3r\, V(\mathbf{r})\mathcal{S}_{\mathbf{s}_0} U(\mathbf{r};\mathbf{r}_0)e^{-ik_0\mathbf{s}\cdot\mathbf{r}} \\ &= \mathcal{S}_{\mathbf{s}_0}\left\{\frac{-1}{4\pi}\int d^3r\, V(\mathbf{r})U(\mathbf{r};\mathbf{r}_0)e^{-ik_0\mathbf{s}\cdot\mathbf{r}}\right\} = \mathcal{S}_{\mathbf{s}_0} f(\mathbf{s},\mathbf{r}_0), \end{aligned} \tag{6.93a}$$

where

$$f(\mathbf{s},\mathbf{r}_0) = \frac{-1}{4\pi}\int d^3r\, V(\mathbf{r})U(\mathbf{r};\mathbf{r}_0)e^{-ik_0\mathbf{s}\cdot\mathbf{r}} \tag{6.93b}$$

is the *spherical-wave scattering amplitude*. This quantity is simply the scattering amplitude observed in a scattering experiment employing an incident spherical wave rather than an incident plane wave and is a special case of the *generalized scattering amplitude* defined in Section 6.4. Equation (6.93a) allows the (plane-wave) scattering amplitude to be computed via a slant stack of the spherical-wave scattering amplitude.

Further reading

Non-relativistic quantum scattering theory, upon which much of the material presented in this chapter is based, is treated in Newton (1982), Taylor (1972) and Sitenko (1991). Discussions of the classical use of the Born approximation in X-ray crystallography can be found in Vainshtein (1974) and Lipson and Cochran (1966). A comparison of the Born and Rytov approximations in one space dimension is presented in Keller (1969) and in three space dimensions for transmission and reflection at an interface in Oristaglio (1985) and, more generally, by Weston (Weston, 1985). A computer study comparing the approximations with exact acoustic scattering from cylinders was performed by Robinson and Greenleaf (Robinson and Greenleaf, 1986). Good overall treatments of the Rytov approximation can be found in Ishimaru (1999), Chernov (1967) and Tatarski (1961).

Problems

6.1 Derive the following alternative form of the Lippmann–Schwinger equation Eq. (6.7a):

$$U(\mathbf{r}, \nu) = U^{(\text{in})}(\mathbf{r}, \nu) + \int d^3r'\, G_+(\mathbf{r}, \mathbf{r}')V(\mathbf{r}')U^{(\text{in})}(\mathbf{r}', \nu),$$

where G_+ is the full Green function of the background with an embedded scatterer.

6.2 Prove that the full outgoing-wave Green function $G_+(\mathbf{r}, \mathbf{r}_0)$ is a symmetric function of its arguments.

6.3 Use Theorem 6.2 to compute the scattering amplitude of a scattering potential of the general form

$$V(\mathbf{r}) = \sum_{m=1}^{M} V_m(\mathbf{r} - \mathbf{X}_m)$$

in terms of the scattering amplitudes of the component potentials $V_m(\mathbf{r})$.

6.4 Use the angular-spectrum expansion of the scattered field given in Eq. (6.33) and the angular-spectrum expansion of the outgoing-wave multipole fields given in Eq. (3.49) of Chapter 3 to derive a multipole expansion of the scattered field including expressions for the multipole moments in terms of the scattering amplitude.

6.5 Use the scattering amplitude of a homogeneous sphere in the angular-spectrum expansion given in Section 6.5 to compute the multipole expansion of the scattered field. You will need to make use of the angular-spectrum expansions of the multipole fields given in Section 3.4.2. Verify that the expansion you obtained agrees with the one obtained in Section 6.3.

6.6 Use the scattering amplitude of a homogeneous cylinder in the 2D angular-spectrum expansion to compute the multipole expansion of the scattered field. You will need to make use of the angular-spectrum expansions of the 2D multipole fields found in Problem 4.13 of Chapter 4. Verify that the expansion you obtained agrees with the one obtained in Section 6.3.

6.7 Express the multipole moments of the scattered field in terms of its boundary value over a sphere that completely surrounds the scattering volume.

6.8 Compute the 2D Born approximation of the scattering amplitude of a homogeneous scatterer with wavenumber k_1 having a radius a_0 and being centered at $\mathbf{X}_0$. Verify that this scattering amplitude is in agreement with Theorems 6.1 and 6.2 but does not satisfy the optical theorem, Theorem 6.3.

6.9 Repeat Problem 6.8 for the 3D case of a sphere of radius a_0 centered at $\mathbf{X}_0$.

6.10 Compute the generalized scattering amplitude of a homogeneous sphere for the case of an incident free multipole field $j_l(kr)Y_l^m(\hat{\mathbf{r}})$ by using the technique given in Section 6.5.

6.11 Compute the field scattered from an infinite Dirichlet plane (a plane over which the field vanishes) located at $z = 0$ due to an incident wavefield radiated by a source $Q(\mathbf{r})$ located in the l.h.s. $z < 0$. Express your answer in terms of the outgoing-wave Green function $G_+(\mathbf{r} - \mathbf{r}')$.

6.12 Use the result obtained in the previous problem to derive the so-called "law of reflection," which states that a plane wave incident from the left half-space with unit propagation vector $\mathbf{s}_0$ onto an infinite plane Dirichlet surface located on the (x, y) plane will generate a reflected plane wave that propagates into the left half-space with unit wave vector $\widetilde{\mathbf{s}_0} = (s_{0_x}, s_{0_y}, -s_{0_z})$.

6.13 Express the scattered (reflected) wavefield found in Problem 6.11 in an angular-spectrum expansion and interpret your result in terms of the law of reflection stated in the previous problem.

6.14 Derive the Ricatti equation Eq. (6.72a) from the Helmholtz equation.

6.15 Derive the form of the Ricatti equation given in Eq. (6.75) from Eq. (6.72a).

6.16 Derive Eq. (6.83).

6.17 Derive Eq. (6.85).

7 Surface scattering and diffraction

In this chapter we turn our attention to scattering from non-penetrable objects, or "surface scattering," and "diffraction" from planar apertures.[1] As was mentioned in the introduction to the previous chapter, the interaction of an incident wave with a non-penetrable scatterer occurs over the surface of the scattering obstacle and is thus defined by some type of boundary condition over this surface. In a similar vein diffraction of an incident wave from apertures cut into non-penetrable surfaces is also defined by some type of boundary condition over the aperture plus surface and thus can, in a certain sense, be considered to be a type of surface scattering. The formal solution to both types of problems is thus obtained in an identical fashion by converting the problem into a boundary-value problem, which is then easily solved using the theory developed in Chapter 2.

The above prescription for "solving" surface scattering and aperture diffraction problems has one missing ingredient: determination of the boundary values required in the solution of the scattering or diffraction problem. This is the ingredient that distinguishes a scattering or diffraction problem from the purely mathematical boundary-value problem. In this chapter we will restrict our attention to non-penetrable objects over which the total field (incident plus scattered) satisfies homogeneous Dirichlet or Neumann conditions. By invoking this condition it is possible to represent the scattered field in terms of either the value of the normal derivative of the total field (the homogeneous Dirichlet case) or the total field itself (the homogeneous Neumann case) over the scatterer surface. Unfortunately, neither of these quantities is known a priori and we will find that they can be easily determined only for scattering objects whose surface coincides with a separable surface for the scalar wave Helmholtz equation. In such cases the unknown boundary condition can be computed and the scattered field represented via an eigenfunction expansion using the theory developed in Chapter 3.

Such exact solutions are not easily attained for non-penetrable scatterers of general shape or for aperture diffraction problems and we will need to resort to approximate solutions based on our intuitive understanding of how waves interact with material bodies. The approximate solutions developed in this chapter include the *physical-optics* approximation (PO approximation) for surface scattering and the *Kirchhoff* approximation in the case of aperture diffraction. These two approximations are closely related and both are short-wavelength approximations; i.e., require that the wavelength be small compared with the scatterer or aperture size.

[1] The term "diffraction" is usually associated with forward scattering resulting from an incident wavefield impinging on an aperture contained in an otherwise impenetrable surface. Unfortunately, as we will find in the following chapter, the term has been used also to describe forward scattering from penetrable objects, as in "diffraction" tomography.

Besides developing approximate solutions to (forward) aperture diffraction problems we will also formulate and obtain approximate solutions to *inverse diffraction problems* for such apertures. Just as our treatment of forward diffraction problems is based on solutions to (forward) boundary-value problems, our treatment of inverse diffraction will be based on solutions to *inverse* boundary-value problems. We have already encountered the inverse boundary-value problem (IBVP) for plane boundaries in Section 2.11 of Chapter 2 and in our development of field back propagation in Section 4.3 of Chapter 4. In the IBVP treated in those sections we considered a source supported within some strip $z^- \leq z \leq z^+$ and radiating a field $U_+(\mathbf{r})$ that was measured over an infinite plane $z = z_0$ lying outside the source strip. The goal of the IBVP was then to compute the radiated field within the interior strip(s) $z_0 \leq z \leq z^-$ and/or $z^+ \leq z \leq z_0$ from Dirichlet or Neumann data over the measurement plane z_0. We showed that field back propagation as developed in Section 4.3 provided an exact, yet unstable, solution to the IBVP, while the Green-function-based approach developed in Section 2.11 provided an approximate solution that was a regularized form of the exact solution obtained via field back propagation. In this chapter we will revisit the IBVP and show how it can be used as a basis for solving inverse diffraction problems from finite apertures. Whereas the IBVP is a purely formal mathematical problem and has limited practical use, the inverse diffraction problem has numerous applications in optics, acoustics and electromagnetics.

Finally, we will also present a brief treatment of inverse scattering from non-penetrable scatterers, where the goal is the determination of the scatterer shape (boundary) from measurements of its scattering amplitude. We will consider methods based on approximate scattering models such as the PO approximation mentioned above as well as methods that are independent of the scattering model, requiring only that the field satisfy homogeneous boundary conditions on the scatterer surface.

7.1 Formulation of the scattering problem for non-penetrable scatterers

The mathematical model for scattering from non-penetrable obstacles embedded in a uniform background medium is the homogeneous Helmholtz equation together with appropriate boundary conditions. These conditions include the far-field behavior as defined by Eq. (6.4) as well as homogeneous Dirichlet or Neumann conditions over the surface $\partial\tau_0$ of the scattering object. On applying standard Green-function techniques to the homogeneous Helmholtz equation satisfied by the total field outside the support of the scatterer and the Helmholtz equation Eq. (6.2) satisfied by the outgoing-wave Green function we find that

$$U(\mathbf{r}', \nu)\nabla_{r'}^2 G_{0+}(\mathbf{r} - \mathbf{r}') - G_{0+}(\mathbf{r} - \mathbf{r}')\nabla_{r'}^2 U(\mathbf{r}', \nu) = \delta(\mathbf{r} - \mathbf{r}')U(\mathbf{r}', \nu),$$

where ν is a parameter defining the incident wave employed in the scattering experiment and G_{0+} is the outgoing-wave Green function to the Helmholtz equation. We now integrate both sides of the above equation over all of space exterior to the scattering volume and use Green's theorem to find that, if $\mathbf{r} \notin \tau_0$,

$$U(\mathbf{r},\nu) = \int_{\Sigma_\infty} dS'\,\hat{\mathbf{n}}'\cdot\{U(\mathbf{r}',\nu)\nabla_{r'}G_{0+}(\mathbf{r}-\mathbf{r}') - G_{0+}(\mathbf{r}-\mathbf{r}')\nabla_{r'}U(\mathbf{r}',\nu)\}$$
$$- \int_{\partial\tau_0} dS'\,\hat{\mathbf{n}}'\cdot\{U(\mathbf{r}',\nu)\nabla_{r'}G_{0+}(\mathbf{r}-\mathbf{r}') - G_{0+}(\mathbf{r}-\mathbf{r}')\nabla_{r'}U(\mathbf{r}',\nu)\},$$

where Σ_∞ is the surface of a sphere of radius $R\to\infty$, $\partial\tau_0$ is the surface of the scattering object and $\hat{\mathbf{n}}'$ is the outward-directed unit vector to these two surfaces. If, on the other hand, $\mathbf{r}\in\tau_0$, the r.h.s. of the above equation is zero.

If we use the asymptotic boundary condition Eq. (6.4) satisfied by the total field at infinity we find that the first term in the above equation becomes (cf. Section 2.4 of Chapter 2)

$$\int_{\Sigma_\infty} dS'\,\hat{\mathbf{n}}'\cdot\{U(\mathbf{r}',\nu)\nabla_{r'}G_{0+}(\mathbf{r}-\mathbf{r}') - G_{0+}(\mathbf{r}-\mathbf{r}')\nabla_{r'}U(\mathbf{r}',\nu)\}$$
$$= \int_{\Sigma_\infty} dS'\,\hat{\mathbf{n}}'\cdot\{U^{(\mathrm{in})}(\mathbf{r}',\nu)\nabla_{r'}G_{0+}(\mathbf{r}-\mathbf{r}') - G_{0+}(\mathbf{r}-\mathbf{r}')\nabla_{r'}U^{(\mathrm{in})}(\mathbf{r}',\nu)\}$$
$$= U^{(\mathrm{in})}(\mathbf{r},\nu).$$

The total field outside of the scattering volume then reduces to

$$U(\mathbf{r},\nu) = U^{(\mathrm{in})}(\mathbf{r},\nu) - \int_{\partial\tau_0} dS'\,U(\mathbf{r}',\nu)\frac{\partial}{\partial n'}G_{0+}(\mathbf{r}-\mathbf{r}') \tag{7.1a}$$

or

$$U(\mathbf{r},\nu) = U^{(\mathrm{in})}(\mathbf{r},\nu) + \int_{\partial\tau_0} dS'\,G_{0+}(\mathbf{r}-\mathbf{r}')\frac{\partial}{\partial n'}U(\mathbf{r}',\nu), \tag{7.1b}$$

where $\partial/\partial n'$ denotes the outward-directed normal derivative to the surface $\partial\tau_0$. Equation (7.1a) applies to Neumann scatterers where the normal derivative of the field vanishes over their surface and Eq. (7.1b) applies to Dirichlet scatterers where the field vanishes over the scatterers' surface. If the field point $\mathbf{r}$ is interior to the scattering volume τ_0 the r.h.s. of both equations must vanish, which can be formally viewed as a requirement that the total interior field to the scattering volume must vanish.

7.1.1 The scattering amplitude

The scattering amplitude is defined in Eq. (6.4) to be the coefficient of the outgoing spherical wave in the asymptotic expansion of the scattered field as $k_0r\to\infty$ in the direction of the unit vector $\mathbf{s}$ resulting from an incident plane wave having unit propagation vector $\mathbf{s}_0$. On making use of the definition of G_{0_+} from Eq. (6.6b) we find that

$$G_{0_+}(r\mathbf{s}-\mathbf{r}') \sim -\frac{1}{4\pi}e^{-ik_0\mathbf{s}\cdot\mathbf{r}'}\frac{e^{ik_0r}}{r}, \quad k_0r\to\infty,$$

and

$$\frac{\partial}{\partial n'}G_{0+}(r\mathbf{s}-\mathbf{r}') \sim \frac{ik_0\mathbf{s}\cdot\hat{\mathbf{n}}'}{4\pi}e^{-ik_0\mathbf{s}\cdot\mathbf{r}'}\frac{e^{ik_0r}}{r}, \quad k_0r\to\infty.$$

If we substitute the above expressions into Eqs. (7.1) we obtain the following expressions for the scattering amplitudes for Neumann and Dirichlet non-penetrable scatterers:

$$f(\mathbf{s},\mathbf{s}_0) = -\frac{ik_0}{4\pi}\int_{\partial\tau_0} dS'\,\mathbf{s}\cdot\hat{\mathbf{n}}'\,U(\mathbf{r}';\mathbf{s}_0)e^{-ik_0\mathbf{s}\cdot\mathbf{r}'}, \quad \text{Neumann surface,} \tag{7.2a}$$

$$f(\mathbf{s},\mathbf{s}_0) = -\frac{1}{4\pi}\int_{\partial\tau_0} dS'\,e^{-ik_0\mathbf{s}\cdot\mathbf{r}'}\frac{\partial}{\partial n'}U(\mathbf{r}',\mathbf{s}_0), \quad \text{Dirichlet surface.} \tag{7.2b}$$

7.1.2 Liouville–Neumann expansion

The set of Eqs. (7.1) play for non-penetrable scatterers the role played by the Lippmann–Schwinger integral equation Eq. (6.7a) for penetrable scatterers. As was the case in potential scattering, we can employ a perturbation expansion in the form of a Liouville–Neumann (Born) series to obtain formal solutions to these equations. A formal iteration yields the results

$$U_{\mathrm{N}_\nu} = U_\nu^{(\mathrm{in})} - \int_{\partial\tau_0} dS'\,\frac{\partial}{\partial n'}G_{0+}U_\nu^{(\mathrm{in})} + \int_{\partial\tau_0} dS'\,\frac{\partial}{\partial n'}G_{0+}\int_{\partial\tau_0} dS''\,\frac{\partial}{\partial n''}G_{0+}U_\nu^{(\mathrm{in})} - \cdots \tag{7.3a}$$

and

$$U_{\mathrm{D}_\nu} = U_\nu^{(\mathrm{in})} + \int_{\partial\tau_0} dS'\,G_{0+}\frac{\partial}{\partial n'}U_\nu^{(\mathrm{in})} + \int_{\partial\tau_0} dS'\,G_{0+}\frac{\partial}{\partial n'}\int_{\partial\tau_0} dS''\,G_{0+}\frac{\partial}{\partial n''}U_\nu^{(\mathrm{in})} + \cdots, \tag{7.3b}$$

where the subscripts N and D are self-explanatory. It is clear from these equations that the mapping from the incident field to the scattered field is linear, whereas the mapping from the scatterer's surface $\partial\tau_0$ to the scattered field is non-linear. We found a similar non-linear mapping from the scattering potential to the scattered field in Chapter 6, where we remarked that this made the inverse scattering problem of determining the scattering potential from scattered field data extremely difficult, which caused us to develop linear approximate mappings such as the Born and Rytov approximations. It might be expected that a similar situation arises in the inverse scattering problem of determining the shape of a non-penetrable object from scattered field data. However, there is one big difference between potential scattering and surface scattering: the dimensionality of the problems. In particular, a scattering potential has three degrees of freedom, whereas a Dirichlet or Neumann surface has only two degrees of freedom. Because of this the inverse scattering problems are vastly different and, as we will show later in this chapter, there are various approaches to inverse surface scattering that can be used to develop inverse scattering algorithms that are valid within the exact scattering models as defined in Eqs. (7.2). On the other hand, linearized versions of these equations are extremely useful in the forward scattering problem and have also been used in surface shape reconstruction as will be described later in the chapter.

Lowest-order approximation

One such linear model is the analog of the Born approximation for potential scattering. This model results from dropping all terms in the Liouville expansions except the leading two terms and is thus given by

$$U^{\rm B}_{{\rm N}_\nu} = U^{\rm (in)}_\nu - \int_{\partial\tau_0} dS' \frac{\partial}{\partial n'} G_{0+} U^{\rm (in)}_\nu, \qquad U^{\rm B}_{D_\nu} = U^{\rm (in)}_\nu + \int_{\partial\tau_0} dS'\, G_{0+} \frac{\partial}{\partial n'} U^{\rm (in)}_\nu. \tag{7.4}$$

Unfortunately, experience has shown that the above Born models for non-penetrable scatterers are not good approximations and it is necessary to resort to ad-hoc (non-perturbational) approximations such as the *physical-optics approximation*, which will be developed later in the chapter.

7.2 Scattering from simple shapes

In cases in which the scatterer surface $\partial\tau_0$ coincides with a separable surface for the scalar Helmholtz equation either a Green function or an eigenfunction expansion of the type developed in Chapter 3 can be employed to exactly solve the forward scattering problem. The Green-function solution employs a Green function satisfying the same homogeneous condition over the surface of the scatterer (either homogeneous Dirichlet or Neumann) as is satisfied by the total field, whereas in the eigenfunction approach the incident and scattered fields are represented via eigenfunction expansions appropriate to the specific separable system and the boundary conditions over $\partial\tau_0$ can be applied term by term to these series. This approach is similar to that employed in Section 6.3 of Chapter 6 to compute the scattered field for penetrable objects whose scattering potentials are piecewise constant within regions bounded by separable surfaces. The simplest example of this type is scattering from an infinite plane surface over which the field satisfies homogeneous Dirichlet or Neumann conditions. This problem was given in the form of Problems 2.16 and 2.17 at the end of Chapter 2 and in the form of Problems 4.9 and 4.10 at the end of Chapter 4. The problems posed in Chapter 2 sought Green-function-based expressions for the scattered field from a Dirichlet plane, whereas those posed in Chapter 4 sought angular-spectrum expansions for the scattered field. Here, we will briefly review the angular-spectrum solution to the problem of an incident wavefield scattering from a Dirichlet or Neumann plane and then look at scattering from spheres and cylinders.

7.2.1 Scattering from an infinite Dirichlet or Neumann plane

We consider an infinite plane surface located at $z = 0$ and illuminated by an incident wavefield propagating into the plane from the l.h.s. $z < 0$. We will represent a general field point $\mathbf{r}$ in terms of its transverse component $\boldsymbol{\rho} = (x, y)$ lying on the plane surface and its longitudinal component z. We can represent the incident wavefield via an angular-spectrum expansion of the general form

$$U^{(\mathrm{in})}(\mathbf{r}) = \int_{K_\rho < k_0} d^2K_\rho\, A(\mathbf{K}_\rho) e^{i\mathbf{K}_\rho \cdot \boldsymbol{\rho}} e^{i\gamma z}, \tag{7.5a}$$

where $\gamma = \sqrt{k_0^2 - K_\rho^2}$ and $A(\mathbf{K}_\rho)$ is an arbitrary plane-wave amplitude ("angular spectrum") of the incident wave. We have assumed for simplicity that the incident wave possesses no evanescent plane-wave components, although this is not a necessary assumption in the theory. Equation (7.5a) is seen to consist of a superposition of homogeneous plane waves whose wave vectors $\mathbf{K}_\rho + \gamma\hat{z}$ all have positive z components and, hence, all of which propagate in the positive-z direction (toward the plane $z = 0$) from the l.h.s.

It is not difficult to show that if the total field satisfies homogeneous Dirichlet conditions over the plane at $z = 0$ then the reflected (scattered) wave propagates into the l.h.s. ($z < 0$) according to the angular-spectrum expansion

$$U_{\mathrm{D}}^{(\mathrm{s})}(\mathbf{r}) = -\int_{K_\rho < k_0} d^2K_\rho\, A(\mathbf{K}_\rho) e^{i\mathbf{K}_\rho \cdot \boldsymbol{\rho}} e^{-i\gamma z}, \tag{7.5b}$$

where we have used the subscript D to denote that this is the field scattered by a Dirichlet plane surface. Equation (7.5b) is seen to be identical to the expansion of the incident wave except for the presence of the minus sign *and* the replacement of $+\gamma$ by $-\gamma$ in the component plane waves comprising the expansion. The expansion of the scattered wave thus consists of a superposition of homogeneous plane waves, all of which propagate in the negative-z direction (away from the plane boundary) with the same angular spectrum (plane-wave amplitude) as the incident wave.

A completely parallel development applies if the total wavefield satisfies homogeneous Neumann conditions over the plane. In this case the normal derivative of the total field must vanish at $z = 0$ so that the scattered field admits the plane-wave expansion

$$U_{\mathrm{N}}^{(\mathrm{s})}(\mathbf{r}) = \int_{K_\rho < k_0} d^2K_\rho\, A(\mathbf{K}_\rho) e^{i\mathbf{K}_\rho \cdot \boldsymbol{\rho}} e^{-i\gamma z}. \tag{7.5c}$$

It is clear that the total field $U_{\mathrm{D}} = U^{(\mathrm{in})} + U_{\mathrm{D}}^{(\mathrm{s})}$ vanishes over the Dirichlet plane and the normal derivative of $U_{\mathrm{N}} = U^{(\mathrm{in})} + U_{\mathrm{N}}^{(\mathrm{s})}$ vanishes over the Neumann plane. We also note the interesting result that the normal derivative of the total field over the Dirichlet plane is given by

$$\frac{\partial}{\partial z} U_{\mathrm{D}}(\mathbf{r})|_{z=0} = 2\int_{K_\rho < k_0} d^2K_\rho\, i\gamma A(\mathbf{K}_\rho) e^{i\mathbf{K}_\rho \cdot \boldsymbol{\rho}} = 2\frac{\partial}{\partial z} U^{(\mathrm{in})}(\mathbf{r})|_{z=0}, \tag{7.6a}$$

while the total field over the Neumann plane is

$$U_{\mathrm{N}}(\mathbf{r})|_{z=0} = 2\int_{K_\rho < k_0} d^2K_\rho\, A(\mathbf{K}_\rho) e^{i\mathbf{K}_\rho \cdot \boldsymbol{\rho}} = 2U^{(\mathrm{in})}(\mathbf{r})|_{z=0}. \tag{7.6b}$$

Equations (7.6) form the foundation of the so-called "physical-optics approximation," which applies to scattering from Dirichlet or Neumann surfaces whose curvatures are smooth relative to the wavelength λ. We will review this approximation in some detail later in this chapter.

7.2.2 Scattering from a sphere

A more interesting example of a simple shape for which the surface scattering problem can be exactly solved is provided by a sphere over which the field satisfies homogeneous Dirichlet or Neumann conditions. We will first solve the problem for a sphere of radius a_0 that is centered at the origin and over which the (total) field must vanish (the Dirichlet problem for a sphere). In this case the appropriate eigenfunctions are the multipole fields into which we can expand both the incident and the scattered fields using the results obtained in Examples 3.5 and 3.4 of Chapter 3:

$$U^{(\mathrm{in})}(\mathbf{r},\nu)=\sum_{l=0}^{\infty}\sum_{m=-l}^{l}a_l^m(\nu)j_l(k_0r)Y_l^m(\hat{\mathbf{r}}), \tag{7.7a}$$

$$U^{(\mathrm{s})}(\mathbf{r},\nu)=\sum_{l=0}^{\infty}\sum_{m=-l}^{l}b_l^m(\nu)h_l^+(k_0r)Y_l^m(\hat{\mathbf{r}}), \tag{7.7b}$$

where the scattered-field expansion is valid outside the sphere; i.e., for $r > a_0$. In these equations Y_l^m are the spherical harmonics of order m and degree l, j_l and h_l^+ are the spherical Bessel and Hankel functions of the first kind of order l, respectively, and we have denoted the angular arguments of the spherical harmonics by the unit vector $\hat{\mathbf{r}}$. The quantities $a_l^m(\nu)$ are the so-called *multipole moments* of the incident wave (parameterized by ν and assumed specified in the scattering problem) and $b_l^m(\nu)$ are the multipole moments of the scattered wave.

The multipole moments $b_l^m(\nu)$ of the scattered field are the unknowns that must be determined using the condition that the total field vanishes over the surface of the sphere. Since the spherical harmonics are orthonormal over the unit sphere the vanishing of the total field (incident plus scattered) then yields the requirement

$$a_l^m(\nu)j_l(k_0a_0)+b_l^m(\nu)h_l^+(k_0a_0)=0 \Rightarrow b_l^m(\nu)=-\frac{j_l(k_0a_0)}{h_l^+(k_0a_0)}a_l^m(\nu), \tag{7.8}$$

where a_0 is the radius of the sphere. Using this result in Eq. (7.7b), we then obtain the following (exact) solution for the scattered field:

$$U^{(\mathrm{s})}(\mathbf{r},\nu)=-\sum_{l=0}^{\infty}\sum_{m=-l}^{l}\frac{j_l(k_0a_0)}{h_l(k_0a_0)}a_l^m(\nu)h_l^+(k_0r)Y_l^m(\hat{\mathbf{r}}). \tag{7.9}$$

The solution Eq. (7.9) is, in principle, exact. However, in practice it would, of course, be necessary to approximate the infinite sum via a finite sum. This is easily accomplished with small error due to the fact that the multipole moments $b_l^m(\mathbf{s}_0)$ decay exponentially fast with index l when l exceeds the cutoff value of $L=[k_0a_0]$, where the bracket denotes the next higher integer.[2] This is illustrated in Fig. 7.1, where we plot the ratio $|j_l(x)/h_l(x)|$ as a

[2] This assumes that the multipole moments $a_l^m(\nu)$ are all bounded, a condition that is guaranteed so long as the incident wave is everywhere bounded (see the problems at the end of this chapter).

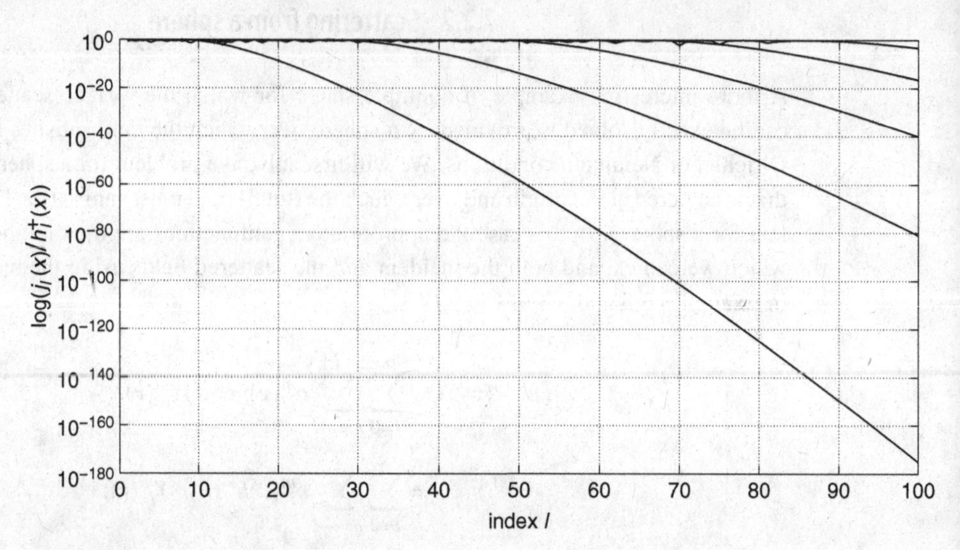

Fig. 7.1 The ratios $|j_l(x)/h_l^+(x)|$ plotted logarithmically as a function of l for values of $x = 10, 30, 50, 70$ and $x = 90$. The break points at $l = x$ where these ratios begin to decay exponentially fast with increasing l are clear in the plots.

function of l for various values of x. It follows from this figure that the multipole expansion Eq. (7.9) can be approximated with small error if we terminate the series at an l value that exceeds the cutoff value of $L = [k_0a_0]$.

The Neumann sphere

The same general approach as was used for the Dirichlet sphere can be employed for a sphere over which the normal derivative of the field must vanish (the Neumann sphere). In place of Eq. (7.8) we now require

$$j_l'(k_0a_0)a_l^m(\nu) + b_l^m(\nu)h_l^{+'}(k_0a_0) = 0 \Rightarrow b_l^m(\nu) = -\frac{j_l'(k_0a_0)}{h_l^{+'}(k_0a_0)}a_l^m(\nu), \tag{7.10}$$

where the prime on the spherical Bessel and Hankel functions denotes a first derivative with respect to their arguments. Using this result in Eq. (7.7b) then yields the following expression for the scattered field from a Neumann sphere:

$$U^{(s)}(\mathbf{r},\nu) = -\sum_{l=0}^{\infty}\sum_{m=-l}^{l}\frac{j_l'(k_0a_0)}{h_l^{+'}(k_0a_0)}a_l^m(\nu)h_l^+(k_0r)Y_l^m(\hat{\mathbf{r}}). \tag{7.11}$$

As in the case of the Dirichlet sphere shown in Fig. 7.1, the multipole moments $b_l^m(\nu)$ of the scattered wave decay exponentially fast when l exceeds the cutoff value of $L = [k_0a_0]$ so that the series Eq. (7.11) can be terminated at any l value that exceeds L.

The scattering amplitude

The (generalized) scattering amplitude is obtained by making use of the asymptotic formula

$$h_l^+(k_0 r) \sim (-i)^{l+1} \frac{e^{ik_0 r}}{k_0 r}, \quad \text{as } k_0 r \to \infty.$$

We then find using Eq. (7.7b) that

$$U^{(s)}(r\mathbf{s}, \nu) \sim \left\{ \sum_{l=0}^{\infty} \sum_{m=-l}^{l} \frac{(-i)^{l+1}}{k_0} b_l^m(\nu) Y_l^m(\mathbf{s}) \right\} \frac{e^{ik_0 r}}{r},$$

as $k_0 r \to \infty$ along the direction of the unit vector $\mathbf{s} = \mathbf{r}/r$. The generalized scattering amplitude is the quantity in brackets:

$$f(\mathbf{s}, \nu) = \sum_{l=0}^{\infty} \sum_{m=-l}^{l} \frac{(-i)^{l+1}}{k_0} b_l^m(\nu) Y_l^m(\mathbf{s}), \tag{7.12}$$

where the multipole moments $b_l^m(\nu)$ are given in Eq. (7.8) for a Dirichlet sphere and by Eq. (7.10) for a Neumann sphere.

7.2.3 Scattering of a plane wave from a cylinder

An especially interesting case that is ideally suited for computer simulations is scattering from a circular cylinder whose axis is aligned along the z axis of a Cartesian coordinate system when illuminated by a plane wave whose unit propagation vector $\mathbf{s}_0$ lies in the (x, y) plane. In this case the problem reduces to two space dimensions and we can use the 2D multipole expansions developed in Section 3.6 of Chapter 3. It was shown in that section that the incident and scattered waves can be expressed in 2D multipole expansions of the form

$$e^{ik_0 \mathbf{s}_0 \cdot \mathbf{r}} = e^{ik_0 r \cos(\phi - \phi_0)} = \sum_{n=-\infty}^{\infty} i^n J_n(k_0 r) e^{in(\phi - \phi_0)}, \tag{7.13a}$$

$$U^{(s)}(\mathbf{r}, \phi_0) = \sum_{n=-\infty}^{\infty} b_n(\phi_0) H_n^+(k_0 r) e^{in\phi}, \tag{7.13b}$$

where r, ϕ are the polar coordinates in an (x, y) plane perpendicular to the cylinder axis, with ϕ_0 being the polar angle of the incident plane wave as measured relative to the positive-x axis. Here, J_n and H_n^+ are the Bessel and Hankel functions of the first kind of order n. For a Dirichlet cylinder of radius a_0 the total field must vanish over the cylinder surface (at $r = a_0$), yielding the result

$$b_n(\phi_0) = -i^n \frac{J_n(k_0 a_0)}{H_n^+(k_0 a_0)} e^{-in\phi_0}, \tag{7.14}$$

which is the 2D version of Eq. (7.8). Substituting the expression for $b_n(\phi_0)$ into Eq. (7.13b) then yields the following expression for the scattered field from a Dirichlet cylinder:

$$U^{(s)}(\mathbf{r},\phi_0) = -\sum_{n=-\infty}^{\infty} i^n \frac{J_n(k_0a_0)}{H_n^+(k_0a_0)} H_n^+(k_0r)e^{in(\phi-\phi_0)}. \tag{7.15}$$

Using an entirely parallel development we find that for a Neumann cylinder

$$b_n(\phi_0) = -i^n \frac{J_n'(k_0a_0)}{H_n^{+'}(k_0a_0)} e^{-in\phi_0}, \tag{7.16}$$

from which we obtain the following expression for the field scattered by a Neumann cylinder:

$$U^{(s)}(\mathbf{r},\phi_0) = -\sum_{n=-\infty}^{\infty} i^n \frac{J_n'(k_0a_0)}{H_n^{+'}(k_0a_0)} H_n^+(k_0r)e^{in(\phi-\phi_0)}. \tag{7.17}$$

We show in Fig. 7.2 the magnitudes of the scattered and total wavefields from a Dirichlet cylinder having a radius of $a_0 = 5\lambda$ when illuminated by a plane wave propagating in the direction of the positive-x axis, i.e., at angle $\phi_0 = 0$. We have, of course, plotted these quantities only outside the cylinders, since the interior fields must vanish. The fields were computed using the multipole expansions given above with a cutoff l value of $[ka_0]$, where $[\cdot]$ stands for the next greatest integer. It is interesting to note that the maximum value of the magnitude of the scattered field is in the forward direction. Although this seems counter-intuitive for a Dirichlet cylinder, it makes sense when it is realized that this is due to the almost total blockage of the incident plane wave in the forward direction so that the scattered wave in this direction must be nearly equal to the negative of the incident plane wave. This is also apparent from the plot of the magnitude of the total (incident plus scattered waves) in Fig. 7.2, where this magnitude assumes a minimum value in the forward direction.

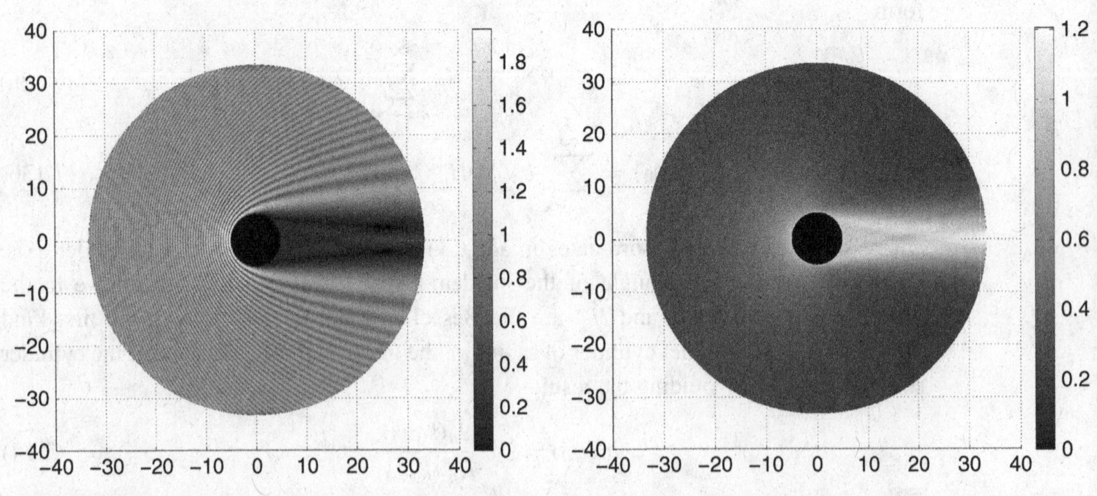

Fig. 7.2 Magnitudes of the total (left) and scattered (right) wavefields for an incident plane wave incident on a Dirichlet cylinder of radius $a_0 = 5\lambda$. The incident plane wave is propagating from left to right in the figures.

The scattering amplitude

The scattering amplitude for a Dirichlet or Neumann cylinder is obtained directly from Eq. (7.13b) upon making use of the asymptotic expression for the Hankel functions given in Eq. (6.28) of Chapter 6:

$$H_n^+(k_0 r) \sim \sqrt{\frac{2}{\pi k_0}} e^{-i(n+\frac{1}{2})\frac{\pi}{2}} \frac{e^{ik_0 r}}{\sqrt{r}}, \quad k_0 r \to \infty. \tag{7.18}$$

On substituting this expression into Eq. (7.13b) we obtain the result

$$U^{(s)}(r\mathbf{s}, \phi_0) \sim \left\{ \sqrt{\frac{2}{\pi k_0}} e^{-i\frac{\pi}{4}} \sum_{n=-\infty}^{\infty} (-i)^n b_n(\phi_0) e^{in\phi} \right\} \frac{e^{ik_0 r}}{\sqrt{r}}, \tag{7.19}$$

as $k_0 r \to \infty$ along the direction of the unit vector $\mathbf{s} = \mathbf{r}/r$ and where $b_n(\phi_0)$ is given in Eq. (7.14) for a Dirichlet cylinder and in Eq. (7.16) for a Neumann cylinder. The scattering amplitude in two space dimensions is the coefficient of the 2D outgoing spherical wave $\exp(ik_0 r)/\sqrt{r}$, from which we conclude that the scattering amplitudes measured at angle ϕ for a Dirichlet and a Neumann cylinder are given by

$$f_D(\phi, \phi_0) = -\sqrt{\frac{2}{\pi k_0}} e^{-i\frac{\pi}{4}} \sum_{n=-\infty}^{\infty} \frac{J_n(k_0 a_0)}{H_n^+(k_0 a_0)} e^{in(\phi-\phi_0)} \tag{7.20a}$$

and

$$f_N(\phi, \phi_0) = -\sqrt{\frac{2}{\pi k_0}} e^{-i\frac{\pi}{4}} \sum_{n=-\infty}^{\infty} \frac{J_n'(k_0 a_0)}{H_n^{+\prime}(k_0 a_0)} e^{in(\phi-\phi_0)}. \tag{7.20b}$$

We show plots of the scattering amplitude for both Dirichlet and Neumann cylinders in Figs. 7.3 and 7.4 for cylinder radii of $a_0 = \lambda, 2\lambda, 3\lambda$ and 4λ. Those in Fig. 7.3 are polar plots of the magnitude of the scattering amplitudes, whereas those in Fig. 7.4 are linear plots of the magnitudes of the scattering amplitudes as a function of the polar angle ϕ for an angle of incidence $\phi_0 = 0$. The polar plots indicate that there is a great deal of commonality in the magnitude of the scattering amplitudes when $a_0 > \lambda$ except in the immediate vicinity of the forward direction $\phi = 0$. Similar results are obtained for larger radii. The plots in Fig. 7.4 show that while the magnitudes of the scattering amplitudes for the two cases are similar if $a_0 > \lambda$, the two scattering amplitudes differ in detail outside of the main lobe.

7.3 The physical-optics approximation

The linearized Born-type approximations Eqs. (7.4) for Neumann or Dirichlet non-penetrable scatterers are not accurate due to the fact that we can expect the wavefield incident on a non-penetrable surface to be partially blocked by the surface, causing the wavefield in the region immediately behind the surface to be small. The linearized approximations, on the other hand, assume that the total wave over the entire surface is equal to

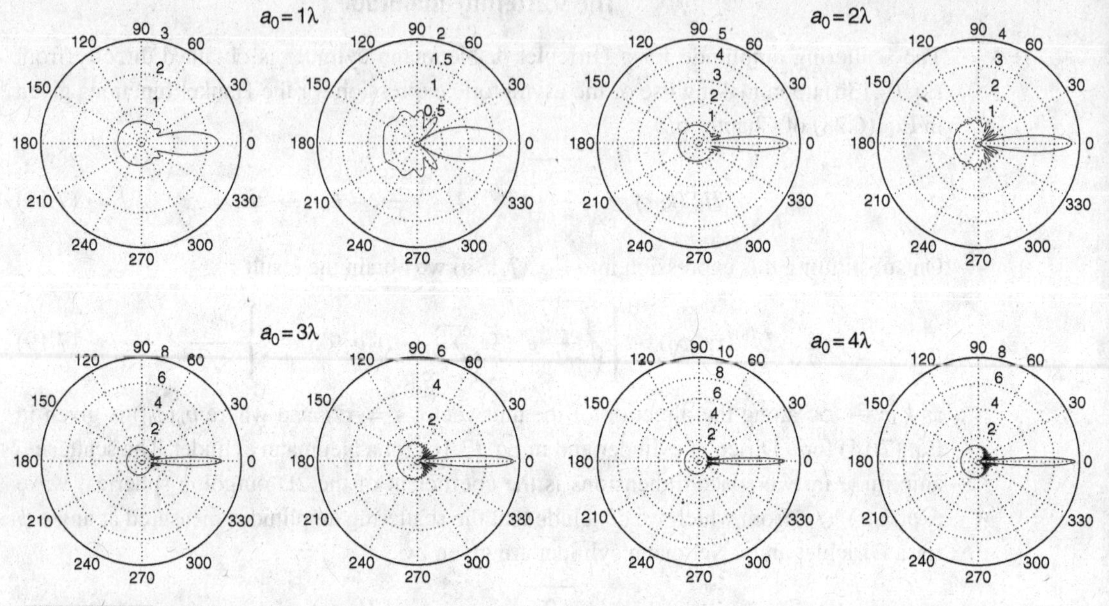

Fig. 7.3 Polar plots of the magnitude of the scattering amplitudes for Dirichlet (left member of each pair of plots) and Neumann cylinders (right member of each pair of plots).

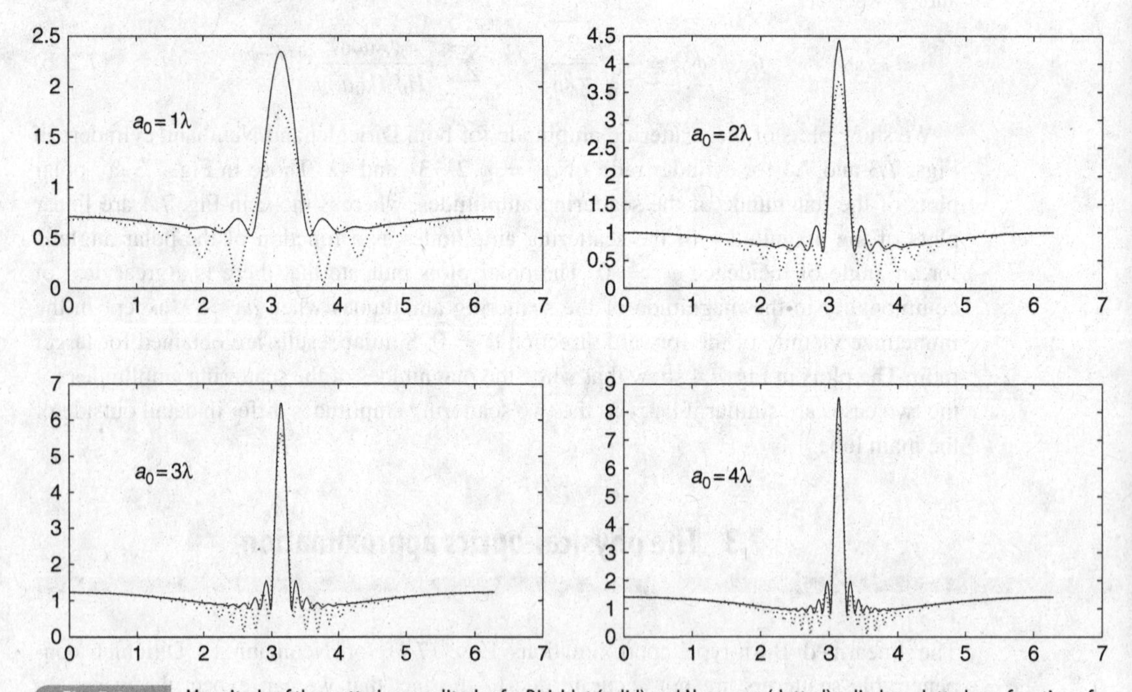

Fig. 7.4 Magnitude of the scattering amplitudes for Dirichlet (solid) and Neumann (dotted) cylinders plotted as a function of polar angle with $\phi = 0$ being the forward scattering direction.

the (unblocked) incident wave, which, although reasonable for a weak penetrable scatterer, will certainly not be the case for a non-penetrable scatterer. A more accurate approximation is possible if the scattering surface is *convex*[3] and varies slowly relative to the wavelength λ of the incident wave; i.e., such that the curvature of the surface varies slowly at the scale of the wavelength. Examples of such a surface would be spheres or cylinders whose radii are large relative to the wavelength. We show a more general example of such a surface in Fig. 7.5.

A Neumann or Dirichlet surface that satisfies the conditions outlined above has the property that we can attach a Neumann or Dirichlet plane at each point of the surface and approximately model the scattering of an incident wave in the immediate vicinity of that point as that of scattering from the locally attached plane. We saw in Section 7.2 that over such a plane the total Neumann field is simply twice the wavefield incident on the plane surface, while the normal derivative of the Dirichlet field is twice the normal derivative of the incident wave. *It is important to note that the development presented in Section 7.2 and these conclusions are based on the assumption that the field incident on the locally attached plane is propagating inwards toward the plane and, in particular, does not include any outward-propagating plane waves.* A general incident wave to a compactly supported scattering structure would, at each point of the surface, contain both inward-propagating plane waves and outward-propagating plane waves and the approximation outlined above requires that only the portion of the incident wavefield that contains the inward-propagating

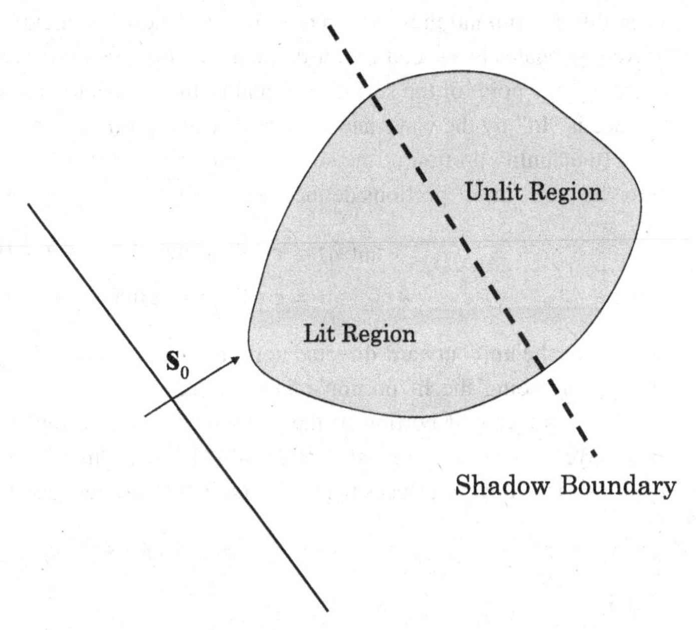

Fig. 7.5 A convex surface suitable for the physical-optics approximation and the decomposition of the surface into "lit" and "unlit" components.

[3] A convex surface is one for which a tangent plane can be attached at each point of the surface such that it doesn't intersect any other points of the surface.

plane waves be included. For example, an incident plane wave on a Dirichlet sphere of radius $a \gg \lambda$ would be mostly blocked by the sphere so that the field on the side of the sphere that does not face the incident plane wave would be essentially zero.

If we define that portion of the incident wave that propagates inward toward the scattering object by $U^{(\mathrm{in})}_{<}(\mathbf{r})$ we then define the *physical-optics approximation* to the scattered wave via the equations

$$U^{(\mathrm{s})}_{\mathrm{PO}}(\mathbf{r}) = -2\int_{\partial\tau_0} dS'\, U^{(\mathrm{in})}_{<}(\mathbf{r}')\frac{\partial}{\partial n'}G_{0+}(\mathbf{r}-\mathbf{r}'), \quad \text{Neumann surface} \tag{7.21a}$$

and

$$U^{(\mathrm{s})}_{\mathrm{PO}}(\mathbf{r}) = 2\int_{\partial\tau_0} dS'\, G_{0+}(\mathbf{r}-\mathbf{r}')\frac{\partial}{\partial n'}U^{(\mathrm{in})}_{<}(\mathbf{r}'), \quad \text{Dirichlet surface,} \tag{7.21b}$$

where we have used the subscript PO to denote the physical-optics (PO) approximation.

7.3.1 Plane-wave incidence

An obvious problem with the PO approximation is the determination of that component of the incident wave that is propagating inward to the surface at any given point. However, this determination is easy in the case of plane-wave incidence, for which the incident wave propagates in a specific well-defined direction. For this case the "inward-propagating wave" at any point of the surface is equal to the incident plane wave if that portion of the surface is "lit" by the wave and is zero over that portion that is in the shadow of the scatterer (the "unlit" portion of the surface). We accordingly decompose the total surface $\partial\tau_0$ into "lit" and "unlit" portions defined according to the equations

$$\partial\tau_{0l}(\mathbf{s}_0) = \mathbf{r} \in \partial\tau_0 \text{ such that } \mathbf{s}_0 \cdot \hat{\mathbf{n}} \le 0,$$
$$\partial\tau_{0l}^{\perp}(\mathbf{s}_0) = \mathbf{r} \in \partial\tau_0 \text{ such that } \mathbf{s}_0 \cdot \hat{\mathbf{n}} > 0,$$

where $\hat{\mathbf{n}}$ is the unit outward-directed normal to the surface at $\mathbf{r}$. The above equations define $\partial\tau_{0l}(\mathbf{s}_0)$ as being the lit portion of the surface that faces the incident plane wave and $\partial\tau_{0l}^{\perp}(\mathbf{s}_0)$ as the unlit portion of the surface that faces away from the incoming incident wave. We illustrate the decomposition into "lit" and "unlit" components in Fig. 7.5. It is clear that for convex surfaces the two components are related via the equation

$$\partial\tau_{0l}^{\perp}(\mathbf{s}_0) = \partial\tau_{0l}(-\mathbf{s}_0) \tag{7.22a}$$

and that

$$\partial\tau_0 = \partial\tau_{0l}(\mathbf{s}_0) \cup \partial\tau_{0l}^{\perp}(\mathbf{s}_0). \tag{7.22b}$$

The PO approximations to the scattered fields for plane-wave incidence are obtained from Eqs. (7.21) by setting

$$U^{(\mathrm{in})}_{<}(\mathbf{r};\mathbf{s}_0) = e^{ik_0\mathbf{s}_0\cdot\mathbf{r}}, \quad \mathbf{r} \in \partial\tau_{0l}(\mathbf{s}_0), \qquad U^{(\mathrm{in})}_{<}(\mathbf{r};\mathbf{s}_0) = 0, \quad \mathbf{r} \in \partial\tau_{0l}^{\perp}(\mathbf{s}_0). \tag{7.23}$$

We then obtain the results

$$U_{\mathrm{PO}}^{(\mathrm{s})}(\mathbf{r}, \mathbf{s}_0) = -2 \int_{\partial\tau_{0l}(\mathbf{s}_0)} dS'\, e^{ik_0\mathbf{s}_0\cdot\mathbf{r}'} \frac{\partial}{\partial n'} G_{0+}(\mathbf{r}-\mathbf{r}') \tag{7.24a}$$

for the Neumann surface and

$$U_{\mathrm{PO}}^{(\mathrm{s})}(\mathbf{r}, \mathbf{s}_0) = 2ik_0 \int_{\partial\tau_{0l}(\mathbf{s}_0)} dS'\, G_{0+}(\mathbf{r}-\mathbf{r}')\mathbf{s}_0 \cdot \hat{\mathbf{n}}' e^{ik_0\mathbf{s}_0\cdot\mathbf{r}'} \tag{7.24b}$$

for the Dirichlet surface.

The PO approximation to the scattering amplitude

The scattering amplitude within the PO approximation for plane-wave incidence is found from Eqs. (7.2) upon replacing the total surface $\partial\tau_0$ by $\partial\tau_{0l}(\mathbf{s}_0)$ and substituting from Eqs. (7.23) for the boundary-value fields over the lit surface. We obtain the results

$$f_{\mathrm{PO}}(\mathbf{s}, \mathbf{s}_0) = -\frac{ik_0}{2\pi} \int_{\partial\tau_{0l}(\mathbf{s}_0)} dS'\, \mathbf{s} \cdot \hat{\mathbf{n}}' e^{-ik_0(\mathbf{s}-\mathbf{s}_0)\cdot\mathbf{r}'}, \quad \text{Neumann surface} \tag{7.25a}$$

and

$$f_{\mathrm{PO}}(\mathbf{s}, \mathbf{s}_0) = -\frac{ik_0}{2\pi} \int_{\partial\tau_{0l}(\mathbf{s}_0)} dS'\, \mathbf{s}_0 \cdot \hat{\mathbf{n}}' e^{-ik_0(\mathbf{s}-\mathbf{s}_0)\cdot\mathbf{r}'}, \quad \text{Dirichlet surface.} \tag{7.25b}$$

7.3.2 Simulation

We compared the PO approximation for scattering from a Dirichlet cylinder with the exact field computation using the multipole expansion developed in Section 7.2.3. The PO computation was implemented using Eq. (7.24b) specialized to the case of an incident plane wave propagating in the positive-x direction in the (x, y) plane. In this case the problem reduces to two space dimensions, where the Green function admits the expansion given in Eq. (3.61) of Chapter 3:

$$G_{0+}(\mathbf{r}-\mathbf{r}') = \frac{i}{4} \sum_{n=-\infty}^{\infty} J_n(k_0 r') H_n^+(k_0 r) e^{in(\phi-\phi')},$$

where J_n and H_n^+ are the Bessel and Hankel functions of the first kind, r, ϕ and r', ϕ' are the polar coordinates of the field point $\mathbf{r}$ and the source point $\mathbf{r}'$, and it is assumed that $r > r'$. The PO approximation for a Dirichlet cylinder then takes the form

$$\begin{aligned} U_{\mathrm{PO}}^{(\mathrm{s})}(r, \phi) &= 2ik_0 a_0 \int_{\frac{\pi}{2}}^{\frac{3}{2}\pi} d\phi' \overbrace{\frac{i}{4} \sum_{n=-\infty}^{\infty} J_n(k_0 a_0) H_n^+(k_0 r) e^{in(\phi-\phi')}}^{G_{0+}(\mathbf{r}-\mathbf{r}')} \cos\phi'\, e^{ik_0 a_0 \cos\phi'} \\ &= -\frac{k_0 a_0}{2} \sum_{n=-\infty}^{\infty} C_n J_n(k_0 a_0) H_n^+(k_0 r) e^{in\phi}, \end{aligned} \tag{7.26a}$$

where the expansion coefficients C_n are given by

$$C_n = \int_{\frac{\pi}{2}}^{\frac{3}{2}\pi} d\phi' \cos\phi' \, e^{ik_0 a_0 \cos\phi'} e^{-in\phi'}.$$

It is not difficult to show that the expansion coefficients satisfy the condition $C_{-n} = C_n$ so that Eq. (7.26a) assumes the simplified form

$$U_{\rm PO}^{(s)}(r,\phi) = -\frac{k_0 a_0}{2}\left[C_0 J_0(k_0 a_0) H_0(k_0 r) + 2\sum_{n=1}^{\infty} C_n J_n(k_0 a_0) H_n^+(k_0 r)\cos(n\phi)\right], \tag{7.26b}$$

which is much simpler to implement in code than Eq. (7.26a).

We show in Figs. 7.6 and 7.7 comparisons of the exact values and PO approximations of the wavefields from a Dirichlet cylinder of radius $a_0 = 13.33\lambda$ illuminated by a plane wave propagating perpendicular to the cylinder axis. The exact results were obtained using the multipole expansion developed in Section 7.2.3, while the PO approximation was computed using Eq. (7.26b). The mesh plots shown in Fig. 7.6 cover a radial range of roughly 90 wavelengths outside the cylinder, while the plots in Fig. 7.7 are over a circle one wavelength outside the cylinder boundary. The PO approximation is seen to be excellent except around 90° and 270°, which correspond to the transition region between the lit and unlit portions of the cylinder. Plots comparing the PO approximation with the exact scattering amplitudes are shown in Fig. 7.8. The exact scattering amplitudes were computed using the expansion developed in Section 7.2.3, while the PO approximation was computed using the expansion

$$f_{\rm PO}(\phi_0) = \sqrt{\frac{k_0 a_0}{2\pi}} e^{-i\frac{\pi}{4}}\left[C_0 J_0(k_0 a_0) H_0(k_0 r) + 2\sum_{n=1}^{\infty}(-i)^n C_n J_n(k_0 a_0)\cos(n\phi)\right],$$

which results from Eq. (7.26b) after making use of Eq. (7.18). The plots shown in Fig. 7.8 illustrate the very close agreement between the exact values and the PO approximations of the magnitudes of the scattering amplitudes for four different cylinder radii. Overall, it is safe to say that the PO approximation to the scattering amplitude is excellent for cylinders whose radii exceed a few wavelengths.

7.4 The Bojarski transformation and linearized inverse surface scattering

N. Bojarski (Bojarski, 1982b) proposed an interesting transformation that can be employed for Dirichlet or Neumann scatterers that are accurately modeled by the PO approximation. The so-called "Bojarski transformation" which is useful in certain inverse scattering applications that we will discuss below is applied to the sum

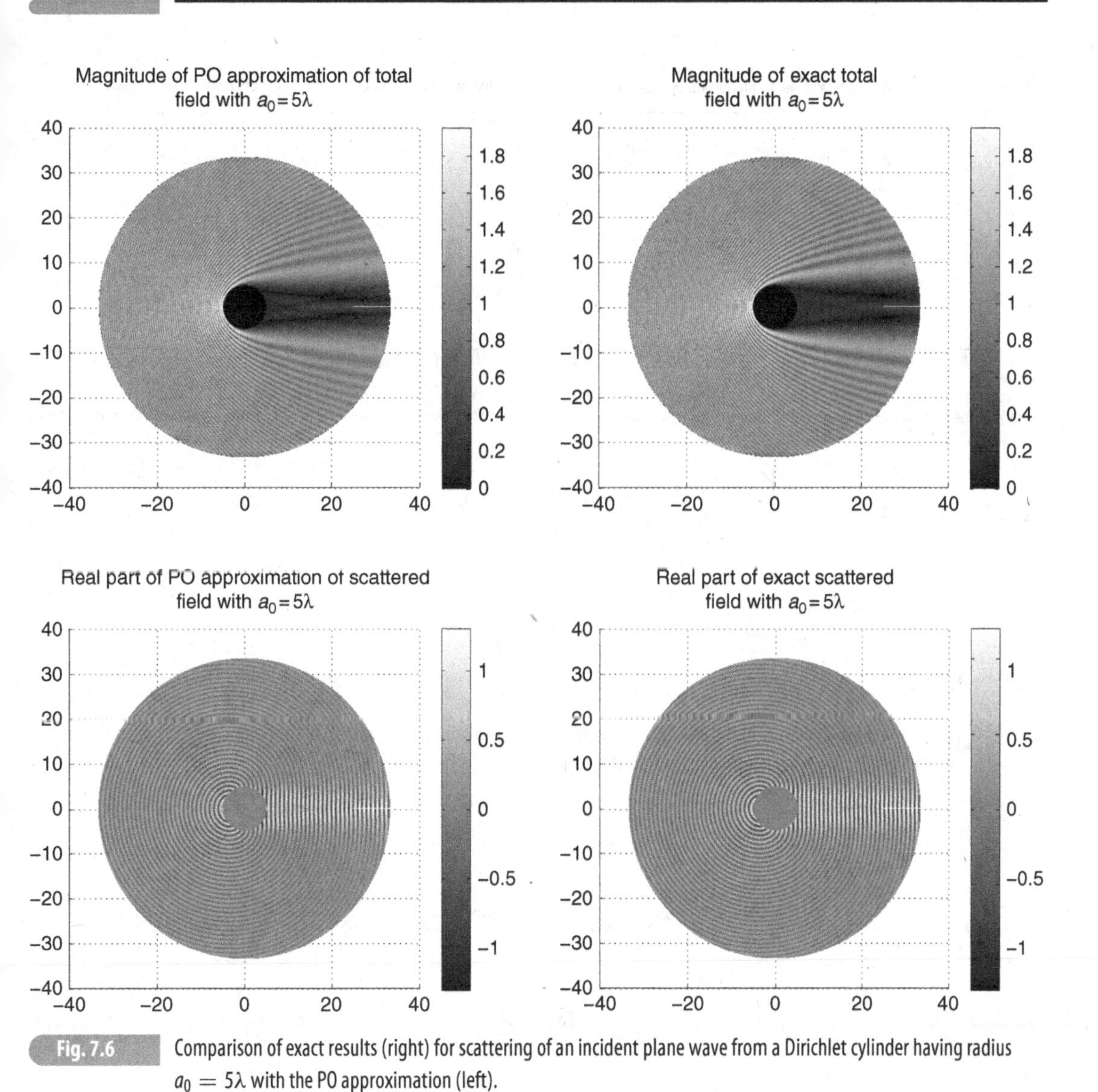

Fig. 7.6 Comparison of exact results (right) for scattering of an incident plane wave from a Dirichlet cylinder having radius $a_0 = 5\lambda$ with the PO approximation (left).

$$F(k_0\mathbf{s}_0) = f_{\mathrm{PO}}(-\mathbf{s}_0, \mathbf{s}_0) + f^*_{\mathrm{PO}}(\mathbf{s}_0, -\mathbf{s}_0). \tag{7.27}$$

The quantity $f_{\mathrm{PO}}(-\mathbf{s}_0, \mathbf{s}_0)$ is the scattering amplitude within the PO approximation computed in the *back-scattering* direction; i.e., for the observation vector $\mathbf{s} = -\mathbf{s}_0$. Equation (7.27) thus computes the sum of the back-scattering amplitude for some arbitrary incident wave direction $\mathbf{s}_0$ plus the complex conjugate of the back-scattering amplitude for the opposite incident wave direction $-\mathbf{s}_0$.

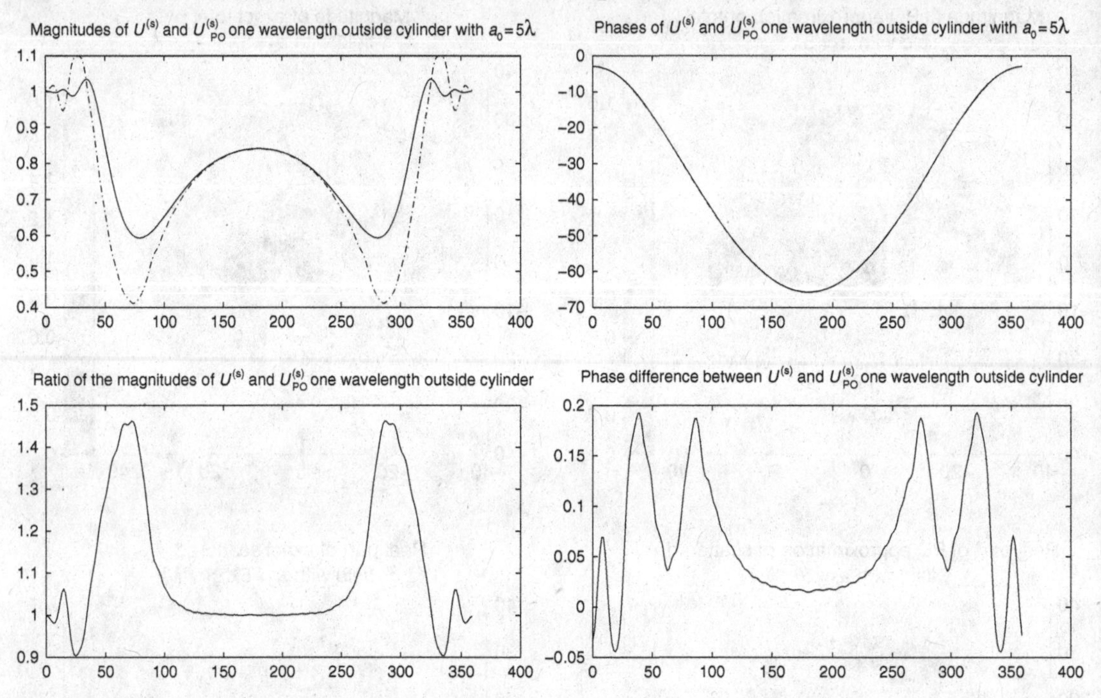

Fig. 7.7 Cross-sectional cuts of the exact values (solid) and PO approximations (dotted) of the field scattered from a Dirichlet cylinder one wavelength outside the cylinder boundary.

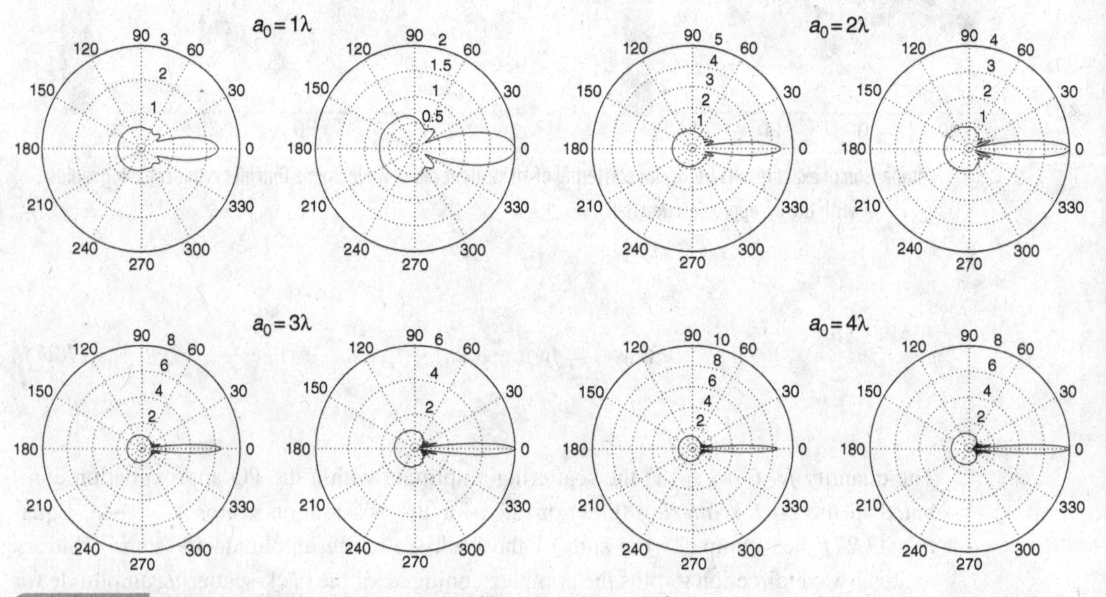

Fig. 7.8 Polar plots comparing the magnitudes of the exact values and PO approximations of the scattering amplitudes from Dirichlet cylinders having different radii.

The back-scattering amplitudes within the PO approximation are readily found from Eqs. (7.25) to be given by

$$f_{\rm PO}(-\mathbf{s}_0,\mathbf{s}_0)=\pm\frac{ik_0}{2\pi}\int_{\partial\tau_{0l}(\mathbf{s}_0)} dS'\,\mathbf{s}_0\cdot\hat{\mathbf{n}}' e^{i2k_0\mathbf{s}_0\cdot\mathbf{r}'},$$

where the top (plus) sign applies for Neumann scatterers and the bottom (minus) sign for Dirichlet scatterers. For convex surfaces the quantity $f_{\rm PO}(\mathbf{s}_0,-\mathbf{s}_0)$ is obtained by replacing $\mathbf{s}_0$ in the above equation by $-\mathbf{s}_0$ *and* by replacing $\partial\tau_{0l}(\mathbf{s}_0)$ by $\partial\tau_{0l}(-\mathbf{s}_0)$, which, according to Eq. (7.22a), is equal to its complement $\partial\tau_{0l}^{\perp}(\mathbf{s}_0)$. We then find that

$$f_{\rm PO}(\mathbf{s}_0,-\mathbf{s}_0)=\mp\frac{ik_0}{2\pi}\int_{\partial\tau_{0l}^{\perp}(\mathbf{s}_0)} dS'\,\mathbf{s}_0\cdot\hat{\mathbf{n}}' e^{-i2k_0\mathbf{s}_0\cdot\mathbf{r}'},$$

where the top sign is for Neumann scatterers and the bottom for Dirichlet scatterers. If we now compute the quantity $F(k_0\mathbf{s}_0)$ defined in Eq. (7.27) we obtain

$$\begin{aligned}F(k_0\mathbf{s}_0)&=\pm\frac{ik_0}{2\pi}\int_{\partial\tau_{0l}(\mathbf{s}_0)} dS'\,\mathbf{s}_0\cdot\hat{\mathbf{n}}' e^{i2k_0\mathbf{s}_0\cdot\mathbf{r}'}\pm\frac{ik_0}{2\pi}\int_{\partial\tau_{0l}^{\perp}(\mathbf{s}_0)} dS'\,\mathbf{s}_0\cdot\hat{\mathbf{n}}' e^{i2k_0\mathbf{s}_0\cdot\mathbf{r}'}\\&=\pm\frac{ik_0}{2\pi}\int_{\partial\tau_0} dS'\,\mathbf{s}_0\cdot\hat{\mathbf{n}}' e^{i2k_0\mathbf{s}_0\cdot\mathbf{r}'},\end{aligned}\qquad(7.28)$$

where we have used Eq. (7.22b) so that now the integration is over the entire surface $\partial\tau_0$.

The Bojarski transformation is completed by using the divergence theorem on Eq. (7.28) and simplifying the result. On applying this theorem we find that

$$\int_{\partial\tau_0} dS'\,\mathbf{s}_0\cdot\hat{\mathbf{n}}' e^{i2k_0\mathbf{s}_0\cdot\mathbf{r}'}=\int_{\tau_0} d^3r'\,\nabla_{r'}\cdot[\mathbf{s}_0 e^{i2k_0\mathbf{s}_0\cdot\mathbf{r}'}]=2ik_0\int_{\tau_0} d^3r'\,e^{i2k_0\mathbf{s}_0\cdot\mathbf{r}'},\qquad(7.29)$$

where τ_0 is the support volume of the scatterer. If we now define the *characteristic function* $\Gamma(\mathbf{r})$ of the scatterer via the equation

$$\Gamma(\mathbf{r})=\begin{cases}1 & \text{if } \mathbf{r}\in\tau_0\\ 0 & \text{else}\end{cases}$$

we can express Eq. (7.29) in the simplified form

$$\int_{\partial\tau_0} dS'\,\mathbf{s}_0\cdot\hat{\mathbf{n}}' e^{i2k_0\mathbf{s}_0\cdot\mathbf{r}'}=2ik_0\int d^3r'\,\Gamma(\mathbf{r}')e^{i2k_0\mathbf{s}_0\cdot\mathbf{r}'}=2ik_0\tilde{\Gamma}(-2k_0\mathbf{s}_0),\qquad(7.30)$$

where

$$\tilde{\Gamma}(\mathbf{K})=\int d^3r'\,\Gamma(\mathbf{r}')e^{-i\mathbf{K}\cdot\mathbf{r}'}.$$

Finally, on substituting Eq. (7.30) into Eq. (7.28) we obtain the desired result

$$F(k_0\mathbf{s}_0)=\mp\frac{k_0^2}{\pi}\tilde{\Gamma}(-2k_0\mathbf{s}_0),\qquad(7.31)$$

where the top sign applies to Neumann scatterers and the bottom to Dirichlet scatterers.

7.4.1 The generalized Bojarski transformation

A generalized form of the Bojarski transformation can be derived that is not limited to backscatter data and that is in the form of the Born approximation of penetrable scatterers obtained in the previous chapter. This generalized form is particularly useful in inverse scattering applications involving Dirichlet or Neumann scatterers since it allows the inversion algorithms developed in the following chapter for Born inversion to be applied to the surface-reconstruction problem discussed below and at the end of the chapter.

The derivation of the transformation follows almost identical lines to those employed above for the classical Bojarski transformation but where we do not require the scattering vector $\mathbf{s}$ to be the negative of the incident wave vector $\mathbf{s}_0$. Thus, in place of Eq. (7.27) we define the quantity

$$F_{\mathrm{g}}(\mathbf{s},\mathbf{s}_0)=f_{\mathrm{PO}}(\mathbf{s},\mathbf{s}_0)+f^*_{\mathrm{PO}}(-\mathbf{s},-\mathbf{s}_0), \tag{7.32}$$

which reduces to $F(k_0\mathbf{s}_0)$ under the replacement of $\mathbf{s}=-\mathbf{s}_0$. Restricting our attention for the moment to Dirichlet scatterers we then find in place of Eq. (7.28) the result

$$\begin{aligned}F_{\mathrm{g}}(\mathbf{s},\mathbf{s}_0)&=-\frac{ik_0}{2\pi}\int_{\partial\tau_{0I}(\mathbf{s}_0)}dS'\,\mathbf{s}_0\cdot\hat{\mathbf{n}}'e^{-ik_0(\mathbf{s}-\mathbf{s}_0)\cdot\mathbf{r}'}\\&\quad-\frac{ik_0}{2\pi}\int_{\partial\tau_{0I}^{\perp}(\mathbf{s}_0)}dS'\,\mathbf{s}_0\cdot\hat{\mathbf{n}}'e^{-ik_0(\mathbf{s}-\mathbf{s}_0)\cdot\mathbf{r}'}\\&=-\frac{ik_0}{2\pi}\int_{\partial\tau_0}dS'\,\mathbf{s}_0\cdot\hat{\mathbf{n}}'e^{-ik_0(\mathbf{s}-\mathbf{s}_0)\cdot\mathbf{r}'}.\end{aligned} \tag{7.33}$$

The transformation is completed by applying the divergence theorem to Eq. (7.33):

$$\begin{aligned}-\frac{ik_0}{2\pi}\int_{\partial\tau_0}dS'\,\mathbf{s}_0\cdot\hat{\mathbf{n}}'e^{-ik_0(\mathbf{s}-\mathbf{s}_0)\cdot\mathbf{r}'}&=-\frac{ik_0}{2\pi}\int_{\tau_0}d^3r'\,\nabla_{r'}\cdot[\mathbf{s}_0e^{-ik_0(\mathbf{s}-\mathbf{s}_0)\cdot\mathbf{r}'}]\\&=\frac{k_0^2}{2\pi}(1-\mathbf{s}_0\cdot\mathbf{s})\int d^3r'\,\Gamma(\mathbf{r}')e^{-ik_0(\mathbf{s}-\mathbf{s}_0)\cdot\mathbf{r}'},\end{aligned} \tag{7.34}$$

where $\Gamma(\mathbf{r}')$ is the *characteristic function* $\Gamma(\mathbf{r})$ of the scatterer. A completely parallel development yields the same result with a minus sign rather than a plus sign for the case of Neumann scatterers. We then obtain the generalized form of Eq. (7.31)

$$F_{\mathrm{g}}(\mathbf{s},\mathbf{s}_0)=\mp\frac{k_0^2}{2\pi}(1-\mathbf{s}_0\cdot\mathbf{s})\tilde{\Gamma}[k_0(\mathbf{s}-\mathbf{s}_0)], \tag{7.35a}$$

where the top sign applies to Neumann scatterers and the bottom to Dirichlet scatterers. We can also write the above result in the alternative form

$$F_{\mathrm{g}}(\mathbf{s},\mathbf{s}_0)=\mp\frac{1}{4\pi}|k_0(\mathbf{s}-\mathbf{s}_0)|^2\tilde{\Gamma}[k_0(\mathbf{s}-\mathbf{s}_0)]. \tag{7.35b}$$

The results obtained above are identical in form to the Born approximation of penetrable scatterers obtained in Section 6.7, where we found that the scattering amplitude of a penetrable scatterer was proportional to the spatial Fourier transform of the scattering potential over sets of Ewald spheres; i.e., over the sets of spatial frequencies defined by $\mathbf{K} = k_0(\mathbf{s} - \mathbf{s}_0)$. Here, the same general result is obtained, where the quantity $F_g(\mathbf{s}, \mathbf{s}_0)$ is seen from Eq. (7.35b) to be proportional to $K^2\tilde{\Gamma}(\mathbf{K})$ over a set of Ewald spheres. Since this transform is the transform of $-\nabla^2\Gamma(\mathbf{r})$, it then follows that within the PO approximation $F_g(\mathbf{s}, \mathbf{s}_0)$ plays the role of the scattering amplitude within the Born approximation and $-\nabla^2\Gamma(\mathbf{r})$ plays the role of the scattering potential. This conclusion has important consequences in the inverse scattering problem for Dirichlet and Neumann scatterers (see below).

7.4.2 Inverse scattering within the PO approximation

The quantities $F(k_0\mathbf{s}_0)$ and $F_g(\mathbf{s}, \mathbf{s}_0)$ can be employed in certain inverse scattering problems involving Dirichlet or Neumann convex-shaped scatterers. In particular, if the wavelength λ is much smaller than the minimum curvature of the (convex) scattering object we can expect the PO approximation to be valid and, hence, Eqs. (7.31) and (7.35) will provide good linearized scattering models that relate the shape of the scatterer (via the characteristic function) to scattered field data in the form of the scattering amplitude. It then follows that these models can be approximately inverted to obtain the shape from knowledge of the scattering amplitude specified over some set of incident- and scattered-wave directions $\mathbf{s}_0$ and $\mathbf{s}$ and some set of wavenumbers k_0. For example, if the forward- and back-scattering amplitudes are specified for all unit vectors $\mathbf{s}_0$ lying on the unit sphere and over some band of frequencies corresponding to wavenumbers $k_0 \in [k_0^-, k_0^+]$ then the characteristic function is approximately given by

$$\Gamma(\mathbf{r}) \approx \frac{1}{(2\pi)^3}\int_{\mathbf{K}\in\mathcal{D}} d^3K\,\tilde{\Gamma}(\mathbf{K})e^{i\mathbf{K}\cdot\mathbf{r}} = \frac{1}{(2\pi)^3}\int_{2k_0^-}^{2k_0^+} K^2\,dK \int_{4\pi} d\Omega_s\,\tilde{\Gamma}(K\mathbf{s})e^{iK\mathbf{s}\cdot\mathbf{r}}$$
$$= \mp\frac{1}{2\pi^2}\int_{k_0^-}^{k_0^+} dk_0 \int_{4\pi} d\Omega_{s_0}\,F^*(k_0\mathbf{s}_0)e^{ik_0\mathbf{s}_0\cdot\mathbf{r}}, \tag{7.36}$$

where the top sign applies for Neumann scatterers and the bottom for Dirichlet scatterers and we have used the result that

$$F(-k_0\mathbf{s}_0) = F^*(k_0\mathbf{s}_0).$$

The actual surface can then be obtained from the above image of the characteristic function by applying the gradient operator and finding the maximum of the magnitude of this quantity:

$$\partial\tau_0 \approx \max\left|\frac{1}{2\pi^2}\int_{k_0^-}^{k_0^+} dk_0 \int_{4\pi} d\Omega_{s_0}\,\mathbf{s}_0F^*(k_0\mathbf{s}_0)e^{2ik_0\mathbf{s}_0\cdot\mathbf{r}}\right|. \tag{7.37}$$

Another case of interest is when the scattering amplitude is specified at a single frequency but over some set of incident and scattered field directions. In this case the inversion algorithms developed in the following chapter for penetrable scatterers within the

Born approximation can be employed to yield an approximate estimate of $\nabla^2\Gamma(\mathbf{r})$ from $F_g(\mathbf{s}, \mathbf{s}_0)$. This inversion is of special interest since it will generate a doublet pulse over the surface of the scatterer, from which its shape will be clearly evident. We will return to the inverse problem for surface scattering in Section 7.7, where we develop this inversion scheme and present other approaches to the surface-reconstruction problem, including one method that does not rely on linearized approximations such as the PO approximation.

7.5 Kirchhoff diffraction theory

The boundary-value problem is a purely mathematical problem associated with the homogeneous Helmholtz equation in some finite or semi-infinite region of space. In this section we will see how the solution to this purely mathematical problem can be employed to obtain (approximate) solutions to practical problems involving the interaction of scalar wavefields with certain types of physical bodies and, in particular, to the problem of wavefield *diffraction*. In the classical diffraction problem a known wavefield $U^{(\text{in})}(\mathbf{r}, \nu)$ is incident onto an aperture located in the plane $z = 0$. The incident wave interacts with the aperture and the result of this interaction is a *diffracted wave* $U^{(\text{d})}(\mathbf{r}, \nu)$ that propagates away from the aperture into the right half-space $z > 0$. The *diffraction problem* consists of determining the diffracted wave from the known incident wavefield and the properties of the diffracting aperture.[4]

A key ingredient of the above description of an aperture diffraction problem is the phrase "properties of the diffracting aperture." This phrase sets the diffraction problem apart from the boundary-value problem since it requires knowledge in the form of a mathematical model for the interaction of the aperture with the incident wave. This additional information cannot come from the Helmholtz equation itself and must, instead, come from the physics describing the interaction of the particular wavefield with the aperture. However, it is possible to obtain a heuristic model for this interaction that applies in many practical applications, by appealing to our intuitive understanding of how waves interact with material bodies. For example, we know that there exist certain material bodies that are almost transparent to an incident wavefield, whereas other types of material are essentially opaque to an incident wave; i.e., essentially block the transmission of an incident wave.

Kirchhoff used such intuitive ideas to obtain an approximate solution of the scalar wave-diffraction problem for the case of a wave incident onto an aperture consisting of a perfectly transparent hole in a perfectly opaque screen.[5] In his solution of the problem Kirchhoff made the seemingly obvious assumption that the field passed unattenuated through the hole in the aperture but was completely blocked by the opaque material of the screen at points not within the hole. Thus, assuming that the incident wave is propagating from

[4] A reflected wave that propagates back into the left half-space will also be generated in the process but is of no interest in classical diffraction theory and is not considered to be a part of the actual diffracted wavefield.

[5] The solution of the dual problem of a perfectly opaque hole in a perfectly transparent screen can be obtained from the solution of this problem using Babinet's principle (Born and Wolf, 1999).

left to right,[6] Kirchhoff hypothesized that the diffracted wave and its normal derivative to the immediate right of the hole were equal to the value of the incident field and its normal derivative over the hole and that the diffracted wave not within the area of the hole vanished identically. Since both the field and its normal derivative were then specified over the entire aperture plane and the wavefield had to be outgoing into the r.h.s. and, hence, had to satisfy the Sommerfeld radiation condition (SRC) in that half-space, Kirchhoff was able to use the first Helmholtz identity (cf. Eq. (2.41a) in Section 2.8 of Chapter 2) to obtain the following expression for the diffracted field in the half-space $z > 0$:

$$U^{(d)}(\mathbf{r}, \nu) = \int_{\mathcal{A}} d^2\rho' \left[G_+(\mathbf{r} - \mathbf{r}'_0) \frac{\partial}{\partial z'} U^{(in)}(\mathbf{r}'_0, \nu) - U^{(in)}(\mathbf{r}'_0, \nu) \frac{\partial}{\partial z'} G_+(\mathbf{r} - \mathbf{r}'_0) \right], \tag{7.38}$$

where $\mathcal{A}$ denotes that area of the hole in the aperture plane here assumed to be $z = 0$ and $\mathbf{r}'_0 = (\boldsymbol{\rho}', z = +\epsilon)$ denotes the position vector on the plane $z = +\epsilon$ with $\epsilon > 0$ arbitrarily small so that $U^{(in)}(\mathbf{r}'_0, \nu)$ is the incident field immediately to the right of the aperture.

The assumptions behind the Kirchhoff solution Eq. (7.38) appear to be very reasonable and the inquisitive reader may wonder why this solution is not, in principle, exact. There are two basic reasons why the Kirchhoff solution is only an approximation to the true diffracted field. First, because the diffracted wave obeys the outgoing-wave radiation condition in the r.h.s. $z > 0$ this field and its normal derivative over the diffraction plane must be related via the second Helmholtz identity (cf. Eq. (2.41b) in Section 2.8). If we then assume that these boundary values are given according to Kirchhoff's assumption we require that

$$\int_{\mathcal{A}} d^2\rho' \left[G_+(\mathbf{r} - \mathbf{r}'_0) \frac{\partial}{\partial z'} U^{(in)}(\mathbf{r}'_0, \nu) - U^{(in)}(\mathbf{r}'_0, \nu) \frac{\partial}{\partial z'} G_+(\mathbf{r} - \mathbf{r}'_0) \right] = 0$$

if the field point $\mathbf{r}$ lies in the left half-space $\tau_- = \{\mathbf{r} : z < 0\}$. On the other hand, the incident wave has to be generated by a source located in the l.h.s. $z < 0$ and, hence, it, by itself, must satisfy the second Helmholtz identity on the l.h.s.; i.e.,

$$\int_{z=0} d^2\rho' \left[G_+(\mathbf{r} - \mathbf{r}'_0) \frac{\partial}{\partial z'} U^{(in)}(\mathbf{r}'_0, \nu) - U^{(in)}(\mathbf{r}'_0, \nu) \frac{\partial}{\partial z'} G_+(\mathbf{r} - \mathbf{r}'_0) \right] = 0,$$

where the integration is over the entire aperture plane $z = 0$. It is clear that these two equations cannot both be satisfied except for the uninteresting case of an infinite aperture.

A second objection to the Kirchhoff solution is that it ignores completely the physics of the interaction of the incident wave with the material of the aperture screen. The Kirchhoff solution Eq. (7.38) is identical for screens made of *any* material so long as the material fits the criterion of being "opaque" to the incident wave. Thus, in the case of an incident optical wave the theory predicts no difference in the diffracted wave irrespective of whether

[6] Kirchhoff's original application was optics and, as any optical physicist or engineer will tell you, light always propagates from left to right.

the material is a perfect conductor or a perfect absorber. In both cases the material fits the loose definition of being opaque, since in the case of an infinite conducting screen all of the incident radiation is reflected, while in the case of an infinite absorbing screen all of the incident radiation is absorbed.

Of the two objections to the Kirchhoff theory given above, the second is, by far, the most serious. Indeed, we will show below that it is possible to convert the diffraction problem into a Rayleigh–Sommerfeld (RS) boundary-value problem so that the diffracted field can be computed entirely in terms of either its boundary value or the boundary value of its normal derivative over the aperture plane. If we then invoke Kirchhoff's assumption that the diffracted field to the immediate right of the aperture is equal to the incident wave within the aperture and zero outside the aperture we are able to arrive at two alternative formulas for the diffracted wave; one involving only the incident wave (the Dirichlet solution to the RS boundary-value problem) and a second involving only the normal derivative of the incident wave (the Neumann solution to the RS problem). Unfortunately, both "solutions" are also not exact and, in particular, are insensitive to the precise nature of the material of the aperture material. Thus, the only real objection with the Kirchhoff theory is that his ansatz for constructing the diffracted field boundary conditions on the aperture plane from the incident wavefield is only approximately correct and its domain of validity must be determined from more fundamental considerations or from actual experimental data.

7.5.1 The Rayleigh–Sommerfeld alternative to the Kirchhoff diffraction formula

We can compute the wavefield diffracted by an aperture located in the plane $z = 0$ in terms of its Dirichlet or Neumann boundary values on this plane using the RS solution to the boundary-value problem given in Eqs. (2.48) of Chapter 2. In particular, we have that

$$U^{(d)}(\mathbf{r},\nu) = \begin{cases} -2\int_{z=z_0} d^2\rho'\, U^{(d)}(\mathbf{r}_0',\nu)(\partial/\partial z')G_+(\mathbf{r}-\mathbf{r}_0'), & \text{Dirichlet conditions,} \\ +2\int_{z=z_0} d^2\rho'\, (\partial/\partial z')U^{(d)}(\mathbf{r}_0',\nu)G_+(\mathbf{r}-\mathbf{r}_0'), & \text{Neumann conditions,} \end{cases}$$

where $U^{(d)}(\mathbf{r}_0',\nu)$ and $(\partial/\partial z')U^{(d)}(\mathbf{r}_0',\nu)$ denote the boundary value of the diffracted wavefield and its normal derivative to the immediate right of the aperture. Note that the integral in the above equations is over the *entire* boundary plane, not just over the aperture region as is the case in the Kirchhoff solution Eq. (7.38). The problem that we face is that we don't know the boundary value of the diffracted field or of its normal derivative over the aperture plane and we must, again, invoke Kirchhoff's ansatz or make some other assumption about the required boundary values in order to proceed.

If we invoke the Kirchhoff ansatz we obtain the results

$$U^{(d)}(\mathbf{r},\nu) = -2\int_A d^2\rho'\, U^{(in)}(\mathbf{r}_0',\nu)\frac{\partial}{\partial z'}G_+(\mathbf{r}-\mathbf{r}_0') \tag{7.39a}$$

and

$$U^{(d)}(\mathbf{r},\nu) = +2\int_A d^2\rho'\, \frac{\partial}{\partial z'}U^{(in)}(\mathbf{r}_0',\nu)G_+(\mathbf{r}-\mathbf{r}_0'), \tag{7.39b}$$

where now the integral is only over the aperture $\mathcal{A}$ as in the Kirchhoff solution Eq. (7.38). Although we have denoted the diffracted wavefields generated by Dirichlet and Neumann boundary conditions using the same symbol U, it is important to realize that the wavefields generated by the two formulas will be different. We will refer to the above two approximate solutions to the diffraction problem as the *Rayleigh–Sommerfeld diffraction formulas*. We might expect either of the above two diffraction formulas to be better or more accurate than Kirchhoff's formula Eq. (7.38), but evidence suggests that this is not the case. Indeed, certain experimental studies of optical diffraction from holes in conducting plates (Wolf and Marchand, 1964) have shown that the opposite is true, especially in the immediate vicinity of the aperture. In this connection we note that the original Kirchhoff solution is actually the arithmetic average of the two RS formulas:

$$U_{\mathrm{K}} = \frac{1}{2}U_{\mathrm{RS}_1} + \frac{1}{2}U_{\mathrm{RS}_2},$$

where U_{K} denotes the original Kirchhoff approximation to the diffracted field and U_{RS_1} and U_{RS_2} are the two Rayleigh–Sommerfeld approximations to the diffracted field.

7.5.2 More general diffraction problems

The Kirchhoff and Rayleigh–Sommerfeld diffraction formulas apply to the case of a hole in a perfectly opaque screen and it is, of course, of interest to generalize these formulas to more general aperture distributions. In between the two extremes of perfectly transparent and perfectly opaque bodies are those bodies that will partially transmit and partially reflect an incident wave. If we then construct a very thin (compared with the wavelength $\lambda = 2\pi/k$ of the wavefield) aperture from such material it is reasonable to assume a model for the aperture–wavefield interaction of the form

$$U_{\mathrm{t}}(\mathbf{r}_0, \nu) = \mathcal{T}(\boldsymbol{\rho})U^{(\mathrm{in})}(\mathbf{r}_0, \nu), \tag{7.40}$$

where $U_{\mathrm{t}}(\mathbf{r}_0, \nu)$ denotes the diffracted wavefield immediately to the right of the aperture and $\mathcal{T}(\boldsymbol{\rho})$ is a transmission coefficient or "transmittance function" that depends on the aperture material and geometry *but is independent of the incident wavefield.*

The above model for the interaction of the incident wave with a general aperture might appear, at first glance, to be, in principle, exact. In particular, the reader may wonder why the wavefield immediately to the right of a thin aperture located on the plane $z = 0$ cannot always be written in the form of Eq. (7.40) for *some* choice of the transmission coefficient $\mathcal{T}$. While it is certainly true that the (exact) boundary-value field can always be so expressed, it is not true that the transmission coefficient will be independent of the incident wavefield. Thus, the model as defined in Eq. (7.40) assumes a *linear* interaction between the aperture and the incident wave. Upon reflection the reader will realize that this assumption is also hidden in the Kirchhoff ansatz that is the limiting case of the model Eq. (7.40) when the transmission coefficient is unity within the aperture (hole) and zero outside this aperture.

If we assume a model of the form given in Eq. (7.40) the diffracted wavefield is then given by the Kirchhoff or RS formulas with the incident wave replaced by the aperture field $U_t(\mathbf{r}_0, \nu)$. For example, using the first RS formula Eq. (7.39a) we obtain the following expression for the diffracted field:

$$U^{(d)}(\mathbf{r}, \nu) = -2 \int_{z=z_0} d^2\rho' \, T(\boldsymbol{\rho}') U^{(in)}(\mathbf{r}_0, \nu) \frac{\partial}{\partial z'} G_+(\mathbf{r} - \mathbf{r}_0'). \tag{7.41a}$$

If we compute the derivative of the Green function appearing in Eq. (7.41a) we can write this formula in the form (cf. Section 4.7 of Chapter 4)

$$U^{(d)}(\mathbf{r}, \nu) = \frac{1}{2\pi} \int_{z=z_0} d^2\rho' \, T(\boldsymbol{\rho}') U^{(in)}(\mathbf{r}_0, \nu) \frac{e^{ik_0|\mathbf{r}-\mathbf{r}_0'|}}{|\mathbf{r} - \mathbf{r}_0'|} \left[-ik_0 \frac{z}{|\mathbf{r} - \mathbf{r}_0'|} + \frac{z}{|\mathbf{r} - \mathbf{r}_0'|^2} \right]. \tag{7.41b}$$

7.5.3 Algorithmic implementation of the diffraction formulas

Although the diffraction formulas Eqs. (7.38), (7.39a), (7.39b) and, more generally, Eqs. (7.41) are computationally simple by modern standards, they were highly intractable in the early and middle parts of the last century when they were first devised and used. Thus a good deal of effort was expended obtaining approximate forms of these formulas that could be used to obtain analytic expressions for the field diffracted from various simple apertures. This was especially true in optical applications, where the distance of the field point $r = |\mathbf{r}|$ can be expected to be large compared with the wavelength λ of the incident wavefield. Depending on the size of the distance r relative to the effective radius a of the aperture, these approximate theories are known as the *Fresnel approximation* and the *Fraunhofer approximation*. These approximate schemes, together with exact eigenfunction implementations of these formulas such as the angular-spectrum expansion developed in Chapter 4, are still important today and are often used in place of the exact formulas because of their simplicity and computational efficiency. Indeed, all of these alternatives to the basic diffraction formulas are Fourier-based and easily implemented using the fast Fourier transform (FFT).

We have already developed the Fresnel approximation in Section 4.7 of Chapter 4, where we compared its performance with both the RS solution to the boundary-value problem and the angular-spectrum implementation of this solution. We found that the Fresnel approximation was excellent at large propagation distances but degraded significantly in the near field of the boundary. On the other hand, the angular-spectrum implementation of the RS formulas, while, in principle, exact, degraded significantly for large propagation distances due to aliasing in the spatial frequency domain. Thus, the computer implementation of boundary-value and aperture diffraction problems is best treated using the angular-spectrum expansion for short propagation distances and the Fresnel approximation for moderate propagation distances. The Fraunhofer approximation is a further approximation to the Fresnel approximation and is valid only at very large propagation distances. We will briefly review these three alternative implementations of the diffraction formulas in this section. For the sake of simplicity we consider only the general formulas Eqs. (7.41) for an aperture located in the plane $z = 0$, with the understanding that implementations of the other diffraction formulas are easily obtained using entirely parallel developments. To obtain the formulas for an aperture located at $z = a_0$ simply replace z by $|z - a_0|$ in the formulas presented below.

The angular-spectrum expansion

The angular-spectrum expansion of the diffracted field is obtained by employing the Weyl expansion of the outgoing-wave Green function given in Eq. (4.4a) of Section 4.1 of Chapter 4. Using this expansion we find that

$$\frac{\partial}{\partial z'}G_+(\mathbf{r}-\mathbf{r}'_0)=\frac{-1}{8\pi^2}\int_{-\infty}^{\infty}d^2K_\rho\, e^{i\mathbf{K}_\rho\cdot(\boldsymbol{\rho}-\boldsymbol{\rho}')+i\gamma z},$$

where $\mathbf{r}'_0=\boldsymbol{\rho}'$ denotes a point on the aperture plane located at $z=0$. On substituting the above expansion into Eqs. (7.41a) we obtain

$$U^{(d)}(\mathbf{r},\nu)=\frac{1}{(2\pi)^2}\int_{-\infty}^{\infty}d^2K_\rho\,\widetilde{\mathcal{T}U^{(in)}}(\mathbf{K}_\rho,\nu)e^{i\mathbf{K}_\rho\cdot\boldsymbol{\rho}+i\gamma z},\tag{7.42a}$$

where

$$\widetilde{\mathcal{T}U^{(in)}}(\mathbf{K}_\rho,\nu)=\int_{z=0}d^2\rho'\,\mathcal{T}(\boldsymbol{\rho}')U^{(in)}(\mathbf{r}'_0,\nu)e^{-i\mathbf{K}_\rho\cdot\boldsymbol{\rho}'}$$

is the spatial Fourier transform of the product of the transmittance function $\mathcal{T}(\boldsymbol{\rho}')$ with the incident wavefield. The above expansion is particularly simple if $U^{(in)}(\mathbf{r},\nu=\hat{\mathbf{z}})=\exp(ik_0\hat{\mathbf{z}}\cdot\mathbf{r})$ is a unit-amplitude plane wave normally incident to the aperture. In this case we find that

$$\widetilde{\mathcal{T}U^{(in)}}(\mathbf{K}_\rho,\hat{\mathbf{z}})=\int_{z'=0}d^2\rho'\,\mathcal{T}(\boldsymbol{\rho}')e^{-i\mathbf{K}_\rho\cdot\boldsymbol{\rho}'}=\tilde{\mathcal{T}}(\mathbf{K}_\rho)$$

is equal to the spatial Fourier transform of the transmittance function and Eq. (7.42a) reduces to

$$U^{(d)}(\mathbf{r},\hat{\mathbf{z}})=\frac{1}{(2\pi)^2}\int_{-\infty}^{\infty}d^2K_\rho\,\tilde{\mathcal{T}}(\mathbf{K}_\rho)e^{i\mathbf{K}_\rho\cdot\boldsymbol{\rho}+i\gamma z}.\tag{7.42b}$$

The angular-spectrum expansion of the diffracted field is efficiently implemented in code using FFTs, as is apparent from the form of Eqs. (7.42).

The Fresnel approximation

The Fresnel approximation to the solution of the RS boundary-value problem was developed in Section 4.7 of Chapter 4, where it is given by Eq. (4.41). Using this approximation for the Kirchhoff diffracted field we obtain the result

$$U^{(d)}(\mathbf{r},\nu)=-\frac{ik_0}{2\pi}\int_{z=0}d^2\rho'\,U^{(d)}(\mathbf{r}',\nu)|_{z'=0}\frac{e^{ik_0z+ik_0\frac{(x-x')^2+(y-y')^2}{2z}}}{z}.$$

On setting the boundary value $U^{(\mathrm{d})}(\mathbf{r}', \nu)|_{z'=0}$ equal to the product of the incident wave and the aperture transmittance function and performing some simplification we then obtain the following Fresnel approximation to the (Kirchhoff-approximated) diffracted wavefield:

$$U^{(\mathrm{d})}(\mathbf{r}, \nu) \approx -\frac{ik_0}{2\pi}\frac{e^{ik_0 z}}{z}\int_{z'=0} d^2\rho'\, T(\boldsymbol{\rho}')U^{(\mathrm{in})}(\mathbf{r}_0', \nu)e^{ik_0 \frac{|\boldsymbol{\rho}-\boldsymbol{\rho}'|^2}{2z}}, \tag{7.43}$$

where $\boldsymbol{\rho} = (x, y)$ and $\boldsymbol{\rho}' = (x', y')$. As discussed in Chapter 4, a sufficient condition for the validity of the Fresnel approximation is that

$$z^3 \gg \frac{\pi}{4\lambda}\,\delta\rho^4|_{\max},$$

where $\delta\rho = |\boldsymbol{\rho} - \boldsymbol{\rho}'|$ and the maximum is to be taken relative to all source points in the diffracting aperture ($z = 0$) and all field points on the plane z for which the field is to be computed.

The Fresnel transform

At first glance it would appear that there is no significant computational advantage of the Fresnel approximation to the exact Kirchhoff formula Eqs. (7.41). However, if we expand the quadratic term in the exponential in Eq. (7.43) we can write the approximate expression in the form

$$U^{(\mathrm{d})}(\mathbf{r}, \nu) \approx -\frac{ik_0}{2\pi}\frac{e^{ik_0 z}}{z}e^{ik_0\frac{\rho^2}{2z}}\int_{z'=0} d^2\rho'\, T(\boldsymbol{\rho}')e^{ik_0\frac{\rho'^2}{2z}}U^{(\mathrm{in})}(\mathbf{r}_0', \nu)e^{-ik_0\frac{\boldsymbol{\rho}\cdot\boldsymbol{\rho}'}{z}}. \tag{7.44}$$

Equation (7.44) represents the Fresnel diffracted field in terms of the spatial Fourier transform of the product of the incident field with the aperture transmission function modulated by $\exp[ik_0\rho'^2/(2z)]$. This shows that the paraxial form of the Fresnel approximation can be computed using an FFT with some additional overhead computations. In this form the Fresnel approximation is sometimes referred to as the *Fresnel transform*.

The Fresnel transform is treated in some detail in Section 4.7 of Chapter 4, where it is compared with the angular-spectrum expansion. As discussed in that section, the Fresnel transform has the advantage over the angular-spectrum expansion in that it is *self-scaling*. By this we mean that the sample spacing $\delta\rho$ in an FFT implementation of this transform increases linearly with the propagation distance z. On the other hand, FFT implementation of the angular-spectrum expansion needs to employ a fixed sample spacing for all propagation distances. Since the diffracted field will expand in the (x, y) plane, with increasing propagation distance the angular-spectrum expansion requires significant buffering in the FFT in order to avoid aliasing in the spatial frequency domain. This is not true for the Fresnel transform, so the diffracted field can be computed very efficiently at large propagation distances using the Fresnel transform rather than the angular-spectrum expansion. Of course, the Fresnel transform is only approximate and requires the propagation distance to be large, whereas the angular-spectrum expansion is, in principle, exact at any propagation distance.

The Fraunhofer approximation

The Fraunhofer approximation is simply a further simplification of the Fresnel approximation. In particular, in the Fraunhofer approximation it is assumed that the Fresnel approximation is valid and, in addition, the term $\exp[ik_0\rho'^2/(2z)] \approx 1$ so that Eq. (7.44) further simplifies to become

$$U^{(\mathrm{d})}(\mathbf{r},\nu) \approx -\frac{ik_0}{2\pi}\frac{e^{ik_0 z}}{z}e^{ik_0\frac{\rho^2}{2z}}\int_{z=0} d^2\rho'\, \mathcal{T}(\boldsymbol{\rho}')U^{(\mathrm{in})}(\mathbf{r}_0',\nu)e^{-ik_0\frac{\boldsymbol{\rho}\cdot\boldsymbol{\rho}'}{z}}. \tag{7.45}$$

A sufficient condition for the validity of the Fraunhofer approximation is that the maximum value of the quadratic phase term $k\rho'^2/(2z)$ across the diffracting aperture be less than a radian. Thus, we require that

$$z \gg \frac{\pi}{\lambda}\rho'^2|_{\max},$$

where the maximum is taken relative to all source points in the diffracting aperture.

From a computational point of view the Fraunhofer approximation offers little advantage over the Fresnel approximation. In particular, an FFT is used in both approximations, where, however, the Fresnel approximation requires also a multiplication of the transmission function $\mathcal{T}$ by the factor $\exp[ik_0\rho'^2/(2z)]$. However, this multiplication has only to be performed once for any given observation plane z, so it does not significantly increase the computational burden. However, the Fraunhofer approximation offers considerable advantages over the Fresnel approximation for computing the diffraction of simple waveforms such as plane waves from simple apertures since the analytic form of the spatial Fourier transform is known for a number of such cases. In this connection we note that

$$\frac{e^{ik_0 z}}{z}e^{ik_0\frac{\rho^2}{2z}} \approx \frac{e^{ik_0 r}}{r},$$

which is accurate to the same level as approximation as the Fresnel approximation. We can then write Eq. (7.45) in the form

$$U_{\mathrm{d}}(\mathbf{r},\nu) \approx f_{\mathrm{d}}(\mathbf{u},\nu)\frac{e^{ik_0 r}}{r},$$

where $\mathbf{u} = \boldsymbol{\rho}/z$ and

$$f_{\mathrm{d}}(\mathbf{u},\nu) = -\frac{ik_0}{2\pi}\int_{z=0} d^2\rho'\, \mathcal{T}(\boldsymbol{\rho}')U^{(\mathrm{in})}(\mathbf{r}_0',\nu)e^{-ik_0\mathbf{u}\cdot\boldsymbol{\rho}'} \tag{7.46a}$$

is called the *diffraction pattern* of the aperture. The diffraction pattern plays the same role for diffraction problems as does the *scattering amplitude* (cf. Section 6.4 of Chapter 6) for scattering problems. Indeed, the process of diffraction is another form of wavefield scattering and can be treated as a limiting form of potential scattering within the Born approximation (see the problems at the end of the chapter).

The diffraction pattern within the Kirchhoff approximation as given in Eq. (7.46a) is seen to be the 2D spatial Fourier transform of the product of the transmittance function with the incident wave to the aperture. If the incident wave is a plane wave then this expression becomes particularly simple and can be computed in closed form for a number of simple apertures. This is one reason for the historical popularity of the Kirchhoff approximation. However, with the advent of modern digital computers and the FFT it has no computational advantages over either the angular-spectrum expansion or the Fresnel approximation and the Fresnel transform. On the other hand, the fact that the diffraction pattern is related to the transmittance function via a spatial Fourier transform makes it an ideal mathematical model to be used in *inverse diffraction*, as will be discussed below.

The diffraction pattern given in Eq. (7.46a) is a function of the coordinate vector $\boldsymbol{\rho}$ over an observation plane located at a fixed distance z from the diffracting aperture. It is useful to have a formula that gives this pattern as a function of the unit observation vector $\mathbf{s} = \mathbf{r}/r$ over a hemisphere in complete analogy with the scattering amplitude of a penetrable or impenetrable scatterer and the radiation pattern of a primary or induced source. Such a formula is easily derived within the same level of approximation as Eq. (7.46a) directly from Eq. (7.41b) and one finds that

$$f_{\mathrm{d}}(\mathbf{s}, \nu) = -\frac{ik_0}{2\pi}\int_{z=0} d^2\rho'\, T(\boldsymbol{\rho}')U^{(\mathrm{in})}(\mathbf{r}_0', \nu)e^{-ik_0\mathbf{s}\cdot\boldsymbol{\rho}'}. \tag{7.46b}$$

We emphasize that the two expressions for "the" diffraction pattern differ fundamentally in that Eq. (7.46a) gives the pattern as function of position on a fixed plane $z \gg \lambda$ lying parallel to the diffraction aperture, whereas Eq. (7.46b) gives this pattern as a function of the unit vector $\mathbf{s}$ over a spherical surface located at fixed radial distance $r \gg \lambda$. We note that the diffraction pattern as given in Eq. (7.46b) can be interpreted as being the radiation pattern generated by the induced source

$$Q(\mathbf{r}) = 2ik_0 T(\boldsymbol{\rho}')U^{(\mathrm{in})}(\mathbf{r}_0', \nu)\delta(z)$$

supported on the $z = 0$ plane. Sources of this type were investigated in connection with the inverse source problem (ISP) in Chapter 5 and were employed in connection with so-called "wavelet" fields in Section 4.5 of Chapter 4.

7.6 Inverse diffraction

Inverse diffraction within the Kirchhoff approximation consists of determining the aperture transmittance function $T(\boldsymbol{\rho})$ from knowledge of the incident wave to the diffracting aperture and the diffracted field or its normal derivative specified over some "measurement plane" located at some distance $z_0 > 0$ from the diffracting aperture. Mathematically, this reduces to solving the integral equation Eqs. (7.41) for T with $U^{(\mathrm{in})}$ known and $U^{(\mathrm{d})}$ or $(\partial/\partial z)U^{(\mathrm{d})}$ specified over a plane $z_0 > 0$. Formally, this problem is identical to the *inverse boundary-value problem* (IBVP) first introduced in Section 2.11 of Chapter 2, where it was approximately solved using an incoming-wave Green function solution to a (forward)

Rayleigh–Sommerfeld (RS) boundary-value problem. It is solved exactly (but unstably) using field back propagation implemented via the angular-spectrum expansion as developed in Section 4.3 of Chapter 4 and in Eq. (7.42b) above. Here we will briefly review those earlier approaches to inverse diffraction and then formulate and solve the problem using the singular value decomposition (SVD) that was developed in Chapter 5. For the sake of simplicity of notation we will restrict our attention to an aperture located on the plane $z = 0$ and illuminated by a normally incident plane wave.

7.6.1 Inverse diffraction using back propagation

The angular-spectrum expansion of the Kirchhoff diffracted field over a plane $z = z_0 > 0$ for a normally incident, unit-amplitude plane wave to an aperture located in the plane $z = 0$ is given by Eq. (7.42b). Spatially Fourier transforming both sides of this equation and solving for $\tilde{T}(\mathbf{K}_\rho)$ then yields the result

$$\tilde{T}(\mathbf{K}_\rho) = \widetilde{U^{(d)}}(\mathbf{K}_\rho, z_0, \hat{\mathbf{z}})e^{-i\gamma z_0}, \tag{7.47a}$$

where

$$\widetilde{U^{(d)}}(\mathbf{K}_\rho, z_0, \hat{\mathbf{z}}) = \int_{z_0} d^2\rho\, U^{(d)}(\boldsymbol{\rho}, z_0, \hat{\mathbf{z}})e^{-i\mathbf{K}_\rho\cdot\boldsymbol{\rho}} \tag{7.47b}$$

is the spatial Fourier transform of the diffracted field over the plane z_0. Equation (7.47a) is the spatial frequency-domain statement of (exact) field back propagation from the diffraction plane at z_0 back to the aperture plane at $z = 0$ that was presented in Section 4.3 of Chapter 4. We obtain the space-domain back-propagated field by simply inverse transforming Eq. (7.47a):

$$T(\boldsymbol{\rho}) = \frac{1}{(2\pi)^2}\int d^2K_\rho\, \widetilde{U^{(d)}}(\mathbf{K}_\rho, z_0, \hat{\mathbf{z}})e^{-i\gamma z_0}e^{i\mathbf{K}_\rho\cdot\boldsymbol{\rho}}, \tag{7.47c}$$

which is a formal solution to the inverse Kirchhoff diffraction problem.

The process of field back propagation was discussed at some length in Chapter 4, where it was pointed out that the process, although in principle exact, is unstable due to the exponential growth of the factor $\exp(-i\gamma z_0)$, $z_0 > 0$, in Eqs. (7.47a) and (7.47c) over the evanescent region $K_\rho > k_0$. We discussed stabilized field back propagation in Section 4.4 of that chapter, where the stabilization was achieved by damping the contribution of the evanescent plane waves in the back-propagation integral Eq. (7.47c). The two schemes that were introduced in that section were hard limiting the integration region in the integral to the homogeneous region $K_\rho < k_0$ and replacing γ by its complex conjugate γ^*. Over the homogeneous region in a non-dispersive medium $\gamma^* = \gamma$, while over the evanescent region $\gamma^* = -\gamma$, so this stabilization scheme returns a result almost identical to that obtained from the hard limiting scheme so long as z is much larger than the wavelength.

The second stabilization scheme outlined above was shown in Section 4.4 to correspond to using the incoming-wave Dirichlet Green function in the approximate solution of the inverse RS boundary-value problem obtained in Section 2.11 of Chapter 2. The use of the incoming-wave Green function in place of the outgoing-wave Green function with boundary-value data was shown in that section to correspond to a stabilized form of field back propagation. Thus, we can also express the stabilized approximate solution of the inverse Kirchhoff diffraction problem directly in the space domain in the form

$$\hat{\mathcal{T}}(\boldsymbol{\rho}) = 2\int_{z_0} d^2\rho'\, U^{(\mathrm{d})}(\mathbf{r}', \hat{\mathbf{z}})\frac{\partial}{\partial z'}G_-(\mathbf{r}-\mathbf{r}'),$$

where $G_- = G_+^*$ is the incoming-wave Green function in a non-dispersive medium and $\mathbf{r}' = (\boldsymbol{\rho}', z_0)$ denotes a point on the diffraction plane.

7.6.2 The SVD formulation of the inverse diffraction problem

We again consider the Kirchhoff diffraction problem formulated for an aperture located in the plane $z = 0$ and a normally incident plane wave via Eq. (7.42b). The inverse Kirchhoff diffraction problem consists of inverting this equation for the transmittance function $\mathcal{T}(\boldsymbol{\rho})$ given Dirichlet data $U^{(\mathrm{d})}(\mathbf{r}, \hat{\mathbf{z}})$ specified over some plane $z = z_0 > 0$. The spatial Fourier transform of this equation over the data plane $z = z_0$ yields $\tilde{\mathcal{T}}(\mathbf{K}_\rho)$ according to Eq. (7.47a), which reduces this problem to the classical problem of inverting the Fourier transform of a space-limited function $\mathcal{T}(\boldsymbol{\rho})$ that we considered in Example 5.7 of Chapter 5. In principle the transform is specified over the entire $\mathbf{K}_\rho$ plane, but in practice it will be limited to the homogeneous region $K_\rho < k$ as discussed above. Thus, a more accurate formulation of the problem is that of inverting the Fourier transform of a space-limited function from frequency-limited data; i.e., to the classical Slepian–Pollak problem discussed in Example 5.8 of Chapter 5.

We will first address the problem using the formulation outlined in Example 5.7, which will yield an exact inversion (within the Kirchhoff approximation) if the spatial Fourier transform of the diffracted field is exactly specified over the entire $\mathbf{K}_\rho$ plane and a least-squares approximate inversion otherwise. We then will briefly outline the use of the Slepian–Pollak theory developed in Example 5.8, which yields an exact solution given exact data only over the homogeneous region of the spectra $K_\rho < k$ and a least-squares solution given noisy data over $K_\rho < k$.

7.6.3 The full data case

For simplicity we will consider the 2D version of the Kirchhoff inverse diffraction problem, in which the aperture is a slit of known width $2a_0$ centered along the x axis of the aperture plane $z = 0$ and the incident wave is a unit-amplitude plane wave having unit propagation vector $\mathbf{s}_0 = \hat{\mathbf{z}}$. The extension to 2D rectangular apertures and arbitrary incident waves is straightforward, while the extension to circular and other separable apertures can be treated using eigenfunctions of the homogeneous Helmholtz equation such as those presented in Chapter 3.

The 2D version of Eq. (7.47a) for a normally incident plane wave can be expressed in the simplified form

$$\frac{1}{\sqrt{2\pi}}\int_{-a_0}^{a_0} dx\, \mathcal{T}(x)e^{-iKx} = f(K), \tag{7.48a}$$

where

$$f(K) = \frac{1}{\sqrt{2\pi}}\widetilde{U^{(d)}}(K, z_0)e^{-i\gamma z_0}, \tag{7.48b}$$

with

$$\gamma = \begin{cases} \sqrt{k^2 - K^2} & K < k, \\ +i\sqrt{K^2 - k^2} & K > k. \end{cases}$$

We now define the Hilbert spaces $\mathcal{H}_{\mathcal{T}}$ of square-integrable functions over $(-a_0, +a_0)$ and $\mathcal{H}_f$ of square-integrable functions of the spatial frequency variable K over $(-\infty, \infty)$ and introduce the operator $\hat{T} : \mathcal{H}_{\mathcal{T}} \to \mathcal{H}_f$,

$$\hat{T} = \frac{1}{\sqrt{2\pi}}\int_{-a_0}^{a_0} dx\, e^{-iKx},$$

used in Example 5.7 of Chapter 5. In terms of this operator, which we will refer to here as the "diffraction operator," we can write Eq. (7.48a) in the abstract form

$$\hat{T}\mathcal{T}(K) = f(K). \tag{7.49a}$$

To invert Eq. (7.48b) for the transmittance function $\mathcal{T}(x)$ we will employ the singular value decomposition (SVD) of $\hat{T}$ that was derived in Example 5.7 of Chapter 5. The SVD was reviewed in some detail in that chapter, where it was employed to solve the inverse source problem (ISP) for the scalar Helmholtz equation in a generally dispersive medium. The advantage of the SVD is that it is a "tried and true" scheme for solving a host of linear inverse problems and can be used here to solve the inverse diffraction problem in a very simple and elegant manner.

The SVD of the operator $\hat{T}$ consists of the set $\{v_p(x) \in \mathcal{H}_{\mathcal{T}}, u_p(K) \in \mathcal{H}_f, \sigma_p \geq 0\}$ of singular functions v_p, u_p and associated singular values σ_p that are indexed by the integer $p = -\infty, \ldots, +\infty$ and satisfy the set of equations

$$\hat{T}v_p = \sigma_p u_p, \qquad \hat{T}^\dagger u_p = \sigma_p v_p. \tag{7.50}$$

In these equations $\hat{T}^\dagger : \mathcal{H}_f \to \mathcal{H}_{\mathcal{T}}$ is the adjoint operator

$$\hat{T}^\dagger = \frac{1}{\sqrt{2\pi}}\mathcal{M}_{a_0}\int_{-\infty}^{+\infty} dK\, e^{iKx},$$

where

$$\mathcal{M}_{a_0} = \begin{cases} 1 & -a_0 \leq x \leq a_0 \\ 0 & \text{else} \end{cases}$$

is a masking operator that space limits the result of the integral transform defining $\hat{T}^\dagger$ to the aperture interval $(-a_0, +a_0)$. The steps involved in computing an adjoint operator are discussed in detail in Chapter 5. The singular functions each comprise an orthonormal basis in their respective Hilbert spaces so that

$$\mathcal{T}(x) = \sum_{p=-\infty}^{\infty} \langle v_p, \mathcal{T}\rangle_{\mathcal{H}_\mathcal{T}} v_p(x), \qquad f(K) = \sum_p \langle u_p, f\rangle_{\mathcal{H}_f} u_p(K), \tag{7.51}$$

where $\langle \cdot, \cdot \rangle_\mathcal{H}$ stands for the standard inner product in the Hilbert space $\mathcal{H}$.

The solution to the inverse diffraction problem is obtained by substituting the above expansions into Eq. (7.49a), making use of Eq. (7.50) and solving for the (unknown) expansion coefficients $\langle v_p, \mathcal{T}\rangle_{\mathcal{H}_\mathcal{T}}$ in terms of the known coefficients $\langle u_p, f\rangle_{\mathcal{H}_f}$. We obtain the result

$$\hat{T}\left[\overbrace{\sum_p \langle v_p, \mathcal{T}\rangle_{\mathcal{H}_\mathcal{T}} v_p(x)}^{\mathcal{T}(x)}\right] = \overbrace{\sum_p \langle u_p, f\rangle_{\mathcal{H}_f} u_p(K)}^{f(K)}$$

$$\Downarrow$$

$$\langle v_p, \mathcal{T}\rangle_{\mathcal{H}_\mathcal{T}} = \frac{\langle u_p, f\rangle_{\mathcal{H}_f}}{\sigma_p}, \qquad \sigma_p > 0.$$

The transmittance function is thus given by

$$\mathcal{T}(x) = \sum_{\sigma_p > 0} \frac{\langle u_p, f\rangle_{\mathcal{H}_f}}{\sigma_p} v_p(x). \tag{7.52}$$

The solution is completed by employing the SVD of $\hat{T}$ that was obtained in Example 5.7. In that example we showed that

$$v_p(x) = \mathcal{M}_{a_0} \frac{e^{i\frac{\pi}{a_0}px}}{\sqrt{2a_0}}, \qquad u_p(K) = \frac{\mathrm{sinc}[(a_0/\pi)(K - (\pi/a_0)p)]}{\sqrt{\pi/a_0}}, \qquad \sigma_p = 1,$$

where

$$\mathrm{sinc}\, x = \frac{\sin(\pi x)}{\pi x}$$

is the "sinc function." On substituting the above into Eq. (7.52) we obtain

$$\mathcal{T}(x) = \mathcal{M}_{a_0} \sum_{p=-\infty}^{\infty} \langle u_p, f\rangle_{\mathcal{H}_f} \frac{e^{i\frac{\pi}{a_0}px}}{\sqrt{2a_0}},$$

where

$$\langle u_p, f\rangle_{\mathcal{H}_f} = \int_{-\infty}^{\infty} dK f(K) \frac{\mathrm{sinc}[(a_0/\pi)(K - (\pi/a_0)p)]}{\sqrt{\pi/a_0}}.$$

The above equations yield an exact (within the L^2 norm of the Hilbert space $\mathcal{H}_\mathcal{T}$) solution of the Kirchhoff inverse diffraction problem. However, it requires that the Fourier transform of the diffracted field be specified over the entire line $-\infty < K < \infty$, which is unrealistic in practice. However, the set of orthonormal singular functions

$u_p(K)$, $p = -\infty, \ldots, +\infty$, forms a basis for expanding the diffraction pattern into the classical Whittaker–Shannon sinc-function expansion of band-limited functions (Arsac, 1984). In particular, on substituting the singular functions $u_p(K)$ into the expansion of the diffracted field transform in Eq. (7.51) we find that

$$f(K) = \sum_p \langle u_p, f \rangle_{\mathcal{H}_f} \frac{\text{sinc}[(a_0/\pi)(K - (\pi/a_0)p)]}{\sqrt{\pi/a_0}}, \tag{7.53}$$

which is the Whittaker–Shannon sinc-function expansion of the band-limited function $f(K)$. This expansion allows us to identify the expansion coefficients $\langle u_p, f \rangle_{\mathcal{H}_f}$ not only as the projections of the transform onto the u_p but also as the samples of the $f(K)$ at the discrete sample points $K_p = (\pi/a_0)p$. By this means we can then obtain a least-squares approximation to the transmission function from computations of the diffracted field transform over some set of points, say K_p, $p = -P, -P+1, \ldots, P-1, P$, which then yields the least-squares approximation

$$\mathcal{T}(x) \approx \mathcal{M}_{a_0} \sum_{p=-P}^{P} f(K_p) e^{i\frac{\pi}{a_0}px}. \tag{7.54}$$

The problem with the solution of the Kirchhoff diffraction problem obtained above is that the Fourier transform $\widetilde{U^{(d)}}(K, z_0)$ of the diffracted field over the line z_0 will decay exponentially fast while the factor $\exp(-i\gamma z_0)$ will grow exponentially fast when $|K| > k$ so that the sample values

$$f(K_p) = \frac{1}{\sqrt{2\pi}} \widetilde{U^{(d)}}(K_p, z_0) e^{-i\gamma_p z_0}$$

will be unreliable over this range; i.e., when γ_p is purely positive imaginary. The sampling series Eq. (7.54) can then only return a least-squares solution with the maximum value of the index P in Eq. (7.54) determined from the condition $|K_P| = k$. A way around this difficulty is provided by the so-called Slepian–Pollak theory, which we will review next.

7.6.4 The Slepian–Pollak theory

Slepian and Pollak (Slepian and Pollak, 1961) were able to obtain the SVD of the frequency-limited Fourier transform of a space (or time)-limited function in a famous paper published in the *Bell System Technical Journal.* In particular, they obtained the SVD of the linear transform

$$\hat{T} = \frac{1}{\sqrt{2\pi}} \mathcal{M}_{K_0} \int_{-a_0}^{a_0} dx\, e^{-iKx},$$

where $\mathcal{M}_{K_0}$ is the masking operator that frequency limits the transform to $-K_0 \leq K \leq K_0$. In inverse diffraction theory we can only expect to determine the transform of the diffracted field over the homogeneous region of the spectra corresponding to $K_0 = k_0 = 2\pi/\lambda$, so this theory is ideally suited to realistically treating the Kirchhoff inverse diffraction problem.

Using results from Sturm–Liouville theory for the homogeneous Helmholtz equation Slepian and Pollak showed that the SVD of the operator $\hat{T}$ is given in terms of the *angular prolate-spheroidal wavefunctions* $S_{0,n}(c,\xi)$ (cf. Example 5.8 of Chapter 5) by

$$v_p(x) = \mathcal{M}_{a_0} \frac{i^{-p/2} S_{0p}(c, x/a_0)}{\sqrt{a_0}||S_{0p}||}, \qquad u_p(K) = \mathcal{M}_{K_0} \frac{i^{p/2} S_{0p}(c, K/K_0)}{\sqrt{K_0}||S_{0p}||},$$

$$\sigma_p = \sqrt{K_0 a_0}|K_p|,$$

where c is a constant parameter and

$$K_n = \frac{2i^n}{\sqrt{2\pi}} R^{(1)}_{on}(c, 1),$$

with $R^{(1)}_{on}(c,1)$ being the radial prolate-spheroidal wavefunctions. Moreover, and most importantly, these functions are orthogonal over the intervals $-1 < \xi < +1$ and $-1 < \omega < +1$ with norm

$$||S_{0,n}||^2 = \int_{-1}^{+1} d\xi\, |S_{0,n}(c,\xi)|^2.$$

The singular functions $\{v_p(x)\}$ form an orthonormal basis in $L^2(-a_0, +a_0)$, while the singular functions $\{u_p(K)\}$ form an orthonormal basis in $L^2(-K_0, +K_0)$. These two sets then can be used to (exactly) solve the inverse diffraction problem within the Kirchhoff approximation. We will not delve further into this solution of the inverse diffraction problem in this book since this approach and solution are very well documented in the open literature both by Slepian and Pollak in their original paper (Slepian and Pollak, 1961) and by many other workers who have extended their work in various directions.

7.7 Determining the shape of a surface scatterer

A problem of some interest in acoustic and electromagnetic scattering is that of determining the shape of an obstacle from measurements of scattered field data generated in a suite of scattering experiments. In the classic form of this problem the scattering process is governed by the Helmholtz equation, the field satisfies homogeneous Dirichlet or Neumann conditions over the scatterer surface and the data are values of the scattering amplitude specified over a set of incident and scattering directions and/or over a band of frequencies. The inverse (scattering) problem then consists of determining the surface $\partial\tau_0$ from this data set. We briefly discussed this problem within the linearized physical-optics (PO) approximation in Section 7.4 for the case in which the data consist of the back-scattering amplitude $f(-\mathbf{s}_0, \mathbf{s}_0)$ specified over a set of wavelengths (frequencies) for incident wave directions $\mathbf{s}_0$ spanning the unit sphere and for a data set consisting of the scattering amplitude specified over a set of incident and scattering directions at a single frequency. These approaches are based on the use of an approximate linearized forward-scattering model that related the surface (via the characteristic function) to the scattering amplitude in much the same way as the Born or Rytov linearized scattering models developed in the previous chapter relate the scattering potential of a penetrable scatterer to the scattering amplitude. An approximate Fourier inversion of these models then yields an approximate characteristic function from which the surface can be estimated.

In this section we will develop the inversion scheme based on the generalized Bojarski transformation obtained in Section 7.4 that employs the scattering amplitude specified over a set of incident and scattering wave directions at a single frequency. We will also examine alternative approaches to the surface-reconstruction problem that do not rely on a linearized scattering model and are based on the simple requirement that the sum of the incident and scattered fields satisfy homogeneous Dirichlet conditions on the surface $\partial\tau_0$. One algorithm for accomplishing this is to compute the scattered field over all of space from the scattering amplitude and determine $\partial\tau_0$ from the requirement that $U^{(\text{in})}(\mathbf{r}) + U^{(\text{s})}(\mathbf{r}) = 0$ when $\mathbf{r} \in \partial\tau_0$. To implement this procedure it is, of course, necessary to compute the scattered field from the scattering amplitude or, in other words, to back propagate the scattered field into the region occupied by the scatterer. As we showed in Chapter 4, evanescent components of the scattered field cannot be reliably determined from field measurements performed more than a few wavelengths from the scatterer, so this procedure will have an inherent error associated with it that can possibly result in large errors for the surface reconstruction. However, in many cases it can yield good results and, most importantly, does not rely on a linearized model for the scattering amplitude. We will review the procedure in the following section.

An alternative approach to the surface reconstruction process that is also model-independent is presented in Section 7.7.2. In that approach we still use the requirement that the field satisfy homogeneous Dirichlet conditions on $\partial\tau_0$ as our basic algorithm for determining the object shape but avoid the problem of determining the scattered field over $\partial\tau_0$ by *selecting an incident field that results in a scattered field having small magnitude outside and on the object's surface.* The required incident field is shown to be readily computed from the SVD of the scattering amplitude and the surface is then (approximately) determined by the condition that the magnitude of the computed incident field $|U^{(\text{in})}(\mathbf{r})|$ be small when the field point $\mathbf{r} \to \partial\tau_0$. A simple example illustrating the method developed here is presented and compared with results obtained by use of the earlier schemes developed in Section 7.7.1.

7.7.1 Surface reconstruction via back propagation

This approach to the surface-reconstruction problem is based on the requirement that the total field on the surface of a Dirichlet scatterer must vanish. If then the total field is back propagated from some arbitrary surface surrounding the scatterer its surface can be estimated from the locus of points over which the back-propagated field vanishes or achieves a minimum. Alternatively, the scattered field can be back propagated and the scatterer surface estimated from the sum of the known incident wave and the back-propagated scattered wavefield. The scheme is not exact, of course, due to the fact that evanescent components of the scattered field cannot be obtained at more than a wavelength from the scatterer, but can be expected to be accurate so long as the scatterer surface does not vary significantly on the scale of the wavelength. Note also that, unlike the PO approximation, the scattering model on which the method is based is exact so that the procedure seeks an approximate solution to an exactly posed problem. This is in contrast to many of the linearized inverse scattering schemes that rely on approximate forward-scattering models such as the PO, Born and Rytov approximations and seek exact inversions of approximate scattering models.

To test the algorithm, we considered a circular Dirichlet cylinder with plane-wave illumination such that the scattered field data are given by the scattering amplitude of the cylinder. The back propagation was implemented using the angular-spectrum expansion limited to homogeneous waves as developed in Section 4.3 of Chapter 4. We used the angle-variable form of the expansion presented for 2D wavefields in Section 4.6, which, for scattered fields, assumes the form

$$U^{(s)}(\mathbf{r}, \mathbf{s}_0) = \sqrt{\frac{k}{2\pi}} e^{i\frac{\pi}{4}} \int_{C_\pm^{(h)}} d\alpha\, f(\mathbf{s}, \mathbf{s}_0) e^{ik_0 \mathbf{s}\cdot\mathbf{r}},$$

where $f(\mathbf{s}, \mathbf{s}_0)$ is the scattering amplitude and the integration contours $C_+^{(h)}$ and $C_-^{(h)}$ are the α-integration contours shown in Fig. 4.9 of Section 4.6 limited to the homogeneous region. In the treatment presented here we assume that the cylinder axis is aligned along the z axis of a Cartesian system and the incident plane wave propagates along the positive-x axis in the (x, y) plane. The angle α in the above expansion is thus the polar angle measured relative to the x axis and the expansion reduces to

$$U^{(s)}(x, y) = \sqrt{\frac{k}{2\pi}} e^{i\frac{\pi}{4}} \int_{-\frac{\pi}{2}}^{\frac{\pi}{2}} d\alpha\, f(\alpha) e^{ik_0(x\cos\alpha + y\sin\alpha)},$$

where $f(\alpha)$ is the scattering amplitude $f(\alpha, \alpha_0)$ with the polar angle of the incident plane wave $\alpha_0 = 0$ and we have restricted our attention to the forward-scattered field propagating into the r.h.s. $x > 0$.

We show in Fig. 7.9 the magnitude of the back-propagated total and scattered fields from a Dirichlet cylinder having a radius of 5λ and evaluated over a square 32 wavelengths on a side. The back-propagated fields were generated from the scattering amplitude over the right half-plane and so the back-propagated fields can be expected to be valid only outside the line $x = 5\lambda$ defined by the edge of the cylinder where the (exact) angular-spectrum expansion converges. However, on comparison of this figure with the exact total and scattered fields presented in Fig. 7.2 it is seen that the back-propagated fields appear reasonably accurate for x values lying within the strip $-5\lambda < x < +5\lambda$ occupied by the cylinder. Note also that the shape of the cylinder is clearly defined by the minimum of the back-propagated total field.

A better estimate of the shape of the cylinder is obtained by employing a number of back-propagated fields from incident plane waves whose propagation directions vary over the unit circle. The simplest such algorithm is to simply incoherently sum the set of back-propagated images of the magnitude of the total back-propagated fields resulting from the different incident plane waves. Alternatively, if we note that the magnitude squared of the total back-propagated field reduces to

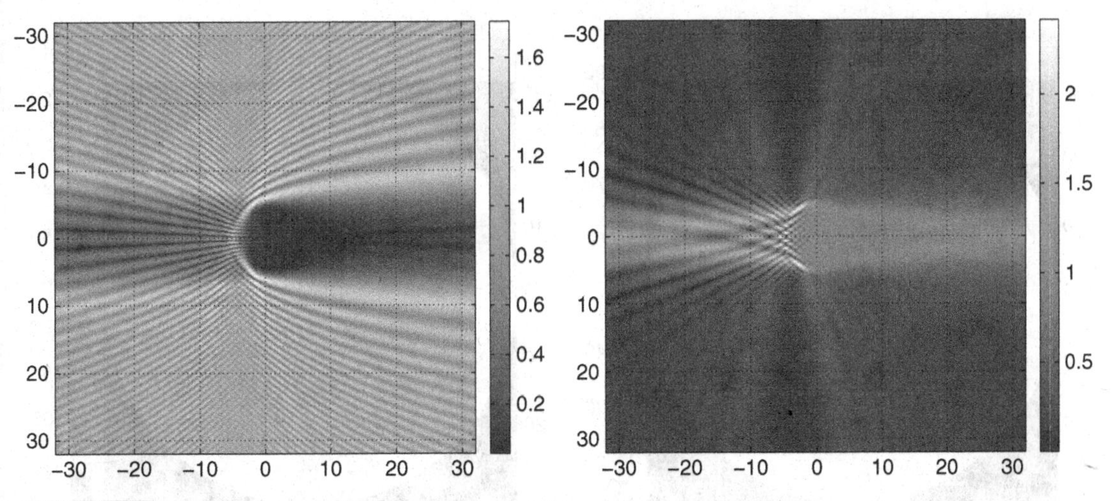

Fig. 7.9 Mesh plots of the absolute values of the total (left) and scattered (right) back-propagated fields from a circular cylinder having a radius of $a_0 = 5\lambda$.

$$|U_{\mathrm{bp}}|^2 = |U_{\mathrm{bp}}^{(\mathrm{s})}|^2 + 1 + 2\Re[e^{-ik_0z}U_{\mathrm{bp}}^{(\mathrm{s})}]$$

we see that the presence of the square of the magnitude of the scattered field will blur out the sharpness of the image of the shape of the scatterer so that a superior estimate is obtained by summing the quantity $\Re[e^{-ik_0z}U_{\mathrm{bp}}^{(\mathrm{s})}]$ over the different incident propagation directions. Because the cylinder is circularly symmetric, these algorithms are implemented by rotating either the magnitude of the total field or $\Re[e^{-ik_0z}U_{\mathrm{bp}}^{(\mathrm{s})}]$ obtained for a single incident wave direction through a set of angles angle $\phi_n = n\,\delta\phi$, $n = 1, 2, \ldots, N$, spanning 2π radians and then summing the rotated images. The resulting images for a cylinder having a radius of $a_0 = 5\lambda$ are shown in Fig. 7.10 for $N = 100$ and radial cuts through these images are shown in Fig. 7.11. It is clear from the radial cuts that the back-propagated demodulated images yield an excellent image of the cylinder and of the cylinder radius.

Inversion based on the PO approximation

The inversion scheme developed above is very similar to the inversion of the PO model Eq. (7.35b) via the *filtered back-propagation* (FBP) algorithm developed in the following chapter for linearized inverse scattering of penetrable scatterers. In particular, it follows from that equation that the spatial Fourier transform of $-\nabla^2\Gamma(\mathbf{r})$ is proportional to $F_{\mathrm{g}}(\mathbf{s}, \mathbf{s}_0)$, which is obtained from the scattering amplitude via the equation

$$F_{\mathrm{g}}(\mathbf{s}, \mathbf{s}_0) = f(\mathbf{s}, \mathbf{s}_0) + f^*(-\mathbf{s}, -\mathbf{s}_0). \tag{7.55}$$

If $\mathbf{s}$ and $\mathbf{s}_0$ are allowed to vary over the entire unit sphere then $F_{\mathrm{g}}(\mathbf{s}, \mathbf{s}_0)$ specifies the transform over the interior of the Ewald limiting sphere shown in Fig. 6.3 within the context of the Born approximation of penetrable scatterers. The Ewald limiting sphere is centered

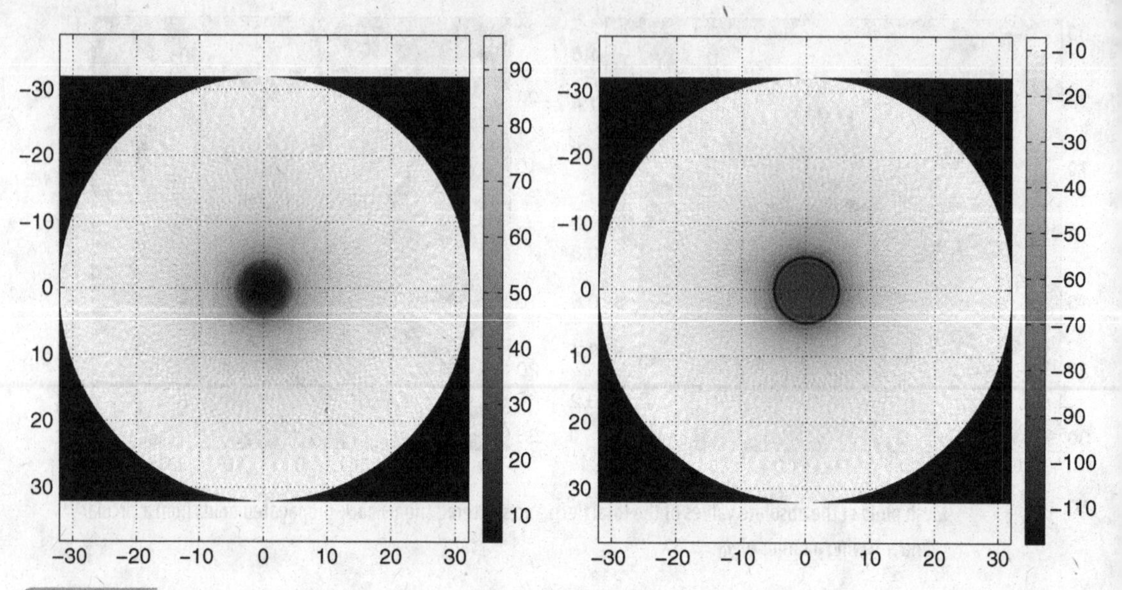

Fig. 7.10 Images of the superposition of the magnitudes of 100 back-propagated total fields from a circular cylinder having a radius of $a_0 = 5\lambda$ (left) and of the demodulated back-propagated scattered fields $\Re[e^{-ik_0 z}U^{(s)}_{bp}]$ (right).

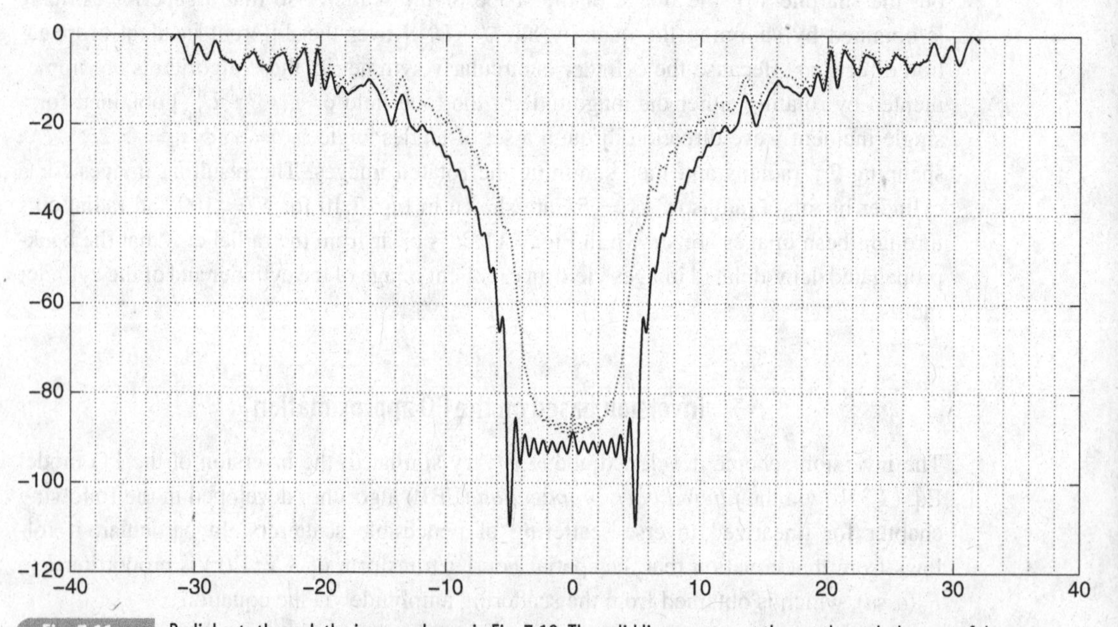

Fig. 7.11 Radial cuts through the images shown in Fig. 7.10. The solid lines represent the cut through the sum of the back-propagated demodulated fields and the dashed lines represent the cut through the sum of the magnitudes of the total back-propagated fields. The radial cuts have been normalized in amplitude for comparison purposes.

at the origin and has a radius of $2k_0$ in Fourier space. For the surface scattering and classical *diffraction tomography* treated in the following chapter it is preferable to limit the scattering directions $\mathbf{s}$ to be in the forward direction $\mathbf{s}\cdot\mathbf{s}_0 > 0$, in which case this quantity specifies the transform over the interior of a smaller sphere having radius $\sqrt{2}k_0$.

For 2D objects the FBP algorithm derived in the following chapter generates an image of $-\nabla^2\Gamma(\mathbf{r})$ via the algorithm

$$I(\mathbf{r}) = \mp\frac{k_0^2}{8\pi^2}\int_{-\pi}^{\pi} d\alpha_0 \int_{-\frac{\pi}{2}}^{\frac{\pi}{2}} d\alpha \sqrt{1-(\mathbf{s}\cdot\mathbf{s}_0)^2}F_g(\mathbf{s},\mathbf{s}_0)e^{ik_0(\mathbf{s}-\mathbf{s}_0)\cdot\mathbf{r}}, \tag{7.56}$$

where, as above, the minus sign is used for a Neumann scatterer and the plus for a Dirichlet scatterer, the angle α is the polar angle of $\mathbf{s}$ measured relative to the unit propagation vector $\mathbf{s}_0$ and α_0 is the polar angle of $\mathbf{s}_0$ measured relative to the positive-x axis. It then follows that $\mathbf{s}\cdot\mathbf{s}_0 = \cos\alpha$ so that $\sqrt{1-(\mathbf{s}\cdot\mathbf{s}_0)^2} = |\sin\alpha|$. Moreover,

$$(\mathbf{s}-\mathbf{s}_0)\cdot\mathbf{r} = \xi\cos\alpha + \eta(\sin\alpha - 1), \tag{7.57}$$

where (ξ,η) are the Cartesian coordinates in the (x,y) plane rotated through the angle α_0 so that $\hat{\eta}_0 = \mathbf{s}_0$ as illustrated in Fig. 8.3 in the following chapter. On substituting for F_g from Eq. (7.55) and making use of Eq. (7.57) we find that Eq. (7.56) simplifies to

$$I(\mathbf{r}) = \mp\frac{k_0^2}{4\pi^2}\Re\left\{\int_{-\pi}^{\pi} d\alpha_0 \int_{-\frac{\pi}{2}}^{\frac{\pi}{2}} d\alpha|\sin\alpha| f(\alpha,\alpha_0)e^{ik_0[\xi\sin\alpha+\eta(\cos\alpha-1)]}\right\}. \tag{7.58}$$

We implemented the above algorithm for the Dirichlet cylinder considered in the inversions given above, where $f(\alpha,\alpha_0) = f(\alpha)$ is independent of the incident wave direction and depends only on the angle α measured relative to the η axis. For this circularly symmetric case the algorithm is implemented exactly along the lines used in those earlier inversions. In particular, except for the presence of the filter $|\sin\alpha|$ in the FBP algorithm (hence the adjective *filtered*) this algorithm consists of back propagation followed by demodulation and is implemented in the same manner as is used in the filtered back-propagation algorithms that will be employed in the following chapter.

The results of the simulation are presented in Fig. 7.12. For a Dirichlet cylinder the characteristic function is given by $\Gamma(r) = 1-\theta(r-a_0)$, where $\theta(\cdot)$ is the unit step function and a_0 is the radius of the cylinder. The FBP algorithm should then generate a filtered version of

$$-\nabla^2[1-\theta(r-a_0)] = \frac{1}{a_0}\delta(r-a_0) + \delta'(r-a_0),$$

where $\delta'(\cdot)$ is the first derivative of the Dirac delta function (the doublet). The presence of the delta function and doublet on the surface of the cylinder is clearly evident in this figure, and an excellent estimate of the cylinder's surface is obtained.

7.7.2 The SVD approach to surface reconstruction

To develop this approach to surface inverse scattering we first have to review some basic material that we developed in earlier chapters. We showed in Section 3.2.1 of Chapter 3

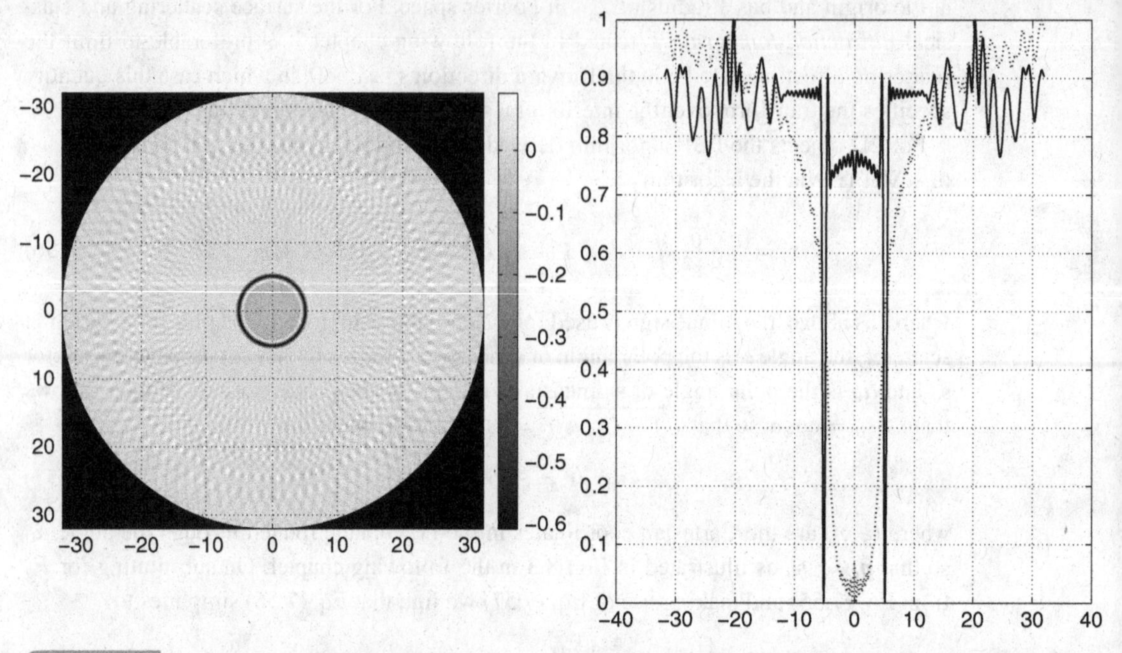

Fig. 7.12 An image generated using the filtered back-propagation algorithm (left) with 100 views of the scatterer and a radial cut through the center of the image (solid line in the plot on the right). Also shown for comparison (dotted line in the plot on the right) is a radial cut through the sum of the back-propagated images shown in Fig. 7.11. The radial cuts have been normalized in amplitude for comparison purposes.

that an arbitrary incident wave satisfying the homogeneous Helmholtz equation over all of space can be represented via the plane-wave expansion

$$U^{(\mathrm{in})}(\mathbf{r},\nu) = \int_0^{2\pi} d\beta_0 \int_0^{\pi} \sin\alpha_0\, d\alpha_0\, A(\mathbf{s}_0,\nu) e^{ik_0\mathbf{s}_0\cdot\mathbf{r}}, \tag{7.59a}$$

in which $\mathbf{s}_0$ is a unit propagation vector having direction cosines $\sin\alpha_0\cos\beta_0$, $\sin\alpha_0\sin\beta_0$, $\cos\alpha_0$ and the plane-wave amplitude $A(\mathbf{s}_0,\nu)$ is an arbitrary square-integrable function on the unit sphere $\mathbf{s}_0 \in \Omega_{s_0}$. We have included the parameter ν in the arguments of the incident wave and its plane-wave amplitude to keep track of possible multiple-scattering experiments using different incident waves. We also showed in Section 6.5 that the scattered field from a compactly supported scattering object resulting from the above incident field can also be expressed in a plane-wave (angular-spectrum) expansion of the general form

$$U^{(\mathrm{s})}(\mathbf{r},\nu) = \frac{ik_0}{2\pi} \int_0^{2\pi} d\beta \int_0^{\pi/2-i\infty} \sin\alpha\, d\alpha\, f(\mathbf{s},\nu) e^{ik_0\mathbf{s}\cdot\mathbf{r}}, \tag{7.59b}$$

where $f(\mathbf{s}, \nu)$ is the analytic continuation of the generalized scattering amplitude of the scattered field, i.e.,

$$U^{(s)}(r\mathbf{s}, \nu) \sim f(\mathbf{s}, \nu)\frac{e^{ik_0 r}}{r},$$

as $k_0 r \to \infty$ in the direction $\mathbf{s}$. The plane-wave expansion Eq. (7.59b) is valid for any field point $\mathbf{r} = (x, y, z > z_0)$, where z_0 is the z coordinate of the closest (x, y) bounding plane to the object surface $\partial\tau_0$. Since the orientation of the Cartesian (x, y, z) coordinate system is arbitrary, an expansion of the form Eq. (7.59b) is possible for every point lying outside the smallest convex surface that completely encloses the scattering volume τ_0.

The detailed form of the α contour of integration in Eq. (7.59b) is unimportant due to the known analyticity of $f(\mathbf{s}, \nu)$. Thus, this contour can be selected to run along the real-α axis from $\alpha = 0$ to $\alpha = \pi/2$ and then parallel to the imaginary axis from $\alpha = \pi/2$ to $\alpha = \pi/2 - i\infty$, corresponding to the decomposition of the scattered field into a *homogeneous part* $U_{\text{h}}^{(s)}$ and an *evanescent part* $U_{\text{e}}^{(s)}$. The homogeneous part of the scattered field is a superposition of unattenuated *homogeneous* plane waves that all propagate into the half-space $z > 0$, while the evanescent part is a superposition of plane waves that propagate perpendicular to the z axis and attenuate exponentially with propagation distance z from the boundary plane $z = z_0$. For this reason, as we discussed in Chapter 4, only the homogeneous part of the scattered field can be reliably computed from scattered field data collected at more than a few wavelengths from the scattering volume τ_0.

We also showed in Section 6.5 that the scattered-field plane-wave amplitude (angular spectra) $f(\mathbf{s}, \nu)$ is related to the incident-field plane-wave amplitude $A(\mathbf{s}_0, \nu)$ through the integral transform

$$f(\mathbf{s}, \nu) = \int_0^{2\pi} d\beta_0 \int_0^{\pi} \sin\alpha_0 \, d\alpha_0 f(\mathbf{s}, \mathbf{s}_0)A(\mathbf{s}_0, \nu), \tag{7.59c}$$

where $f(\mathbf{s}, \mathbf{s}_0)$ is the (plane-wave) scattering amplitude of the scatterer. The above mapping holds for all real and complex unit vectors $\mathbf{s}$, so knowledge of the incident-field plane-wave amplitude and the scattering amplitude can, in principle, uniquely and completely determine both the homogeneous part and the evanescent part of the scattered field. However, this would require an analytic continuation of the scattering amplitude onto complex unit vectors $\mathbf{s}$, which is highly ill-posed and not at all practical in any real-world application. The basic underlying reason for the ill-posed nature of this operation goes back to our earlier remarks regarding the irrecoverable loss of information when the scattered field measurements are performed at more than a few wavelengths from the scattering volume.

SVD of the scattering amplitude and modal expansion of the fields

The transform Eq. (7.59c) defines a linear mapping $\hat{T} : L^2(\Omega_{s_0}) \to L^2(\Omega_s)$ that admits the SVD (see Section 5.4.2 of Chapter 5)

$$\hat{T} v_{\mathbf{p}} = \sigma_p u_{\mathbf{p}}, \qquad \hat{T}^\dagger u_{\mathbf{p}} = \sigma_p v_{\mathbf{p}},$$

where $p = |\mathbf{p}|$ and the singular values $\sigma_p \geq 0$, $p = 0, 1, \ldots$, are ordered from largest ($p = 0$) to smallest and have a limit point of zero at $p = \infty$; i.e., $\lim_{p\to\infty} \sigma_p = 0$. We have indexed the singular vectors $\{v_{\mathbf{p}}(\mathbf{s}_0), u_{\mathbf{p}}(\mathbf{s})\}$ with the vector $\mathbf{p}$ since, in general, each singular value will be degenerate.

Associated with each pair of plane-wave amplitudes (singular vectors) $\{v_{\mathbf{p}}(\mathbf{s}_0), u_{\mathbf{p}}(\mathbf{s})\}$ is a pair of incident and scattered fields $\{V_{\mathbf{p}}(\mathbf{r}), \sigma_p U_{\mathbf{p}}(\mathbf{r})\}$ computed from the plane-wave expansions Eqs. (7.59); i.e.,

$$V_{\mathbf{p}}(\mathbf{r}) = \int_0^{2\pi} d\beta_0 \int_0^{\pi} \sin\alpha_0 \, d\alpha_0 \, v_{\mathbf{p}}(\mathbf{s}_0) e^{ik_0 \mathbf{s}_0 \cdot \mathbf{r}}, \tag{7.60a}$$

$$U_{\mathbf{p}}(\mathbf{r}) = \frac{ik_0}{2\pi} \int_0^{2\pi} d\beta \int_0^{\pi - i\infty} \sin\alpha \, d\alpha \, u_{\mathbf{p}}(\mathbf{s}) e^{ik_0 \mathbf{s} \cdot \mathbf{r}} \sim u_{\mathbf{p}}(\hat{\mathbf{r}}) \frac{e^{ik_0 r}}{r}, \quad k_0 r \to \infty, \tag{7.60b}$$

such that $\sigma_p U_{\mathbf{p}}(\mathbf{r})$ is the scattered field generated from the incident field $V_{\mathbf{p}}(\mathbf{r})$. Moreover, since the singular sets $\{v_{\mathbf{p}}, p = 0, 1, \ldots\}$ and $\{u_{\mathbf{p}}, p = 0, 1, \ldots\}$ are complete sets of orthonormal functions in Ω_{s_0} and Ω_s, respectively, it follows that any set of plane-wave amplitudes $A(\mathbf{s}_0, \nu)$ and $f(\mathbf{s}, \nu)$ can be represented in the forms

$$A(\mathbf{s}_0, \nu) = \sum_{p=0}^{\infty} A_{\mathbf{p}}(\nu) v_{\mathbf{p}}(\mathbf{s}_0), \qquad f(\mathbf{s}, \nu) = \sum_{p=0}^{\infty} \sigma_p A_{\mathbf{p}}(\nu) u_{\mathbf{p}}(\mathbf{s}),$$

from which we conclude that the incident field and the associated scattered field can be represented via the expansions

$$V^{(\mathrm{in})}(\mathbf{r}) = \sum_{p=0}^{\infty} A_{\mathbf{p}}(\nu) V_{\mathbf{p}}(\mathbf{r}), \qquad U^{(\mathrm{s})}(\mathbf{r}) = \sum_{p=0}^{\infty} \sigma_p A_{\mathbf{p}}(\nu) U_{\mathbf{p}}(\mathbf{r}), \tag{7.61a}$$

where the expansion coefficients $\{A_{\mathbf{p}}\}$ must satisfy the condition

$$\sum_{p=0}^{\infty} |A_{\mathbf{p}}|^2 < \infty$$

but are otherwise arbitrary.

Determining a Dirichlet surface from the modal expansions

The total field (incident plus scattered) from a Dirichlet surface must satisfy homogeneous boundary conditions on the surface $\partial\tau_0$ and this includes the modal pair $\{V_{\mathbf{p}}(\mathbf{r}), \sigma_p U_{\mathbf{p}}(\mathbf{r})\}$; i.e.,

$$V_{\mathbf{p}}(\mathbf{r}) + \sigma_p U_{\mathbf{p}}(\mathbf{r}) = 0, \quad \mathbf{r} = \mathbf{r}_0.$$

It then follows that

$$\lim_{\mathbf{r}\to\mathbf{r}_0} \sum_{p=0}^{\infty} A_{\mathbf{p}}(\nu) F(V_{\mathbf{p}}(\mathbf{r}) + \sigma_p U_{\mathbf{p}}(\mathbf{r})) = 0, \tag{7.62}$$

where $F(x)$ is an arbitrary non-negative function that is zero at $x = 0$. Strictly speaking, the above results apply only if the surface $\partial\tau_0$ is convex (see the discussion under

Eq. (7.59b)). However, the plane-wave expansions of the individual modal fields $U_{\mathbf{p}}(\mathbf{r})$ can be expected to converge everywhere outside the origin so that any finite series approximations of Eq. (7.62) should apply to general, non-convex boundaries.

As discussed earlier, only the homogeneous part of the scattered field can be reliably computed from far-field data, and this applies also to the modal fields $U_{\mathbf{p}}(\mathbf{r})$. Thus, the algorithm Eq. (7.62) will, in general, lead to errors in determining the boundary $\partial\tau_0$ because of inaccuracies in the computation of these modal fields. However, we can avoid this problem by selecting the expansion coefficients $A_{\mathbf{p}}$ so as to minimize the scattered-field contribution to these equations (Luke and Devaney, 2007). This is easily accomplished by setting a cutoff singular index P_0 such that $\sigma_p < \epsilon$, $\forall p > P_0$, where ϵ is some small parameter, and then demanding that $A_{\mathbf{p}} = 0$, $\forall p < P_0$. Equation (7.62) then reduces to the form

$$\lim_{\mathbf{r}\to\mathbf{r}_0} \sum_{p=P_0}^{P_1} A_{\mathbf{p}}(\nu)F(V_{\mathbf{p}}(\mathbf{r}) + \sigma_p U_{\mathbf{p}}(\mathbf{r})) \approx \lim_{\mathbf{r}\to\mathbf{r}_0} \sum_{p=P_0}^{P_1} A_{\mathbf{p}}(\nu)F(V_{\mathbf{p}}(\mathbf{r})) \approx 0, \tag{7.63}$$

where P_1 is some maximum index value that depends on the specific application.

The success of the above algorithm hinges on the facts that (1) the singular values $\sigma_p < \epsilon$, $\forall p \geq P_0$, and (2) the amplitude of each of the modal fields $U_{\mathbf{p}}(\mathbf{r})$ remains bounded in the vicinity of the boundary $\partial\tau_0$. To examine this second condition in more depth we define the scattered-field energy radiated out of the scattering volume τ_0 in any of these modal fields in the usual manner as

$$E_{\mathbf{p}} = \frac{1}{k}\Im \int_{\partial\tau_0} dS_0\, U_{\mathbf{p}}{}^*(\mathbf{r}_0)\frac{\partial}{\partial n_0}U_{\mathbf{p}}(\mathbf{r}_0) = \int_{\Omega_s} d\Omega_s |u_p(\mathbf{s})|^2 = 1,$$

where $\mathbf{r}_0$ is a general field point on the scattering surface $\partial\tau_0$ and we have made use of Eq. (7.60b). It then follows that the individual scattered-field modes must remain bounded as $\mathbf{r} \to \mathbf{r}_0 \in \partial\tau_0$ so that the algorithm Eq. (7.63) will be valid.

Example 7.1 As an example we consider 2D scattering from a cylinder over which the field satisfies homogeneous Dirichlet conditions. The scattering amplitude for a Dirichlet cylinder centered on the origin and having radius a_0 was found in Section 7.2.3 to be given by

$$f(\alpha, \alpha_0) = -\sqrt{\frac{2}{\pi k_0}} e^{-i\frac{\pi}{4}} \sum_{n=-\infty}^{\infty} \frac{J_n(k_0 a_0)}{H_n^+(k_0 a_0)} e^{in(\alpha-\alpha_0)}. \tag{7.64}$$

It is easy to verify that the singular system $\{v_{\mathbf{p}}, u_{\mathbf{p}}, \sigma_p\}$ is, in this case, given by

$$v_{\mathbf{p}}(\alpha_0) = \frac{1}{\sqrt{2\pi}} e^{\pm ip\alpha_0}, \qquad u_{\mathbf{p}}(\alpha) = \frac{e^{i\phi_p}}{\sqrt{2\pi}} e^{\pm ip\alpha}, \qquad \sigma_p = \sqrt{\frac{8\pi}{k_0}} \left| \frac{J_p(k_0 a_0)}{H_p(k_0 a_0)} \right|, \tag{7.65a}$$

where

$$\phi_p = \frac{3\pi}{4} + \text{Arg}[J_p(k_0 a_0)/H_p(k_0 a_0)] \tag{7.65b}$$

and the plus sign gives one of the two singular vectors and the minus sign the second for each singular value σ_p.

The modal fields $V_\mathbf{p}$ and $U_\mathbf{p}$ are constructed from the 2D versions of the plane-wave expansions Eq. (7.60) with the plane-wave amplitudes set equal to $v_p(\mathbf{s}_0)$ and $u_p(\mathbf{s})$. We showed in Chapters 3 and 4 that these 2D versions assume the form

$$V_\mathbf{p}(\mathbf{r}) = \int_{-\pi}^{\pi} d\alpha_0\, v_\mathbf{p}(\alpha_0)e^{ik_0\mathbf{s}_0\cdot\mathbf{r}}, \qquad U_\mathbf{p}(\mathbf{r}) = \sqrt{\frac{k}{2\pi}}e^{i\frac{\pi}{4}}\int_{C_\pm} d\alpha\, u_\mathbf{p}(\alpha)e^{ik_0\mathbf{s}\cdot\mathbf{r}}, \tag{7.66}$$

where α_0 and α are, respectively, the polar angles of the unit propagation vectors $\mathbf{s}_0$ and $\mathbf{s}$. On substituting for the singular vectors from Eqs. (7.65a) we find that

$$V_\mathbf{p}(r,\theta) = \sqrt{2\pi}\, i^p J_p(k_0 r)e^{\pm ip\theta}, \qquad U_\mathbf{p}(r,\theta) = \sqrt{2\pi}\, e^{i\phi_p} i^p H_p^+(k_0 r)e^{\pm ip\theta}, \tag{7.67}$$

where (r,θ) are the cylindrical polar coordinates of the field vector $\mathbf{r}$. We selected the functional $F(x)$ in Eqs. (7.63) to be $F(x) = |x|^2$ and the expansion coefficients $A_\mathbf{p} = 1$ so that this equation reduces to

$$\lim_{r\to a}\sum_{p=P_0}^{P_1}\left|J_p(k_0 r) - \frac{J_p(k_0 a_0)}{H_p(k_0 a_0)}H_p(k_0 r)\right|^2 \approx \lim_{r\to a}\sum_{p=P_0}^{P_1}|J_p(k_0 r)|^2 = 0, \tag{7.68}$$

with $P_1 > P_0$ and P_0 such that $\sigma_p < \epsilon$, $\forall p > P_0$.

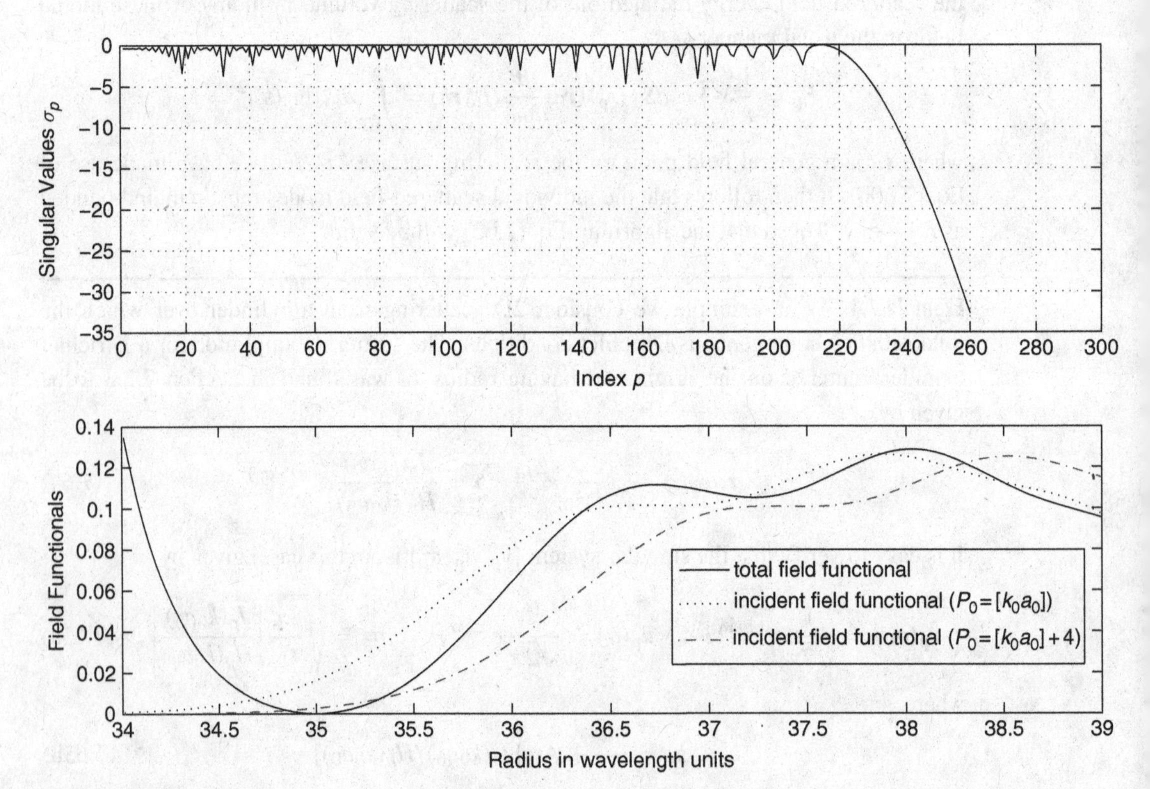

Fig. 7.13 The top plot shows the singular values $\sigma_p = |J_p(k_0 a_0)/H_p(k_0 a_0)|$ for a cylinder having radius $a = 35$. In the bottom part we show plots of the l.h.s. of Eq. (7.68) and the r.h.s. of this equation for $P_0 = [k_0 a_0]$ and for $P_1 = P_0 + 20$, for $a = 35$. We also show a plot of the r.h.s. of this equation for the case in which $P_0 = [k_0 a_0] + 4$.

We present a plot of the singular values $\sigma_p = |J_p(k_0 a_0)/H_p(k_0 a_0)|$ using unit wavelength ($k = 2\pi$) and cylinder radius $a = 35\lambda = 35$ in the top part of Fig. 7.13. It is clear from this figure that the cutoff $P_0 \approx [k_0 a_0] \approx 220$, where $[x]$ indicates the nearest integer approximation of x. In the bottom part of Fig. 7.13 we show plots of the sums on the left- and right-hand sides of Eq. (7.68) for the cases in which $P_0 = [k_0 a_0]$ and $P_0 = [k_0 a_0] + 4$ and $P_1 = P_0 + 20$. We obtained similar results for other choices of the cylinder radius a and index limits P_0 and P_1 so long as $P_0 > [k_0 a_0]$ is not significantly greater than $[k_0 a_0]$.

Further reading

An excellent historical account of diffraction theory is given in Baker and Copson (1950). More modern treatments are available in Born and Wolf (1999), Cowley (1966) and Arsac (1984). The earliest papers on inverse scattering from surfaces include those by Lewis (Lewis, 1969), which was based on the physical-optics approximation, and Imbriale and Mittra (Imbriale and Mittra, 1970), who employed field back propagation to generate the inversions. A comprehensive and well-researched and -tested procedure for inverse scattering from surfaces within the framework of exact scattering theory is the so-called "linear sampling method" developed by Kirsch, Colton and co-workers (Kirsch, 1998; Kirsch and Ritter, 2000; Haddar and Piana, 2003). Anyone interested in this subject should certainly consult their work before beginning any research in the area; see also Colton and Kress (1992, 1991). Cheney (Cheney, 2001) has related this method to the MUSIC algorithm which will be presented in Chapter 10, while direct application of the MUSIC algorithm to the problem of inverse surface scattering was presented in Solna *et al.* (2006) and Marengo *et al.* (2007). Pierri and co-workers have used the SVD to determine the shape of metallic scatterers (Pierri *et al.*, 2006) and also investigated the shape-reconstruction problem using optical scattered intensities (Soldovieri and Pierri, 2008). Langenberg and co-workers have developed a number of algorithms for non-destructive evaluation that are based on inverse diffraction and inverse scattering, which are summarized in Langenberg *et al.* (1999). A completely different approach to surface-structure determination that we have not discussed is possible using the analytic properties of the scattering amplitude and a generalized form (Damelin and Devaney, 2007; Kusiak and Sylvester, 2005; Hansen *et al.*, 2000) of the famous Paley–Wiener theorem (Paley and Wiener, 1934).

The application of inverse scattering in identifying and tracking targets has, of course, been treated extensively within the military community, mostly using ad-hoc approaches. Methods such as SAR, ISAR, etc. are closely related and can be developed in a rigorous manner from the theories presented in this and the following chapter. Suggested reading on such treatments of these imaging schemes is presented in Borden (1999, 2002) and Cheney and Borden (2008, 2009). See also the tomography-based scheme presented in Das and Boerner (1978).

Problems

7.1 Prove that a square-integrable free field (one that satisfies the homogeneous Helmholtz equation over all of space) has finite multipole moments.

7.2 Derive Eq. (7.5b).

7.3 Use the angular-spectrum expansions developed in Section 7.2.1 to compute the outgoing-wave Green functions in the half-space $z < 0$ that satisfy homogeneous Dirichlet and Neumann conditions on the plane $z = 0$.

7.4 Generalize the angular-spectrum expansions developed in Section 7.2.1 to the case of an incident wave radiated by a source in the right half-space $z > z_0$ and reflecting from the plane $z = z_0$, where it satisfies homogeneous Dirichlet or Neumann conditions.

7.5 Derive the 2D versions of Eqs. (7.25).

7.6 Derive Eq. (7.26a).

7.7 Derive the generalized scattering amplitudes for Dirichlet and Neumann surfaces within the PO approximation

(1) from the definition of the PO scattered fields given in Eqs. (7.21) and

(2) from their plane-wave scattering amplitudes and the relationship Eq. (6.35b).

(3) Verify that they are the same and reduce to the plane-wave scattering amplitudes for plane-wave incidence.

7.8 Compute the scattered field and scattering amplitude for plane-wave incidence within the PO approximation for a Dirichlet sphere.

7.9 Use the scattering amplitude for a Dirichlet cylinder within the PO approximation given in Section 7.3 to verify the 2D version of Eq. (7.31).

7.10 Use the scattering amplitude for a Dirichlet cylinder within the PO approximation given in Section 7.3 to verify the 2D version of Eq. (7.35).

7.11 Derive the expression for the Kirchhoff diffraction pattern given in Eq. (7.46b).

7.12 Compute the Kirchhoff diffraction pattern of a circular disk as a function of the transverse coordinates in an observation plane located at distance z from the center of the disk.

7.13 Compute the diffraction pattern of the disk in Problem 7.12 in the far field as a function of the unit vector $\mathbf{s}$.

7.14 Give an argument for why the field diffracted by a circular aperture that subtends the solid angle Ω_0 from a source located in the l.h.s. is approximately given by

$$U^{(d)}(\mathbf{r}) \approx \int_{\Omega_0} d\Omega_s A^{(in)}(\mathbf{s}) e^{ik_0 \mathbf{s}\cdot\mathbf{r}},$$

where $A(\mathbf{s})$ is the angular spectra of the incident wave to the aperture. State the requirements that must hold in order for the approximation to be accurate.

7.15 Derive Eq. (7.58) from Eq. (7.56).

7.16 Generalize the formulation developed in Example 7.1 to the case of a sphere.

8 Classical inverse scattering and diffraction tomography

The "direct" or "forward" scattering problem was treated in the preceding two chapters, where the goal was the computation of a scattered field given knowledge of the scattering object and the incident wavefield. In the "inverse scattering problem" (ISCP) the goal is the determination of the scattering object given knowledge of the incident wave and the scattered wave over some restricted region of space. In Chapter 6 we treated so-called "penetrable" scatterers, where the incident wave penetrates into the interior of the obstacle, thus creating an "induced volume source" that then radiates as a conventional volume source of the type treated in earlier chapters. In Chapter 7 we treated non-penetrable scatterers, where the interaction of the incident wave with the obstacle occurs only over the object's surface. We also treated certain inverse problems associated with non-penetrable scatterers in that chapter that included inverse diffraction and the ISCP of determining the shape of a Dirichlet or Neumann scatterer from its scattering amplitude. In this chapter we will treat the ISCP for penetrable scatterers. We will also make the simplifying assumption that the scattering object is embedded in a uniform *lossless* medium. This assumption will be discarded in the next chapter, where we will treat scatterers embedded in non-uniform and dispersive media.

We pointed out in Chapter 5 that the difficulty of the "inverse source problem" (ISP) lies in the fact that the radiated field from which the source is to be determined is known only over space points that lie in some restricted region of space that is *outside* the support of the (unknown) source. The same is true of the ISCP: if the scattered field is known over all of space the scattering structure is easily determined. For example, for the potential scattering treated in Chapter 6 the total (incident plus scattered) field satisfies the equation

$$[\nabla^2 + k_0^2 - V(\mathbf{r})]U(\mathbf{r}, \nu) = 0, \tag{8.1}$$

where k_0 is the background wavenumber, ν is a parameter that parameterizes the incident wave and $V(\mathbf{r})$ is the unknown scattering potential. If the scattered field is known over all of space (including the interior of the scattering volume) then the scattering potential is easily computed via the equation

$$V(\mathbf{r}) = \frac{[\nabla^2 + k_0^2]U(\mathbf{r}, \nu)}{U(\mathbf{r}, \nu)}.$$

In any real experiment the scattered field can be measured only at space points that are exterior to the scattering volume, so the above simple solution to the ISCP is not possible.

However, it is possible to estimate the interior field from exterior field measurements from which an estimate of the scattering potential can be computed. Thus, in essence the ISCP for penetrable scatterers reduces to the determination of the field interior to the scattering volume from exterior scattered field measurements.

A similar situation was encountered in the ISP treated in Chapter 5. In that problem measurements of the field radiated by an unknown source are performed in space regions exterior to the source and are to be employed to estimate the unknown source. In the first part of Chapter 5, where we treated the ISP for the wave equation, we employed *field back propagation* to estimate the interior field from which we obtained an estimate of the source. In the later part of Chapter 5, where we considered dispersive media, we employed a formal mathematical approach to the problem that allowed us to estimate the source directly from the exterior field without first estimating the interior field. We will employ both of these approaches in this chapter, although, in fact, they are fundamentally the same.

One major difference between the ISP and the ISCP is that in the ISCP we are allowed the luxury of performing more than one scattering experiment using different incident wavefields. Because of this we obtain multiple "looks" or "views" of the scattering potential. On the other hand, in the ISP the source is fixed and we get only a single view of the source provided by its (single) radiated field. We will show later in this chapter that these multiple views of a scattering potential provided by the scattered field data acquired in multiple scattering experiments are analogous to the multiple views that are generated in a suite of tomographic experiments in X-ray tomography. Indeed, we will show that the ISCP can be cast in an entirely parallel form to that of computed tomography (CT), in which form it becomes known as *diffraction tomography* (DT). We will examine both formulations of the ISCP in this chapter.

Linearization of the inverse scattering problem

We showed in Chapters 6 and 7 that the process of scattering is *linear* when it is considered to be a transformation from the incident wave to the scattered wave but *non-linear* when it is considered to be a transformation from the scattering object to the scattered wave. It then follows that the ISCP, which consists of deducing the scatterer from scattered field data, reduces to the inversion of a set of non-linear transformations obtained over the suite of scattering experiments. An obvious approach to this problem is to linearize this set of transformations, thus reducing the problem to a form that can be solved using well-established mathematical theory. In the case of potential scattering there are two standard linearization schemes that were developed in Chapter 6: the Born approximation and the Rytov approximation (Tatarski, 1961; Chernov, 1967; Ishimaru, 1999). We will employ both linearization schemes in this chapter and, in the process, obtain the two formulations of linearized inverse scattering alluded to above: (i) Born-based inverse scattering and (ii) Rytov-based inverse scattering, which is commonly known as diffraction tomography. In this chapter we will limit our attention to weakly scattering objects embedded in a lossless and constant (uniform) background medium but generalize our results to a more general class of linearized models in the following chapter.

8.1 Born inverse scattering from far-field data

An important class of incident wavefields is constituted by the plane waves

$$U^{(\mathrm{in})}(\mathbf{r},\mathbf{s}_0)=e^{ik_0\mathbf{s}_0\cdot\mathbf{r}},$$

where the parameter ν is now the unit propagation vector $\mathbf{s}_0$ of the incident plane wave. The plane waves are important for a number of reasons, not the least of which is that, as shown in Chapter 3, they form a complete set of elementary solutions to the homogeneous Helmholtz equation into which any incident wavefield can be expanded. However, they are also important in that certain controlled scattering experiments can directly employ a set of incident plane waves in order to generate a suite of scattered field data for the ISCP. A classic example of this occurs in structure determination in X-ray crystallography, where a set of incident plane waves with unit propagation vectors $\mathbf{s}_0$ is employed and the scattered field measurements are performed many wavelengths from the scattering object. Under these conditions the scattered field is well approximated by its far-field expression

$$U^{(\mathrm{s})}(r\mathbf{s};\mathbf{s}_0)\sim f(\mathbf{s};\mathbf{s}_0)\frac{e^{ik_0r}}{r},\quad k_0r\to\infty,$$

where $f(\mathbf{s};\mathbf{s}_0)$ is the *scattering amplitude* of the scattering object with $\mathbf{s}=\mathbf{r}/r$ the unit vector along the direction of the field point $\mathbf{r}$. Plane waves and far-field measurements are also used in certain inverse scattering applications in electromagnetics, ultrasound and optics. In all of these applications the ISCP reduces to the determination of a scattering potential $V(\mathbf{r})$ from specification of its scattering amplitude[1] $f(\mathbf{s},\mathbf{s}_0)$ over some set of scattering directions $\mathbf{s}$ and some set of incident-wave directions $\mathbf{s}_0$.

The scattering amplitude within the Born approximation was derived in Section 6.7.2 of Chapter 6, where it was shown to be related to the scattering potential via the equation

$$f_{\mathrm{B}}(\mathbf{s},\mathbf{s}_0)=-\frac{1}{4\pi}\tilde{V}[k_0(\mathbf{s}-\mathbf{s}_0)],\tag{8.2}$$

where

$$\tilde{V}(\mathbf{K})=\int_{\tau_0}d^3r\,V(\mathbf{r})e^{-i\mathbf{K}\cdot\mathbf{r}}$$

is the spatial Fourier transform of the scattering potential and the subscript B on the scattering amplitude denotes the Born approximation. For fixed incident-wave direction $\mathbf{s}_0$ the locus of points

$$\mathbf{K}=k_0(\mathbf{s}-\mathbf{s}_0)$$

[1] A complication occurs in X-ray and optical scattering, where only intensity field measurements can be performed. This problem, called the *phase problem*, can be solved in certain cases by employing holographic measurement systems (Yamaguchi and Zhang, 1997) or by use of so-called *phase-retrieval algorithms* (Taylor, 1981; Maleki *et al.*, 1992; Fienup, 1982; Guizar-Sicairos and Fienup, 2006; Gerchberg and Saxton, 1972; Gonsalves, 1976). See also Devaney (1989).

lies on the surface of a sphere of radius k_0 centered at the point $\mathbf{K} = -k_0\mathbf{s}_0$, which is known as an *Ewald sphere* (Lipson and Cochran, 1966; Born and Wolf, 1999) (see Figure 6.3 of Section 6.7.2). For any given incident-plane-wave direction $\mathbf{s}_0$ the scattering amplitude specified over all scattering directions $\mathbf{s} \in \Omega$ thus determines the scattering potential's spatial Fourier transform over the surface of a single Ewald sphere. Here, and from this point onward, we denote the entire (real) unit sphere by Ω (i.e., $\Omega = 4\pi$ steradians) and will denote limited regions of the unit sphere by Ω_s or Ω_{s_0}, etc. If a complete suite of scattering experiments is performed, using all incident-wave directions $\mathbf{s}_0 \in \Omega$, the totality of resulting Ewald-sphere surfaces will sweep out the interior of a sphere of radius $2k_0$ called the *Ewald limiting sphere*. In this ideal case $\tilde{V}(\mathbf{K})$ is then determined throughout the interior of the Ewald limiting sphere and a *low-pass-filtered approximation* is obtained via an inverse spatial Fourier transform:

$$V_{\mathrm{LP}}(\mathbf{r}) = \frac{1}{(2\pi)^3} \int_{|\mathbf{K}|\leq 2k_0} d^3K \;\overbrace{\tilde{V}(\mathbf{K})}^{-4\pi f_{\mathrm{B}}(\mathbf{s},\mathbf{s}_0)}\; e^{i\mathbf{K}\cdot\mathbf{r}}. \tag{8.3}$$

We note that the spatial Fourier transform $\tilde{V}(\mathbf{K})$ of a compactly supported scattering potential is an entire analytic function of the spatial frequency vector $\mathbf{K}$ so that specification of the Fourier transform throughout the interior of the Ewald limiting sphere allows the transform to be determined over all of $\mathbf{K}$ space via the process of analytic continuation. Within the linearized Born scattering model an exact and complete solution to the ISCP can then be obtained via an inverse Fourier transform of the analytically continued transform. However, as discussed in Section 6.7.2, such a procedure is computationally unstable and the best that can be expected in practice is the low-pass-filtered approximation defined in Eq. (8.3). Indeed, in any real experimental situation the scattering amplitude will be determined only over some finite discrete set of scattering directions $\mathbf{s} \in \Omega_{\mathbf{s}}$ and for a finite number of incident-wave directions $\mathbf{s}_0 \in \Omega_{\mathbf{s}_0}$. One method of obtaining an approximate solution to the linearized ISCP in such cases is then to interpolate the transform data specified over the set of Ewald spheres onto a regular cubic grid in $\mathbf{K}$ space from which $V_{\mathrm{LP}}(\mathbf{r})$ can then be approximated using an inverse discrete Fourier transform. This method has been employed extensively in the ISCP, especially, within the discipline of *X-ray crystal-structure determination*. We will treat this so-called "limited-view problem" later in this and the following chapter.

8.1.1 Born inversion from ideal data

Equation (8.3) by itself is not actually an inversion algorithm for computing the low-pass-filtered approximation to the scattering potential since it requires first an interpolation from the Ewald-spherical surfaces $\mathbf{K} = k_0(\mathbf{s} - \mathbf{s}_0)$ over which $\tilde{V}(\mathbf{K})$ is specified from the scattering amplitude onto a regular grid in $\mathbf{K}$ space over which the integral can then be numerically approximated. Alternatively, one can transform the 3D integral in Eq. (8.3) into a 4D integral over the two unit vectors $\mathbf{s}$ and $\mathbf{s}_0$, in which case no interpolation or other scheme is required in order to generate the low-pass-filtered approximation directly from the scattering-amplitude data. In this section we will employ this second procedure and derive a one-step inversion algorithm that employs the "natural" coordinate vectors **s** and $\mathbf{s}_0$ rather than the spatial frequency vector **K**. We will assume that the scattering amplitude is modeled exactly by the Born approximation according to Eq. (8.2) and that it is specified over all scattering $\mathbf{s} \in \Omega$ and incident $\mathbf{s}_0 \in \Omega$. Later we will relax these conditions and examine situations of non-ideal data not exactly modeled by the Born approximation and specified over limited sets of scattering and incident wave directions.

To obtain the desired inversion algorithm we will employ a generalized form of the scheme that was used in Chapter 5 to solve the ISP for the wave equation. In the case of the ISP we first back propagated the field data to generate an "image" of the source that was related to the exact source by an integral equation (the Porter–Bojarski integral equation), which was then solved to yield an estimate of the source. In the ISCP we perform a suite of scattered-field experiments using different incident wavefields each of which generates a scattered wave via the process of radiation by a different "induced source." For plane-wave incidence and within the Born approximation we showed in Chapter 6 that these induced sources are related to the scattering potential and incident plane wave via the equation

$$Q(\mathbf{r}; \mathbf{s}_0) = V(\mathbf{r})e^{ik_0\mathbf{s}_0\cdot\mathbf{r}}.$$

The scattered fields radiated by this set of induced sources gives us a set of separate "partial views" of the scattering potential and we thus have to modify the scheme employed in the ISP to accommodate this entire set of scattered (radiated) fields.

The required modification consists of first back propagating each of the separate scattered-field data sets (each view of the scattering potential) to generate a set of intermediate images of the induced sources for the various experiments. The next step consists of "demodulating" each induced source image by multiplication by $\exp(-ik_0\mathbf{s}_0 \cdot \mathbf{r})$ to obtain a partial image of the scattering potential and then summing the entire set of partial view images. This summed image is found to be related to the scattering potential via an integral equation that is then solved to yield an estimate of the scattering potential. In the case of complete ideal Born data under consideration here, the estimate we generate is precisely the low-pass-filtered version of the scattering potential defined in Eq. (8.3) and the solution we obtain is in the form of an integral over the unit vectors $\mathbf{s}_0$ and **s** as desired.

In the case of far-field data and plane-wave incidence the back-propagated field $\Phi(\mathbf{r}; \mathbf{s}_0)$ generated in each experiment is given by Eq. (5.7a) of Example 5.2 of Chapter 5 with the radiation pattern $f(\mathbf{s})$ replaced by the scattering amplitude $f(\mathbf{s}, \mathbf{s}_0)$:

$$\Phi(\mathbf{r}; \mathbf{s}_0) = \frac{ik_0}{2\pi} \int d\Omega_s f_{\mathrm{B}}(\mathbf{s}, \mathbf{s}_0)e^{ik_0\mathbf{s}\cdot\mathbf{r}},$$

where the integral is performed over the entire unit sphere Ω. It is mathematically convenient to use the quantity

$$\Theta(\mathbf{r}; \mathbf{s}_0) = \frac{i}{2k_0}\Phi(\mathbf{r}; \mathbf{s}_0) = -\frac{1}{4\pi} \int d\Omega_s f_{\mathrm{B}}(\mathbf{s}, \mathbf{s}_0)e^{ik_0\mathbf{s}\cdot\mathbf{r}}$$

rather than $\Phi(\mathbf{r};\mathbf{s}_0)$ in the following development. The demodulated and summed set of partial images is then found to be

$$F(\mathbf{r}) = \int d\Omega_{s_0}\, e^{-ik_0\mathbf{s}_0\cdot\mathbf{r}}\Theta(\mathbf{r};\mathbf{s}_0) = -\frac{1}{4\pi}\int d\Omega_{s_0}\int d\Omega_s f_{\mathrm{B}}(\mathbf{s},\mathbf{s}_0)e^{ik_0(\mathbf{s}-\mathbf{s}_0)\cdot\mathbf{r}}. \tag{8.4a}$$

On making use of Eq. (8.2) we find that the above equation can be written in the form

$$F(\mathbf{r}) = -\frac{1}{4\pi}\int d\Omega_{s_0}\int d\Omega_s \overbrace{\left(-\frac{1}{4\pi}\int d^3r'\, V(\mathbf{r}')e^{-ik_0(\mathbf{s}-\mathbf{s}_0)\cdot\mathbf{r}'}\right)}^{f_{\mathrm{B}}(\mathbf{s},\mathbf{s}_0)} e^{ik_0(\mathbf{s}-\mathbf{s}_0)\cdot\mathbf{r}}$$
$$= \int d^3r'\, V(\mathbf{r}')H(\mathbf{r}-\mathbf{r}'), \tag{8.4b}$$

where

$$H(\mathbf{R}) = \frac{1}{(4\pi)^2}\int d\Omega_{s_0}\int d\Omega_s\, e^{ik_0(\mathbf{s}-\mathbf{s}_0)\cdot\mathbf{R}}, \tag{8.4c}$$

with $\mathbf{R} = \mathbf{r}-\mathbf{r}'$. Equation (8.4b) is an integral equation for the scattering potential $V(\mathbf{r})$ in terms of the demodulated and summed set of partial images $F(\mathbf{r})$ that is directly computed from the scattering amplitude f_{B} via Eq. (8.4a). We will refer to Eq. (8.4b) as the *ISCP integral equation*.

The kernel $H(\mathbf{R})$ of the ISCP integral equation Eq. (8.4b) can be computed in closed form by making use of the expansion

$$j_0(k_0R) = \frac{1}{4\pi}\int d\Omega_s\, e^{ik_0\mathbf{s}\cdot\mathbf{R}},$$

where j_0 is the zeroth-order spherical Bessel function of the first kind and $R = |\mathbf{R}|$. On making use of the above result in Eq. (8.4c) we find that

$$H(\mathbf{R}) = \left|\frac{1}{4\pi}\int d\Omega_s\, e^{ik_0\mathbf{s}\cdot\mathbf{R}}\right|^2 = j_0^2(k_0R). \tag{8.5}$$

The above closed-form expression for the kernel $H(\mathbf{R})$ of the ISCP integral equation allows us to compute its spatial Fourier transform $\tilde{H}(\mathbf{K})$ and, hence, solve the integral equation using standard Fourier-transform techniques. In particular, we have that[2]

$$\tilde{H}(\mathbf{K}) = \int d^3R\, j_0^2(k_0R)e^{-i\mathbf{K}\cdot\mathbf{R}} = \int_0^\infty R^2\, dR\, j_0^2(k_0R)\overbrace{\int d\Omega_R\, e^{-i\mathbf{K}\cdot\mathbf{R}}}^{4\pi j_0(k_0R)}$$
$$= 4\pi\int_0^\infty R^2\, dR\, j_0^2(k_0R)j_0(k_0R) = \begin{cases}\pi^2/(k_0^2K) & \text{if } K\le 2k_0,\\ 0 & \text{else.}\end{cases} \tag{8.6}$$

If we Fourier transform both sides of the ISCP integral equation Eq. (8.4b) we then obtain the result

$$\tilde{F}(\mathbf{K}) = \tilde{H}(\mathbf{K})\tilde{V}(\mathbf{K}) = \begin{cases}\pi^2/(k_0^2K)\tilde{V}(\mathbf{K}) & \text{if } K\le 2k_0,\\ 0 & \text{else.}\end{cases} \tag{8.7}$$

[2] The final expression Eq. (8.6) is obtained using contour integration and the calculus of residues.

It can be seen from Eq. (8.7) that the transform of the inhomogeneous term $F(\mathbf{r})$ of the ISCP integral equation defined in Eq. (8.4a) is band-limited to the Ewald limiting sphere $|\mathbf{K}| = K \leq 2k_0$. It then follows that F merely determines the transform of the scattering potential within the Ewald limiting sphere or, equivalently, merely determines the low-pass-filtered version V_{LP} of the scattering potential defined in Eq. (8.3). In particular, for $K \leq 2k_0$ we conclude from Eq. (8.7) that

$$\tilde{V}_{\text{LP}}(\mathbf{K}) = \frac{k_0^2 K}{\pi^2} \tilde{F}(\mathbf{K}),$$

where

$$\tilde{V}_{\text{LP}}(\mathbf{K}) = \int d^3r \, V_{\text{LP}}(\mathbf{r}) e^{-i\mathbf{K}\cdot\mathbf{r}}$$

is the spatial Fourier transform of the low-pass-filtered scattering potential.

The low-pass-filtered version of the scattering potential V_{LR} is then found to be

$$V_{\text{LP}}(\mathbf{r}) = \frac{1}{(2\pi)^3} \int d^3K \, \overbrace{\frac{k_0^2 K}{\pi^2} \tilde{F}(\mathbf{K})}^{\tilde{V}_{\text{LP}}(\mathbf{K})} e^{i\mathbf{K}\cdot\mathbf{r}}. \tag{8.8}$$

We can express the transform $\tilde{F}(\mathbf{K})$ using Eq. (8.4a) in the form

$$\begin{aligned}
\tilde{F}(\mathbf{K}) &= \int d^3r \overbrace{\left(-\frac{1}{4\pi} \int d\Omega_{s_0} \int d\Omega_s f_{\text{B}}(\mathbf{s}, \mathbf{s}_0) e^{ik_0(\mathbf{s}-\mathbf{s}_0)\cdot\mathbf{r}}\right)}^{F(\mathbf{r})} e^{-i\mathbf{K}\cdot\mathbf{r}} \\
&= -2\pi^2 \int d\Omega_{s_0} \int d\Omega_s f_{\text{B}}(\mathbf{s}, \mathbf{s}_0) \delta(\mathbf{K} - k_0(\mathbf{s} - \mathbf{s}_0)),
\end{aligned}$$

which, when substituted into Eq. (8.8), yields the result

$$V_{\text{LP}}(\mathbf{r}) = -\frac{k_0^2}{4\pi^3} \int d^3K \, K \left\{ \int d\Omega_{s_0} \int d\Omega_s f_{\text{B}}(\mathbf{s}, \mathbf{s}_0) \delta(\mathbf{K} - k_0(\mathbf{s} - \mathbf{s}_0)) \right\} e^{i\mathbf{K}\cdot\mathbf{r}}$$

$$\Downarrow$$

$$V_{\text{LP}}(\mathbf{r}) = -\frac{k_0^3}{4\pi^3} \int d\Omega_{s_0} \int d\Omega_s |\mathbf{s} - \mathbf{s}_0| f_{\text{B}}(\mathbf{s}, \mathbf{s}_0) e^{ik_0(\mathbf{s}-\mathbf{s}_0)\cdot\mathbf{r}}, \tag{8.9}$$

which is the sought-after inversion algorithm.

8.1.2 The filtered back-propagation algorithm

The inversion algorithm Eq. (8.9) is one version of the *filtered back-propagation algorithm* (FBP algorithm) of linearized inverse scattering. We can decompose this algorithm into the following three steps:

- filtering of the scattering amplitude $f_B(\mathbf{s}, \mathbf{s}_0)$ implemented by the multiplication

$$\overline{f_B}(\mathbf{s}, \mathbf{s}_0) = \frac{ik_0^2}{2\pi^2}|\mathbf{s} - \mathbf{s}_0| f_B(\mathbf{s}, \mathbf{s}_0),$$

- back propagation of the filtered scattering amplitude at fixed incident-wave direction followed by demodulation to yield the *partial reconstructions*

$$V_{LP}(\mathbf{r}; \mathbf{s}_0) = \overbrace{e^{-ik_0\mathbf{s}_0\cdot\mathbf{r}}}^{\text{demodulation}} \overbrace{\frac{ik_0}{2\pi}\int d\Omega_s \overline{f_B}(\mathbf{s}, \mathbf{s}_0) e^{ik_0\mathbf{s}\cdot\mathbf{r}}}^{\text{back propagation}},$$

- summation of the partial reconstructions over all incident-wave directions $\mathbf{s}_0$ to yield the low-pass-filtered version of the scattering potential

$$V_{LP}(\mathbf{r}) = \int d\Omega_0\, V_{LP}(\mathbf{r}; \mathbf{s}_0).$$

We should note that the FBP algorithm employed with ideal (Born) data will generate the theoretical ideal reconstruction

$$V_{LP}(\mathbf{r}) = \frac{1}{(2\pi)^3}\int_{K\le 2k_0} d^3K\, \tilde{V}(\mathbf{K}) e^{i\mathbf{K}\cdot\mathbf{r}},$$

which is $V(\mathbf{r})$ band-limited to the Ewald limiting sphere. If the transform $\tilde{V}(\mathbf{K})$ extends outside this sphere the reconstruction can have pronounced ripples (the Gibbs phenomenon) introduced by the hard band-limiting caused by the abrupt cutoff of the transform at $K = 2k_0$. To minimize these ripples a smoothing filter such as a Blackman, Hamming or wavelet filter can be employed in the FBP algorithm. This then introduces one more filter in the filtering process given above.

8.1.3 Inverse scattering identity

The inversion formula Eq. (8.9) is actually a mathematical identity between the low-pass-filtered version $V_{LP}(\mathbf{r})$ of *any* function $V(\mathbf{r})$ and its spatial Fourier transform defined over the set of Ewald spheres. Indeed, if we replace $f_B(\mathbf{s}, \mathbf{s}_0)$ in Eq. (8.4a) by $-\tilde{V}[k_0(\mathbf{s}-\mathbf{s}_0)]/(4\pi)$ we find that the inversion formula Eq. (8.9) yields the result

$$V_{LP}(\mathbf{r}) = \frac{k_0^3}{(2\pi)^4}\int d\Omega_{s_0}\int d\Omega_s |\mathbf{s} - \mathbf{s}_0| \tilde{V}[k_0(\mathbf{s} - \mathbf{s}_0)] e^{ik_0(\mathbf{s}-\mathbf{s}_0)\cdot\mathbf{r}}, \tag{8.10a}$$

where $\tilde{V}(\mathbf{K})$ is the spatial Fourier transform of an arbitrary function $V(\mathbf{r})$ and V_{LP} is its low-pass-filtered version band-limited to the Ewald limiting sphere $K \le 2k_0$. We will refer to Eq. (8.10a) as the *inverse scattering identity*.

On setting $\mathbf{r} = 0$ in Eq. (8.10a) and using the definition Eq. (8.3) of a low-pass-filtered function we obtain the result

$$\int_{K\leq 2k_0} d^3K\,\tilde{V}(\mathbf{K}) = \frac{k_0^3}{2\pi}\int d\Omega_{s_0}\int d\Omega_s |\mathbf{s}-\mathbf{s}_0|\tilde{V}[k_0(\mathbf{s}-\mathbf{s}_0)]. \tag{8.10b}$$

Equation (8.10b) is an identity that relates the integral of the transform of an arbitrary function $\tilde{V}(\mathbf{K})$ over the Ewald limiting sphere to the integral of this transform expressed as a function of the difference vector $k_0(\mathbf{s}-\mathbf{s}_0)$ and weighted by $k_0^3|\mathbf{s}-\mathbf{s}_0|/(2\pi)$. This identity will be used later to evaluate integrals of the form appearing on the r.h.s. of Eq. (8.10b).

It is apparent that the inverse scattering identity Eq. (8.10a) and the identity Eq. (8.10b) result from the change of integration variables

$$\mathbf{K} = k_0(\mathbf{s}-\mathbf{s}_0).$$

Unfortunately, in three space dimensions this encompasses a change from three variables (the Cartesian components of $\mathbf{K}$) to four variables (the polar and azimuthal angles of the unit vectors $\mathbf{s}$ and $\mathbf{s}_0$) and there does not appear to be any straightforward method for performing this via simple change-of-variable techniques (see, however, Burridge and Beylkin (1988)). However, the situation is quite different in two space dimensions, as the following example illustrates.

Example 8.1 The inversion formula Eq. (8.9) and, in particular, the inverse scattering identity Eq. (8.10a) can be derived in the 2D case by a simple change of variables. In particular, in two space dimensions we have that

$$V_{\mathrm{LP}}(\boldsymbol{\rho}) = \frac{1}{(2\pi)^2}\int_{K\leq 2k_0} d^2K\,\tilde{V}(\mathbf{K})e^{i\mathbf{K}\cdot\boldsymbol{\rho}}, \tag{8.11}$$

where $\boldsymbol{\rho} = x\hat{\mathbf{x}} + y\hat{\mathbf{y}}$ and $\mathbf{K} = K_x\hat{\mathbf{x}} + K_y\hat{\mathbf{y}}$ are vectors on the (x, y) plane. We now make the change of integration variable

$$\begin{aligned}\mathbf{K} &= k_0(\mathbf{s}-\mathbf{s}_0)\\ &= \overbrace{k_0(\cos\alpha - \cos\alpha_0)}^{K_x}\,\hat{\mathbf{x}} + \overbrace{k_0(\sin\alpha - \sin\alpha_0)}^{K_y}\,\hat{\mathbf{y}},\end{aligned} \tag{8.12}$$

where α and α_0 are, respectively, the polar angles of the unit vectors $\mathbf{s}$ and $\mathbf{s}_0$. The Jacobian of the above transformation is found to be

$$\begin{aligned}J &= \frac{\partial(K_x, K_y)}{\partial(\alpha, \alpha_0)}\\ &= \begin{vmatrix} -k_0\sin\alpha & k_0\sin\alpha_0 \\ k_0\cos\alpha & -k_0\cos\alpha_0 \end{vmatrix}\\ &= k_0^2|\sin\alpha\cos\alpha_0 - \cos\alpha\sin\alpha_0|\\ &= k_0^2\sin(\alpha-\alpha_0) = k_0^2\sqrt{1-\cos^2(\alpha-\alpha_0)}\\ &= k_0^2\sqrt{1-(\mathbf{s}\cdot\mathbf{s}_0)^2}.\end{aligned}$$

On making the change of variables Eq. (8.12) in Eq. (8.11) we then find that

$$V_{\mathrm{LP}}(\boldsymbol{\rho}) = \frac{k_0^2}{2(2\pi)^2}\int_{-\pi}^{\pi} d\alpha_0\int_{-\pi}^{\pi} d\alpha\sqrt{1-(\mathbf{s}\cdot\mathbf{s}_0)^2}\,\tilde{V}[k_0(\mathbf{s}-\mathbf{s}_0)]e^{ik_0(\mathbf{s}-\mathbf{s}_0)\cdot\boldsymbol{\rho}}, \tag{8.13}$$

which is the inverse scattering identity in two space dimensions. Note that the extra factor of $1/2$ appears due to double coverage in the integration on going from Eq. (8.11) to Eq. (8.13).

8.1.4 The FBP algorithm in two space dimensions

We showed in Section 6.4 that in two space dimensions (2D) the scattering amplitude is related to the scattering potential via the equation

$$f(\mathbf{s}, \mathbf{s}_0) = -\sqrt{\frac{1}{8\pi k_0}} e^{i\frac{\pi}{4}} \int d^2\rho\, V(\boldsymbol{\rho}) U(\boldsymbol{\rho}; \mathbf{s}_0) e^{-ik_0 \mathbf{s}\cdot\boldsymbol{\rho}}. \tag{8.14}$$

Within the Born approximation this then yields the result

$$\tilde{V}[k_0(\mathbf{s} - \mathbf{s}_0)] = -\sqrt{8\pi k_0} e^{-i\frac{\pi}{4}} f_B(\mathbf{s}, \mathbf{s}_0), \tag{8.15}$$

which, when employed in the 2D inverse scattering identity derived in the above example, yields the 2D FBP algorithm

$$V_{LP}(\boldsymbol{\rho}) = \sqrt{\frac{k_0^5}{(2\pi)^3}} e^{i\frac{3\pi}{4}} \int_{-\pi}^{\pi} d\alpha_0 \int_{-\pi}^{\pi} d\alpha \sqrt{1 - (\mathbf{s}\cdot\mathbf{s}_0)^2} f_B(\mathbf{s}, \mathbf{s}_0) e^{ik_0(\mathbf{s}-\mathbf{s}_0)\cdot\boldsymbol{\rho}}. \tag{8.16}$$

As mentioned above, it is advisable to employ a smoothing filter in addition to the filter function $\sqrt{1 - (\mathbf{s}\cdot\mathbf{s}_0)^2}$ in the above algorithm.

Circularly symmetric scatterers

An especially interesting case is that of a circularly symmetric scatterer such as a cylinder when the incident $\mathbf{s}_0$ and scattered $\mathbf{s}$ wave directions cover the unit circle. For this scenario the scattering amplitude $f(\mathbf{s}, \mathbf{s}_0) = f(\alpha)$, where the angle α is the polar angle of $\mathbf{s}$ measured relative to the direction of the incident wave vector $\mathbf{s}_0$ and varies from $-\pi$ to $+\pi$. The 2D FBP algorithm Eq. (8.16) then becomes

$$V_{LP}(\boldsymbol{\rho}) = \sqrt{\frac{k_0^5}{(2\pi)^3}} e^{i\frac{3\pi}{4}} \int_{-\pi}^{\pi} d\alpha_0 \int_{-\pi}^{\pi} d\alpha |\sin\alpha| f_B(\alpha) e^{ik_0(\mathbf{s}-\mathbf{s}_0)\cdot\boldsymbol{\rho}}. \tag{8.17}$$

Since the integral over α is fixed and independent of α_0 we can perform the α_0 integration to obtain

$$\int_{-\pi}^{\pi} d\alpha_0\, e^{ik_0(\mathbf{s}-\mathbf{s}_0)\cdot\boldsymbol{\rho}} = \int_{-\pi}^{\pi} d\alpha_0\, e^{i\sqrt{2}k_0\rho\sqrt{1-\cos\alpha}\cos\alpha_0} = 2\pi J_0(\sqrt{2}k_0\rho\sqrt{1-\cos\alpha}),$$

where we have used the result that

$$k_0(\mathbf{s} - \mathbf{s}_0)\cdot\boldsymbol{\rho} = k_0|\mathbf{s} - \mathbf{s}_0|\rho\cos\chi = \sqrt{2}k_0\rho\sqrt{1 - \cos\alpha}\cos\chi,$$

where χ is the angle between the vectors $(\mathbf{s} - \mathbf{s}_0)$ and $\boldsymbol{\rho}$, and have then replaced χ by α_0 since the integrand in the above integral is periodic in χ. Equation (8.17) then becomes

$$V_{\rm LP}(\rho) = \sqrt{\frac{k_0^5}{2\pi}} e^{i\frac{3\pi}{4}} \int_{-\pi}^{\pi} d\alpha |\sin\alpha| f_{\rm B}(\alpha) J_0(\sqrt{2}k_0\rho\sqrt{1-\cos\alpha}). \tag{8.18}$$

Equation (8.18) is easily implemented in code and is ideally suited for computer simulations testing the performance of the FBP algorithm on exact and idealized (Born) data from circularly symmetric scatterers. We performed such simulations using exact and Born scattering amplitudes computed as described in Section 6.8 of Chapter 6 for single cylinders and pairs of concentric cylinders. In that section we compared the Born and exact scattering amplitudes for cylinders having indices of refraction varying from $n_{\rm r} = 1.01$ to $n_{\rm r} = 1.07$ corresponding to scattering potentials $V = k_0^2(1 - n_{\rm r}^2)$ varying from -0.8 to -5.7. These are rather strong scatterers, which were found to result in large errors between the Born and exact scattering amplitudes and, consequently, would result in large errors in the reconstructions generated using the FBP algorithm with exact scattered-field data. Consequently we employed lower index values in the reconstructions we show below, with $n_{\rm r}$ varying between $n_{\rm r} = 1.005$ and $n_{\rm r} = 1.02$ in steps of $\delta n_{\rm r} = 0.005$. This range then corresponds to scattering potentials varying from -0.4 to -1.6.

The results of the simulations for a single penetrable cylinder having a radius of $a_0 = 4\lambda$ and various index values are shown in Fig. 8.1, where the Born data employed in the FBP algorithm generated the theoretical optimum reconstruction defined in Eq. (8.11), while the exact data generated least-squares inversions (see the following section). Both the ideal and the approximate solutions exhibit pronounced Gibbs effects as described earlier, so we

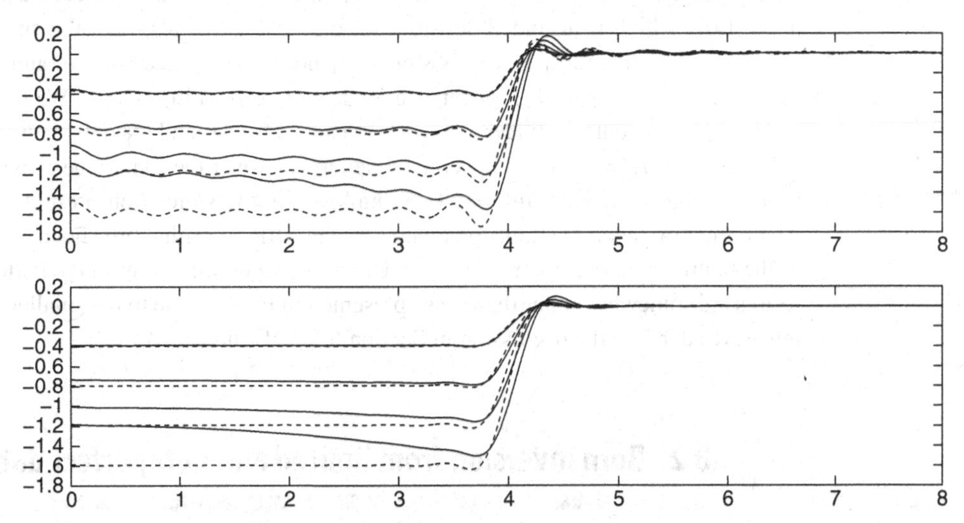

Fig. 8.1 Comparison of reconstructions obtained using the FBP algorithm on exact (solid) and Born (dashed) data for a single cylinder of radius $a_0 = 4\lambda$ and index values ranging from $n_{\rm r} = 1.005$ corresponding to $V = -0.4$ (best reconstruction) to $n_{\rm r} = 1.02$ corresponding to $V = -1.6$ (worst reconstruction) in steps of $\delta n_{\rm r} = 0.005$. The reconstructions in the bottom part included the use of a standard Hamming window in the filtering operation.

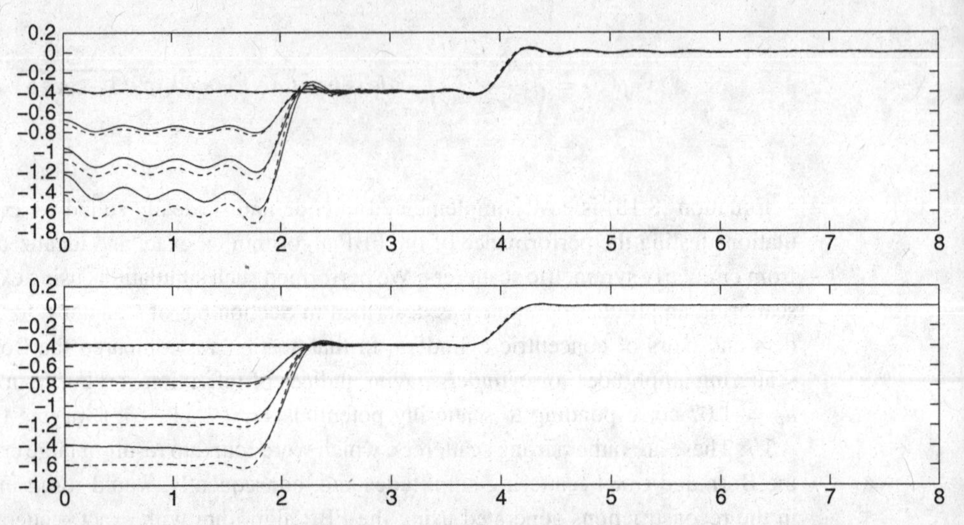

Fig. 8.2 Comparison of reconstructions obtained using the FBP algorithm on exact (solid) and Born (dashed) data for a pair of concentric cylinders. The outer cylinder had a fixed index of $n_r = 1.005$ and radius $a_2 = 4\lambda$ and the inner cylinder had a radius of $a_1 = 2\lambda$ and index values ranging from $n_r = 1.005$ (best reconstruction) to $n_r = 1.02$ (worst reconstruction) in steps of $\delta n_r = 0.005$. The reconstructions in the bottom part included the use of a standard Hamming window in the filtering operation.

also generated inversions that employed a Hamming filter to smooth the ripples introduced by the hard band-limiting. It can be seen from Fig. 8.1 that the FBP algorithm with exact data degrades monotonically with increasing n_r and would be considered unreliable after $n_r = 1.02$, which is an index deviation of only 2%. Note also that the smoothing filter reduces the rippling due to the Gibbs phenomenon by an appreciable amount.

The results for a pair of concentric cylinders are shown in Fig. 8.2. We fixed the index and radius of the outer cylinder at $n_r = 1.005$ and $a_2 = 4\lambda$, and fixed the radius of the inner cylinder to be $a_1 = 2\lambda$ and allowed its index to vary over the four values from $n_r = 1.005$ to $n_r = 1.02$ in steps of $\delta n_r = 0.005$ employed in the simulation presented in Fig. 8.1. Again the raw reconstructions generated by the FBP algorithm are shown in the top part of the figure and those obtained using a Hamming smoothing filter in the bottom part. The results are somewhat better than those presented in Fig. 8.1 due to the smaller radius of the inner cylinder (see the discussion in Section 6.7.2 of Chapter 6).

8.2 Born inversion from limited and non-perfect data

The FBP algorithm that we have just derived is a solution to the integral equation

$$f(\mathbf{s}, \mathbf{s}_0) = -\frac{1}{4\pi}\int d^3r\, V(\mathbf{r})e^{-ik_0(\mathbf{s}-\mathbf{s}_0)\cdot\mathbf{r}} \tag{8.19}$$

in the special case in which the scattering amplitude is *exactly* modeled by the Born approximation and, hence, is a function of the difference vector $\mathbf{s} - \mathbf{s}_0$ and f is specified for all unit vectors $\mathbf{s}_0$ and $\mathbf{s}$ on the unit sphere Ω. We now consider the general case in which the observed scattering amplitude $f(\mathbf{s}, \mathbf{s}_0)$ need not be a function of the difference vector $\mathbf{s} - \mathbf{s}_0$ and is specified only over limited sets of incident and scattered field directions $\mathbf{s}_0 \in \Omega_{s_0}$, $\mathbf{s} \in \Omega_s$. We first consider the case in which the scattering amplitude is not exactly modeled by the Born approximation but where $\Omega_{s_0} = \Omega_s = \Omega$ and then treat the limited-data case.

8.2.1 Non-perfect data

When the scattering amplitude is not exactly modeled by the Born approximation the integral equation Eq. (8.19) will not possess a solution $V_{\mathrm{LP}}(\mathbf{r})$ due to the fact that the scattering amplitude is not in the range of the integral transform defined by the r.h.s. of this equation. Although we cannot exactly solve this equation we can obtain a *least-squares solution* $\hat{V}$ that minimizes the integral squared error between the observed scattering amplitude f and the Born scattering amplitude f_{B} defined by Eq. (8.19) with $V = \hat{V}$. The least-squares solution is thus defined to be that approximate solution $\hat{V}$ that minimizes the squared error

$$\mathcal{L} = \int d\Omega_{s_0} \int d\Omega_s \left| f(\mathbf{s}, \mathbf{s}_0) - \overbrace{\frac{1}{4\pi} \int d^3r' \, \hat{V}(\mathbf{r}') e^{-ik_0(\mathbf{s}-\mathbf{s}_0)\cdot\mathbf{r}'}}^{f_{\mathrm{B}}(\mathbf{s},\mathbf{s}_0)} \right|^2 \tag{8.20}$$

The minimization of Eq. (8.20) is performed in the usual way by requiring that the first variation with respect to $\hat{V}$ vanish. The first variation yields the result

$$\begin{aligned}
\delta\mathcal{L} &= \frac{1}{2\pi} \int d\Omega_{s_0} \int d\Omega_s \left\{ \left[f(\mathbf{s}, \mathbf{s}_0) + \frac{1}{4\pi} \int d^3r' \, \hat{V}(\mathbf{r}') e^{-ik_0(\mathbf{s}-\mathbf{s}_0)\cdot\mathbf{r}'} \right] \right. \\
&\qquad \left. \times \int d^3r \, \delta\hat{V}^*(\mathbf{r}) e^{ik_0(\mathbf{s}-\mathbf{s}_0)\cdot\mathbf{r}} \right\} + \text{c.c.} \\
&= \frac{1}{2\pi} \int d^3r \, \delta\hat{V}^*(\mathbf{r}) \int d\Omega_{s_0} \\
&\quad \times \int d\Omega_s \left[f(\mathbf{s}, \mathbf{s}_0) e^{ik_0(\mathbf{s}-\mathbf{s}_0)\cdot\mathbf{r}} + \frac{1}{4\pi} \int d^3r' \, \hat{V}(\mathbf{r}') e^{ik_0(\mathbf{s}-\mathbf{s}_0)\cdot(\mathbf{r}-\mathbf{r}')} \right] + \text{c.c.},
\end{aligned}$$

where c.c. stands for the complex conjugate of the first term on the r.h.s of the equation. On setting the variation equal to zero and performing some minor manipulations we obtain the result

$$-\frac{1}{4\pi} \int d\Omega_{s_0} \int d\Omega_s f(\mathbf{s}, \mathbf{s}_0) e^{ik_0(\mathbf{s}-\mathbf{s}_0)\cdot\mathbf{r}} = \int d^3r' \, \hat{V}(\mathbf{r}') H(\mathbf{r} - \mathbf{r}'), \tag{8.21}$$

where H is the kernel defined in Eq. (8.4c). Equation (8.21) is recognized as the ISCP integral equation Eq. (8.4b) which has its solution in the form of the filtered back-propagation algorithm Eq. (8.9).

We conclude from the above that the filtered back-propagation algorithm can be applied to non-perfect data, in which case it generates the least-squares solution $\hat{V}$. The least-squares solution has the property that it minimizes the integral squared error defined in

Eq. (8.20) between the measured scattered amplitude and a Born scattered amplitude that is related to the scattering potential $\hat{V}$ via the integral equation Eq. (8.19). We will find in Section 8.4 that this property follows automatically from the properties of the SVD presented in Section 5.4 of Chapter 5.

8.2.2 Limited-data case I

The above treatment can also be employed to determine the least-squares solution to Eq. (8.20) for the case in which the scattering amplitude is specified over limited (but dense) sets of incident and scattered field directions $\mathbf{s}_0 \in \Omega_{\mathbf{s}_0}$ and $\mathbf{s} \in \Omega_{\mathbf{s}}$. In this case the integrals in Eq. (8.20) extend only over the regions $\Omega_{\mathbf{s}_0}$ and $\Omega_{\mathbf{s}}$. However, other than this change, the above treatment goes through without modification, with the result that the least-squares solution again satisfies the integral equation Eq. (8.21) with the integrals over the incident and scattered field directions limited to $\Omega_{\mathbf{s}_0}$ and $\Omega_{\mathbf{s}}$ and where the filter function is now given by

$$H(\mathbf{R}) = \frac{1}{(4\pi)^2} \int_{\Omega_{\mathbf{s}_0}} d\Omega_{s_0} \int_{\Omega_{\mathbf{s}}} d\Omega_s \, e^{ik_0(\mathbf{s}-\mathbf{s}_0)\cdot\mathbf{R}}. \tag{8.22}$$

Equation (8.21) with H given by Eq. (8.22) and $\mathbf{s}_0$ and $\mathbf{s}$ restricted to the solid angles $\Omega_{\mathbf{s}_0}$ and $\Omega_{\mathbf{s}}$ can still be inverted, and one finds that it is again given by the FBP algorithm with the integrals restricted to $\Omega_{\mathbf{s}_0}$ and $\Omega_{\mathbf{s}}$.

8.2.3 Limited-data case II

A more realistic limited-data case occurs when the scattering amplitude is specified over limited discrete sets of incident and scattered field directions $\mathbf{s}_{0n}$, $n = 1, 2, \ldots, N$, and $\mathbf{s}_m$, $m = 1, 2, \ldots, M$. We will address this problem using iterative algorithms in Section 8.6.1 and also in Section 9.10 of Chapter 9, where we also incorporate a support constraint into the formulation of the problem. For now we outline a formulation along the lines of the FBP algorithm.

A natural approximation to the solution to the ISCP in the limited-data case would be to simply replace the FBP algorithm Eq. (8.9) by its discrete-sum approximation:

$$V_{\mathrm{LP}}(\mathbf{r}) \approx -\frac{k_0^3}{4\pi^3} \sum_{n=1}^{N} \delta\Omega_{s_{0n}} \sum_{m=1}^{M} \delta\Omega_{s_m} |\mathbf{s}_m - \mathbf{s}_{0n}| f_{\mathrm{B}}(\mathbf{s}_m, \mathbf{s}_{0n}) e^{ik_0(\mathbf{s}_m-\mathbf{s}_{0n})\cdot\mathbf{r}}. \tag{8.23}$$

This approximation is adequate if the sample spacings $\delta\Omega_{s_{0n}}$ and $\delta\Omega_{s_m}$ are sufficiently small and if the incident and scattered wave directions span the entire unit sphere. A more systematic approach to this limited-data case is to compute the least-squares solution to the problem; i.e., determine the scattering potential that minimizes the summed squared error between the measured scattered amplitude and the Born model for this scattering amplitude.

In this limited-data case the integrated squared error defined in Eq. (8.20) for data specified over dense sets of incident and scattered field directions is replaced by

$$\mathcal{L} = \sum_{n=1}^{N}\sum_{m=1}^{M} |f(\mathbf{s}_m, \mathbf{s}_{0n}) + \frac{1}{4\pi}\int d^3r'\ \hat{V}(\mathbf{r}')e^{-ik_0(\mathbf{s}_m-\mathbf{s}_{0n})\cdot\mathbf{r}'}|^2.$$

By again following steps almost identical to those employed in the complete-data case we find that the least-squares solution $\hat{V}$ must satisfy the integral equation

$$-\frac{1}{4\pi}\sum_{n=1}^{N}\sum_{m=1}^{M} f(\mathbf{s}_m, \mathbf{s}_{0n})e^{ik_0(\mathbf{s}_m-\mathbf{s}_{0n})\cdot\mathbf{r}} = \int d^3r'\ \hat{V}(\mathbf{r}')H(\mathbf{r}-\mathbf{r}'),$$

where H is the filter

$$H(\mathbf{R}) = \frac{1}{(4\pi)^2}\sum_{n=1}^{N}\sum_{m=1}^{M} e^{ik_0(\mathbf{s}_m-\mathbf{s}_{0n})\cdot\mathbf{R}}.$$

It should be clear that this solution will converge to the FBP algorithm in the limit where the two sets $\mathbf{s}_m \to \mathbf{s} \in \Omega$ and $\mathbf{s}_{0n} \to \mathbf{s}_0 \in \Omega$.

8.3 Non-uniqueness and non-scattering scatterers

In our derivation of the inverse scattering identity and of the FBP algorithm we assumed that the scattering amplitude $f(\mathbf{s}, \mathbf{s}_0)$ is specified over a continuum of incident $\mathbf{s}_0$ and scattering $\mathbf{s}$ directions completely covering the entire unit sphere Ω and was exactly modeled by the Born scattering model. In this case the scattering potential is uniquely determined up to a high-pass function

$$F_{\mathrm{HP}}(\mathbf{r}) = \frac{1}{(2\pi)^3}\int_{K>2k_0} d^3K\ \tilde{V}(\mathbf{K})e^{i\mathbf{K}\cdot\mathbf{r}},$$

where $\tilde{V}(\mathbf{K})$ is an arbitrary square-integrable function of the spatial frequency vector $\mathbf{K}$. Thus, in such cases the ISCP is non-unique only up to components of the scattering potential that are band-limited to spatial frequencies that lie outside the Ewald limiting sphere. Such components limit the spatial resolution of Born inversion but do not pose a serious problem in most applications where the wavelength λ is small compared with the sought-after resolution of the inversion.

A more serious issue arises in all practical applications in which the data are specified only over a finite set of incident and/or scattered wave directions. The reason for this is that *within the linearized Born scattering model* there exists a countably infinite number of scattering potentials $V_{\mathrm{ns}}^{\mathrm{B}}(\mathbf{r})$ whose scattering amplitude $f(\mathbf{s}, \mathbf{s}_0)$ vanishes identically over any finite sets Ω_{s_0} and/or Ω_s of incident and/or scattering wave directions. These scattering potentials, which play in the ISCP the role which the non-radiating sources Q_{nr} play in the ISP, are called *non-scattering potentials* and increase immensely the non-uniqueness of the linearized ISCP. Non-scattering potentials were covered in some detail in Section 6.9 of Chapter 6 and here we present an overview of the main results obtained in that section.

8.3.1 Non-scattering potentials within the Born approximation

We showed in Section 6.9 that within the Born approximation any scattering potential of the general form

$$V_{\rm ns}(\mathbf{r};\mathbf{s}_0) = (\nabla^2 + 2ik_0\mathbf{s}_0 \cdot \nabla)\chi(\mathbf{r}),$$

where $\chi(\mathbf{r})$ is compactly supported within the scattering volume τ_0 and continuously differentiable but otherwise arbitrary, will generate a zero scattered field outside τ_0 from an incident plane wave with unit propagation vector $\mathbf{s}_0$. Such a scattering potential will thus lie in the null space of the integral transform Eq. (8.19) for that *particular value* of $\mathbf{s}_0$. More generally, we showed that a scattering potential of the general form

$$V_{\rm ns}^{\rm B}(\mathbf{r}) = \prod_{m=1}^{M}(\nabla^2 + 2ik_0\mathbf{s}_{0m} \cdot \nabla)\chi(\mathbf{r}), \tag{8.24}$$

where $\chi(\mathbf{r})$ is assumed to be continuously differentiable up to order $2M-1$ and compactly supported within a scattering volume τ_0, will not scatter in any of the M scattering experiments employing incident plane waves having the M unit propagation vectors $\mathbf{s}_{0m}$, $m = 1, 2, \ldots, M$. This scattering potential thus lies in the null space of Eq. (8.19) for this entire set of incident plane-wave directions and any scattering direction $\mathbf{s}$.

Consider now a scattering potential of the general form

$$V(\mathbf{r}) = V_0(\mathbf{r}) + V_{\rm ns}^{\rm B}(\mathbf{r}),$$

where V_0 is an arbitrary potential supported within τ_0 and $V_{\rm ns}^{\rm B}$ is a Born non-scattering potential also supported within τ_0 and of the general form given in Eq. (8.24). Because the Born approximation is a linear mapping between the scattering potential and the scattered field, it follows that V and V_0 will generate identical scattered fields outside τ_0 over the set of M plane-wave scattering experiments employing incident plane waves having the M unit propagation vectors $\mathbf{s}_{0m}$, $m = 1, 2, \ldots, M$. The ISCP will thus not have a unique solution for this data set.

The above argument shows that the ISCP does not possess a unique solution from scattered field data collected in a finite set of scattering experiments employing incident plane waves. Moreover, we also showed in Section 6.9 of Chapter 6 that the potential defined in Eq. (8.24) with $\mathbf{s}_{0m}$ replaced by $\mathbf{s}_m$ will generate a zero scattering amplitude $f(\mathbf{s}_m, \mathbf{s}_0)$ over the set of M scattering directions $\mathbf{s}_m$, $m = 1, 2, \ldots, M$, for any incident plane wave having an arbitrary unit propagation vector $\mathbf{s}_0$. We also showed that this scattering potential generated a zero generalized scattering amplitude $f(\mathbf{s}_m; \nu)$ over the same set of scattering directions for arbitrary incident waves. We thus conclude that the ISCP will also not have a unique solution for scattered-field data equal to the generalized scattering amplitudes measured over any finite set of scattering directions in any finite set of scattering experiments employing arbitrary incident waves.

8.4 Hilbert-space formulation of Born inverse scattering

We reviewed the basics of Hilbert-space theory in Chapter 5, where we employed this theory to solve the inverse source problem (ISP). Here, we employ the same theory to solve the ISCP within the Born approximation. By analogy with our treatment of the ISP in Chapter 5 we define the Hilbert space $\mathcal{H}_V$ of complex-valued scattering potentials $V(\mathbf{r})$ supported in a (possibly infinite)[3] scattering volume τ_0. This space has the standard inner product and norm

$$\langle V_1, V_2\rangle_{\mathcal{H}_V} = \int d^3r\, V_1^*(\mathbf{r})V_2(\mathbf{r}), \tag{8.25a}$$

$$||V|| = \sqrt{\langle V, V\rangle_{\mathcal{H}_V}} = \sqrt{\int d^3r |V(\mathbf{r})|^2}, \tag{8.25b}$$

where the integrals are over all of space. We emphasize that, unlike for the Hilbert space $L^2(\tau_0)$ of sources employed in Chapter 5, the scattering potentials contained in $\mathcal{H}_V$ need not be compactly supported. However, we will require that all scattering potentials within the space $\mathcal{H}_V$ possess finite norm $||V||$.

In addition to the Hilbert space $\mathcal{H}_V$ of scattering potentials we also define a Hilbert space $\mathcal{H}_f$ of scattering amplitudes $f(\mathbf{s}, \mathbf{s}_0)$. The Hilbert space $\mathcal{H}_f$ has the standard inner product and norm

$$\langle f_1, f_2\rangle_{\mathcal{H}_f} = \int d\Omega_{s_0} \int d\Omega_s f_1^*(\mathbf{s}, \mathbf{s}_0) f_2(\mathbf{s}, \mathbf{s}_0), \tag{8.26a}$$

$$||f|| = \sqrt{\langle f, f\rangle_{\mathcal{H}_f}} = \sqrt{\int d\Omega_{s_0} \int d\Omega_s |f(\mathbf{s}, \mathbf{s}_0)|^2}, \tag{8.26b}$$

where the integrals are over the entire unit sphere Ω.

The Hilbert-space setting provided above is ideally suited for solving the ISCP within the Born approximation. As we have shown earlier, this problem consists of inverting the integral equation Eq. (8.19)

$$f(\mathbf{s}, \mathbf{s}_0) = -\frac{1}{4\pi} \int d^3r\, V(\mathbf{r}) e^{-ik_0(\mathbf{s}-\mathbf{s}_0)\cdot\mathbf{r}} \tag{8.27a}$$

for the scattering potential V in terms of the scattering amplitude f. The integral equation Eq. (8.27a) is a *linear mapping* from the Hilbert space of scattering potentials $\mathcal{H}_V$ into the Hilbert space $\mathcal{H}_f$. We will write Eq. (8.27a) in the symbolic form

$$\hat{T}V = f, \tag{8.27b}$$

[3] Throughout this section we will not constrain the solutions to the ISCP to have compact support. Such constraints can be easily incorporated into the iterative algorithms treated in Section 8.6.1 and will be incorporated into our more general formulation of the ISCP in an inhomogeneous background in Chapter 9.

where $\hat{T}: \mathcal{H}_V \to \mathcal{H}_f$ is the linear operator

$$\hat{T} = -\frac{1}{4\pi}\int d^3r\, e^{-ik_0(\mathbf{s}-\mathbf{s}_0)\cdot\mathbf{r}}. \tag{8.27c}$$

Comparison with the ISP

The Hilbert-space formulation of the ISP with far-field data that was employed in Chapter 5 was based on the integral transform

$$f(\mathbf{s}) = -\frac{1}{4\pi}\int_{\tau_0} d^3r\, Q(\mathbf{r})e^{-ik_0\mathbf{s}\cdot\mathbf{r}}, \tag{8.28}$$

where we have suppressed the frequency variable ω in the arguments of the radiation pattern $f(\mathbf{s})$ and source $Q(\mathbf{r})$, and the source volume τ_0 is assumed to be finite. On comparison with Eq. (8.27a) we see that the ISCP and ISP have very similar mathematical structures and, indeed, become identical if just a single scattering experiment is employed in the ISCP and if the constraint that the source be compactly supported in τ_0 is dropped. In the latter case the scattered field is equal to the field radiated by the *induced source*

$$Q(\mathbf{r}, \mathbf{s}_0) = V(\mathbf{r})e^{ik\mathbf{s}_0\cdot\mathbf{r}}$$

and the radiation pattern $f(\mathbf{s})$ of this induced source is the scattering amplitude $f(\mathbf{s}, \mathbf{s}_0)$.

8.4.1 Adjoint and composite operators

We define the *adjoint operator* (cf. Section 5.4.1) in the usual way:

$$\langle f, \hat{T}V\rangle_{\mathcal{H}_f} = \langle \hat{T}^\dagger f, V\rangle_{\mathcal{H}_V}.$$

Using the definition Eq. (8.27c) we find that

$$\begin{aligned}\langle f, \hat{T}V\rangle_{\mathcal{H}_f} &= \int d\Omega_{s_0}\int d\Omega_s f^*(\mathbf{s}, \mathbf{s}_0)\overbrace{-\frac{1}{4\pi}\int d^3r\, V(\mathbf{r})e^{-ik_0(\mathbf{s}-\mathbf{s}_0)\cdot\mathbf{r}}}^{\hat{T}V}\\ &= \int d^3r\left\{-\frac{1}{4\pi}\int d\Omega_{s_0}\int d\Omega_s f(\mathbf{s}, \mathbf{s}_0)e^{ik_0(\mathbf{s}-\mathbf{s}_0)\cdot\mathbf{r}}\right\}^* V(\mathbf{r}),\end{aligned}$$

from which we conclude that

$$\hat{T}^\dagger = -\frac{1}{4\pi}\int d\Omega_{s_0}\int d\Omega_s\, e^{ik_0(\mathbf{s}-\mathbf{s}_0)\cdot\mathbf{r}}, \tag{8.29}$$

where, again, we have assumed that the background medium is lossless so that the wavenumber k_0 is real-valued.

We will also have need of the composite operators $\hat{T}^\dagger\hat{T} : \mathcal{H}_V \to \mathcal{H}_V$ and $\hat{T}\hat{T}^\dagger : \mathcal{H}_f \to \mathcal{H}_f$. These composite operators can be constructed directly from the definitions of $\hat{T}$ and $\hat{T}^\dagger$ given in Eqs. (8.27c) and (8.29). First considering $\hat{T}^\dagger\hat{T}$ we find that

$$\hat{T}^\dagger\hat{T}V = \int d^3r'\, H(\mathbf{r}-\mathbf{r}')V(\mathbf{r}'), \tag{8.30}$$

where

$$H(\mathbf{r}-\mathbf{r}') = j_0^2(k_0|\mathbf{r}-\mathbf{r}'|)$$

is the kernel defined in Eq. (8.5).

Consider now the composite operator $\hat{T}\hat{T}^\dagger : \mathcal{H}_{\mathcal{H}_f} \to \mathcal{H}_{\mathcal{H}_f}$. We have that

$$\hat{T}\hat{T}^\dagger f = \frac{\pi}{2}\int d\Omega_{s_0'}\int d\Omega_{s'}\,\delta[k_0(\mathbf{s}-\mathbf{s}_0) - k_0(\mathbf{s}'-\mathbf{s}_0')]f(\mathbf{s}',\mathbf{s}_0'), \tag{8.31}$$

where $\delta(\cdot)$ is the 3D Dirac delta function.

We can obtain a closed-form expression for action of the composite operator $\hat{T}\hat{T}^\dagger$ on scattering amplitudes $f_B \in \mathcal{H}_f$ within the Born approximation by making use of the identity Eq. (8.10b). In particular, we find that

$$\hat{T}\hat{T}^\dagger f_B = -\frac{\pi^2}{k_0^3}\frac{f_B(\mathbf{s},\mathbf{s}_0)}{|\mathbf{s}-\mathbf{s}_0|}. \tag{8.32}$$

8.4.2 Singular value decomposition

The singular value decomposition (SVD) was reviewed in Section 5.4 of Chapter 5 and employed in later sections of that chapter to solve the ISP. For the ISCP formulated within the Born approximation the SVD takes the form

$$\hat{T}v_p = \sigma_p u_p(\mathbf{s},\mathbf{s}_0), \qquad \hat{T}^\dagger u_p = \sigma_p v_p(\mathbf{r}), \tag{8.33}$$

where $\hat{T}$ is defined in Eqs. (8.27), $\sigma_p \geq 0$ are the singular values and $v_p \in \mathcal{H}_{\mathcal{H}_V}$ and $u_p \in \mathcal{H}_{\mathcal{H}_f}$ are the singular functions. As in the ISP, we employ an integer p to index the singular system. However, this is done merely for ease and simplicity of notation and, indeed, we will find in the Born scattering model that the index is, in fact, a continuous variable in R^3.

Normal equations

The normal equations satisfied by the singular functions v_p and u_p are found to be

$$\hat{T}^\dagger\hat{T}v_p(\mathbf{r}) = \sigma_p^2 v_p(\mathbf{r}), \qquad \hat{T}\hat{T}^\dagger u_p(\mathbf{s},\mathbf{s}_0) = \sigma_p^2 u_p(\mathbf{s},\mathbf{s}_0),$$

where $\hat{T}^\dagger\hat{T}$ and $\hat{T}\hat{T}^\dagger$ are the composite operators defined in Eqs. (8.30) and (8.31). On making use of these definitions we find that

$$\int d^3r'\, H(\mathbf{r}-\mathbf{r}')v_p(\mathbf{r}') = \sigma_p^2 v_p(\mathbf{r}), \tag{8.34a}$$

$$\frac{\pi}{2}\int d\Omega_{s_0'}\int d\Omega_{s'}\,\delta[k_0(\mathbf{s}-\mathbf{s}_0) - k_0(\mathbf{s}'-\mathbf{s}_0')]u_p(\mathbf{s}',\mathbf{s}_0') = \sigma_p^2 u_p(\mathbf{s},\mathbf{s}_0). \tag{8.34b}$$

Equations (8.34) constitute the normal equations for the SVD of the far-field ISCP problem within the Born approximation. They are standard eigenfunction equations satisfied by the two singular functions v_p and u_p.

The singular functions v_p

We first consider Eq. (8.34a) satisfied by the singular functions v_p. It is seen that this equation is the eigenfunction equation associated with the ISCP integral Eq. (8.4b). We can solve this equation by noting that since the kernel H is convolutional a general solution will be of the form

$$v_p(\mathbf{r}) = A_p e^{i\boldsymbol{\kappa}\cdot\mathbf{r}},$$

where $\boldsymbol{\kappa}$ is a free parameter and A_p is a normalization constant. On substituting the above expression into the l.h.s. of Eq. (8.34a) we obtain the result

$$\int d^3r'\, H(\mathbf{r}-\mathbf{r}')v_p(\mathbf{r}') = A_p \int d^3r'\, H(\mathbf{r}-\mathbf{r}')e^{i\boldsymbol{\kappa}\cdot\mathbf{r}'}$$

$$= A_p e^{i\boldsymbol{\kappa}\cdot\mathbf{r}} \overbrace{\int d^3r'\, H(\mathbf{r}')e^{-i\boldsymbol{\kappa}\cdot\mathbf{r}'}}^{\tilde{H}(\boldsymbol{\kappa})}$$

$$= A_p e^{i\boldsymbol{\kappa}\cdot\mathbf{r}} \begin{cases} \pi^2/(k_0^2\kappa) & \text{if } \kappa \le 2k_0, \\ 0 & \text{else,} \end{cases} \tag{8.35a}$$

where we have made use of Eq. (8.6).

We conclude from Eq. (8.35a) that the index p is actually the parameter $\boldsymbol{\kappa} \in R^3$ and that the singular functions $v_p(\mathbf{r}) = v(\mathbf{r};\boldsymbol{\kappa})$ have a continuous spectrum; their singular values $\sigma(\boldsymbol{\kappa})$ are a function of the continuous variable $\boldsymbol{\kappa} \in R^3$. It then follows that they possess a delta-function normalization; i.e., they are orthogonal with delta-function weighting

$$\int d^3\kappa\, v^*(\mathbf{r},\boldsymbol{\kappa})v(\mathbf{r},\boldsymbol{\kappa}') = \delta(\boldsymbol{\kappa}-\boldsymbol{\kappa}'),$$

which then leads to the requirement that

$$|A_p|^2 = |A(\boldsymbol{\kappa})|^2 = \frac{1}{(2\pi)^3} \to A(\boldsymbol{\kappa}) = \frac{1}{(2\pi)^{3/2}}. \tag{8.35b}$$

On making use of Eqs. (8.35) we then conclude that

$$v_p(\mathbf{r}) = v(\mathbf{r},\boldsymbol{\kappa}) = \frac{e^{i\boldsymbol{\kappa}\cdot\mathbf{r}}}{(2\pi)^{3/2}}, \tag{8.36a}$$

$$\sigma_p^2 = \sigma^2(\boldsymbol{\kappa}) = \begin{cases} \pi^2/(k_0^2\kappa) & \text{if } \kappa \le 2k_0, \\ 0 & \text{else.} \end{cases} \tag{8.36b}$$

The singular functions u_p

The singular functions u_p corresponding to non-zero singular values $\sigma_p = \sigma(\boldsymbol{\kappa}) > 0$ are obtained directly from Eq. (8.33):

$$\sigma(\boldsymbol{\kappa})u(\mathbf{s}, \mathbf{s}_0; \boldsymbol{\kappa}) = \hat{T}v(\mathbf{s}, \mathbf{s}_0; \boldsymbol{\kappa})$$
$$= -\frac{1}{4\pi}\int d^3r\, e^{-ik_0(\mathbf{s}-\mathbf{s}_0)\cdot\mathbf{r}}\left\{\frac{1}{(2\pi)^{3/2}}e^{i\boldsymbol{\kappa}\cdot\mathbf{r}}\right\}$$
$$= -\sqrt{\frac{\pi}{2}}\delta(k_0(\mathbf{s}-\mathbf{s}_0)-\boldsymbol{\kappa}). \tag{8.37}$$

We conclude that the singular functions $v(\mathbf{s}, \mathbf{s}_0; \boldsymbol{\kappa})$ corresponding to non-zero singular values $\kappa > 0$ are given by

$$u(\mathbf{s}, \mathbf{s}_0; \boldsymbol{\kappa}) = \frac{\hat{T}v(\mathbf{s}, \mathbf{s}_0; \boldsymbol{\kappa})}{\sigma(\boldsymbol{\kappa})} = -\sqrt{\frac{\kappa}{2\pi}}k_0\,\delta(k_0(\mathbf{s}-\mathbf{s}_0)-\boldsymbol{\kappa}). \tag{8.38}$$

8.4.3 Solution to the inverse scattering problem

The integral equation Eq. (8.27a) will possess a solution only if the scattering amplitude $f \in \mathcal{H}_f$ is in the range of the operator $\hat{T}$. Moreover, the solution, if it exists, will be non-unique due to the fact that the operator $\hat{T}$ has a null space consisting of all functions band-limited to spatial frequencies that lie outside the Ewald limiting sphere. However, we showed in our review of basic Hilbert-space theory in Section 5.4 of Chapter 5 that the SVD-based solution that we obtained in Eq. (5.48b) and that we will employ here to solve the linearized ISCP actually generates a *least-squares pseudo-inverse* of any linear mapping of the general form of Eq. (8.27b). This means that if the scattering amplitude $f(\mathbf{s}, \mathbf{s}_0)$ is not in the range of the operator $\hat{T}$ the inversion that we will derive below will be a scattering potential that minimizes the mean-squared error between the actual scattering amplitude (data) and that generated by our solution. This also means that the reconstructed scattering potential has the minimum L^2 norm among all least-squares solutions to the linearized inverse problem. The least-squares pseudo-inverse solution to the ISCP is obtained using the singular system $\sigma_p = \sigma(\boldsymbol{\kappa}), v_p = v(\mathbf{r}; \boldsymbol{\kappa}), u_p = u(\mathbf{s}, \mathbf{s}_0; \boldsymbol{\kappa})$ following the same procedure as that employed in Chapter 5 for the ISP. In particular, in analogy with Eq. (5.48b) of Section 5.4 we can represent the solution in the form

$$\hat{V}(\mathbf{r}) = \Sigma_{\sigma_p>0}\frac{\langle u_p, f\rangle_{\mathcal{H}_f}}{\sigma_p}v_p(\mathbf{r})$$
$$= \int_{\kappa\le 2k_0} d^3\kappa\,\frac{k_0\sqrt{\kappa}}{\pi}\langle u_p, f\rangle_{\mathcal{H}_f}v(\mathbf{r}; \boldsymbol{\kappa}), \tag{8.39a}$$

where

$$\langle u_p, f\rangle_{\mathcal{H}_f} = \int d\Omega_{s0}\int d\Omega_s\, v^*(\mathbf{s}, \mathbf{s}_0; \boldsymbol{\kappa})f(\mathbf{s}, \mathbf{s}_0)$$
$$= \int d\Omega_{s0}\int d\Omega_s\left[-\sqrt{\frac{\kappa}{2\pi}}k_0\delta[\boldsymbol{\kappa}-k_0(\mathbf{s}-\mathbf{s}_0)]f(\mathbf{s}, \mathbf{s}_0)\right], \tag{8.39b}$$

and where we have denoted the least-squares pseudo-inverse by $\hat{V}$. If we then substitute Eq. (8.39b) into Eq. (8.39a) we obtain

$$\hat{V}(\mathbf{r}) = \int_{\kappa \leq 2k_0} d^3\kappa \frac{k_0\sqrt{\kappa}}{\pi} \left\{ \int d\Omega_{s_0} \int d\Omega_s \left[-\sqrt{\frac{\kappa}{2\pi}} k_0 \delta[\boldsymbol{\kappa} - k_0(\mathbf{s} - \mathbf{s}_0)] f(\mathbf{s}, \mathbf{s}_0) \right] \right\}$$
$$\times \frac{e^{i\boldsymbol{\kappa}\cdot\mathbf{r}}}{(2\pi)^{3/2}}$$
$$= -\frac{k_0^3}{4\pi^3} \int d\Omega_{s_0} \int d\Omega_s |\mathbf{s} - \mathbf{s}_0| f(\mathbf{s}, \mathbf{s}_0) e^{ik_0(\mathbf{s}-\mathbf{s}_0)\cdot\mathbf{r}},$$

which is recognized as the filtered back-propagation (FBP) algorithm Eq. (8.9).

Because of the general property that the SVD solution generates a minimum-norm least-squares solution it follows immediately that the FBP algorithm has this property. Thus, if the scattering amplitude is not a function of the difference $\mathbf{s} - \mathbf{s}_0$ and thus is not an ideal Born scattering amplitude, the FBP algorithm will generate a least-squares solution that will minimize the integrated squared error between the scattering amplitude generated by the scattering potential reconstruction and the actual scattering amplitude (data). This property applies also if the scattering amplitude is specified only over a region of either or both of the unit spheres Ω_s and Ω_{s_0}. We are also guaranteed that the FBP algorithm generates a minimum-norm solution; i.e., that $|\hat{V}|$ will be minimum among all least-squares solution to the linearized ISCP.

8.5 Born inversion using non-plane-wave probes and arbitrary measurement surfaces

The solution to the Born ISCP obtained in the preceding sections assumes knowledge of the scattering amplitude $f(\mathbf{s}, \mathbf{s}_0)$ which is directly measured in the far field of the scattering potential V. In this section we consider the case in which the field measurements are performed on an arbitrary boundary $\partial\tau$ surrounding the scattering potential and incident waves other than plane waves are employed.

8.5.1 Data collected on arbitrary surfaces

If we make use of the first Helmholtz identity as given in Eq. (2.41a) of Chapter 2 we find that at all points outside an arbitrary surface $\partial\tau$ surrounding the scattering potential the scattered field $U^{(s)}(\mathbf{r}; \nu)$ generated from an arbitrary incident wave parameterized by ν can be expressed in the form

$$U^{(s)}(\mathbf{r}; \nu) = \int_{\partial\tau} dS' \left[G_{0_+}(\mathbf{r} - \mathbf{r}') \frac{\partial}{\partial n'} U^{(s)}(\mathbf{r}'; \nu) - U^{(s)}(\mathbf{r}'; \nu) \frac{\partial}{\partial n'} G_{0_+}(\mathbf{r} - \mathbf{r}') \right], \quad (8.40)$$

where the partial derivatives are with respect to the outward-directed normals to the surface $\partial\tau$. The generalized scattering amplitude is defined to be the radiation pattern associated with the scattered field $U^{(s)}(\mathbf{r}; \nu)$ and can be computed from the above equation by making use of the asymptotic form for the outgoing-wave Green function

$$G_{0+}(\mathbf{r}-\mathbf{r}') \sim -\frac{1}{4\pi} e^{-ik_0\mathbf{s}\cdot\mathbf{r}'} \frac{e^{ik_0 r}}{r}, \quad \text{as} \quad r \to \infty, \tag{8.41}$$

where $\mathbf{s} = \mathbf{r}/r$ is the unit vector along the direction of the asymptotic field point $\mathbf{r}$. On making use of Eq. (8.41) in Eq. (8.40) we obtain

$$U^{(s)}(\mathbf{r};\nu) \sim \overbrace{-\frac{1}{4\pi}\int_{\partial\tau} dS' \left[\frac{\partial}{\partial n'} U^{(s)}(\mathbf{r}';\nu) + ik_0(\mathbf{s}\cdot\hat{\mathbf{n}}')U^{(s)}(\mathbf{r}';\nu)\right] e^{-ik_0\mathbf{s}\cdot\mathbf{r}'}}^{f(\mathbf{s},\nu)} \frac{e^{ik_0 r}}{r},$$

where

$$f(\mathbf{s},\nu) = -\frac{1}{4\pi}\int_{\partial\tau} dS' \left[\frac{\partial}{\partial n'} U^{(s)}(\mathbf{r}';\nu) + ik_0(\mathbf{s}\cdot\hat{\mathbf{n}}')U^{(s)}(\mathbf{r}';\nu)\right] e^{-ik_0\mathbf{s}\cdot\mathbf{r}'} \tag{8.42}$$

is the generalized scattering amplitude.

If the incident wave is a plane wave with unit propagation vector $\mathbf{s}_0$ then Eq. (8.42) can be used to compute the conventional scattering amplitude $f(\mathbf{s},\mathbf{s}_0)$ from the field and its normal derivative over $\partial\tau$, which can then be used in the FBP algorithm. It is also possible to generate a reconstruction from non-plane-wave data using either the method of *slant stacking* (see discussion below) or the algorithms presented in the following chapter.

Plane measurement boundaries

Equation (8.42) is an expression for the generalized scattering amplitude in terms of the scattered field and its normal derivative measured over an arbitrary surface $\partial\tau$ surrounding the scattering potential V. As we showed in Chapters 1 and 2, the field and its normal derivative are not independent and it is also possible to compute the scattered field exterior to the surface $\partial\tau$ in terms of either the scattered field or its normal derivative over $\partial\tau$ by using the solutions to the exterior boundary-value problem given in Eqs. (2.42). However, these solutions require knowledge of the Dirichlet or Neumann Green function, which can be easily computed only for separable boundaries. One important example of a separable boundary is the union of a set of parallel planes located at $z = z_< \le z^-$ and $z = z_> \ge z^+$ lying on either side of the scattering potential assumed to be localized to the strip $z^- \le z \le z^+$. For this case we can employ the solution to the Rayleigh–Sommerfeld boundary-value problems presented in Chapter 2 to express the scattered field throughout the two half-spaces $z < z_<$ and $z > z_>$ in terms of the value of the scattered field (the Dirichlet solution) or its normal derivative (the Neumann solution) over the two planes $z = z_<$ and $z = z_>$. In particular, for the case of Dirichlet conditions we find that

$$U^{(s)}(\mathbf{r};\nu) = \begin{cases} 2\int_{z_<} dS'\, U^{(s)}_{z_<}(\boldsymbol{\rho}';\nu)(\partial/\partial z')G_{0+}(\mathbf{r}-\mathbf{r}_0'), & \text{if } z < z_<, \quad (8.43a) \\ -2\int_{z_>} dS'\, U^{(s)}_{z_>}(\boldsymbol{\rho}';\nu)(\partial/\partial z')G_{0+}(\mathbf{r}-\mathbf{r}_0'), & \text{if } z > z_>, \quad (8.43b) \end{cases}$$

where $U^{(s)}_{z_<}(\boldsymbol{\rho}';\nu)$ and $U^{(s)}_{z_>}(\boldsymbol{\rho}';\nu)$ denote the values of the scattered field over the planes $z = z_<$ and $z = z_>$, respectively, and a similar result is obtained for the case of Neumann conditions over these planes.

We can employ the same general procedure as was used above to derive Eq. (8.42) to compute the generalized scattering amplitude from Eqs. (8.43). In particular, by making use of the asymptotic form for the outgoing-wave Green function given in Eqs. (8.41) we have that

$$\frac{\partial}{\partial n'}G_{0+}(\mathbf{r}-\mathbf{r}') = \hat{\mathbf{z}}'\cdot\nabla_{r'}G_{0+}(\mathbf{r}-\mathbf{r}') \sim \frac{ik}{4\pi}s_z e^{-ik_0\mathbf{s}\cdot\mathbf{r}'}\frac{e^{ik_0 r}}{r},$$

where $s_z = \hat{\mathbf{z}}\cdot\mathbf{s}$ is the z component of the unit observation vector $\mathbf{s} = \mathbf{r}/r$. If we substitute the above result into Eqs. (8.43) we find that

$$U^{(s)}(\mathbf{r};\nu) \sim \begin{cases} \left\{\dfrac{ik_0 s_z}{2\pi}\displaystyle\int_{z_<} dS'\, U^{(s)}_{z_<}(\boldsymbol{\rho}';\nu)e^{-ik_0\mathbf{s}\cdot\mathbf{r}'}\right\}\dfrac{e^{ik_0 r}}{r}, & \text{if } z < z_<, \\ \left\{-\dfrac{ik_0 s_z}{2\pi}\displaystyle\int_{z_>} dS'\, U^{(s)}_{z_>}(\boldsymbol{\rho}';\nu)e^{-ik_0\mathbf{s}\cdot\mathbf{r}'}\right\}\dfrac{e^{ik_0 r}}{r}, & \text{if } z > z_>. \end{cases}$$

We conclude from the above that the generalized scattering amplitude is given by the equations

$$f(\mathbf{s},\nu) = \begin{cases} \dfrac{ik_0 s_z}{2\pi}\displaystyle\int_{z_<} dS'\, U^{(s)}_{z_<}(\boldsymbol{\rho}';\nu)e^{-ik_0\mathbf{s}\cdot\mathbf{r}'}, & \text{if } s_z < 0, & (8.44a) \\ -\dfrac{ik_0 s_z}{2\pi}\displaystyle\int_{z_>} dS'\, U^{(s)}_{z_>}(\boldsymbol{\rho}';\nu)e^{-ik_0\mathbf{s}\cdot\mathbf{r}'}, & \text{if } s_z > 0. & (8.44b) \end{cases}$$

The above equations, which were derived in Section 6.4 using the angular-spectrum expansion of the scattered field, allow the generalized scattering amplitude to be determined for all values of the unit vector $\mathbf{s}$ from measurements of the scattered field over any two parallel bounding planes of the scattering potential. In the case of incident plane waves the parameter ν is the unit wavenumber $\mathbf{s}_0$ of the incident plane waves and the generalized scattering amplitude reduces to the conventional scattering amplitude $f(\mathbf{s},\mathbf{s}_0)$. In this case the filtered back-propagation algorithm Eq. (8.9) can be employed to generate the low-pass-filtered approximation V_{LP} to the scattering potential V from scattered-field measurements over the two planes $z = z_<$ and $z = z_>$. We will find in Section 8.7 that a variant of this procedure forms the basis of diffraction tomography (DT).

Locally plane measurement surfaces

Equation (8.42) applies to arbitrary surfaces and, in addition, can be approximately evaluated in terms of the scattered field (or its normal derivative) alone if the curvature of the surface $\partial\tau$ is small relative to the wavelength. In such cases the surface can be locally approximated as a plane surface and the Rayleigh–Sommerfeld Green function can be employed, thus yielding the result

$$f(\mathbf{s},\nu) \approx -\frac{ik}{2\pi}\int_{\partial\tau} dS'(\mathbf{s}\cdot\hat{\mathbf{n}}')U^{(s)}(\mathbf{r}';\nu)e^{-ik_0\mathbf{s}\cdot\mathbf{r}'}, \tag{8.45}$$

with an analogous result holding for Neumann data on $\partial\tau$. When the incident waves are plane waves so that $\nu = \mathbf{s}_0$, Eqs. (8.42) and (8.45) yield expressions for the scattering amplitude $f(\mathbf{s},\mathbf{s}_0)$ that can then be employed in the filtered back-propagation algorithm Eq. (8.9) to generate the low-pass-filtered approximation V_{LP} to the scattering potential V.

8.5.2 Incident spherical waves

In the case in which spherical incident waves are employed in a suite of scattering experiments the slant-stack algorithm discussed in Section 6.11 can be employed to convert the spherical-wave scattering amplitude $f(\mathbf{s},\mathbf{r}_0)$ to the conventional (plane-wave) scattering amplitude. In particular, we have from Eq. (6.93a)

$$f(\mathbf{s};\mathbf{s}_0) = \mathcal{S}_{\mathbf{s}_0} f(\mathbf{s},\mathbf{r}_0),$$

where $\mathcal{S}_{\mathbf{s}_0}$ is the slant-stack operator defined by its property of converting a suite of spherical waves covering a closed surface $\mathbf{r}_0 \in \partial\tau_0$ into a plane wave having unit propagation vector $\mathbf{s}_0$; i.e.,

$$e^{ik_0\mathbf{s}_0\cdot\mathbf{r}} = \mathcal{S}_{\mathbf{s}_0} G(\mathbf{r},\mathbf{r}_0).$$

By applying the slant-stack operator to the expression for the generalized scattering amplitude given in Eq. (8.42) we obtain the result

$$f(\mathbf{s};\mathbf{s}_0) = \mathcal{S}_{\mathbf{s}_0}\left\{\overbrace{-\frac{1}{4\pi}\int_{\partial\tau} dS' \left[\frac{\partial}{\partial n'}U^{(s)}(\mathbf{r}';\mathbf{r}_0) + ik_0(\mathbf{s}\cdot\hat{\mathbf{n}}')U^{(s)}(\mathbf{r}';\mathbf{r}_0)\right] e^{-ik_0\mathbf{s}\cdot\mathbf{r}'}}^{f(\mathbf{s},\mathbf{r}_0)}\right\}$$

$$= -\frac{1}{4\pi}\int_{\partial\tau} dS' \left[\frac{\partial}{\partial n'}\left\{\mathcal{S}_{\mathbf{s}_0}U^{(s)}(\mathbf{r}';\mathbf{r}_0)\right\} + ik_0(\mathbf{s}\cdot\hat{\mathbf{n}}')\left\{\mathcal{S}_{\mathbf{s}_0}U^{(s)}\right\}(\mathbf{r}';\mathbf{r}_0)\right] e^{-ik_0\mathbf{s}\cdot\mathbf{r}'}.$$

The above equation allows the (conventional) scattering amplitude to be determined from measurements of the scattered waves generated by a suite of spherical incident waves. The scattered-field measurements are performed over any closed surface $\partial\tau$ that surrounds the scattering potential and the incident spherical waves are distributed uniformly over another surface $\partial\tau_0$ that also surrounds the scattering potential.

8.6 Iterative algorithms

We begin with the defining equation for the least-squares minimum-norm solution $\hat{V}$:

$$\hat{T}^\dagger\hat{T}\hat{V} = \hat{T}^\dagger f. \tag{8.46}$$

As discussed earlier, the solution $\hat{V}$ will be equal to the low-pass-filtered approximation V_{LP} to the scattering potential in the case in which the data are in the range of $\hat{T}$ and will minimize the integral squared error between the scattering amplitude generated by $\hat{V}$ and the data in cases in which the data are not perfect and, in addition, will have the minimum L^2 norm among all such reconstructions. The least-squares minimum-norm solution is generated by the filtered back-propagation (FBP) algorithm Eq. (8.9). However, in some instances it is desirable to employ an iterative algorithm rather than the FBP algorithm. This occurs, for example, in cases of limited data and in cases in which we wish to incorporate constraints into the solution. For example, the FBP algorithm generates a scattering potential having infinite support, which is unrealistic. In many cases we know or have a good idea of the scattering potential support and it is desirable to build this prior knowledge into the reconstruction algorithm.

8.6.1 Limited-data problems

We consider here the important case in which the scattering amplitude $f(\mathbf{s}, \mathbf{s}_0)$ is known only over a set of discrete scattering directions $\mathbf{s}_j$, $j = 1, 2, \ldots, N_s$, and only for a finite set of incident-wave directions $\mathbf{s}_{0j}$, $j = 1, 2, \ldots, N_{s_0}$. The scattering operator $\hat{T}$ is still defined by Eq. (8.27c), with $\mathbf{s}$ and $\mathbf{s}_0$ replaced by $\mathbf{s}_j$ and $\mathbf{s}_{0j}$, respectively. The Hilbert space of scattering potentials $\mathcal{H}_V$ remains the same, with the same inner product and norm as defined in Eqs. (8.25). However, the Hilbert space of scattering amplitudes $\mathcal{H}_f$ is now replaced by the Hilbert space of complex numbers $f_{j,j'} = f(\mathbf{s}_j, \mathbf{s}_{0j'})$ having the standard inner product and norm

$$\langle f_1, f_2 \rangle_{\mathcal{H}_f} = \sum_{j'=1}^{N_{s_0}} \sum_{j=1}^{N_s} f_{1\,j,j'}^* f_{2\,j,j'},$$

$$||f|| = \sqrt{\langle f, f \rangle_{\mathcal{H}_f}} = \sqrt{\sum_{j'=1}^{N_{s_0}} \sum_{j=1}^{N_s} |f_{j,j'}|^2}.$$

It is easy to show that the adjoint operator $\hat{T}^\dagger$ is now given by

$$\hat{T}^\dagger = -\frac{1}{4\pi} \sum_{j'=1}^{N_{s_0}} \sum_{j=1}^{N_s} e^{ik_0(\mathbf{s}_j - \mathbf{s}_{0j'})\cdot\mathbf{r}},$$

which is simply the full-data adjoint operator defined in Eq. (8.29) with the integrals over $\mathbf{s}_0$ and $\mathbf{s}$ replaced by discrete sums over $\mathbf{s}_{0j'}$ and $\mathbf{s}_j$.

The composite operators $\hat{T}^\dagger\hat{T}$ and $\hat{T}\hat{T}^\dagger$ are also easily computed. These operators, like the adjoint $\hat{T}^\dagger$, are obtained from the full-data operators defined in Eqs. (8.30) and (8.31) by replacing the integrals over $\mathbf{s}_0$ and $\mathbf{s}$ by discrete sums over $\mathbf{s}_{0j'}$ and $\mathbf{s}_j$. Of particular interest here is the composite operator $\hat{T}^\dagger\hat{T}$ which is defined exactly by Eq. (8.30) but where the kernel H is now given by

$$H(\mathbf{R}) = \frac{1}{(4\pi)^2} \sum_{j'=1}^{N_{s_0}} \sum_{j=1}^{N_s} e^{ik_0(\mathbf{s}_j - \mathbf{s}_{0j'})\cdot\mathbf{R}}.$$

We can develop an iterative algorithm for the limited-data ISCP directly from Eq. (8.46) by rewriting this equation in the form

$$\hat{V} = \hat{V} + \hat{T}^\dagger[f - \hat{T}\hat{V}],$$

where $\hat{T}$ and $\hat{T}^\dagger$ are the limited-data operators defined above. We now define a sequence of scattering potentials $\hat{V}^n$, $n = 0, 1, 2, \ldots$, that are required to satisfy the equation

$$\hat{V}^{n+1} = \hat{V}^n + \hat{T}^\dagger[f - \hat{T}\hat{V}^n], \quad n = 0, 1, 2, \ldots. \tag{8.47}$$

It is clear that if the sequence $\hat{V}^n$ converges then it converges to the solution of the ISCP; i.e., if

$$\lim_{n\to\infty} \hat{V}^n \to \hat{V}$$

then $\hat{V}$ satisfies Eq. (8.46) and, hence, is the least-squares solution to the limited-data ISCP.

The SIRT algorithm

The iteration Eq. (8.47) is in the form of a SIRT (Andersen and Kak, 1984) (for simultaneous iterative reconstruction technique) algorithm that is similar to the well-known ART (Bender *et al.*, 1970) (for algebraic reconstruction technique) algorithm of computed tomography (see also Gilbert (1972)). It can be interpreted as consisting of three steps.

1. Computation of the scattering amplitude $f^n = \hat{T}\hat{V}^n$ corresponding to the nth approximation $\hat{V}^n$ of the scattering potential.
2. Back propagation of the residual error $f - f^n$ between the actual scattering amplitude and the nth approximation of this quantity implemented via the adjoint operator $\hat{T}^\dagger$.
3. Addition of the back-propagated residual error to the nth approximation $\hat{V}^n$ to obtain the $(n+1)$th approximation $\hat{V}^{n+1}$.

8.6.2 Incorporation of constraints

We can employ the SIRT algorithm defined above both in full- and in limited-data problems that incorporate constraints by making use of a *constraint operator* $\mathcal{C}$. The constraint operator has the property of projecting any scattering potential into the subspace $\mathcal{H}_{V_C} \subset \mathcal{H}_V$ of scattering potentials that satisfy the various constraints imposed on the scattering potential V. In order for this scheme to work it is required, of course, that the constraints be of such a nature that the constrained scattering potentials form a proper subspace of the Hilbert space $\mathcal{H}_V$. Examples of such constraints include support constraints, positivity (or negativity) constraints and bandwidth constraints. Thus, for example, a support constraint operator is defined by the property

$$CV = \begin{cases} V & \text{if } \mathbf{r} \in \tau_0, \\ 0 & \text{else,} \end{cases}$$

where τ_0 is the known support of the scattering potential V.

One form of a modified SIRT algorithm that incorporates the constraint operator is given by

$$\hat{V}^{n+1} = C\hat{V}^n + \hat{T}^\dagger[f - \hat{T}C\hat{V}^n], \quad n = 0, 1, 2, \ldots, \tag{8.48}$$

although other forms of the algorithm are also possible. The constrained SIRT algorithm defined in Eq. (8.48) is equivalent to the usual SIRT algorithm Eq. (8.47) followed by application of the constraint operator at each step of the iteration. The sequence now converges only if both $\hat{V}^n \to \hat{V}$ and $C\hat{V} = \hat{V}$.

8.7 Tomographic formulation of inverse scattering

In this section we will again employ a linearized version of the inverse scattering problem (ISCP) but will base the theory on the *Rytov approximation* developed in Section 6.10 of Chapter 6. The Rytov approximation is obtained as a linearization of the non-linear Ricatti differential equation satisfied by the (complex) space-dependent *phase* of the total field $U = U^{(\mathrm{in})} + U^{(s)}$ rather than of the Helmholtz equation Eq. (8.1) satisfied by the total field U. We showed in Chapter 6 that the two formulations of the (forward) scattering problem (the Helmholtz equation for U and the Ricatti equation for the complex phase of U) are mathematically equivalent and, hence, yield the same (exact) forward solution for the field. However, the linearized versions of the two formulations are different and the Rytov approximation presented in Section 6.10 of Chapter 6 is generally more accurate and has a greater domain of applicability than the Born approximation. Although the two linearized formulations of the forward-scattering problem have different domains of applicability we showed in Section 6.10 that the underlying mathematical structure of the ISCP is identical within the two formulations so that many of the results obtained earlier in this chapter can again be employed within the Rytov approximation to the ISCP presented here.

The reformulated theory of inverse scattering based on the Rytov approximation employs the complex phase of the field rather than the scattered field as the measured field quantity upon which the ISCP is based. The complex phase describes the shape and amplitude of the wavefronts of the field diffracted by the scatterer and for this reason the ISCP formulated using the Rytov approximation has become known as *diffraction tomography* (DT).[4] The "tomography" part of its name derives from the fact that in the special cases in which the field measurements are performed in the *forward-scattering* direction the theory resembles the classical theory of *computed tomography* and, indeed, reduces to the latter theory in the limit where the wavelength λ tends to zero.

It is customary in DT to employ the *complex index of refraction* $n(\mathbf{r})$ in place of the scattering potential $V(\mathbf{r})$. These two quantities are related according to the equation

$$V(\mathbf{r}) = k_0^2[1 - n^2(\mathbf{r})], \tag{8.49}$$

where, as usual in this chapter, we have suppressed the frequency variable ω in the arguments of the scattering potential and index of refraction. If this relationship is used in the Helmholtz equation Eq. (8.1) we obtain

$$[\nabla^2 + k_0^2 n^2(\mathbf{r})]U(\mathbf{r}; \nu) = 0,$$

which will be employed in place of Eq. (8.1) from this point on in this chapter. The goal of DT is then the determination of the complex index of refraction $n(\mathbf{r})$ from measurements of the complex phase of the field outside the scattering volume τ_0.

[4] The term diffraction tomography has also been applied to the Born ISCP due to the mathematical similarity of the inversion algorithms. However, we will reserve the term "diffraction tomography" for the Rytov formulation of the ISCP presented in this section.

8.7.1 The Rytov approximation

The Rytov approximation is based on a representation of the total field U in terms of a complex-valued phase W via the equation

$$U(\mathbf{r};\nu) = e^{ik_0 W(\mathbf{r};\nu)}.$$

The complex-valued phase function $W(\mathbf{r};\nu)$ can be decomposed into the sum of the phase W_0 of the incident wave and a perturbation component δW that arises due to the scattering (diffraction) of the incident wave from the deviation of the complex index of refraction $n(\mathbf{r})$ from its background value of unity. The perturbation component of the phase is then found to satisfy a non-linear differential equation whose solution can be expanded into a Liouville–Neumann expansion of the general form (cf. Section 6.10.2)

$$\delta W = \sum_{n=1}^{\infty} \delta W^{(n)} \epsilon^n,$$

where ϵ denotes the strength of the deviation of the index of refraction from unity. The Rytov approximation $\delta W_R = \epsilon\, \delta W^{(1)}$ to the phase δW results from retaining only the first term in the above expansion and is found to satisfy the *linear* differential equation

$$ik_0\, \nabla^2 \delta W_R - 2k_0^2\, \nabla W_0 \cdot \nabla \delta W_R + k_0^2(n^2 - 1) = 0.$$

In the important case of incident plane waves the phase $W_0 = \mathbf{s}_0 \cdot \mathbf{r}$ and this equation then assumes the form

$$ik_0\, \nabla^2 \delta W_R - 2k_0^2 \mathbf{s}_0 \cdot \nabla \delta W_R = -k_0^2(n^2 - 1). \tag{8.50}$$

Classical DT assumes incident plane-wave fields and, hence, is governed by Eq. (8.50). Although the theory can be extended to arbitrary incident waves, historically plane waves have played a dominant role in applications and will be assumed throughout the development in the following exposition. In any case, the inversion algorithms developed in classical DT that are based on the assumption of incident plane waves can be employed for other types of incident waves such as spherical waves through the process of *slant stacking* as discussed in Section 8.5.2.

8.7.2 The short-wavelength limit of DT

In the limiting case in which the wavelength $\lambda \to 0$ ($k_0 \to \infty$) the first term in Eq. (8.50) is negligible in comparison with the other two terms and we obtain the result

$$\mathbf{s}_0 \cdot \nabla \delta w_R = \frac{n^2 - 1}{2} \approx \delta n, \tag{8.51}$$

where we have denoted the short-wavelength limit of δW_R by the lower-case symbol δw_R and

$$\delta n(\mathbf{r}) = n(\mathbf{r}) - 1.$$

In making the above approximation in Eq. (8.51) we have used the fact that the Rytov approximation requires that $n^2 - 1$ be small.

Equation (8.51) is the basic model of diffraction tomography of weakly inhomogeneous objects within the short-wavelength limit. This equation is equivalent to the simple differential equation

$$\frac{\partial}{\partial \eta}\delta w_{\mathrm{R}} = \delta n,$$

where $\eta = \mathbf{s}_0 \cdot \mathbf{r}$ is the position variable along the Cartesian coordinate axis defined by the direction of propagation of the incident plane wave. The above equation can be immediately integrated to yield

$$\delta w_{\mathrm{R}}(\mathbf{r})\big|_{\eta=a}^{\eta=b} = \int_a^b d\eta\, \delta n(\mathbf{r}),$$

where the integration has been performed along the *straight-line ray path* that extends from $\eta = a$ to $\eta = b$. If we take the points $\eta = a$ and $\eta = b = l_0$ to lie outside the object then $\delta w_{\mathrm{R}}|_{\eta=a} = 0$ and the integration on the r.h.s. of the above equation can be extended from $-\infty$ to ∞ and we obtain the standard tomographic model

$$\delta w_{\mathrm{R}}(\mathbf{r})|_{\eta=l_0} = \int_{-\infty}^{\infty} d\eta\, \delta n(\mathbf{r}). \tag{8.52}$$

Equation (8.52) is the underlying mathematical model for computed tomography (CT) and appears here as the limiting model for diffraction tomography (DT) within the Rytov approximation in the short-wavelength limit.

8.7.3 Computed tomography

We consider the experimental setup illustrated in Fig. 8.3, which shows a plane wave with unit propagation vector $\mathbf{s}_0$ incident on an object having a complex index of refraction $n(\mathbf{r})$. We define Cartesian coordinate systems (x, y, z) and (ξ, η, z) whose z axes coincide and are perpendicular to the direction of propagation of the incident plane wave. The (x, y, z) coordinate system is assumed to be fixed in space while the (ξ, η, z) system rotates about the z axis with the direction of propagation of the incident plane wave in such a way that the η axis always remains aligned along this direction as indicated in the figure; i.e., $\mathbf{s}_0 = \hat{\boldsymbol{\eta}}$. If the unit wave vector $\mathbf{s}_0$ then makes the angle α_0 with the positive y axis then the ξ, η system is simply the x, y system rotated in a counter-clockwise direction by the angle α_0 about the z axis. For future reference we note that the x, y and ξ, η coordinate systems are related by the orthogonal transformations

$$x = \xi \cos\alpha_0 - \eta \sin\alpha_0, \qquad y = \xi \sin\alpha_0 + \eta \cos\alpha_0, \tag{8.53a}$$

$$\eta = -x \sin\alpha_0 + y \cos\alpha_0, \qquad \xi = x \cos\alpha_0 + y \sin\alpha_0. \tag{8.53b}$$

Within the short-wavelength limit of the Rytov approximation the complex phase perturbation measured along the line $\eta = l_0$ which is assumed to lie outside the object is then approximately given by Eq. (8.52), which we can express in the form

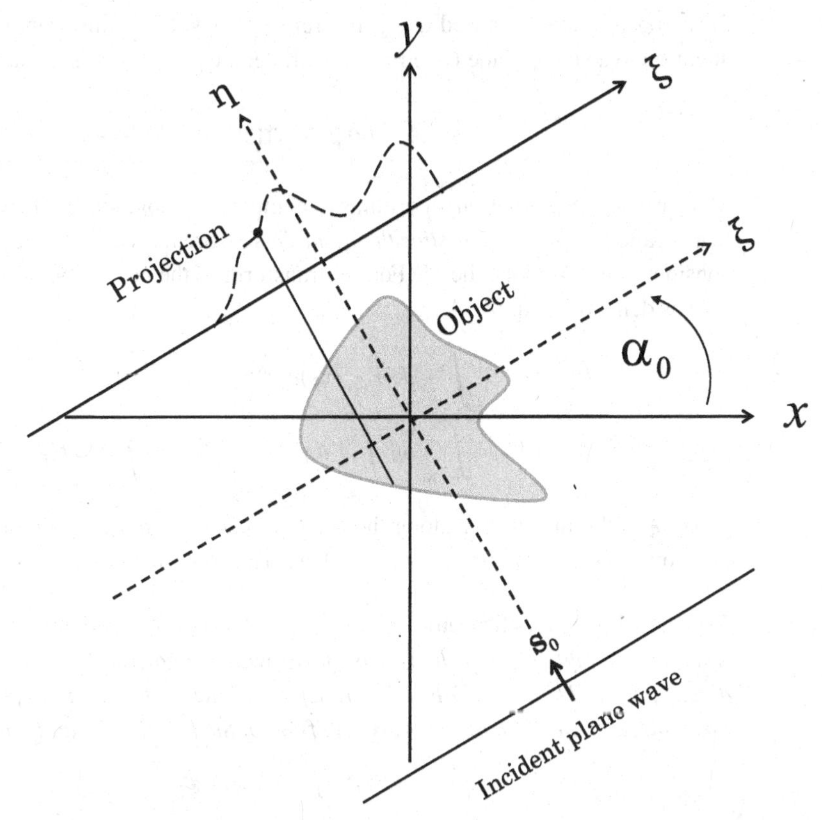

Fig. 8.3 The geometry for computed tomography. The solid line running through the object in the figure denotes a typical line over which the straight-line integral defined in Eq. (8.52) is performed, generating the so-called "projection" of the object onto the ξ axis.

$$\delta w_{\mathrm{R}}(\mathbf{r})|_{\eta=l_0} = P_{\alpha_0}\,\delta n(\xi) = \int_{-\infty}^{\infty} d\eta\,\delta n(x,y), \tag{8.54}$$

where the integration is performed along the line ξ = constant and $\delta n(x,y) = \delta n(\mathbf{r})$ evaluated on the plane $z = z_0$. We will not, at this point, discuss the details of how the phase perturbation over the measurement line is determined from the wavefield over this line but will return to this issue later in our treatment of diffraction tomography below. However, we should point out that this computation is not trivial, due to the phase-unwrapping problem, and is one of the major road blocks to the successful implementation of DT in ultrasound and optical tomography.

The quantity $P_{\alpha_0}\,\delta n(\xi)$ is called a *projection* of the two-dimensional function $\delta n(x,y)$ and can be interpreted roughly as being a "shadow" of the index perturbation $\delta n(x,y) = \delta n(x,y,z_0)$ generated by the incident plane wave along the line $\eta = l_0$. The goal of computed tomography is to *reconstruct* the function $\delta n(x,y)$ from a set of projections $P_{\alpha_0}\,\delta n(\xi)$ taken at various *viewing angles* α_0. Computed tomography then generates a reconstruction of the index perturbation over the plane $z = z_0$, but, since this plane is arbitrary, an approximate 3D reconstruction of the index perturbation can be obtained using a set of

2D reconstructions acquired using different z planes. From this point on we will limit our attention to a single plane (fixed z) and will denote points in this plane by $\boldsymbol{\rho} = (x, y)$.

8.7.4 The projection-slice theorem

Many of the reconstruction algorithms of computed tomography (CT) depend on a theorem that is called the *projection-slice theorem*. This theorem is easily proved from the following considerations. We take the 1D Fourier transform of the projection of an arbitrary function $f(\boldsymbol{\rho})$ as defined in Eq. (8.54):

$$\widetilde{P_{\alpha_0} f}(K) = \int_{-\infty}^{\infty} d\xi \, P_{\alpha_0} f(\xi) e^{-iK\xi}$$
$$= \int_{-\infty}^{\infty} d\xi \int_{-\infty}^{\infty} d\eta \, f(x, y) e^{-iK\xi} = \int d^2\rho \, f(\boldsymbol{\rho}) e^{-iK\hat{\boldsymbol{\xi}}\cdot\boldsymbol{\rho}},$$

where $\hat{\boldsymbol{\xi}}$ is the unit vector along the ξ coordinate axis and $\boldsymbol{\rho} = (x, y)$ denotes the vector position of a general point in the (x, y) plane. The above result yields the following theorem.

Theorem 8.1 (the projection-slice theorem) *Let $P_{\alpha_0} f(\xi)$ be a projection of a function $f(x, y)$ taken at the angle α_0 that the ξ axis makes with the positive x axis of the fixed x, y coordinate system. Then the 1D Fourier transform of the projection $P_{\alpha_0} f(\xi)$ is equal to a slice taken through the 2D Fourier transform $\tilde{f}(\mathbf{K})$ of the function f along the line $\mathbf{K} = K\hat{\boldsymbol{\xi}}$; i.e.,*

$$\widetilde{P_{\alpha_0} f}(K) = \tilde{f}(K\hat{\boldsymbol{\xi}}).$$

The projection-slice theorem leads directly to a popular method of tomographic reconstruction; namely, the *filtered back-projection algorithm*, which we will now discuss.

8.7.5 The filtered back-projection algorithm

We begin by representing δn using a 2D Fourier integral:

$$\delta n(\boldsymbol{\rho}) = \frac{1}{(2\pi)^2} \int d^2K \, \widetilde{\delta n}(\mathbf{K}) e^{i\mathbf{K}\cdot\boldsymbol{\rho}}.$$

If we make a change in integration variables from Cartesian to cylindrical in the above equation we find that

$$\delta n(\boldsymbol{\rho}) = \frac{1}{(2\pi)^2} \int_{-\pi}^{\pi} d\alpha_0 \int_0^{\infty} K \, dK \, \widetilde{\delta n}(K\hat{\mathbf{K}}) e^{iK\hat{\mathbf{K}}\cdot\boldsymbol{\rho}}$$
$$= \frac{1}{2(2\pi)^2} \int_{-\pi}^{\pi} d\alpha_0 \int_{-\infty}^{\infty} |K| dK \, \widetilde{\delta n}(K\hat{\mathbf{K}}) e^{iK\hat{\mathbf{K}}\cdot\boldsymbol{\rho}}, \tag{8.55}$$

where $\hat{\mathbf{K}}$ is the unit vector along the $\mathbf{K}$ direction. If we now simply align the $\mathbf{K}$ direction with the ξ axis so that $\hat{\mathbf{K}} = \hat{\boldsymbol{\xi}}$ and make use of the projection-slice theorem we can write

$$\widetilde{\delta n}(K\hat{\mathbf{K}}) = \widetilde{P_{\alpha_0} \delta n}(K),$$

which when used in Eq. (8.55) yields the filtered back-projection (FBP) algorithm:

$$\delta n(\boldsymbol{\rho}) = \frac{1}{2(2\pi)^2}\int_{-\pi}^{\pi} d\alpha_0 \int_{-\infty}^{\infty} |K| dK\, \widetilde{P_{\alpha_0}\,\delta n}(K) e^{iK\hat{\boldsymbol{\xi}}\cdot\boldsymbol{\rho}}. \tag{8.56}$$

The FBP algorithm as defined in Eq. (8.56) owes its name to the fact that it can be interpreted as the composite operation of 1D filtering of the projections followed by the process of "back projection." To see this, we make the substitution

$$\xi = \hat{\boldsymbol{\xi}}\cdot\boldsymbol{\rho} = x\cos\alpha_0 + y\sin\alpha_0,$$

which follows directly from Eqs. (8.53b). On substituting the above into Eq. (8.56) we obtain

$$\begin{aligned}\delta n(\boldsymbol{\rho}) &= \frac{1}{2(2\pi)^2}\int_{-\pi}^{\pi} d\alpha_0 \int_{-\infty}^{\infty} |K| dK\, \widetilde{P_{\alpha_0}\,\delta n}(K) e^{iK(x\cos\alpha_0 + y\sin\alpha_0)} \\ &= \frac{1}{4\pi}\int_{-\pi}^{\pi} d\alpha_0\, \overline{P_{\alpha_0}\,\delta n}(x\cos\alpha_0 + y\sin\alpha_0),\end{aligned} \tag{8.57a}$$

where

$$\begin{aligned}\overline{P_{\alpha_0}\,\delta n}(\xi) &= \frac{1}{2\pi}\int_{-\infty}^{\infty} |K| dK\, \widetilde{P_{\alpha_0}\,\delta n}(K) e^{iK\xi} \\ &= h \circ P_{\alpha_0}\,\delta n(\xi)\end{aligned} \tag{8.57b}$$

is the *filtered projection*. In this equation h denotes the 1D convolutional filter

$$h(\xi) = \frac{1}{2\pi}\int_{-\infty}^{\infty} |K| dK\, e^{iK\xi} \tag{8.58}$$

and $\circ$ denotes the 1D convolution operation

$$h \circ g(\xi) = \int_{-\infty}^{\infty} d\xi'\, h(\xi - \xi') g(\xi')$$

for any function $g(\xi)$.

The FBP algorithm as given in Eq. (8.57a) is thus seen to consist of three steps.

1. Convolutional filtering of the projections with the filter h to obtain $\overline{P_{\alpha_0}\,\delta n}(\xi)$.
2. Replacement of the argument τ of the filtered projections via the formula $\xi = x\cos\alpha_0 + y\sin\alpha_0$.
3. Summation of all of the quantities obtained in step 2 over all view angles α_0.

The second and third steps, together, constitute the operation of "back projection," thus leading to the name *filtered back-projection algorithm*.

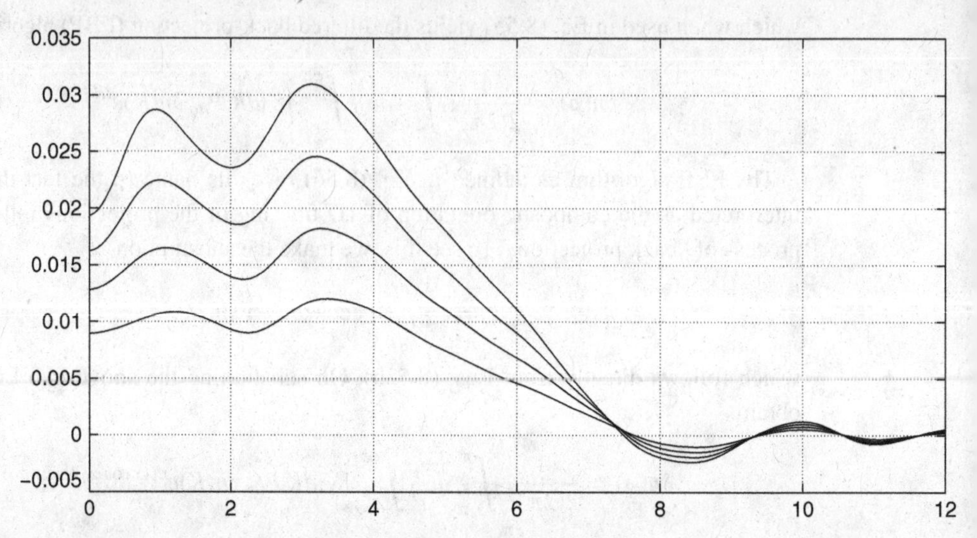

Fig. 8.4 Reconstructions obtained using the CT filtered back-projection algorithm with exactly computed phase data from four cylinders all having a radius of 6λ and having relative indices varying from 1.01 to 1.025 in steps of $\delta n_r = 0.005$. The reconstructions are in the form of radial cuts through the center of the cylinders.

8.7.6 Computed tomography of circularly symmetric objects

The reconstruction algorithms of CT are especially simple for circularly symmetric objects centered at the origin where the spatial Fourier transform $\tilde{\delta n}(\mathbf{K}) = \tilde{\delta n}(K)$ is a function only of the magnitude of the 2D spatial frequency vector $\mathbf{K}$ and the projections $P_{\alpha_0}\,\delta n(\xi)$ are independent of the incident-wave direction α_0. In that case we can simplify the FBP algorithm as given in Eq. (8.56) to obtain

$$\delta n(\boldsymbol{\rho}) = \frac{1}{2(2\pi)^2}\int_{-\pi}^{\pi} d\alpha_0 \int_{-\infty}^{\infty} |K| dK\, \widetilde{P\,\delta n}(K) e^{iK\hat{\boldsymbol{\xi}}\cdot\boldsymbol{\rho}}$$
$$= \frac{1}{2\pi}\int_0^{\infty} K\, dK\, \widetilde{P\,\delta n}(K) J_0(K\rho). \tag{8.59}$$

We implemented the reconstruction algorithm Eq. (8.59) using scattered-field data from penetrable cylinders having different indices of refraction and show the results in Fig. 8.4. The reconstructions shown in the figure are radial cuts through the center of single cylinders having a radius $a_0 = 6\lambda$ and four relative indices varying from 1.01 to 1.025 in steps of $\delta n_r = 0.005$ and employed scattered-field data computed using the multipole expansion developed in Section 6.3 of Chapter 6. The algorithm employed the phase perturbation of the total field (incident plus scattered) over a line two wavelengths outside the cylinder surface. Ideally, the reconstructions should be step functions with the value of the cylinder index over the range $0 \le \rho \le a_0$ and zero for $\rho > a_0$. The extremely bad reconstructions generated by the algorithm are due to the fact that the scattered field does not propagate along the straight-ray paths assumed in CT but rather diffracts and expands as it propagates through the cylinder. This diffraction is taken into account at least to lowest order in DT, as will be developed in the following section.

8.8 Diffraction tomography

The CT model described in the previous section requires that the wavelength λ be sufficiently small that the first term in the Rytov model given in Eqs. (8.50) can be neglected in comparison with the remaining two terms in this equation. It is apparent from the results presented in Fig. 8.4 that this requirement can fail if the incident wave suffers significant diffraction in passing through the object.[5] In this section we will discuss the general case in which λ need not be small and so will examine the ISCP formulated within the Rytov approximation and valid for any wavelength. We will again assume plane-wave incidence with the understanding that the plane-wave data can be obtained from a full suite of spherical-wave data using the slant-stack algorithm described in Chapter 6. A generalized form of inverse scattering that can directly employ arbitrary incident wavefields is developed in Chapter 9.

We showed in Section 6.10.5 that potential scattering theory formulated using the Rytov approximation can be formally converted to the scattering theory formulated within the Born approximation. In particular, we showed that the Rytov approximation to the phase perturbation is *mathematically* related to the Born approximation to the scattered field via the equation

$$\delta W_{\mathrm{R}} = -\frac{i}{k_0} e^{-ik_0 \mathbf{s}_0 \cdot \mathbf{r}} U_{\mathrm{B}}^{(\mathrm{s})}(\mathbf{r}). \tag{8.60}$$

Although the two approximations (Born and Rytov) are mathematically related, it is important to note that the Rytov approximation is an approximation to the complex phase of the field, whereas the Born approximation is an approximation to the scattered field. Because of this they have different domains of validity; i.e., they are not equally good approximations. In particular, the Rytov approximation is generally a more accurate approximation to the complex phase than the Born approximation is to the scattered field if the scattering object is large compared with the wavelength. On the other hand, as we showed in Section 6.10.5, the Rytov approximation degrades as the distance between the scattering volume τ_0 and the field observation point $\mathbf{r}$ increases and becomes identical to the Born approximation in the far-field limit (Colton and Kress, 1992).

8.8.1 Hybrid formulation

It is possible to employ diffraction tomography as formulated via the Rytov approximation with scattered-field data acquired at any distance from the scattering potential by first back propagating the field data to the near field of the scattering potential, after which the Rytov-based FBP algorithm is employed. This scheme, which has been called the "hybrid method" (Johansen *et al.*, 1990), is easy, in principle, to implement but suffers from the

[5] In X-ray CT the wavelength of the X-rays is much smaller than the scale at which the object varies so that very little diffraction occurs and the FBP algorithm and other CT-based algorithms can be successfully employed.

difficulty of phase unwrapping the back-propagated field. The back propagation of the scattered field can be performed using any of the algorithms presented in Chapter 4, including Green-function-based methods as well as Fourier-based methods such as the angular-spectrum expansion and the Fresnel transform. An alternative approach (He and Greenleaf, 1986) is to back propagate the complex phase directly using the non-linear Ricatti equation Eq. (6.72a) which it satisfies. This approach avoids the issue of phase unwrapping but is computationally demanding and there can be problems associated with instabilities in regions where caustics introduce phase discontinuities.

8.8.2 Reduction to a set of 2D inverse scattering problems

The fact that the Rytov approximation to the complex phase of the field is mathematically identical to the Born approximation to the scattered field allows us to employ all of the results obtained earlier in this chapter to the ISCP formulated within the Rytov approximation. In particular, the FBP algorithm can be directly employed within the Rytov approximation to yield an inversion algorithm employing the complex phase perturbation δW_R of the (total) field rather than the (complex) amplitude of the scattered field $U_B^{(s)}$. However, in diffraction tomography, as opposed to the classical ISCP, we employ only *forward-scattered field data* (similarly to the CT case) and employ *near-field* measurements (or measurements of the back-propagated field in the hybrid method) as opposed to the far-field measurements of the scattering amplitude $f(\mathbf{s}, \mathbf{s}_0)$ used in the classical ISCP problem and entering into the FBP algorithm as defined in Eq. (8.9). Thus, in order to transform the results obtained earlier in the chapter to the case of DT it is necessary to specialize these results to be applicable to near-field measurements of forward-scattered-field data.

Although the necessary modifications to the earlier Born-based theory can be made directly in the full 3D case, it is preferable to first develop the inversion algorithms for planar projections of the 3D index perturbation δn onto specified planes and then later extend the theory to the full 3D case. The reduction of the 3D case to planar projections of the index perturbation over planes is possible because of the following theorem.

Theorem 8.2 *Let the unit vectors $\mathbf{s}_0$ of a set of incident plane waves all lie in a plane perpendicular to the z axis of a Cartesian coordinate system $(x, y, z) = (\boldsymbol{\rho}, z)$ and let $U^{(s)}(\mathbf{r}; \mathbf{s}_0)$ be the field scattered by the scattering potential V for any of these incident plane waves. Then the planar projections*

$$P_z U_B^{(s)}(\boldsymbol{\rho}; \mathbf{s}_0) = \int_{-\infty}^{\infty} dz\, U_B^{(s)}(\mathbf{r}; \mathbf{s}_0) \tag{8.61a}$$

of the Born approximation $U_B^{(s)}(\mathbf{r}; \mathbf{s}_0)$ to the scattered fields projected onto this plane are equal to the Born approximation to the 2D fields scattered by the planar projection $P_z V(\boldsymbol{\rho})$ of the scattering potential; i.e.,

$$P_z U_B^{(s)}(\boldsymbol{\rho}; \mathbf{s}_0) = \int d^2\rho'\, P_z V(\boldsymbol{\rho}') e^{ik_0 \mathbf{s}_0 \cdot \boldsymbol{\rho}'} G_{0+}(\boldsymbol{\rho} - \boldsymbol{\rho}'), \tag{8.61b}$$

where

$$P_z V(\boldsymbol{\rho}) = \int_{-\infty}^{\infty} dz\, V(\mathbf{r})$$

are the projections of the scattering potential onto the (x, y) *plane and*

$$G_{0_+}(\boldsymbol{\rho}) = -\frac{i}{4} H_0(k\rho)$$

is the outgoing-wave Green function to the 2D Helmholtz equation.

The above theorem effectively reduces the ISCP within the Born approximation to a sequence of 2D inverse scattering problems. In particular, by projecting the scattered field onto the (x, y) plane according to Eq. (8.61a) and then solving the 2D ISCP corresponding to a set of incident plane waves all lying in this plane the above theorem states that the resulting 2D scattering potential generated by the inversion is equal to a planar projection of the 3D scattering potential onto the (x, y) plane. By repeating the procedure for a set of such planes it is then possible to determine the projections of the 3D potential onto all of these planes from which the potential can then be determined via a 3D CT algorithm. Moreover, in many applications so-called "out-of-plane scattering" will be small so that on employing an incident plane-wave beam that is essentially zero outside some small region of the z axis the resulting 2D reconstruction will be a good approximation to the cross-section of the potential over that wedge region. It is then possible to obtain planar cross-sectional images of a 3D structure by this means.

To prove the theorem, we first note that the Born approximation to the scattered field satisfies the equation

$$[\nabla^2 + k_0^2] U_{\mathrm{B}}^{(\mathrm{s})}(\mathbf{r}) = V(\mathbf{r}) e^{ik_0 \mathbf{s}_0 \cdot \mathbf{r}},$$

where

$$V(\mathbf{r}) = k_0^2 [1 - n^2(\mathbf{r})].$$

If we now take the planar projection of both sides of the above equation satisfied by $U_{\mathrm{B}}^{(\mathrm{s})}$ onto the plane containing the unit propagation vectors $\mathbf{s}_0$ we find that

$$\begin{aligned}
\int_{-\infty}^{\infty} dz [\nabla^2 + k_0^2] U_{\mathrm{B}}^{(\mathrm{s})}(\mathbf{r}) &= \int_{-\infty}^{\infty} dz\, V(\mathbf{r}) e^{ik_0 \mathbf{s}_0 \cdot \boldsymbol{\rho}} \frac{\partial}{\partial z} U_{\mathrm{B}}^{(\mathrm{s})}\Big|_{-\infty}^{\infty} + [\nabla_\rho^2 + k_0^2] \int_{-\infty}^{\infty} dz\, U_{\mathrm{B}}^{(\mathrm{s})}(\mathbf{r}) \\
&= \int_{-\infty}^{\infty} dz\, V(\mathbf{r}) e^{ik_0 \mathbf{s}_0 \cdot \boldsymbol{\rho}} [\nabla_\rho^2 + k_0^2] P_z U_{\mathrm{B}}^{(\mathrm{s})}(\boldsymbol{\rho}) \\
&= P_z V(\boldsymbol{\rho}) e^{ik_0 \mathbf{s}_0 \cdot \boldsymbol{\rho}},
\end{aligned}$$

which establishes the theorem.

The above theorem is stated for the Born approximation to the scattered field but is trivially extended to the Rytov approximation and, hence, to diffraction tomography within this approximation. This, of course, follows from the linear relationship Eq. (8.60) between the Born and Rytov scattering models. Within the Rytov approximation the above theorem takes the following form.

Theorem 8.3 *Let the unit vectors* $\mathbf{s}_0$ *of a set of incident plane waves all lie in a plane perpendicular to the z axis of a Cartesian coordinate system* $(x, y, z) = (\boldsymbol{\rho}, z)$ *and let* $\delta W(\mathbf{r}; \mathbf{s}_0)$ *be the complex phase of the total (incident plus scattered) field generated by the scattering potential V for any of these incident plane waves. Then the planar projections*

$$P_z\, \delta W_{\rm R}(\boldsymbol{\rho}; \mathbf{s}_0) = \int_{-\infty}^{\infty} dz\, \delta W_{\rm R}(\mathbf{r}; \mathbf{s}_0)$$

of the Rytov approximation $\delta W_{\rm R}(\mathbf{r}; \mathbf{s}_0)$ *to the phase projected onto this plane are equal to the Rytov approximation phase of the 2D fields scattered by the planar projection* $P_z V(\boldsymbol{\rho})$ *of the scattering potential; i.e.,*

$$P_z\, \delta W_{\rm R}(\boldsymbol{\rho}; \mathbf{s}_0) = -\frac{i}{k_0} e^{-ik_0 \mathbf{s}_0 \cdot \boldsymbol{\rho}} \int d^2\rho'\, P_z V(\boldsymbol{\rho}') e^{ik_0 \mathbf{s}_0 \cdot \boldsymbol{\rho}'} G_{0+}(\boldsymbol{\rho} - \boldsymbol{\rho}'), \tag{8.62}$$

where $P_z V(\boldsymbol{\rho})$ *are the projections of the scattering potential onto the* (x, y) *plane and* $G_{0+}(\boldsymbol{\rho})$ *is the outgoing-wave Green function to the 2D Helmholtz equation.*

The proof of the theorem follows immediately from Eq. (8.60) and Theorem 8.2.

8.9 Diffraction tomography in two space dimensions

The above two theorems allow us to solve the full 3D inverse scattering problem within either the Born or the Rytov approximation using a sequence of 2D inversions that we will find are generalizations of the FBP algorithm of CT. In the following discussion we will develop the theory for 2D diffraction tomography without reference to the 3D case and will, in particular, employ a 2D scattering potential $V(\boldsymbol{\rho})$ and 2D scattered field $U^{(\rm s)}(\boldsymbol{\rho}, \mathbf{s}_0)$ and phase perturbation $\delta W(\boldsymbol{\rho}, \mathbf{s}_0)$ and restrict our attention to near-field data acquired using the "classical" tomographic geometry illustrated in Fig. 8.3. We will also make the approximation

$$V(\boldsymbol{\rho}) = k_0^2[1 - n^2(\boldsymbol{\rho})] \approx -2k_0^2\, \delta n(\boldsymbol{\rho}), \tag{8.63}$$

which is required in order for the Rytov approximation to be valid. As illustrated in Fig. 8.3, the (2D) index perturbation δn is illuminated by an incident plane wave whose unit propagation vector $\mathbf{s}_0$ lies in the $\boldsymbol{\rho}$ plane, making the angle α_0 with respect to the positive-y axis, and the complex phase of the total field is measured outside the scattering volume τ_0 along the line ξ in the rotated ξ, η coordinate system illustrated in Fig. 8.3. The geometry is thus identical to that used in classical CT and all of the discussion relative to that figure given earlier applies here.

The field data consist of the complex phase perturbation of the total (incident plus scattered) field over the measurement line ξ, namely the total (measured) phase minus the constant and known phase of the incident plane wave on this measurement line. In two space dimensions the Rytov phase perturbation for plane-wave incidence is given by Eq. (8.62) with $P_z\, \delta W_{\rm R}$ replaced by $\delta W_{\rm R}$ and $P_z V$ replaced by $-2k_0^2\, \delta n$ according to Eq. (8.63). For the classical scan geometry illustrated in Fig. 8.3 we have that $\mathbf{s}_0 = \hat{\boldsymbol{\eta}}$ and $\mathbf{s}_0 \cdot \boldsymbol{\rho} = \eta$, so the Rytov phase perturbation on the measurement line is given by

$$\delta W_{\mathrm{R}}(\xi;\alpha_0) = 2ik_0 e^{-ik_0 l_0} \int d^2\rho'\, \delta n(\boldsymbol{\rho}') e^{ik_0\eta'} G_{0_+}(\boldsymbol{\rho} - \boldsymbol{\rho}'), \tag{8.64}$$

where we used α_0 to denote the dependence of the phase on $\mathbf{s}_0$ and $\boldsymbol{\rho} = \xi\hat{\boldsymbol{\xi}} + l_0\hat{\boldsymbol{\eta}}$ denotes a point on the measurement line, which we assume is at a constant distance $\eta = l_0$ from the center of rotation of the tomographic system.

We will refer to the mapping $\delta n(\boldsymbol{\rho}) \rightarrow \delta W_{\mathrm{R}}(\xi;\alpha_0)$ defined in Eq. (8.64) as a *generalized projection* of the index perturbation δn onto the measurement line ξ and denote it, in analogy to the CT case, by $\mathcal{P}_{\alpha_0}\,\delta n(\xi)$; i.e.,

$$\delta W_{\mathrm{R}}(\xi;\alpha_0) = \mathcal{P}_{\alpha_0}\,\delta n(\xi) = 2ik_0 e^{-ik_0 l_0} \int d^2\rho'\, \delta n(\boldsymbol{\rho}') e^{ik_0\eta'} G_{0_+}(\boldsymbol{\rho} - \boldsymbol{\rho}'). \tag{8.65}$$

The DT reconstruction problem then reduces to that of estimating (reconstructing) the index perturbation $\delta n(\boldsymbol{\rho})$ from a set of generalized projections $\mathcal{P}_{\alpha_0}\,\delta n(\xi)$ taken at various *viewing angles* α_0 contained in some set $\alpha_0 \in S_{\alpha_0}$.

8.9.1 The generalized projection-slice theorem

We have already seen that in the short-wavelength limit the above model reduces to conventional CT, which is grounded in the projection-slice theorem relating the 1D transform of the phase perturbation to a slice of the 2D transform of the index perturbation. It is then natural to take a spatial Fourier transform of both sides of the defining equation Eq. (8.65) of the generalized projection with respect to ξ in the hope of obtaining a generalization of this theorem that will lead to an inversion algorithm within the Rytov approximation that applies at all wavelengths. In computing this transform we will make use of the Weyl expansion of the 2D outgoing-wave Green function given in Section 4.6 of Chapter 4. For the geometry in hand this expansion takes the form

$$G_{0_+}(\boldsymbol{\rho} - \boldsymbol{\rho}') = \frac{-i}{4\pi} \int_{-\infty}^{\infty} \frac{d\kappa}{\gamma}\, e^{i[\kappa(\xi-\xi') + \gamma(\eta-\eta')]}, \tag{8.66}$$

where

$$\gamma = \begin{cases} \sqrt{k_0^2 - \kappa^2} & \kappa < k_0, \\ i\sqrt{\kappa^2 - k_0^2} & \kappa > k_0. \end{cases}$$

On substituting Eq. (8.66) into Eq. (8.64) with $\eta = l_0$ and performing some simple algebra we obtain

$$\mathcal{P}_{\alpha_0}\delta n(\xi) = \frac{k_0 e^{-ik_0 l_0}}{2\pi} \int_{-\infty}^{\infty} \frac{d\kappa}{\gamma}\, \widetilde{\delta n}(\kappa, \gamma - k_0) e^{i(\kappa\xi + \gamma l_0)},$$

where

$$\widetilde{\delta n}(\kappa,\gamma-k_0)=\int d^2\rho\,\delta n(\rho)e^{-i\mathbf{K}\cdot\rho}\big|_{\mathbf{K}=\kappa\hat{\boldsymbol{\xi}}+(\gamma-k_0)\hat{\boldsymbol{\eta}}},$$

is the 2D spatial Fourier transform $\widetilde{\delta n}(\mathbf{K})$ of the 2D index perturbation over the arc $\mathbf{K}=\kappa\hat{\boldsymbol{\xi}}+(\gamma-k_0)\hat{\boldsymbol{\eta}}$. The above two equations encompass the so-called *generalized projection-slice theorem*, which we state formally as follows.

Theorem 8.4 (the generalized projection-slice theorem) *Let $P_{\alpha_0}f(\xi)$ be a generalized projection of a function $f(x,y)$ taken at the angle α_0 that the ξ axis makes with the positive-x axis of the fixed x, y coordinate system. Then the 1D Fourier transform of the generalized projection $\mathcal{P}_{\alpha_0}f(\xi)$ is proportional to a slice taken through the 2D Fourier transform $\tilde{f}(\mathbf{K})$ of the function f along the arc $\mathbf{K}=\kappa\hat{\boldsymbol{\xi}}+(\gamma-k_0)\hat{\boldsymbol{\eta}}$; i.e.,*

$$\widetilde{\mathcal{P}_{\alpha_0}f}(\kappa)=\frac{k_0e^{i(\gamma-k_0)l_0}}{\gamma}\tilde{f}(\kappa,\gamma-k_0),\tag{8.67a}$$

where

$$\widetilde{\mathcal{P}_{\alpha_0}f}(\kappa)=\int_{-\infty}^{\infty}d\xi\,\mathcal{P}_{\alpha_0}f(\xi)e^{-i\kappa\xi},\tag{8.67b}$$

is the 1D spatial Fourier transform of the generalized projection of the function $f(x,y)$ over the measurement line $\eta=l_0$.

In the limit $\lambda\to 0$ we have that $\gamma\to k_0$ so that the above theorem reduces to the usual projection-slice theorem, Theorem 8.1. When the wavelength is finite the straight-line slice occurring in the projection-slice theorem deforms to a semicircular arc that corresponds to the inner half of an Ewald circle and the linear transformation from the 2D function $f(x,y)$ to the 1D generalized projection includes the multiplying factor

$$H(\kappa)=\frac{k_0}{\gamma}e^{i(\gamma-k_0)l_0},$$

which acts as a filter in this transformation (Goodman, 1968). Over the homogeneous region of the spectrum ($\kappa<k_0$) γ is real-valued and this filter has a magnitude k_0/γ that increases with spatial frequency and a non-linear phase $(\gamma-k_0)l_0$ that oscillates non-linearly with κ with the instantaneous frequency of oscillation increasing as a function of κ^2. The spatial frequency $\kappa=k_0$ is a cutoff frequency after which the filter decreases exponentially fast with increasing κ. Moreover, using an argument identical to that employed in Section 4.2 of Chapter 4, it is easily established that the product $H(\kappa)\tilde{f}(\kappa,\gamma-k_0)$ also decreases exponentially fast after this cutoff frequency so long as the radius of the support of the function $f(\rho)$ is many wavelengths less than the radius l_0 of the measurement circle. This, of course, is due to the fact that the evanescent components of the scattered field decay exponentially fast outside the support of the scattering potential. In this case, which we will assume here and in the following, the generalized projections are band-limited to the homogeneous region of the spectrum $\kappa<k_0$.

8.9.2 The filtered back-propagation algorithm

The reconstruction problem for the classical scan geometry reduces to estimating the index perturbation $\delta n(\boldsymbol{\rho})$ from a set of generalized projections related to the data (phase perturbations) via the equations

$$\mathcal{P}_{\alpha_0}\,\delta n(\xi) = \delta W_R(\xi;\alpha_0),$$

and to the index perturbation via the generalized projection-slice theorem. According to this theorem a given generalized projection determines the 2D spatial Fourier transform of the index perturbation over the arc

$$\mathbf{K} = \kappa\hat{\boldsymbol{\xi}} + (\gamma - k_0)\hat{\boldsymbol{\eta}}. \tag{8.68}$$

As discussed in conjunction with the generalized projection-slice theorem, the spatial frequencies lying outside the band $-k_0 < \kappa < +k_0$ correspond to evanescent components of the scattered field and hence will be unusable if the measurement circle lies many wavelengths distant from the support of $\delta n(\boldsymbol{\rho})$, which we will assume is the case here.[6]

The unit vector $\hat{\boldsymbol{\eta}}$ is equal to the unit propagation vector $\mathbf{s}_0$ of the incident plane wave and if we define the unit vector $\mathbf{s}$ according to the equation

$$k_0\mathbf{s} = \kappa\hat{\boldsymbol{\xi}} + \gamma\hat{\boldsymbol{\eta}}, \quad -k_0 < \kappa < +k_0,$$

then we conclude that the arc defined in Eq. (8.68) over which the generalized projections specify the 2D transform of the index perturbations corresponds to the Ewald semicircle

$$\mathbf{K} = k_0(\mathbf{s} - \mathbf{s}_0), \tag{8.69}$$

with $\mathbf{s}_0 \cdot \mathbf{s} \geq 0$. We thus conclude that

$$\widetilde{\mathcal{P}_{\alpha_0}\,\delta n}(\kappa) = \delta\widetilde{W}_R(\kappa,\alpha_0) = \frac{k_0 e^{i(\gamma - k_0)l_0}}{\gamma}\widetilde{\delta n}[k_0(\mathbf{s} - \mathbf{s}_0)], \tag{8.70}$$

with $\mathbf{s}$ defined in Eq. (8.69) and with $\mathbf{s} \cdot \mathbf{s}_0 \geq 0$.

By varying the unit propagation vector $\mathbf{s}_0$ over the unit circle we conclude from Eq. (8.70) that $\widetilde{\delta n}(\mathbf{K})$ can be determined throughout the interior of the circle $K = |\mathbf{K}| \leq \sqrt{2}k_0$, which is analogous to the interior of the Ewald limiting sphere that we encountered in our treatment of the ISCP formulated within the Born approximation in Section 8.1. In that earlier treatment we found that the scattering amplitude specified for all incident and scattering directions uniquely determined the transform of the scattering potential throughout the interior of the Ewald limiting sphere, which has a radius of $2k_0$ rather than $\sqrt{2}k_0$. The reason for the difference is that in the classical scan configuration we use only forward-scattered data corresponding to $\mathbf{s}_0 \cdot \mathbf{s} \geq 0$, unlike in the Born inversion from the scattering amplitude, which employs scattered-field data in all directions.

The inversion algorithm for the ISCP within the Born approximation from the scattering amplitude was based on the so-called *inverse scattering identity* derived in Section 8.1.3 for the 3D case and in Example 8.1 of that section for the 2D case. In those treatments it was assumed that the scattered-field data were obtained for all scattering directions $\mathbf{s}$,

[6] Evanescent plane waves can be included in the data using the Green-function-based inversion algorithms developed in the following chapter.

whereas for the classical scan configuration under consideration here we are restricted to forward-scattered field data corresponding to the semicircular arcs defined in Eq. (8.69) with $\mathbf{s}_0 \cdot \mathbf{s} \geq 0$. Using a development completely parallel to that employed in Example 8.1, we then find that the 2D inverse scattering identity modified to use only forward-scattering data is given by

$$\delta n_{\mathrm{LP}}(\boldsymbol{\rho}) = \frac{k_0^2}{2(2\pi)^2} \int_{-\pi}^{\pi} d\alpha_0 \int_0^{\pi} d\alpha \sqrt{1-(\mathbf{s}\cdot\mathbf{s}_0)^2}\, \widetilde{\delta n}[k_0(\mathbf{s}-\mathbf{s}_0)] e^{ik_0(\mathbf{s}-\mathbf{s}_0)\cdot\boldsymbol{\rho}}, \tag{8.71a}$$

where α is the angle that $\mathbf{s}$ makes relative to the ξ axis and

$$\delta n_{\mathrm{LP}}(\boldsymbol{\rho}) = \frac{1}{(2\pi)^2} \int_{K \leq \sqrt{2}k_0} d^2K\, \widetilde{\delta n}(\mathbf{K}) e^{i\mathbf{K}\cdot\boldsymbol{\rho}}, \tag{8.71b}$$

is the index perturbation band limited to the circle $K \leq \sqrt{2}k_0$.

Our final step is to substitute for $\widetilde{\delta n}[k_0(\mathbf{s}-\mathbf{s}_0)]$ from the generalized projection-slice theorem and convert Eq. (8.71a) into a more usable form that parallels the FBP algorithm Eq. (8.57a) of CT. This form follows from the change of integration variable $\alpha \to \kappa$ according to the equations

$$\kappa = k_0 \cos\alpha, \qquad \gamma = k_0 \sin\alpha, \qquad d\kappa = -\gamma\, d\alpha, \qquad \mathbf{s}_0 \cdot \mathbf{s} = \frac{\gamma}{k_0},$$

from which we find that Eq. (8.71a) becomes

$$\delta n_{\mathrm{LP}}(\boldsymbol{\rho}) = \frac{1}{2(2\pi)^2} \int_{-\pi}^{\pi} d\alpha_0 \int_{-k_0}^{k_0} d\kappa |\kappa| \widetilde{\delta W}_{\mathrm{R}}(\kappa, \alpha_0) e^{i[\kappa\xi + (\gamma - k_0)(\eta - l_0)]}. \tag{8.72}$$

The inversion algorithm Eq. (8.72) is the "classical" *filtered back-propagation algorithm* of diffraction tomography within the Rytov approximation. This algorithm, which is the DT version of the FBP algorithm presented in Section 8.1.2 can be decomposed into three steps.

- Filtering of the data (complex phase perturbations) $\delta W_{\mathrm{R}}(\kappa, \alpha_0)$ acquired at angle α_0 implemented in the spatial frequency domain via
$$\widetilde{\widetilde{\delta W}}_{\mathrm{R}}(\kappa, \alpha_0) = |\kappa| \widetilde{\delta W}_{\mathrm{R}}(\kappa, \alpha_0).$$
- Back propagation of the filtered data to yield the *partial reconstructions*
$$\delta n_{\mathrm{LP}}(\boldsymbol{\rho}, \alpha_0) = \frac{1}{2\pi} \int_{-k_0}^{k_0} d\kappa\, \widetilde{\widetilde{\delta W}}_{\mathrm{R}}(\kappa, \alpha_0) e^{i[\kappa\xi + (\gamma - k_0)(\eta - l_0)]}.$$
- Summation of the partial reconstructions over all incident-wave directions (angles) α_0
$$\delta n_{\mathrm{LP}}(\boldsymbol{\rho}) = \frac{1}{4\pi} \int_{-\pi}^{\pi} d\alpha_0\, \delta n_{\mathrm{LP}}(\boldsymbol{\rho}, \alpha_0).$$

It is easily verified that the DT FBP algorithm reduces to the classical FBP algorithm of CT in the limit $\lambda \to 0$ $(k_0 \to \infty)$.

8.9.3 Diffraction tomography of circularly symmetric objects

As was the case in CT, the FBP algorithm of DT is especially simple for circularly symmetric objects centered at the origin, where the spatial Fourier transform $\widetilde{\delta n}(\mathbf{K}) = \widetilde{\delta n}(K)$ is a function only of the magnitude of the 2D spatial frequency vector $\mathbf{K}$ and the generalized projections $P_{\alpha_0}\delta n(\xi)$ are independent of the incident-wave direction α_0. In that case we can simplify the FBP algorithm as given in Eq. (8.72) to obtain

$$\delta n_{\mathrm{LP}}(\boldsymbol{\rho}) = \frac{1}{2(2\pi)^2}\int_{-\pi}^{\pi} d\alpha_0 \int_{-k_0}^{k_0} d\kappa |\kappa| \widetilde{\delta W}_{\mathrm{R}}(\kappa) e^{i[\kappa\xi+(\gamma-k_0)(\eta-l_0)]}$$
$$= \frac{1}{2\pi}\int_0^{k_0} d\kappa |\kappa| \widetilde{\delta W}_{\mathrm{R}}(\kappa) e^{-i(\gamma-k_0)l_0} J_0\left(\sqrt{\kappa^2+(\gamma-k_0)^2}\rho\right). \qquad (8.73a)$$

It is easily verified that Eq. (8.73a) reduces to its CT equivalent Eq. (8.59) in the limit $k_0 \to \infty$. We can also write Eq. (8.73a) in the alternative form

$$\delta n_{\mathrm{LP}}(\boldsymbol{\rho}) = \frac{1}{2\pi}\int_0^{k_0} d\kappa |\kappa| \widetilde{\delta W}_{\mathrm{R}}(\kappa) e^{-i(\gamma-k_0)l_0} J_0(\sqrt{2}k_0\rho\sqrt{1-\gamma/k_0}).$$

Exact and Rytov data for four penetrable cylinders having a radius $a_0 = 6\lambda$ and real-valued indices of refraction ranging from $n_r = 1.01$ to $n_r = 1.025$ in steps of $\delta n_r = 0.005$ are shown in Fig. 8.5. These plots are of the real and imaginary parts of the phase deviation

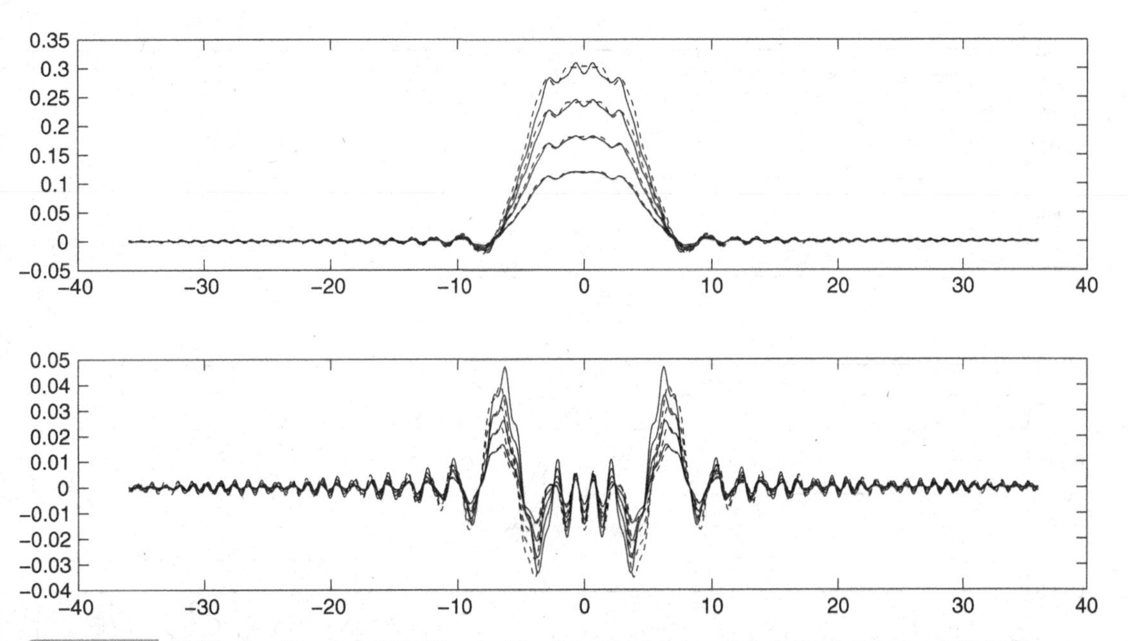

Fig. 8.5 Real (top) and imaginary (bottom) parts of the phase deviation generated over the line $l_0 = a_0 + 2\lambda$ by four penetrable cylinders having a radius $a_0 = 6\lambda$ and indices of refraction ranging from $n_r = 1.01$ to $n_r = 1.025$ in steps of $\delta n_r = 0.005$. The solid lines are the exactly computed phases and the dashed lines are the Rytov phases.

$\delta W(\xi)$ along the measurement line $\eta = 8\lambda$, which is two wavelengths removed from the cylinder boundary. The exact phases were computed using the algorithm

$$\delta W(\xi) = \frac{\ln U}{ik_0} - \frac{\ln e^{ik_0 l_0}}{ik_0}, \tag{8.74}$$

where U is the total exact field (incident plus scattered). The Rytov phases were computed directly using Eq. (8.60) with the Born scattered field computed using the multipole expansion of this field developed in Section 6.8. It can be seen from Fig. 8.5 that the agreement between the Rytov and exact phase data is excellent and that no phase wrapping occurs with this particular data set since the radius and indices of the cylinders are sufficiently small that the phase delay introduced by the passage of the incident wave through the cylinders is smaller than 2π. Reconstructions obtained using this data set in the FBP algorithm in the form given in Eqs. (8.73) are shown in Fig. 8.6. As expected because of the close agreement between the exact and Rytov data the reconstructions are excellent.

We employed the same set of cylinder indices but with a larger cylinder radius equal to 12λ in the simulations shown in Figs. 8.7 and 8.8. In this case the phase delay introduced by the larger cylinders results in a phase wrap in the data for the largest cylinder in the exact data. No wrapping occurs in the Rytov phase since it is computed directly from the Born scattered field using Eq. (8.60) and, hence, is automatically guaranteed to be free of phase wraps. The phase wrap is catastrophic in the reconstructions of this cylinder's radial profile from the exact data, as can be seen in Fig. 8.8. Moreover, the use of the standard Matlab unwrap algorithm fails on this data set. Note, however, that the reconstructions of

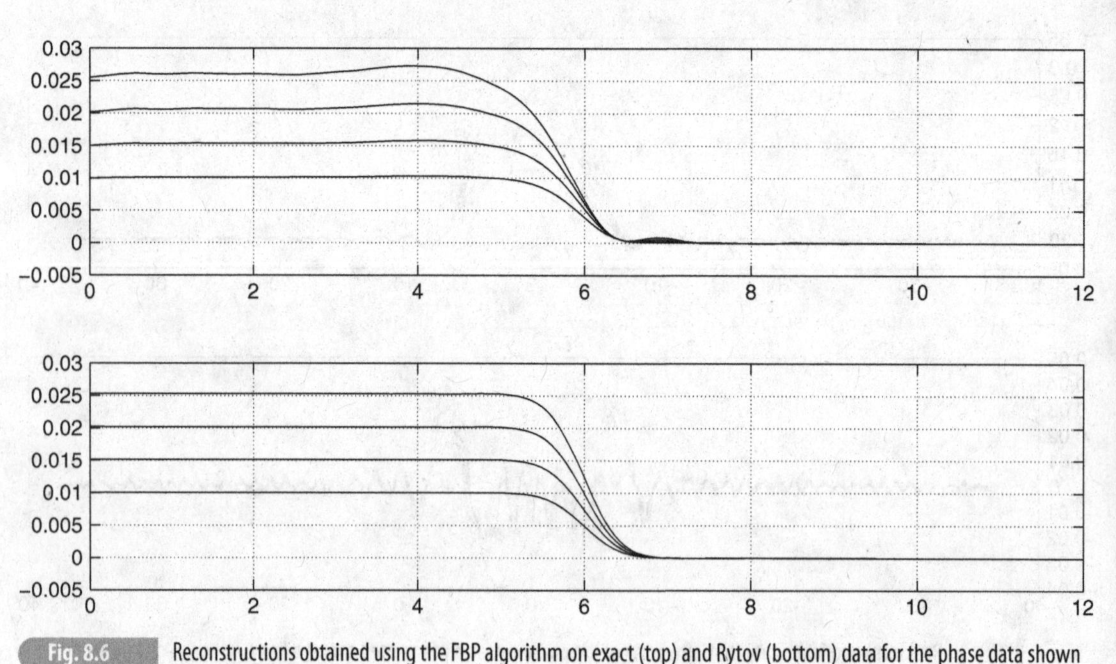

Fig. 8.6 Reconstructions obtained using the FBP algorithm on exact (top) and Rytov (bottom) data for the phase data shown in Fig. 8.5. The reconstructions were obtained using the FBP algorithm with a standard Hamming window included in the filtering operation.

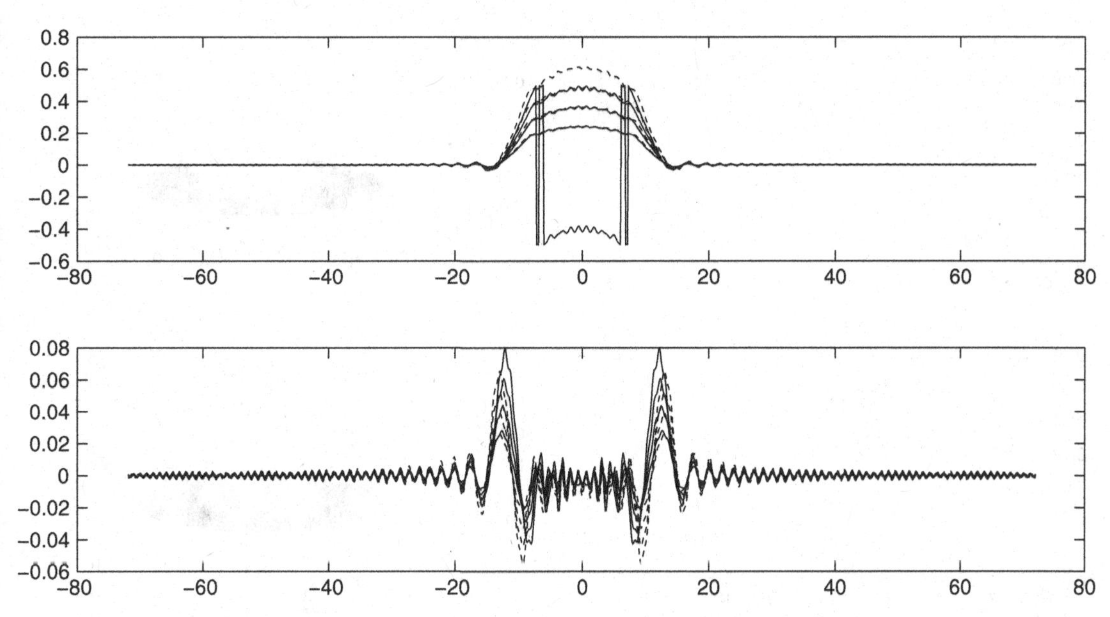

Fig. 8.7 Real (top) and imaginary (bottom) parts of the phase deviation generated over the line $l_0 = a_0 + 2\lambda$ by four penetrable cylinders having a radius $a_0 = 12\lambda$ and indices of refraction ranging from $n_r = 1.01$ to $n_r = 1.025$ in steps of $\delta n_r = 0.005$. The solid lines are the exactly computed phases and the dashed lines are the Rytov phases. It is seen that the exact phase computed for the largest-index cylinder is not properly unwrapped and so generates a severely distorted reconstruction.

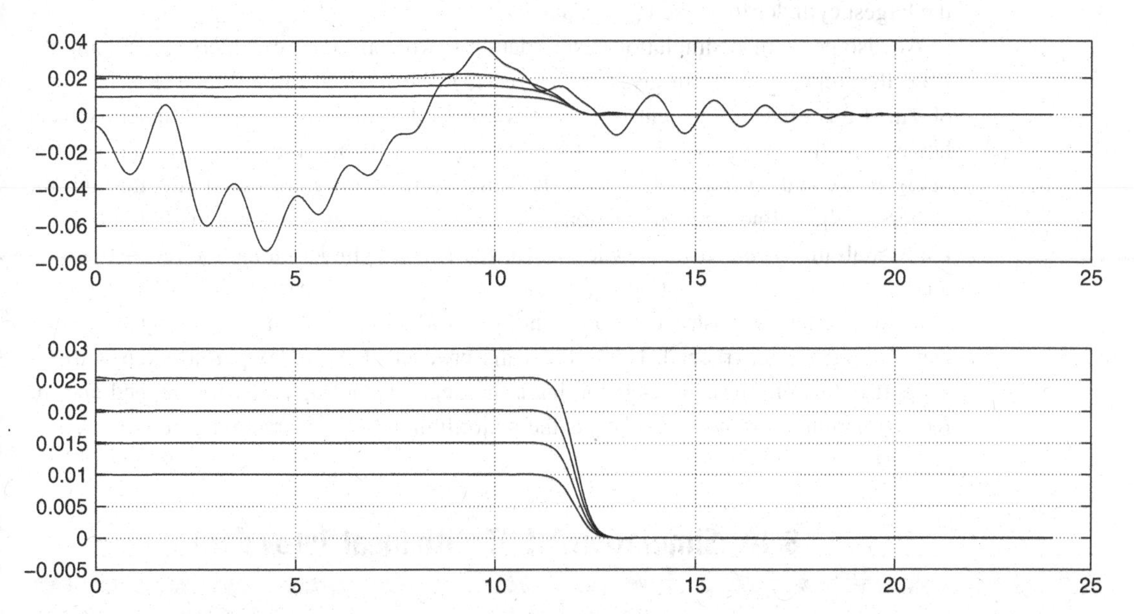

Fig. 8.8 Reconstructions obtained using the FBP algorithm on exact (top) and Rytov (bottom) data for the phase data shown in Fig. 8.7. The reconstructions were obtained using the FBP algorithm with a standard Hamming window included in the filtering operation. It is seen that the failure to have a properly unwrapped phase led to a terrible reconstruction of the index profile of the cylinder having the largest index.

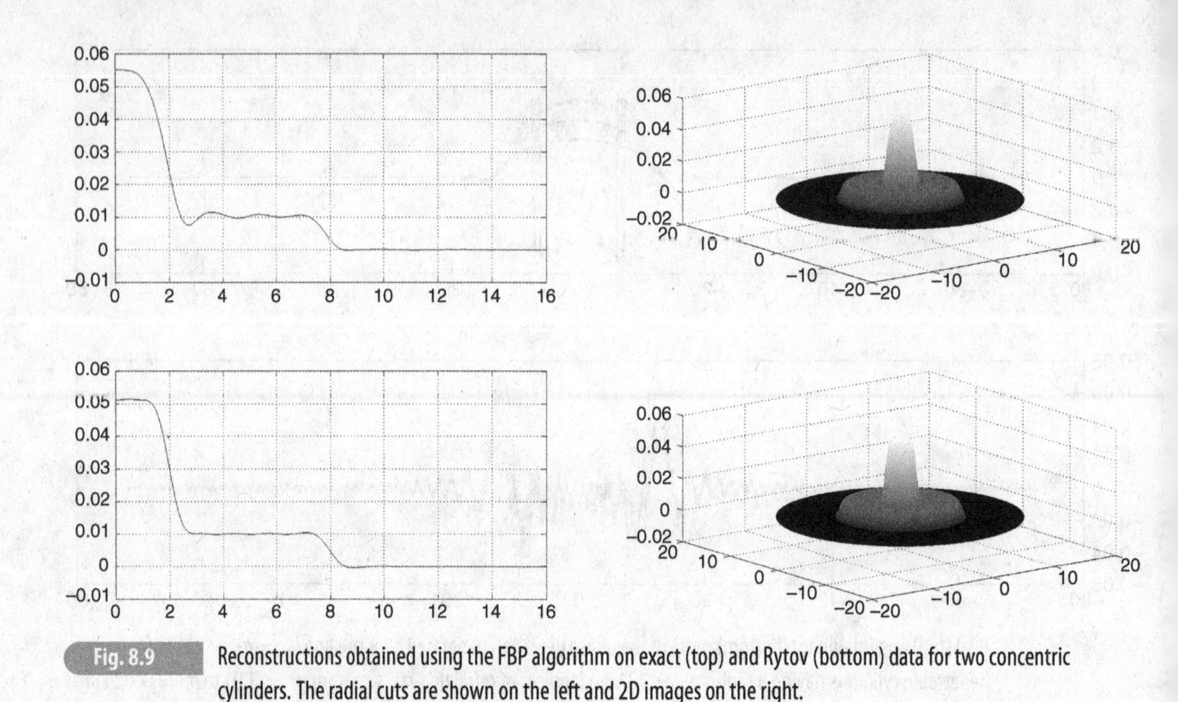

Fig. 8.9 Reconstructions obtained using the FBP algorithm on exact (top) and Rytov (bottom) data for two concentric cylinders. The radial cuts are shown on the left and 2D images on the right.

the cylinder profiles having smaller index values are excellent, as is the reconstruction of the largest cylinder from the Rytov data.

We also performed simulations using data for two concentric cylinders again using the FBP algorithm in the form given in Eqs. (8.73). We show in Fig 8.9 the reconstructions obtained for a pair of concentric cylinders centered at the origin with the inner cylinder having a radius of $a_1 = 2\lambda$ and an index of $n_r = 1.01$ and the outer cylinder a radius of $a_2 = 8\lambda$ and an index of $n_r = 1.05$. This figure shows both radial cuts through the centers of the cylinder reconstructions and 2D plots generated from the radial cuts using the Matlab mfile *pol2cart.m*. As it was for the single-cylinder cases, the reconstructions are excellent. Other simulations of single and concentric cylinders yield similar results: good to excellent reconstructions up to index perturbations of about 5% so long as the data can be properly unwrapped. However, if the product of the index perturbation with the support radius of the scatterer is too large the phase cannot be easily unwrapped and the reconstruction algorithms fail. This is the major limitation of diffraction tomography.

8.10 Simulations of DT with ideal Rytov data

Here we present simulations of CT and DT using ideal Rytov data generated using the generalized projection-slice theorem via Eq. (8.70). Our goal is to acquaint the reader with the steps required in the processing of real data in wavefield tomography and to compare the quality of the DT reconstructions generated via the FBP algorithm with that of those generated from the same data set using the standard CT FBP algorithm without regard to the accuracy of the data set; i.e., we tacitly assume that the scattering object is such that the Rytov data are accurate and that no phase wrapping has occurred. We select scattering objects (phantoms) that are particularly simple, consisting of a set of circular cylinders having differing indices and radii and, hence, of the general form

$$\delta n(\boldsymbol{\rho}) = \sum_{j=1}^{N} \delta n_j \operatorname{circ}[a_j(\boldsymbol{\rho} - \boldsymbol{\rho}_j)], \tag{8.75a}$$

where circ is the circ function having radius a_j and center $\boldsymbol{\rho}_j$ and δn_j are a set of real constants. The 2D spatial Fourier transform of $\delta n(\boldsymbol{\rho})$ is readily found to be given by

$$\widetilde{\delta n}(\mathbf{K}) = 2\pi \sum_{j=1}^{N} \delta n_j \frac{J_1(Ka_j)}{K} e^{-i\mathbf{K}\cdot\boldsymbol{\rho}_j}. \tag{8.75b}$$

The ideal Rytov data are thus given by Eq. (8.70) with $\widetilde{\delta n}[k_0(\mathbf{s} - \mathbf{s}_0)]$ given by Eq. (8.75b) with

$$\mathbf{K} = k_0(\mathbf{s} - \mathbf{s}_0) = \kappa\hat{\boldsymbol{\xi}} + (\gamma - k_0)\hat{\boldsymbol{\eta}}.$$

We thus obtain

$$\widetilde{\delta W}_R(\kappa, \alpha_0) = \frac{2\pi k_0 e^{i(\gamma - k_0)l_0}}{\gamma} \sum_{j=1}^{N} \delta n_j \frac{J_1(\sqrt{2k_0}a_j\sqrt{k_0 - \gamma})}{\sqrt{2k_0}\sqrt{k_0 - \gamma}} e^{-i[\kappa\xi_j + (\gamma - k_0)\eta_j]}, \tag{8.76}$$

where $\boldsymbol{\rho}_j = \xi_j\hat{\boldsymbol{\xi}} + \eta_j\hat{\boldsymbol{\eta}}$. The actual phase data $\delta W(\xi, \alpha_0)$ are, of course, given by the 1D inverse spatial Fourier transform of $\widetilde{\delta W}(\kappa, \alpha_0)$ computed via the algorithm Eq. (8.74).

We take as a general phantom a circular version of the famous Shepp and Logan head phantom (Shepp and Logan, 1974). The parameters for this phantom are given in Table 8.1 and an image of the phantom itself is shown in the left-hand part of Fig. 8.13 later.

Shown in Fig. 8.10 are reconstructions of the head phantom obtained using 3, 6, 10, and 31 equally spaced view angles over 2π radians with the wavelength $\lambda = 1$, which is roughly 1/80 of the outer diameter of the phantom. The increasing clarity and accuracy of the reconstruction with increasing view angles is apparent. The focusing effect of the back-propagation step of the filtered phase data into the image space is also clear, especially in the first two reconstructions. The reconstructions are all band-limited to $K \leq \sqrt{2}k_0$ and are thus somewhat blurred by the low-pass filtering inherent in the FBP algorithm. We also used a Hamming filter on the phase data to remove Gibbs ringing in the reconstructions. We show the same set of reconstructions in Fig. 8.11, but for a much smaller wavelength equal to $\lambda = 0.1$. The increased resolution is apparent, as is the change in the focusing in each of the partial views. This is due to the fact that as the wavelength decreases the back propagations tend toward back projections, which are defined by parallel lines with no focusing. Because of this we can expect the standard CT FBP algorithm to yield a fairly good reconstruction at this shorter wavelength, which is indeed the case, as shown by the CT reconstruction in Fig. 8.12.

Table 8.1 Head phantom parameters (all lengths are in wavelength units)

x_j	y_j	a_j	δn_j
0	0	40.25	0.5
0	−0.92	38.41	−0.5
11	0	10.5	−0.2
−11	0	14.25	−0.2
0	17.5	11.5	0.1
0	5	2.3	0.15
0	−5	2.3	0.15
−4	−30.25	2.3	0.15
0	−30.25	1.15	0.15
3	−30.25	1.725	0.15

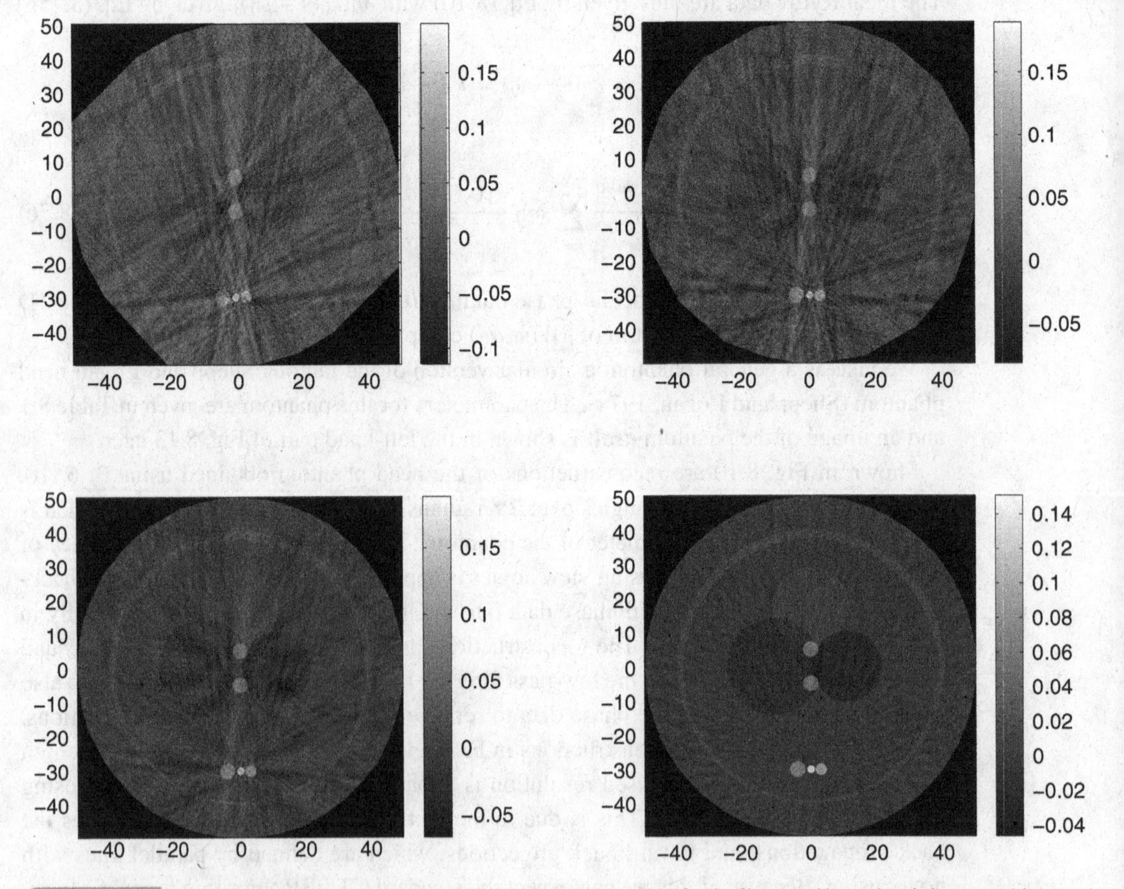

Fig. 8.10 Reconstructions obtained using the FBP algorithm on Rytov data for the Shepp and Logan head phantom with $\lambda = 1$, for 3 views (top left), 6 views (top right), 10 views (bottom left) and 31 views (bottom right), all equally spaced.

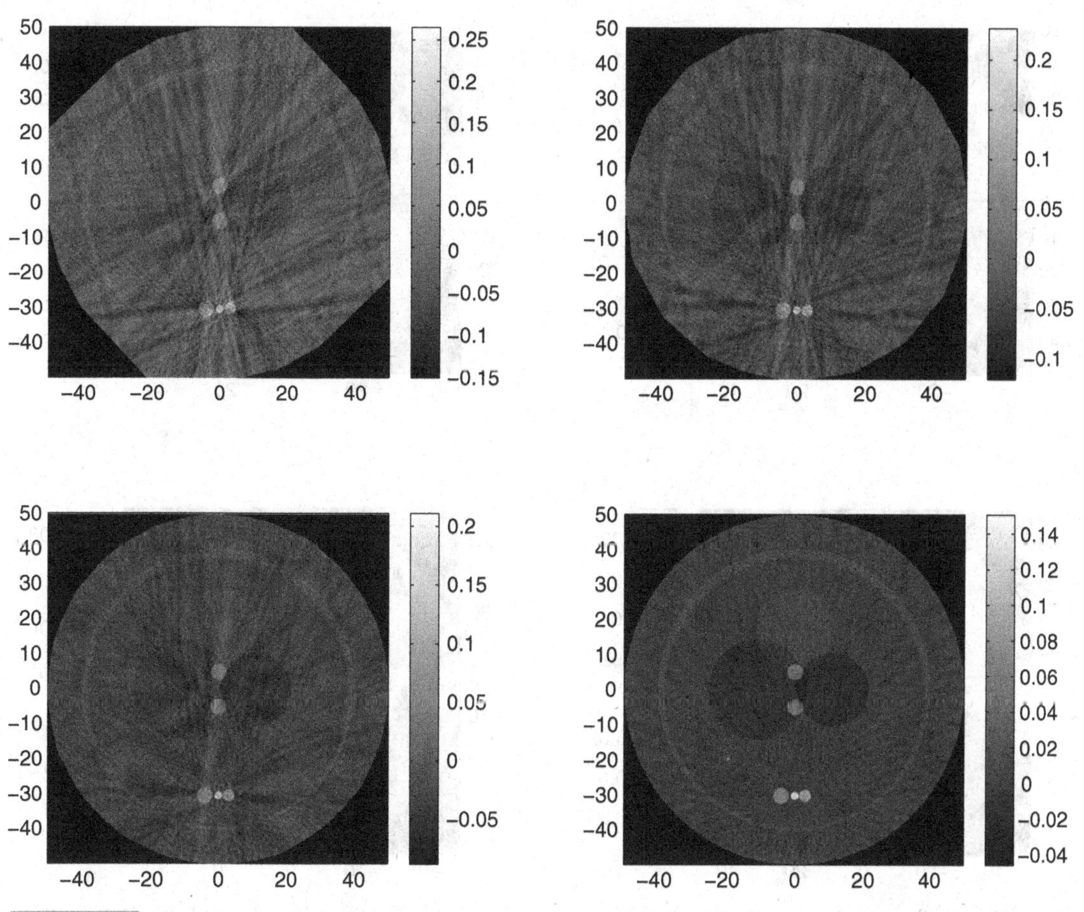

Fig. 8.11 Reconstructions obtained using the FBP algorithm on Rytov data for the Shepp and Logan head phantom with $\lambda = 0.1$, for 3 views (top left), 6 views (top right), 10 views (bottom left) and 31 views (bottom right), all equally spaced.

As a last example we show in the right-hand part of Fig. 8.13 the reconstruction obtained at $\lambda = 1$ for 109 equally spaced view angles. Shown in the left-hand part of Fig. 8.13 is an ideal image of the head phantom obtained using an inverse 2D spatial Fourier transform (2D IFFT) of $\widetilde{\delta n}(\mathbf{K})$ as given in Eq. (8.75b) but not band-limited to $K \leq \sqrt{2}k_0$. The reconstruction is seen to be in good agreement with the ideal image both in shape and in gray level. Even greater agreement can be obtained using larger numbers of view angles in the FBP reconstruction.

8.11 Three-dimensional diffraction tomography

It is possible to directly obtain a complete 3D reconstruction for the classical scan system without first obtaining a set of 2D reconstructions of planar projections as developed

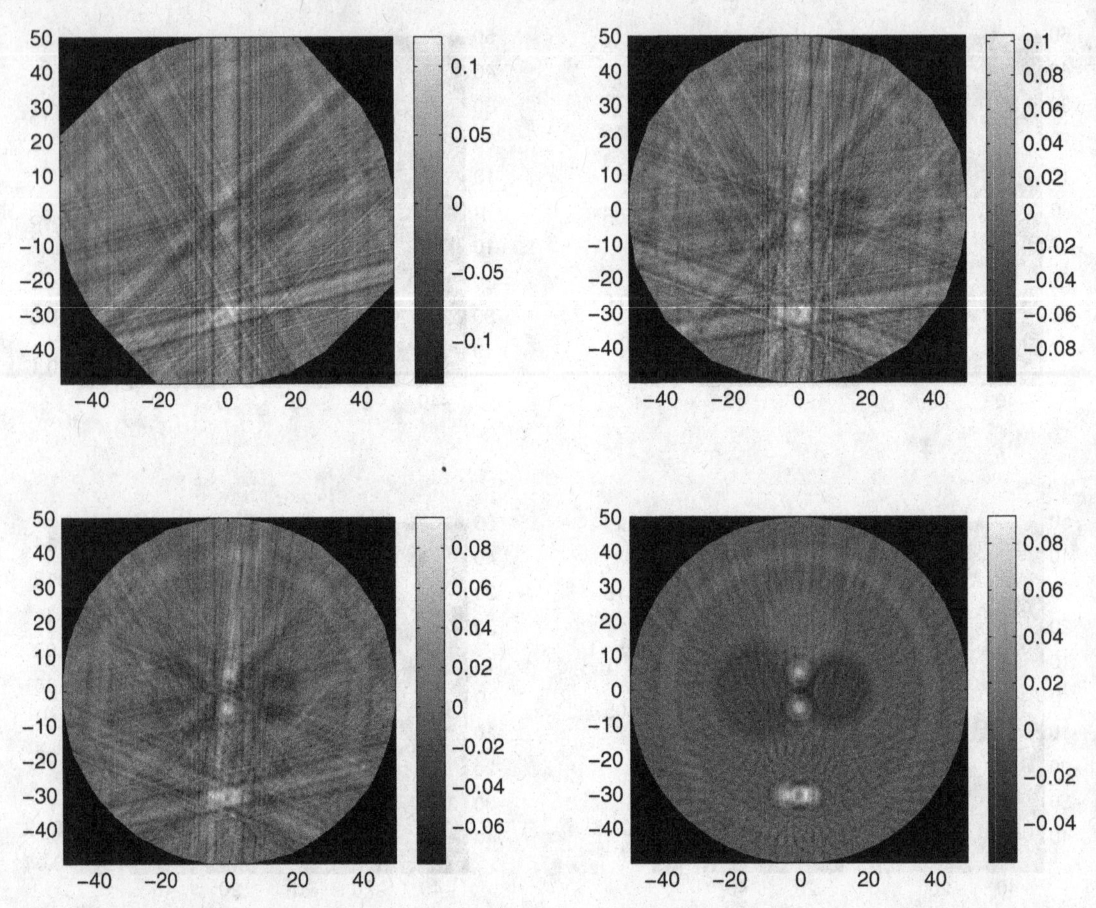

Fig. 8.12 Reconstructions obtained using the CT FBP algorithm on Rytov data for the Shepp and Logan head phantom with $\lambda = 0.1$, for 3 views (top left), 6 views (top right), 10 views (bottom left) and 31 views (bottom right), all equally spaced.

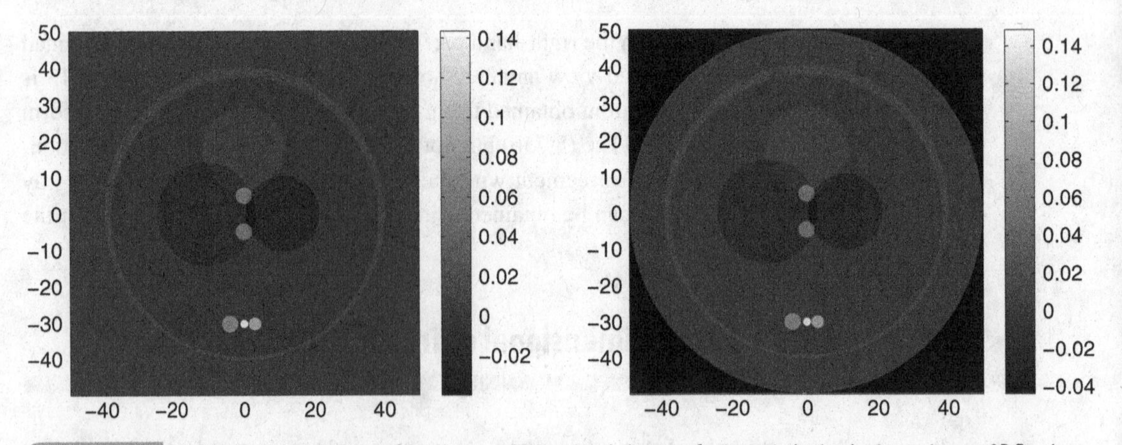

Fig. 8.13 (Left) Ideal reconstruction of the Shepp and Logan head phantom for $\lambda = 1$ obtained using an inverse 2D Fourier transform of $\widetilde{\delta n}(\mathbf{K})$ band-limited to $K \leq \sqrt{2}k_0$. (Right) FBP-generated reconstruction from Rytov phase data using 109 view angles.

in the previous section. The key to this is the 3D inverse scattering identity developed in Section 8.1. Again we will work with DT within the Rytov approximation for the classical scan system, where now the data will be the complex phase perturbation of the total field specified over a set of planes whose unit normals are the unit propagation vectors $\mathbf{s}_0$ of the incident plane waves. We will denote position on these planes by the vector $\boldsymbol{\rho}$ whose Cartesian components are ξ and χ, and, as in the 2D case, use η to denote the third Cartesian component, whose positive axis is aligned along the unit propagation vector $\mathbf{s}_0$ of the incident plane wave. The Rytov model for the phase perturbations measured over a given measurement plane is given in Eq. (8.60), which can be expressed in the form

$$\delta W_{\mathrm{R}}(\boldsymbol{\rho}, \mathbf{s}_0) = 2ik_0 e^{-ik_0 l_0} \int_{\tau_0} d^3r'\, G_{0_+}(\mathbf{r} - \mathbf{r}')\delta n(\mathbf{r}')e^{ik_0\mathbf{s}_0\cdot\mathbf{r}}, \tag{8.77}$$

where $\mathbf{r} = \boldsymbol{\rho} + l_0\mathbf{s}_0$ denotes a point on the measurement plane, with $\boldsymbol{\rho}$ being the projection of $\mathbf{r}$ onto this plane.

Following our treatment of the 2D case, we expand the 3D Green function in the Weyl (angular-spectrum) expansion relative to the measurement plane and find that (cf. Sections 4.1 and 4.2 of Chapter 4)

$$\delta W_{\mathrm{R}}(\boldsymbol{\rho}, \mathbf{s}_0) = \frac{k_0 e^{-ik_0 l_0}}{(2\pi)^2} \int_{K_\rho \le k_0} \frac{d^2K_\rho}{\gamma} \tilde{\delta n}[k_0(\mathbf{s} - \mathbf{s}_0)]e^{ik_0\mathbf{s}\cdot\mathbf{r}}, \tag{8.78}$$

where $\mathbf{K}_\rho = (K_\xi, K_\chi)$ is the Fourier spatial frequency vector conjugate to $\boldsymbol{\rho}$ and

$$k_0\mathbf{s} = \mathbf{K}_\rho + \gamma\mathbf{s}_0, \tag{8.79}$$

with $\gamma = \sqrt{k_0^2 - K_\rho^2}$. Note that in Eq. (8.78) we have assumed that the radius of the measurement sphere l_0 is many wavelengths larger than the radius of the support volume τ_0 of the scattering potential. This then limits the integration to the homogeneous region of the spectra, as discussed in our treatment of the 2D case in the previous section, and requires that $\mathbf{s}_0 \cdot \mathbf{s} \ge 0$.

Again, following our treatment of the 2D case, we transform both sides of Eq. (8.78) and use the definition of $\mathbf{s}$ given above to find that

$$\widetilde{\delta W}_{\mathrm{R}}(\mathbf{K}_\rho, \mathbf{s}_0) = \frac{k_0 e^{i(\gamma - k_0)l_0}}{\gamma} \tilde{\delta n}[k_0(\mathbf{s} - \mathbf{s}_0)], \tag{8.80a}$$

where

$$\widetilde{\delta W}_{\mathrm{R}}(\mathbf{K}_\rho, \mathbf{s}_0) = \int d^2\rho\, \delta W_{\mathrm{R}}(\boldsymbol{\rho}, \mathbf{s}_0)e^{-i\mathbf{K}_\rho\cdot\boldsymbol{\rho}} \tag{8.80b}$$

is the 2D spatial Fourier transform of the phase perturbation over the measurement plane. Equation (8.80a) is the generalization of Eq. (8.70) to the 3D case, and much of the discussion following that earlier equation applies here. In particular, due to the requirement that $\mathbf{s}_0 \cdot \mathbf{s} \ge 0$, a full set of data corresponding to $\delta W_{\mathrm{R}}(\boldsymbol{\rho}, \mathbf{s}_0)$ being specified for $\mathbf{s}_0$ lying over the entire unit sphere results only in specification of $\tilde{\delta n}(\mathbf{K})$ over a sphere having radius $\sqrt{2}k_0$ rather than over the entire Ewald limiting sphere. The 3D inverse scattering identity given in Section 8.1.3 then yields a low-pass-filtered version of δn band-limited to this sphere.

The inverse scattering identity modified to use only forward-scattered field data corresponding to $\mathbf{s}_0 \cdot \mathbf{s}$ yields the result

$$\delta n_{\mathrm{LP}}(\mathbf{r}) = \frac{k_0^3}{(2\pi)^4} \int d\Omega_{s_0} \int_{\mathbf{s}_0\mathbf{s}\geq 0} d\Omega_s |\mathbf{s}-\mathbf{s}_0| \widetilde{\delta n}[k_0(\mathbf{s}-\mathbf{s}_0)] e^{ik_0(\mathbf{s}-\mathbf{s}_0)\cdot\mathbf{r}}, \tag{8.81a}$$

where

$$\delta n_{\mathrm{LP}}(\mathbf{r}) = \frac{1}{(2\pi)^3} \int_{K\leq\sqrt{2}k_0} d^3K\, \widetilde{\delta n}(\mathbf{K}) e^{i\mathbf{K}\cdot\mathbf{r}}. \tag{8.81b}$$

Following our treatment of the 2D case we substitute for $\widetilde{\delta n}[k_0(\mathbf{s}-\mathbf{s}_0)]$ from Eq. (8.80a) and convert Eq. (8.81a) using the change of integration variable according to Eq. (8.79). We find that

$$d\Omega_s = \frac{d^2K_\rho}{\gamma}, \qquad |\mathbf{s}-\mathbf{s}_0| = \sqrt{2}\sqrt{1-\gamma},$$

which then yields

$$\delta n_{\mathrm{LP}}(\mathbf{r}) = \frac{k_0^2}{(2\pi)^4} \int d\Omega_{s_0} \int d^2K_\rho\, \widetilde{\delta W}_{\mathrm{R}}(\mathbf{K}_\rho, \mathbf{s}_0) e^{i[\mathbf{K}_\rho\cdot\boldsymbol{\rho}+(\gamma-k_0)(\eta-l_0)]}. \tag{8.82}$$

Further reading

The Born approximation is used extensively in X-ray crystallography (Vainshtein, 1974; Lipson and Cochran, 1966; Cowley, 1966), while one of the first, and most often-quoted, papers in Born inverse scattering theory in optics is Wolf (1969). The use of the Born approximation using broad-band data in the time domain is given in Norton and Linzer (1981). The first use of the SVD in Born inverse scattering is due to B. DeFacio (Brander and DeFacio, 1986), while Nachman and Waag (Nachman *et al.*, 1997) used the SVD to decompose the scattering amplitude (see our treatment of MUSIC in Chapter 10). The "inverse scattering identity" Eq. (8.9) was first derived in Devaney (1982) along the lines of the derivation presented here and later in Beylkin (1983) using a completely different approach. Fiddy and co-workers have done a great deal of work on the inverse scattering problem within the field of optics (Byrne and Fiddy, 1987; Fiddy and Testorf, 2006; Ross *et al.*, 1979), where the phase problem is of major importance. Various approaches to DT and linearized inverse scattering have been developed by a number of workers (Maleki *et al.*, 1992; Devaney, 1989; Gbur and Wolf, 2002).

Exact (non-linear) inverse scattering for the 1D Schrödinger equation has been treated using a number of different approaches (Newton, 1980a; Sylvester and Winebrenner, 1998; Sylvester and Gylys-Colwell, 1996), while the treatment of exact inverse scattering in two or more space dimensions has had very little success. Important contributions, however, include those by Prosser (Prosser, 1969), who was an early investigator of multi-dimensional non-linear inverse scattering, and Roger Newton (Newton, 1980b, 1982) and the work of Adrian Nachman and co-workers, who have established uniqueness theorems for a number of inverse scattering problems (Nachman, 1996; Isakov and Nachman, 1995). Many of the mathematical issues that are encountered in inverse scattering theory were outlined in an important paper by Sabatier (Sabatier, 1983). Non-linear-based inverse scattering in the form of non-linear diffraction tomography has been developed using Volterra series in Tsihrintzis and Devaney (2000a, 2000b, 2000c) while the effect of multiple scattering on linearized DT was investigated in Belkebir *et al.* (2006).

The Radon transform, which underlies much of computed tomography, first appeared in Radon (1917). Computed tomography (CT) was first introduced by Nobel prize winner Hounsfield (Hounsfield, 1973). The book by Kak and Slaney (Kak and Slaney, 1988) contains an easily read presentation of both classical CT and diffraction tomography (DT), while Born and Wolf (1999) give a nice and concise treatment of DT. Other excellent treatments of CT are given in the books by Natterer (Natterer, 1986) and Gabor Herman (Herman, 1980). The famous Shepp and Logan head phantom first appeared in Shepp and Logan (1974) and so-called "ghost images" (the CT analog of non-scattering scatterers) are treated in Louis (1981).

Two of the earliest papers on the Rytov-based ISCP are Iwata and Nagata (1974) and Mueller *et al.* (1979). The use of DT in geophysics was championed by Alan Witten and co-workers (Witten *et al.*, 1992; Levy and Witten, 1996), while Weglein (Stolt and Weglein, 1985) and his co-workers have done an immense amount of work on Born-based geophysical inversion within the context of the oil industry. Twomey's book (Twomey, 2002) is a wonderful and thorough treatment of inverse methods especially useful in geophysics applications. Greenleaf and co-workers (Greenleaf, 1980, 1983; Greenleaf and Bahn, 1981; He and Greenleaf, 1986) pioneered the use of inverse scattering and DT in medical ultrasound tomography, while the first application in geophysics is in Devaney (1984). Wolf has reviewed the use of DT in optical applications (Wolf, 1996) and has developed the theory for random scatterers (Fischer and Wolf, 1997). Diffraction tomography has also been formulated for random media in Tsihrintzis and Devaney (1993) and specialized to geophysical applications in Tsihrintzis and Devaney (1994). Pan has made a number of contributions within the field of diffraction tomography, many of which are summarized in Pan (1998). Pan and Kak (Pan and Kak, 1983) compared the FBP algorithm with Fourier interpolation in **K** space.

Stamnes and his students have done extensive work in all phases of DT but especially in optical tomography (Wedberg and Stamnes, 1996a, 1996b) while Lauer (Lauer, 2002) has developed an optical microscope based on DT inversion. Time-domain versions of DT (Melamed *et al.*, 1996; Melamed and Heyman, 1997) and inverse scattering (Bolomey *et al.*, 1981) that allow much of the theory presented in this chapter to be applied with wide-band data have been developed. Schotland and co-workers have developed a form of DT for diffusive imaging (Markel and Schotland, 2001; Schotland and Markel, 2001) and have also applied DT to near-field microscopy (Carney and Schotland, 2000). Finally, it is of interest to note that the former president of MIT Charles Vest was one of the earliest researchers to use a form of DT in optical imaging applications (Cha and Vest, 1979). Tabbara and co-workers have applied diffraction tomography both in biomedical and in non-destructive-testing applications (Tabbara *et al.*, 1988).

Problems

8.1 Compute the coherent point-spread function of the low-pass-filtered approximation generated via Eq. (8.3).

8.2 Complete the derivation of Eq. (8.6) using contour-integration techniques.

8.3 Fill in the steps in the derivation of Eq. (8.18) from Eq. (8.17).

8.4 Verify that the FBP algorithm with $\mathbf{s}_0$ and $\mathbf{s}$ restricted to the solid angles Ω_{s_0} and Ω_s satisfies the integral equation Eq. (8.19).

8.5 Complete the derivation of Eq. (8.21).

8.6 Derive the expressions for $\hat{T}\hat{T}^\dagger$ given in Eq. (8.30) and for $\hat{T}^\dagger\hat{T}$ given in Eq. (8.31) and Eq. (8.32).

8.7 Complete the derivation of Eq. (8.42).

8.8 Express the generalized scattering amplitude given in Eq. (8.42) in terms of Dirichlet data over a sphere that completely encloses the scattering volume τ_0. Use this result to express the generalized scattering amplitude in terms of the multipole moments of the scattered field.

8.9 Derive Eqs. (8.44) directly from Eqs. (8.42). Hint: use the relationship between the spatial Fourier transforms of the scattered field and its normal derivative over a plane surface derived in Example 4.4 of Chapter 4.

8.10 Derive the second line in Eq. (8.55) from the first line.

8.11 Complete the derivation of Eq. (8.59).

8.12 Derive Eq. (8.71a).

8.13 Derive Eq. (8.72).

8.14 Complete the derivation of Eq. (8.73a); i.e., show that

$$\frac{1}{2\pi}\int_{-\pi}^{\pi} d\alpha_0\, e^{i[\kappa\xi+(\gamma-k_0)\eta]} = J_0(\sqrt{\kappa^2+(\gamma-k_0)^2}\rho),$$

where α_0 is the angle formed by the η coordinate axis with the fixed x axis. (Hint: write $\kappa\xi+(\gamma-k_0)\eta$ as the dot product of two vectors.)

8.15 Derive Eq. (8.75b).

9 Waves in inhomogeneous media

In this chapter we generalize the theory presented in Chapters 1–8 to inhomogeneous background media where the wavenumber is a known function of position $\mathbf{r}$ and temporal frequency[1] ω. As in most of the book we will work strictly in the frequency domain, with the understanding that the time-domain results, if desired, can be obtained via an inverse temporal Fourier transform. The basic wave model that we employ is the inhomogeneous Helmholtz equation

$$[\nabla^2 + k_0^2(\mathbf{r}) - V(\mathbf{r})]U(\mathbf{r}) = Q(\mathbf{r}), \tag{9.1a}$$

where $k_0(\mathbf{r})$ is the space-varying wavenumber of a "background" medium in which are embedded a scattering potential V and primary source Q. We assume throughout the chapter that the background wavenumber $k_0(\mathbf{r})$ is a known quantity that is at least piecewise continuous and satisfies the asymptotic condition

$$k_0(\mathbf{r}) \to k_0, \quad r \to \infty, \tag{9.1b}$$

where k_0 is a real (possibly frequency-dependent) constant. Both the background wavenumber $k_0(\mathbf{r})$ and the scattering potential $V(\mathbf{r})$ can be complex quantities, although, for the sake of simplicity, we will sometimes restrict the background wavenumber to being real-valued.

The background medium in this chapter corresponds to the uniform (homogeneous) background assumed in earlier chapters. In regions exterior to the support regions of primary sources or scatterers (where $Q = V = 0$) radiated or scattered waves in the inhomogeneous background satisfy the homogeneous Helmholtz equation

$$[\nabla^2 + k_0^2(\mathbf{r})]U_0(\mathbf{r}) = 0.$$

If we define the *background scattering potential*

$$V_0(\mathbf{r}) = k_0^2 - k_0^2(\mathbf{r}) \tag{9.2a}$$

we can express the Helmholtz equation in the form

[1] For the sake of economy we will not explicitly display the frequency variable ω in the arguments of the various field quantities.

$$[\nabla^2 + k_0^2 - V_0(\mathbf{r})]U_0(\mathbf{r}) = 0, \tag{9.2b}$$

which is precisely the model we employed in Chapter 6 in our treatment of classical scattering theory. Thus many of the results obtained in that chapter carry over and will be used in the current chapter. However, a major difference between our treatment in Chapter 6 and that employed here is emphasis. In that earlier chapter our goal was to obtain scattered-field models that could be used in inverse scattering applications for scatterers embedded in a uniform background. Consequently we were mainly concerned with the properties of the scattered field *outside* of the support volume of the scattering potential. In the current chapter our goal is to obtain models of background waves *interior* to the support of the background scattering potential that can then be used in inverse source and scattering problems for sources and scatterers embedded in an inhomogeneous medium. We will often make use of the Helmholtz equation in the form of Eq. (9.2b) in the following developments, with the understanding that it now plays the role of the (classical) homogeneous Helmholtz equation in a uniform medium having a constant wavenumber.

9.1 Background-medium Green functions

A Green function for an inhomogeneous background satisfies the Helmholtz equation

$$[\nabla^2 + k_0^2(\mathbf{r})]G_0(\mathbf{r}, \mathbf{r}') = \delta(\mathbf{r} - \mathbf{r}') \tag{9.3}$$

and boundary conditions appropriate to the particular problem that is to be solved. In the case of radiation and scattering problems in unbounded backgrounds the appropriate boundary condition is causality in the time domain, which translates into the *Sommerfeld radiation condition* (SRC) or "outgoing-wave condition" in the frequency domain. Denoting the outgoing-wave background Green function by $G_{0_+}(\mathbf{r}, \mathbf{r}')$, the SRC is defined by

$$G_{0_+}(\mathbf{r}, \mathbf{r}') \sim g_+(\mathbf{s}, \mathbf{r}')\frac{e^{ik_0 r}}{r} + O\left(\frac{1}{r^2}\right), \quad r \to \infty, \tag{9.4a}$$

where $g_+(\mathbf{s}, \mathbf{r}')$ is the *radiation pattern* for the outgoing-wave Green function along the direction defined by the unit vector $\mathbf{s} = \mathbf{r}/r$ and we have used the condition that $k_0(\mathbf{r}) \to k_0$ as $r \to \infty$. Two other Green functions that will be used are the so-called "incoming-wave" Green function which satisfies the incoming-wave radiation condition

$$G_{0_-}(\mathbf{r}, \mathbf{r}') \sim g_-(\mathbf{s}, \mathbf{r}')\frac{e^{-ik_0 r}}{r} + O\left(\frac{1}{r^2}\right), \quad r \to \infty, \tag{9.4b}$$

where $g_-(\mathbf{s}, \mathbf{r}')$ is the incoming-wave Green function's radiation pattern, and the "conjugate-wave Green function" which is simply the complex conjugate of G_{0_+}. If the background wavenumber $k_0(\mathbf{r})$ is real-valued the conjugate-wave Green function reduces to the incoming-wave Green function $G_{0_-}(\mathbf{r}, \mathbf{r}')$ and $g_- = g_+^*$. The conjugate-wave and incoming-wave Green functions for homogeneous backgrounds were defined in Section 2.2 of Chapter 2 and were used throughout many of the earlier chapters.

In the case of homogeneous backgrounds for which $k_0(\mathbf{r}) = k_0$ we were able to compute the outgoing- and incoming-wave Green functions $G_{0\pm}$ in closed form to find that[2]

$$G_{0+}(\mathbf{r}-\mathbf{r}') = -\frac{1}{4\pi}\frac{e^{ik_0|\mathbf{r}-\mathbf{r}'|}}{|\mathbf{r}-\mathbf{r}'|} \sim \overbrace{-\frac{1}{4\pi}e^{-ik_0\mathbf{s}\cdot\mathbf{r}'}}^{g_+(\mathbf{s},\mathbf{r}')}\frac{e^{ik_0 r}}{r}, \tag{9.5a}$$

$$G_{0-}(\mathbf{r}-\mathbf{r}') = -\frac{1}{4\pi}\frac{e^{-ik_0|\mathbf{r}-\mathbf{r}'|}}{|\mathbf{r}-\mathbf{r}'|} \sim \overbrace{-\frac{1}{4\pi}e^{ik_0\mathbf{s}\cdot\mathbf{r}'}}^{g_-(\mathbf{s},\mathbf{r}')}\frac{e^{-ik_0 r}}{r}. \tag{9.5b}$$

A closed-form or analytical solution for the inhomogeneous background Green functions or their radiation patterns is not possible except for very simple backgrounds such as those consisting of piecewise-constant components with separable boundaries. However, many important theoretical results can still be obtained for general backgrounds that are important in imaging and inverse source and scattering problems.

9.1.1 The reciprocity condition for the Green functions

We have already encountered the outgoing-wave Green function G_{0_+} for a non-uniform medium in Chapter 6 where we studied the scattering problem for a scattering potential embedded in a uniform background medium having (constant) wavenumber k_0. Indeed, if we express the scattering potential in that chapter by its form $V_0(\mathbf{r}) = k_0^2 - k_0^2(\mathbf{r})$ given in Eq. (9.2a) we find that the background Green function G_{0_+} corresponds to what we referred to as the "full Green function" in Chapter 6. One of the properties of this Green function that was stated there was the reciprocity property

$$G_{0_+}(\mathbf{r},\mathbf{r}') = G_{0_+}(\mathbf{r}',\mathbf{r}). \tag{9.6}$$

We didn't actually prove the reciprocity property in Chapter 6 but simply stated that its proof followed along parallel lines to those used in Section 2.8.4 of Chapter 2 to show that the uniform-medium Green functions satisfying homogeneous Dirichlet or Neumann conditions on any closed surface $\partial\tau$ surrounding two field points $\mathbf{r}_1$ and $\mathbf{r}_2$ will be symmetric functions of $\mathbf{r}_1$ and $\mathbf{r}_2$. For the sake of completeness we supply a proof of the reciprocity property here both for the incoming- and for the outgoing-wave Green function.

As was done in Section 2.8.4 of Chapter 2, we begin with the two equations satisfied by the Green functions computed for two source points located at $\mathbf{r}_1$ and $\mathbf{r}_2$:

$$[\nabla_{r'}^2 + k_0^2(\mathbf{r}')]G_{0\pm}(\mathbf{r}',\mathbf{r}_j) = \delta(\mathbf{r}'-\mathbf{r}_j), \quad j = 1,2,$$

[2] We will also denote the (infinite) uniform-medium outgoing- and incoming-wave Green functions using the subscript "0." No confusion should arise since the uniform-medium Green functions will be functions of the difference $\mathbf{r}-\mathbf{r}'$ and so can easily be identified by their arguments.

with $\mathbf{r}_1$ and $\mathbf{r}_2$ both contained within some volume τ, which we will take to be an infinite sphere with surface $\partial\tau = \Sigma_\infty$ and radius $r' \to \infty$. Using by-now-familiar manipulations we find that

$$\int_{\Sigma_\infty} dS' \left[G_{0\pm}(\mathbf{r}', \mathbf{r}_2)\frac{\partial}{\partial n'}G_{0\pm}(\mathbf{r}', \mathbf{r}_1) - G_{0\pm}(\mathbf{r}', \mathbf{r}_1)\frac{\partial}{\partial n'}G_{0\pm}(\mathbf{r}', \mathbf{r}_2)\right]$$
$$= G_{0\pm}(\mathbf{r}_1, \mathbf{r}_2) - G_{0\pm}(\mathbf{r}_2, \mathbf{r}_1),$$

where we have used Green's theorem and the above holds $\forall\, \mathbf{r}_1, \mathbf{r}_2$. Now, since $G_{0\pm}$ must satisfy the boundary conditions given in Eq. (9.4) (with $\mathbf{r}'$ and $\mathbf{r}$ interchanged), we have that

$$\begin{aligned} G_{0\pm}(\mathbf{r}_1, \mathbf{r}_2) - G_{0\pm}(\mathbf{r}_2, \mathbf{r}_1) &= \int_{\Sigma_\infty} dS' \left[G_{0\pm}(\mathbf{r}', \mathbf{r}_2)\frac{\partial}{\partial n'}G_{0\pm}(\mathbf{r}', \mathbf{r}_1) \right. \\ &\qquad \left. - G_{0\pm}(\mathbf{r}', \mathbf{r}_1)\frac{\partial}{\partial n'}G_{0\pm}(\mathbf{r}', \mathbf{r}_2)\right] \\ &\sim \int_{4\pi} d\Omega' \left[\pm ik_0 g_\pm(\hat{\mathbf{r}}', \mathbf{r}_2) g_\pm(\hat{\mathbf{r}}', \mathbf{r}_1) e^{\pm 2ik_0 r'} \right. \\ &\qquad \left. \mp ik_0 g_\pm(\hat{\mathbf{r}}', \mathbf{r}_1) g_\pm(\hat{\mathbf{r}}', \mathbf{r}_2) e^{\pm 2ik_0 r'}\right] + O\left(\frac{1}{r'^2}\right) \\ &= 0 \end{aligned}$$

in the limit $r' \to \infty$, which establishes the desired result.

9.1.2 Plane-wave scattering states

We were able to easily obtain the radiation patterns of the outgoing- and incoming-wave Green functions for homogeneous backgrounds in Section 1.5.1 of Chapter 1 by setting $|\mathbf{r} - \mathbf{r}'| \sim r - \mathbf{s}\cdot\mathbf{r}'$ as $r \to \infty$ to yield the results given in Eqs. (9.5). Unfortunately, we cannot use a similar approach for inhomogeneous backgrounds since the Green functions are not known in closed form. However, we can obtain an analogous result using the "plane-wave scattering states[3]" $\psi_\pm(\mathbf{r}, k_0\mathbf{s})$ of the background medium in place of the plane waves $\exp(\pm ik_0\mathbf{s}\cdot\mathbf{r}')$ that occur in $g_\pm(\mathbf{s}, \mathbf{r}')$ for the homogeneous-background case. The plane-wave scattering states are defined to be the outgoing- and incoming-wave solutions to the non-uniform-medium Helmholtz equation which we can write in the form of Eq. (9.2b),

$$[\nabla^2 + k_0^2 - V_0(\mathbf{r})]\psi_\pm(\mathbf{r}, k_0\mathbf{s}_0) = 0, \tag{9.7a}$$

which satisfy the boundary conditions

$$\psi_\pm(\mathbf{r}, k_0\mathbf{s}_0) \sim e^{ik_0\mathbf{s}_0\cdot\mathbf{r}} + f_\pm(\mathbf{s}, \mathbf{s}_0)\frac{e^{\pm ik_0 r}}{r}. \tag{9.7b}$$

[3] In the literature of quantum scattering theory these states are known as the *stationary scattering states*.

The plane-wave scattering state $\psi_+(\mathbf{r}, k_0\mathbf{s}_0)$ is recognized as being the total (incident plus scattered) field resulting from an incident plane wave propagating in the $\mathbf{s}_0$ direction in a homogeneous medium having wavenumber k_0 in which is embedded the background scattering potential $V_0(\mathbf{r}) = k_0^2 - k_0^2(\mathbf{r})$ defined in Eq. (9.2a) and the far-field amplitude $f_+(\mathbf{s}, \mathbf{s}_0)$ of the scattered wave is the *scattering amplitude* of this scattering potential. The plane-wave scattering states satisfy the Lippmann–Schwinger (LS) equations that were derived in Section 6.2 of Chapter 6, which can be written in either of the two forms

$$\psi_\pm(\mathbf{r}, k_0\mathbf{s}_0) = e^{ik_0\mathbf{s}_0\cdot\mathbf{r}} + \int d^3r'\, G_{0_\pm}(\mathbf{r}, \mathbf{r}')V_0(\mathbf{r}')e^{ik_0\mathbf{s}_0\cdot\mathbf{r}'} \tag{9.8a}$$

and

$$\psi_\pm(\mathbf{r}, k_0\mathbf{s}_0) = e^{ik_0\mathbf{s}_0\cdot\mathbf{r}} + \int d^3r'\, G_{0\pm}(\mathbf{r} - \mathbf{r}')V_0(\mathbf{r}')\psi_\pm(\mathbf{r}', k_0\mathbf{s}_0), \tag{9.8b}$$

with $V_0(\mathbf{r}) = k_0^2 - k_0^2(\mathbf{r})$ and where $G_{0\pm}(\mathbf{r}-\mathbf{r}')$ are the Green functions of the *homogeneous medium* with wavenumber k_0.

To determine the radiation pattern of the two Green functions $G_{0_\pm}(\mathbf{r}, \mathbf{r}')$ we use the LS equations satisfied by these two Green functions that were also derived in Section 6.2 of Chapter 6:

$$G_{0_+}(\mathbf{r}, \mathbf{r}') = G_{0\pm}(\mathbf{r} - \mathbf{r}') + \int d^3r''\, G_{0\pm}(\mathbf{r} - \mathbf{r}'')V_0(\mathbf{r}'')G_{0_\pm}(\mathbf{r}'', \mathbf{r}').$$

If we now let $r \to \infty$ and make use of Eqs. (9.5) we find that

$$G_{0_\pm}(\mathbf{r}, \mathbf{r}') \sim -\frac{1}{4\pi}e^{\mp ik_0\mathbf{s}\cdot\mathbf{r}'}\frac{e^{\pm ik_0r}}{r} + \int d^3r'' \left[-\frac{1}{4\pi}e^{\mp ik_0\mathbf{s}\cdot\mathbf{r}''}\frac{e^{\pm ik_0r}}{r}\right] V_0(\mathbf{r}'')G_{0_\pm}(\mathbf{r}'', \mathbf{r}')$$

$$= -\frac{1}{4\pi}\frac{e^{\pm ik_0r}}{r}\overbrace{\left\{e^{\mp ik_0\mathbf{s}\cdot\mathbf{r}'} + \int d^3r''\, e^{\mp ik_0\mathbf{s}\cdot\mathbf{r}''}V_0(\mathbf{r}'')G_{0_\pm}(\mathbf{r}'', \mathbf{r}')\right\}}^{\psi_\pm(\mathbf{r}', \mp k_0\mathbf{s})}.$$

We thus conclude that

$$G_{0_\pm}(\mathbf{r}, \mathbf{r}') \sim -\frac{1}{4\pi}\frac{e^{\pm ik_0r}}{r}\psi_\pm(\mathbf{r}', \mp k_0\mathbf{s}), \quad r \to \infty, \tag{9.9a}$$

which yields the result

$$g_\pm(\mathbf{s}, \mathbf{r}') = -\frac{1}{4\pi}\psi_\pm(\mathbf{r}', \mp k_0\mathbf{s}). \tag{9.9b}$$

In words, *the radiation pattern $g_\pm(\mathbf{s}, \mathbf{r}')$ of the outgoing (incoming)-wave Green function in a background medium with wavenumber $k_0(\mathbf{r}')$ is $-1/(4\pi)$ times the outgoing (incoming) plane-wave scattering state for this medium for a unit-amplitude incident plane wave with propagation vector $\mp k_0\mathbf{s}$.* Note that in the special case in which the background medium is uniform with wavenumber k_0 the radiation pattern reduces to the uniform-medium case defined in Eqs. (9.5).

Example 9.1 If we take the inner product of both sides of the LS equation Eq. (9.8b) with the spherical harmonics $Y_l^m(\hat{\mathbf{r}})$ and make use of the multipole expansions of the incident plane wave and the outgoing-wave Green function G_{0+} given in Eqs. (3.35) and (3.38a) of Chapter 3 we obtain the result

$$\langle Y_l^m, \psi_\pm \rangle = 4\pi i^l Y_l^{m*}(\mathbf{s}_0) j_l(k_0 r)$$
$$\mp ik_0 \int d^3r' \, j_l(k_0 r_<) h_l^\pm(k_0 r_>) Y_l^{m*}(\hat{\mathbf{r}}') V_0(\mathbf{r}') \psi_\pm(\mathbf{r}', k_0 \mathbf{s}_0),$$

where $r_< = \min r, r'$ and $r_> = \max r, r'$, and

$$\langle Y_l^m, \psi_\pm \rangle = \int d\Omega \, Y_l^{m*}(\hat{\mathbf{r}}) \psi_\pm(\mathbf{r}, k_0 \mathbf{s}_0)$$

denotes the inner product over the unit sphere. If the background scattering potential is spherically symmetric ($V_0(\mathbf{r}) = V_0(r)$) the above equation becomes

$$\langle Y_l^m, \psi_\pm \rangle = 4\pi i^l Y_l^{m*}(\mathbf{s}_0) j_l(k_0 r) \mp ik_0 \int_0^\infty r'^2 \, dr' \, j_l(k_0 r_<) h_l^\pm(k_0 r_>) V_0(r') \langle Y_l^m, \psi_\pm \rangle.$$

We thus conclude that for spherically symmetric backgrounds the plane-wave scattering states admit the expansion

$$\psi_\pm(\mathbf{r}, k_0 \mathbf{s}_0) = 4\pi \sum_{l,m} i^l g_l^\pm(r) Y_l^{m*}(\mathbf{s}_0) Y_l^m(\hat{\mathbf{r}}), \tag{9.10a}$$

where $g_l^\pm(r)$ satisfies the integral equation

$$g_l^\pm(r) = j_l(k_0 r) \mp ik_0 \int_0^\infty r'^2 \, dr' \, j_l(k_0 r_<) h_l^\pm(k_0 r_>) V_0(r') g_l^\pm(r'). \tag{9.10b}$$

The result obtained in the above example is a consequence of the fact that the Helmholtz equation is separable in spherical coordinates if the background scattering potential $V_0(\mathbf{r}) = V_0(r)$ is spherically symmetric. Indeed, in this case the homogeneous Helmholtz equation satisfied by the plane-wave scattering states can be expressed in spherical coordinates in the form (cf. Section 3.3 of Chapter 3)

$$\left[\frac{1}{r^2} \frac{\partial}{\partial r} \left(r^2 \frac{\partial}{\partial r} \right) - \frac{L^2}{r^2} + k_0^2 - V_0(r) \right] \psi_\pm(\mathbf{r}, k_0 \mathbf{s}_0) = 0, \tag{9.11}$$

where L^2 is the square of the angular-momentum operator whose eigenfunctions are the spherical harmonics Y_l^m with eigenvalue $l(l+1)$. It follows from Eq. (9.11) that the plane-wave scattering states can be expressed in the form of Eq. (9.10a) as a superposition of the product of the spherical harmonics with the functions $g_l^\pm(r)$ that then satisfy the integral equation Eq. (9.10b). On substituting this expansion into Eq. (9.11) we find that the radial functions also satisfy the differential equation

$$\left\{ \frac{d}{dr} \left(r^2 \frac{d}{dr} \right) + r^2 [k_0^2 - V_0(r)] - l(l+1) \right\} g_l^\pm(r) = 0, \tag{9.12}$$

and behave as the sum of $j_l(k_0 r)$ and a scattered wave component $h_l^{\pm}(k_0 r)$ as $r \to \infty$, which follows from the boundary condition Eq. (9.7b) satisfied by $\psi_{\pm}(\mathbf{r}, k_0\mathbf{s}_0)$. In the limit where $V_0(r) \to 0$ Eq. (9.12) reduces to the differential equation satisfied by the spherical Bessel functions $g_l^{\pm}(r) = j_l(k_0 r)$, as is also found directly from Eq. (9.10b).

9.2 The radiation problem in non-uniform backgrounds

The field radiated by a source $Q(\mathbf{r})$ in a non-uniform background medium satisfies Eq. (9.1a) with the scattering potential $V(\mathbf{r})$ set equal to zero. We will assume that the source is causal and supported within some finite spatial volume τ_0 and that the wavenumber $k_0(\mathbf{r})$ satisfies the asymptotic condition Eq. (9.1b). The physical model thus corresponds to a causal source localized within a finite and causal non-uniform background that eventually becomes uniform with wavenumber k_0 sufficiently far from the source volume τ_0. We can thus regard the radiated field $U_{0_+}(\mathbf{r})$ from Q to be the field that is radiated by a *composite* causal source consisting of Q embedded in the non-uniform causal background. It then follows that U_{0_+} must behave as an outgoing spherical wave as $r \to \infty$ and satisfy the Sommerfeld–Radiation condition (SRC) in both of the two equivalent forms

$$\lim_{r\to\infty} r\left[\frac{\partial U_{0_+}(\mathbf{r})}{\partial r} - ik_0 U_{0_+}(\mathbf{r})\right] \to 0$$

and

$$U_{0_+}(\mathbf{r}) \sim f(\mathbf{s})\frac{e^{ik_0 r}}{r} + O\left(\frac{1}{r^2}\right), \quad r \to \infty,$$

where $f(\mathbf{s})$ is the "radiation pattern" of the field along the direction $\mathbf{s} = \mathbf{r}/r$.

9.2.1 The Green-function solution to the radiation problem

Following identical steps to those employed in the uniform-background case in Section 2.4 of Chapter 2 we start with the pair of equations

$$[\nabla_{r'}^2 + k_0^2(\mathbf{r}')]G_{0_+}(\mathbf{r}, \mathbf{r}') = \delta(\mathbf{r} - \mathbf{r}'),$$
$$[\nabla_{r'}^2 + k_0^2(\mathbf{r}')]U_{0_+}(\mathbf{r}') = Q(\mathbf{r}'),$$

where we require that both the Green function and the field satisfy the SRC and have made use of the fact that the outgoing-wave Green function G_{0_+} is a symmetric function of its arguments. On multiplying the top equation by U_{0_+} and the bottom by G_{0_+} and subtracting one of the resulting two equations from the other we obtain

$$U_{0_+}(\mathbf{r}')\nabla_{r'}^2 G_{0_+}(\mathbf{r}, \mathbf{r}') - G_{0_+}(\mathbf{r}, \mathbf{r}')\nabla_{r'}^2 U_{0_+}(\mathbf{r}') = U_{0_+}(\mathbf{r}')\delta(\mathbf{r} - \mathbf{r}') - G_{0_+}(\mathbf{r}, \mathbf{r}')Q(\mathbf{r}').$$

By integrating the above equation over a volume $\tau \supset \tau_0$ containing the source space region τ_0 and having closed boundary $\partial\tau$ we obtain

$$\chi(\mathbf{r}) + \int_{\tau_0} d^3r'\, G_{0_+}(\mathbf{r}, \mathbf{r}')Q(\mathbf{r}') = U_{0_+}(\mathbf{r}), \quad \mathbf{r} \in \tau, \tag{9.13a}$$

and

$$\chi(\mathbf{r})+\int_{\tau_0} d^3r'\, G_{0_+}(\mathbf{r},\mathbf{r}')Q(\mathbf{r}')=0, \quad \mathbf{r}\in\tau^{\perp}, \tag{9.13b}$$

where $\tau^{\perp}$ denotes the infinite region exterior to τ (the complement of τ). In deriving the above we have used the fact that the source Q vanishes outside the space region τ_0 and have defined

$$\chi(\mathbf{r})=\int_{\partial\tau} dS'\left[U_{0_+}(\mathbf{r}')\frac{\partial}{\partial n'}G_{0_+}(\mathbf{r},\mathbf{r}')-G_{0_+}(\mathbf{r},\mathbf{r}')\frac{\partial}{\partial n'}U_{0_+}(\mathbf{r}')\right]. \tag{9.14}$$

We have also made use of Green's theorem

$$\int_{\tau} d^3r'[U_{0_+}\nabla^2_{\mathbf{r}'}G_{0_+}-G_{0_+}\nabla^2_{\mathbf{r}'}U_{0_+}]=\int_{\partial\tau} dS'\left[U_{0_+}\frac{\partial}{\partial n'}G_{0_+}-G_{0_+}\frac{\partial}{\partial n'}U_{0_+}\right],$$

where the partial derivatives are taken with respect to the outward-directed normal to $\partial\tau$. The field χ can be shown to vanish in the limit where the volume τ tends to an infinite sphere Σ_∞ with radius $r'\to\infty$ so long as we require the Green function and field U_{0_+} to satisfy the SRC (see the problems at the end of the chapter) and we obtain

$$U_{0_+}(\mathbf{r})=\int_{\tau_0} d^3r'\, G_{0_+}(\mathbf{r},\mathbf{r}')Q(\mathbf{r}'). \tag{9.15}$$

9.2.2 The Kirchhoff–Helmholtz representation of the radiated field

We return to the set of Eqs. (9.13), where we now take the volume $\tau\supset\tau_0$ to be arbitrary. If we now make use of the Green-function solution Eq. (9.15) in these equations we conclude that

$$\chi(\mathbf{r})=0, \quad \mathbf{r}\in\tau, \qquad \chi(\mathbf{r})+U_{0_+}(\mathbf{r})=0, \quad \mathbf{r}\in\tau^{\perp}.$$

On substituting the expression for χ given in Eq. (9.14) these two equations yield the results

$$\int_{\partial\tau} dS'\left[G_{0_+}(\mathbf{r},\mathbf{r}')\frac{\partial}{\partial n'}U_{0_+}(\mathbf{r}')-U_{0_+}(\mathbf{r}')\frac{\partial}{\partial n'}G_{0_+}(\mathbf{r},\mathbf{r}')\right]=U_{0_+}(\mathbf{r}), \quad \mathbf{r}\in\tau^{\perp}, \tag{9.16a}$$

and

$$\int_{\partial\tau} dS'\left[G_{0_+}(\mathbf{r},\mathbf{r}')\frac{\partial}{\partial n'}U_{0_+}(\mathbf{r}')-U_{0_+}(\mathbf{r}')\frac{\partial}{\partial n'}G_{0_+}(\mathbf{r},\mathbf{r}')\right]=0, \quad \mathbf{r}\in\tau. \tag{9.16b}$$

Equations (9.16) are a generalization to the case of inhomogeneous backgrounds of the famous *Kirchhoff–Helmholtz theorem* for homogeneous backgrounds that was established in Section 2.5 of Chapter 2 and, in the time domain, for the wave equation in Chapter 1. As mentioned in those earlier treatments, they are sometimes referred to as the *Helmholtz identities*. The first Helmholtz identity, Eq. (9.16a), is a formal solution to the *exterior boundary-value problem*; i.e., it is the solution to the homogeneous Helmholtz equation throughout $\tau^{\perp}$ that achieves specified boundary conditions in the form of the field $U_{0_+}(\mathbf{r}')$

and $\partial U_{0_+}(\mathbf{r}')/\partial n'$ at all points $\mathbf{r}'$ on the boundary $\partial\tau$ and satisfies the SRC. However, this boundary-value problem is not well posed in the sense that the boundary values $U_{0_+}(\mathbf{r}')$ and $\partial U_{0_+}(\mathbf{r}')/\partial n'$ are not independent and therefore cannot be assigned arbitrarily. As was the case for homogeneous backgrounds, these two quantities are coupled by the second Helmholtz identity, which implies that Dirichlet, Neumann or mixed conditions are necessary and sufficient in order to yield a unique solution to a properly posed boundary-value problem both in homogeneous and in inhomogeneous background media. We will not go into a detailed treatment of these boundary-value problems since it would follow identical lines to that developed in Chapter 2.

9.2.3 The Porter–Bojarski integral equation

The Porter–Bojarski (PB) integral equation was derived in Chapter 2 and used in Chapter 5 to solve the inverse source problem (ISP) in a homogeneous and lossless background medium. This equation relates the source of a radiated or scattered wave to the wavefield back propagated from over-specified boundary-value data over a closed surface surrounding the source support region. The equation plays a fundamental role in time-reversal imaging of sources embedded in lossless media and will be used in the following chapter in our treatment of imaging of systems of discrete scatterers.

We can derive the PB integral equation in a general inhomogeneous medium by following the same steps as used above in deriving the solution to the radiation problem, where, however, we replace the outgoing-wave Green function by the incoming-wave Green function $G_{0_-}(\mathbf{r},\mathbf{r}')$. Under this replacement Eq. (9.13a) becomes

$$\Phi(\mathbf{r}) + \int_{\tau_0} d^3r'\, G_{0_-}(\mathbf{r},\mathbf{r}')Q(\mathbf{r}') = U_{0_+}(\mathbf{r}),$$

where

$$\Phi(\mathbf{r}) = \int_{\partial\tau} dS' \left[U_{0_+}(\mathbf{r}')\frac{\partial}{\partial n'}G_{0_-}(\mathbf{r},\mathbf{r}') - G_{0_-}(\mathbf{r},\mathbf{r}')\frac{\partial}{\partial n'}U_{0_+}(\mathbf{r}')\right]. \tag{9.17a}$$

If we now make use of Eq. (9.15) we obtain the PB integral equation

$$\Phi(\mathbf{r}) = \int_{\tau_0} d^3r'\, G_{0_f}(\mathbf{r},\mathbf{r}')Q(\mathbf{r}'), \tag{9.17b}$$

where

$$G_{0_f}(\mathbf{r},\mathbf{r}') = G_{0_+}(\mathbf{r},\mathbf{r}') - G_{0_-}(\mathbf{r},\mathbf{r}') \tag{9.18}$$

is the free-space propagator of the medium. This quantity is the difference between two Green functions and, hence, satisfies the homogeneous Helmholtz equation over all of space.

As mentioned above, the PB integral equation forms the foundation of one formulation of the ISP and, as we will find in the following chapter, plays a key role in time-reversal imaging in lossless media. In such media $G_{0_-} = G^*_{0_+}$, so the complex conjugate of Φ defined in Eq. (9.17a) can be interpreted as being the field radiated into the interior region τ by surface sources equal to the time-reversed radiated (or scattered) field. This field is thus an "image" of the source formed by field data over $\partial\tau$ and, as we saw in Chapter 5, is, in some cases, a good approximation to the solution of the ISP.

9.2.4 The radiation pattern of the field

If we substitute the asymptotic expression Eq. (9.4a) for the outgoing-wave Green function into Eq. (9.15) we obtain

$$U_{0_+}(\mathbf{r}) \sim \left\{\int_{\tau_0} d^3r'\, g_+(\mathbf{s},\mathbf{r}')Q(\mathbf{r}')\right\}\frac{e^{ik_0r}}{r},$$

from which we conclude that the radiation pattern of the field is given by

$$f(\mathbf{s}) = -\frac{1}{4\pi}\int_{\tau_0} d^3r'\, \psi_+(\mathbf{r}',-k_0\mathbf{s})Q(\mathbf{r}'), \tag{9.19a}$$

where we have used the expression for the outgoing-wave Green-function radiation pattern $g_+(\mathbf{s},\mathbf{r}')$ given in Eq. (9.9b). This result reduces to the homogeneous background radiation pattern

$$f(\mathbf{s}) = -\frac{1}{4\pi}\int_{\tau_0} d^3r'\, e^{-ik_0\mathbf{s}\cdot\mathbf{r}'}Q(\mathbf{r}') = -\frac{1}{4\pi}\tilde{Q}(k_0\mathbf{s}),$$

where

$$\tilde{Q}(\mathbf{K}) = \int_{\tau_0} d^3r'\, e^{-i\mathbf{K}\cdot\mathbf{r}'}Q(\mathbf{r}')$$

is the spatial Fourier transform of the source when $k_0(\mathbf{r}')$ is constant and equal to k_0.

We can also express the radiation pattern in terms of boundary-value data over any surface $\partial\tau$ that completely surrounds the source volume τ_0. In particular, on making use of Eqs. (9.16a) and (9.9b) we find that

$$U_{0_+}(\mathbf{r}) \sim \int_{\partial\tau} dS'\left[g_+(\mathbf{s},\mathbf{r}')\frac{e^{ik_0r}}{r}\frac{\partial}{\partial n'}U_{0_+}(\mathbf{r}') - U_{0_+}(\mathbf{r}')\frac{\partial}{\partial n'}g_+(\mathbf{s},\mathbf{r}')\frac{e^{ik_0r}}{r}\right]$$
$$= \left\{-\frac{1}{4\pi}\int_{\partial\tau} dS'\left[\psi_+(\mathbf{r}',-k_0\mathbf{s})\frac{\partial}{\partial n'}U_{0_+}(\mathbf{r}') - U_{0_+}(\mathbf{r}')\frac{\partial}{\partial n'}\psi_+(\mathbf{r}',-k_0\mathbf{s})\right]\right\}\frac{e^{ik_0r}}{r},$$

which yields the result

$$f(\mathbf{s}) = -\frac{1}{4\pi}\int_{\partial\tau} dS'\left[\psi_+(\mathbf{r}',-k_0\mathbf{s})\frac{\partial}{\partial n'}U_{0_+}(\mathbf{r}') - U_{0_+}(\mathbf{r}')\frac{\partial}{\partial n'}\psi_+(\mathbf{r}',-k_0\mathbf{s})\right]. \tag{9.19b}$$

9.3 Generalized plane-wave expansions

The plane waves $\exp(ik_0\mathbf{s}_0 \cdot \mathbf{r})$ played a dominant role in radiation and scattering problems in homogeneous background media and we might expect that the plane-wave scattering states $\psi_\pm(\mathbf{r}, k_0\mathbf{s}_0)$ should play a similar role in these problems in inhomogeneous backgrounds. These quantities reduce to plane waves in the limit where $k_0(\mathbf{r}) \to k_0$ and are, thus, a natural generalization of the pure plane waves to inhomogeneous background media. Although these "generalized plane waves" are, indeed, useful in a number of applications, their utility is limited due to the fact that they are not, except in certain special cases, associated with any geometry that forms a separable system for the non-uniform-medium Helmholtz equation Eq. (9.7a). As discussed in Chapter 3, the conventional Helmholtz equation is separable in 11 coordinate systems that include the Cartesian system. Because of this the plane waves form a complete set of functions for representing solutions to the homogeneous Helmholtz equation *and* have the property that they can be used to fit data specified over all of space, as in the initial-value problem for the wave equation, or over infinite plane surfaces, as in the Rayleigh–Sommerfeld boundary-value problem. The Helmholtz equation in a non-uniform medium will not generally be separable in any coordinate system and, although the plane-wave scattering states form a complete set for expanding wavefields propagating in the background media, the expansion coefficients in these expansions cannot be easily determined from boundary-value data. However, as mentioned above, the plane-wave scattering states and the generalized plane-wave expansions employing these states have some utility in certain applications and for this reason, and for the sake of completeness, we include a treatment of such expansions in this section.

9.3.1 Generalized plane-wave expansions to the homogeneous Helmholtz equation in a non-uniform medium

We consider a wavefield $U_0(\mathbf{r})$ satisfying the homogeneous Helmholtz equation Eq. (9.7a) over all of space

$$[\nabla_{r'}^2 + k_0^2(\mathbf{r}')]U_0(\mathbf{r}') = 0.$$

If we couple the above equation with the Helmholtz equation for the outgoing-wave Green function $G_{0_+}(\mathbf{r}, \mathbf{r}')$ we obtain

$$G_{0_+}(\mathbf{r}, \mathbf{r}')\nabla_{r'}^2 U_0(\mathbf{r}') - U_0(\mathbf{r}')\nabla_{r'}^2 G_{0_+}(\mathbf{r}, \mathbf{r}') = -U_0(\mathbf{r}')\delta(\mathbf{r} - \mathbf{r}').$$

If we now integrate the above equation over the interior of an infinite sphere with radius $r' \to \infty$ and use Green's theorem we obtain

$$U_0(\mathbf{r}) = \lim_{r' \to \infty} r'^2 \int_{4\pi} d\Omega_{r'} \left\{ U_0(\mathbf{r}')\frac{\partial}{\partial r'}G_{0_+}(\mathbf{r}, \mathbf{r}') - G_{0_+}(\mathbf{r}, \mathbf{r}')\frac{\partial}{\partial r'}U_0(\mathbf{r}') \right\}.$$

As a final step we make use of the asymptotic expansions of the field $U_0(\mathbf{r}')$ and the outgoing-wave Green function $G_{0_+}(\mathbf{r}, \mathbf{r}')$ as $r' \to \infty$ with $\mathbf{r}$ finite. Since the field is a solution to the *homogeneous* Helmholtz equation it will not satisfy a radiation condition but will, instead, have both outgoing- and incoming-wave components at infinity, i.e.,

$$U_0(\mathbf{r}') \sim u_+(\hat{\mathbf{r}}')\frac{e^{ik_0 r'}}{r'} - u_-(\hat{\mathbf{r}}')\frac{e^{-ik_0 r'}}{r'} + O\left(\frac{1}{r'^2}\right), \quad r' \to \infty. \tag{9.20}$$

On making use of this result and the far-field expression of the Green function given in Eq. (9.9a) (with $\mathbf{r}$ and $\mathbf{r}'$ exchanged) we find after a bit of algebra that

$$U_0(\mathbf{r}) = \frac{ik_0}{2\pi}\int_{4\pi} d\Omega_s\, u_-(\mathbf{s})\psi_+(\mathbf{r}, -k_0\mathbf{s}). \tag{9.21a}$$

A completely parallel development using G_{0_-} in place of G_{0_+} yields the alternative expansion

$$U_0(\mathbf{r}) = \frac{ik_0}{2\pi}\int_{4\pi} d\Omega_s\, u_+(\mathbf{s})\psi_-(\mathbf{r}, k_0\mathbf{s}). \tag{9.21b}$$

The above (generalized) plane-wave expansions are precisely of the form of the plane-wave expansions of solutions to the homogeneous Helmholtz equation in a uniform background that were derived in Section 3.2 of Chapter 3. In both cases they consist of superpositions of plane waves or plane-wave scattering states with an arbitrary plane-wave amplitude. Note, however, that the derivation of the plane-wave expansions given above not only yields the general form of the expansion but also relates the plane-wave amplitudes to the far-field amplitudes of the incoming- and outgoing-wave components of the field! We employed a different derivation of the plane-wave expansion in Chapter 3 that did not yield this result.

Example 9.2 If we specialize the general plane-wave expansions given in Eqs. (9.21) to the case of a homogeneous background with constant wavenumber k_0 then the plane-wave scattering states $\psi_\pm(\mathbf{r}, k_0\mathbf{s})$ become the plane wave $\exp(ik_0\mathbf{s}\cdot\mathbf{r})$ and the expansions become

$$U_0(\mathbf{r}) = \frac{ik_0}{2\pi}\int_{4\pi} d\Omega_s\, u_\pm(\mathbf{s})e^{\pm ik_0\mathbf{s}\cdot\mathbf{r}}. \tag{9.22}$$

As an example we consider the free-field propagator

$$G_{0_f}(\mathbf{r}-\mathbf{r}') = G_{0_+}(\mathbf{r}-\mathbf{r}') - G_{0_-}(\mathbf{r}-\mathbf{r}') = -\frac{i}{2\pi}\frac{\sin(k_0|\mathbf{r}-\mathbf{r}'|)}{|\mathbf{r}-\mathbf{r}'|}, \tag{9.23}$$

which played a major role in the ISP for the wave equation treated at the beginning of Chapter 5. The free-field propagator is the difference between two Green functions and, hence, satisfies the homogeneous Helmholtz equation over all of space. It can thus be represented via the plane-wave expansion Eq. (9.22) with $\mathbf{r}'$ a free parameter. The far-field amplitudes $u_\pm(\mathbf{s})$ are readily obtained from Eq. (9.23) and we find that

$$u_\pm(\mathbf{s}) = -\frac{1}{4\pi}e^{\mp ik_0\mathbf{s}\cdot\mathbf{r}'},$$

which, when substituted into Eq. (9.22), yields the result

$$G_{0_f}(\mathbf{r}-\mathbf{r}') = -\frac{ik_0}{8\pi^2}\int_{4\pi} d\Omega_s\, e^{ik_0\mathbf{s}\cdot(\mathbf{r}-\mathbf{r}')}.$$

The plane-wave expansion of the free-field propagator found above was previously derived in Example 4.2 of Chapter 4 by making use of the angular-spectrum expansions of the outgoing- and incoming-wave Green functions in the definition Eq. (9.23) of the free-field propagator. The above derivation based on the general expressions for the plane-wave amplitudes given in Eq. (9.20) is much simpler and more elegant than the previous derivation and allows solutions of the homogeneous Helmholtz equation both in uniform and in non-uniform media to be computed given only the far-field behavior of the fields.

Example 9.3 As a second example we consider the plane-wave expansion of the free-field propagator in an inhomogeneous medium. This quantity which previously arose as the kernel of the PB integral equation in Section 9.2.3 is the natural generalization of the free-field propagator in a homogeneous medium, being defined as the difference between the outgoing- and incoming-wave Green functions of the medium. The free-field propagator behaves asymptotically as

$$\begin{aligned} G_{0_f}(\mathbf{r},\mathbf{r}') &= G_{0_+}(\mathbf{r},\mathbf{r}') - G_{0_-}(\mathbf{r},\mathbf{r}') \\ &\sim g_+(\mathbf{r},\hat{\mathbf{r}}')\frac{e^{ik_0r'}}{r'} - g_-(\mathbf{r},\hat{\mathbf{r}}')\frac{e^{-ik_0r'}}{r'} + O\left(\frac{1}{r'^2}\right), \quad r'\to\infty, \end{aligned}$$

with $g_\pm(\mathbf{r},\hat{\mathbf{r}}';k_0)$ defined in Eq. (9.9b). Because it is the difference between two Green functions, the free-field propagator satisfies the homogeneous Helmholtz equation with respect to both $\mathbf{r}$ and $\mathbf{r}'$ and thus admits a plane-wave expansion of the form given in Eqs. (9.21) with either $\mathbf{r}$ or $\mathbf{r}'$ being a free parameter. On making use of Eq. (9.3) and the definition of $u_\pm$ given in Eq. (9.20) we find that

$$u_\pm(\mathbf{s}) = -\frac{1}{4\pi}\psi_\pm(\mathbf{r},\mp k_0\mathbf{s}),$$

where we have selected $\mathbf{r}$ to be the free parameter. The plane-wave expansion of the free-field propagator is then obtained from either of Eqs. (9.21) and found to be

$$G_{0_f}(\mathbf{r},\mathbf{r}') = -\frac{ik_0}{8\pi^2}\int_{4\pi} d\Omega_s\, \psi_+(\mathbf{r}',-k_0\mathbf{s})\psi_-(\mathbf{r},k_0\mathbf{s}), \tag{9.24a}$$

which, due to symmetry of the free-field propagator, can also be written in the form

$$G_{0_f}(\mathbf{r},\mathbf{r}') = -\frac{ik_0}{8\pi^2}\int_{4\pi} d\Omega_s\, \psi_+(\mathbf{r},-k_0\mathbf{s})\psi_-(\mathbf{r}',k_0\mathbf{s}). \tag{9.24b}$$

It is easily verified that this plane-wave expansion of G_{0_f} reduces to the plane-wave expansion of the free-field propagator given in Example 9.2 when $k_0(\mathbf{r})\to k_0$.

9.3.2 Generalized angular-spectrum expansions

The plane-wave expansions that we obtained in the preceding section apply to wavefields that propagate in the background medium without the presence of sources or scatterers and employ plane-wave scattering states $\psi_+(\mathbf{r}, k_0\mathbf{s})$ that have purely real unit propagation vectors $\mathbf{s}$ lying on the surface of the (real) unit sphere. We saw in the case of uniform backgrounds in earlier chapters that these types of plane-wave expansions do not suffice for representing either radiated or scattered fields and it is necessary to include so-called *evanescent plane waves* that are characterized by complex unit propagation vectors in the expansions of these types of wavefields.

For example, the plane-wave expansion of the outgoing-wave Green function in a uniform background having wavenumber k_0, called the *Weyl expansion*, that was derived in Section 4.1 of Chapter 4 is given by

$$G_{0_+}(\mathbf{r}-\mathbf{r}') = -\frac{ik_0}{8\pi^2}\int_{-\pi}^{\pi} d\beta \int_{C_\pm} \sin\alpha \, d\alpha \, e^{ik_0\mathbf{s}\cdot(\mathbf{r}-\mathbf{r}')},$$

where $\mathbf{s} = \sin\alpha\cos\beta\,\hat{\mathbf{x}} + \sin\alpha\sin\beta\,\hat{\mathbf{y}} + \cos\alpha\,\hat{\mathbf{z}}$ and the contours $C_\pm$ are displayed in Fig. 4.1 of Chapter 4; C_+ is used if $z > z'$ and C_- if $z < z'$. These types of plane-wave expansions that can represent radiated and scattered fields are referred to as *angular-spectrum expansions* and played an important role in the theory developed in previous chapters. For example, the angular-spectrum expansion of outgoing-wave solutions to the Helmholtz equation formed the basis in Chapter 4 for the important concept of field *back propagation* that was used extensively in our solutions of the ISP and ISCP in later chapters.

Although it is possible to derive a generalized type of Weyl expansion for the outgoing-wave Green function $G_{0_+}(\mathbf{r}, \mathbf{r}')$ the details of the derivation are quite involved, especially in the case of lossy backgrounds where the background scattering potential V_0 is complex. The derivation is a bit more manageable in the case of lossless backgrounds where V_0 is real-valued and the Helmholtz operator is Hermitian. In this case standard results from quantum collision theory can be employed and we show in Appendix B that

$$G_{0_+}(\mathbf{r}, \mathbf{r}') = -\frac{ik_0}{8\pi^2}\int_{-\pi}^{\pi} d\beta \int_{C_\pm} \sin\alpha \, d\alpha \, \psi_+(\mathbf{r}, k_0\mathbf{s})\psi_-^*(\mathbf{r}', -k_0\mathbf{s}), \tag{9.25a}$$

where, as in the homogeneous-background case, the contour C_+ is used if $z > z'$ and the contour C_- is used if $z < z'$. It is evident that Eq. (9.25a) reduces to the conventional Weyl expansion in the homogeneous-background case. In their angle-variable forms given above the Weyl expansion for both homogeneous and inhomogeneous backgrounds does not decompose into evanescent plane waves that decay in a specific direction but generally includes plane waves that decay or grow in more than one direction. However, the plane waves and plane-wave scattering states are entire analytic functions of the unit propagation vector **s**, so the α contours of integration can be transformed to lie along the real-α axis and the line $\Re\alpha = \pi/2$ in the complex-α plane, which correspond exactly to a decomposition into homogeneous and evanescent plane waves (see the discussion in Section 3.2.2 of Chapter 3).

The Weyl expansion can be written in an alternative form by making use of the reciprocity condition Eq. (9.6). In particular, we find that

$$G_{0_+}(\mathbf{r},\mathbf{r}') = -\frac{ik_0}{8\pi^2}\int_{-\pi}^{\pi} d\beta \int_{C_\mp} \sin\alpha\, d\alpha\, \psi_+(\mathbf{r}', k_0\mathbf{s})\psi_-(\mathbf{r}, -k_0\mathbf{s}).$$

The α contour of integration is changed in the above expansion so that C_- is now used if $z > z'$ and C_+ if $z < z'$. However, if we now make the change of integration variables $\beta \to \beta + \pi$ and $\alpha \to \pi - \alpha$ resulting in $\mathbf{s} \to -\mathbf{s}$ then $C_\mp \to C_\pm$ and the expansion assumes the form (cf. Section 4.1 of Chapter 4)

$$G_{0_+}(\mathbf{r},\mathbf{r}') = -\frac{ik_0}{8\pi^2}\int_{-\pi}^{\pi} d\beta \int_{C_\pm} \sin\alpha\, d\alpha\, \psi_-(\mathbf{r}, k_0\mathbf{s})\psi_+(\mathbf{r}', -k_0\mathbf{s}), \tag{9.25b}$$

where now C_+ is again used if $z > z'$ and C_- if $z < z'$.

Although we have derived the generalized Weyl expansion in Appendix B under the assumption of a lossless medium where the Helmholtz operator is Hermitian, Eqs. (9.25) actually hold more generally in cases in which V_0 is complex and the background is lossy. However, the derivation of the expansions in this more general scenario is quite involved and requires the use of biorthogonal expansions (Morse and Feshbach, 1953), so it will not be presented here.

9.3.3 Angular-spectrum expansion of the radiated field in non-uniform media

We now assume that the source is confined to a strip $z^- \leq z \leq z^+$ and restrict our attention to field points **r** lying outside this strip. Then, using arguments completely parallel to those employed in Section 4.2 of Chapter 4, we substitute the generalized Weyl expansion of the Green function in Eq. (9.25b) into Eq. (9.15) to obtain the angular-spectrum expansion of the radiated field valid everywhere outside the source strip $z^- \leq z \leq z^+$:

$$U_{0_+}(\mathbf{r}) = \int_{\tau_0} d^3r' \left\{-\frac{ik_0}{8\pi^2}\int_{-\pi}^{\pi} d\beta \int_{C_\pm} \sin\alpha\, d\alpha\, \psi_-(\mathbf{r}, k_0\mathbf{s})\psi_+(\mathbf{r}', -k_0\mathbf{s})\right\} Q(\mathbf{r}')$$

$$\Downarrow$$

$$U_{0_+}(\mathbf{r}) = \frac{ik_0}{2\pi}\int_{-\pi}^{\pi} d\beta \int_{C_\pm} \sin\alpha\, d\alpha\, A(k_0\mathbf{s})\psi_-(\mathbf{r}, k_0\mathbf{s}), \tag{9.26a}$$

where the angular spectrum $A(k_0\mathbf{s})$ is given by

$$A(k_0\mathbf{s}) = -\frac{1}{4\pi}\int_{\tau_0} d^3r'\, Q(\mathbf{r}')\psi_+(\mathbf{r}', -k_0\mathbf{s}) \tag{9.26b}$$

and the α integration contour C_+ is used in the r.h.s. $z > z^+$ and the contour C_- in the l.h.s. $z < z^-$. We note that the orientation of the Cartesian coordinate system is arbitrary, so such an expansion can be used throughout any two parallel half-spaces that lie outside the source volume τ_0.

On making use of Eq. (9.19a) we can express the angular-spectrum expansion in terms of the radiation pattern of the field[4]

$$U_{0_+}(\mathbf{r}) = \frac{ik_0}{2\pi}\int_{-\pi}^{\pi} d\beta \int_{C_\pm} \sin\alpha\, d\alpha\, f(\mathbf{s})\psi_-(\mathbf{r}, k_0\mathbf{s}). \tag{9.27}$$

The above expansion performs the operation of field *back propagation* from the radiation pattern of the source that was developed for homogeneous backgrounds in Section 4.3 of Chapter 4. All of the discussion of field back propagation that was presented in that section also applies here for inhomogeneous backgrounds, including and most importantly, the instability of this process due to the inclusion of *evanescent waves* in the angular-spectrum expansion. Since the evanescent plane waves *decay* exponentially fast with distance from the source boundary planes $z = z^\pm$ they *grow* exponentially fast as the source boundary planes are approached from more distant boundaries. It then follows that any small errors in the (analytically continued) radiation pattern $f(\mathbf{s})$ in the evanescent region will be amplified exponentially fast as the source region is approached from the far field (see the discussion in Section 4.3 of Chapter 4); i.e., exact field back propagation is mathematically unstable. Despite being unstable it is, nevertheless, mathematically exact and allows us to prove a number of general results and theorems that are important for a complete understanding of inverse problems related to the wave and Helmholtz equations.

9.4 Non-radiating sources in non-uniform media

The basic definition and all of the results pertaining to frequency-domain non-radiating (NR) sources embedded in uniform media developed in Sections 1.7.1 and 2.7 of Chapters 1 and 2 carry over to inhomogeneous backgrounds. In particular, such sources are still characterized by Definition 1.1, which gives as necessary and sufficient conditions for a source to be NR at some given frequency ω that its radiated field $U_{0_+}(\mathbf{r})$ as given

[4] Note that we require that the radiation pattern be extended into an entire function of the unit vector $\mathbf{s}$ in Eq. (9.27). That this is possible follows from the assumption that the source support τ_0 is compact and the plane-wave scattering states are entire functions of $\mathbf{s}$.

by Eq. (9.15) must vanish everywhere outside its spatial support τ_0 at that particular frequency. Piecewise-continuous compactly supported NR sources are constructed using the recipe (cf. Eq. (1.57))

$$Q_{\rm nr}(\mathbf{r}) = [\nabla^2 + k_0^2(\mathbf{r})]\Pi(\mathbf{r}), \tag{9.28}$$

where $\Pi(\mathbf{r})$ is a function that is compactly supported in the spatial volume τ_0 at any given frequency ω and possesses continuous first partial derivatives throughout this volume but is otherwise arbitrary. The field radiated by the NR source defined above is found using Eq. (9.15):

$$U_{\rm nr}(\mathbf{r}) = \int_{\tau_0} d^3r'\, G_{0_+}(\mathbf{r},\mathbf{r}')\overbrace{\{[\nabla_{r'}^2 + k_0^2(\mathbf{r}')]\Pi(\mathbf{r}')\}}^{Q_{\rm nr}(\mathbf{r}')}$$
$$= \int_{\tau_0} d^3r'\, \overbrace{[\nabla_{r'}^2 + k_0^2(\mathbf{r}')]G_{0_+}(\mathbf{r},\mathbf{r}')}^{\delta(\mathbf{r}-\mathbf{r}')}\,\Pi(\mathbf{r}') = \Pi(\mathbf{r}),$$

where we have twice integrated by parts and dropped the surface terms due to the assumption that Π is continuously differentiable throughout τ_0. The field $U_{\rm nr}$ vanishes outside τ_0, which then establishes that the source defined via Eq. (9.28) is an NR source at frequency ω and, moreover, that the field it generates is the compactly supported function $\Pi(\mathbf{r})$.

The question arises as to whether an NR source in a uniform medium will also be NR when embedded in a non-uniform medium. The answer is definitely no, due to the fact that the radiated field within the interior of the source will interact with the background in which it is embedded and thus generate an outgoing (scattered) wave that will not vanish outside the source's support. In particular, the field radiated by such a source when embedded in a non-uniform background is given by

$$U_{0_+}(\mathbf{r}) = \int_{\tau_0} d^3r'\, \overbrace{[\nabla_{r'}^2 + k_0^2]\Pi(\mathbf{r}')}^{Q_{0_{\rm NR}}(\mathbf{r}')}\, G_{0_+}(\mathbf{r},\mathbf{r}'), \tag{9.29a}$$

where $Q_{0_{\rm NR}}(\mathbf{r}')$ is an NR source in the uniform medium having wavenumber k_0. If we now integrate by parts and drop the surface terms due to the continuity and differentiability conditions on $\Pi(\mathbf{r}')$ we find that at all points outside τ_0

$$U_{0_+}(\mathbf{r}) = \int_{\tau_0} d^3r'\, \Pi(\mathbf{r}')[\nabla_{r'}^2 + k_0^2]G_{0_+}(\mathbf{r},\mathbf{r}') = \int_{\tau_0} d^3r'\, \Pi(\mathbf{r}')V_0(\mathbf{r}')G_{0_+}(\mathbf{r},\mathbf{r}'), \tag{9.29b}$$

which is the field scattered from the background scattering potential $V_0(\mathbf{r}')$ confined to τ_0 by the interior field $\Pi(\mathbf{r}')$ radiated by the source $Q_{0_{\rm NR}}(\mathbf{r}')$ within the source region.

9.4.1 Non-radiating sources and the radiation pattern

In our treatment of NR sources for the Helmholtz equation in Chapter 2 we established Theorem 2.1 that, in effect, states that the necessary and sufficient condition for a compactly supported source in a uniform medium to be NR is that its radiation pattern vanish. The same theorem holds for compactly supported sources in non-uniform media, which we state formally as follows.

Theorem 9.1 *Let $Q(\mathbf{r})$ be a piecewise-continuous source compactly supported in τ_0 in a background medium with wavenumber $k_0(\mathbf{r})$. Then the necessary and sufficient condition for Q to be NR at any given frequency ω is that the source's radiation pattern $f(\mathbf{s})$ vanish at that frequency.*

The necessary part of the theorem is obvious from the definition of NR sources. A constructive proof of the sufficiency condition can be established in two steps as follows. First we use the angular-spectrum expansion given in Eq. (9.27) to construct the field everywhere outside the convex hull of the source support τ_0. This then shows that the vanishing of the radiation pattern guarantees that the radiated field vanishes everywhere outside the convex hull of τ_0 and, thus, establishes the theorem for source supports τ_0 whose surfaces are convex. Now assume that the source surface is not convex. We then follow a procedure entirely analogous to that used in Section 4.8.3 to continue the known field outside the convex hull of the source support to points that lie outside τ_0 but interior to the convex hull. Since the field vanishes outside the convex hull, the continued field will also vanish, which then establishes the theorem.

9.5 The inverse source problem

In this section we will generalize the treatment of the ISP presented in Chapter 5 to inhomogeneous background media. As in that chapter, we will employ a Hilbert-space formulation of the problem that makes use of the singular value decomposition (SVD) to generate a least-squares minimum norm (pseudo-inverse) solution of the ISP. We will then specialize this general formulation to two specific cases that are easily solved. More general source geometries and backgrounds can be treated using the developed theory but will, in general, require a numerically based solution.

The question naturally arises as to what applications require a solution to the ISP for a source embedded in an inhomogeneous medium. As mentioned at the beginning of Chapter 5, there are basically two applications for the ISP, irrespective of whether it is posed for sources in uniform backgrounds or for sources in inhomogeneous backgrounds. The first of these is that of 3D (or 2D) imaging, where the goal is to estimate the interior of a radiating source from observations of the radiated field. This problem is very closely tied to the inverse scattering problem (ISCP) that was treated in Chapter 8 and that will be generalized to inhomogeneous backgrounds later in this chapter. As was the case for homogeneous backgrounds, the formulation of the ISP for inhomogeneous backgrounds will lay the foundation for the ISCP in inhomogeneous backgrounds and will be useful in its own right for imaging coherent sources embedded in known backgrounds.

The second application is that of source, or antenna, design, where the goal is to design a source to radiate a prescribed field (radiation pattern). This second application for the ISP in inhomogeneous media is intriguing since it offers the possibility of designing novel antennas by embedding a conventional antenna in an inhomogeneous medium. The idea is that the *induced source* (antenna in background) will somehow outperform the antenna when radiating in a uniform background. We will investigate this aspect of the ISP in some detail below, where we will find that fundamental limitations allow, at best, only moderate performance increases to be obtained by the use of such schemes (Devaney *et al.*, 2007).

9.5.1 General formulation

We showed in Chapter 5 that in the case of homogeneous backgrounds the ISP can be directly posed in terms of the radiation pattern with no loss of generality. The same applies to inhomogeneous backgrounds, where it follows from Eq. (9.27) that the radiation pattern uniquely specifies the radiated field everywhere outside the source support volume τ_0. In view of this we can pose the ISP directly in terms of the radiation pattern without any loss of generality. Indeed, we showed in Section 9.2.4 that the radiation pattern can be computed from field data specified over any closed surface surrounding the source support volume so that, if necessary, the ISP can be solved in two steps for data specified over such surfaces.

Following our treatment of the ISP presented in Chapter 5, we define the operator

$$\hat{T} = -\frac{1}{4\pi}\int_{\tau_0} d^3r'\,\psi_+(\mathbf{r}', -k_0\mathbf{s}), \tag{9.30a}$$

so that Eq. (9.19a) assumes the form

$$\hat{T}Q = f. \tag{9.30b}$$

Equation (9.30b) is identical in form to the defining equation Eq. (5.29) for the ISP in a homogeneous background and virtually all of the discussion presented in Chapter 5 regarding this equation applies here. In particular, we will assume that the source Q is contained in the Hilbert space $\mathcal{H}_Q = L^2(\tau_0)$ of square-integrable functions in τ_0 and that the radiation pattern $f(\mathbf{s})$ is in the Hilbert space $\mathcal{H}_f = L^2(\Omega)$ of square-integrable functions on the unit sphere. Both of these spaces are equipped with the standard inner products

$$\langle Q_1, Q_2\rangle_{\mathcal{H}_Q} = \int_{\tau_0} d^3r\, Q_1^*(\mathbf{r})Q_2(\mathbf{r}),$$

$$\langle f_1, f_2\rangle_{\mathcal{H}_f} = \int_{4\pi} d\Omega_s f_1^*(\mathbf{s})f_2(\mathbf{s}),$$

with induced norms $||Q|| = \sqrt{\langle Q, Q\rangle_{\mathcal{H}_Q}}$ and $||f|| = \sqrt{\langle f, f\rangle_{\mathcal{H}_f}}$.

The adjoint operator $\hat{T}^\dagger$ is obtained in the usual manner from its defining equation

$$\langle f, \hat{T}Q\rangle_{\mathcal{H}_f} = \langle \hat{T}^\dagger f, Q\rangle_{\mathcal{H}_Q},$$

from which we find that

$$\hat{T}^\dagger f = -\frac{1}{4\pi}\mathcal{M}_{\tau_0}\int_{4\pi} d\Omega\, \psi_+^*(\mathbf{r}', -k_0\mathbf{s})f(\mathbf{s}), \tag{9.31a}$$

where

$$\mathcal{M}_{\tau_0} = \begin{cases} 1 & \text{if } \mathbf{r} \in \tau_0 \\ 0 & \text{else} \end{cases} \tag{9.31b}$$

is a masking operator.

Following almost identical steps to those employed in Examples 5.3 and 5.4 of Chapter 5 and using the homogeneous plane-wave expansions developed in Section 9.3, it is not difficult to show that the operator $\hat{T}$ and its adjoint are compact.

9.5.2 Singular value decomposition

The fact that the operators $\hat{T}$ and $\hat{T}^\dagger$ are compact allows us to employ the SVD to solve the ISP. Following the treatment employed in Section 5.4.2 of Chapter 5 we thus define the system $\{v_p, u_p, \sigma_p\}$ via the set of equations

$$\hat{T}v_p = \sigma_p u_p, \qquad \hat{T}^\dagger u_p = \sigma_p v_p, \tag{9.32a}$$

where the *singular values* σ_p are a discrete set of non-negative constants and $p = 0, 1, \ldots$ is an integer index that labels the singular set. The singular functions $v_p \in \mathcal{H}_Q$ and $u_p \in \mathcal{H}_f$ are orthonormal and complete in their respective Hilbert spaces and can thus represent the operators $\hat{T}$ and $\hat{T}^\dagger$ in the expansions

$$\hat{T} = \int_{\tau_0} d^3r \sum_p \sigma_p u_p(\mathbf{s})v_p^*(\mathbf{r}), \qquad \hat{T}^\dagger = \int_{4\pi} d\Omega \sum_p \sigma_p v_p(\mathbf{r})u_p^*(\mathbf{s}), \tag{9.32b}$$

which are called the "singular value decompositions" of the two operators. It follows from the defining equations Eqs. (9.32a) that the singular functions v_p and u_p satisfy the *normal equations*

$$\hat{T}^\dagger\hat{T}v_p = \sigma_p^2 v_p, \qquad \hat{T}\hat{T}^\dagger u_p = \sigma_p^2 u_p. \tag{9.32c}$$

The composite operator $\hat{T}^\dagger \hat{T}$ is easily found using the definitions given in Eq. (9.30a) and (9.31a). In particular, we have that

$$\hat{T}^\dagger \hat{T} = \frac{1}{(4\pi)^2}\mathcal{M}_{\tau_0}\int_{4\pi} d\Omega\, \psi_+^*(\mathbf{r}, -k_0\mathbf{s}) \int_{\tau_0} d^3r'\, \psi_+(\mathbf{r}', -k_0\mathbf{s})$$
$$= \frac{1}{(4\pi)^2}\mathcal{M}_{\tau_0}\int_{\tau_0} d^3r' \left\{\int_{4\pi} d\Omega\, \psi_+^*(\mathbf{r}, -k_0\mathbf{s})\psi_+(\mathbf{r}', -k_0\mathbf{s})\right\}. \tag{9.33}$$

Example 9.4 If we substitute the representation for $\hat{T}^\dagger \hat{T}$ given in Eq. (9.33) into the normal equation Eq. (9.32c) we find that the singular functions $v_p(\mathbf{r})$ having singular values $\sigma_p > 0$ satisfy the homogeneous Helmholtz equation *with background wavenumber* $k_0^*(\mathbf{r})$ within the source support volume τ_0; i.e.,

$$[\nabla^2 + k_0^{2*}(\mathbf{r})]\sigma_p^2 v_p(\mathbf{r}) = [\nabla^2 + k_0^{2*}(\mathbf{r})]\hat{T}^\dagger \hat{T} v_p(\mathbf{r})$$
$$= [\nabla^2 + k_0^{2*}(\mathbf{r})]\frac{1}{(4\pi)^2}\mathcal{M}_{\tau_0}$$
$$\times \int_{\tau_0} d^3r' \left\{\int_{4\pi} d\Omega\, \psi_+^*(\mathbf{r}, -k_0\mathbf{s})\psi_+(\mathbf{r}', -k_0\mathbf{s})\right\} v_p(\mathbf{r}')$$
$$= \frac{1}{(4\pi)^2}\mathcal{M}_{\tau_0}\int_{4\pi} d\Omega \left\{\overbrace{[\nabla^2 + k_0^{2*}(\mathbf{r})]\psi_+^*(\mathbf{r}, -k_0\mathbf{s})}^{0}\right.$$
$$\left.\times \int_{\tau_0} d^3r'\, v_p(\mathbf{r}')\psi_+(\mathbf{r}', -k_0\mathbf{s})\right\}$$
$$= 0,$$

so long as the field point $\mathbf{r} \in \tau_0$. The above result can be seen to be the generalization of the corresponding result established for homogeneous backgrounds in Example 5.6.

9.5.3 The least-squares pseudo-inverse solution of the ISP

The singular system $\{v_p, u_p, \sigma_p\}$ can, in principle, be computed for any given background wavenumber and source geometry from the defining equations Eq. (9.32a) or from the normal equations Eqs. (9.32c). Since the singular functions $\{v_p(\mathbf{r})\}$ are complete and orthonormal in $\mathcal{H}_Q$ and $\{u_p(\mathbf{s})\}$ are complete and orthonormal in $\mathcal{H}_f$ we can expand the unknown source $Q(\mathbf{r})$ and (known) data $f(\mathbf{s})$ into these sets to obtain

$$Q(\mathbf{r}) = \sum_p \langle v_p, Q\rangle_{\mathcal{H}_Q} v_p(\mathbf{r}), \tag{9.34a}$$

$$f(\mathbf{s}) = \sum_p \langle u_p, f\rangle_{\mathcal{H}_f} u_p(\mathbf{s}). \tag{9.34b}$$

On substituting these expansions into Eq. (9.30b) and making use of Eq. (9.32a) we then obtain the result

$$\sum_p \sigma_p \langle v_p, Q \rangle_{\mathcal{H}_Q} u_p(\mathbf{s}) = \sum_p \langle u_p, f \rangle_{\mathcal{H}_f} u_p(\mathbf{s}),$$

from which we conclude that

$$\langle v_p, Q \rangle_{\mathcal{H}_Q} = \frac{\langle u_p, f \rangle_{\mathcal{H}_f}}{\sigma_p}, \quad \sigma_p > 0. \tag{9.35}$$

As a final step we substitute the above result into the expansion Eq. (9.34a) to obtain the following formal solution to the ISP

$$\hat{Q}(\mathbf{r}) = \sum_{\sigma_p > 0} \frac{\langle u_p, f \rangle_{\mathcal{H}_f}}{\sigma_p} v_p(\mathbf{r}). \tag{9.36}$$

The above "solution" of the ISP is identical in form to that obtained in Section 5.4.4 of Chapter 5 for the case of a source embedded in a homogeneous background and all of the discussion presented in that section applies here. In particular, we showed that, because the above expansion extends only over non-zero singular values, $\hat{Q}$ has no components in the null space of the operator $\hat{T}$ and is, thus, a minimum-norm or "pseudo-inverse" solution of the ISP; i.e., among all solutions to Eqs. (9.30b) $\hat{Q}$ is that particular solution whose norm $||\hat{Q}||$ is smallest. Moreover, in the case of non-perfect data, it is, at best, a least-squares solution to Eq. (9.30b). We say "at best" because the above expansion will not even converge in $\mathcal{H}_Q$ unless the data $f(\mathbf{s})$ satisfy the *Picard condition*

$$\sum_{\sigma_p > 0} \left| \frac{\langle u_p, f \rangle_{\mathcal{H}_f}}{\sigma_p} \right|^2 < \infty, \tag{9.37}$$

which is simply a guarantee that $\hat{Q}$ have finite norm in $\mathcal{H}_Q$. We will not delve further into these issues here, but simply refer the reader to the extensive treatment of this material presented earlier in Section 5.4.4 for further details.

9.6 Solution of the ISP for spherically symmetric backgrounds

In order to construct $\hat{Q}$ it is necessary to compute the singular system for singular values $\sigma_p > 0$. Even for the case of a homogeneous background this can be done analytically only for source geometries whose boundaries coincide with a surface in one of the 11 separable coordinate systems for the Helmholtz operator in a uniform medium. In the case of non-homogeneous backgrounds the same requirement applies, but where the Helmholtz operator is now defined for the non-homogeneous medium; i.e., the constant background wavenumber k_0 is replaced by $k_0(\mathbf{r})$. Because of this the number of coordinate systems in which this operator is separable is much reduced and will, in general, be zero. One case of importance that can be easily solved is that of a source compactly supported within a spherical volume $\tau_0 : r < a_0$ in a spherically symmetric background characterized by a scattering potential $V_0(r)$.[5]

The plane-wave scattering states $\psi_\pm(\mathbf{r}, k_0\mathbf{s}_0)$ for spherically symmetric backgrounds were found in Example 9.1. If in that example we replace $\mathbf{s}_0$ by $-\mathbf{s}$ and use the fact that

$$\sum_m Y_l^{m*}(-\mathbf{s})Y_l^m(\hat{\mathbf{r}}) = \sum_m Y_l^m(-\mathbf{s})Y_l^{m*}(\hat{\mathbf{r}})$$

we find that

$$\psi_+(\mathbf{r}, -k_0\mathbf{s}) = 4\pi \sum_{l,m} i^l g_l^+(r) Y_l^m(-\mathbf{s}) Y_l^{m*}(\hat{\mathbf{r}}), \tag{9.38a}$$

where the radial functions $g_l^+(r)$ satisfy the integral equation

$$g_l^+(r) = j_l(k_0 r) - ik_0 \int_0^\infty r'^2\, dr'\, j_l(k_0 r_<) h_l^+(k_0 r_>) V_0(r') g_l^+(r'). \tag{9.38b}$$

On substituting Eq. (9.38a) into the r.h.s. of Eq. (9.33) we then find that

$$\begin{aligned}\hat{T}^\dagger \hat{T} &= \frac{1}{(4\pi)^2} \mathcal{M}_{\tau_0} \int_{\tau_0} d^3r' \left\{ \int_{4\pi} d\Omega_s\, \psi_+^*(\mathbf{r}, -k_0\mathbf{s}) \psi_+(\mathbf{r}', -k_0\mathbf{s}) \right\} \\ &= \mathcal{M}_{\tau_0} \sum_{l,m} \int_{\tau_0} d^3r'\, g_l^{+*}(r) g_l^+(r') Y_l^m(\hat{\mathbf{r}}) Y_l^{m*}(\hat{\mathbf{r}}') v_p(\mathbf{r}').\end{aligned}$$

The normal equation Eq. (9.32c) satisfied by the singular functions $v_p(\mathbf{r})$ then becomes

$$\mathcal{M}_{\tau_0} \sum_{l,m} \int_{\tau_0} d^3r'\, g_l^{+*}(r) g_l^+(r') Y_l^m(\hat{\mathbf{r}}) Y_l^{m*}(\hat{\mathbf{r}}') v_p(\mathbf{r}') = \sigma_p^2 v_p(\mathbf{r}). \tag{9.39}$$

The kernel on the l.h.s. of Eq. (9.39) is separable and, using arguments identical to those employed in Section 5.5 of Chapter 5 in our solution of the ISP for a source in a spherical support volume in a uniform background, we find that the solution of the above equation is given by

$$v_{l,m}(\mathbf{r}) = \begin{cases} -(i^l/\sigma_l) g_l^{+*}(r) Y_l^m(\hat{\mathbf{r}}) & \text{if } r < a_0, \\ 0 & \text{if } r > a_0, \end{cases} \tag{9.40a}$$

where the index p is the doublet $p = \{l, m\}$ and

$$\sigma_l = \sqrt{\int_0^{a_0} r^2\, dr |g_l^+(r)|^2}. \tag{9.40b}$$

Note that, in agreement with Example 9.4, the singular functions defined in Eq. (9.40a) satisfy the homogeneous Helmholtz equation with wavenumber $k_0^*(\mathbf{r})$.

[5] We emphasize that the source need not be spherically symmetric, only the background medium in which it is embedded.

The singular functions $\{u_p, \sigma_p > 0\}$ can be computed from the singular functions $\{v_p\}$ using the defining equations Eq. (9.32a) for the SVD. Again, following an identical procedure to that employed in Section 5.5, we find that

$$u_{l,m}(\mathbf{s}) = Y_l^m(\mathbf{s}), \qquad \sigma_l > 0, \tag{9.40c}$$

which are identical to the singular functions $u_{l,m}$ found in Section 5.5.

Solution of the ISP

On making use of the SVD obtained above in the least-squares pseudo-inverse $\hat{Q}$ given in Eq. (9.36) we obtain the result

$$\hat{Q}(\mathbf{r}) = \sum_{l=0}^{\infty}\sum_{m=-l}^{l} \frac{\langle Y_l^m, f\rangle_{\mathcal{H}_f}}{\sigma_l} v_{l,m}(\mathbf{r}) = \sum_{l=0}^{\infty}\sum_{m=-l}^{l} \frac{f_l^m}{\sigma_l} v_{l,m}(\mathbf{r})$$

$$\Downarrow$$

$$\hat{Q}(\mathbf{r}) = -\sum_{l=0}^{\infty}\sum_{m=-l}^{l} \frac{i^l f_l^m}{\sigma_l^2} g_l^{+*}(r) Y_l^m(\hat{\mathbf{r}}), \quad r < a_0, \tag{9.41a}$$

where

$$f_l^m = \langle Y_l^m, f\rangle_{\mathcal{H}_f} = \int d\Omega_s f(\mathbf{s}) Y_l^{m*}(\mathbf{s}), \tag{9.41b}$$

are the radiation-pattern Fourier coefficients first introduced in Section 4.8.4 of Chapter 4.

The Picard condition

On making use of Eq. (9.41b) we find that the Picard condition Eq. (9.37) assumes the form

$$\sum_{\sigma_p > 0} \left| \frac{f_l^m}{\sigma_p} \right|^2 < \infty, \tag{9.42}$$

which is a condition that must be satisfied by the expansion coefficients of the radiation pattern of a source of a given (specified) radius. We showed in Section 5.6 of Chapter 5 that this condition places a lower bound on the size of a source to radiate a given field and the two theorems proven in that section carry over to the current application of sources embedded in non-homogeneous backgrounds. The only difference between the two cases is that the singular values σ_p will no longer be those that we computed for homogeneous backgrounds, so the details of the theorem proofs will be different.

9.6.1 Solution of the ISP for a piecewise-constant spherically symmetric background

In order to complete the solution of the ISP in a spherically symmetric background it is necessary to compute the radial functions $g_l^+(r)$ within the interior of the source support region τ_0. These functions satisfy the LS equation Eq. (9.38b) and also the differential equation Eq. (9.12) and are the radial components of the field scattered by the background scattering potential $V_0(r)$ for an incident wave equal to the free-space multipole field

$$U^{(\text{in})}(\mathbf{r}) = j_l(k_0 r) Y_l^m(\hat{\mathbf{r}}).$$

One case in which the radial functions are easily computed is for a piecewise-constant spherically symmetric background whose scattering potential is of the form

$$V_0(r) = \begin{cases} k_0^2[1 - n_\text{r}^2] & r \leq R_0, \\ 0 & r > R_0, \end{cases} \tag{9.43}$$

where $n_\text{r} = k/k_0$ is the relative index of refraction of the background region to that at infinity. We computed the scattered field from such a scattering potential in Section 6.3 of Chapter 6 and can use that result here to compute the required radial functions $g_l^+(r)$ entering into the pseudo-inverse solution to the ISP found above.

We only need the radial functions within the support region τ_0 of the source, which is assumed to be a sphere of radius $a_0 < R_0$ concentric to the background spherical support region. On making use of the results obtained in Section 6.3 we find that the total field (incident plus scattered) within this region for an incident wave equal to the free-space multipole field defined above is given by

$$g_l^+(r) Y_l^m(\hat{\mathbf{r}}) = a_l^m(\nu) j_l(k_0 n_\text{r} r) Y_l^m(\hat{\mathbf{r}}), \quad r \leq a_0, \tag{9.44a}$$

where

$$a_l^m = \frac{i/(k_0 R_0)^2}{j_l(k_0 n_\text{r} R_0) h_l^{+\prime}(k_0 R_0) - n_\text{r} j_l'(k_0 n_\text{r} R_0) h_l^+(k_0 R_0)}. \tag{9.44b}$$

The radial functions $g_l^+(r)$ within the source support region are thus proportional to the spherical Bessel functions $j_l(k_0 n_\text{r} r)$ with proportionality constant equal to a_l^m. The singular values are then found to be

$$\sigma_l = |a_l^m| \overbrace{\sqrt{\int_0^{a_0} r^2\, dr |j_l(k_0 n_\text{r} R_0)|^2}}^{\mu_l(k_0 n_\text{r} a_0)} = |a_l^m| \mu_l(k_0 n_\text{r} a_0). \tag{9.44c}$$

9.6.2 Super-resolution

As we mentioned at the beginning of the previous section, the question arises as to whether a primary source embedded in a non-uniform medium can, in some sense, outperform the same source embedded in a uniform background. One measure of performance is the minimum source size required in order to radiate a given radiation pattern, which for a source embedded in a uniform background is governed by Theorem 5.1. Unfortunately, it is not possible to "beat" this theorem and obtain "super-resolution" since a composite source consisting of a primary source Q_0 embedded in an inhomogeneous background is, in fact, an induced source embedded in a homogeneous medium and thus also subject to Theorem 5.1! In particular, the induced source is given by

$$Q_0' = Q_0 + V_0 U,$$

where U is the wave radiated by the primary source embedded in the background scattering potential V_0. For a proper comparison the induced (composite) source must have the same support as the primary source, which requires that the background potential have this support. The composite source radiates into the homogeneous medium having wavenumber k_0 with the same support as Q_0 and is thus subject to Theorem 5.1. This conclusion holds *regardless of the background potential!* Thus, although, as we will show below, it is possible to obtain certain moderate gains (Devaney *et al.*, 2007) by embedding a source in a non-uniform background, Theorem 5.1 strictly applies and limits the achievable gains and benefits of immersing primary sources in inhomogeneous backgrounds no matter how they are constructed.

Although the performance of the *composite* (induced) source consisting of a primary source Q_0 embedded in a non-uniform background is identical to that of a (different) primary source ($Q_0' = Q_0 + V_0 U$) embedded in a uniform background, the primary source Q_0 embedded in the background will, for some backgrounds, have better characteristics than that of Q_0 embedded in a uniform background. The reason for this is that the rates of decay and cutoff values of the singular values σ_p for Q_0 embedded in a non-uniform medium can, in some cases, be better than those for the same source in a uniform medium so that in this sense moderate gains can be achieved. For the case considered above, namely that of a uniform spherically symmetric background, this can be easily seen by comparing the singular values $\mu_l(k_0 a_0)$ for a primary source embedded in a uniform background with wavenumber k_0 with the singular values $\sigma_l = a_l^m \mu(k_0 n_r a_0)$ for a primary source embedded in a uniform spherically symmetric background. At first glance the situation looks promising since the quantities $\mu_l(x)$ begin their exponential decay with index l at $l = [x]$, where the bracket $[\cdot]$ stands for the next higher integer. Thus, for a source embedded in a background with relative index of refraction $n_r > 1$ the factor $\mu_l(k_0 n_r a_0)$ will begin its exponential decay at $l = n_r[k_0 a_0]$, whereas $\mu_l(k_0 a_0)$ begins its decay at $[k_0 a_0]$. The embedded primary source will thus have a smaller L^2 norm than that of the non-embedded one for a given radiation pattern so long as $n_r > 1$. Unfortunately, the multiplying factor a_l^m offsets this gain somewhat, since it also decays exponentially fast with index l. We show in Fig. 9.1 semi-log plots of $\mu_l(k_0 a_0)$, $\mu_l(n_r k_0 a_0)$ and σ_l for a background relative index $n_r = 1.5$ and for $[k_0 a_0] = 26$ and $n_r[k_0 a_0] = 39$. The singular values $\mu_l(k_0 a_0)$ begin their exponential decay at $l \approx 26$, whereas the $\mu_l(n_r k_0 a_0)$ begin theirs at $l \approx 39$. The cutoff point for the singular values σ_l of the composite source lies between these two values at $l \approx 34$.

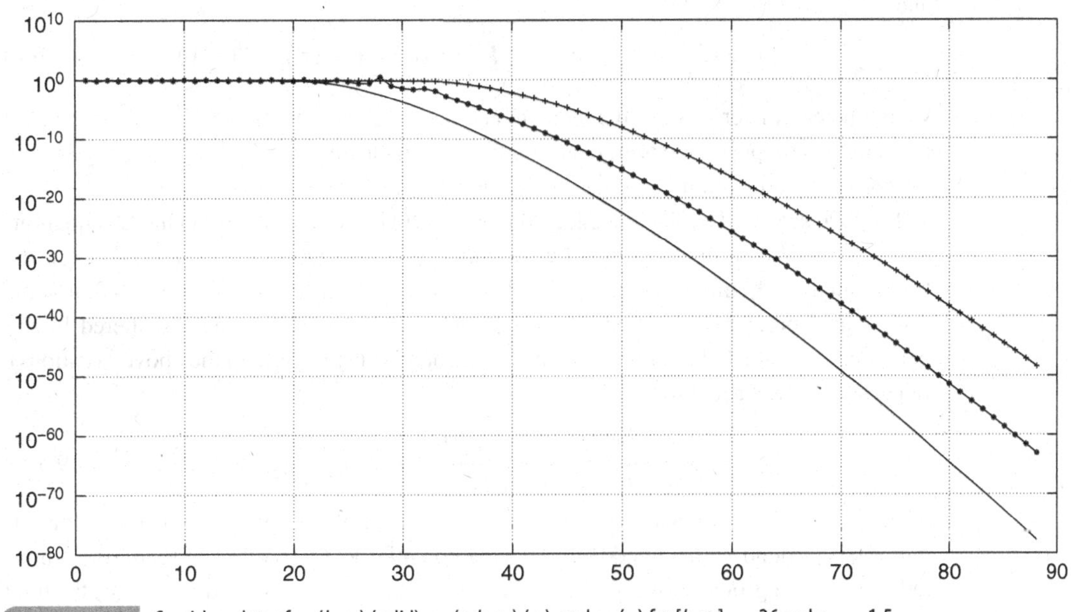

Fig. 9.1 Semi-log plots of $\mu_l(k_0a_0)$ (solid), $\mu_l(n_rk_0a_0)$ (+), and σ_l (•) for $[k_0a_0] = 26$ and $n_r = 1.5$.

9.7 Scattering in a non-uniform background medium

In this section we consider the scattering of an incident wave propagating in a non-uniform background by an embedded penetrable scatterer characterized by the scattering potential $V(\mathbf{r})$. Virtually all of the theory and results established in our treatment of scattering in a uniform background established in Chapter 6 apply here, with the uniform-medium Green function $G_{0_+}(\mathbf{r}-\mathbf{r}')$ replaced by the non-uniform-medium background Green function $G_{0_+}(\mathbf{r},\mathbf{r}')$. The Green function for a composite medium having scattering potential $V_0 + V$ satisfies the Lippmann–Schwinger (LS) equation in either of the two forms

$$G_+(\mathbf{r},\mathbf{r}') = G_{0_+}(\mathbf{r},\mathbf{r}') + \int d^3r''\, G_{0_+}(\mathbf{r},\mathbf{r}'')V(\mathbf{r}'')G_+(\mathbf{r}'',\mathbf{r}') \tag{9.45a}$$

and

$$G_+(\mathbf{r},\mathbf{r}') = G_{0_+}(\mathbf{r},\mathbf{r}') + \int d^3r''\, G_+(\mathbf{r},\mathbf{r}'')V(\mathbf{r}'')G_{0_+}(\mathbf{r}'',\mathbf{r}'), \tag{9.45b}$$

where we have denoted the composite (total) medium Green function with the subscript +. An incident wave propagating in the background will scatter from the embedded scattering potential V and generate a total wave (incident plus scattered) that satisfies the LS equation in either of the two forms

$$U_+(\mathbf{r},\nu) = U_0^{(\mathrm{in})}(\mathbf{r},\nu) + \int d^3r'\, G_{0_+}(\mathbf{r},\mathbf{r}')V(\mathbf{r}')U_+(\mathbf{r}',\nu) \tag{9.46a}$$

and

$$U_+(\mathbf{r},\nu) = U_0^{(\text{in})}(\mathbf{r},\nu) + \int d^3r'\, G_+(\mathbf{r},\mathbf{r}')V(\mathbf{r}')U_0^{(\text{in})}(\mathbf{r}',\nu), \tag{9.46b}$$

where the parameter ν labels the incident and resulting total fields and $U_+(\mathbf{r},\nu)$ is the total wave in the composite medium resulting from the incident wave $U_0^{(\text{in})}(\mathbf{r},\nu)$ propagating in the background medium.

The field scattered by the embedded potential $V(\mathbf{r})$ is the second term in the LS equations and is formally the field radiated by the induced source $Q(\mathbf{r},\nu) = V(\mathbf{r})U_+(\mathbf{r},\nu)$ in the background medium. However, the total scattered field that would be observed at some distant field point $\mathbf{r} = r\mathbf{s}$ along the direction $\mathbf{s}$ is the sum of the field scattered by the potential V and the field scattered by the background (the first term in the above equations). In particular, we have that

$$U_+(\mathbf{r},\nu) \sim f_+(\mathbf{s},\nu)\frac{e^{ik_0r}}{r} = f_{0_+}(\mathbf{s},\nu)\frac{e^{ik_0r}}{r} + \delta f(\mathbf{s},\nu)\frac{e^{ik_0r}}{r}, \quad r\to\infty, \tag{9.47a}$$

where $f_+(\mathbf{s},\nu)$ is the (generalized) scattering amplitude of the composite medium consisting of background with embedded scattering potential V, $f_{0_+}(\mathbf{s},\nu)$ is the scattering amplitude of the background wave alone (cf. Eq. (9.7b)) and $\delta f(\mathbf{s},\nu)$ is the deviation of the total scattering amplitude from the scattering amplitude of the background alone. The last of these three quantities is simply the radiation pattern of the induced source $V(\mathbf{r}')U_+(\mathbf{r}',\nu)$ derived in Section 9.2.4 and is thus given by

$$\delta f(\mathbf{s},\nu) = -\frac{1}{4\pi}\int_{\tau_0} d^3r'\, \psi_+(\mathbf{r}',-k_0\mathbf{s})\, \overbrace{V(\mathbf{r}')U_+(\mathbf{r}',\nu)}^{Q(\mathbf{r}',\nu)}, \tag{9.47b}$$

where $\psi_+(\mathbf{r}',-k_0\mathbf{s})$ is the background plane-wave scattering state with unit propagation vector $-\mathbf{s}$. On making use of Eq. (9.47b) in Eq. (9.47a) we obtain the following expression for the total scattering amplitude of the composite medium:

$$f_+(\mathbf{s},\nu) = f_{0_+}(\mathbf{s},\nu) + \overbrace{-\frac{1}{4\pi}\int_{\tau_0} d^3r'\, \psi_+(\mathbf{r}',-k_0\mathbf{s})V(\mathbf{r}')U_+(\mathbf{r}',\nu)}^{\delta f_+(\mathbf{s},\nu)}. \tag{9.47c}$$

The mappings from the incident wave propagating in the background to the total and scattered fields according to the LS equations above are linear functionals of the incident waves but non-linear functionals of the embedded scattering potential $V(\mathbf{r})$. Both of these properties follow directly from a Liouville–Neumann expansion of the LS equations along the lines employed in Chapter 6 for scattering potentials embedded in a uniform background medium. The non-linearity of these equations and of the scattering amplitudes with respect to the embedded scattering potential makes the inverse scattering problem (ISCP) extremely difficult, as was discussed extensively in Chapters 6 and 8. As in those earlier chapters, we are thus led to linearization of these equations by dropping higher-order terms in the Liouville–Neumann expansion, thus yielding the Born approximation in the uniform-medium case and the so-called *distorted-wave* Born approximation (DWBA) for non-uniform backgrounds.

9.8 The distorted-wave Born approximation

An important procedure in many inverse scattering applications is the process of *index matching* whereby an object (scatterer) of interest is embedded in a known scattering background whose index of refraction (background scattering potential) is closely matched to the average index of refraction of the scattering object. The background is generally selected to be simple so that its exact scattering amplitude can be easily computed. The ISCP then reduces to estimating the scattering potential (index-of-refraction profile) of the object from the observed scattering amplitude of the composite scatterer consisting of the object embedded in the known scattering background. The key points are that the scattering from the embedded scatterer will be weak and the scattering amplitude of the background is known so that a linearized version of Eq. (9.47c) can be employed to compute the embedded scattering potential.

The scattering model that employs the known scattering amplitude of a background in which is embedded a weak scatterer of interest is known as the distorted-wave Born approximation (DWBA). The DWBA of the scattering amplitudes defined in Eqs. (9.47) results from approximating the total exact wave $U_+(\mathbf{r}', \nu)$ by the total wave propagating in the background medium $U_0^{(\text{in})}(\mathbf{r}', \nu)$. We then obtain

$$\delta f_{\text{DB}}(\mathbf{s}, \nu) = -\frac{1}{4\pi}\int_{\tau_0} d^3r'\, \psi_+(\mathbf{r}', -k_0\mathbf{s})\, \overbrace{V(\mathbf{r}')U_0^{(\text{in})}(\mathbf{r}', \nu)}^{Q_{\text{DB}}(\mathbf{r}',\nu)}, \tag{9.48a}$$

for a general incident wave, and

$$\delta f_{\text{DB}}(\mathbf{s}, \mathbf{s}_0) = -\frac{1}{4\pi}\int_{\tau_0} d^3r'\, \psi_+(\mathbf{r}', -k_0\mathbf{s})\, \overbrace{V(\mathbf{r}')\psi_+(\mathbf{r}', k_0\mathbf{s}_0)}^{Q_{\text{DB}}(\mathbf{r}',\mathbf{s}_0)}, \tag{9.48b}$$

for an incident plane-wave scattering state, where we have denoted the DWBA to the various quantities with the "DB" subscript. The DWBA scattering amplitudes are linear functionals of the scattering potential $V(\mathbf{r})$ and thus can form the basis of linearized inverse scattering in inhomogeneous media, which we shall discuss in the next section. They can, of course, also be used as simple forward-scattering models to compute approximations to the scattering amplitude of weakly scattering potentials embedded in a known background medium.

9.8.1 The DWBA for a pair of concentric homogeneous cylinders

As an example we compute the DWBA of the scattering amplitude $f_+(\mathbf{s}, \mathbf{s}_0)$ for a pair of concentric homogeneous penetrable cylinders for the case in which the incident- and scattered-field unit vectors $\mathbf{s}_0$ and $\mathbf{s}$ both lie in the plane perpendicular to the axis of the cylinders. In this case the background consists of the larger cylinder with scattering amplitude $f_{0_+}(\mathbf{s}, \mathbf{s}_0)$ and

$$\delta f_{\mathrm{DB}}(\mathbf{s}, \mathbf{s}_0) = -\sqrt{\frac{1}{8\pi k_0}} e^{i\frac{\pi}{4}} \int_{\tau_0} d^2r' \, \psi_{0_+}(\mathbf{r}', -k_0\mathbf{s}) V(\mathbf{r}') \psi_{0_+}(\mathbf{r}', k_0\mathbf{s}_0) \tag{9.49}$$

is the scattering amplitude of the smaller cylinder when embedded in the larger cylinder, where all quantities are defined on the (x, y) plane. We take the background cylinder to have a radius a_2 and index n_2 and the interior cylinder to have a radius $a_1 < a_2$ and index n_1. The scattering amplitude $f_{0_+}(\mathbf{s}, \mathbf{s}_0)$ of the background cylinder was computed in Example 6.2 of Chapter 6, where it was found to be given by the 2D multipole expansion

$$f_{0_+}(\mathbf{s}, \mathbf{s}_0) = \sqrt{\frac{2}{\pi k_0}} e^{-i\frac{\pi}{4}} \sum_{l=-\infty}^{\infty} R_l e^{il(\alpha - \alpha_0)}, \tag{9.50}$$

where R_l are the generalized reflection coefficients for the outer cylinder defined in Eq. (6.17c) of Chapter 6.

The interior wavefield to the background cylinder for a general incident wave was computed in Section 6.3 of Chapter 6, where it was found to be given by the multipole expansion

$$U_+(\mathbf{r}, \nu) = \sum_{l=-\infty}^{\infty} a_l(\nu) J_l(k_0 n_2 r) e^{il\phi}, \quad r \le a_2, \tag{9.51}$$

where r, ϕ are the polar coordinates of the field point $\mathbf{r}$ in the x, y system centered along the axis of the cylinder and

$$a_l(\nu) = T_l a_{0_l}(\nu), \tag{9.52a}$$

where $a_{0_l}(\nu)$ are the multipole moments of the incident wave to the cylinder and, for an incident plane wave propagating with polar angle ν, are given by

$$a_{0_l}(\nu) = i^l e^{-il\nu}, \tag{9.52b}$$

and

$$T_l = \frac{2i/(\pi k_0 a_2)}{J_l(k_0 n_2 a_2) H_l^{+\prime}(k_0 a_2) - n_2 J_l'(k_0 n_2 a_2) H_l^+(k_0 a_2)}$$

are the generalized transmission coefficients of the background cylinder. The plane-wave scattering states for this background are thus given by Eq. (9.51), with $\nu = \alpha + \pi$ for $\psi_+(\mathbf{r}', -k_0\mathbf{s})$ and $\nu = \alpha_0$ for $\psi_+(\mathbf{r}', k_0\mathbf{s}_0)$.

The DWBA for a circular cylinder is given by Eq. (9.49), with the scattering potential

$$V(\mathbf{r}) = \delta V = k_0^2(n_2^2 - n_1^2), \quad r < a_1, \tag{9.53}$$

and zero otherwise. Because of the circular symmetry of the scattering potential we can perform the integration over the polar angle in the expression for $\delta f_{\mathrm{DB}}(\mathbf{s},\mathbf{s}_0)$ to find that

$$\begin{aligned}
\delta f_{\mathrm{DB}}(\mathbf{s},\mathbf{s}_0) &= -\sqrt{\frac{1}{8\pi k_0}}e^{i\frac{\pi}{4}}\,\delta V\int_0^{a_1} r'\,dr'\int_0^{2\pi} d\phi'\,\psi_+(\mathbf{r}',-k_0\mathbf{s})\psi_+(\mathbf{r}',k_0\mathbf{s}_0)\\
&= -\sqrt{\frac{\pi}{2k_0}}e^{i\frac{\pi}{4}}\,\delta V\sum_{l=-\infty}^{\infty}(-1)^l\int_0^{a_1} r'\,dr'|J_l(k_0n_2r')|^2 a_{-l}(\alpha+\pi)a_l(\alpha_0)\\
&= -\sqrt{\frac{\pi}{2k_0}}e^{i\frac{\pi}{4}}\,\delta V\sum_{l=-\infty}^{\infty}(-1)^l\mu_l^2(k_0n_2a_1)a_{-l}(\alpha+\pi)a_l(\alpha_0),
\end{aligned} \tag{9.54}$$

where we have used the result that $J_{-l}(x)=(-1)^lJ_l(x)$ and where $\mu_l^2(ka)$ is our old friend (cf. Section 6.8 of Chapter 6)

$$\mu_l^2(ka)=\int_0^a r\,dr\,J_l^2(k_0r)=\frac{a^2}{2}[J_l^2(ka)-J_{l-1}(ka)J_{l+1}(ka)].$$

On making use of Eqs. (9.52) we conclude that

$$a_l(\alpha_0)=T_li^le^{-il\alpha_0},\qquad a_{-l}(\alpha+\pi)=T_{-l}i^{-l}e^{il(\alpha+\pi)}=(-1)^lT_li^{-l}e^{il\alpha},$$

where we have made use of the fact that $T_l=T_{-l}$. On making use of the above results in Eq. (9.54) we then obtain the final expression

$$\delta f_{\mathrm{DB}}(\mathbf{s},\mathbf{s}_0)=-\sqrt{\frac{\pi}{2k_0}}e^{i\frac{\pi}{4}}\,\delta V\sum_{l=-\infty}^{\infty}T_l^2\mu_l^2(k_0n_2a_1)e^{i(\alpha-\alpha_0)}, \tag{9.55a}$$

and

$$\begin{aligned}
f_T(\mathbf{s},\mathbf{s}_0) &= \sqrt{\frac{2}{\pi k_0}}e^{-i\frac{\pi}{4}}\sum_{l=-\infty}^{\infty}R_le^{i(\alpha-\alpha_0)}\\
&\quad-\sqrt{\frac{\pi}{2k_0}}e^{i\frac{\pi}{4}}\,\delta V\sum_{l=-\infty}^{\infty}T_l^2\mu_l^2(k_0n_2a_1)e^{i(\alpha-\alpha_0)}.
\end{aligned} \tag{9.55b}$$

We note that in the limit where the background cylinder has index $n_2=n_0=1$ we have $T_l\to 1$, $R_l\to 0$, and Eq. (9.55b) reduces to the Born approximation for scattering from a single cylinder with radius a_1 embedded in an infinite homogeneous medium with wavenumber k_0 found in Section 6.8 of Chapter 6. On the other hand, if $\delta V\to 0$, f_T reduces to the exact expression for scattering from a single cylinder of radius a_2 embedded in an infinite homogeneous medium with wavenumber k_0.

We compared the DWBA of the scattering amplitude of two concentric cylinders computed above with the exact scattering amplitude found in Example 6.3 of Section 6.4 of Chapter 6 and with the Born approximation for these two cylinders given in Section 6.8 of that chapter. The exact scattering amplitude is also given by a multipole expansion of the form of Eq. (9.50) but with the generalized reflection coefficients now given by $R_l=b_l(\mathbf{s}_0)/a_{0_l}(\mathbf{s}_0)$, where the $b_l(\mathbf{s}_0)$ are obtained as the solutions to the matrix equation

Eq. (6.19) of Section 6.3 of Chapter 6 and with $a_{0_l}(\mathbf{s}_0) = i^l e^{-il\alpha_0}$ given by Eq. (9.52a). The standard Born approximation was found to be given by

$$f_B(\mathbf{s}, \mathbf{s}_0) = -\sqrt{\frac{\pi}{2k_0}} e^{i\frac{\pi}{4}} \sum_{l=-\infty}^{\infty} [\delta V \mu_l^2(k_0 a_1) + V_2 \mu_l^2(k_0 a_2)] e^{il(\alpha-\alpha_0)}, \tag{9.56a}$$

where δV is defined in Eq. (9.53) and

$$V_2 = k_0^2(1 - n_2^2), \tag{9.56b}$$

is the amplitude of the scattering potential of the outer cylinder alone.

We compared the exact, Born and distorted-wave Born approximations to the scattering amplitudes of two concentric cylinders having varying radii and relative indices of refraction. We show in Figs. 9.2 and 9.3 the magnitude and phase of the three scattering amplitudes computed for the pair of concentric cylinders previously considered in Section 6.8. The two cylinders have radii $a_1 = 2\lambda$ and $a_2 = 4\lambda$ and relative indices of refraction of 1.03 and 1.07. The various combinations of indices are thus (1.03, 1.03) (top left in the figures), which reduces to a single cylinder having index 1.03 and radius $a_2 = 4\lambda$, (1.03, 1.07) (top right in the figures), (1.07, 1.03) (bottom left in the figures) and (1.07, 1.07) (bottom right in the figures), which reduces to a single cylinder having index 1.07 and radius $a_2 = 4\lambda$. Note that, as expected, the DWBA of the two concentric cylinders having the same relative index of refraction is exact. It is also clear from the figures that the DWBA outperforms

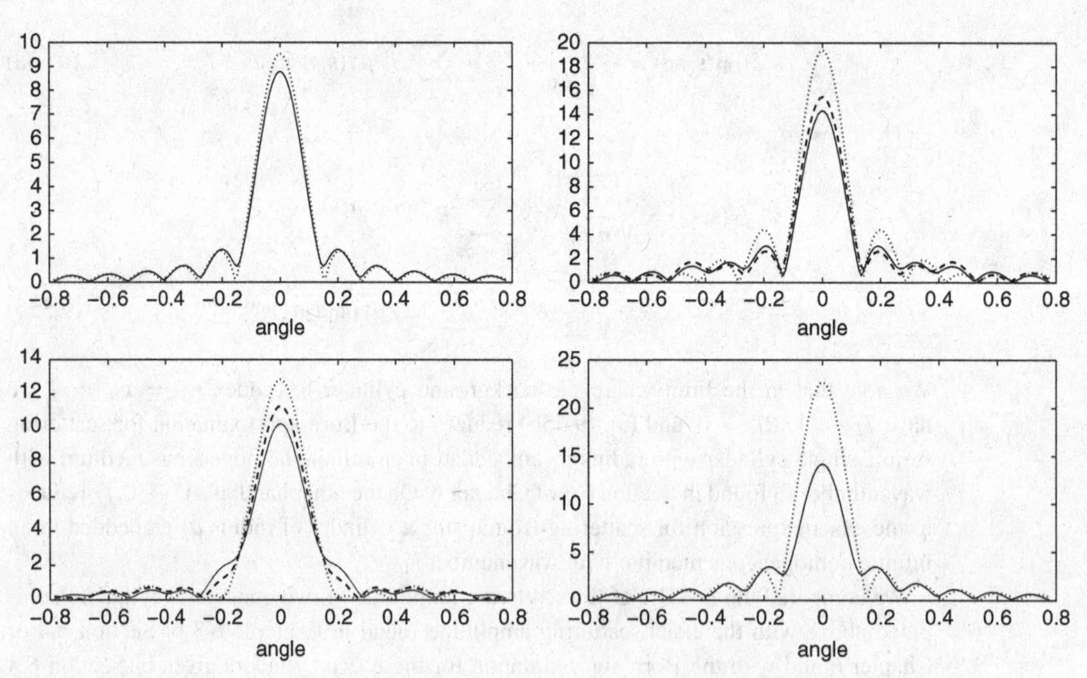

Fig. 9.2 Magnitudes of the exact values (solid) and of the Born approximation (dotted) and DWBA (dashed) to the scattering amplitudes of two concentric cylinders having radii of $a_1 = 2\lambda$ and $a_2 = 4\lambda$ and relative indices of refraction of $(n_1, n_2) = (1.03, 1.03)$ (top left), (1.03, 1.07) (top right), (1.07, 1.03) (bottom left) and (1.07, 1.07) (bottom right).

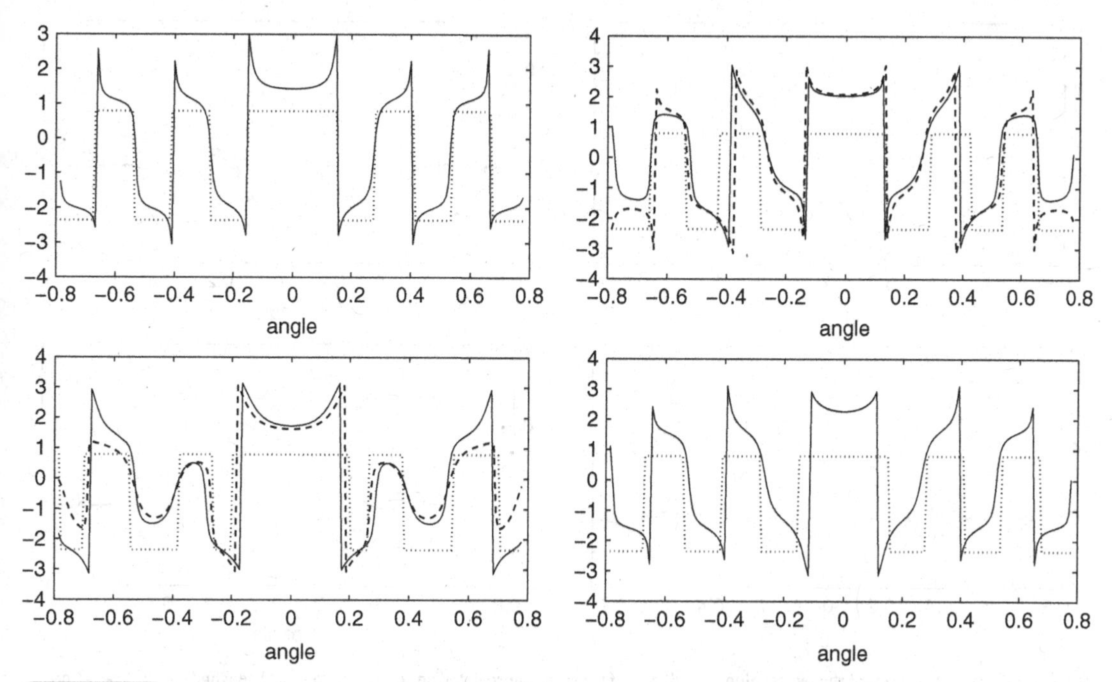

Fig. 9.3 Phase plots of the exact values (solid) and of the Born approximation (dotted) and DWBA (dashed) to the scattering amplitudes of two concentric cylinders having radii of $a_1 = 2\lambda$ and $a_2 = 4\lambda$ and relative indices of refraction of $(n_1, n_2) = (1.03, 1.03)$ (top left), (1.03, 1.07) (top right), (1.07, 1.03) (bottom left) and (1.07, 1.07) (bottom right).

the Born approximation in all cases, especially as regards the (real) phase of the scattering amplitudes shown in Fig. 9.3. The phase of the scattering amplitude plays a dominant role in inverse scattering applications that rely heavily on field back propagation, which employs the real and imaginary parts of the scattering amplitude in the back-propagation process. We show in Figs. 9.4 and 9.5 the real and imaginary parts of the scattering amplitudes displayed in Figs. 9.2 and 9.3. It is readily apparent from these two figures that the DWBA is much superior to the Born approximation in this example.

As a second example we selected the background cylinder to have a radius $a_2 = 8\lambda$ and the inner cylinder the same radius as used above, namely $a_1 = 2\lambda$. We selected the indices to be $n_1 = (1.1, 1.2)$ and $n_2 = (1.05, 1.08)$. We display only the DWBA and exact scattering amplitudes in Figs. 9.6 and 9.7 since the Born results were very poor and not worth showing. We also only display the real and imaginary parts of the scattering amplitudes since, as illustrated above, they are much more indicative of the accuracy of the approximations than are the magnitude and phase. It can be seen from the figures that the DWBA results in almost perfect agreement with the exact scattering amplitudes for $n_1 = 1.1$ and $n_2 = 1.08$ (top right in the figures) and does reasonably well for $n_1 = 1.1$ and $n_2 = 1.05$. On the other hand, the results for $n_1 = 1.2$ (bottom parts of the figures) are not accurate and a better index match between the two cylinders would be required in any inverse scattering application.

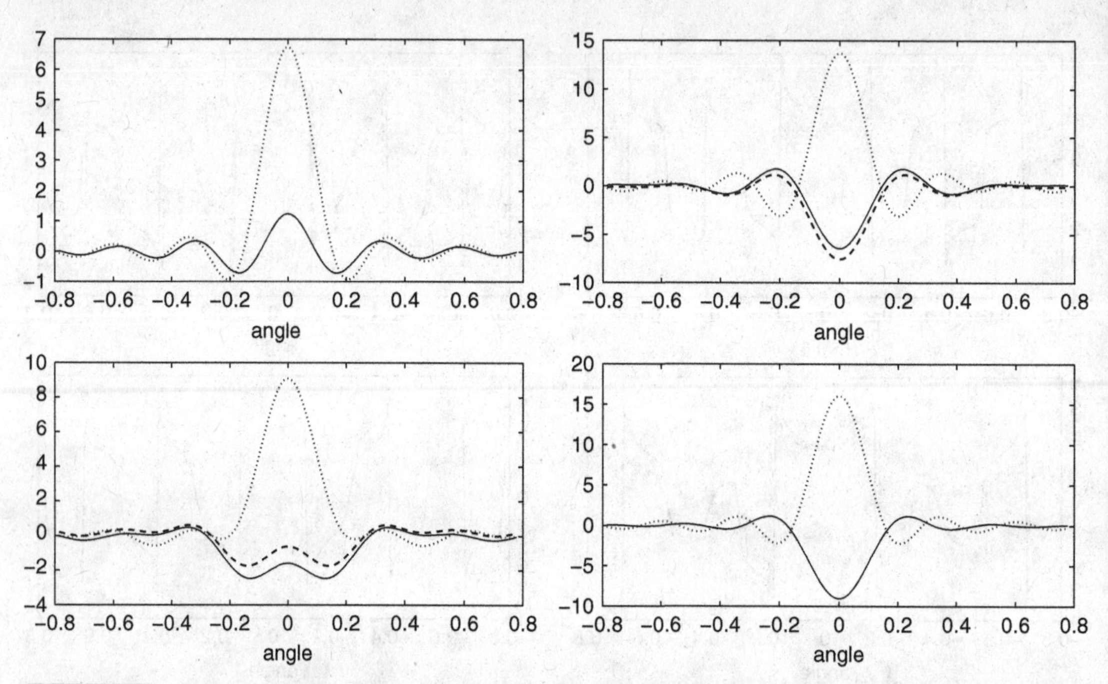

Fig. 9.4 Real parts of the exact values (solid) and of the Born approximation (dotted) and DWBA (dashed) to the scattering amplitudes of two concentric cylinders having radii of $a_1 = 2\lambda$ and $a_2 = 4\lambda$ and relative indices of refraction of $(n_1, n_2) = (1.03, 1.03)$ (top left), (1.03, 1.07) (top right), (1.07, 1.03) (bottom left) and (1.07, 1.07) (bottom right).

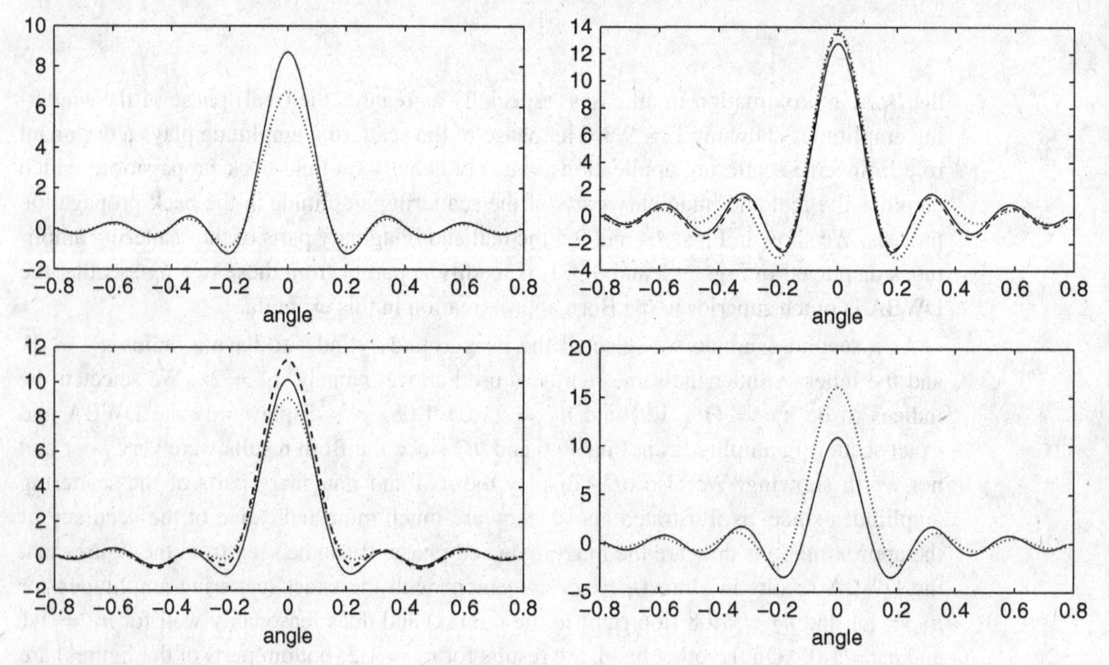

Fig. 9.5 Imaginary parts of the exact values (solid) and of the Born approximation (dotted) and DWBA (dashed) to the scattering amplitudes of two concentric cylinders having radii of $a_1 = 2\lambda$ and $a_2 = 4\lambda$ and relative indices of refraction of $(n_1, n_2) = (1.03, 1.03)$ (top left), (1.03, 1.07) (top right), (1.07, 1.03) (bottom left) and (1.07, 1.07) (bottom right).

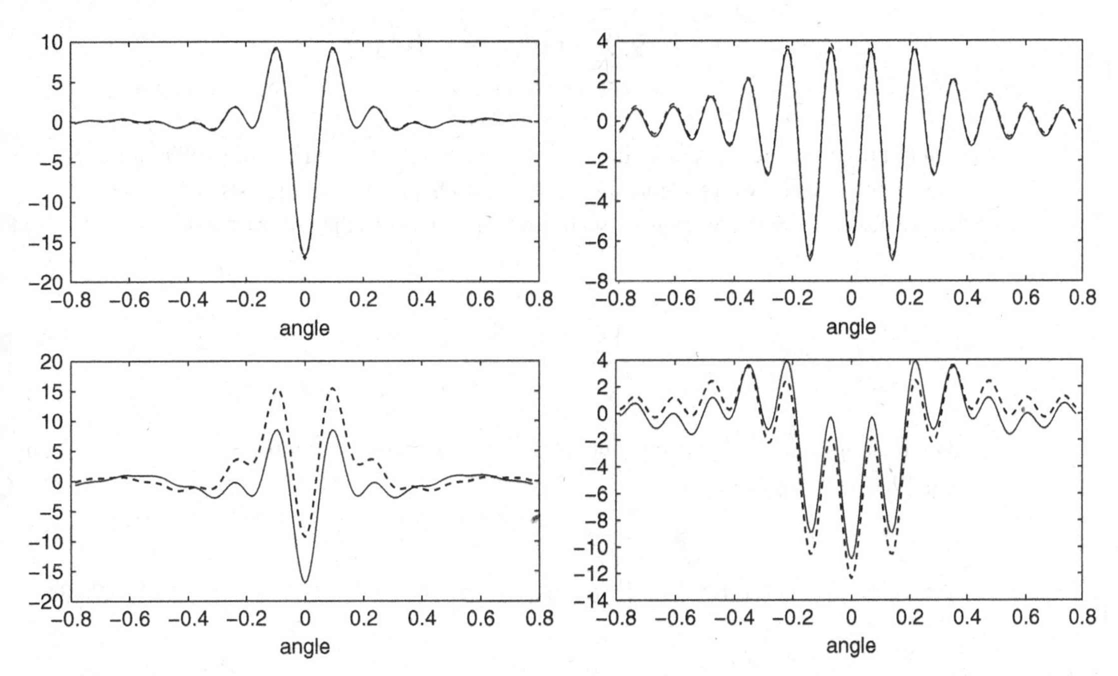

Fig. 9.6 Real parts of the exact values (solid) and of the DWBA (dashed) to the scattering amplitudes of two concentric cylinders having radii of $a_1 = 2\lambda$ and $a_2 = 8\lambda$ and relative indices of refraction of $(n_1, n_2) = (1.1, 1.05)$ (top left), (1.1, 1.08) (top right), (1.2, 1.05) (bottom left) and (1.2, 1.08) (bottom right).

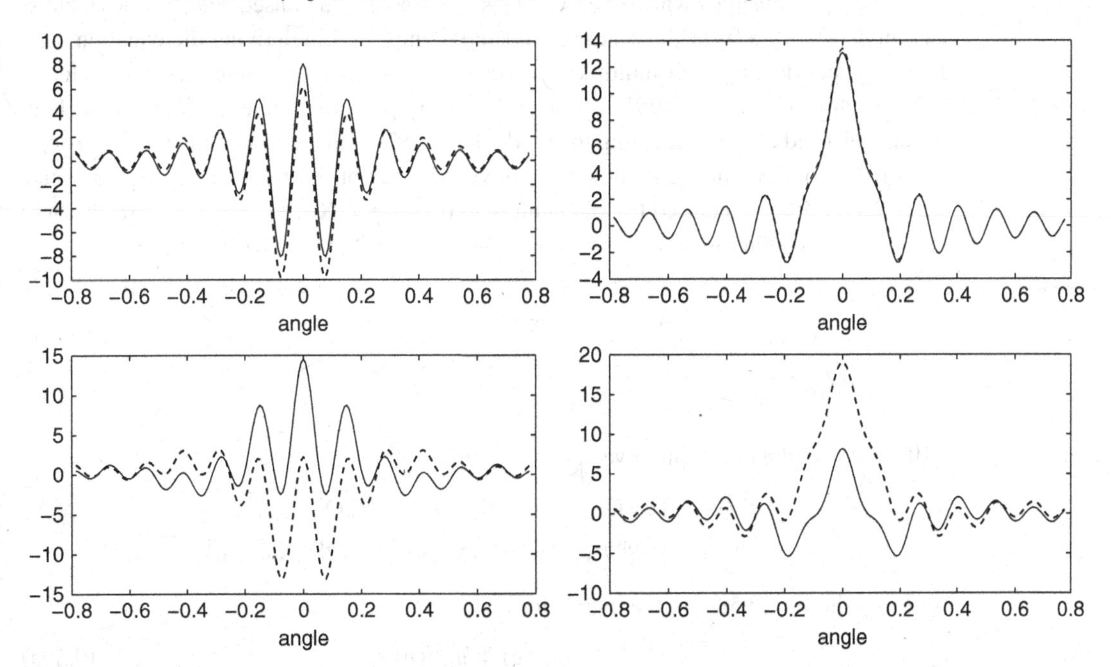

Fig. 9.7 Imaginary parts of the exact values (solid) and of the DWBA (dashed) to the scattering amplitudes of two concentric cylinders having radii of $a_1 = 2\lambda$ and $a_2 = 8\lambda$ and relative indices of refraction of $(n_1, n_2) = (1.1, 1.05)$ (top left), (1.1, 1.08) (top right), (1.2, 1.05) (bottom left) and (1.2, 1.08) (bottom right)..

9.9 Foldy–Lax theory

An important scattering model in a number of applications and that we will employ in our treatment of time-reversal imaging in the next chapter is that of a set of idealized point (delta-function) scatterers whose scattering potential is of the general form

$$V(\mathbf{r}) = \sum_{m=1}^{M} \mathcal{V}_m \delta(\mathbf{r} - \mathbf{X}_m). \tag{9.57}$$

On making use of the above model for the scattering potential in the LS equation Eq. (9.46a) we obtain

$$U_+(\mathbf{r}, \nu) = U_0^{(\text{in})}(\mathbf{r}, \nu) + \sum_{m=1}^{M} \mathcal{V}_m G_{0_+}(\mathbf{r}, \mathbf{X}_m) U_+(\mathbf{X}_m, \nu). \tag{9.58a}$$

The above equation runs into immediate difficulty if we take the field point $\mathbf{r}$ to be any of the scattering centers $\mathbf{X}_m$ due to the fact that the background Green function occurring in the equation diverges when $\mathbf{r} = \mathbf{X}_m$. This is, of course, a consequence of the unphysical model Eq. (9.57) for the scattering potential employed in deriving the equation. In a famous paper dealing with multiple scattering in a system of discrete random scatterers Foldy (Foldy, 1945; Lax, 1951; Ishimaru, 1999) proposed a simple fix for this problem, which consisted of supplementing the model Eq. (9.57) with the requirement that the scattered field from any of the point scatterers vanish back at their scattering centers. Thus, while Eq. (9.58a) is required to hold at all field points $\mathbf{r} \neq \mathbf{X}_m$, $m = 1, 2, \ldots, M$, the field at any of the scatterer locations is required to satisfy the "renormalized" LS equation

$$U_+(\mathbf{X}_m, \nu) = U_0^{(\text{in})}(\mathbf{X}_m, \nu) + \sum_{m' \neq m} \mathcal{V}_{m'} G_{0_+}(\mathbf{X}_m, \mathbf{X}_{m'}) U_+(\mathbf{X}_{m'}, \nu). \tag{9.58b}$$

If we define the two column vectors

$$u_+(\nu) = [U_+(\mathbf{X}_1, \nu), U_+(\mathbf{X}_2, \nu), \ldots, U_+(\mathbf{X}_M, \nu)]^{\mathrm{T}},$$
$$u_0^{(\text{in})}(\nu) = [U_0^{(\text{in})}(\mathbf{X}_1, \nu), U_0^{(\text{in})}(\mathbf{X}_2, \nu), \ldots, U_0^{(\text{in})}(\mathbf{X}_M, \nu)]^{\mathrm{T}},$$

we can write Eq. (9.58b) in the matrix form

$$H u_+(\nu) = u_0^{(\text{in})}(\nu), \tag{9.59a}$$

where H is the $M \times M$ matrix with elements

$$H(m, m') = \delta_{m,m'} - \mathcal{V}'_m [\delta_{m,m'} - G_{0_+}(\mathbf{X}_m, \mathbf{X}_{m'})]. \tag{9.59b}$$

Since we know the background Green function, the incident wave and the scattering centers $\mathbf{X}_m$ and scattering strengths $\mathcal{V}_m$ the Foldy–Lax formulation and, in particular, Eq. (9.58b) reduces the forward-scattering problem[6] to the *linear* problem of inverting the matrix equation Eq. (9.59b). Once the field amplitudes $U_+(\mathbf{X}_m, \nu)$ at the various scattering centers have been computed, the field $U_+(\mathbf{r}, \nu)$ is then easily computed using Eq. (9.58a).

Example 9.5 The Foldy–Lax formulation given above can be used to compute the composite-medium Green function $G_+(\mathbf{r}, \mathbf{r}')$ by selecting the incident wave $U_0^{(\text{in})}(\mathbf{r}, \nu) = G_{0_+}(\mathbf{r}, \mathbf{r}')$. Equations (9.58) then become

$$G_+(\mathbf{r}, \mathbf{r}') = G_{0_+}(\mathbf{r}, \mathbf{r}') + \sum_{m=1}^{M} \mathcal{V}_m G_{0_+}(\mathbf{r}, \mathbf{X}_m) G_+(\mathbf{X}_m, \mathbf{r}'),$$

$$G_+(\mathbf{X}_m, \mathbf{r}') = G_{0_+}(\mathbf{X}_m, \mathbf{r}') + \sum_{m' \neq m} \mathcal{V}_{m'} G_{0_+}(\mathbf{X}_m, \mathbf{X}_{m'}) G_+(\mathbf{X}_{m'}, \mathbf{r}'),$$

where $\mathbf{r}' \neq \mathbf{X}_m$, $m = 1, 2, \ldots, M$. We can cast the second equation in the form of the matrix equation Eq. (9.59a), where now $\nu \to \mathbf{r}'$ and u_+ and $u^{(\text{in})}$ are replaced by

$$g_+(\mathbf{r}') = [G_+(\mathbf{X}_1, \mathbf{r}'), G_+(\mathbf{X}_2, \mathbf{r}'), \ldots, G_+(\mathbf{X}_M, \mathbf{r}')]^{\mathrm{T}},$$

$$g_{0_+}(\mathbf{r}') = [G_{0_+}(\mathbf{X}_1, \mathbf{r}'), G_{0_+}(\mathbf{X}_2, \mathbf{r}'), \ldots, G_{0_+}(\mathbf{X}_M, \mathbf{r}')]^{\mathrm{T}}.$$

9.10 Inverse scattering within the DWBA

We will address only the ISCP formulated within the linearized DWBA presented in the last section. We will also limit our presentation to the limited-view problem in which one performs only a limited number of scattering experiments and field measurements. We will first develop the theory for the ISCP formulated in terms of the scattering amplitude $f_+(\mathbf{s}^{(j)}, \mathbf{s}_0^{(k)})$ specified over a set of N_j scattering directions $\mathbf{s}^{(j)}$ and N_k incident plane-wave propagation vectors $\mathbf{s}_0^{(k)}$ but will also later consider the more general scenario in which the incident waves are generated by a set of N_k transmitting antennas and the measurements are performed by a set of N_j receiving antennas, both distributed over an arbitrary set of locations outside the scattering volume τ_0. Our formulation of the limited-view ISCP mirrors closely our treatment of the limited-view ISP presented in Section 5.8 of Chapter 5.

As usual we will cast the problem in Hilbert space such that the unknown scattering potential $V(\mathbf{r})$ is contained in the Hilbert space $\mathcal{H}_V = L^2(\tau_0)$ of square-integrable functions compactly supported in τ_0 and the field data are in the space $\mathcal{H}_f$. In the limited-view ISCP under consideration here $\mathcal{H}_f$ is the space of complex N-tuples with $N = N_j \times N_k$ and with standard inner product defined by

[6] We will find in the following chapter that the Foldy–Lax formulation, coupled with the DORT and/or MUSIC algorithms, also reduces the ISCP of computing the scattering strengths $\mathcal{V}_m$ from scattered-field data to a linear problem.

$$\langle f_1, f_2 \rangle_{\mathcal{H}_f} = \sum_{j=1}^{N_j} \sum_{k=1}^{N_k} f_{1_{j,k}}^* f_{2_{j,k}}. \tag{9.60a}$$

The Hilbert space $\mathcal{H}_V$ of scattering potentials is assumed to possess the standard inner product

$$\langle V_1, V_2 \rangle_{\mathcal{H}_V} = \int_{\tau_0} d^3r\, V_1^*(\mathbf{r}) V_2(\mathbf{r}). \tag{9.60b}$$

The two spaces have their respective induced norms $||f||_{\mathcal{H}_f} = \sqrt{\langle f, f \rangle_{\mathcal{H}_f}}$ and $||V||_{\mathcal{H}_V} = \sqrt{\langle V, V \rangle_{\mathcal{H}_V}}$.

9.10.1 The far-field limited-view ISCP

We can cast the far-field ISCP directly in terms of the scattering-potential perturbation $\delta f_{\rm DB}(\mathbf{s}, \mathbf{s}_0)$ specified over the set of scattering and incident plane-wave directions $\mathbf{s}^{(j)}, \mathbf{s}_0^{(k)}$ in an obvious generalization of the statement of the ISCP for the constant-background case considered in Chapter 8. In particular, in analogy with Eqs. (8.27) of that chapter we have that

$$\hat{T} V = \delta f, \tag{9.61}$$

where $\hat{T} : \mathcal{H}_V \to \mathcal{H}_f$ and $\hat{T}^\dagger : \mathcal{H}_f \to \mathcal{H}_V$ are the linear operators

$$\hat{T} = -\frac{1}{4\pi} \int_{\tau_0} d^3r'\, \psi_+(\mathbf{r}', -k_0\mathbf{s}^{(j)}) \psi_+(\mathbf{r}', k_0\mathbf{s}_0^{(k)}), \tag{9.62a}$$

$$\hat{T}^\dagger = -\frac{1}{4\pi} \mathcal{M}_{\tau_0} \sum_{j=1}^{N_j} \sum_{k=1}^{N_k} \psi_+^*(\mathbf{r}, -k_0\mathbf{s}^{(j)}) \psi_+^*(\mathbf{r}, k_0\mathbf{s}_0^{(k)}), \tag{9.62b}$$

where $\mathcal{M}_{\tau_0}$ is the masking operator defined in Eq. (9.31b).

We now introduce, in analogy with our treatment of the limited-view ISP in Section 5.8, the functions

$$\chi_n(\mathbf{r}) = -\frac{1}{4\pi} \mathcal{M}_{\tau_0} \psi_+^*(\mathbf{r}, -k_0\mathbf{s}^{(j)}) \psi_+^*(\mathbf{r}, k_0\mathbf{s}_0^{(k)}), \tag{9.63}$$

where $n = j + (k-1)N_j$ ordered such that, for each $j = 1, 2, \ldots, N_j$, $k = 1, 2, \ldots, N_k$. In terms of the functions χ_n we find that

$$\hat{T} = \int d^3r'\, \chi_n^*(\mathbf{r}'), \qquad \hat{T}^\dagger = \sum_n \chi_n(\mathbf{r}),$$

so that the normal equations for the singular functions $v_p(\mathbf{r}) \in \mathcal{H}_V$ and singular vectors $u_p \in \mathcal{H}_f$ are given by

$$\hat{T}^{\dagger}\hat{T}v_p = \sum_{n=1}^{N} \langle \chi_n, v_p \rangle_{\mathcal{H}_V} \chi_n(\mathbf{r}) = \sigma_p^2 v_p(\mathbf{r}), \tag{9.64a}$$

$$\hat{T}\hat{T}^{\dagger}u_p = \int d^3r\, \chi_n^*(\mathbf{r}) \langle \chi_{n'}^*(\mathbf{r}), u_p \rangle_{\mathcal{H}_f} = \sigma_p^2 u_p. \tag{9.64b}$$

It follows from Eq. (9.64a) that the singular functions v_p corresponding to non-zero singular values $\sigma_p > 0$ are linear combinations of the functions χ_n. By then following the same steps as used in Section 5.8 one finds that

$$v_p(\mathbf{r}) = \sum_{n=1}^{N} u_p(n)\chi_n(\mathbf{r})$$
$$= -\frac{1}{4\pi}\mathcal{M}_{\tau_0} \sum_{j=1}^{N_j}\sum_{k=1}^{N_k} u_p(j,k)\psi_+^*(\mathbf{r}, -k_0\mathbf{s}^{(j)})\psi_+^*(\mathbf{r}, k_0\mathbf{s}_0^{(k)}), \quad \sigma_p > 0, \tag{9.65}$$

where the expansion coefficients $u_p(j,k) = u_p(n)$, $n = j + (k-1)N_j$, are the components of the singular vectors u_p and are the eigenvectors of the Hermitian matrix $\langle \chi_n, \chi_{n'} \rangle_{\mathcal{H}_V}$ whose eigenvalues are the square of the singular values σ_p^2; i.e.,

$$\sum_{n'=1}^{N} \langle \chi_n, \chi_{n'} \rangle_{\mathcal{H}_V} u_p(n') = \sigma_p^2 u_p(n). \tag{9.66}$$

The solution to the limited-view far-field ISCP is given by the usual expression

$$\hat{V}(\mathbf{r}) = \sum_{\sigma_p > 0} \frac{\langle u_p, \delta f \rangle_{\mathcal{H}_f}}{\sigma_p} v_p(\mathbf{r}), \tag{9.67a}$$

where

$$\langle u_p, \delta f \rangle_{\mathcal{H}_f} = \sum_{j=1}^{N_j}\sum_{k=1}^{N_k} u_p^*(j,k)\delta f_+(\mathbf{s}^{(j)}, \mathbf{s}_0^{(k)}), \tag{9.67b}$$

with the singular functions $v_p(\mathbf{r})$, $\sigma_p > 0$ given in Eq. (9.65).

It is important to note that the SVD $\{v_p, u_p, \sigma_p\}$ is totally independent of the scattered-field data and, hence, can be computed once and for all for any given experimental arrangement. Moreover, $\langle \chi_n, \chi_{n'} \rangle_{\mathcal{H}_V}$ will be rank deficient or nearly singular, so only a relatively small number of terms will need to be included in the inversion Eq. (9.67a). This means, of course, that only this number of singular functions will need to be stored in memory, which is an important consideration since the full number of terms will be $N_j \times N_k \times N^3$, where N^3 is the number of pixels in a given 3D image space.

9.10.2 Back-propagation imaging

The reconstruction algorithm Eq. (9.67a) is in the form of the *filtered back-propagation algorithms* discussed extensively in Chapters 5 and 8. In particular, the expansion coefficients

$$\frac{\langle u_p, \delta f\rangle_{\mathcal{H}_f}}{\sigma_p}$$

can be viewed as being the result of linear filtering of the raw data δf in the subspace spanned by the singular vectors u_p. These filtered data are then back-propagated into τ_0 by summation over the singular functions $v_p(\mathbf{r})$. If we replace the filters $1/\sigma_p$ by σ_p we obtain the algorithm

$$\tilde{V}(\mathbf{r}) = \sum_p \sigma_p \langle u_p, \delta f\rangle_{\mathcal{H}_f} v_p(\mathbf{r}) = \hat{T}^\dagger \, \delta f, \tag{9.68a}$$

where we have used the result that the adjoint operator has the representation

$$\hat{T}^\dagger = \sum_p \sigma_p v_p(\mathbf{r}) u_p^*(j,k).$$

As discussed in those earlier chapters, Eq. (9.68a) corresponds to simple back propagation of the data and can be implemented directly by making use of the definition of $\hat{T}^\dagger$ given in Eq. (9.62b):

$$\tilde{V}(\mathbf{r}) = -\frac{1}{4\pi}\mathcal{M}_{\tau_0} \sum_{j=1}^{N_j}\sum_{k=1}^{N_k} \psi_+^*(\mathbf{r}, -k_0\mathbf{s}^{(j)})\psi_+^*(\mathbf{r}, k_0\mathbf{s}_0^{(k)})\delta f_+(\mathbf{s}^{(j)}, \mathbf{s}_0^{(k)}). \tag{9.68b}$$

The process of computing the reconstruction of the scattering potential by means of the back-propagation algorithm as given in Eq. (9.68b) is seen to be simple and requires absolutely no work. One simply multiplies the raw data by the conjugates of the plane-wave scattering states and sums over the sets of incident and observation directions. In cases in which the singular values σ_p are essentially constant over a pass band, as is often the case, this raw image of the scattering potential can be quite accurate, as we have found in our treatments of the ISP and ISCP in earlier chapters. In any case the back-propagation algorithm allows one to compute a quick approximation to the scattering potential directly from the data and the plane-wave scattering states of the background.

9.10.3 The limited-view problem in a homogeneous background

In this section we specialize the general formulation presented above to the ISCP for a 2D scattering potential compactly supported within a circle of radius a_0 in a 2D homogeneous lossless background having wavenumber k_0. The plane-wave scattering states are, in this case, the usual plane waves

$$\psi_+(\mathbf{r}, k_0\mathbf{s}) = e^{ik_0\mathbf{s}\cdot\mathbf{r}},$$

where both $\mathbf{s}$ and $\mathbf{r}$ are vectors on the (x, y) plane. The far-field ISCP within the DWBA then reduces to the far-field ISCP within the normal Born approximation which consists of inverting the mapping (cf. Section 8.1.3)

$$f(\mathbf{s}, \mathbf{s}_0) = -\sqrt{\frac{1}{8\pi k_0}} e^{i\frac{\pi}{4}} \int_{\tau_0} d^2r \, V(\mathbf{r}) e^{-ik_0\mathbf{s}\cdot\mathbf{r}} e^{ik_0\mathbf{s}_0\cdot\mathbf{r}},$$

where τ_0 is a circle of radius a_0 centered at the origin.

We can cast the 2D limited-view problem within the Hilbert-space formulation given in Eq. (9.61) by simply replacing δf in that equation by

$$\delta f(\mathbf{s}^j, \mathbf{s}_0^k) = \sqrt{\frac{k_0}{2\pi}} e^{-i\frac{\pi}{4}} f(\mathbf{s}^j, \mathbf{s}_0^k),$$

where, as above, $\mathbf{s}^j, j = 1, 2, \ldots, N_j$, are the set of scattering unit vectors and $\mathbf{s}_0^k$, $k = 1, 2, \ldots, N_k$, the set of incident plane-wave unit vectors. The operators $\hat{T}$ and $\hat{T}^\dagger$ defined in Eqs. (9.62) then become

$$\hat{T} = -\frac{1}{4\pi} \int_{\tau_0} d^2r' \, e^{ik_0 \delta\mathbf{s}^{(j,k)} \cdot \mathbf{r}'}, \qquad \hat{T}^\dagger = -\frac{1}{4\pi} \mathcal{M}_{\tau_0} \sum_{j,k} e^{ik_0 \delta\mathbf{s}^{(j,k)} \cdot \mathbf{r}},$$

where now $\mathcal{M}_{\tau_0}$ is the masking operator for a circle of radius a_0 and centered at the origin of the (x, y) plane and $\delta\mathbf{s}^{(j,k)} = \mathbf{s}^{(j)} - \mathbf{s}_0^{(k)}$. The quantities χ_n defined in Eq. (9.63) are given by

$$\chi_n(\mathbf{r}) = -\frac{1}{4\pi} \mathcal{M}_{\tau_0} e^{ik_0 \delta\mathbf{s}^{(j,k)} \cdot \mathbf{r}}, \tag{9.69}$$

where $n = j + (k-1)N_j$.

The singular vectors $v_p(\mathbf{r})$ are given by the expansion Eq. (9.65) with the χ_n now defined in Eq. (9.69) and the expansion coefficients $u_p(n)$ are again the components of the singular vectors u_p and satisfy the matrix equation Eq. (9.66). The matrix $\langle \chi_n, \chi_{n'} \rangle_{\mathcal{H}_V}$ is particularly simple in this case and can be found in closed form:

$$\begin{aligned} \langle \chi_n, \chi_{n'} \rangle_{\mathcal{H}_V} &= \frac{1}{(4\pi)^2} \int_0^{a_0} r \, dr \int_0^{2\pi} d\phi \, e^{-ik_0(\delta\mathbf{s}^{(j,k)} - \delta\mathbf{s}^{(j',k')}) \cdot \mathbf{r}} \\ &= \frac{a_0^2}{8\pi} \frac{J_1(k_0 a_0 |\delta\mathbf{s}^{(j,k)} - \delta\mathbf{s}^{(j',k')}|)}{k_0 a_0 |\delta\mathbf{s}^{(j,k)} - \delta\mathbf{s}^{(j',k')}|}. \end{aligned} \tag{9.70}$$

As a final step the pseudo-inverse is given by Eq. (9.67a), with the expansion coefficients $\langle u_p, \delta f \rangle_{\mathcal{H}_f}$ obtained from the solutions of Eq. (9.66) using the matrix computed from Eq. (9.70).

Example 9.6 We performed a simulation of the 2D limited-view problem for a scattering potential consisting of two circular disks embedded in a uniform background:

$$V(\mathbf{r}) = \sum_{j=1}^{2} V_j \, \mathrm{Circ}[a_j(\mathbf{r} - \mathbf{R}_j)],$$

where Circ is the "Circ function" having radius a_j and center $\mathbf{R}_j$ and V_j, $j = 1, 2$, are two real constants. The 2D scattering potential is then found to be (cf. Eqs. (8.75) of Chapter 8)

$$f(\mathbf{s}, \mathbf{s}_0) = -\sqrt{\frac{1}{8\pi k_0}} e^{i\frac{\pi}{4}} \tilde{V}(\mathbf{K}) = -\sqrt{\frac{\pi}{2k_0}} e^{i\frac{\pi}{4}} \sum_{j=1}^{N} V_j \frac{J_1(Ka_j)}{K} e^{-i\mathbf{K}\cdot\mathbf{R}_j},$$

where $\mathbf{K} = k_0(\mathbf{s} - \mathbf{s}_0)$.

We took the overall scattering-potential support radius to be $a_0 = 5\lambda$ and first used 13 incident-wave directions $\mathbf{s}_0^j$ uniformly distributed over the unit circle and 17 scattered-wave directions $\mathbf{s}^k$ uniformly distributed between $-\pi/2$ and $\pi/2$ relative to the incident-wave

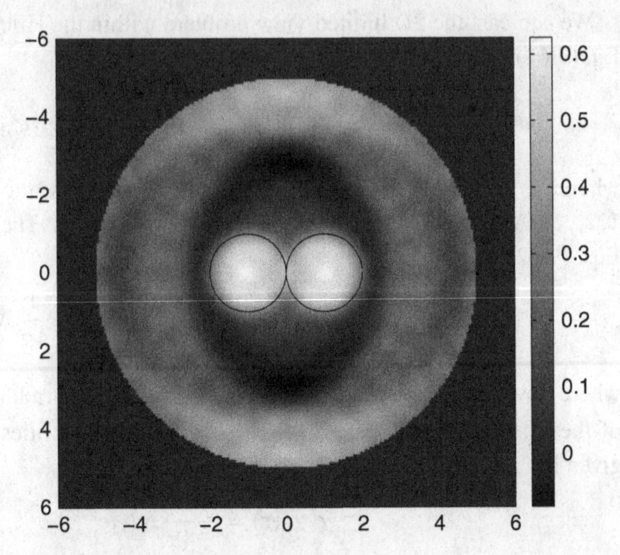

Fig. 9.8 Reconstruction of two circular scatterers using 13 incident plane waves and 17 scattering directions. Two circles indicating the actual supports of the two scatterers are drawn.

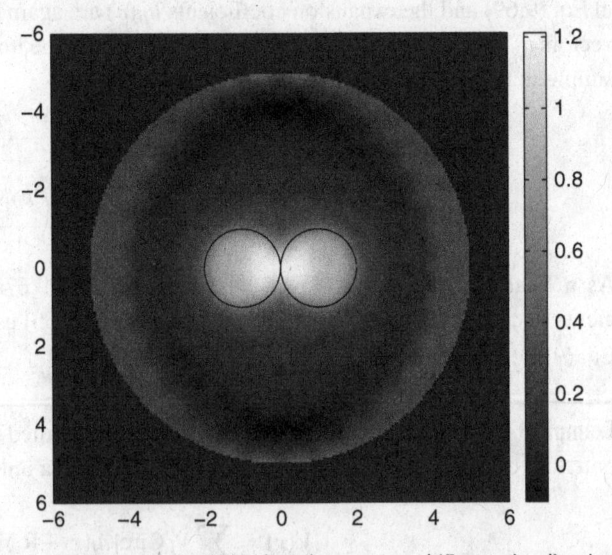

Fig. 9.9 Reconstruction of two circular scatterers using 17 incident plane waves and 17 scattering directions. Two circles indicating the actual supports of the two scatterers are drawn.

direction; i.e., over 180° in the forward-scattering direction for each incident plane wave. We limited the scattered-field directions to the forward direction since we are using the Born approximation in a uniform background where backscattering will be small for scattering potentials more than a wavelength in radius. The 221 singular values σ_p varied from a maximum of 2.16 to zero, with a significant drop off at $p = 189$, which we selected as a cutoff value in the reconstruction algorithm.

We centered one of the two scatterers one wavelength to the left of the origin at $x = -\lambda$ and $y = 0$ and the second at $x = \lambda$ and $y = 0$. The two scatterers were given the same radius of $a = \lambda$ and scattering strengths of unit amplitude. The reconstruction is shown in Fig. 9.8, where we have added the circles to indicate the supports of the scatterers. The supports and locations of the two scatterers are clearly correct in the reconstruction, but it has limited overall quality due to the limited number of incident and scattering directions employed. Increasing the number of incident and scattering directions increases the quality of the reconstruction, but at the cost of computer memory. We show in Fig. 9.9 a reconstruction obtained using the same number of scattering directions but 17 incident-wave directions. The quality of the reconstruction clearly improves, as is evident from the more accurate scattering strengths and the decrease in the "halo" around the two scatterers.

9.11 The ISCP using data generated and acquired by sets of antennas

The formulation and SVD-based solution of the far-field ISCP within the DWBA presented in the last section is easily generalized to be applicable to experiments employing transmitter and receiver elements (antennas) distributed over an arbitrary set of points outside of the scattering volume τ_0. In the simplest case we can model the antennas as point sources or sinks, in which case the incident wave generated from one of the transmitting antennas is the background Green function $G_{0_+}(\mathbf{r}, \boldsymbol{\alpha}_k)$, where $\boldsymbol{\alpha}_k$, $k = 1, 2, \ldots N_\alpha$, is the location of the antenna. The field measured by a point antenna located at $\boldsymbol{\beta}_j$, $j = 1, 2, \ldots, N_\beta$, is then the total Green function $G_+(\boldsymbol{\beta}_j, \boldsymbol{\alpha}_k)$ of the composite medium consisting of the scatterer with unknown scattering potential $V(\mathbf{r})$ embedded in the background with known scattering potential $V_0(\mathbf{r})$. The scattering model for the ISCP is then found from the LS Eqs. (9.45) to be given by

$$G_+(\boldsymbol{\beta}_j, \boldsymbol{\alpha}_k) = G_{0_+}(\boldsymbol{\beta}_j, \boldsymbol{\alpha}_k) + \int d^3r' \, G_+(\boldsymbol{\beta}_j, \mathbf{r}')V(\mathbf{r}')G_{0_+}(\mathbf{r}', \boldsymbol{\alpha}_k). \tag{9.71}$$

The quantity

$$K_{j,k} = G_+(\boldsymbol{\beta}_j, \boldsymbol{\alpha}_k) - G_{0_+}(\boldsymbol{\beta}_j, \boldsymbol{\alpha}_k)$$

is known as the *multistatic data matrix* and is the obvious generalization of the conventional scattering amplitude to near-field applications. It follows from Eq. (9.71) that the multistatic data matrix is related to the unknown scattering potential via the equation

$$K_{j,k} = \int d^3r' \, G_+(\boldsymbol{\beta}_j, \mathbf{r}')V(\mathbf{r}')G_{0_+}(\mathbf{r}', \boldsymbol{\alpha}_k), \tag{9.72a}$$

and within the DWBA by

$$K_{j,k} \approx \int d^3r' \, G_{0_+}(\boldsymbol{\beta}_j, \mathbf{r}')V(\mathbf{r}')G_{0_+}(\mathbf{r}', \boldsymbol{\alpha}_k). \tag{9.72b}$$

We can generalize the above development to extended transmitting and receiving antennas by modeling the transmitting antennas as surface sources consisting of singlet and doublet components as described in Sections 1.8 and 2.12 for homogeneous backgrounds. These models are easily generalized to inhomogeneous backgrounds by simple replacement of the constant-background Green functions by non-uniform-background Green functions. The corresponding models for receiving antennas are then obtained by simple arguments of reciprocity. The simplest such model, and the one we will employ here, radiates a field within the background according to the equation

$$\psi_t(\mathbf{r}, \boldsymbol{\alpha}_k) = \int d^2r' \, \mathcal{R}_t(\mathbf{r}', \boldsymbol{\alpha}_k) G_{0_+}(\mathbf{r}, \mathbf{r}'), \tag{9.73a}$$

where, as above, $\boldsymbol{\alpha}_k$ is the location of the antenna element and $\mathcal{R}_t$ is its "transmission function," which is essentially the singlet component of a finite-sized source distributed over the surface of the antenna. The quantity $\psi_t(\mathbf{r}, \boldsymbol{\alpha}_k)$ is the transmitting antenna's "response function" and is the field radiated by the antenna and, thus, the generalization of $G_{0_+}(\mathbf{r}, \boldsymbol{\alpha}_k)$ to the case in which the field is radiated by a finite-sized antenna.

The model for the receiving antenna located at $\boldsymbol{\beta}_j$ is given by

$$\psi_r(\boldsymbol{\beta}_j, \mathbf{r}) = \int d^2r' \, \mathcal{R}_r(\boldsymbol{\beta}_j, \mathbf{r}') G_+(\mathbf{r}', \mathbf{r}), \tag{9.73b}$$

where $\mathcal{R}_r$ is the receiving antenna's "transmission function." The quantity $\psi_r(\boldsymbol{\beta}_j, \mathbf{r})$ is the receiving antenna's response function and is the measured response at an extended antenna element located at $\boldsymbol{\beta}_j$ to the field radiated by a point source at $\mathbf{r}$ and is thus the generalization of $G_+(\boldsymbol{\beta}_j, \mathbf{r})$ to the case in which the field is measured by a finite-sized antenna. Within the DWBA $\psi_r(\boldsymbol{\beta}_j, \mathbf{r})$ becomes

$$\psi_r(\boldsymbol{\beta}_j, \mathbf{r}) \approx \int d^2r' \, \mathcal{R}_r(\boldsymbol{\beta}_j, \mathbf{r}') G_{0_+}(\mathbf{r}', \mathbf{r}). \tag{9.73c}$$

The generalization of the exact and DWBA models for the multistatic data matrix given in Eqs. (9.72) to finite-sized antennas is found to be

$$K_{j,k} = \int d^3r' \, \psi_r(\boldsymbol{\beta}_j, \mathbf{r}') V(\mathbf{r}') \psi_t(\mathbf{r}', \boldsymbol{\alpha}_k), \tag{9.74}$$

where $\psi_r(\boldsymbol{\beta}_j, \mathbf{r})$ is given by Eq. (9.73b) within the exact scattering model and by Eq. (9.73c) for scattering within the DWBA.

Solution of the limited-view ISCP

We now consider the limited-view ISCP formulated in terms of the DWBA of the multistatic data matrix as given by Eqs. (9.74) with $\psi_r(\boldsymbol{\beta}_j, \mathbf{r})$ defined in Eq. (9.73c). We can express this equation in the standard form

$$\hat{T}V = K,$$

where $K = K_{j,k} \in \mathcal{H}_f = l^2(C^N)$, $N = N_\alpha \times N_\beta$, is the multistatic data matrix, $V \in \mathcal{H}_V = L^2(\tau_0)$ is the unknown scattering potential and $\hat{T} : \mathcal{H}_V \to \mathcal{H}_f$ is the linear operator

$$\hat{T} = \int d^3r' \, \psi_r(\boldsymbol{\beta}_j, \mathbf{r}')\psi_t(\mathbf{r}', \boldsymbol{\alpha}_k).$$

The Hilbert space $\mathcal{H}_f$ has the same inner product as in the far-field limited-view problem defined in Eq. (9.60a), from which it follows that

$$\hat{T}^\dagger = \mathcal{M}_{\tau_0} \sum_{j=1}^{N_\beta} \sum_{k=1}^{N_\alpha} \psi_r^*(\boldsymbol{\beta}_j, \mathbf{r})\psi_t^*(\mathbf{r}, \boldsymbol{\alpha}_k).$$

Following identical lines as employed in the far-field limited-view ISCP we now introduce the functions

$$\chi_n(\mathbf{r}) = \mathcal{M}_{\tau_0} \psi_r^*(\boldsymbol{\beta}_j, \mathbf{r})\psi_t^*(\mathbf{r}, \boldsymbol{\alpha}_k), \tag{9.75}$$

where $\mathcal{M}_{\tau_0}$ is the masking operator defined in Eq. (9.31b) and $n = j + (k-1)N_\beta$ ordered such that, for each $j = 1, 2, \ldots, N_\beta, k = 1, 2, \ldots, N_\alpha$. The normal equations for the singular functions $v_p(\mathbf{r}) \in \mathcal{H}_V$ and singular vectors $u_p \in \mathcal{H}_f$ are then found to be again given by Eq. (9.64), with the functions χ_n now defined according to Eqs. (9.75). The solution of the normal equations are given by

$$v_p(\mathbf{r}) = \mathcal{M}_{\tau_0} \sum_{j=1}^{N_\beta} \sum_{k=1}^{N_\alpha} u_p(j,k)\psi_r^*(\boldsymbol{\beta}_j, \mathbf{r})\psi_t^*(\mathbf{r}, \boldsymbol{\alpha}_k), \quad \sigma_p > 0, \tag{9.76}$$

where the expansion coefficients $u_p(j,k) = u_p(n)$, $n = j + (k-1)N_\beta$, and the squares of the singular values σ_p^2 are the eigenvectors and eigenvalues of the Hermitian matrix

$$\langle \chi_n, \chi_{n'} \rangle_{\mathcal{H}_V} = \int_{\tau_0} d^3r \, \psi_r(\boldsymbol{\beta}_j, \mathbf{r})\psi_t(\mathbf{r}, \boldsymbol{\alpha}_k)\psi_r^*(\boldsymbol{\beta}_{j'}, \mathbf{r})\psi_t^*(\mathbf{r}, \boldsymbol{\alpha}_{k'}). \tag{9.77}$$

Finally, the solution of the ISCP is given by

$$\hat{V}(\mathbf{r}) = \sum_{\sigma_p > 0} \frac{\langle u_p, K \rangle_{\mathcal{H}_f}}{\sigma_p} v_p(\mathbf{r}),$$

where

$$\langle u_p, K \rangle_{\mathcal{H}_f} = \sum_{j=1}^{N_\beta} \sum_{k=1}^{N_\alpha} u_p^*(j,k)K_{j,k},$$

with the singular functions $v_p(\mathbf{r})$, $\sigma_p > 0$ given in Eq. (9.76).

The back-propagation algorithm

As was the case for far-field data the reconstruction algorithm obtained above is a filtered back-propagation algorithm and can be approximated by the back-propagation algorithm obtained by replacing the inverse filters $1/\sigma_p$ by σ_p. One then obtains

$$\tilde{V}(\mathbf{r}) = \sum_p \sigma_p \langle u_p, K \rangle_{\mathcal{H}_f} v_p(\mathbf{r}) = \hat{T}^\dagger K$$

$$= \mathcal{M}_{\tau_0} \sum_{j=1}^{N_\beta} \sum_{k=1}^{N_\alpha} \psi_{\mathrm{r}}^*(\boldsymbol{\beta}_j, \mathbf{r}) \psi_{\mathrm{t}}^*(\mathbf{r}, \boldsymbol{\alpha}_k) K_{j,k}, \tag{9.78}$$

where we have denoted the back-propagated reconstruction by $\tilde{V}(\mathbf{r})$. We will find in the next chapter in our treatment of time-reversal imaging of systems of discrete scatterers that the back-propagation algorithm as given above generates a raw image from which the locations of the various scatterers can be estimated.

Further reading

A time-domain version of inverse scattering in inhomogeneous media is presented in Melamed *et al.* (1999a, 1990b). Ramm (Ramm, 1990) has studied the completeness property of products of solutions to partial differential equations, which play a crucial role in the solution of the limited-view problem presented in Section 9.10. An alternative computationally based scheme for inverse scattering within the DWBA is presented by Chew (Chew, 1990) and an iterative method is given in Chew and Wang (1990), while Heyman and colleagues have formulated inverse scattering in inhomogeneous media in the time domain (Melamed *et al.*, 1996). The treatment of the limited-view problem presented in Section 9.10 is similar to the inversion schemes employed by Twomey (Twomey, 2002). Hansen and Yaghjian (Hansen and Yaghjian, 1999) provide a complete treatment of antenna theory that is especially appropriate for the material covered in this book.

Problems

9.1 Show that in the special case in which the background is lossless so that the scattering potential V_0 is real-valued we have that

$$\psi_+^*(\mathbf{r}, k_0\mathbf{s}_0) = e^{-ik_0\mathbf{s}_0\cdot\mathbf{r}} + \int d^3r'\, G_-(\mathbf{r}, \mathbf{r}') V_0(\mathbf{r}') e^{-ik_0\mathbf{s}_0\cdot\mathbf{r}'},$$

from which we conclude that for such backgrounds

$$\psi_+^*(\mathbf{r}', k_0\mathbf{s}_0) = \psi_-(\mathbf{r}', -k_0\mathbf{s}_0).$$

9.2 Compute the plane-wave scattering states in two space dimensions for an inhomogeneous medium consisting of a homogeneous plane-parallel slab with constant wavenumber $k \neq k_0$.

9.3 Prove that the plane-wave scattering states satisfy the relationship

$$\psi_+(\mathbf{r}, k_0\mathbf{s}_0; \mathbf{X}_0) = e^{ik_0\mathbf{s}_0\cdot\delta\mathbf{X}_0} \psi_+(\mathbf{r}, k_0\mathbf{s}_0; \mathbf{X}_0'),$$

where $\delta\mathbf{X}_0 = \mathbf{X}_0 - \mathbf{X}_0'$ with $\mathbf{X}_0$ and $\mathbf{X}_0'$ being any two central locations of the background scattering potential.

9.4 Fill in the missing steps in the derivation of Eqs. (9.10) of Example 9.1.

9.5 Fill in the missing steps in the derivation of Eq. (9.15).

9.6 Complete the derivation of Eqs. (9.21).

9.7 Complete the derivation of Eqs. (9.24).

9.8 Derive Eq. (9.25b) from Eq. (9.25a) and the reciprocity condition satisfied by the background Green functions.

9.9 Prove that the operators $\hat{T}$ and $\hat{T}^\dagger$ defined in Eqs. (9.30a) and (9.30b) are compact.

9.10 Complete the derivation of Eq. (9.55).

9.11 Prove the inhomogeneous-medium "field uniqueness theorem," which states that the field radiated by a source compactly supported in a space region τ_0 within a non-uniform medium is uniquely determined over all space points lying outside τ_0 by the radiated field or its normal derivative (Dirichlet or Neumann conditions) over any closed surface $\partial\tau$ that completely surrounds τ_0.

9.12 Show that for a lossless inhomogeneous background

$$\hat{T}^\dagger\hat{T} = \frac{i}{2k_0}\mathcal{M}_{\tau_0}\int_{\tau_0} d^3r'\, G_{0_f}(\mathbf{r},\mathbf{r}'),$$

where $\hat{T}$ is the operator defined in Eq. (9.30a).

9.13 Formulate and solve the 2D ISP for a source compactly supported within a homogeneous plane-parallel slab with constant wavenumber $k \neq k_0$ and Dirichlet data over two bounding planes. Compare and contrast your solution with that obtained in Section 5.3 of Chapter 5.

9.14 Compute the scattering amplitude for a set of discrete point scatterers embedded in an inhomogeneous medium using the Foldy–Lax scattering model.

9.15 Complete the derivation of Eq. (9.70) in Section 9.10.3.

9.16 Derive Eq. (9.66).

9.17 Derive Eq. (9.74).

10 Time-reversal imaging for systems of discrete scatterers

In this chapter we turn our attention to systems of discrete scatterers embedded in an inhomogeneous background medium characterized by a background scattering potential $V_0(\mathbf{r})$. The system of scatterers is assumed to have a scattering potential of the general form

$$V(\mathbf{r}) = \sum_{m=1}^{M} V_m(\mathbf{r}), \tag{10.1a}$$

where the component scattering potentials $V_m(\mathbf{r})$ are each compactly supported within non-overlapping regions $\tau_m \subset \tau_0$ with centers $\mathbf{X}_m$. In the simplest, but important, case in which the supports τ_m of the component scatterers are all much smaller than the wavelength of the wavefields of interest we can make the approximation

$$V_m(\mathbf{r}) \approx \mathcal{V}_m \delta(\mathbf{r} - \mathbf{X}_m),$$

where the scattering strengths $\mathcal{V}_m$, $m = 1, 2, \ldots, M$, are a set of real or complex constants. The scattering potential of the system of scatterers then becomes

$$V(\mathbf{r}) = \sum_{m=1}^{M} \mathcal{V}_m \delta(\mathbf{r} - \mathbf{X}_m). \tag{10.1b}$$

We will employ the simple model Eq. (10.1b) throughout most of this chapter. Generalizations of this model to extended discrete scatterers have been developed by Chambers and co-workers (Chambers and Berryman, 2004; Chambers and Gautesen, 2001; Chambers, 2002) and Zhao (Zhao, 2004). As usual, we work in the frequency domain but will not explicitly display the frequency ω unless necessary.

Our main topics in the chapter are time-reversal imaging and inverse scattering from systems having the general model Eq. (10.1a) or the simplified model Eq. (10.1b). Our major emphasis is in so-called "computational time-reversal imaging," where the goal is to determine the scattering centers $\mathbf{X}_m$ and, possibly, the scattering strengths $\mathcal{V}_m$ in the case of the model Eq. (10.1b) from measurements of the multistatic data matrix introduced at the end of the previous chapter. We will also, however, briefly review so-called "experimental time reversal" using a sequence of actual (physical) experiments to focus an incident wave on one or more of the various discrete scatterers. Within the area of computational time-reversal imaging we review the DORT (Prada and Fink, 1994, 1995) (standing for "décomposition de l'opérateur de retournement temporel") and time-reversal MUSIC (Lev-Ari and Devaney, 2000) algorithms and will also show how these algorithms are related to the solution of the limited-view ISCP as formulated in the last chapter in terms of the multistatic data matrix. We will begin our treatment with a brief review of experimental time-reversal imaging as developed more than 20 years ago by the famous French group headed by M. Fink.

10.1 Experimental time-reversal imaging

We consider a system of discrete scatterers and an antenna system that lies outside the support τ_0 of the system and has the ability both to generate an incident wave propagating into the system and to measure the resulting scattered wave over some set of points exterior to τ_0. Our goal is to develop an experimental procedure that is able to generate a set of incident waves having the property that they each focus on a different scatterer in the system *without prior knowledge of the scatterer locations or of the background medium in which they are embedded!* In the ideal and simplest case we can imagine the scattering system to be embedded in a homogeneous and non-dispersive background with wave propagation and scattering governed by the wave equation. We will also assume that the system is completely surrounded by a closed surface $\partial\tau$ over which time-domain surface singlet and doublet sources as described in Section 1.8 can be deployed and that time-domain measurements of both the scattered field $u^{(s)}(\mathbf{r},t)$ and its normal derivative $\partial u^{(s)}(\mathbf{r},t)/\partial n$ can be performed.

Restricting our attention, for the moment, to the ideal situation described above, we now imagine that an arbitrary wave is incident to the scattering system and that both the resulting scattered field and its normal derivative are measured over $\partial\tau$. We showed in Section 5.1.3 of Chapter 5 that if these field data are time-reversed and used as singlet and doublet sources on $\partial\tau$ the resulting incident wave to the scattering system will be a singularity-free version of $u^{(s)}(\mathbf{r},-t)$; i.e., these surface sources will generate an incident wave that closely resembles the time-reversed scattered wave from the system of scatterers. Since the discrete scatterers comprising the scattering system will generate outgoing expanding spherical scattered waves centered at their respective scattering centers, the time-reversed scattered field and, hence, the new incident wave, will consist of a set of incoming contracting spherical waves with focal points at the various scattering centers.[1] If this process of illumination using the time-reversed scattered-wave data from the previous experiment is repeated in a sequence of experiments one can expect that eventually the sequence will converge, yielding an incident wave $u_1^{(in)}$ that will focus on the strongest scatterer in the system.[2] The entire process can then be repeated, where, however, the incident wave used in any experiment is processed to remove the component lying along $u_1^{(in)}$.

[1] These incoming spherical waves are the free-space propagators $g_f(\mathbf{r}-\mathbf{X}_m,t)$ equal to the difference between the retarded and advanced Green functions to the wave equation (see Section 5.1.3) that focus at $\mathbf{X}_m$ but have no singularity.

[2] The incident wave in any given experiment is assumed to be normalized over the measurement boundary $\partial\tau$. Otherwise the sequence of scattered waves would converge to zero due to the inherent loss of energy experienced by the incident field in the scattering process.

This new sequence can then be expected to converge to focus on the second strongest scatterer. Again the process is repeated, where now the incident waves are processed to remove the components lying both along $u_1^{(\text{in})}$ and along $u_2^{(\text{in})}$. By this means then it is possible to selectively focus on all of the scatterers in the system.

10.1.1 Experimental time-reversal imaging in non-uniform media

The above discussion was certainly "suggestive" but did not detail the mathematics behind the iterative procedure and also assumed a homogeneous and non-dispersive background medium with wave propagation and scattering governed by the wave equation. Here we will generalize the above treatment to a system of discrete scatterers embedded in an inhomogeneous but still lossless background having scattering potential $V_0(\mathbf{r})$ and will examine the iterative procedure in the frequency domain, where we can get a better understanding of the underlying mathematics behind the procedure. As above, we assume that both the scattered field and its normal derivative are measurable over a closed surface $\partial\tau$ surrounding the scattering volume τ_0 and that singlet and doublet sources can be deployed over the surface to radiate an incident wave into τ_0.

We begin by assuming that an arbitrary but unfocused wave $U_1^{(\text{in})}$ is incident to the system of discrete scatterers and that both the resulting scattered field and its normal derivative are measured over $\partial\tau$. If we now normalize and time reverse (complex conjugate) these boundary-value data and use them as surface sources that radiate into the interior $\tau \supset \tau_0$ we generate the new incident wave

$$U_2^{(\text{in})}(\mathbf{r}) = \int_{\partial\tau} dS' \left[\psi^*(\mathbf{r}') \frac{\partial}{\partial n'} G_{0+}(\mathbf{r}, \mathbf{r}') - G_{0+}(\mathbf{r}, \mathbf{r}') \frac{\partial}{\partial n'} \psi^*(\mathbf{r}') \right],$$

where ψ is the scattered wave generated in the first experiment normalized over the measurement boundary and G_{0+} is the outgoing-wave Green function in the background medium. If we now make use of the Porter–Bojarski (PB) integral equation derived in Section 9.2.3 and the assumption that the background medium is lossless so that $G_{0+}^* = G_{0-}$ we find that

$$U_2^{(\text{in})}(\mathbf{r}) = \Phi^*(\mathbf{r}) = -\int_{\tau_0} d^3r' \, G_{0_f}(\mathbf{r}, \mathbf{r}') Q^*(\mathbf{r}'), \tag{10.2}$$

where Q is the induced source and

$$G_{0_f}(\mathbf{r}, \mathbf{r}') = G_{0+}(\mathbf{r}, \mathbf{r}') - G_{0-}(\mathbf{r}, \mathbf{r}')$$

is the free-space propagator of the background medium and

$$\Phi(\mathbf{r}) = \int_{\partial\tau} dS' \left[\psi(\mathbf{r}') \frac{\partial}{\partial n'} G_{0-}(\mathbf{r}, \mathbf{r}') - G_{0-}(\mathbf{r}, \mathbf{r}') \frac{\partial}{\partial n'} \psi(\mathbf{r}') \right]$$

is the field back propagated into τ from the boundary-value data.

Assume, for the moment, that the wavelength is sufficiently large that the approximate form of the scattering potential given in Eq. (10.1b) can be employed. The induced source is then given by

$$Q(\mathbf{r}) = V(\mathbf{r})U_1(\mathbf{r}) = \sum_{m=1}^{M} V_m U_1(\mathbf{X}_m)\delta(\mathbf{r} - \mathbf{X}_m),$$

where U_1 is the sum of the incident and scattered wavefields from the first experiment. If we substitute this equation into Eq. (10.2) and perform some simple algebra we obtain

$$U_2^{(\mathrm{in})}(\mathbf{r}) = -\sum_{m=1}^{M} V_m^* U_1^*(\mathbf{X}_m) G_{0_\mathrm{f}}(\mathbf{r}, \mathbf{X}_m). \tag{10.3}$$

The free space propagator $G_{0_\mathrm{f}}(\mathbf{r}, \mathbf{X}_m)$ is a regularized version of the background Green function and will peak at the various scattering centers $\mathbf{X}_m$ so that $U_2^{(\mathrm{in})}(\mathbf{r})$ will focus on the scattering centers of the system. The strength of the focusing is proportional to the product $|V_m U_1(\mathbf{X}_m)|$, which, for an unfocused incident wave, can be expected to be maximized at the location of the strongest scatterer so that the overall wave scattered from the system will then be dominated by the component from this strongest scatterer. The process is then repeated using the normalized new scattered wave to generate an even stronger focused wave at the location of the strongest scatterer and so on until the iterative process converges to a pair of incident and scattered waves that satisfies the equation

$$U^{(\mathrm{s})}(\mathbf{r}) = \sigma U^{(\mathrm{in})*}(\mathbf{r}), \qquad \frac{\partial}{\partial n} U^{(\mathrm{s})}(\mathbf{r}) = \sigma \frac{\partial}{\partial n} U^{(\mathrm{in})*}(\mathbf{r}), \quad \mathbf{r} \in \partial\tau, \tag{10.4}$$

where σ is a constant.

Assume now that the normalized final incident wave

$$\psi_1 = \frac{U^{(\mathrm{in})}(\mathbf{r})}{||U^{(\mathrm{in})}(\mathbf{r})||}$$

and its normal derivative from this first sequence of experiments is stored and a new sequence of experiments begun, whereby the first and subsequent incident waves are processed to remove the components along ψ_1 and $\partial\psi_1/\partial n$; i.e.,

$$U^{(\mathrm{in})} \to U^{(\mathrm{in})} - \langle \psi_1, U^{(\mathrm{in})} \rangle \psi_1,$$

$$\frac{\partial}{\partial n} U^{(\mathrm{in})} \to \frac{\partial}{\partial n} U^{(\mathrm{in})} - \left\langle \frac{\partial}{\partial n}\psi_1, \frac{\partial}{\partial n} U^{(\mathrm{in})} \right\rangle \frac{\partial}{\partial n}\psi_1,$$

where the brackets $\langle \cdot \rangle$ denote the L^2 inner product over the measurement surface $\partial\tau$. Since the mapping from incident to scattered field is linear, the set of incident waves processed as above will not generate a focused component on the strongest scatterer but rather will converge to an incident wave that focuses on the second-strongest scatterer. This final incident wave will again satisfy Eq. (10.4) with generally a different σ and will, by virtue of the way it was constructed, be orthogonal to ψ_1. It then follows that the entire sequence of experiments can again be repeated, where now the components along both ψ_1 and ψ_2 are removed from each iteration of the sequence. The process can then be repeated until all of the scatterers in the system have sequentially been focused on.

10.2 Time-reversal imaging using a finite set of antennas

Our discussion of experimental time-reversal imaging presented above is, of course, oversimplified in that it assumes that continuously distributed field transmission and receiver systems are employed over the surface $\partial\tau$ surrounding the system of scatterers. Any real experiment must use finite sets of transmitting antennas and receiving antennas of the type employed in our treatment of the limited-view inverse scattering problem (ISCP) in Section 9.11 in the last chapter. In this section we will present a brief review of experimental time-reversal imaging implemented with two finite sets of co-located transmit and receive antenna arrays that we will model as point (delta-function) sources that radiate the background Green function. We use such simplified ideal point antenna elements only for the sake of simplicity and the development is easily generalized to more realistic antenna models as described in Section 9.11. We also mention that the two arrays need not be regularly and/or densely spaced but must, of course, be co-located, for otherwise it would not be possible to determine the required (time-reversed) driving signals to the transmitter array from measurements of the scattered field over the (separate) receiver array.

10.2.1 Experimental time-reversal imaging

We again consider the scattering system defined in Eq. (10.1b) embedded in a lossless inhomogeneous background but where the experimental system now consists of a finite set of transmit and receive antennas, which, as mentioned above, we will model as ideal point sources and sinks. We assume that the antennas are co-located over an arbitrary set of points $\boldsymbol{\alpha}_k, k = 1, 2, \ldots, N_\alpha$, outside of the overall scattering volume τ_0. The set of transmitting antennas is excited by frequency-domain signals $t_k(\omega)$, $k = 1, 2, \ldots, N_\alpha$ and the set of co-located N_α receiving antennas output the corresponding frequency-domain signals $r_j(\omega)$, $j = 1, 2, \ldots, N_\alpha$. Our goal, as above, is to employ a sequence of scattering experiments that results in a set of radiated (incident) waves that selectively focus on the various discrete scatterers comprising the scattering system.

We consider a particular experiment in which a set of signals $t_k(\omega)$ is applied to the transmitting antennas, generating an incident wave $U^{(\text{in})}(\mathbf{r})$ in the background medium. The incident wave will be of the general form

$$U^{(\text{in})}(\mathbf{r}) = \sum_{k=1}^{N_\alpha} t_k(\omega) G_{0+}(\mathbf{r}, \boldsymbol{\alpha}_k), \tag{10.5}$$

where $G_{0+}(\mathbf{r}, \boldsymbol{\alpha}_k)$ is the background Green function radiated by the kth transmitter element. Using the simplified model Eq. (10.1b) for the scattering potential, the scattered wave will be given by the Lippmann–Schwinger equation Eq. (9.46b) as

$$U^{(\text{s})}(\mathbf{r}) = \sum_{m=1}^{M} V_m G_+(\mathbf{r}, \mathbf{X}_m) U^{(\text{in})}(\mathbf{X}_m),$$

where G_+ is the Green function of the composite medium consisting of the scattering potential $V(\mathbf{r})$ embedded in the background medium. Owing to our assumption of ideal point antenna elements the output from any given receiving antenna $r_j(\omega)$ is equal to the scattered field evaluated at the antenna location[3] $\boldsymbol{\alpha}_j$, which then yields the equation

$$r_j(\omega) = \sum_{k=1}^{N_\alpha} \left\{ \sum_{m=1}^{M} V_m G_+(\boldsymbol{\alpha}_j, \mathbf{X}_m) G_{0_+}(\mathbf{X}_m, \boldsymbol{\alpha}_k) \right\} t_k(\omega). \tag{10.6}$$

The quantity

$$K_{j,k}(\omega) = \sum_{m=1}^{M} V_m G_+(\boldsymbol{\alpha}_j, \mathbf{X}_m) G_{0_+}(\mathbf{X}_m, \boldsymbol{\alpha}_k) \tag{10.7}$$

is recognized as being the *multistatic data matrix* first introduced in the last section of the previous chapter. This quantity is the response at a receiving antenna located at $\boldsymbol{\alpha}_j$ due to an input signal $t_k(\omega) = 1$ at a transmitting antenna located at $\boldsymbol{\alpha}_k$ and is *independent of the actual driving signals $t_k(\omega)$ applied in any given experiment.* Since we are using ideal point sources and sinks, the K matrix is actually the (scattered-field component) of the composite-medium Green function $G_+(\boldsymbol{\alpha}_j, \boldsymbol{\alpha}_k)$, from which it follows from reciprocity that the matrix K is symmetric. We can then write Eq. (10.6) in the matrix form

$$r(\omega) = Kt(\omega),$$

where $K = K^{\mathrm{T}}$ is the $N_\alpha \times N_\alpha$ multistatic data matrix and $r(\omega)$ and $t(\omega)$ are column vectors with components $r_j(\omega)$ and $t_k(\omega)$, respectively.

Now consider a sequence of experiments in which, as earlier, we employ normalized time-reversed scattered-field data to generate the incident wave in each successive experiment. In place of the boundary-value fields used in the previous section we now use the input and output signals $t_j(\omega)$ and $r_k(\omega)$ to and from the various antennas distributed over the set of points $\boldsymbol{\alpha}_j$, $j = 1, 2, \ldots, N_\alpha$. If the sequence converges we obtain in place of Eq. (10.4) the requirement

$$Kt(\omega) = \sigma t^*(\omega), \tag{10.8a}$$

where again σ is a constant. The above equation is easily converted to an eigen-equation that must be satisfied by the transmission signals by pre-multiplying both sides with the adjoint matrix $K^\dagger = K^{*\mathrm{T}} = K^*$:

$$K^\dagger K t(\omega) = |\sigma|^2 t(\omega). \tag{10.8b}$$

[3] The antenna elements will, of course, measure the total field (incident plus scattered), but we assume that the incident-wave component is known and removed by suitable means.

The $N_\alpha \times N_\alpha$ Hermitian matrix

$$\mathcal{T} = K^\dagger K$$

is called the *time-reversal matrix* and has a complete set of N_α orthonormal eigenvectors μ_p and real non-negative eigenvalues $\lambda_p \geq 0$, $p = 1, 2, \ldots, N_\alpha$. We conclude from the above analysis that the sequence of time-reversal experiments will converge to an input signal $t(\omega)$ that is one of the eigenvectors of $\mathcal{T}$. Indeed, it is not difficult to show that this sequence corresponds to solving for the eigenvectors using the so-called "power method," which will converge to the eigenvector having the largest eigenvalue. Whether or not this largest eigenvector when applied as input to the antenna system focuses on the strongest scatterer depends on the total number of scatterers M and the number of transmit/receive antennas N_α and also depends on the separation of the scattering centers relative to the wavelength of the radiation.

10.2.2 Eigenvectors of the time-reversal matrix

In order to simplify the analysis it is convenient to define the N_α-dimensional *background Green-function vectors*

$$g_0(\mathbf{r}) = [G_{0_+}(\boldsymbol{\alpha}_1, \mathbf{r}), G_{0_+}(\boldsymbol{\alpha}_2, \mathbf{r}), \ldots, G_{0_+}(\boldsymbol{\alpha}_{N_\alpha}, \mathbf{r})]^{\mathrm{T}}, \tag{10.9a}$$

which are simply the column vectors formed from the N_α background Green functions radiated by the antenna elements. We also introduce the associated *composite-medium Green-function vectors*

$$g(\mathbf{r}) = [G_+(\boldsymbol{\alpha}_1, \mathbf{r}), G_+(\boldsymbol{\alpha}_2, \mathbf{r}), \ldots, G_+(\boldsymbol{\alpha}_{N_\alpha}, \mathbf{r})]^{\mathrm{T}}, \tag{10.9b}$$

which are the column vectors formed from the composite-medium Green function. On making use of Eqs. (10.7) we can then express the multistatic data matrix as the weighted sum of outer products of the Green-function vectors evaluated at the various scatterer locations

$$K = \sum_{m=1}^{M} \mathcal{V}_m g(\mathbf{X}_m) g_0^{\mathrm{T}}(\mathbf{X}_m). \tag{10.10}$$

We will refer to the Green-function vectors $g_0(\mathbf{X}_m)$ and $g(\mathbf{X}_m)$ evaluated at the scatterer locations as the background and composite-medium *antenna vectors*.

The rank $\mathcal{R}(K)$ of the multistatic data matrix is clearly less than or equal to the smaller of the number M of antenna vectors and the number N_α of antenna elements. If $M \leq N_\alpha$ it is likely that the antenna vectors will be linearly independent except, possibly, for some very specialized antenna configurations and scatterer locations that are very unlikely to occur,[4] so we will make this assumption throughout this chapter. We then have

$$\text{Rank(K)} = \mathcal{R}(K) = \min(M, N_\alpha), \tag{10.11}$$

a result that will be important in the analysis presented below.

In terms of the antenna vectors the time-reversal matrix then becomes

$$\mathcal{T} = K^\dagger K = \sum_{m=1}^{M} \sum_{m'=1}^{M} g_0^*(\mathbf{X}_m) \Lambda(m, m') g_0^{\mathrm{T}}(\mathbf{X}_{m'}), \tag{10.12a}$$

where

$$\Lambda(m, m') = \mathcal{V}_m^* \mathcal{V}_{m'} g^\dagger(\mathbf{X}_m) g(\mathbf{X}_{m'}). \tag{10.12b}$$

It is clear that the rank of the time-reversal matrix is equal to the rank of the multistatic data matrix $\mathcal{R}(K)$ which satisfies Eq. (10.11).

If we expand the inner product in Eq. (10.12b) we obtain

$$\Lambda(m, m') = \mathcal{V}_m^* \mathcal{V}_{m'} \sum_{j=1}^{N_\alpha} G_+^*(\mathbf{X}_m, \boldsymbol{\alpha}_j) G_+(\boldsymbol{\alpha}_j, \mathbf{X}_{m'}).$$

We can interpret the quantity

$$H(\mathbf{r}, \mathbf{r}') = \sum_{j=1}^{N_\alpha} G_+^*(\mathbf{r}, \boldsymbol{\alpha}_j) G_+(\boldsymbol{\alpha}_j, \mathbf{r}') = g^\dagger(\mathbf{r}) g(\mathbf{r}') \tag{10.13}$$

as being the *coherent point-spread function (CPSF) of the antenna system embedded in the composite medium.* In particular, this quantity represents an "image" of a source point $\mathbf{r}'$ in the composite medium formed by the antenna system from measurement of the outgoing-wave Green function $G_+(\mathbf{r}, \mathbf{r}')$ at the various antenna elements. Mathematically this equation represents the CPSF as the back propagation of the outgoing-wave Green function of the composite medium from its measured values across the antenna system into the scatterer region τ_0. It follows that the inner product occurring in Eq. (10.12b) is the antenna-system point-spread function evaluated at the image and source points $\mathbf{r} = \mathbf{X}_m$ and $\mathbf{r}' = \mathbf{X}_{m'}$, and Λ can be expressed in terms of the CPSF as

$$\Lambda(m, m') = \mathcal{V}_m^* \mathcal{V}_{m'} H(\mathbf{X}_m, \mathbf{X}_{m'}). \tag{10.14}$$

[4] We will assume throughout this chapter that the antenna vectors are linearly independent. This can be easily established for the special case of an infinite homogeneous background (Devaney, 2000).

Well-resolved scatterers

For large, closely spaced, antenna systems and sparse distributions of scatterers the CPSF $H(\mathbf{r}, \mathbf{X}_{m'})$ will have a maximum at the source point location $\mathbf{r} = \mathbf{X}_{m'}$ and will decay in amplitude for points removed from this image point. For sparse systems or for systems having few elements the CPSF will have a complicated structure that consists of ridges and valleys (so-called "grating modes") that converge to the source point at which the CPSF achieves a local maximum or, at least, a maximum in some plane or line that contains the source point. The effective spatial extent of the point-spread function is determined by the geometry of the antenna system, the wavelength of the radiation and the density of the scatterers. For example, for a densely spaced (less than a half-wavelength spacing) planar square antenna array in a uniform background, the *transverse* effective spatial region of support of the CPSF is roughly a square having sides of length $z\lambda/a$, where z is the distance of the array from the system and $a = N_\alpha\delta$ is the length of the side of the array, with δ being the spacing between array elements. The extent of the region of the CPSF in the longitudinal direction (z direction) is considerably larger and, again, depends on the wavelength and array geometry.

We conclude from the above discussion that if a scatterer located at $\mathbf{X}_m$ is separated by distances larger than the effective spatial extent of the CPSF from *all other scatterers* in the system then $H(\mathbf{X}_m, \mathbf{X}_{m'})$ will approximately vanish for $m' \neq m$ and $\Lambda(m, m')$ reduces approximately to

$$\Lambda(m, m') = \mathcal{V}_m^* \mathcal{V}_{m'} g^\dagger(\mathbf{X}_m) g(\mathbf{X}_{m'}) \approx |\mathcal{V}_m|^2 \rho_m \delta_{m,m'}, \tag{10.15a}$$

where

$$\rho_m = H(\mathbf{X}_m, \mathbf{X}_m) = ||g(\mathbf{X}_m)||^2 \tag{10.15b}$$

is the squared norm of the mth antenna vector of the composite medium and $\delta_{m,m'}$ is the Kronecker delta function. If this situation occurs we say that the mth scatterer is *well resolved* by the array.

By making use of Eqs. (10.12a) and (10.15a) we find that if the members of some subset $m \in \mathcal{M}$ of scatterers are well resolved the time-reversal matrix will decompose into the two components

$$\mathcal{T} \approx \sum_{m \in \mathcal{M}} |\mathcal{V}_m|^2 \rho_m g_0^*(\mathbf{X}_m) g_0^{\mathrm{T}}(\mathbf{X}_m) + \sum_{m,m' \notin \mathcal{M}} \Lambda(m, m') g_0^*(\mathbf{X}_m) g_0^{\mathrm{T}}(\mathbf{X}_{m'}), \tag{10.16a}$$

from which it follows that the eigenvectors μ_m, $m \in \mathcal{M}$ are approximately equal to the normalized complex conjugates of the background antenna vectors $g_0(\mathbf{X}_m)$, $m \in \mathcal{M}$ and the eigenvalues are equal to $|\mathcal{V}_m|^2 \rho_{0_m} \rho_m$, $m \in \mathcal{M}$, i.e.,

$$\mu_m = \frac{g_0^*(\mathbf{X}_m)}{\sqrt{\rho_{0_m}}}, \qquad \lambda_m = |\mathcal{V}_m|^2 \rho_{0_m} \rho_m, \quad m \in \mathcal{M}, \tag{10.16b}$$

where

$$\rho_{0_m} = ||g_0(\mathbf{X}_m)||^2$$

are the squared norms of the background antenna vectors. The total number of well-resolved scatterers $N_{\mathcal{M}}$ must clearly be less than or equal to the rank $\mathcal{R}(K)$ of the time-reversal matrix.

Non-resolved scatterers

Those scatterers that do not satisfy the condition of being well resolved, Eq. (10.15a), are said to be "non-resolved" by the antenna array. These scatterers are then represented by the second sum in the decomposition of the time-reversal matrix in Eq. (10.16a) and are associated with eigenvectors μ_m that are linear combinations of the complex conjugates of the antenna vectors entering this summation. Since the rank of the time-reversal matrix is $\mathcal{R}(K) = \min(M, N_\alpha)$ there will be a total of $\mathcal{R}(K)$ eigenvectors having non-zero eigenvalues. These $\mathcal{R}(K)$ eigenvectors are divided between the well-resolved and non-resolved targets so that

$$N_{\mathcal{M}} + N_{\mathcal{M}^\perp} = \mathcal{R}(K), \tag{10.17}$$

where $N_{\mathcal{M}}$ and $N_{\mathcal{M}^\perp}$ are, respectively, the numbers of resolved and unresolved scatterers. The remaining (if any) N_α eigenvectors have zero eigenvalue and are not associated with any scatterers. These eigenvectors will be found to play a key role in time-reversal MUSIC, which will be developed in the following section.

10.2.3 Focusing with the eigenvectors of the time-reversal matrix

We are now in a position to interpret the results obtained in a convergent sequence of time-reversal imaging experiments. As we found earlier, any such sequence will result in an antenna input signal $t = \{t_k(\omega)\}$, $k = 1, 2, \ldots, N_\alpha$, that is an eigenvector μ_m of the time-reversal matrix. In the case of a well-resolved scatterer located at $\mathbf{X}_m$ the eigenvector μ_m will be given by Eq. (10.16b), which then results in the incident wave

$$U_m(\mathbf{r}) = \sum_{k=1}^{N_\alpha} \overbrace{\frac{G^*_{0+}(\mathbf{X}_m, \boldsymbol{\alpha}_k)}{\sqrt{\rho_{0_m}}}}^{t_k(\omega)=\mu_m(k)} G_{0+}(\mathbf{r}, \boldsymbol{\alpha}_k) = \frac{g_0^\dagger(\mathbf{X}_m)}{\sqrt{\rho_{0_m}}} g_0(\mathbf{r}) = \frac{H_0^*(\mathbf{r}, \mathbf{X}_m)}{\sqrt{\rho_{0_m}}}, \tag{10.18}$$

where

$$H_0(\mathbf{r}, \mathbf{r}') = \sum_{j=1}^{N_\alpha} G^*_{0+}(\mathbf{r}, \boldsymbol{\alpha}_j) G_{0+}(\boldsymbol{\alpha}_j, \mathbf{r}') = g_0^\dagger(\mathbf{r}) g_0(\mathbf{r}')$$

is the CPSF of the background medium. The wavefields radiated by a set of eigenvectors μ_m, $m \in \mathcal{M}$ belonging to well-resolved scatterers then individually focus on the different scatterers with image fields proportional to the complex conjugate of the CPSF of the array in the background medium evaluated at the scatterer location.

The eigenvectors not belonging to the set of well-resolved scatterers are associated with the non-resolved scatterers and are of the general form

$$\mu_m = \sum_{m' \notin \mathcal{M}} \left\langle \frac{g_0^*(\mathbf{X}_{m'})}{\sqrt{\rho_{0_{m'}}}}, \mu_m \right\rangle \frac{g_0^*(\mathbf{X}_{m'})}{\sqrt{\rho_{0_{m'}}}}. \tag{10.19}$$

The input signal $t_k(\omega)$ at the kth antenna is then

$$t_k(\omega) = \sum_{m' \notin \mathcal{M}} \left\langle \frac{g_0^*(\mathbf{X}_{m'})}{\sqrt{\rho_{0'_m}}}, \mu_m \right\rangle \frac{G^*_{0_+}(\mathbf{X}_{m'}, \boldsymbol{\alpha}_k)}{\sqrt{\rho_{0_{m'}}}}$$

and the resulting incident wave generated at the conclusion of the sequence of experiments is found to be

$$U_m(\mathbf{r}) = \sum_{k=1}^{N_\alpha} \left\{ \sum_{m' \notin \mathcal{M}} \left\langle \frac{g_0^*(\mathbf{X}_{m'})}{\sqrt{\rho_{0_{m'}}}}, \mu_m \right\rangle \frac{G^*_{0_+}(\mathbf{X}_{m'}, \boldsymbol{\alpha}_k)}{\sqrt{\rho_{0_{m'}}}} \right\} G_{0_+}(\mathbf{r}, \boldsymbol{\alpha}_k)$$
$$= \sum_{m' \notin \mathcal{M}} \left\langle \frac{g_0^*(\mathbf{X}_{m'})}{\sqrt{\rho_{0_{m'}}}}, \mu_m \right\rangle \frac{H_0^*(\mathbf{r}, \mathbf{X}_{m'})}{\sqrt{\rho_{0_{m'}}}}. \tag{10.20}$$

We thus conclude that *for scatterers that are not resolved the sequence of iterative time-reversal experiments will result in an incident wave that focuses on groups of scatterers rather than a single scatterer.*

10.3 Computational time-reversal imaging

It follows from the linearity of the mapping from incident to scattered wavefield that the output signals $r_j(\omega)$, $j = 1, 2, \ldots, N_\alpha$, generated in any physical experiment are completely specified by the multistatic data matrix $K = \{K_{j,k}\}$ and the set of input signals $t_k(\omega)$, $k = 1, 2, \ldots, N_\alpha$. Moreover, if we know the background scattering potential $V_0(\mathbf{r})$ we can compute the background Green function $G_{0_+}(\mathbf{r}, \mathbf{r}')$ and eigenvectors of the time-reversal matrix and thus also compute the actual incident wave $U^{(\text{in})}(\mathbf{r})$ generated from the eigenvectors of the time-reversal matrix. In other words we can *synthetically generate the time-reversal images of the system of scatterers that are generated in a sequence of actual time-reversal experiments!* We call this procedure of *computing* the time-reversal images of the scattering system *computational time-reversal imaging.*

We should note that experimental time-reversal imaging possesses one big advantage over computational time-reversal imaging: one need not know the background scattering potential (or Green function) in order to perform experimental time-reversal imaging. On the other hand, it has a disadvantage in that the actual scattering centers are not determined in the process. To actually determine the scattering centers and, possibly, their strengths it is necessary to have knowledge of the background Green function with which computer algorithms can be devised to perform this determination. Moreover, for experimental time-reversal imaging[5] to work it is necessary to have the transmitter and receiver antenna arrays co-located. Otherwise it would not be possible to determine the required (time-reversed) driving signals to the transmitter array from measurements of the scattered field over the

[5] We emphasize that by "experimental" time-reversal imaging we mean using a sequence of experiments as described earlier to generate focused waves into a medium. One or more experiments must, of course, be performed also in computational time-reversal imaging, but they differ from the sequence used in what we have termed experimental time-reversal imaging.

(separate) receiver array. However, in computational time-reversal imaging each antenna in the system can be regarded as being *both* a transmitter *and* a receiver due to wavefield reciprocity, so time-reversal imaging can be implemented both for co-located and for non-co-located sets of transmitter and receiver arrays. We will develop computational time-reversal imaging in this section for the general case of non-co-located transmit and receive antenna arrays. Our development follows closely that presented in Lehman and Devaney (2003).

We assume that the transmitting antennas are located at $\boldsymbol{\alpha}_k$, $k = 1, 2, \ldots, N_\alpha$, and the set of receiving antennas at $\boldsymbol{\beta}_j$, $j = 1, 2, \ldots, N_\beta$, and that a set of experiments yielding the $N_\beta \times N_\alpha$ multistatic data matrix will be performed. This quantity is the received signal at antenna element $\boldsymbol{\beta}_j$ due to a unit-amplitude applied signal at antenna element $\boldsymbol{\alpha}_k$ and, due to reciprocity, can also be viewed as being the received signal at antenna element $\boldsymbol{\alpha}_k$ due to a unit-amplitude applied signal at antenna element $\boldsymbol{\beta}_j$.

Using the Lippmann–Schwinger equations derived in Section 9.7 of the last chapter we can express the multistatic data matrix as

$$K_{j,k}(\omega) = \int_{\tau_0} G_+(\boldsymbol{\beta}_j, \mathbf{r}')V(\mathbf{r}')G_{0_+}(\mathbf{r}', \boldsymbol{\alpha}_k) = \int_{\tau_0} G_{0_+}(\boldsymbol{\beta}_j, \mathbf{r}')V(\mathbf{r}')G_+(\mathbf{r}', \boldsymbol{\alpha}_k),$$

where $G_{0_+}(\mathbf{r}, \mathbf{r}')$ is the background Green function and $G_+(\mathbf{r}, \mathbf{r}')$ the composite-medium Green function. We can interpret the first of the above two representations as resulting from using the α array as the transmitting array and the β array as the receiving array, while the second representation has the two arrays reversed. We emphasize that the above equations are exact in that they include all of the multiple scattering between the scatterers and background. The distorted-wave Born approximation (DWBA) (see Section 9.8) results from approximating the composite-medium Green function in these expressions by the background Green function. However, we will develop the theory using the exact multiple-scattering formulation as defined by the above expressions for the K matrix.

We will employ the simplified model for the scattering potential given in Eq. (10.1b), for which the above expressions for the K matrix reduce to

$$K_{j,k}(\omega) = \sum_{m=1}^{M} V_m G_+(\boldsymbol{\beta}_j, \mathbf{X}_m)G_{0_+}(\mathbf{X}_m, \boldsymbol{\alpha}_k) = \sum_{m=1}^{M} V_m G_{0_+}(\boldsymbol{\beta}_j, \mathbf{X}_m)G_+(\mathbf{X}_m, \boldsymbol{\alpha}_k),$$

which we can express in the compact forms

$$K(\omega) = \sum_{m=1}^{M} V_m g_\beta(\mathbf{X}_m) g_{0_\alpha}^{\mathrm{T}}(\mathbf{X}_m) = \sum_{m=1}^{M} V_m g_{0_\beta}(\mathbf{X}_m) g_\alpha^{\mathrm{T}}(\mathbf{X}_m), \tag{10.21a}$$

where $g_{0_\alpha}(\mathbf{r})$ and $g_{0_\beta}(\mathbf{r})$ are the Green-function vectors

$$g_{0_\alpha}(\mathbf{r}) = [G_{0_+}(\boldsymbol{\alpha}_1, \mathbf{r}), G_{0_+}(\boldsymbol{\alpha}_2, \mathbf{r}), \ldots, G_{0_+}(\boldsymbol{\alpha}_{N_\alpha}, \mathbf{r})]^{\mathrm{T}}, \tag{10.21b}$$

$$g_{0_\beta}(\mathbf{r}) = [G_{0_+}(\boldsymbol{\beta}_1, \mathbf{r}), G_{0_+}(\boldsymbol{\beta}_2, \mathbf{r}), \ldots, G_{0_+}(\boldsymbol{\beta}_{N_\beta}, \mathbf{r})]^{\mathrm{T}}, \tag{10.21c}$$

and g_α and g_β are the Green-function vectors obtained by replacing the background Green function G_{0_+} in these expressions by the composite-medium Green function G_+. As in our treatment of experimental time-reversal imaging presented in the previous section we will refer to the various Green-function vectors evaluated at a scatterer location $\mathbf{X}_m$ as the *antenna vectors.*

10.3.1 Singular value decomposition of the multistatic data matrix

The multistatic data matrix $K(\omega)$ defines a linear transformation between the space C^{N_α} of antenna-α excitation signals and the space C^{N_β} of antenna-β received signals and admits a singular value decomposition (SVD) defined by the set of equations

$$Kv_p = \sigma_p u_p, \qquad K^\dagger u_p = \sigma_p v_p, \tag{10.22a}$$

where $v_p \in C^{N_\alpha}$, $u_p \in C^{N_\beta}$ are the singular (column) vectors and $\sigma_p \geq 0$ are the set of singular values, where p is a positive integer that indexes the system. The singular vectors satisfy the normal equations

$$K^\dagger K v_p = \sigma_p^2 v_p, \qquad KK^\dagger u_p = \sigma_p^2 u_p, \tag{10.22b}$$

from which it follows that there will a total of $N_{\sigma>0} = \mathcal{R}(K) = \min(M, N_\alpha, N_\beta)$ non-zero singular values $\sigma_p > 0$ and a total of $N_\sigma = \min(N_\alpha, N_\beta)$ singular values σ_p with associated singular vectors $\{v_p, u_p\}$, $p = 1, 2, \ldots, N_\sigma$. We tacitly assume that these N_σ singular vectors are extended, if necessary, to include the kernel (null-space vectors) of $K^\dagger K$ and $KK^\dagger$ so that there will be a grand total of N_α singular vectors v_p and N_β singular vectors u_p. The members of the (extended) set of singular vectors v_p are orthonormal and complete in C^{N_α}, while those of the set u_p are orthonormal and complete in C^{N_β}; i.e.,

$$v_p^\dagger v_{p'} = \delta_{p,p'}, \quad p, p' = 1, 2, \ldots, N_\alpha, \qquad u_p^\dagger u_{p'} = \delta_{p,p'}, \quad p, p' = 1, 2, \ldots, N_\beta.$$

The SVDs of the matrices K and $K^\dagger$ are then expressed in terms of the singular system via the equations

$$K = \sum_{p=1}^{N_\sigma} \sigma_p u_p v_p^\dagger, \qquad K^\dagger = \sum_{p=1}^{N_\sigma} \sigma_p v_p u_p^\dagger, \tag{10.22c}$$

where $v_p^\dagger = v_p^{*\mathrm{T}}$ and similarly for $u_p^\dagger$.

The normal equation satisfied by the singular vectors v_p is seen to be identical to the eigen-equation Eq. (10.8b) satisfied by the excitation signals in a convergent sequence of experimental time-reversal imaging experiments employing the α array as both a transmit and a receive array. Indeed, we can interpret the quantity $K^\dagger K$ as being the time-reversal matrix associated with the α antenna array and the singular vectors v_p are then the required excitation signals that, when applied to this array, ensure that it radiates a wave that will focus on one or more of the scatterers. Similarly, we can interpret the quantity $KK^\dagger$ as being the (complex conjugate) of the time-reversal matrix K^*K^T associated with the β

antenna array and the complex conjugates of the singular vectors u_p^* are then the required excitation signals that, when applied to the β array, ensure that it radiates a wave that will also focus on one or more of the scatterers. Thus, each separate antenna array can act as both a transmitting array and a receiving array and possesses its own time-reversal matrix and its own set of eigenvectors that can be associated in a one-to-one manner with individual or groups of scatterers. When excited by an eigenvector of their respective time-reversal matrices they will each generate an incident wave that focuses on one or more of the scatterers. However, we must be careful to note that these two focused waves might not focus on the same scatterer or groups of scatterers![6] Thus, although for given p the singular vectors v_p and u_p are each (separately) associated with specific scatterers or group of scatterers, they need not be the *same* scatterer or group of scatterers. The only guaranteed exception to this possibility is when the two arrays are coincident and $u_p = v_p^*$.

Although we cannot in practice perform the conceptual set of experiments leading to the above results, we can perform an actual set of experiments to measure the multistatic data matrix K. If the singular vectors v_p are then computed from this measured matrix and the actual (physical) transmit array is excited by one of the singular vectors $t_\alpha = v_p$ the resulting incident wave will focus on a single scatterer or group of scatterers. Similarly, if the receiver array, used as a transmitting array, is excited by one of the singular vectors u_p it will also generate an incident wave that will focus on a (possibly different) set of scatterers. Thus, in this way we can extend the application of experimental time-reversal imaging to non-co-located transmit and receiver arrays. More importantly, we can also use these conclusions in computational time-reversal imaging to generate separate images using focused wavefields *both* from the transmit *and* from the receiver arrays. Thus, in particular, knowing the background Green functions, we can compute these focused waves from the set of singular vectors $v_p, u_p, p = 1, 2, \ldots, N_{\sigma>0}$, of the K matrix. However, as noted in the previous paragraph we must beware of the possibility that these two focused waves might not focus on the same scatterer or set of scatterers.

10.3.2 DORT

The conventional DORT images are computed from the singular vectors via the generalization of Eq. (10.18) according to the two equations

$$U_p(\mathbf{r};\alpha) = \sum_{k=1}^{N_\alpha} v_p(k) G_{0_+}(\mathbf{r}, \boldsymbol{\alpha}_k) = v_p^{\mathrm{T}} g_{0_\alpha}(\mathbf{r}), \qquad (10.23a)$$

$$U_p(\mathbf{r};\beta) = \sum_{j=1}^{N_\beta} u_p^*(k) G_{0_+}(\mathbf{r}, \boldsymbol{\beta}_j) = g_{0_\beta}^{\mathrm{T}}(\mathbf{r}) u_p^*, \qquad (10.23b)$$

with $p = 1, 2, \ldots, N_\alpha$ for $U_p(\mathbf{r};\alpha)$ and $p = 1, 2, \ldots, N_\beta$ for $U_p(\mathbf{r};\beta)$; i.e., each array

6 The reason for this is that the eigenvalues of the two separate arrays depend on both the scatterer strengths as well as the CPSFs of the separate arrays evaluated at the scatterer locations. Thus, a given eigenvalue σ_p^2 common to both arrays can correspond to different sets of scatterers for each array.

can generate from its singular vectors a number of images equal to the number of elements in the array. Of these sets of images only a total of $N_{\sigma>0} \leq \min(M, N_\alpha, N_\beta)$ will be associated with non-zero singular values and, hence, be images of single (well-resolved) scatterers or groups of non-resolved scatterers. Because the images generated from the two arrays might not be of the same scatterer or group of scatterers, we cannot, in general, combine them to form a single composite image of the *same* scatterer or group of scatterers. We can, of course, form a composite image by simply adding the two separate images to obtain

$$\phi_p(\mathbf{r}) = g_{0_\beta}^{\mathrm{T}}(\mathbf{r})u_p^* + v_p^{\mathrm{T}} g_{0_\alpha}(\mathbf{r}), \quad p = 1, 2, \ldots, N_{\sigma>0}, \tag{10.24a}$$

which will focus on the scatterers associated with both arrays. We can also multiply the two separate images to obtain

$$\chi_p(\mathbf{r}) = g_{0_\beta}^{\mathrm{T}}(\mathbf{r})u_p^* v_p^{\mathrm{T}} g_{0_\alpha}(\mathbf{r}), \quad p = 1, 2, \ldots, N_{\sigma>0}, \tag{10.24b}$$

which will then focus on the scatterer or set of scatterers common to the two array images. Various other ways to combine the two images are, of course, possible.

Example 10.1 As an example we consider two coincident linear arrays with element-to-element spacing of $\delta = 2\lambda$ in 2D space. Each array has 11 elements and is distributed above a set of three point (line) scatterers. The 2D outgoing-wave Green function is given by Eq. (2.19a) of Chapter 2

$$G_+(\mathbf{R}, \omega) = \frac{-i}{4} H_0^+(kR),$$

where $H_0^+(\cdot)$ is the zeroth-order Hankel function of the first kind. The multistatic data matrix and Green-function and antenna vectors are then constructed from Eqs. (10.21) using the above 2D Green function.

In this case we have $M = 3$ and $N_\alpha = N_\beta = 11$ so that the rank $\mathcal{R}(K) = 3$, thus yielding three non-zero singular values σ_p, $p = 1, 2, 3$, with associated singular vectors $(v_p, u_p = v_p^*)$, $p = 1, 2, 3$. Since the two arrays are coincident they generate identical DORT images given by

$$\phi_p(\mathbf{r}) = g_{0_\beta}^{\mathrm{T}}(\mathbf{r})u_p^* = v_p^{\mathrm{T}} g_{0_\alpha}(\mathbf{r}), \quad p = 1, 2, 3.$$

We show in Fig. 10.1 the results of a computer simulation for two different sets of scatterers. The top images correspond to three well-resolved scatterers, such that each image is that of a single scatterer. The three singular values are seen to be comparable in the top images. The bottom images correspond to three scatterers that are aligned perpendicular to the antenna arrays and, hence, cannot be resolved due to the poor longitudinal resolution of a linear array. This is indicated by the magnitudes of the second and third singular values, which are much smaller than that of the first.

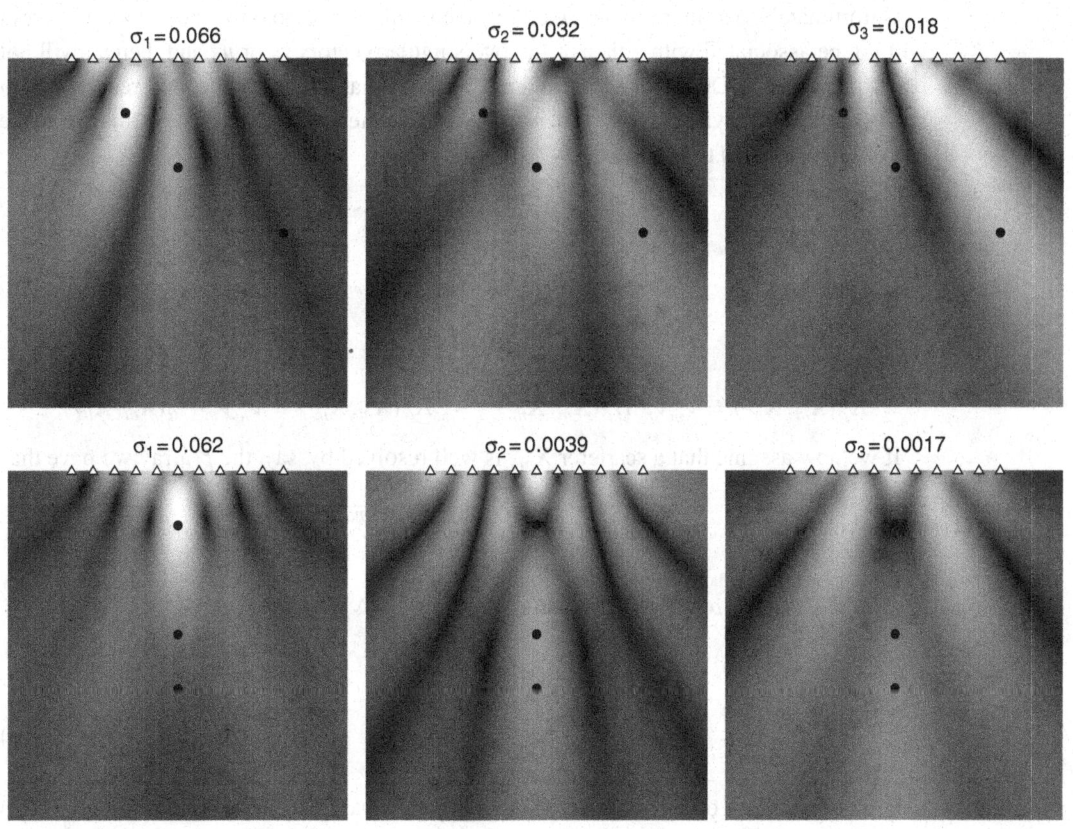

Fig. 10.1 DORT images of three well-resolved scatterers (top) and three unresolved scatterers (bottom). The positions of the antennas are shown at the top of each image and the locations of the scattering centers are displayed as black dots in the images. The singular values associated with each image are indicated above the images.

Well-resolved scatterers

The background and composite-medium CPSFs of the α and β arrays are defined in the usual way via the equations

$$H_{0_\alpha}(\mathbf{r},\mathbf{r}') = g^\dagger_{0_\alpha}(\mathbf{r})g_{0_\alpha}(\mathbf{r}'), \qquad H_{0_\beta}(\mathbf{r},\mathbf{r}') = g^\dagger_{0_\beta}(\mathbf{r})g_{0_\beta}(\mathbf{r}'),$$
$$H_\alpha(\mathbf{r},\mathbf{r}') = g^\dagger_\alpha(\mathbf{r})g_\alpha(\mathbf{r}'), \qquad H_\beta(\mathbf{r},\mathbf{r}') = g^\dagger_\beta(\mathbf{r})g_\beta(\mathbf{r}').$$

A specific scatterer located at $\mathbf{X}_m$ is then said to be well resolved with respect to the α and/or β array if

$$H_\alpha(\mathbf{X}_m,\mathbf{X}_{m'}) \approx \rho_{\alpha_m}\delta_{m,m'}, \qquad H_\beta(\mathbf{X}_m,\mathbf{X}_{m'}) \approx \rho_{\beta_m}\delta_{m,m'},$$

where

$$\rho_{\alpha_m} = ||g_\alpha(\mathbf{X}_m)||^2, \qquad \rho_{\beta_m} = ||g_\beta(\mathbf{X}_m)||^2.$$

Unfortunately, a scatterer that is well resolved with respect to only one of the two arrays will not be associated with either of the two singular vectors v_p or u_p and, hence, will not result in a focused DORT image. Rather, it requires that a scatterer be well resolved with respect to both arrays for DORT focusing on the scatterer to occur. To see this we make use of Eqs. (10.21a) to find that

$$K^\dagger K = \sum_{m,m'} \Lambda_\beta(\mathbf{X}_m, \mathbf{X}_{m'}) g^*_{0_\alpha}(\mathbf{X}_m) g^{\mathrm{T}}_{0_\alpha}(\mathbf{X}_{m'}), \tag{10.25a}$$

$$KK^\dagger = \sum_{m,m'} \Lambda^*_\alpha(\mathbf{X}_m, \mathbf{X}_{m'}) g_{0_\beta}(\mathbf{X}_m) g^\dagger_{0_\beta}(\mathbf{X}_{m'}), \tag{10.25b}$$

where

$$\Lambda_\beta(\mathbf{X}_m, \mathbf{X}_{m'}) = \mathcal{V}^*_m \mathcal{V}_{m'} H_\beta(\mathbf{X}_m, \mathbf{X}_{m'}), \qquad \Lambda_\alpha(\mathbf{X}_m, \mathbf{X}_{m'}) = \mathcal{V}^*_m \mathcal{V}_{m'} H_\alpha(\mathbf{X}_m, \mathbf{X}_{m'}).$$

If we now assume that a scatterer $\mathbf{X}_{m_0}$ is well resolved by, say, the β array we have that

$$H_\beta(\mathbf{X}_{m_0}, \mathbf{X}_{m'}) \approx \rho_{\beta_{m_0}} \delta_{m_0,m'}$$

so that

$$K^\dagger K \approx |\mathcal{V}_{m_0}|^2 \rho_{\beta_{m_0}} g^*_{0_\alpha}(\mathbf{X}_{m_0}) g^{\mathrm{T}}_{0_\alpha}(\mathbf{X}_{m_0}) + \sum_{m,m' \neq m_0} \Lambda_\beta(\mathbf{X}_m, \mathbf{X}_{m'}) g^*_{0_\alpha}(\mathbf{X}_m) g^{\mathrm{T}}_{0_\alpha}(\mathbf{X}_{m'}).$$

For

$$v_{m_0} = \frac{g^*_{0_\alpha}(\mathbf{X}_{m_0})}{\sqrt{\rho_{\alpha_{m_0}}}} \tag{10.26a}$$

to be an eigenvector of the normal equations $K^\dagger K v_{m_0} = \sigma^2_{m_0} v_{m_0}$ then requires that $g^*_{0_\alpha}(\mathbf{X}_{m_0})$ be orthogonal to the second term in the expansion of $K^\dagger K$ given above and this requires that

$$g^{\mathrm{T}}_{0_\alpha}(\mathbf{X}_{m'}) g^*_{0_\alpha}(\mathbf{X}_{m_0}) = H_{0_\alpha}^*(\mathbf{X}_{m'}, \mathbf{X}_{m_0}) = 0, \quad m' \neq m_0,$$

and thus requires that the scatterer at $\mathbf{X}_{m_0}$ also be well resolved[7] by the α array! In a similar fashion we conclude that for

$$u_{m_0} = \frac{g_{0_\beta}(\mathbf{X}_{m_0})}{\sqrt{\rho_{\beta_{m_0}}}} \tag{10.26b}$$

to be an eigenvector of the normal equations $KK^\dagger u_{m_0} = \sigma^2_{m_0} u_{m_0}$ requires that

$$g^\dagger_{0_\beta}(\mathbf{X}_{m'}) g_{0_\beta}(\mathbf{X}_{m_0}) = H_{0_\beta}(\mathbf{X}_{m'}, \mathbf{X}_{m_0}) = 0, \quad m' \neq m_0,$$

and thus requires that the scatterer at $\mathbf{X}_{m_0}$ also be well resolved by the β array!

In summary, we see that a scatterer located at $\mathbf{X}_m$ that is well resolved by both the α and β arrays is associated with singular vectors v_m and u_m that will generate a composite DORT image

$$\phi_m(\mathbf{r}) = g^{\mathrm{T}}_{0_\beta}(\mathbf{r}) u^*_m + v^{\mathrm{T}}_m g_{0_\alpha}(\mathbf{r}) = \frac{H^*_\beta(\mathbf{X}_m, \mathbf{r}) + H^*_\alpha(\mathbf{X}_m, \mathbf{r})}{\sqrt{\rho_{\beta_m} \rho_{\alpha_m}}}.$$

[7] In this case the scatterer is well resolved by the array in the background medium rather than in the composite medium.

Example 10.2 As an example of non-coincident arrays we consider two linear arrays that are oriented perpendicular to each other. As in Example 10.1, each array has 11 elements, with element-to-element spacing of $\delta = 2\lambda$ in 2D space and we again take three scatterers so that $M = 3$ and $N_\alpha = N_\beta = 11$ and the rank $\mathcal{R}(K) = 3$, thus yielding three non-zero singular values σ_p, $p = 1, 2, 3$, with associated singular vectors (v_p, u_p), $p = 1, 2, 3$. Since the two arrays are not coincident, they separate DORT images given by

$$\phi_{\alpha_p}(\mathbf{r}) = v_p^{\mathrm{T}} g_{0_\alpha}(\mathbf{r}), \qquad \phi_{\beta_p}(\mathbf{r}) = g_{0_\beta}^{\mathrm{T}}(\mathbf{r}) u_p^*, \quad p = 1, 2, 3.$$

We show in Fig. 10.2 the results of a computer simulation for a single set of scatterers that are well resolved by both arrays. The top images are those generated by the singular vectors v_p and are thus associated with the α array and the bottom are those generated by the singular vectors u_p and are thus associated with the β array. In this particular example

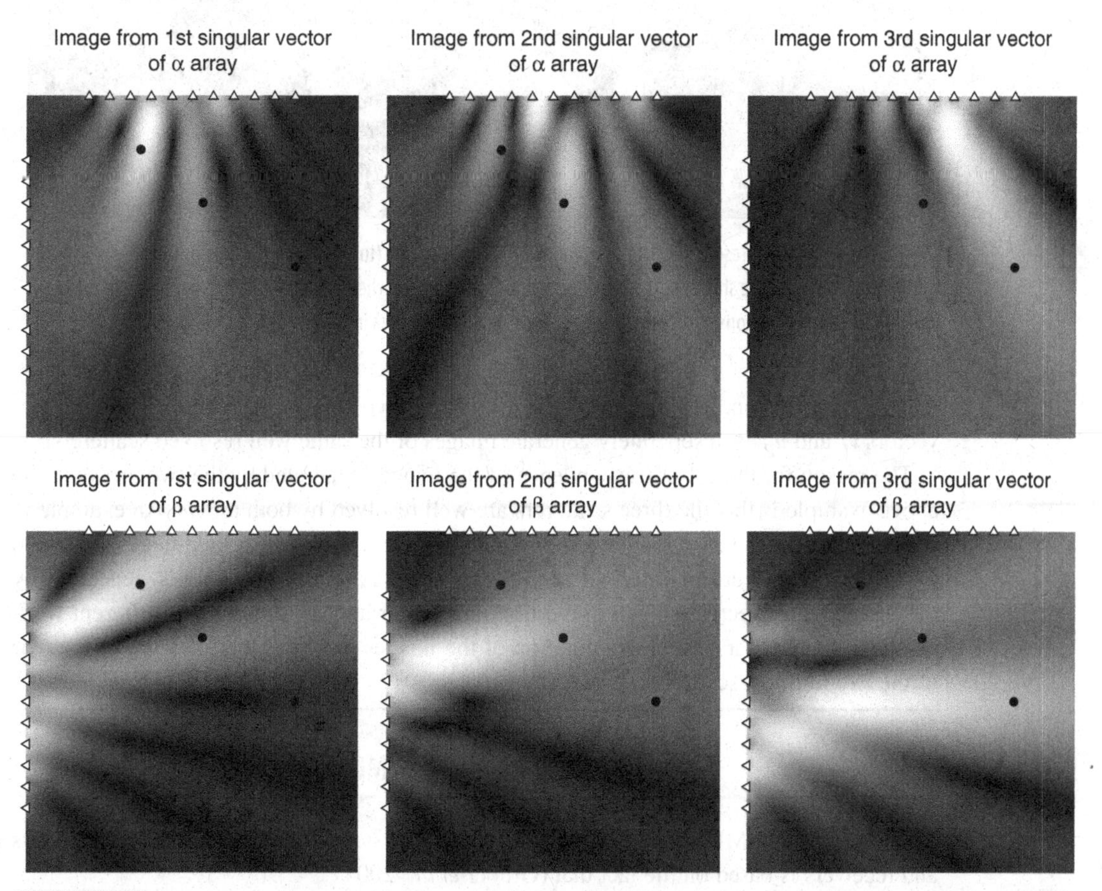

Fig. 10.2 DORT images of three well-resolved scatterers obtained from the α array (top) and β array (bottom). The positions of the α-array antennas are shown at the top of each figure and those of the β array on the left-hand side of the images. The actual locations of the scatterers are displayed as black dots in the images.

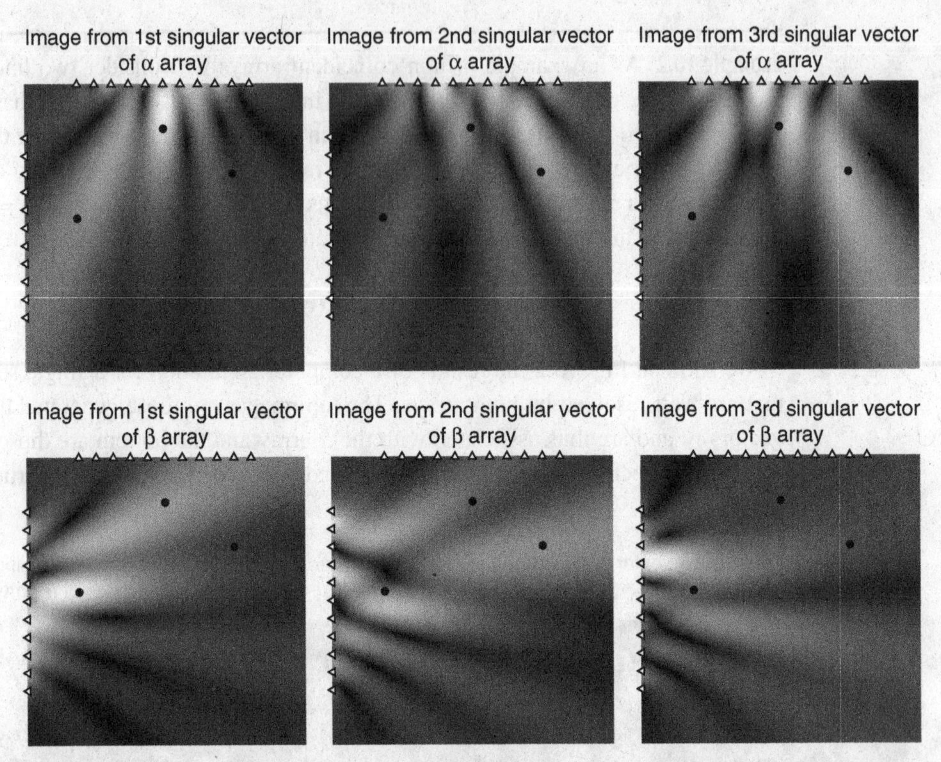

Fig. 10.3 DORT images of three unresolved scatterers obtained from the α array (top) and β array (bottom). The positions of the α-array antennas are shown at the top of each figure and those of the β array on the left-hand side of the images. The actual locations of the scatterers are displayed as black dots in the images.

there is a one-to-one correspondence between the two sets of images; i.e., the singular vectors v_p and u_p each separately generate images of the same well resolved scatterers.

The reason for the 1:1 correspondence of the images generated by the two arrays in the above example is that the three scatterers are well resolved by both arrays. An example in which this is not the case is shown in Fig. 10.3, where it is clear that the images generated by the two arrays are of different sets of scatterers. Although the images generated by the first singular vectors shown in the first image pair appear to be of two different well-resolved scatterers a careful examination of the images indicates that they are of the pair of the two leftmost scatterers.

10.3.3 Time-reversal MUSIC

The time-reversal MUSIC algorithm for a general system of non-coincident transmitters and receivers is based on the fact that (Gruber *et al.*, 2004)

$$\eta(K)^{\perp} = \text{Span}(v_p, \sigma_p > 0) = \text{Span}(g^*_{0_\alpha}(\mathbf{X}_m)),$$
$$\eta(K^{\dagger})^{\perp} = \text{Span}(u_p, \sigma_p > 0) = \text{Span}(g_{0_\beta}(\mathbf{X}_m)),$$

where $\eta(K) = \text{Span}(v_p, \sigma_p = 0)$ is the null space of K and $\eta(K^\dagger) = \text{Span}(u_p, \sigma_p = 0)$ the null space of $K^\dagger$. Moreover, if $M < (N_\alpha, N_\beta)$ then the two null spaces are not empty so that

$$v_p^{\mathrm{T}} g_{0_\alpha}(\mathbf{r}) = 0, \qquad g_{0_\beta}^{\mathrm{T}}(\mathbf{r}) u_p^* = 0, \quad \mathbf{r} = \mathbf{X}_m$$

when $v_p \in \eta(K)$ and $u_p \in \eta(K^\dagger)$ or when $p = M+1, \ldots, N_\alpha$ in the first case and $p = M+1, \ldots, N_\beta$ in the second case. We can then form the pseudo-spectrum

$$\Phi(\mathbf{r}) = \frac{1}{\sum_{p=N_{\sigma>0}+1}^{N_\alpha} |v_p^{\mathrm{T}} g_{0_\alpha}(\mathbf{r})| * \sum_{p=N_{\sigma>0}+1}^{N_\beta} |g_{0_\beta}^{\mathrm{T}}(\mathbf{r}) u_p^*|}, \tag{10.27}$$

which will peak, ideally to infinity, at the various scatterer locations $\mathbf{X}_m$. In practice it is necessary to include a regularization parameter $\epsilon > 0$ in the denominator to broaden the width around the peak values of $\Phi(\mathbf{r})$ at the various scatterer locations. Otherwise they are difficult to visually detect from a grayscale image.

As in defining the DORT images according to Eqs. (10.24), there exist other possible ways to compute a pseudo-spectrum. We have chosen to employ the products of the sums of the magnitudes of the components $v_p^{\mathrm{T}} g_{0_\alpha}(\mathbf{r})$ and $g_{0_\beta}^{\mathrm{T}}(\mathbf{r}) u_p^*$ rather than summing the complex amplitudes in order to avoid the possibility of false maxima in $\Phi(\mathbf{r})$ introduced by destructive interference of the various terms in the denominator.

Example 10.3 As an example we consider time-reversal MUSIC in two space dimensions with the two coincident arrays employed in Example 10.1. The pseudo-spectrum in this case reduces to

$$\Phi(\mathbf{r}) = \frac{1}{\sum_{\sigma_p=0} |v_p^{\mathrm{T}} g_{0_\alpha}(\mathbf{r})|}.$$

We took three scatterers with scattering locations varying from well removed from each other to closely spaced. The results of the simulation are shown in Fig. 10.4. It can be seen that even for the closely spaced scatterers there is relatively good separation of the focus point on the individual scatterers.

As a second example we applied the MUSIC algorithm to the same scattering system but using the two non-coincident arrays employed in Example 10.2. The results of the simulation are shown in Fig. 10.5. It is seen that the resulting images have better resolution than those shown in Fig. 10.4 along the vertical (y) axis due to the good transverse resolution of the β array.

10.3.4 Filtered DORT and multiple-frequency algorithms

Unlike DORT, which has the capability (for well-resolved scatterers) to generate separate images of individual scatterers, the time-reversal MUSIC algorithm generates an image that has all scatterers included. If we are willing to give up the advantage of having images of individual scatterers then there are alternatives to MUSIC that can be employed and that,

Fig. 10.4 MUSIC images of three scatterers in two space dimensions obtained using the coincident linear arrays employed in Example 10.1. The actual locations of the scatterers are displayed as black dots in the images.

in many cases, work as well as or better than the MUSIC algorithm. We will refer to these alternative algorithms as "filtered DORT algorithms" since they generate their images by weighting the component DORT images with filters $F_\alpha(p)$ and $F_\beta(p)$ and summing the components over the index p. Thus, on making use of Eq. (10.24a) we obtain the filtered DORT image

$$\overline{\phi}(\mathbf{r}) = \sum_{\sigma_p>0} F_\beta(p) g_{0_\beta}^{\mathrm{T}}(\mathbf{r}) u_p^* + \sum_{\sigma_p>0} F_\alpha(p) v_p^{\mathrm{T}} g_{0_\alpha}(\mathbf{r}). \tag{10.28a}$$

If we select the filters $F_\beta(p) = F_\alpha(p) = \delta_{p,m}$ we clearly obtain the standard composite DORT image corresponding to the mth singular vectors of the K matrix.

In a similar fashion we can use the basic DORT image defined in Eq. (10.24b) to generate the filtered DORT image

$$\overline{\chi}(\mathbf{r}) = \sum_{\sigma_p>0} F(p) g_{0_\beta}^{\mathrm{T}}(\mathbf{r}) u_p^* * v_p^{\mathrm{T}} g_{0_\alpha}(\mathbf{r}). \tag{10.28b}$$

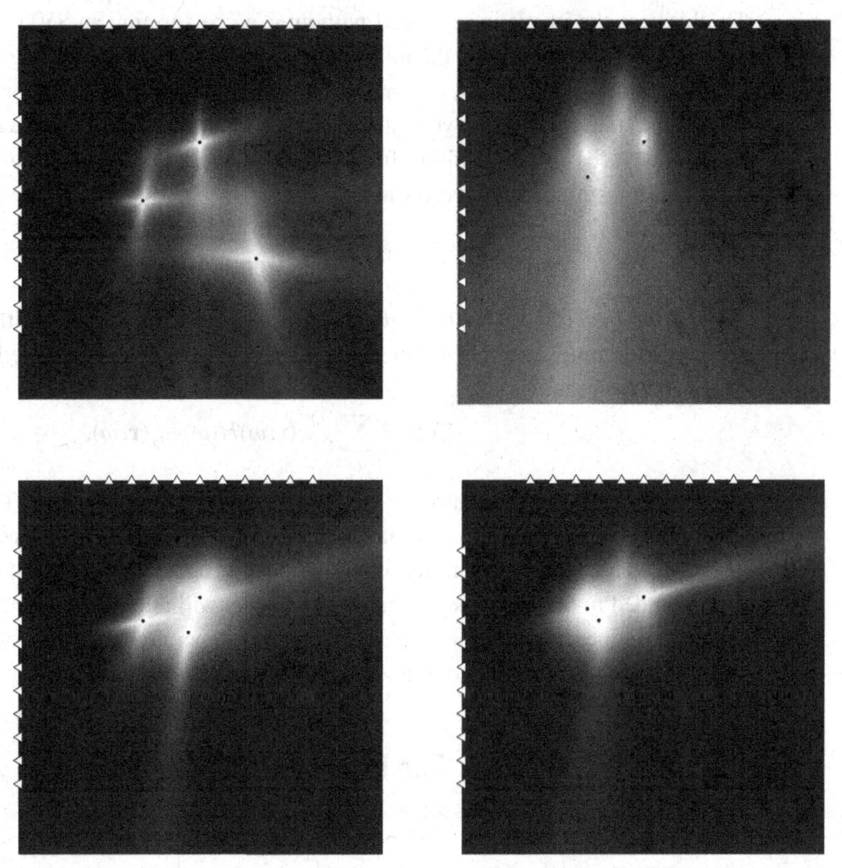

Fig. 10.5 MUSIC images of three scatterers in two space dimensions obtained using the non-coincident linear arrays employed in Example 10.2. The actual locations of the scatterers are displayed as black dots in the images.

By summing over the singular values in the above equation we obtain

$$\overline{\chi}(\mathbf{r}) = g_{0_\beta}^{\mathrm{T}}(\mathbf{r})\mathcal{H}g_{0_\alpha}(\mathbf{r}), \tag{10.29a}$$

where $\mathcal{H}$ is the $N_\beta \times N_\alpha$ matrix

$$\mathcal{H} = \sum_{\sigma_p>0} F(p)u_p^* v_p^{\mathrm{T}}. \tag{10.29b}$$

An important special choice that is very easily implemented results from selecting $F(p) = \sigma_p$, i.e., equal to the singular values of the SVD of the K matrix. In this case we have that

$$\mathcal{H} = \sum_{\sigma_p>0} \sigma_p u_p^* v_p^{\mathrm{T}} = K^*,$$

so that

$$\overline{\chi}(\mathbf{r}) = g_{0_\beta}^{\mathrm{T}}(\mathbf{r})K^* g_{0_\alpha}(\mathbf{r}) = \sum_{j,k} G_{0_+}(\mathbf{r}, \boldsymbol{\beta}_j)K_{j,k}^* G_{0_+}(\mathbf{r}, \boldsymbol{\alpha}_k). \tag{10.30}$$

In this case the image is generated without even performing an SVD of the K matrix and can be interpreted as being the image generated by first time reversing the multistatic data matrix and then letting the time-reversed data radiate into the interior region τ_0 containing the scatterers. Alternatively, it can also be interpreted as being the complex-conjugate (time-reversed) image resulting from the back propagation of the multistatic data matrix from all transmitter and receiver locations.

Multiple frequencies

All of the material covered in this chapter is easily extended to multiple frequencies by simply summing over the frequency band. For example, the filtered DORT algorithm Eq. (10.29a) becomes

$$\overline{\chi}(\mathbf{r}) = \sum_{\omega} g_{0_\beta}^{\mathrm{T}}(\mathbf{r}, \omega)\mathcal{H}(\omega) g_{0_\alpha}(\mathbf{r}, \omega), \tag{10.31}$$

where $g_{0_\beta}, g_{0_\alpha}$ and the matrix $\mathcal{H}$ now depend on the frequency ω. The use of multiple frequencies will improve performance in all of the algorithms but especially in DORT and MUSIC. The filtering operation can also benefit from multiple frequencies since it can be extended to the frequency domain and, hence, will allow a frequency-domain matched filter to be incorporated into the overall filtering operation. The use of matched filters in DORT imaging is discussed in a number of papers in the literature.

10.4 The inverse scattering problem

The ISCP formulated for a set of discrete scatterers consists of estimating their scattering strengths $\mathcal{V}_p$ and scattering centers $\mathbf{X}_p$ from knowledge of the scattered field over some set of points that lies outside of the scattering volume $\mathcal{V}$ (Devaney *et al.*, 2005). Here we will restrict our attention to scattering experiments employing a finite set of ideal point (delta-function) transmitting and receiving antennas as employed in our treatment of computational time-reversal imaging in Section 10.3 and data consisting of the multistatic data matrix $K_{j,k}(\omega)$. Our results are easily extended to more realistic antenna models as described in Section 9.11 and to the classical case of data consisting of the scattering amplitude $f(\mathbf{s}, \mathbf{s}_0)$ specified over some set of incident $\mathbf{s}_0$ and scattering $\mathbf{s}$ directions by letting the antenna locations recede to infinity.

The basic scattering model is given by Eq. (10.21a), which we will write in the form

$$K_{j,k}(\omega) = \sum_{m=1}^{M} \mathcal{V}_m G_{0_+}(\boldsymbol{\beta}_j, \mathbf{X}_m) G_+(\mathbf{X}_m, \boldsymbol{\alpha}_k). \tag{10.32a}$$

The above equation defines a non-linear transformation from the (unknown) scattering strengths $\mathcal{V}_m$ and scattering centers $\mathbf{X}_m$ to the K matrix and the ISCP consists of estimating the $\mathcal{V}_m$ and $\mathbf{X}_m$ from $K_{j,k}$, $j = 1, 2, \ldots, N_\beta$, $k = 1, 2, \ldots, N_\alpha$.

We can remove the non-linearity associated with the unknown scattering centers $\mathbf{X}_m$ by first estimating these quantities using DORT, MUSIC, or one of the filtered DORT imaging schemes described in Section 10.3. However, even with these quantities known it *appears* that the presence of the composite-medium Green function in Eq. (10.32a) also introduces a non-linearity because of its dependence on the scattering strengths $\mathcal{V}_m$. However, the Foldy–Lax formulation of the forward-scattering problem presented in Section 9.9 of the last chapter gets around this problem by renormalizing the Lippmann–Schwinger equation satisfied by the $G_+(\mathbf{X}_m, \alpha_k)$ in such a way that these quantities depend only on $\mathcal{V}_{m'}$ with $m' \neq m$. In particular, we showed in Example 9.5 of Section 9.9 that according to this model

$$G_+(\mathbf{X}_m, \alpha_k) = G_{0_+}(\mathbf{X}_m, \alpha_k) + \sum_{m' \neq m} \mathcal{V}_{m'} G_{0_+}(\mathbf{X}_m, \mathbf{X}_{m'}) G_+(\mathbf{X}_{m'}, \alpha_k). \qquad (10.32b)$$

It follows from Eq. (10.32b) that $G_+(\mathbf{X}_m, \alpha_k)$ is linearly related to the set of scattering strengths $\mathcal{V}_m$ so that, under the assumption that the scattering centers $\mathbf{X}_m$ are known, the unknown scattering strengths are related to the multistatic data matrix via a linear transform that is readily inverted.

Further reading

It is impossible to list even a small percentage of the many papers that have been published in the general area of time-reversal imaging. Many of them are from the famous French group headed by M. Fink (Thomas *et al.*, 1995; Spoliansky *et al.*, 1996; Prada and Fink, 1994; Lerosey *et al.*, 2007), who have developed most of the practical applications of this theory. An interesting paper connecting time-reversal imaging to chaos and using the Foldy–Lax formulation of multiple scattering outlined in the previous chapter is Snieder and Scales (1998). Larry Carin and his group at Duke have also been very productive in theoretical as well as experimental studies of time reversal in electromagnetic systems (Liu *et al.*, 2005, 2007), while Ishimaru and his group have investigated the use of time reversal in enhanced back scattering (Ishimaru *et al.*, 2007). The MUSIC time-reversal algorithm was first presented in Lev-Ari and Devaney (2000) and in an unpublished manuscript (Devaney, 2000), which contains a proof of the linear independence of the Green-function vectors for a homogeneous background. The algorithm is an adaptation of the algorithm with the same name first developed by R. Schmidt in his Ph.D. thesis (Schmidt, 1981). The algorithm is also used within the signal-processing community for resolving narrow spectral peaks (Therrien, 1992; Stoica and Moses, 1997) and in angle-of-arrival estimation in radar (Schmidt, 1986).

Problems

10.1 Prove that $G^*_{0_f} = -G_{0_f}$.
10.2 Derive Eqs. (10.2) and (10.3).

10.3 Use the Lippmann–Schwinger equations satisfied by the composite-medium Green function to prove that the multistatic data matrix given in Eq. (10.7) is symmetric.

10.4 Determine the asymptotic form of the multistatic data matrix given in Eq. (10.7) in the limit when the transmit and receive locations lie on the surface of an infinite sphere.

10.5 Prove that, for the simple scattering potential model given in Eq. (10.1b) and coincident point transmitter and receiver arrays, $K_{j,k}(\omega)$ is the scattered field component of the composite-medium Green function $G_+(\alpha_j, \alpha_k)$.

10.6 Prove that the antenna vectors $g(\mathbf{X}_1)$ and $g(\mathbf{X}_2)$ for the case of two co-located transmit and receive elements in a homogeneous background are linearly independent except for certain special scatterer and antenna locations. Determine these special situations in which linear independence breaks down.

10.7 Derive the representations of the K matrix given in Eq. (10.21) starting from the Lippmann–Schwinger equations from the previous chapter.

10.8 Formulate the SVD for the case of far-field transmit and receive co-located antenna elements considered in Problem 10.4.

10.9 Show that the singular values for the far-field SVD considered in the previous problem are invariant under a finite translation of the scattering system.

10.10 Show that the singular vectors for the far-field SVD considered in the previous two problems merely suffer a phase shift under a finite translation of the scattering system and determine what that phase shift is.

10.11 Show that the Green-function vectors $g(\mathbf{r})$ and $g_0(\mathbf{r})$ are related via the equation

$$g(\mathbf{r}) = g_0(\mathbf{r}) + \sum_m V_m G_{0_+}(\mathbf{r}, \mathbf{X}_m) g(\mathbf{X}_m).$$

10.12 Show that in place of Eqs. (10.25) we can also represent the matrices $K^\dagger K$ and $KK^\dagger$ in the alternative form

$$K^\dagger K = \sum_{m,m'} \Lambda_{0_\beta}(\mathbf{X}_m, \mathbf{X}_{m'}) g_\alpha^*(\mathbf{X}_m) g_\alpha^{\mathrm{T}}(\mathbf{X}_{m'}), \tag{10.33a}$$

$$KK^\dagger = \sum_{m,m'} \Lambda_{0_\alpha}^*(\mathbf{X}_m, \mathbf{X}_{m'}) g_\beta(\mathbf{X}_m) g_\beta^\dagger(\mathbf{X}_{m'}), \tag{10.33b}$$

where

$$\Lambda_{0_\beta}(\mathbf{X}_m, \mathbf{X}_{m'}) = V_m^* V_{m'} H_{0_\beta}(\mathbf{X}_m, \mathbf{X}_{m'}),$$
$$\Lambda_{0_\alpha}(\mathbf{X}_m, \mathbf{X}_{m'}) = V_m^* V_{m'} H_{0_\alpha}(\mathbf{X}_m, \mathbf{X}_{m'}).$$

10.13 Discuss the implications of using the representations Eqs. (10.33) rather than those in Eqs. (10.25) in the definition of a scatterer being well resolved and also in the actual generation of the DORT image of a well-resolved scatterer.

10.14 Discuss the implications of using the representations Eqs. (10.33) rather than those in Eqs. (10.25) in constructing the SVD of the K matrix and in time-reversal MUSIC.

10.15 Express the filtered DORT and MUSIC algorithms using multiple frequencies.

11 The electromagnetic field

11.1 Maxwell equations

We work in the frequency domain where the "electromagnetic" (EM) field consists of the electric $\mathbf{E}(\mathbf{r}, \omega)$ and magnetic $\mathbf{H}(\mathbf{r}, \omega)$ field vectors and the electric $\mathbf{D}(\mathbf{r}, \omega)$ and magnetic $\mathbf{B}(\mathbf{r}, \omega)$ flux vectors, where ω is the temporal frequency. The time-dependent fields and fluxes are obtained, in the usual way, via an inverse temporal Fourier transform so that, for example,

$$\mathbf{e}(\mathbf{r}, t) = \frac{1}{2\pi} \int_{-\infty}^{\infty} d\omega\, \mathbf{E}(\mathbf{r}, \omega) e^{-i\omega t},$$

where, of course,

$$\mathbf{E}(\mathbf{r}, \omega) = \int_{-\infty}^{\infty} dt\, \mathbf{e}(\mathbf{r}, t) e^{i\omega t}.$$

From this point onward we will not include the temporal frequency ω in the arguments of the various field quantities except in special cases where its exclusion can result in confusion or we wish to emphasize frequency dependence. We emphasize, however, that all of the EM field quantities depend to some extent on ω and this dependence must be accounted for except in narrow-band applications such as occur in optics, where the use of lasers is common.

The four field vectors are coupled by the famous Maxwell equations, which, in the SI system of units, assume the form

$$\begin{aligned}
\nabla \cdot \mathbf{D}(\mathbf{r}) &= \rho(\mathbf{r}), \\
\nabla \cdot \mathbf{B}(\mathbf{r}) &= 0, \\
\nabla \times \mathbf{E}(\mathbf{r}) &= i\omega \mathbf{B}(\mathbf{r}), \\
\nabla \times \mathbf{H}(\mathbf{r}) &= -i\omega \mathbf{D}(\mathbf{r}) + \mathbf{J}(\mathbf{r}),
\end{aligned}$$

where $\mathbf{J}$ and ρ are the current density and charge density, respectively. These two quantities are coupled via the charge–current conservation equation

$$\nabla \cdot \mathbf{J}(\mathbf{r}) = i\omega\rho(\mathbf{r}), \tag{11.1}$$

which can be inferred directly from the Maxwell equations.

The four field vectors are also coupled via the so-called *constitutive relations* which, in the case of *linear and isotropic* media (which we will assume throughout this chapter) are given by

$$\mathbf{D}(\mathbf{r}) = \epsilon(\mathbf{r})\mathbf{E}(\mathbf{r}), \qquad \mathbf{B}(\mathbf{r}) = \mu(\mathbf{r})\mathbf{H}(\mathbf{r}),$$

where ϵ and μ are the dielectric "constant" and permitivity of the medium, respectively. These two parameters are generally complex and are directly related to the complex index of refraction $n(\mathbf{r})$ of the medium via the equation

$$n(\mathbf{r}) = c\sqrt{\mu(\mathbf{r})\epsilon(\mathbf{r})},$$

where c is the velocity of light in vacuum. The wavenumber $k_0(\mathbf{r})$ of the medium is then given by the usual equation

$$k_0(\mathbf{r}) = n(\mathbf{r})\frac{\omega}{c} = \omega\sqrt{\mu(\mathbf{r})\epsilon(\mathbf{r})}.$$

On making use of Eq. (11.1) in the set of Maxwell equations we obtain a set of equations involving only the electric and magnetic field vectors

$$\begin{aligned}
\nabla \cdot \epsilon(\mathbf{r})\mathbf{E}(\mathbf{r}) &= \rho(\mathbf{r}),\\
\nabla \cdot \mu(\mathbf{r})\mathbf{H}(\mathbf{r}) &= 0,\\
\nabla \times \mathbf{E}(\mathbf{r}) &= i\omega\mu(\mathbf{r})\mathbf{H}(\mathbf{r}),\\
\nabla \times \mathbf{H}(\mathbf{r}) &= -i\omega\epsilon(\mathbf{r})\mathbf{E}(\mathbf{r}) + \mathbf{J}(\mathbf{r}).
\end{aligned}$$

11.1.1 Maxwell equations for a homogeneous isotropic medium

If the medium in which the charge and current densities are embedded is uniform (properties independent of $\mathbf{r}$) then the Maxwell equations simplify to become

$$\epsilon_0 \nabla \cdot \mathbf{E}(\mathbf{r}) = \rho(\mathbf{r}), \tag{11.2a}$$

$$\nabla \cdot \mathbf{H}(\mathbf{r}) = 0, \tag{11.2b}$$

$$\nabla \times \mathbf{E}(\mathbf{r}) = i\omega\mu_0\mathbf{H}(\mathbf{r}), \tag{11.2c}$$

$$\nabla \times \mathbf{H}(\mathbf{r}) = -i\omega\epsilon_0\mathbf{E}(\mathbf{r}) + \mathbf{J}(\mathbf{r}). \tag{11.2d}$$

The above set of equations can be easily uncoupled and lead to the pair of vector Helmholtz equations

$$\nabla \times \nabla \times \mathbf{E}(\mathbf{r}) - k_0^2\mathbf{E}(\mathbf{r}) = i\omega\mu_0\mathbf{J}(\mathbf{r}), \tag{11.3a}$$

$$\nabla \times \nabla \times \mathbf{H}(\mathbf{r}) - k_0^2\mathbf{H}(\mathbf{r}) = \nabla \times \mathbf{J}(\mathbf{r}), \tag{11.3b}$$

where $k_0 = k_0(\omega)$ is now a frequency-dependent complex constant given by

$$k_0 = \frac{\omega}{c}n_0 = \omega\sqrt{\mu_0\epsilon_0}. \tag{11.4}$$

The set of Maxwell equations Eqs. (11.2) or the two vector Helmholtz equations Eqs. (11.3) together with the first Maxwell equation Eq. (11.2a) and the charge–current conservation equation Eq. (11.1) form the basis for the radiation and inverse source problems for electromagnetic fields in linear isotropic media.

11.1.2 Maxwell equations in the spatial frequency domain

We define the spatial Fourier transform pair $\mathbf{E}(\mathbf{r}) \Longleftrightarrow \tilde{\mathbf{E}}(\mathbf{K})$

$$\tilde{\mathbf{E}}(\mathbf{K}) = \int d^3r\, \mathbf{E}(\mathbf{r})e^{-i\mathbf{K}\cdot\mathbf{r}}, \qquad \mathbf{E}(\mathbf{r}) = \frac{1}{(2\pi)^3}\int d^3K\, \tilde{\mathbf{E}}(\mathbf{K})e^{i\mathbf{K}\cdot\mathbf{r}},$$

with similar definitions for the magnetic field vector and the current and charge densities. As usual we assume that

$$\frac{\partial^n}{\partial x_j^n}\mathbf{E}(\mathbf{r}) \Longleftrightarrow (iK_j)^n\tilde{\mathbf{E}}(\mathbf{K}), \tag{11.5}$$

at least up to order $n = 2$, where x_j is any Cartesian coordinate of the position vector $\mathbf{r}$ and K_j is the associated component of the spatial frequency vector $\mathbf{K}$.

On taking the spatial Fourier transform of the set of Maxwell equations Eqs. (11.2) we obtain the result

$$i\epsilon_0\mathbf{K}\cdot\tilde{\mathbf{E}}(\mathbf{K}) = \tilde{\rho}(\mathbf{K}), \tag{11.6a}$$

$$\mathbf{K}\cdot\tilde{\mathbf{H}}(\mathbf{K}) = 0, \tag{11.6b}$$

$$\mathbf{K}\times\tilde{\mathbf{E}}(\mathbf{K}) = \omega\mu_0\tilde{\mathbf{H}}(\mathbf{K}), \tag{11.6c}$$

$$i\mathbf{K}\times\tilde{\mathbf{H}}(\mathbf{K}) = -i\omega\epsilon_0\tilde{\mathbf{E}}(\mathbf{K}) + \tilde{\mathbf{J}}(\mathbf{K}). \tag{11.6d}$$

The charge–current conservation equation Eq. (11.1) yields the result

$$\mathbf{K}\cdot\tilde{\mathbf{J}}(\mathbf{K}) = \omega\tilde{\rho}(\mathbf{K}). \tag{11.7}$$

The spatial Fourier transform of the vector Helmholtz equations Eqs. (11.3) are easily obtained directly from Eqs. (11.6). We find that

$$\mathbf{K}\times\mathbf{K}\times\tilde{\mathbf{E}}(\mathbf{K}) + k_0^2\tilde{\mathbf{E}}(\mathbf{K}) = -i\omega\mu_0\tilde{\mathbf{J}}(\mathbf{K}), \tag{11.8a}$$

$$\mathbf{K}\times\mathbf{K}\times\tilde{\mathbf{H}}(\mathbf{K}) + k_0^2\tilde{\mathbf{H}}(\mathbf{K}) = -i\mathbf{K}\times\tilde{\mathbf{J}}(\mathbf{K}). \tag{11.8b}$$

11.2 The Helmholtz theorem

The Helmholtz theorem states that any suitably well-behaved vector field $\mathbf{V}(\mathbf{r})$ can be uniquely decomposed into the sum of a longitudinal part $\mathbf{V}_{\mathrm{L}}(\mathbf{r})$ that has zero curl and a transverse part $\mathbf{V}_{\mathrm{T}}(\mathbf{r})$ that has zero divergence; i.e.,

$$\mathbf{V}(\mathbf{r}) = \mathbf{V}_{\mathrm{L}}(\mathbf{r}) + \mathbf{V}_{\mathrm{T}}(\mathbf{r}),$$

where

$$\nabla\times\mathbf{V}_{\mathrm{L}}(\mathbf{r}) = 0, \qquad \nabla\cdot\mathbf{V}_{\mathrm{T}}(\mathbf{r}) = 0. \tag{11.9}$$

The proof of the theorem is somewhat tedious in the space domain but there is a simple and elegant proof for vector-valued fields that admit a spatial Fourier decomposition for which the relationship Eq. (11.5) holds and that we will assume in the following treatment.

To establish the theorem we begin by using the vector identity

$$\mathbf{K} \times \mathbf{K} \times \tilde{\mathbf{V}}(\mathbf{K}) \equiv \mathbf{K}[\mathbf{K} \cdot \tilde{\mathbf{V}}(\mathbf{K})] - K^2 \tilde{\mathbf{V}}(\mathbf{K}), \tag{11.10}$$

where $\tilde{\mathbf{V}}(\mathbf{K})$ is the spatial Fourier transform of the field $\mathbf{V}(\mathbf{r})$. It then follows that this transform admits the decomposition

$$\tilde{\mathbf{V}}(\mathbf{K}) = \mathbf{K}\frac{\mathbf{K} \cdot \tilde{\mathbf{V}}(\mathbf{K})}{K^2} - \frac{\mathbf{K} \times \mathbf{K} \times \tilde{\mathbf{V}}(\mathbf{K})}{K^2}.$$

We can then define longitudinal and transverse components via the equations

$$\tilde{\mathbf{V}}_{\mathrm{L}}(\mathbf{K}) = \mathbf{K}\frac{\mathbf{K} \cdot \tilde{\mathbf{V}}(\mathbf{K})}{K^2}, \qquad \tilde{\mathbf{V}}_{\mathrm{T}}(\mathbf{K}) = -\frac{\mathbf{K} \times \mathbf{K} \times \tilde{\mathbf{V}}(\mathbf{K})}{K^2}, \tag{11.11}$$

where the space-dependent inverse transforms $V_{\mathrm{L}}(\mathbf{r})$ and $V_{\mathrm{T}}(\mathbf{r})$ satisfy Eqs. (11.9) on account of our assumption that the transform $\tilde{\mathbf{V}}(\mathbf{K})$ satisfies Eq. (11.5).

The space-dependent field components $\mathbf{V}_{\mathrm{L}}(\mathbf{r})$ and $\mathbf{V}_{\mathrm{T}}(\mathbf{r})$ are found by taking an inverse Fourier transform of Eqs. (11.11). We find that

$$\mathbf{V}_{\mathrm{L}}(\mathbf{r}) = \frac{1}{(2\pi)^3}\int d^3K\, \mathbf{K}\frac{\mathbf{K} \cdot \tilde{\mathbf{V}}(\mathbf{K})}{K^2} e^{i\mathbf{K}\cdot\mathbf{r}},$$

$$\mathbf{V}_{\mathrm{T}}(\mathbf{r}) = -\frac{1}{(2\pi)^3}\int d^3K\, \frac{\mathbf{K} \times \mathbf{K} \times \tilde{\mathbf{V}}(\mathbf{K})}{K^2} e^{i\mathbf{K}\cdot\mathbf{r}}.$$

As a final step we note that if we again make use of Eqs. (11.5) we can express the Helmholtz decomposition in the form

$$\mathbf{V}(\mathbf{r}) = \nabla\phi(\mathbf{r}) + \nabla \times \mathbf{A}(\mathbf{r}),$$

where the *scalar potential* ϕ and *vector potential* $\mathbf{A}$ are given by

$$\phi(\mathbf{r}) = -\frac{i}{(2\pi)^3}\int d^3K\, \frac{\mathbf{K} \cdot \tilde{\mathbf{V}}(\mathbf{K})}{K^2} e^{i\mathbf{K}\cdot\mathbf{r}},$$

$$\mathbf{A}(\mathbf{r}) = \frac{i}{(2\pi)^3}\int d^3K\, \frac{\mathbf{K} \times \tilde{\mathbf{V}}(\mathbf{K})}{K^2} e^{i\mathbf{K}\cdot\mathbf{r}}.$$

It should be noted that the decomposition of a given vector field into longitudinal and transverse components is unique, although the scalar and vector potentials are not, and that the theorem does not require the vector field $\mathbf{V}(\mathbf{r})$ to satisfy the vector Helmholtz equation, or any equation for that matter. We also note that it follows from the Maxwell equations that the magnetic field vector has zero divergence and, hence, is totally transverse, while the electric field vector has both longitudinal and transverse components. We will make frequent use of the decomposition of a vector field into longitudinal and transverse components in later sections of this chapter.

Example 11.1 As an example we decompose the transform of a current density $\mathbf{J}(\mathbf{r})$ into longitudinal and transverse parts using the Fourier-based approach used in proving the Helmholtz theorem. On making use of Eqs. (11.11) we find that

$$\tilde{\mathbf{J}}_{\mathrm{L}}(\mathbf{K}) = \mathbf{K}\frac{\mathbf{K}\cdot\tilde{\mathbf{J}}(\mathbf{K})}{K^2} = \mathbf{K}\frac{\omega\tilde{\rho}(\mathbf{K})}{K^2},$$

$$\tilde{\mathbf{J}}_{\mathrm{T}}(\mathbf{K}) = -\frac{\mathbf{K}\times\mathbf{K}\times\tilde{\mathbf{J}}(\mathbf{K})}{K^2}.$$

Since $\tilde{\mathbf{J}}(\mathbf{K}) = \tilde{\mathbf{J}}_{\mathrm{L}}(\mathbf{r}) + \tilde{\mathbf{J}}_{\mathrm{T}}(\mathbf{r})$ we can also express the transform of the transverse part of the current in the form

$$\tilde{\mathbf{J}}_{\mathrm{T}}(\mathbf{K}) = \tilde{\mathbf{J}}(\mathbf{K}) - \mathbf{K}\frac{\omega\tilde{\rho}(\mathbf{K})}{K^2}, \tag{11.12}$$

which also follows directly from expanding the triple cross product in the above equation for the transform of the transverse current using the identity Eq. (11.10).

Example 11.2 As a second example we decompose the transforms of the electric and magnetic field vectors into longitudinal and transverse parts according to the Helmholtz theorem. On making use of the identity Eq. (11.10) we find from Eqs. (11.8) that

$$(-K^2 + k_0^2)\tilde{\mathbf{E}}(\mathbf{K}) = -i\omega\mu_0\tilde{\mathbf{J}}(\mathbf{K}) + \frac{i}{\epsilon_0}\mathbf{K}\tilde{\rho}(\mathbf{K}),$$

$$(-K^2 + k_0^2)\tilde{\mathbf{H}}(\mathbf{K}) = -i\mathbf{K}\times\tilde{\mathbf{J}}(\mathbf{K}) = -i\mathbf{K}\times\tilde{\mathbf{J}}_{\mathrm{T}}(\mathbf{K}),$$

where we have also made use of Eqs. (11.6a) and (11.6b) and the easily proven identity

$$\mathbf{K}\times\tilde{\mathbf{J}}(\mathbf{K}) \equiv \mathbf{K}\times\tilde{\mathbf{J}}_{\mathrm{T}}(\mathbf{K}).$$

Solving for the field transforms we then find that

$$\tilde{\mathbf{E}}(\mathbf{K}) = \frac{-i\omega\mu_0\tilde{\mathbf{J}}(\mathbf{K}) + (i/\epsilon_0)\mathbf{K}\tilde{\rho}(\mathbf{K})}{-K^2 + k_0^2},$$

$$\tilde{\mathbf{H}}(\mathbf{K}) = \frac{-i\mathbf{K}\times\tilde{\mathbf{J}}_{\mathrm{T}}(\mathbf{K})}{-K^2 + k_0^2}.$$

As noted earlier and as follows from the above equations, the magnetic field is transverse and depends only on the transverse component of the current density, while the electric field has both a transverse and a longitudinal component. However, if we recall that $k_0^2 = \omega^2\epsilon_0\mu_0$ we can express the transform of the source term for the electric field in the form

$$-i\omega\mu_0\tilde{\mathbf{J}}(\mathbf{K}) + \frac{i}{\epsilon_0}\mathbf{K}\tilde{\rho}(\mathbf{K}) = -i\omega\mu_0\left[\tilde{\mathbf{J}}(\mathbf{K}) - \mathbf{K}\frac{\omega\tilde{\rho}(\mathbf{K})}{k_0^2}\right],$$

which, on making use of Eq. (11.12) of the previous example, is seen to reduce to $-i\omega\mu_0\tilde{\mathbf{J}}_{\mathrm{T}}(\mathbf{K})$ when $K = k_0$. We will show in a later section that both the electric field and the magnetic field depend only on the Fourier transform of the transverse component of the current on the boundary $K = k_0$ everywhere outside the charge–current source region τ_0.

11.3 The EM radiation problem

The radiation problem for EM fields generated by sources in a uniform isotropic medium consists of solving the vector Helmholtz equations Eqs. (11.3) for the electric and magnetic field vectors subject to causality in the time domain, which, as we showed in Chapter 1, translates into the "outgoing-wave" or Sommerfeld radiation condition (SRC) in the frequency domain. We can write these equations in the form

$$[\nabla^2 + k_0^2]\mathbf{E}(\mathbf{r}) = -i\omega\mu_0\mathbf{J}(\mathbf{r}) + \frac{1}{\epsilon_0}\nabla\rho(\mathbf{r}),$$
$$[\nabla^2 + k_0^2]\mathbf{H}(\mathbf{r}) = -\nabla\times\mathbf{J}(\mathbf{r}),$$

where we have made use of the Maxwell equation Eq. (11.2a). The solutions to the above equations that satisfy the SRC are then immediately obtained using the outgoing scalar-wave Green function

$$G_+(\mathbf{R}) = -\frac{1}{4\pi}\frac{e^{ik_0R}}{R},$$

and we find that

$$\mathbf{E}_+(\mathbf{r}) = -\int_{\tau_0} d^3r' \left[i\omega\mu_0\mathbf{J}(\mathbf{r}') - \frac{1}{\epsilon_0}\nabla_{r'}\rho(\mathbf{r}')\right] G_+(\mathbf{r}-\mathbf{r}'), \tag{11.13a}$$

$$\mathbf{H}_+(\mathbf{r}) = -\int_{\tau_0} d^3r' \, \nabla_{r'}\times\mathbf{J}(\mathbf{r}')G_+(\mathbf{r}-\mathbf{r}'), \tag{11.13b}$$

where τ_0 is the spatial support of the charge and current distributions and we have used a plus-sign subscript to denote that these are the outgoing-wave solutions to the Maxwell equations.

11.3.1 The dyadic Green function

The solution to the EM radiation problem given in Eqs. (11.13) employs a vector source comprised of both the charge and current densities and a scalar Green function. It is sometimes desirable to use a form of this solution in terms of just the current density and a tensor Green function in the form of a *dyadic*. The dyadic is, in fact, a second-order Cartesian tensor expressed in terms of constant unit vectors $\hat{\mathbf{x}}_j$, $j = 1,2,3$, relative to a fixed Cartesian coordinate system labeled by (x_1, x_2, x_3). The dyadic $\overleftrightarrow{\mathbf{A}}$ corresponding to the second-order tensor (orthogonal matrix) a_{ij} is given by

$$\overleftrightarrow{\mathbf{A}} = \sum_{i,j} a_{ij}\hat{\mathbf{x}}_i\hat{\mathbf{x}}_j.$$

The identity dyadic (diagonal matrix) corresponds to $a_{ij} = \delta_{i,j}$ and is thus given by

$$\overleftrightarrow{\mathbf{I}} = \sum_{j=1}^{3} \hat{\mathbf{x}}_j \hat{\mathbf{x}}_j,$$

so that

$$\overleftrightarrow{\mathbf{I}} \cdot \mathbf{V} = \mathbf{V} \cdot \overleftrightarrow{\mathbf{I}} = \mathbf{V},$$

for any vector **V**.

A Green-function dyadic for the electric and magnetic field vectors **E** and **H** is obtained as the solution to the inhomogeneous vector Helmholtz equation

$$\nabla \times \nabla \times \overleftrightarrow{\mathbf{G}}(\mathbf{r}, \mathbf{r}', \omega) - k_0^2 \overleftrightarrow{\mathbf{G}}(\mathbf{r}, \mathbf{r}', \omega) = \overleftrightarrow{\mathbf{I}}\, \delta(\mathbf{r} - \mathbf{r}'), \tag{11.14}$$

which satisfies specified boundary conditions. For radiation and scattering problems in an infinite homogeneous and isotropic medium the appropriate boundary condition is the Sommerfeld radiation condition (SRC) that requires the Green-function dyadic to be outgoing at infinity. By applying standard Green-function techniques to the vector Helmholtz equations Eqs. (11.3) satisfied by the electric or magnetic field vectors with Eq. (11.14) and using the SRC it can be shown that the two field vectors admit the integral representations

$$\mathbf{E}_+(\mathbf{r}) = -i\omega\mu_0 \int_{\tau_0} d^3r'\, \overleftrightarrow{\mathbf{G}}(\mathbf{r}, \mathbf{r}') \cdot \mathbf{J}(\mathbf{r}'), \tag{11.15a}$$

$$\mathbf{H}_+(\mathbf{r}) = -\int_{\tau_0} d^3r'\, \nabla_{r'} \times \mathbf{J}(\mathbf{r}') \cdot \overleftrightarrow{\mathbf{G}}(\mathbf{r}, \mathbf{r}'), \tag{11.15b}$$

where the Green dyadic is given by

$$\overleftrightarrow{\mathbf{G}}(\mathbf{r}, \mathbf{r}') = \left[\overleftrightarrow{\mathbf{I}} + \frac{1}{k_0^2} \nabla_r \nabla_r\right] G_+(\mathbf{r} - \mathbf{r}'). \tag{11.15c}$$

Equations (11.15) can be derived simply by transforming the field representations given in Eqs. (11.13) into the form of Eqs. (11.15). For example, we can express the electric field vector from Eq. (11.13a) in the form

$$\begin{aligned}
\mathbf{E}_+(\mathbf{r}) &= -i\omega\mu_0 \int_{\tau_0} d^3r' \left[\mathbf{J}(\mathbf{r}') - \frac{1}{i\omega\mu_0\epsilon_0} \nabla_{r'} \rho(\mathbf{r}')\right] G_+(\mathbf{r} - \mathbf{r}') \\
&= -i\omega\mu_0 \int_{\tau_0} d^3r' \left[\mathbf{J}(\mathbf{r}') G_+(\mathbf{r} - \mathbf{r}') - \frac{1}{i\omega\mu_0\epsilon_0} \rho(\mathbf{r}') \nabla_r G_+(\mathbf{r} - \mathbf{r}')\right],
\end{aligned}$$

where we have integrated by parts and used the fact that

$$\nabla_{r'} G_+(\mathbf{r} - \mathbf{r}') = -\nabla_r G_+(\mathbf{r} - \mathbf{r}').$$

If we now make use of the charge–current conservation equation Eq. (11.1) and the dispersion relationship Eq. (11.4) we obtain

$$\begin{aligned}\mathbf{E}_+(\mathbf{r}) &= -i\omega\mu_0 \int_{\tau_0} d^3r' \left[\mathbf{J}(\mathbf{r}')G_+(\mathbf{r}-\mathbf{r}') + \frac{\nabla_{r'}\cdot\mathbf{J}(\mathbf{r}')}{k_0^2}\nabla_r G_+(\mathbf{r}-\mathbf{r}')\right] \\ &= -i\omega\mu_0 \int_{\tau_0} d^3r' \left[\mathbf{J}(\mathbf{r}')G_+(\mathbf{r}-\mathbf{r}') + \frac{\mathbf{J}(\mathbf{r}')}{k_0^2}\cdot\nabla_r\nabla_r G_+(\mathbf{r}-\mathbf{r}')\right] \\ &= -i\omega\mu_0 \int_{\tau_0} d^3r'\, \mathbf{J}(\mathbf{r}')\cdot\left[\overleftrightarrow{I} + \frac{1}{k_0^2}\nabla_r\nabla_r\right] G_+(\mathbf{r}-\mathbf{r}'),\end{aligned}$$

which is Eq. (11.15a) with the Green dyadic given in Eq. (11.15c). A similar development can be used to obtain the Green dyadic representation of the magnetic field vector.

11.3.2 The radiation patterns

The radiation patterns of the EM field vectors are obtained using the asymptotic form of the outgoing-wave Green function G_+ (cf. Section 1.5.1 of Chapter 1)

$$G_+(\mathbf{r}-\mathbf{r}') \sim -\frac{1}{4\pi}e^{-ik_0\mathbf{s}\cdot\mathbf{r}'}\frac{e^{ik_0r}}{r}, \quad \text{as } r\to\infty,$$

where $\mathbf{s} = \mathbf{r}/r$ is the unit vector in the direction of the field point $\mathbf{r}$. On substituting the above expression into the set Eqs. (11.13) we obtain the result

$$\mathbf{E}_+(\mathbf{r}) \sim \mathbf{f}_e(\mathbf{s})\frac{e^{ik_0r}}{r}, \qquad \mathbf{H}_+(\mathbf{r}) \sim \mathbf{f}_h(\mathbf{s})\frac{e^{ik_0r}}{r}, \tag{11.16}$$

where the electric and magnetic radiation patterns $\mathbf{f}_e(\mathbf{s})$ and $\mathbf{f}_h(\mathbf{s})$ are given by

$$\begin{aligned}\mathbf{f}_e(\mathbf{s}) &= \frac{1}{4\pi}\int_{\tau_0} d^3r' \left[i\omega\mu_0\mathbf{J}(\mathbf{r}') - \frac{1}{\epsilon_0}\nabla_{r'}\rho(\mathbf{r}')\right] e^{-ik_0\mathbf{s}\cdot\mathbf{r}'}, \\ \mathbf{f}_h(\mathbf{s}) &= \frac{1}{4\pi}\int_{\tau_0} d^3r'\, \nabla_{r'}\times\mathbf{J}(\mathbf{r}')e^{-ik_0\mathbf{s}\cdot\mathbf{r}'}.\end{aligned}$$

We see from the above equations that the radiation patterns are proportional to spatial Fourier transforms of linear combinations of the current and charge distributions evaluated on the surface $\mathbf{K} = k_0\mathbf{s}$. We can then use Eq. (11.5) to simplify these expressions and obtain

$$\mathbf{f}_e(\mathbf{s}) = \frac{1}{4\pi}\left[i\omega\mu_0\tilde{\mathbf{J}}(k_0\mathbf{s}) - \frac{ik_0\mathbf{s}}{\epsilon_0}\tilde{\rho}(\mathbf{k}_0\mathbf{s})\right] = \frac{i\omega\mu_0}{4\pi}\tilde{\mathbf{J}}_{\mathrm{T}}(k_0\mathbf{s}), \tag{11.17a}$$

$$\mathbf{f}_h(\mathbf{s}) = \frac{ik}{4\pi}\mathbf{s}\times\tilde{\mathbf{J}}(k_0\mathbf{s}) = \frac{ik}{4\pi}\mathbf{s}\times\tilde{\mathbf{J}}_{\mathrm{T}}(k_0\mathbf{s}), \tag{11.17b}$$

where, in deriving the expression for $\mathbf{f}_e(\mathbf{s})$, we have used the spatial frequency-domain charge–current conservation equation Eq. (11.7) and the definition of the transverse part of the current given in Eq. (11.12) of Example 11.1. It follows from the above expressions for the radiation patterns that

$$\mathbf{f}_e(\mathbf{s}) = -\frac{\omega\mu_0}{k_0}\mathbf{s}\times\mathbf{f}_h(\mathbf{s}), \qquad \mathbf{f}_h(\mathbf{s}) = \frac{k_0}{\omega\mu_0}\mathbf{s}\times\mathbf{f}_e(\mathbf{s}), \tag{11.18}$$

so that either radiation pattern determines the other and they both depend only on the transverse part of the charge–current distribution. Note also that both patterns are perpendicular (transverse) to the unit observation vector $\mathbf{s}$. This conclusion is obvious from Eqs. (11.17) and is a consequence of the fact that both the electric field and the magnetic field are transverse fields (have zero divergence) outside of the charge–current support volume τ_0.

11.3.3 The Kirchhoff–Helmholtz representation of the radiated field

In the region $\tau^{\perp}$ exterior to a simply connected region $\tau \supset \tau_0$ that contains the source region τ_0 the electric and magnetic field vectors satisfy the homogeneous Helmholtz equations

$$[\nabla^2 + k_0^2]\mathbf{E}_+(\mathbf{r}) = 0, \qquad [\nabla^2 + k_0^2]\mathbf{H}_+(\mathbf{r}) = 0, \quad \mathbf{r}\in\tau^{\perp}.$$

By then following steps identical to those used in Section 2.5 of Chapter 2 we can obtain "Helmholtz identities" and the so-called "Kirchhoff–Helmholtz representation" of the EM fields that are generalizations of those obtained in Section 2.5 for the scalar-wave case. For example, we find for the electric field

$$\mathbf{E}_+(\mathbf{r}) = \int_{\partial\tau} dS' \left[G_+(\mathbf{r}-\mathbf{r}')\frac{\partial}{\partial n'}\mathbf{E}_+(\mathbf{r}') - \mathbf{E}_+(\mathbf{r}')\frac{\partial}{\partial n'}G_+(\mathbf{r}-\mathbf{r}')\right],$$

if $\mathbf{r}\in\tau^{\perp}$ and

$$\int_{\partial\tau} dS' \left[G_+(\mathbf{r}-\mathbf{r}')\frac{\partial}{\partial n'}\mathbf{E}_+(\mathbf{r}') - \mathbf{E}_+(\mathbf{r}')\frac{\partial}{\partial n'}G_+(\mathbf{r}-\mathbf{r}')\right] = 0,$$

if $\mathbf{r}\in\tau$ with an identical result for the magnetic field vector and where the normal derivatives are directed out of τ into the infinite exterior region $\tau^{\perp}$ bounded by $\partial\tau$ and a sphere at infinity. The above two equations are generally referred to as the Helmholtz identities, with the top equation being the Kirchhoff–Helmholtz representation of the electric field in the exterior region $\tau^{\perp}$ and the bottom equation a homogeneous Fredholm integral equation that must be satisfied by the boundary values of the field and its normal derivative.

As discussed in Section 2.5, in the case of scalar wavefields the above integral equation guarantees that the boundary values of the (scalar) field and its normal derivative are not independent and cannot be independently specified so that the Kirchhoff–Helmholtz representation of this field is not a solution to a properly posed boundary value for the (exterior) scalar-wave boundary-value problem. The same holds true in the vector-wave case treated here but in spades. In particular, not only are the boundary values of the EM fields and their normal derivatives not independent, but even the boundary values of the fields themselves are over-specified due to the fact that these fields must have zero divergence outside the source region τ_0. Indeed, as discussed in our treatment of the scalar-wave boundary-value problems in Section 2.8 it is always possible to find outgoing-wave Green functions $G_\mathrm{D}(\mathbf{r},\mathbf{r}')$ and $G_\mathrm{N}(\mathbf{r},\mathbf{r}')$ that satisfy homogeneous Dirichlet (G_D vanishing on $\partial\tau$) or Neumann (normal derivative $(\partial/\partial n)G_\mathrm{N}$ vanishing on $\partial\tau$) and that, when used in the Kirchhoff–Helmholtz field representation yield a proper solution to the exterior boundary-value problem in the scalar-wave case. In the vector-wave case using these Green functions in place of G_+ will remove either the boundary value of the EM field or that of its normal derivative, but we will still end up with an improperly posed vector-wave boundary problem. We will return to this issue in a later section where we will solve EM boundary-value problems using modal expansions of the vector Helmholtz equation for planar and spherical boundaries.

11.4 Angular-spectrum expansions of the radiated field

We can obtain angular-spectrum expansions of the radiated EM fields of the type derived in Section 4.2 of Chapter 4 for the field radiated by scalar sources by making use of the Weyl expansion derived in Section 4.1 of that chapter. This expansion can be expressed in either Cartesian or angular integration variables, and in Cartesian variables assumes the form

$$G_+(\mathbf{R}) = \frac{-i}{8\pi^2}\int_{-\infty}^{\infty}\frac{d^2K_\rho}{\gamma}\,e^{i\mathbf{k}_0^\pm\cdot\mathbf{R}}, \tag{11.19}$$

where the plus sign is used if $Z = \hat{\mathbf{z}}\cdot\mathbf{R} > 0$ and the minus sign if $Z < 0$. Here,

$$\mathbf{k}_0^\pm = \mathbf{K}_\rho \pm \gamma\hat{\mathbf{z}},$$

with $\mathbf{K}_\rho = (K_x, K_y)$ and

$$\gamma = \begin{cases} \sqrt{k_0^2 - K_\rho^2} & \text{if } K_\rho^2 < \Re k_0^2, \\ i\sqrt{K_\rho^2 - k_0^2} & \text{if } K_\rho^2 > \Re k_0^2. \end{cases}$$

As discussed in Section 4.1, the Weyl expansion is in the form of a superposition of plane waves having wave vectors $\mathbf{k}_0^\pm$, which are generally complex due to the fact that in a dispersive medium $\Im k_0 > 0$. These wave vectors also satisfy the requirement that $\mathbf{k}_0^\pm\cdot\mathbf{k}_0^\pm = k_0^2$ as a consequence of the fact that the plane waves satisfy the homogeneous Helmholtz equation. The plane waves for which $K_\rho^2 < \Re k_0^2$ are weakly inhomogeneous plane waves and have a complex wave vector due to the dispersive nature of the medium in which they propagate. If the loss in this medium as characterized by $\Im k_0$ were to vanish these particular plane waves would become homogeneous plane waves and have a unit magnitude over all of space. The plane waves for which $K_\rho^2 > \Re k_0^2$ are *evanescent* plane waves and have a complex wave vector that does not become real in the limit where $\Im k_0 \to 0$. These plane waves derive their inhomogeneous character from the fact that we allow the (K_x, K_y)

components of the wave vectors $\mathbf{k}_0^{\pm} = K_x\hat{\mathbf{x}} + K_y\hat{\mathbf{y}} \pm \gamma\hat{\mathbf{z}}$ to vary over the entire K_x, K_y plane, thus requiring the z component $\pm\gamma$ to be inherently complex when $K_x^2 + K_y^2 > \Re k_0^2$.

The angle-variable form of the Weyl expansion

If we make the transformation from Cartesian (K_x, K_y) to spherical (α, β) integration variables in the Weyl expansion Eq. (11.19) we showed in Section 4.1 that

$$\frac{d^2 K_\rho}{\gamma} \Rightarrow k_0 \sin\alpha \, d\beta \, d\alpha, \tag{11.20a}$$

and

$$\mathbf{k}_0^{\pm} \Rightarrow k_0 \mathbf{s} = k_0 \overbrace{(\sin\alpha\cos\beta\,\hat{\mathbf{x}} + \sin\alpha\sin\beta\,\hat{\mathbf{y}} + \cos\alpha\,\hat{\mathbf{z}})}^{\mathbf{s}}, \tag{11.20b}$$

where the azimuthal angle β varies from 0 to 2π and the polar angle α varies along the contour C_+ in Fig. 3.1 of Chapter 3 for the wave vector $\mathbf{k}_0^+$ and over the contour C_- in this figure for the wave vector $\mathbf{k}_0^-$. We then find that the Weyl expansion assumes the form

$$G_+(\mathbf{R}) = -\frac{ik_0}{8\pi^2} \int_{-\pi}^{\pi} d\beta \int_{C_\pm} \sin\alpha \, d\alpha \, e^{ik_0 \mathbf{s}\cdot\mathbf{R}}, \tag{11.21}$$

where the contour C_+ is used if $Z > 0$ and C_- if $Z < 0$. The integrand in the above angular-spectrum expansion is an entire analytic function of the angles α and β so that the precise shape of the integration contours $C_\pm$ is unimportant and, as discussed in Section 3.2.3, the decomposition of the α contour to lie along the real axis and then along the line $\Re\alpha = \pi/2$ corresponds to a separation of the plane waves in the expansions into weakly inhomogeneous and evanescent plane waves.

11.4.1 The angle-variable form of the angular-spectrum expansion of the EM field

We first consider the angular-spectrum expansion of the EM field in angle variables, where the Weyl expansion is given in Eq. (11.21). We assume that the spatial support volume τ_0 of the charge–current distribution is contained within the strip $z^- \leq z \leq z^+$ and restrict our attention to field points $\mathbf{r} = (\boldsymbol{\rho}, z)$ that lie outside this strip. We note that, since the orientation of the (x, y, z) coordinate system is arbitrary, our results hold outside any such strip containing the source support volume τ_0. If we then substitute the angle-variable form of the Weyl expansion into Eq. (11.13a) we obtain the result

$$\begin{aligned}\mathbf{E}_+(\mathbf{r}) &= \frac{ik_0}{8\pi^2} \int_{-\pi}^{\pi} d\beta \int_{C_\pm} \sin\alpha \, d\alpha \left\{ \int_{\tau_0} d^3r' \left[i\omega\mu_0 \mathbf{J}(\mathbf{r}') - \frac{1}{\epsilon_0}\nabla_{r'}\rho(\mathbf{r}') \right] e^{-ik_0\mathbf{s}\cdot\mathbf{r}'} \right\} e^{ik_0\mathbf{s}\cdot\mathbf{r}} \\ &= \frac{ik_0}{2\pi} \int_{-\pi}^{\pi} d\beta \int_{C_\pm} \sin\alpha \, d\alpha \, \mathbf{A}_e(k_0\mathbf{s}) e^{ik_0\mathbf{s}\cdot\mathbf{r}},\end{aligned} \tag{11.22a}$$

where C_+ is used in $z > z^+$ and C_- in $z < z^-$, and the electric-field plane-wave amplitude (angular spectrum) $\mathbf{A}_e(k_0\mathbf{s})$ is given by

$$\mathbf{A}_e(k_0\mathbf{s}) = \frac{1}{4\pi}\int_{\tau_0} d^3r' \left[i\omega\mu_0\mathbf{J}(\mathbf{r}') - \frac{1}{\epsilon_0}\nabla_{r'}\rho(\mathbf{r}')\right] e^{-ik_0\mathbf{s}\cdot\mathbf{r}'}$$
$$= \frac{1}{4\pi}\left[i\omega\mu_0\tilde{\mathbf{J}}(k_0\mathbf{s}) - \frac{ik_0\mathbf{s}}{\epsilon_0}\tilde{\rho}(k_0\mathbf{s})\right] = \frac{i\omega\mu_0}{4\pi}\tilde{\mathbf{J}}_T(k_0\mathbf{s}). \tag{11.22b}$$

A completely parallel derivation leads to the result that

$$\mathbf{H}_+(\mathbf{r}) = \frac{ik_0}{2\pi}\int_{-\pi}^{\pi} d\beta \int_{C_\pm} \sin\alpha\, d\alpha\, \mathbf{A}_h(k_0\mathbf{s})e^{ik_0\mathbf{s}\cdot\mathbf{r}}, \tag{11.23a}$$

where the magnetic-field angular spectrum $\mathbf{A}_h(k_0\mathbf{s})$ is given by

$$\mathbf{A}_h(k_0\mathbf{s}) = \frac{1}{4\pi}\int_{\tau_0} d^3r'\, \nabla_{r'} \times \mathbf{J}(\mathbf{r}')e^{-ik_0\mathbf{s}\cdot\mathbf{r}'} = \frac{ik_0}{4\pi}\mathbf{s}\times\tilde{\mathbf{J}}_T(k_0\mathbf{s}). \tag{11.23b}$$

As in the case of the electric-field angular-spectrum expansion, the magnetic-field expansion is valid at all field points $\mathbf{r}$ lying outside the source strip $z^- \le z \le z^+$.

The two angular spectra and, hence, the electric and magnetic fields outside the source strip are seen to depend only on the spatial Fourier transform of the transverse part of the current distribution on the boundary $\mathbf{K} = k_0\mathbf{s}$. However, since the orientation of the x, y, z coordinate system is arbitrary, this result, as mentioned in Example 11.2, must apply everywhere outside the smallest convex region that contains the source (the so-called *convex hull* of the source support volume τ_0). We also note that the two angular spectra $\mathbf{A}_e(k_0\mathbf{s})$ and $\mathbf{A}_h(k_0\mathbf{s})$ are entire analytic functions of the (generally complex) unit vector $\mathbf{s}$, so the contours $C_\pm$ can be arbitrarily deformed in Eqs. (11.22a) and (11.23a) so long as they run from $\alpha = 0$ to $\alpha = \pi/2 - i\infty$ in the case of C_+ and from $\alpha = \pi/2 + i\infty$ to $\alpha = \pi$ in the case of C_-. As discussed above, the deformation of these contours to run along the real-α axis and along the line $\Re\alpha = \pi/2$ corresponds to the decomposition of the expansions into weakly inhomogeneous and evanescent plane waves, respectively.

The two angular spectra are functionally identical to the two radiation patterns $\mathbf{f}_e(\mathbf{s})$ and $\mathbf{f}_h(\mathbf{s})$, respectively, defined in Eqs. (11.17) and, hence, must satisfy the relationships in Eqs. (11.18); i.e.,

$$\mathbf{A}_e(k_0\mathbf{s}) = -\frac{\omega\mu_0}{k_0}\mathbf{s}\times\mathbf{A}_h(k_0\mathbf{s}), \qquad \mathbf{A}_h(k_0\mathbf{s}) = \frac{k_0}{\omega\mu_0}\mathbf{s}\times\mathbf{A}_e(k_0\mathbf{s}). \tag{11.24}$$

We should note that, although the angular spectra and radiation patterns are functionally identical, they have the important difference that the radiation patterns are the amplitudes of the electric and magnetic field vectors in the far field along the directions $\mathbf{s} = \mathbf{r}/r$ and, hence, are defined only on the *real* unit sphere, whereas the angular spectra are defined for both real and complex unit vectors $\mathbf{s}$ and, hence, are properly interpreted as being analytic continuations of the radiation patterns (see the discussion below).

11.4.2 Back propagation from the radiation patterns

When the unit vector $\mathbf{s}$ lies on the real unit sphere the angular spectra $\mathbf{A}_e(k_0\mathbf{s})$ and $\mathbf{A}_h(k_0\mathbf{s})$ are seen to be equal to the radiation patterns $\mathbf{f}_e(\mathbf{s})$ and $\mathbf{f}_h(\mathbf{s})$, respectively. This is precisely the same result as we obtained in the scalar-wave case considered in Section 4.2.2

of Chapter 4. As discussed in that section, the radiation patterns are boundary values of entire analytic functions of the unit vector $\mathbf{s}$ and the angular spectra can be considered to be analytic continuations of these boundary values onto complex unit vectors $\mathbf{s}$ having azimuthal angles α lying on the integration contours $C_{\pm}$. This then allows the radiated field to be computed everywhere outside the source strip using *field back propagation* implemented via the angular-spectrum expansions and analytically continued radiation patterns. However, as discussed in Sections 4.2.2 and 4.3 such a process of analytic continuation is not stable and cannot be used in any practical application. However, we can deform the contours $C_{\pm}$ to lie along the real-α axis and along the line $\Re\alpha = \pi/2$ as described above and approximate the radiated fields using only the contributions from the weakly inhomogeneous plane waves whose amplitudes are the observed radiation patterns. Indeed, we employed such approximations in the scalar-wave case in Section 4.4 of Chapter 4, where they were found to be excellent except in the immediate vicinity of the source strip.

A further word of caution is necessary regarding back propagation from the radiation pattern in the case of EM vector wavefields that is not necessary in the scalar-wave case treated in Chapter 4. In particular, it follows from Eqs. (11.18) that the radiation patterns $\mathbf{f}_{\mathrm{e}}(\mathbf{s})$ and $\mathbf{f}_{\mathrm{h}}(\mathbf{s})$ must both be perpendicular to the unit vector $\mathbf{s}$. On the other hand, measurements of these quantities in any physical experiment will not, in general, satisfy this requirement, which will result in back-propagated fields that will not have zero divergence as is required of both EM fields outside of the source support region τ_0. A simple way to correct this problem is to project the measured radiation patterns onto the unit sphere and use the projected patterns in the stabilized angular-spectrum expansions. This is essentially what we will do using the so-called Debye representation of the EM fields that will be presented in a later section of the chapter.

11.4.3 The Cartesian-variable form of the angular-spectrum expansion of the EM field

We now consider the angular-spectrum expansion of the EM field in Cartesian integration variables, where the Weyl expansion is given in Eq. (11.19). As in the angle-variable form of the expansion derived above, we will restrict our attention to field points $\mathbf{r} = (\boldsymbol{\rho}, z)$ that lie outside the source strip $z^- \leq z \leq z^+$. The required expansion can be obtained by employing a parallel derivation to that used above for the angle-variable form, starting with the Cartesian-variable form of the Weyl expansion given in Eq. (11.19). Alternatively, we can simply make the inverse transformation to that defined in Eqs. (11.20) in the angle-variable form of the angular-spectrum expansions in Eqs. (11.22a) and (11.23a). For the sake of simplicity we will use this second scheme and make the transformation in these expansions from angle integration variables to Cartesian integration variables via

$$k_0 \sin\alpha \, d\beta \, d\alpha \Rightarrow \frac{d^2K_\rho}{\gamma},$$

$$k_0\mathbf{s} \Rightarrow \mathbf{k}_0^{\pm} = \mathbf{K}_\rho \pm \gamma\hat{\mathbf{z}}.$$

We find that Eqs. (11.22a) and (11.23a) yield the results

$$\mathbf{E}_+(\mathbf{r}) = \frac{i}{2\pi}\int_{-\infty}^{\infty}\frac{d^2K_\rho}{\gamma}\mathbf{A}_\mathrm{e}(\mathbf{k}_0^\pm)e^{i\mathbf{k}_0^\pm\cdot\mathbf{r}}, \tag{11.25a}$$

$$\mathbf{H}_+(\mathbf{r}) = \frac{i}{2\pi}\int_{-\infty}^{\infty}\frac{d^2K_\rho}{\gamma}\mathbf{A}_\mathrm{h}(\mathbf{k}_0^\pm)e^{i\mathbf{k}_0^\pm\cdot\mathbf{r}}, \tag{11.25b}$$

where $\mathbf{A}_\mathrm{e}(\mathbf{k}_0^\pm)$ and $\mathbf{A}_\mathrm{h}(\mathbf{k}_0^\pm)$ are, respectively, $\mathbf{A}_\mathrm{e}(k_0\mathbf{s})$ and $\mathbf{A}_\mathrm{h}(k_0\mathbf{s})$ under the above transformation and are thus given by

$$\mathbf{A}_\mathrm{e}(\mathbf{k}_0^\pm) = \frac{1}{4\pi}\left[i\omega\mu_0\tilde{\mathbf{J}}(\mathbf{k}_0^\pm) - \frac{i\mathbf{k}_0^\pm}{\epsilon_0}\tilde{\rho}(\mathbf{k}_0^\pm)\right] = \frac{i\omega\mu_0}{4\pi}\tilde{\mathbf{J}}_\mathrm{T}(\mathbf{k}_0^\pm), \tag{11.26a}$$

$$\mathbf{A}_\mathrm{h}(\mathbf{k}_0^\pm) = \frac{i}{4\pi}\mathbf{k}_0^\pm \times \tilde{\mathbf{J}}_\mathrm{T}(\mathbf{k}_0^\pm). \tag{11.26b}$$

Both forms of the expansions require that the z coordinate of the field point $\mathbf{r} = \boldsymbol{\rho} + z\hat{\mathbf{z}}$ lie outside of the source strip $z^- \le z \le z^+$, with $\mathbf{k}_0^+$ used in the expansions if $z > z^+$ and $\mathbf{k}_0^-$ if $z < z^-$.

As was the case with the angle-variable form of the angular-spectrum expansion, the angular spectra $\mathbf{A}_\mathrm{e}(\mathbf{k}_0^\pm)$ and $\mathbf{A}_\mathrm{h}(\mathbf{k}_0^\pm)$ both depend only on the Fourier transform of the transverse part of the current density on the boundary $\sqrt{\mathbf{K}\cdot\mathbf{K}} = k_0$ and are analytically continued radiation patterns and must satisfy Eqs. (11.24) under the replacement of $k_0\mathbf{s}$ with $\mathbf{k}_0^\pm$:

$$\mathbf{A}_\mathrm{e}(\mathbf{k}_0^\pm) = -\frac{1}{\omega\epsilon_0}\mathbf{k}_0^\pm \times \mathbf{A}_\mathrm{h}(\mathbf{k}_0^\pm), \qquad \mathbf{A}_\mathrm{h}(\mathbf{k}_0^\pm) = \frac{1}{\omega\mu_0}\mathbf{k}_0^\pm \times \mathbf{A}_\mathrm{e}(\mathbf{k}_0^\pm). \tag{11.27}$$

Moreover, the radiated field can, in principle, be exactly, but unstably, computed everywhere outside the source strip via field back propagation using the analytically continued radiation patterns in Eqs. (11.25) or can be stably computed by truncating these expansions to weakly inhomogeneous plane waves. However, field back propagation from the radiation patterns is best implemented using the angle-variable forms of the angular-spectrum expansions given in Eqs. (11.22a) and (11.23a) while the Cartesian-variable form of these expansions is best used when back propagating from planar boundary-value data, as we will now describe.

11.4.4 Forward and back propagation from planar boundary-value data

If we take an inverse spatial Fourier transform of the electric and magnetic field vectors as given in the angular-spectrum expansions Eqs. (11.25) over any plane $z = z_0$ that lies outside of the source strip we find that

$$\tilde{\mathbf{E}}_+(\mathbf{K}_\rho, z_0) = \frac{2\pi i}{\gamma}\mathbf{A}_\mathrm{e}(\mathbf{k}_0^\pm)e^{\pm i\gamma z_0}, \qquad \tilde{\mathbf{H}}_+(\mathbf{K}_\rho, z_0) = \frac{2\pi i}{\gamma}\mathbf{A}_\mathrm{h}(\mathbf{k}_0^\pm)e^{\pm i\gamma z_0},$$

where the plus sign is used if $z_0 > z^+$ and the minus sign if $z_0 < z^-$ and

$$\tilde{\mathbf{E}}_+(\mathbf{K}_\rho, z_0) = \int d^2\rho\, \mathbf{E}_+(\boldsymbol{\rho}, z_0)e^{-i\mathbf{K}_\rho\cdot\boldsymbol{\rho}}, \qquad \tilde{\mathbf{H}}_+(\mathbf{K}_\rho, z_0) = \int d^2\rho\, \mathbf{H}_+(\boldsymbol{\rho}, z_0)e^{-i\mathbf{K}_\rho\cdot\boldsymbol{\rho}},$$

are the 2D transforms of the EM fields over the boundary-value plane. The above equations directly relate the angular spectra of the electric and magnetic field vectors to the spatial Fourier transforms of these fields over arbitrary planar surfaces that lie outside the source strip. It then follows that it is possible to compute the angular spectra both over the homogeneous and over the evanescent regions and, hence, the radiated EM field everywhere in the r.h.s. $z > z^+$ from field data acquired over *any* plane in this half-space and throughout the half-space $z < z^-$ from field data acquired over any plane in this half-space.

The above results are identical to those that we obtained for scalar fields in Section 4.3 of Chapter 4, where we discussed field *forward* and *back* propagation from boundary-value data acquired over infinite planes lying outside the source region τ_0. As discussed in that section, if the angular spectrum is acquired from field data over a plane z_0 that lies to the right of the source strip ($z_0 > z^+$) then the resulting angular-spectrum expansion converges throughout the r.h.s. $z > z^+$ and implements field forward propagation from the data if $z > z_0$ and field back propagation if $z < z_0$. On the other hand, if the angular spectrum is acquired from field data over a plane z_0 that lies to the left of the source strip ($z_0 < z^-$) then the resulting angular-spectrum expansion converges throughout the l.h.s. $z < z^-$ and implements field forward propagation from the data if $z < z_0$ and field back propagation if $z > z_0$. In other words, forward propagation is the process of computing the field at points more distant from the source than the data boundary, whereas field back propagation is the process of computing the field at points closer to the source than the data boundary. As discussed extensively in earlier chapters field forward propagation can be formulated as a properly posed *boundary-value problem* and is perfectly stable and well-posed, whereas field back propagation cannot be so formulated and is unstable and ill-posed.

As was the case in EM field back propagation from the radiation patterns, EM field forward and back propagation from boundary-value data is further complicated by the fact that the EM fields must have zero divergence outside the source region. This requires that the angular spectra $\mathbf{A}_e(\mathbf{k}_0^\pm)$ and $\mathbf{A}_h(\mathbf{k}_0^\pm)$ be perpendicular to the wave vectors $\mathbf{k}_0^\pm$ and there is no guarantee that the angular spectra computed from measured boundary value data via the above equations will satisfy this condition. A simple way to insure that this condition is satisfied and that the propagated or back-propagated fields have zero divergence is to project the measured data onto the perpendicular plane to $\mathbf{k}_0^\pm$ and this is, in fact, what is done in deriving the so-called *Whittaker representation* of the fields in the following section.

11.5 The Whittaker representation of the radiated fields

The vector-valued amplitudes (angular spectra) of the plane waves in the Cartesian form of the angular-spectrum expansions of the EM fields derived in the last section are perpendicular to the propagation vectors $\mathbf{k}_0^\pm$ and, hence, can be decomposed into two orthogonal components on the plane perpendicular to these vectors. We can thus express the angular spectra in terms of the two orthogonal vectors $\mathbf{k}_0^\pm \times \hat{\mathbf{z}}$ and $\mathbf{k}_0^\pm \times \mathbf{k}_0^\pm \times \hat{\mathbf{z}}$ in the form

$$\mathbf{A}_{\rm e}(\mathbf{k}_0^{\pm}) = i\mathbf{k}_0^{\pm} \times \hat{\mathbf{z}}\hat{\Pi}_{\rm h}^{\rm w}(\mathbf{k}_0^{\pm}) - \frac{i}{\omega\epsilon_0}\mathbf{k}_0^{\pm} \times \mathbf{k}_0^{\pm} \times \hat{\mathbf{z}}\hat{\Pi}_{\rm e}^{\rm w}(\mathbf{k}_0^{\pm}), \tag{11.28a}$$

$$\mathbf{A}_{\rm h}(\mathbf{k}_0^{\pm}) = i\mathbf{k}_0^{\pm} \times \hat{\mathbf{z}}\hat{\Pi}_{\rm e}^{\rm w}(\mathbf{k}_0^{\pm}) + \frac{i}{\omega\mu_0}\mathbf{k}_0^{\pm} \times \mathbf{k}_0^{\pm} \times \hat{\mathbf{z}}\hat{\Pi}_{\rm h}^{\rm w}(\mathbf{k}_0^{\pm}), \tag{11.28b}$$

where the triple vector product $\mathbf{k}_0^{\pm} \times \mathbf{k}_0^{\pm} \times \hat{\mathbf{z}}$ is taken to mean $\mathbf{k}_0^{\pm} \times [\mathbf{k}_0^{\pm} \times \hat{\mathbf{z}}]$. We will refer to the two scalar quantities $\hat{\Pi}_{\rm e}^{\rm w}$ and $\hat{\Pi}_{\rm h}^{\rm w}$ as the Whittaker electric and magnetic plane-wave amplitudes. It is easy to verify that $\mathbf{A}_{\rm e}(\mathbf{k}_0^{\pm})$ and $\mathbf{A}_{\rm h}(\mathbf{k}_0^{\pm})$ as represented via Whittaker plane-wave amplitudes satisfy the conditions given in Eqs. (11.27).

The Whittaker plane-wave amplitudes are obtained by projecting the electric- and/or magnetic-field plane-wave amplitudes onto the two orthogonal vectors $\mathbf{k}_0^{\pm} \times \hat{\mathbf{z}}$ and $\mathbf{k}_0^{\pm} \times \mathbf{k}_0^{\pm} \times \hat{\mathbf{z}}$. In terms of the electric-field plane-wave amplitude we find, after performing some simple vector algebra and using the fact that $\mathbf{k}_0^{\pm} \cdot \mathbf{k}_0^{\pm} = k_0^2$, that

$$\hat{\Pi}_{\rm h}^{\rm w}(\mathbf{k}_0^{\pm}) = -i\frac{\mathbf{k}_0^{\pm} \times \hat{\mathbf{z}}}{K_\rho^2} \cdot \mathbf{A}_{\rm e}(\mathbf{k}_0^{\pm}), \qquad \hat{\Pi}_{\rm e}^{\rm w}(\mathbf{k}_0^{\pm}) = \frac{i}{\omega\mu_0}\frac{\mathbf{k}_0^{\pm} \times \mathbf{k}_0^{\pm} \times \hat{\mathbf{z}}}{K_\rho^2} \cdot \mathbf{A}_{\rm e}(\mathbf{k}_0^{\pm}).$$

A similar calculation yields the result

$$\hat{\Pi}_{\rm e}^{\rm w}(\mathbf{k}_0^{\pm}) = -i\frac{\mathbf{k}_0^{\pm} \times \hat{\mathbf{z}}}{K_\rho^2} \cdot \mathbf{A}_{\rm h}(\mathbf{k}_0^{\pm}), \qquad \hat{\Pi}_{\rm h}^{\rm w}(\mathbf{k}_0^{\pm}) = -\frac{i}{\omega\epsilon_0}\frac{\mathbf{k}_0^{\pm} \times \mathbf{k}_0^{\pm} \times \hat{\mathbf{z}}}{K_\rho^2} \cdot \mathbf{A}_{\rm h}(\mathbf{k}_0^{\pm}).$$

The Whittaker representation of the EM fields outside the source strip $z^- \leq z \leq z^+$ is obtained by substituting the plane-wave amplitudes from Eqs. (11.28) into the angular-spectrum expansions Eqs. (11.25). We obtain

$$\mathbf{E}_+(\mathbf{r}) = \frac{i}{2\pi}\int_{-\infty}^{\infty}\frac{d^2K_\rho}{\gamma}\left\{i\mathbf{k}_0^{\pm} \times \hat{\mathbf{z}}\hat{\Pi}_{\rm h}^{\rm w}(\mathbf{k}_0^{\pm}) - \frac{i}{\omega\epsilon_0}\mathbf{k}_0^{\pm} \times \mathbf{k}_0^{\pm} \times \hat{\mathbf{z}}\hat{\Pi}_{\rm e}^{\rm w}(\mathbf{k}_0^{\pm})\right\}e^{i\mathbf{k}_0^{\pm}\cdot\mathbf{r}}, \tag{11.29a}$$

$$\mathbf{H}_+(\mathbf{r}) = \frac{i}{2\pi}\int_{-\infty}^{\infty}\frac{d^2K_\rho}{\gamma}\left\{i\mathbf{k}_0^{\pm} \times \hat{\mathbf{z}}\hat{\Pi}_{\rm e}^{\rm w}(\mathbf{k}_0^{\pm}) + \frac{i}{\omega\mu_0}\mathbf{k}_0^{\pm} \times \mathbf{k}_0^{\pm} \times \hat{\mathbf{z}}\hat{\Pi}_{\rm h}^{\rm w}(\mathbf{k}_0^{\pm})\right\}e^{i\mathbf{k}_0^{\pm}\cdot\mathbf{r}}. \tag{11.29b}$$

If we make use of the identities

$$\mathbf{k}_0^{\pm} \times \hat{\mathbf{z}}e^{i\mathbf{k}_0^{\pm}\cdot\mathbf{r}} \equiv -i\nabla \times \hat{\mathbf{z}}e^{i\mathbf{k}_0^{\pm}\cdot\mathbf{r}}, \qquad \mathbf{k}_0^{\pm} \times \mathbf{k}_0^{\pm} \times \hat{\mathbf{z}}e^{i\mathbf{k}_0^{\pm}\cdot\mathbf{r}} \equiv -\nabla \times \nabla \times \hat{\mathbf{z}}e^{i\mathbf{k}_0^{\pm}\cdot\mathbf{r}},$$

we can also write Eqs. (11.29) in the standard form

$$\mathbf{E}_+(\mathbf{r}) = \nabla \times \hat{\mathbf{z}}\Pi_{\rm h}^{\rm w}(\mathbf{r}) + \frac{i}{\omega\epsilon_0}\nabla \times \nabla \times \hat{\mathbf{z}}\Pi_{\rm e}^{\rm w}(\mathbf{r}), \tag{11.30a}$$

$$\mathbf{H}_+(\mathbf{r}) = \nabla \times \hat{\mathbf{z}}\Pi_{\rm e}^{\rm w}(\mathbf{r}) - \frac{i}{\omega\mu_0}\nabla \times \nabla \times \hat{\mathbf{z}}\Pi_{\rm h}^{\rm w}(\mathbf{r}), \tag{11.30b}$$

where

$$\Pi_{\rm e}^{\rm w}(\mathbf{r}) = \frac{i}{2\pi}\int_{-\infty}^{\infty}\frac{d^2K_\rho}{\gamma}\,\hat{\Pi}_{\rm e}^{\rm w}(\mathbf{k}_0^{\pm})e^{i\mathbf{k}_0^{\pm}\cdot\mathbf{R}}, \tag{11.31a}$$

$$\Pi_{\rm h}^{\rm w}(\mathbf{r}) = \frac{i}{2\pi}\int_{-\infty}^{\infty}\frac{d^2K_\rho}{\gamma}\,\hat{\Pi}_{\rm h}^{\rm w}(\mathbf{k}_0^{\pm})e^{i\mathbf{k}_0^{\pm}\cdot\mathbf{R}}. \tag{11.31b}$$

The angular-spectrum expansions Eqs. (11.31) and, hence, the Whittaker representation in either of the two forms given above converge everywhere outside the source strip $z^- \leq z \leq z^+$. However, since the orientation of the (x, y, z) coordinate system is arbitrary, a Whittaker representation can be employed everywhere outside the convex hull of the source support volume τ_0.

The Whittaker representation was first obtained by E. T. Whittaker (Whittaker, 1904) in 1903 and was derived within the context of antenna theory by Borgiotti (Borgiotti, 1962) using the procedure employed above. Borgiotti's work is of special interest since it constitutes a general method that can be extended to spherical geometries and the Debye representation that will be derived below.

Example 11.3 The Whittaker representation is not restricted to the EM field vectors and can be employed to decompose the transverse part of any vector field whose spatial Fourier transform satisfies the conditions Eqs. (11.5). As an example, we consider the transverse part of the current density $\mathbf{J}_{\rm T}(\mathbf{r})$ whose transform we can express in the form

$$\tilde{\mathbf{J}}_{\rm T}(\mathbf{K}) = i\mathbf{K}\times\hat{\mathbf{z}}\tilde{Q}_{\rm h}^{\rm w}(\mathbf{K}) - \frac{i}{\omega\epsilon_0}\mathbf{K}\times\mathbf{K}\times\hat{\mathbf{z}}\tilde{Q}_{\rm e}^{\rm w}(\mathbf{K}), \tag{11.32}$$

where

$$\tilde{Q}_{\rm h}^{\rm w}(\mathbf{K}) = \int_{\tau_0} d^3r\, Q_{\rm h}^{\rm w}(\mathbf{r})e^{-i\mathbf{K}\cdot\mathbf{r}},$$

$$\tilde{Q}_{\rm e}^{\rm w}(\mathbf{K}) = \int_{\tau_0} d^3r\, Q_{\rm e}^{\rm w}(\mathbf{r})e^{-i\mathbf{K}\cdot\mathbf{r}}$$

are the spatial Fourier transforms of the two scalar sources $Q_{\rm h}^{\rm w}(\mathbf{r})$ and $Q_{\rm e}^{\rm w}(\mathbf{r})$, which we will refer to as the Whittaker magnetic and electric scalar source components of the transverse current density. Indeed, on taking an inverse Fourier transform of both sides of Eq. (11.32) and employing steps essentially identical to those used in transforming from Eq. (11.28a) to the standard form of the Whittaker representation of the electric field vector given in Eq. (11.30a) we find that

$$\mathbf{J}_{\rm T}(\mathbf{r}) = \nabla\times\hat{\mathbf{z}}Q_{\rm h}^{\rm w}(\mathbf{r}) + \frac{i}{\omega\epsilon_0}\nabla\times\nabla\times\hat{\mathbf{z}}Q_{\rm e}^{\rm w}(\mathbf{r}).$$

The two scalar source transforms are directly related to the angular spectra of the two Whittaker potentials of the field $\mathbf{E}_+(\mathbf{r})$ radiated by the current density $\mathbf{J}$. In particular, it follows from Eqs. (11.26a), (11.28a) and (11.32) above that

$$\overbrace{i\mathbf{k}_0^{\pm} \times \hat{\mathbf{z}}\hat{\Pi}_{\mathrm{h}}^{\mathrm{w}}(\mathbf{k}_0^{\pm}) - \frac{i}{\omega\epsilon_0}\mathbf{k}_0^{\pm} \times \mathbf{k}_0^{\pm} \times \hat{\mathbf{z}}\hat{\Pi}_{\mathrm{e}}^{\mathrm{w}}(\mathbf{k}_0^{\pm})}^{\mathbf{A}_{\mathrm{e}}(\mathbf{k}_0^{\pm})}$$

$$= \frac{i\omega\mu_0}{4\pi}\left[\overbrace{i\mathbf{k}_0^{\pm} \times \hat{\mathbf{z}}\tilde{Q}_{\mathrm{h}}^{\mathrm{w}}(\mathbf{k}_0^{\pm}) - \frac{i}{\omega\epsilon_0}\mathbf{k}_0^{\pm} \times \mathbf{k}_0^{\pm} \times \hat{\mathbf{z}}\,\tilde{Q}_{\mathrm{e}}^{\mathrm{w}}(\mathbf{k}_0^{\pm})}^{\tilde{\mathbf{J}}_{\mathrm{T}}(\mathbf{k}_0^{\pm})}\right].$$

This then requires that

$$\tilde{Q}_{\mathrm{h}}^{\mathrm{w}}(\mathbf{k}_0^{\pm}) = \frac{4\pi i}{\omega\mu_0}\hat{\Pi}_{\mathrm{h}}^{\mathrm{w}}(\mathbf{k}_0^{\pm}), \qquad \tilde{Q}_{\mathrm{e}}^{\mathrm{w}}(\mathbf{k}_0^{\pm}) = \frac{-4\pi i}{\omega\mu_0}\hat{\Pi}_{\mathrm{e}}^{\mathrm{w}}(\mathbf{k}_0^{\pm}). \tag{11.33}$$

In words, the boundary values of the spatial Fourier transforms of the Whittaker magnetic and electric scalar source components on the boundaries $\mathbf{K} = \mathbf{k}_0^{\pm}$ are proportional to the corresponding angular spectra of the Whittaker magnetic and electric scalar potentials. The Whittaker decomposition of the transverse part of the current density and this later result incorporated in Eqs. (11.33) play an important role in the EM inverse source problem (ISP) for sources compactly supported within the source strip $z^- < z < z^+$.

11.5.1 Boundary-value problems and field back propagation using the Whittaker representation

The Rayleigh–Sommerfeld (RS) boundary-value problem for scalar waves was solved in Section 2.9 of Chapter 2 using Green-function methods and then using the angular-spectrum expansion in Section 4.3 of Chapter 4. Here we will generalize the treatment given in Chapter 4 to the EM case by using the Whittaker representation implemented via angular-spectrum expansions. As in the scalar-wave case the EM RS problem consists of computing an outgoing-wave field throughout a source-free half-space $z > z_0$ or $z < z_0$ from properly specified boundary-value data on the plane $z = z_0$. In the case of scalar waves "properly" specified data consists of the value of the field (Dirichlet), its normal derivative (Neumann) or linear combinations of the two, while in the EM case it is known to consist of specification of the tangential components of the EM fields. We thus address the problem of computing the outgoing-wave electric and magnetic field vectors throughout a source-free half-space from specification of their tangential components on the data plane $z = z_0$. We will show that this solution can be employed to implement field back propagation into the region bounded by the data plane and the planar boundary of the charge–current distribution at $z = z^{\pm}$ that generated the field.

An outgoing EM wave radiated by sources contained within a source strip $z^- < z < z_+$ can be expressed everywhere outside this strip in the Whittaker plane-wave expansions Eq. (11.29). Considering for the moment the electric field vector we find that its tangential component on a plane $z = z_0$ that lies outside the source strip assumes the form

$$\hat{\mathbf{z}} \times \mathbf{E}_{+}(\boldsymbol{\rho}, z_0) = \frac{i}{2\pi} \int_{-\infty}^{\infty} \frac{d^2 K_\rho}{\gamma} \left\{ i\hat{\mathbf{z}} \times \mathbf{k}_0^{\pm} \times \hat{\mathbf{z}} \hat{\Pi}_{\mathrm{h}}^{\mathrm{w}}(\mathbf{k}_0^{\pm}) - \frac{i}{\omega\epsilon_0} \hat{\mathbf{z}} \times \mathbf{k}_0^{\pm} \times \mathbf{k}_0^{\pm} \times \hat{\mathbf{z}} \hat{\Pi}_{\mathrm{e}}^{\mathrm{w}}(\mathbf{k}_0^{\pm}) \right\} e^{i\gamma z_0} e^{i\mathbf{K}_\rho \cdot \boldsymbol{\rho}}$$

$$= \frac{1}{(2\pi)^2} \int_{-\infty}^{\infty} \frac{d^2 K_\rho}{\gamma} \left\{ -2\pi \mathbf{K}_\rho \hat{\Pi}_{\mathrm{h}}^{\mathrm{w}}(\mathbf{k}_0^{\pm}) \mp \frac{2\pi}{\omega\epsilon_0} \gamma \mathbf{K}_\rho \times \hat{\mathbf{z}} \hat{\Pi}_{\mathrm{e}}^{\mathrm{w}}(\mathbf{k}_0^{\pm}) \right\} e^{i\gamma z_0} e^{i\mathbf{K}_\rho \cdot \boldsymbol{\rho}},$$

where we have used the identities

$$\hat{\mathbf{z}} \times \mathbf{k}_0^{\pm} \times \hat{\mathbf{z}} \equiv \mathbf{K}_\rho, \qquad \hat{\mathbf{z}} \times \mathbf{k}_0^{\pm} \times \mathbf{k}_0^{\pm} \times \hat{\mathbf{z}} \equiv \mp\gamma \mathbf{K}_\rho \times \hat{\mathbf{z}}.$$

By performing a 2D inverse Fourier transform over the boundary-value plane we then obtain

$$\widetilde{\mathbf{E}_{\mathrm{T}}}(\mathbf{K}_\rho, z_0) e^{\mp i\gamma z_0} = -\frac{2\pi}{\gamma} \mathbf{K}_\rho \hat{\Pi}_{\mathrm{h}}^{\mathrm{w}}(\mathbf{k}_0^{\pm}) \mp \frac{2\pi}{\omega\epsilon_0} \mathbf{K}_\rho \times \hat{\mathbf{z}} \hat{\Pi}_{\mathrm{e}}^{\mathrm{w}}(\mathbf{k}_0^{\pm}),$$

where

$$\widetilde{\mathbf{E}_{\mathrm{T}}}(\mathbf{K}_\rho, z_0) = \int d^2\rho\, \hat{\mathbf{z}} \times \mathbf{E}(\boldsymbol{\rho}, z_0) e^{-i\mathbf{K}_\rho \cdot \boldsymbol{\rho}}$$

is the 2D spatial Fourier transform of the tangential electric field vector over the boundary-value plane. An entirely parallel computation yields

$$\widetilde{\mathbf{H}_{\mathrm{T}}}(\mathbf{K}_\rho, z_0) e^{\mp i\gamma z_0} = -\frac{2\pi}{\gamma} \mathbf{K}_\rho \hat{\Pi}_{\mathrm{e}}^{\mathrm{w}}(\mathbf{k}_0^{\pm}) \pm \frac{2\pi}{\omega\mu_0} \mathbf{K}_\rho \times \hat{\mathbf{z}} \hat{\Pi}_{\mathrm{h}}^{\mathrm{w}}(\mathbf{k}_0^{\pm}),$$

where

$$\widetilde{\mathbf{H}_{\mathrm{T}}}(\mathbf{K}_\rho, z_0) = \int d^2\rho\, \hat{\mathbf{z}} \times \mathbf{H}(\boldsymbol{\rho}, z_0) e^{-i\mathbf{K}_\rho \cdot \boldsymbol{\rho}}.$$

The two Whittaker plane-wave amplitudes are then found to be

$$\hat{\Pi}_{\mathrm{h}}^{\mathrm{w}}(\mathbf{k}_0^{\pm}) = -\frac{\gamma}{2\pi K_\rho^2} \mathbf{K}_\rho \cdot \widetilde{\mathbf{E}_{\mathrm{T}}}(\mathbf{K}_\rho, z_0) e^{\mp i\gamma z_0}, \tag{11.34a}$$

$$\hat{\Pi}_{\mathrm{e}}^{\mathrm{w}}(\mathbf{k}_0^{\pm}) = \mp \frac{\omega\epsilon_0}{2\pi K_\rho^2} \mathbf{K}_\rho \times \hat{\mathbf{z}} \cdot \widetilde{\mathbf{E}_{\mathrm{T}}}(\mathbf{K}_\rho, z_0) e^{\mp i\gamma z_0}, \tag{11.34b}$$

which can also be expressed in the form

$$\hat{\Pi}_{\mathrm{e}}^{\mathrm{w}}(\mathbf{k}_0^{\pm}) = -\frac{\gamma}{2\pi K_\rho^2} \mathbf{K}_\rho \cdot \widetilde{\mathbf{H}_{\mathrm{T}}}(\mathbf{K}_\rho, z_0) e^{\mp i\gamma z_0}, \tag{11.34c}$$

$$\hat{\Pi}_{\mathrm{h}}^{\mathrm{w}}(\mathbf{k}_0^{\pm}) = \pm \frac{\omega\mu_0}{2\pi K_\rho^2} \mathbf{K}_\rho \times \hat{\mathbf{z}} \cdot \widetilde{\mathbf{H}_{\mathrm{T}}}(\mathbf{K}_\rho, z_0) e^{\mp i\gamma z_0}. \tag{11.34d}$$

This then leads to a solution of the EM RS boundary-value problem for the electric and magnetic field vectors in the form of the Whittaker angular-spectrum representation as given in Eqs. (11.29) or in terms of the standard form of the Whittaker representation given in Eqs. (11.30) with

$$\Pi_{\mathrm{e}}^{\mathrm{w}}(\mathbf{r}) = \mp\frac{i\omega\epsilon_0}{(2\pi)^2}\int_{-\infty}^{\infty}\frac{d^2K_\rho}{\gamma}\frac{\mathbf{K}_\rho\times\hat{\mathbf{z}}}{K_\rho^2}\cdot\widetilde{\mathbf{E}}_{\mathrm{T}}(\mathbf{K}_\rho, z_0)e^{\mp i\gamma z_0}e^{i\mathbf{k}_0^{\pm}\cdot\mathbf{r}}, \tag{11.35a}$$

$$\Pi_{\mathrm{h}}^{\mathrm{w}}(\mathbf{r}) = -\frac{i}{(2\pi)^2}\int_{-\infty}^{\infty}d^2K_\rho\,\frac{\mathbf{K}_\rho}{K_\rho^2}\cdot\widetilde{\mathbf{E}}_{\mathrm{T}}(\mathbf{K}_\rho, z_0)e^{\mp i\gamma z_0}e^{i\mathbf{k}_0^{\pm}\cdot\mathbf{r}}, \tag{11.35b}$$

with the upper signs applying for propagation into the r.h.s. $z > z_0$ with z_0 to the right of the source and the lower signs for propagation into the l.h.s. $z < z_0$ with z_0 lying to the left of the source. Similarly, the solution of the RS boundary-value problem for the magnetic field is also given in terms of the Whittaker representation, where now the plane-wave amplitudes are computed from the boundary values of the magnetic field according to the equations

$$\Pi_{\mathrm{h}}^{\mathrm{w}}(\mathbf{r}) = \pm\frac{i\omega\mu_0}{(2\pi)^2}\int_{-\infty}^{\infty}\frac{d^2K_\rho}{\gamma}\frac{\mathbf{K}_\rho\times\hat{\mathbf{z}}}{K_\rho^2}\cdot\widetilde{\mathbf{H}}_{\mathrm{T}}(\mathbf{K}_\rho, z_0)e^{\mp i\gamma z_0}e^{i\mathbf{k}_0^{\pm}\cdot\mathbf{r}}, \tag{11.36a}$$

$$\Pi_{\mathrm{e}}^{\mathrm{w}}(\mathbf{r}) = -\frac{i}{(2\pi)^2}\int_{-\infty}^{\infty}d^2K_\rho\,\frac{\mathbf{K}_\rho}{K_\rho^2}\cdot\widetilde{\mathbf{H}}_{\mathrm{T}}(\mathbf{K}_\rho, z_0)e^{\mp i\gamma z_0}e^{i\mathbf{k}_0^{\pm}\cdot\mathbf{r}}. \tag{11.36b}$$

Field back propagation

The angular-spectrum expansions for the electric and magnetic field vectors and the two Whittaker potentials vectors in their basic forms given in Eqs. (11.29) and (11.31) converge everywhere outside the source strip $z^- < z < z^+$. It then follows that the above solutions to the EM RS boundary-value problems when implemented in terms of *exact* boundary-value data via Eqs. (11.35) and (11.36) will also converge everywhere outside the source strip and will implement forward propagation in the exterior regions $z^+ < z_0 < z$ and $z < z_0 < z^-$ and back propagation in the interior strips $z^+ < z < z_0$ and $z_0 < z < z^+$. As discussed earlier, forward propagation is a stable process and will continue to be valid even for non-exact boundary-value data, while back propagation is an unstable process and might not converge if the data are not exact due to the exponential growth of the evanescent plane waves in the interior strips. Thus, the process of field back propagation has to be regularized using one of the schemes discussed in Section 4.4 of Chapter 4.

Example 11.4 We showed in Example 11.3 that the transverse part of the current density $\mathbf{J}_{\mathrm{T}}(\mathbf{r})$ admits a Whittaker representation in terms of magnetic and electric scalar source components and that the spatial Fourier transforms of these two source components on the boundaries $\mathbf{K} = \mathbf{k}_0^{\pm}$ are related to the angular spectra of the magnetic and electric Whittaker potentials through the equations

$$\tilde{Q}_{\mathrm{h}}^{\mathrm{w}}(\mathbf{k}_0^{\pm}) = \frac{-4\pi i}{\omega\mu_0}\hat{\Pi}_{\mathrm{h}}^{\mathrm{w}}(\mathbf{k}_0^{\pm}), \qquad \tilde{Q}_{\mathrm{e}}^{\mathrm{w}}(\mathbf{k}_0^{\pm}) = \frac{-4\pi i}{\omega\mu_0}\hat{\Pi}_{\mathrm{e}}^{\mathrm{w}}(\mathbf{k}_0^{\pm}).$$

If we then make use of Eqs. (11.34) we find that

$$\tilde{Q}_{\mathrm{h}}^{\mathrm{w}}(\mathbf{k}_0^{\pm}) = \frac{-4\pi i}{\omega\mu_0}\left[\overbrace{-\frac{\gamma}{2\pi K_\rho^2}\mathbf{K}_\rho\cdot\widetilde{\mathbf{E}}_{\mathrm{T}}(\mathbf{K}_\rho, z_0)e^{\mp i\gamma z_0}}^{\hat{\Pi}_{\mathrm{h}}^{\mathrm{w}}(\mathbf{k}_0^{\pm})}\right],$$

$$\tilde{Q}_{\rm e}^{\rm w}(\mathbf{k}_0^{\pm}) = \frac{-4\pi i}{\omega\mu_0}\Bigg[\overbrace{\mp\frac{\omega\epsilon_0}{2\pi K_\rho^2}\mathbf{K}_\rho \times \hat{\mathbf{z}} \cdot \widetilde{\mathbf{E}_{\rm T}}(\mathbf{K}_\rho, z_0)e^{\mp i\gamma z_0}}^{\hat{\Pi}_{\rm e}^{\rm w}(\mathbf{k}_0^{\pm})}\Bigg],$$

with analogous equations relating the scalar source components to boundary values of the tangential components of the magnetic field vector. The above equations allow the transforms of the two scalar source components on the boundaries $\mathbf{K} = \mathbf{k}_0^{\pm}$ to be determined in terms of the (properly specified) boundary values of the tangential electric field vector over bounding planes that lie outside the source strip. This result can be used to reduce the EM inverse source problem for sources confined to the plane-parallel slab to that of two uncoupled scalar problems such as those treated in Section 5.3 of Chapter 5.

11.6 Debye representation and multipole expansions of radiated fields

Here we will work with the angle-variable form of the angular-spectrum expansions where the angular spectra $\mathbf{A}_{\rm e}(k_0\mathbf{s})$ and $\mathbf{A}_{\rm h}(k_0\mathbf{s})$ are perpendicular to the unit propagation vector $\mathbf{s}$ and thus lie on the surface of the unit sphere where they can be expressed in terms of the polar and azimuthal unit vectors $\hat{\boldsymbol{\alpha}}$ and $\hat{\boldsymbol{\beta}}$. We can write these decompositions in analogy to those used in deriving the Whittaker representation in the form

$$\mathbf{A}_{\rm e}(k_0\mathbf{s})| = -i\mathbf{L}\hat{\Pi}_{\rm h}^{\rm d}(k_0\mathbf{s}) + \frac{ik_0}{\omega\epsilon_0}\mathbf{s} \times \mathbf{L}\hat{\Pi}_{\rm e}^{\rm d}(k_0\mathbf{s}), \tag{11.37a}$$

$$\mathbf{A}_{\rm h}(k_0\mathbf{s}) = -i\mathbf{L}\,\hat{\Pi}_{\rm e}^{\rm d}(k_0\mathbf{s}) - \frac{ik_0}{\omega\mu_0}\mathbf{s} \times \mathbf{L}\hat{\Pi}_{\rm h}^{\rm d}(k_0\mathbf{s}), \tag{11.37b}$$

where $\hat{\Pi}_{\rm e}^{\rm d}$ and $\hat{\Pi}_{\rm h}^{\rm d}$ are the Debye electric and magnetic plane-wave amplitudes and

$$\mathbf{L} = -ik_0\mathbf{s} \times \nabla_{k_0\mathbf{s}} = i\left[\hat{\boldsymbol{\alpha}}\frac{1}{\sin\alpha}\frac{\partial}{\partial\beta} - \hat{\boldsymbol{\beta}}\frac{\partial}{\partial\alpha}\right]$$

is the *angular-momentum operator* first introduced in Section 3.3 of Chapter 3. The validity of the decomposition of the electric and magnetic field plane-wave amplitudes in the form given in Eqs. (11.37) is known as *Hodge's decomposition theorem* and was established by Calvin Wilcox using harmonic analysis (Wilcox, 1957) and by the author in his Ph.D. thesis using simple arguments based on the properties of the angular-momentum operator (Devaney, 1971).

The Debye plane-wave amplitudes $\hat{\Pi}_{\rm e}^{\rm d}(k_0\mathbf{s})$ and $\hat{\Pi}_{\rm h}^{\rm d}(k_0\mathbf{s})$ satisfy the partial differential equations

$$L^2\hat{\Pi}^{\rm d}_{\rm h}(k_0\mathbf{s}) = i\mathbf{L}\cdot\mathbf{A}_{\rm e}(k_0\mathbf{s}) = \frac{i\omega\mu_0}{k_0}\mathbf{s}\times\mathbf{L}\cdot\mathbf{A}_{\rm h}(k_0\mathbf{s}),$$
$$L^2\hat{\Pi}^{\rm d}_{\rm e}(k_0\mathbf{s}) = i\mathbf{L}\cdot\mathbf{A}_{\rm h}(k_0\mathbf{s}) = -\frac{i\omega\epsilon_0}{k_0}\mathbf{s}\times\mathbf{L}\cdot\mathbf{A}_{\rm e}(k_0\mathbf{s}),$$

where $L^2 = \mathbf{L}\cdot\mathbf{L}$ is the square of the angular-momentum operator. The above equations are derived directly from Eqs. (11.37) on making use of the easily proven identities

$$\mathbf{L}\cdot[\mathbf{s}\times\mathbf{L}] = [\mathbf{s}\times\mathbf{L}]\cdot\mathbf{L} = 0, \qquad [\mathbf{s}\times\mathbf{L}]\cdot[\mathbf{s}\times\mathbf{L}] = L^2.$$

It is easy to verify that $\mathbf{A}_{\rm e}(k_0\mathbf{s})$ and $\mathbf{A}_{\rm h}(k_0\mathbf{s})$ as represented in terms of the Debye plane-wave amplitudes satisfy the conditions given in Eqs. (11.18).

The angular-spectrum expansion form of the Debye representation of the EM fields outside the source strip $z^- \le z \le z^+$ is obtained by substituting the plane-wave amplitudes from Eqs. (11.37) into the angular-spectrum expansions Eqs. (11.22a) and (11.23a). We find that

$$\mathbf{E}_+(\mathbf{r}) = -\frac{ik_0}{2\pi}\int_{-\pi}^{\pi}d\beta\int_{C_\pm}\sin\alpha\,d\alpha\left\{i\mathbf{L}\hat{\Pi}^{\rm d}_{\rm h}(k_0\mathbf{s}) - \frac{ik_0}{\omega\epsilon_0}\mathbf{s}\times\mathbf{L}\hat{\Pi}^{\rm d}_{\rm e}(k_0\mathbf{s})\right\}e^{ik_0\mathbf{s}\cdot\mathbf{r}},$$
$$\mathbf{H}_+(\mathbf{r}) = -\frac{ik_0}{2\pi}\int_{-\pi}^{\pi}d\beta\int_{C_\pm}\sin\alpha\,d\alpha\left\{i\mathbf{L}\hat{\Pi}^{\rm d}_{\rm e}(k_0\mathbf{s}) + \frac{ik_0}{\omega\mu_0}\mathbf{s}\times\mathbf{L}\hat{\Pi}^{\rm d}_{\rm h}(k_0\mathbf{s})\right\}e^{ik_0\mathbf{s}\cdot\mathbf{r}}.$$

If we now use the fact that the first-order partial differential operators $\mathbf{L}$ and $\mathbf{s}\times\mathbf{L}$ involve only derivatives with respect to the polar and azimuthal angles α and β of the unit vector $\mathbf{s}$ we can integrate by parts and then make use of the identities

$$\overbrace{k_0\mathbf{s}\times\nabla_{\mathbf{k_0 s}}}^{i\mathbf{L}}\,e^{ik_0\mathbf{s}\cdot\mathbf{r}} = ik_0\mathbf{s}\times\mathbf{r}e^{ik_0\mathbf{s}\cdot\mathbf{r}} = \nabla\times\mathbf{r}e^{ik_0\mathbf{s}\cdot\mathbf{r}}, \tag{11.38a}$$
$$\overbrace{ik^2\mathbf{s}\times\mathbf{s}\times\nabla_{k_0\mathbf{s}}}^{-k_0\mathbf{s}\times\mathbf{L}}\,e^{ik_0\mathbf{s}\cdot\mathbf{r}} = -k_0^2\mathbf{s}\times\mathbf{s}\times\mathbf{r}e^{ik_0\mathbf{s}\cdot\mathbf{r}} = \nabla\times\nabla\times\mathbf{r}e^{ik_0\mathbf{s}\cdot\mathbf{r}}, \tag{11.38b}$$

to express the above angular-spectrum expansion of the EM field in the "standard form" of the Debye representation

$$\mathbf{E}_+(\mathbf{r}) = \nabla\times\mathbf{r}\Pi^{\rm d}_{\rm h}(\mathbf{r}) + \frac{i}{\omega\epsilon_0}\nabla\times\nabla\times\mathbf{r}\Pi^{\rm d}_{\rm e}(\mathbf{r}), \tag{11.39a}$$
$$\mathbf{H}_+(\mathbf{r}) = \nabla\times\mathbf{r}\Pi^{\rm d}_{\rm e}(\mathbf{r}) - \frac{i}{\omega\mu_0}\nabla\times\nabla\times\mathbf{r}\Pi^{\rm d}_{\rm h}(\mathbf{r}), \tag{11.39b}$$

where

$$\Pi^{\rm d}_{\rm e}(\mathbf{r}) = \frac{ik_0}{2\pi}\int_{-\pi}^{\pi}d\beta\int_{C_\pm}\sin\alpha\,d\alpha\,\hat{\Pi}^{\rm d}_{\rm e}(k_0\mathbf{s})e^{ik_0\mathbf{s}\cdot\mathbf{r}}, \tag{11.40a}$$
$$\Pi^{\rm d}_{\rm h}(\mathbf{r}) = \frac{ik_0}{2\pi}\int_{-\pi}^{\pi}d\beta\int_{C_\pm}\sin\alpha\,d\alpha\,\hat{\Pi}^{\rm d}_{\rm h}(k_0\mathbf{s})e^{ik_0\mathbf{s}\cdot\mathbf{r}}. \tag{11.40b}$$

As was the case for the Whittaker representation, the Debye representation as implemented via the angular-spectrum expansions will hold everywhere outside of the source strip but can be employed everywhere outside the *convex hull* of the source support volume τ_0 using appropriate coordinate transformations.

The Debye representation has a long and distinguished history within the field of electromagnetics. It was first obtained by Debye (Debye, 1909) in 1909 and was studied and employed by vast numbers of researchers in the following years. Its main use is in the derivation of the multipole expansion of radiated and scattered EM fields, which we will address in the following section.

Radiation patterns of the Debye potentials

We showed in Section 4.2.2 of Chapter 3 that the radiation pattern of an outgoing scalar wavefield is equal to its angular spectrum for $\mathbf{s}$ lying on the real unit sphere. We thus conclude that

$$\hat{\Pi}^{\rm d}_{\rm e}(k_0\mathbf{s}) = f^{\rm d}_{\rm e}(\mathbf{s}), \qquad \hat{\Pi}^{\rm d}_{\rm h}(k_0\mathbf{s}) = f^{\rm d}_{\rm h}(\mathbf{s}), \tag{11.41a}$$

where

$$\Pi^{\rm d}_{\rm e}(r\mathbf{s}) \equiv f^{\rm d}_{\rm e}(\mathbf{s})\frac{e^{ik_0r}}{r}, \qquad \Pi^{\rm d}_{\rm h}(r\mathbf{s}) \equiv f^{\rm d}_{\rm h}(\mathbf{s})\frac{e^{ik_0r}}{r}, \quad r\to\infty, \tag{11.41b}$$

with $f^{\rm d}_{\rm e}(\mathbf{s})$ and $f^{\rm d}_{\rm h}(\mathbf{s})$ being the radiation patterns of the electric and magnetic Debye potentials, respectively. It then follows that the angular spectra and radiation patterns of the EM fields can be represented in terms of the scalar radiation patterns of the Debye potentials via Eqs. (11.37); i.e.,

$$\mathbf{f}_{\rm e}(\mathbf{s}) = \mathbf{A}_{\rm e}(k_0\mathbf{s}) = -i\mathbf{L}f^{\rm d}_{\rm h}(\mathbf{s}) + \frac{ik_0}{\omega\epsilon_0}\mathbf{s}\times\mathbf{L}f^{\rm d}_{\rm e}(\mathbf{s}), \tag{11.42a}$$

$$\mathbf{f}_{\rm h}(\mathbf{s}) = \mathbf{A}_{\rm h}(k_0\mathbf{s}) = -i\mathbf{L}f^{\rm d}_{\rm e}(\mathbf{s}) - \frac{ik_0}{\omega\mu_0}\mathbf{s}\times\mathbf{L}f^{\rm d}_{\rm h}(\mathbf{s}). \tag{11.42b}$$

Example 11.5 Like the Whittaker representation, the Debye representation is not restricted to the EM field vectors and can be employed to decompose the transverse part of any vector field whose spatial Fourier transform satisfies the conditions Eqs. (11.5). In Example 11.3 we represented the transverse part of the current density $\mathbf{J}_{\rm T}(\mathbf{r})$ in the Whittaker representation and here we represent this quantity in the Debye representation. We begin by expressing the transform of $\mathbf{J}_{\rm T}$ in the form

$$\tilde{\mathbf{J}}_{\rm T}(\mathbf{K}) = -i\mathbf{L}\tilde{Q}^{\rm d}_{\rm h}(\mathbf{K}) + \frac{i}{\omega\epsilon_0}\mathbf{K}\times\mathbf{L}\tilde{Q}^{\rm d}_{\rm e}(\mathbf{K}), \tag{11.43}$$

where

$$\tilde{Q}^{\rm d}_{\rm h}(\mathbf{K}) = \int_{\tau_0} d^3r\, Q^{\rm d}_{\rm h}(\mathbf{r})e^{-i\mathbf{K}\cdot\mathbf{r}}, \qquad \tilde{Q}^{\rm d}_{\rm e}(\mathbf{K}) = \int_{\tau_0} d^3r\, Q^{\rm d}_{\rm e}(\mathbf{r})e^{-i\mathbf{K}\cdot\mathbf{r}},$$

are the spatial Fourier transforms of the magnetic and electric Debye source components of $\mathbf{J}_{\rm T}(\mathbf{r})$. On taking an inverse Fourier transform of both sides of the above equations and employing steps essentially identical to those used in transforming from Eq. (11.37a) to the standard form of the Debye representation of the electric field vector given in Eq. (11.39a) we find that

$$\mathbf{J}_{\mathrm{T}}(\mathbf{r}) = \nabla \times \mathbf{r}Q_{\mathrm{h}}^{\mathrm{d}}(\mathbf{r}) + \frac{i}{\omega\epsilon_0}\nabla \times \nabla \times \mathbf{r}Q_{\mathrm{e}}^{\mathrm{d}}(\mathbf{r}). \tag{11.44}$$

We can relate these two scalar sources to the angular spectra of the Debye potentials as well as to the radiation patterns of the EM field radiated by the current density $\mathbf{J}$. These relationships follow directly from Eq. (11.43) above and from Eqs. (11.22b), (11.23b) and (11.42). Following steps parallel to those used in Example 11.3, we find that

$$\tilde{Q}_{\mathrm{h}}^{\mathrm{d}}(k_0\mathbf{s}) = \frac{-4\pi i}{\omega\mu_0}\hat{\Pi}_{\mathrm{h}}^{\mathrm{d}}(k_0\mathbf{s}) = \frac{-4\pi i}{\omega\mu_0}f_{\mathrm{h}}^{\mathrm{d}}(\mathbf{s}), \tag{11.45a}$$

$$\tilde{Q}_{\mathrm{e}}^{\mathrm{d}}(k_0\mathbf{s}) = \frac{-4\pi i}{\omega\mu_0}\hat{\Pi}_{\mathrm{e}}^{\mathrm{d}}(k_0\mathbf{s}) = \frac{-4\pi i}{\omega\mu_0}f_{\mathrm{e}}^{\mathrm{d}}(\mathbf{s}). \tag{11.45b}$$

The Debye decomposition of $\mathbf{J}_{\mathrm{T}}$ and the above relationships between the scalar source transforms and the Debye potential radiation patterns will play an important role in our solution of the EM ISP for sources compactly supported within a spherical volume.

11.6.1 Multipole expansion of the radiated field

The multipole expansion of the EM field is obtained using the Debye representation and results from expanding the angular spectra of the two Debye potentials into series of spherical harmonics. In particular, we set

$$\hat{\Pi}_{\mathrm{h}}^{\mathrm{d}}(k_0\mathbf{s}) = -\sum_{l=1}^{\infty}\sum_{m=-l}^{l}(-i)^l q_{l,m}^{\mathrm{h}} Y_l^m(\mathbf{s}), \qquad \hat{\Pi}_{\mathrm{e}}^{\mathrm{d}}(k_0\mathbf{s}) = -\sum_{l=1}^{\infty}\sum_{m=-l}^{l}(-i)^l q_{l,m}^{\mathrm{e}} Y_l^m(\mathbf{s}), \tag{11.46a}$$

where the coefficients

$$q_{l,m}^{\mathrm{h}} = -i^l\int_{4\pi} d\Omega_s\, \hat{\Pi}_{\mathrm{h}}^{\mathrm{d}}(k_0\mathbf{s})Y_l^{m*}(\mathbf{s}), \qquad q_{l,m}^{\mathrm{e}} = -i^l\int_{4\pi} d\Omega_s\, \hat{\Pi}_{\mathrm{h}}^{\mathrm{e}}(k_0\mathbf{s})Y_l^{m*}(\mathbf{s}), \tag{11.46b}$$

are referred to, respectively, as the magnetic and electric multipole moments and the factors $-(-i)^l$ are employed for later algebraic simplicity. We note that, since the angular spectra $\hat{\Pi}_{\mathrm{h}}^{\mathrm{d}}(k_0\mathbf{s})$ and $\hat{\Pi}_{\mathrm{e}}^{\mathrm{d}}(k_0\mathbf{s})$ are, respectively, equal to the radiation patterns $f_{\mathrm{h}}^{\mathrm{d}}(\mathbf{s})$ and $f_{\mathrm{e}}^{\mathrm{d}}(\mathbf{s})$ (cf. Eqs. (11.41a)), Eqs. (11.46) also hold, with the angular spectra replaced by the radiation patterns.

On substituting Eqs. (11.46a) into the plane-wave expansions Eqs. (11.40) we obtain

$$\Pi_{\mathrm{e}}^{\mathrm{d}}(\mathbf{r}) = -ik\sum_{l=1}^{\infty}\sum_{m=-l}^{l} q_{l,m}^{\mathrm{e}}\left\{\frac{(-i)^l}{2\pi}\int_{-\pi}^{\pi}d\beta\int_{C_\pm}\sin\alpha\, d\alpha\, Y_l^m(\mathbf{s})e^{ik_0\mathbf{s}\cdot\mathbf{r}}\right\}$$

$$= -ik\sum_{l=1}^{\infty}\sum_{m=-l}^{l} q_{l,m}^{\mathrm{e}} h_l^+(k_0 r)Y_l^m(\hat{\mathbf{r}}), \tag{11.47a}$$

$$\Pi_{\mathrm{h}}^{\mathrm{d}}(\mathbf{r}) = -ik\sum_{l=1}^{\infty}\sum_{m=-l}^{l} q_{l,m}^{\mathrm{h}}\left\{\frac{(-i)^l}{2\pi}\int_{-\pi}^{\pi} d\beta \int_{C_\pm} \sin\alpha\, d\alpha\, Y_l^m(\mathbf{s})e^{ik_0\mathbf{s}\cdot\mathbf{r}}\right\}$$

$$= -ik\sum_{l=1}^{\infty}\sum_{m=-l}^{l} q_{l,m}^{\mathrm{h}} h_l^+(k_0r)Y_l^m(\hat{\mathbf{r}}), \tag{11.47b}$$

where $h_l^+(k_0r)$ are the spherical Hankel functions of the first kind and we have made use of the angular-spectrum expansion of the outgoing-wave scalar multipole fields

$$h_l^+(k_0r)Y_l^m(\hat{\mathbf{r}}) = \frac{(-i)^l}{2\pi}\int_{-\pi}^{\pi} d\beta \int_{C_\pm} \sin\alpha\, d\alpha\, Y_l^m(\mathbf{s})e^{ik_0\mathbf{s}\cdot\mathbf{r}} \tag{11.48}$$

derived in Section 3.4.2 of Chapter 3. The above multipole expansions of the two Debye potentials are identical in form to the multipole expansion of the scalar field developed in Chapter 4 and related to the angular-spectrum expansion of this field in Section 4.10 of that chapter. Like the multipole expansions of the scalar fields considered in Chapter 4, these expansions converge and represent the two Debye potentials everywhere outside the smallest sphere that contains the source support volume τ_0.

As a final step we substitute the above multipole expansions into the Debye representation Eqs. (11.39) to obtain

$$\mathbf{E}_+(\mathbf{r}) = \sum_{l=1}^{\infty}\sum_{m=-l}^{l} q_{l,m}^{\mathrm{h}}\mathbf{E}_{l,m}^{\mathrm{h}}(\mathbf{r}) + \sum_{l=1}^{\infty}\sum_{m=-l}^{l} q_{l,m}^{\mathrm{e}}\mathbf{E}_{l,m}^{\mathrm{e}}(\mathbf{r}), \tag{11.49a}$$

$$\mathbf{H}_+(\mathbf{r}) = \sum_{l=1}^{\infty}\sum_{m=-l}^{l} q_{l,m}^{\mathrm{e}}\mathbf{H}_{l,m}^{\mathrm{e}}(\mathbf{r}) + \sum_{l=1}^{\infty}\sum_{m=-l}^{l} q_{l,m}^{\mathrm{h}}\mathbf{H}_{l,m}^{\mathrm{h}}(\mathbf{r}), \tag{11.49b}$$

where

$$\mathbf{E}_{l,m}^{\mathrm{h}}(\mathbf{r}) = \mathbf{H}_{l,m}^{\mathrm{e}}(\mathbf{r}) = -ik\,\nabla\times\mathbf{r}h_l^+(k_0r)Y_l^m(\hat{\mathbf{r}}), \tag{11.50a}$$

$$\mathbf{E}_{l,m}^{\mathrm{e}}(\mathbf{r}) = -\frac{\mu_0}{\epsilon_0}\mathbf{H}_{l,m}^{\mathrm{h}}(\mathbf{r}) = \sqrt{\frac{\mu_0}{\epsilon_0}}\,\nabla\times\nabla\times\mathbf{r}h_l^+(k_0r)Y_l^m(\hat{\mathbf{r}}), \tag{11.50b}$$

and where the expansions Eqs. (11.49) converge outside the smallest sphere that contains the source support volume τ_0. We will refer to the pair $\mathbf{E}_{l,m}^{\mathrm{h}}(\mathbf{r})$, $\mathbf{H}_{l,m}^{\mathrm{h}}(\mathbf{r})$ as the transverse electric (TE) or magnetic multipole fields and the pair $\mathbf{E}_{l,m}^{\mathrm{e}}(\mathbf{r})$, $\mathbf{H}_{l,m}^{\mathrm{e}}(\mathbf{r})$ as the transverse magnetic (TM) or electric multipole fields. These fields admit angular-spectrum expansions

$$\mathbf{E}_{l,m}^{\mathrm{h}}(\mathbf{r}) = \mathbf{H}_{l,m}^{\mathrm{e}}(\mathbf{r}) = -\frac{(-i)^l k_0}{2\pi}\int_{-\pi}^{\pi} d\beta \int_{C_\pm} \sin\alpha\, d\alpha\, \mathbf{Y}_l^m(\mathbf{s})e^{ik_0\mathbf{s}\cdot\mathbf{r}}, \tag{11.51a}$$

$$\mathbf{E}^{\mathrm{e}}_{l,m}(\mathbf{r}) = -\frac{\mu_0}{\epsilon_0}\mathbf{H}^{\mathrm{h}}_{l,m}(\mathbf{r}) = \sqrt{\frac{\mu_0}{\epsilon_0}}\frac{(-i)^l k_0}{2\pi}\int_{-\pi}^{\pi} d\beta \int_{C_\pm} \sin\alpha\, d\alpha\, \mathbf{s}\times\mathbf{Y}^m_l(\mathbf{s})e^{ik_0\mathbf{s}\cdot\mathbf{r}}, \quad (11.51b)$$

where

$$\mathbf{Y}^m_l(\mathbf{s}) = \mathbf{L}Y^m_l(\mathbf{s}) \quad (11.52)$$

are the vector spherical harmonics first introduced in Chapter 3. The above two expansions are generalizations of the angular-spectrum representation of the scalar multipole field given in Eq. (11.48).

11.7 Vector spherical-harmonic expansion of the radiation pattern

We can obtain an expansion of the EM radiation patterns into the vector spherical harmonics by asymptotically expanding the multipole expansions of the two EM fields or, more simply, by substituting the expansions Eqs. (11.46a) of the angular spectra of the Debye potentials into Eqs. (11.42) for the EM radiation patterns. Using this approach we find that

$$\mathbf{f}_{\mathrm{e}}(\mathbf{s}) = i\sum_{l=1}^{\infty}\sum_{m=-l}^{l}(-i)^l q^{\mathrm{h}}_{l,m}\mathbf{Y}^m_l(\mathbf{s}) - \frac{ik_0}{\omega\epsilon_0}\sum_{l=1}^{\infty}\sum_{m=-l}^{l}(-i)^l q^{\mathrm{e}}_{l,m}\mathbf{s}\times\mathbf{Y}^m_l(\mathbf{s}), \quad (11.53a)$$

$$\mathbf{f}_{\mathrm{h}}(\mathbf{s}) = i\sum_{l=1}^{\infty}\sum_{m=-l}^{l}(-i)^l q^{\mathrm{e}}_{l,m}\mathbf{Y}^m_l(\mathbf{s}) + \frac{ik_0}{\omega\mu_0}\sum_{l=1}^{\infty}\sum_{m=-l}^{l}(-i)^l q^{\mathrm{h}}_{l,m}\mathbf{s}\times\mathbf{Y}^m_l(\mathbf{s}). \quad (11.53b)$$

The $\mathbf{Y}^m_l(\mathbf{s})$ and $\mathbf{s}\times\mathbf{Y}^m_l(\mathbf{s})$ are orthogonal with L^2 norm over the unit sphere of $\sqrt{l(l+1)}$, so the multipole moments can be computed by simply projecting the EM radiation patterns onto these two functions. We find that

$$q^{\mathrm{h}}_{l,m} = \frac{i^{l-1}}{l(l+1)}\int_{4\pi} d\Omega_s\, \mathbf{f}_{\mathrm{e}}(\mathbf{s})\cdot\mathbf{Y}^{m*}_l(\mathbf{s}), \quad (11.54a)$$

$$q^{\mathrm{e}}_{l,m} = -\frac{i^{l-1}\omega\epsilon_0}{l(l+1)k_0}\int_{4\pi} d\Omega_s\, \mathbf{f}_{\mathrm{e}}(\mathbf{s})\cdot\mathbf{s}\times\mathbf{Y}^{m*}_l(\mathbf{s}), \quad (11.54b)$$

with a similar result relating the multipole moments to the radiation pattern of the magnetic field.

The electric and magnetic multipole moments are represented in terms of the angular spectra and radiation patterns of the Debye potentials via Eqs. (11.46b) and in terms of the radiation patterns (and angular spectra) of the EM field vectors via Eqs. (11.54). We can also express the multipole moments in terms of the transverse part of the current density by substituting for $\mathbf{f}_{\mathrm{e}}(\mathbf{s})$ from Eq. (11.17a) into Eqs. (11.54). We obtain the result

$$q^{\mathrm{h}}_{l,m} = -\frac{\omega\mu_0 i^l}{4\pi l(l+1)}\int_{4\pi} d\Omega_s\, \tilde{\mathbf{J}}_{\mathrm{T}}(k_0\mathbf{s})\cdot\mathbf{Y}^{m*}_l(\mathbf{s}), \quad (11.55a)$$

$$q^{\mathrm{e}}_{l,m} = \frac{\omega\mu_0 i^l}{4\pi l(l+1)}\sqrt{\frac{\epsilon_0}{\mu_0}}\int_{4\pi} d\Omega_s\, \tilde{\mathbf{J}}_{\mathrm{T}}(k_0\mathbf{s})\cdot\mathbf{s}\times\mathbf{Y}^{m*}_l(\mathbf{s}). \quad (11.55b)$$

Field back propagation from the radiation patterns

The Debye representation of the EM fields implemented in the form of the multipole expansion forms the basis for properly posed forward- and back-propagation algorithms in spherical coordinate systems and, in particular, for back propagation from the radiation patterns. For example, for back propagation from the electric field radiation pattern the algorithm consists of first computing the multipole moments from Eqs. (11.54) and then using the resulting multipole moments in Eqs. (11.49) to compute the EM fields everywhere outside of the smallest sphere that completely encloses the source support volume τ_0. A similar computation using the magnetic field radiation pattern in place of $\mathbf{f}_e(\mathbf{s})$ can be employed. If required, the field can be further continued into the region between the bounding sphere and τ_0 using a free-field EM multipole expansion such as was discussed in Section 4.8.3 of Chapter 4 for the scalar-wave case.

11.8 The EM inverse source problem

We will restrict our attention to the EM analog of the *antenna-synthesis problem* formulated for scalar sources compactly supported in a spherical domain in Chapter 5, although the EM analog of the ISP for scalar sources confined to plane-parallel strips (parallel piped geometries) presented in Chapter 5 can also be easily treated. The reader may recall that the key to obtaining analytic solutions to the scalar-wave ISP was the use of eigenfunction expansions of solutions to the scalar Helmholtz equation and, likewise, the key to obtaining such solutions to the EM ISP is the use of eigenfunction expansions of solutions to the vector Helmholtz equation satisfied by the electric and magnetic field vectors. While the scalar Helmholtz equation is separable in 11 coordinate systems, leading to analytic solutions of the scalar wave ISP for 11 source geometries, the vector Helmholtz equation is separable in only two basic systems: spherical and Cartesian coordinates, corresponding to the EM ISP for spherical and parallel piped source geometries. Here, we will solve the EM ISP for spherical source geometries in terms of the radiation pattern of the electric or magnetic field vector and leave the problem for sources confined to plane-parallel strips in terms of the tangential electric or magnetic field vectors on two planar boundaries lying outside the source strip to the highly motivated reader (Problem 11.13 at the end of the chapter).

11.8.1 The EM ISP for sources supported in spherical regions

We will base our treatment on the radiation pattern $\mathbf{f}_e(\mathbf{s})$ of the electric field vector, with the understanding that a completely parallel treatment using the magnetic field radiation pattern can be employed. The key to our solution of the EM ISP is the observation that *the Debye representation maps the EM ISP into two uncoupled scalar inverse source problems that we can easily solve using the theory presented in Chapter 5!* In particular, the transverse current density $\mathbf{J}_T(\mathbf{r})$ is mapped into the two uncoupled scalar sources $Q_h^d(\mathbf{r})$

and $Q_e^d(\mathbf{r})$ via Example 11.5 and the EM radiation pattern $\mathbf{f}_e(\mathbf{s})$ is mapped into the two Debye radiation patterns $f_e(\mathbf{s})$ and $f_h(\mathbf{s})$ according to Eqs. (11.42). The EM ISP has thus been reduced to that of determining the scalar source Q_h^d from its corresponding radiation pattern $f_h(\mathbf{s})$ and determining the scalar source Q_e^d from its corresponding radiation pattern $f_e(\mathbf{s})$. The transverse current is then computed using Eq. (11.44) of Example 11.5.

The scalar sources are coupled to their radiation patterns by Eqs. (11.45) of Example 11.5. These two equations are identical in form to those relating the radiation patterns to the sources in our treatment of the scalar ISP in Chapter 5 and all the Hilbert-space machinery that we developed in that chapter applies here. Following the treatment presented in Chapter 5 we thus define the Hilbert space $\mathcal{H}_Q = L^2(\tau_0)$ of square-integrable scalar sources compactly supported within the spatial volume τ_0 and the Hilbert space $\mathcal{H}_f = L^2(\Omega)$ of square-integrable scalar radiation patterns over the unit sphere. The ISP then consists of inverting the mapping $\hat{T} : \mathcal{H}_Q \rightarrow \mathcal{H}_f$ defined by

$$\hat{T}Q = \frac{i}{\omega\mu_0} f, \tag{11.56a}$$

where

$$\hat{T} = -\frac{1}{4\pi}\int_{\tau_0} d^3r\, e^{-ik_0\mathbf{s}\cdot\mathbf{r}}, \tag{11.56b}$$

and where the radiation pattern f and source Q are either the pair $\{f_h(\mathbf{s}), Q_h^d(\mathbf{r})\}$ or the pair $\{f_e(\mathbf{s}), Q_e^d(\mathbf{r})\}$. The extra factor $i/(\omega\mu_0)$, which is not present in the pure scalar-wave case, is required in Eq. (11.56a) due to the presence of the multiplying factor $i\omega\mu_0$ to the current density in the vector Helmholtz equation Eq. (11.3a).

In this section we will address the EM version of the *antenna-synthesis problem* solved in Section 5.5 of Chapter 5. The EM version of this problem consists of determining the transverse current density $\mathbf{J}_T(\mathbf{r})$ whose support volume τ_0 is a sphere of radius a_0 centered on the origin and that radiates a specified radiation pattern $\mathbf{f}_e(\mathbf{s})$. As discussed in Chapter 5 this problem does not possess a unique solution due to the possible presence of non-radiating sources within τ_0 and might not even possess any solution since the data (radiation pattern) might not be in the range of the mapping Eq. (11.56a). We can obtain, however, a *least-squares pseudo-inverse* to the mapping, which yields the source $\hat{Q}(\mathbf{r})$ which is *that particular* least-squares solution to Eq. (11.56a) that possesses minimum L^2 norm in the Hilbert space $\mathcal{H}_Q$. The pseudo-inverse solution found in Section 5.5 translated to the EM case is given by

$$\hat{Q}(\mathbf{r},\omega) = \frac{i}{\omega\mu_0}\sum_{l=0}^{\infty}\sum_{m=-l}^{l}\frac{q_{lm}}{\sigma_l^2(k_0a_0)}j_l^*(k_0r)Y_l^m(\hat{\mathbf{r}}), \quad r < a_0, \tag{11.57a}$$

where $q_{l,m}$ are the EM field multipole moments and

$$\sigma_l^2(k_0a_0) = \int_0^{a_0} r^2\,dr|j_0(k_0r)|^2 \tag{11.57b}$$

are the squares of the singular values of the operator $\hat{T}$ computed in Section 5.5. The above results apply equally both to the electric, $Q_e^d(\mathbf{r})$, and to the magnetic, $Q_h^d(\mathbf{r})$, scalar components of the source (transverse current density). The field multipole moments that enter the above solution are given in terms of the angular spectra or radiation patterns of the Debye potentials in Eqs. (11.46b) or in terms of the EM radiation patterns by Eqs. (11.54).

The pseudo-inverse transverse current density $\hat{\mathbf{J}}_{\mathrm{T}}(\mathbf{r})$ is obtained by substituting the pseudo-inverse scalar sources given above into Eq. (11.44) of Example 11.5. We obtain

$$\mathbf{J}_{\mathrm{T}}(\mathbf{r}) = \frac{i}{\omega\mu_0}\sum_{l=0}^{\infty}\sum_{m=-l}^{l}\frac{q_{lm}^{\mathrm{h}}}{\sigma_l^2(k_0a_0)}\nabla\times\mathbf{r}j_l^*(k_0r)Y_l^m(\hat{\mathbf{r}})$$
$$-\frac{1}{k_0^2}\sum_{l=0}^{\infty}\sum_{m=-l}^{l}\frac{q_{lm}^{\mathrm{e}}}{\sigma_l^2(k_0a_0)}\nabla\times\nabla\times\mathbf{r}j_l^*(k_0r)Y_l^m(\hat{\mathbf{r}}),\qquad r<a_0.$$

Example 11.6 The above treatment allows a transverse current density to be computed from a specified EM radiation pattern. It is also possible to compute an EM radiation pattern and corresponding transverse current density from two pre-specified scalar radiation patterns and their associated scalar sources. As an example we consider the scalar wavelet field from Section 4.5 of Chapter 4 whose radiation pattern is given by

$$f(\mathbf{s}) = e^{k_0a\cos\theta},$$

where θ is the polar angle of the unit vector $\mathbf{s}$ and a is a positive constant parameter of the wavelet field. The generalized Fourier coefficients of this radiation pattern are found to be

$$f_{l,m} = -i^{-l}q_{l,m} = \int d\Omega_s f(\mathbf{s},\omega)Y_l^{m*}(\mathbf{s}) = \int_0^{\pi}\sin\theta\,d\theta\int_{-\pi}^{\pi}d\phi\,e^{k_0a\cos\theta}Y_l^{m*}(\theta,\phi),$$

where $q_{l,m}$ are the multipole moments. It is shown in Example 5.10 that the above integration can be performed, leading to

$$q_{l,m} = -i^lf_{l,m} = -\sqrt{4\pi(2l+1)}j_l(ik_0a).$$

The minimum-norm source that will radiate this scalar wavelet field is then given by Eq. (11.57a) with the $q_{l,m}$ given above:

$$\hat{Q}(\mathbf{r},\omega) = -\frac{\sqrt{4\pi}i}{\omega\mu_0}\sum_{l=0}^{\infty}\sum_{m=-l}^{l}\frac{\sqrt{2l+1}j_l(ik_0a)}{\sigma_l^2(k_0a_0)}j_l^*(k_0r)Y_l^m(\hat{\mathbf{r}}),\qquad r\le a_0,$$

where $\hat{Q}$ can be either the magnetic or the electric scalar source. On taking $\hat{Q}=\hat{Q}_{\mathrm{h}}^{\mathrm{d}}$ we then find that

$$\mathbf{J}_{\mathrm{T}}(\mathbf{r}) = -\frac{\sqrt{4\pi}i}{\omega\mu_0}\sum_{l=0}^{\infty}\sum_{m=-l}^{l}\frac{\sqrt{2l+1}j_l(ik_0a)}{\sigma_l^2(k_0a_0)}\nabla\times\mathbf{r}j_l^*(k_0r)Y_l^m(\hat{\mathbf{r}}),\qquad r\le a_0.$$

11.9 Electromagnetic scattering theory

We now consider a medium consisting of a uniform isotropic background with parameters ϵ_0 and μ_0, in which is embedded a localized inhomogeneity supported within the spatial region τ_0 and having parameters $\epsilon(\mathbf{r}) = \epsilon_0 + \delta\epsilon(\mathbf{r})$ and $\mu(\mathbf{r}) = \mu_0 + \delta\mu(\mathbf{r})$, where $\delta\epsilon$ and $\delta\mu$ both vanish outside of τ_0. The EM field within the composite medium satisfies the Maxwell equations Eqs. (11.2), which can be expressed in the form

$$\epsilon_0 \nabla \cdot \mathbf{E}(\mathbf{r}) = -\nabla \cdot [\delta\epsilon(\mathbf{r})\mathbf{E}(\mathbf{r})] + \rho(\mathbf{r}),$$

$$\mu_0 \nabla \cdot \mathbf{H}(\mathbf{r}) = -\nabla \cdot [\delta\mu(\mathbf{r})\mathbf{H}(\mathbf{r})],$$

$$\nabla \times \mathbf{E}(\mathbf{r}) = i\omega\mu_0\mathbf{H}(\mathbf{r}) + i\omega\,\delta\mu(\mathbf{r})\mathbf{H}(\mathbf{r}),$$

$$\nabla \times \mathbf{H}(\mathbf{r}) = -i\omega\epsilon_0\mathbf{E}(\mathbf{r}) - i\omega\,\delta\epsilon(\mathbf{r})\mathbf{E}(\mathbf{r}) + \mathbf{J}(\mathbf{r}).$$

The above set of equations can be interpreted as describing radiation and scattering in a homogeneous isotropic background medium having parameters ϵ_0 and μ_0 in which are embedded the usual primary electric charge and current densities ρ and $\mathbf{J}$ as well as induced electric *and* magnetic charge and current densities

$$\rho_e(\mathbf{r}) = -\nabla \cdot [\delta\epsilon(\mathbf{r})\mathbf{E}(\mathbf{r})], \qquad \mathbf{J}_e(\mathbf{r}) = -i\omega\,\delta\epsilon(\mathbf{r})\mathbf{E}(\mathbf{r}), \tag{11.58a}$$

$$\rho_h(\mathbf{r}) = -\nabla \cdot [\delta\mu(\mathbf{r})\mathbf{H}(\mathbf{r})], \qquad \mathbf{J}_h(\mathbf{r}) = -i\omega\,\delta\mu(\mathbf{r})\mathbf{H}(\mathbf{r}). \tag{11.58b}$$

It is easily verified that the induced charge–current distributions satisfy the charge–current conservation equation Eq. (11.1).

The primary charge–current distribution $\rho, \mathbf{J}$ radiates an incident EM wavefield $(\mathbf{E}^{(\text{in})}, \mathbf{H}^{(\text{in})})$ that propagates in the background medium and that then interacts with the localized inhomogeneity inducing the above charge–current distributions. These induced sources then re-radiate a *scattered* field $(\mathbf{E}^{(\text{s})}, \mathbf{H}^{(\text{s})})$ that adds to the incident field, creating the net overall electric and magnetic field vectors

$$\mathbf{E}(\mathbf{r}) = \mathbf{E}^{(\text{in})}(\mathbf{r}) + \mathbf{E}^{(\text{s})}(\mathbf{r}), \qquad \mathbf{H}(\mathbf{r}) = \mathbf{H}^{(\text{in})}(\mathbf{r}) + \mathbf{H}^{(\text{s})}(\mathbf{r}).$$

The incident field is radiated by the primary charge–current distribution in the infinite homogeneous isotropic background medium and, hence, obeys the set of Maxwell equations Eqs. (11.2) and the vector Helmholtz equations Eqs. (11.3). By subtracting Eqs. (11.2) with $\mathbf{E} = \mathbf{E}^{(\text{in})}$ and $\mathbf{H} = \mathbf{H}^{(\text{in})}$ from the above set of Maxwell equations we find that the scattered fields satisfy the set of equations

$$\epsilon_0 \nabla \cdot \mathbf{E}^{(\text{s})}(\mathbf{r}) = \rho_e(\mathbf{r}),$$

$$\mu_0 \nabla \cdot \mathbf{H}^{(\text{s})}(\mathbf{r}) = \rho_h(\mathbf{r}),$$

$$\nabla \times \mathbf{E}^{(\text{s})}(\mathbf{r}) = i\omega\mu_0\mathbf{H}^{(\text{s})}(\mathbf{r}) - \mathbf{J}_h(\mathbf{r}),$$

$$\nabla \times \mathbf{H}^{(\text{s})}(\mathbf{r}) = -i\omega\epsilon_0\mathbf{E}^{(\text{s})}(\mathbf{r}) + \mathbf{J}_e(\mathbf{r}),$$

where the fields $\mathbf{E}$ and $\mathbf{H}$ appearing in the induced sources defined in Eqs. (11.58) are the total fields (incident plus scattered). If we then follow steps identical to those employed in deriving the vector Helmholtz equations Eqs. (11.3) we find that the above set of equations reduces to the vector Helmholtz equations

$$\nabla \times \nabla \times \mathbf{E}^{(s)}(\mathbf{r}) - k_0^2 \mathbf{E}^{(s)}(\mathbf{r}) = i\omega\mu_0 \mathbf{J}_e(\mathbf{r}) - \nabla \times \mathbf{J}_h(\mathbf{r}),$$
$$\nabla \times \nabla \times \mathbf{H}^{(s)}(\mathbf{r}) - k_0^2 \mathbf{H}^{(s)}(\mathbf{r}) = i\omega\epsilon_0 \mathbf{J}_h(\mathbf{r}) + \nabla \times \mathbf{J}_e(\mathbf{r}),$$

where $k_0 = \omega\sqrt{\epsilon_0\mu_0}$.

We can convert the above set of vector Helmholtz equations to scalar Helmholtz equations by following a procedure identical to that employed in our treatment of the EM radiation problem in Section 11.3. We obtain

$$[\nabla^2 + k_0^2]\mathbf{E}^{(s)}(\mathbf{r}) = -i\omega\mu_0 \mathbf{J}_e(\mathbf{r}) + \frac{1}{\epsilon_0}\nabla\rho_e(\mathbf{r}) + \nabla \times \mathbf{J}_h(\mathbf{r}), \tag{11.59a}$$

$$[\nabla^2 + k_0^2]\mathbf{H}^{(s)}(\mathbf{r}) = -i\omega\epsilon_0 \mathbf{J}_h(\mathbf{r}) + \frac{1}{\mu_0}\nabla\rho_h(\mathbf{r}) - \nabla \times \mathbf{J}_e(\mathbf{r}). \tag{11.59b}$$

11.9.1 The Lippmann–Schwinger equations

As in our treatment of scalar-wave scattering theory in Chapter 6, we will later have to deal with suites of scattering experiments employing different incident waves, which we will label with the parameter ν. The incident, scattered and total (incident plus scattered) waves will thus be represented as $\mathbf{E}^{(in)}(\mathbf{r}, \nu)$ and $\mathbf{E}^{(s)}(\mathbf{r}, \nu)$, and we have

$$\mathbf{E}(\mathbf{r}, \nu) = \mathbf{E}^{(in)}(\mathbf{r}, \nu) + \mathbf{E}^{(s)}(\mathbf{r}, \nu).$$

The induced charge–current distributions will also, of course, depend on the parameter ν, which will be included in their arguments to emphasize this dependence. The scattered fields satisfy the SRC so that the outgoing-wave solutions to Eqs. (11.59) for the scattered EM fields are identical in form to the solutions to the radiation problem found in Section 11.3. We then find that the total EM fields are given by

$$\mathbf{E}(\mathbf{r}, \nu) = \mathbf{E}^{(in)}(\mathbf{r}, \nu) - \int_{\tau_0} d^3r' \Big[i\omega\mu_0 \mathbf{J}_e(\mathbf{r}', \nu) - \frac{1}{\epsilon_0}\nabla\rho_e(\mathbf{r}', \nu) - \nabla \times \mathbf{J}_h(\mathbf{r}', \nu)\Big] G_+(\mathbf{r} - \mathbf{r}'), \tag{11.60a}$$

$$\mathbf{H}(\mathbf{r}, \nu) = \mathbf{H}^{(in)}(\mathbf{r}, \nu) - \int_{\tau_0} d^3r' \Big[i\omega\epsilon_0 \mathbf{J}_h(\mathbf{r}', \nu) - \frac{1}{\mu_0}\nabla\rho_h(\mathbf{r}', \nu) + \nabla \times \mathbf{J}_e(\mathbf{r}', \nu)\Big] G_+(\mathbf{r} - \mathbf{r}'), \tag{11.60b}$$

where τ_0 is the scattering volume and the second terms in the above two equations are the scattered fields.

Equations (11.60) are not *solutions* to the EM scattering problem since the induced sources depend on the total fields **E** and **H**. They are, in fact, EM versions of the Lippmann–Schwinger integral equation governing potential scattering of scalar waves in Chapter 6. As in the scalar-wave case, the above integral equations cannot, in general, be solved analytically and usually require numerical solutions that will not be pursued here. However, as was the case in scalar-wave scattering theory presented in Chapter 6, they form the basis of EM potential scattering theory and will be employed in the following to develop much of this theory.

11.9.2 Electromagnetic scattering amplitudes

The EM scattering amplitudes are straightforward generalizations of the scalar wave scattering amplitudes defined in Chapter 6. As in that chapter, we define the "generalized" scattering amplitudes to be the radiation patterns of the scattered EM fields resulting from an arbitrary fixed incident wave. The scattering amplitudes are formally identical to the radiation patterns found in Section 11.3, which yield the results

$$\mathbf{f}_{\mathrm{e}}(\mathbf{s},\nu)=\frac{i\omega\mu_0}{4\pi}\tilde{\mathbf{J}}_{\mathrm{eT}}(k_0\mathbf{s},\nu)-\frac{ik_0}{4\pi}\mathbf{s}\times\tilde{\mathbf{J}}_{\mathrm{hT}}(k_0\mathbf{s},\nu),\tag{11.61a}$$

$$\mathbf{f}_{\mathrm{h}}(\mathbf{s},\nu)=\frac{i\omega\epsilon_0}{4\pi}\tilde{\mathbf{J}}_{\mathrm{hT}}(k_0\mathbf{s},\nu)+\frac{ik_0}{4\pi}\mathbf{s}\times\tilde{\mathbf{J}}_{\mathrm{eT}}(k_0\mathbf{s},\nu),\tag{11.61b}$$

where

$$\tilde{\mathbf{J}}_{\mathrm{eT}}(k_0\mathbf{s},\nu)=\tilde{\mathbf{J}}_{\mathrm{e}}(k_0\mathbf{s},\nu)-\frac{\omega\mathbf{s}}{k_0}\tilde{\rho}_{\mathrm{e}}(\mathbf{k}_0\mathbf{s},\nu)=-\mathbf{s}\times\mathbf{s}\times\tilde{\mathbf{J}}_{\mathrm{e}}(k_0\mathbf{s},\nu),$$

$$\tilde{\mathbf{J}}_{\mathrm{hT}}(k_0\mathbf{s},\nu)=\tilde{\mathbf{J}}_{\mathrm{h}}(k_0\mathbf{s},\nu)-\frac{\omega\mathbf{s}}{k_0}\tilde{\rho}_{\mathrm{h}}(\mathbf{k}_0\mathbf{s},\nu)=-\mathbf{s}\times\mathbf{s}\times\tilde{\mathbf{J}}_{\mathrm{h}}(k_0\mathbf{s},\nu)$$

are the spatial Fourier transforms of the transverse parts of the induced current densities on the sphere $\mathbf{K}=k_0\mathbf{s}$. It is easily verified that the EM scattering amplitudes are related via Eqs. (11.18).

11.9.3 angular-spectrum expansions

For a given incident wave the EM field scattered by a localized inhomogeneity is formally identical to the EM field radiated by a primary source and thus admits angular-spectrum expansions such as given in Section 11.4. In particular, for a scatterer localized to the strip $z^-<z<z^+$ we find that

$$\mathbf{E}^{(\mathrm{s})}(\mathbf{r},\nu)=\frac{ik_0}{2\pi}\int_{-\pi}^{\pi}d\beta\int_{C\pm}\sin\alpha\,d\alpha\,\mathbf{A}_{\mathrm{e}}(k_0\mathbf{s},\nu)e^{ik_0\mathbf{s}\cdot\mathbf{r}},$$

$$\mathbf{H}^{(\mathrm{s})}(\mathbf{r},\nu)=\frac{ik_0}{2\pi}\int_{-\pi}^{\pi}d\beta\int_{C\pm}\sin\alpha\,d\alpha\,\mathbf{A}_{\mathrm{h}}(k_0\mathbf{s},\nu)e^{ik_0\mathbf{s}\cdot\mathbf{r}},$$

where C_+ is used in $z > z^+$ and C_- in $z < z^-$, and where the plane-wave amplitudes (angular spectra) $\mathbf{A}_e(k_0\mathbf{s}, \nu)$ and $\mathbf{A}_h(k_0\mathbf{s}, \nu)$ are analytic continuations of the scattering amplitudes $\mathbf{f}_e(\mathbf{s}, \nu)$ and $\mathbf{f}_h(\mathbf{s}, \nu)$ onto the contours $C_\pm$ as was the case for the field radiated by a primary source discussed in Section 11.4. The expansions converge and represent the scattered fields everywhere outside the strip $z^- < z < z^+$.

The Cartesian-variable form of the angular-spectrum expansions of the scattered EM fields are found by extension of the Cartesian-variable form of the expansions of the radiated field given in Section 11.4. Again, for a scatterer localized to the strip $z^- < z < z^+$ we find that

$$\mathbf{E}^{(s)}(\mathbf{r}, \nu) = \frac{i}{2\pi} \int_{-\infty}^{\infty} \frac{d^2 K_\rho}{\gamma} \mathbf{A}_e(\mathbf{k}_0^\pm, \nu) e^{i\mathbf{k}_0^\pm \cdot \mathbf{r}},$$

$$\mathbf{H}^{(s)}(\mathbf{r}, \nu) = \frac{i}{2\pi} \int_{-\infty}^{\infty} \frac{d^2 K_\rho}{\gamma} \mathbf{A}_h(\mathbf{k}_0^\pm, \nu) e^{i\mathbf{k}_0^\pm \cdot \mathbf{r}},$$

where $\mathbf{A}_e(\mathbf{k}_0^\pm, \nu)$ and $\mathbf{A}_h(\mathbf{k}_0^\pm, \nu)$ are, respectively, $\mathbf{A}_e(k_0\mathbf{s}, \nu)$ and $\mathbf{A}_e(k_0\mathbf{s}, \nu)$ under the transformation

$$k_0\mathbf{s} \Rightarrow \mathbf{k}_0^\pm = \mathbf{K}_\rho \pm \gamma \hat{\mathbf{z}},$$

and are thus given by

$$\mathbf{A}_e(\mathbf{k}_0^\pm, \nu) = \frac{i\omega\mu_0}{4\pi} \tilde{\mathbf{J}}_{eT}(\mathbf{k}_0^\pm, \nu) - \frac{i}{4\pi} \mathbf{k}_0^\pm \times \tilde{\mathbf{J}}_{hT}(\mathbf{k}_0^\pm, \nu),$$

$$\mathbf{A}_h(\mathbf{k}_0^\pm, \nu) = \frac{i\omega\epsilon_0}{4\pi} \tilde{\mathbf{J}}_{hT}(\mathbf{k}_0^\pm, \nu) + \frac{i}{4\pi} \mathbf{k}_0^\pm \times \tilde{\mathbf{J}}_{eT}(\mathbf{k}_0^\pm, \nu).$$

As in the case of the angular form of the angular-spectrum expansions the Cartesian forms require that the z coordinate of the field point $\mathbf{r} = \boldsymbol{\rho} + z\hat{\mathbf{z}}$ lie outside of the scatterer strip $z^- \le z \le z^+$, with $\mathbf{k}_0^+$ used in the expansions if $z > z^+$ and $\mathbf{k}_0^-$ if $z < z^-$.

11.10 The Born approximation

The Born approximation results from performing a perturbation expansion of the scattered fields in Eqs. (11.60) in the strength of the perturbations $\delta\epsilon$ and $\delta\mu$ of the material properties of the scatterer and retaining only the lowest order in the expansion. Following lines similar to those employed in Section 6.7, we replace $\delta\epsilon$ and $\delta\mu$ by $\eta\,\delta\epsilon$ and $\eta\,\delta\mu$ and expand the EM scattered fields via

$$\mathbf{E}^{(s)}(\mathbf{r}, \nu) = \eta \mathbf{E}_B^{(s)}(\mathbf{r}, \nu) + O(\eta^2), \qquad \mathbf{H}^{(s)}(\mathbf{r}, \nu) = \eta \mathbf{H}_B^{(s)}(\mathbf{r}, \nu) + O(\eta^2),$$

where the subscript "B" stands for the Born approximation and the perturbation parameter η will eventually be set equal to unity. The Born approximation to the induced charge–current distributions in Eqs. (11.58) results from expanding these quantities in η and retaining only the terms of lowest order. We find that

$$\rho_e^B(\mathbf{r}, \nu) = -\nabla \cdot [\delta\epsilon(\mathbf{r}) \mathbf{E}^{(in)}(\mathbf{r}, \nu)], \qquad \mathbf{J}_e^B(\mathbf{r}, \nu) = -i\omega\,\delta\epsilon(\mathbf{r}) \mathbf{E}^{(in)}(\mathbf{r}, \nu),$$

$$\rho_h^B(\mathbf{r}, \nu) = -\nabla \cdot [\delta\mu(\mathbf{r}) \mathbf{H}^{(in)}(\mathbf{r}, \nu)], \qquad \mathbf{J}_h^B(\mathbf{r}, \nu) = -i\omega\,\delta\mu(\mathbf{r}) \mathbf{H}^{(in)}(\mathbf{r}).$$

Substituting these expressions into the scattered fields in Eqs. (11.60) we then obtain the Born approximations

$$\mathbf{E}^{(s)}_{\mathrm{B}}(\mathbf{r},\nu) = -\int_{\tau_0} d^3r' \left[i\omega\mu_0 \mathbf{J}^{\mathrm{B}}_{\mathrm{e}}(\mathbf{r}',\nu) - \frac{1}{\epsilon_0}\nabla\rho^{\mathrm{B}}_{\mathrm{e}}(\mathbf{r}',\nu) - \nabla\times\mathbf{J}^{\mathrm{B}}_{\mathrm{h}}(\mathbf{r}',\nu)\right] G_+(\mathbf{r}-\mathbf{r}'),$$

$$\mathbf{H}^{(s)}_{\mathrm{B}}(\mathbf{r},\nu) = -\int_{\tau_0} d^3r' \left[i\omega\epsilon_0 \mathbf{J}^{B}_{\mathrm{h}}(\mathbf{r}',\nu) - \frac{1}{\mu_0}\nabla\rho^{\mathrm{B}}_{\mathrm{h}}(\mathbf{r}',\nu) + \nabla\times\mathbf{J}^{B}_{\mathrm{e}}(\mathbf{r}',\nu)\right] G_+(\mathbf{r}-\mathbf{r}').$$

Unlike the EM Lippmann–Schwinger equations Eq. (11.60), the Born approximation to the induced sources does not depend on the scattered field, so the above two equations are the actual solutions for the EM fields within the Born approximation. These solutions are seen to depend linearly both on the incident EM field and on the material perturbations $\delta\epsilon$ and $\delta\mu$. For these reasons, the Born approximation is ideally suited to the inverse scattering problem (ISCP) which we will treat in the following section.

11.10.1 Born scattering amplitudes

The scattering amplitudes within the Born approximation are obtained by substituting the Born induced currents into Eqs. (11.61). We find that

$$\mathbf{f}^{\mathrm{B}}_{\mathrm{e}}(\mathbf{s},\nu) = -\frac{k_0^2}{4\pi\epsilon_0}\mathbf{s}\times\mathbf{s}\times\widetilde{\delta\epsilon\,\mathbf{E}^{(\mathrm{in})}}(k_0\mathbf{s}) - \frac{\omega k_0}{4\pi}\mathbf{s}\times\widetilde{\delta\mu\,\mathbf{H}^{(\mathrm{in})}}(k_0\mathbf{s}), \quad (11.62a)$$

$$\mathbf{f}^{\mathrm{B}}_{\mathrm{h}}(\mathbf{s},\nu) = \frac{k_0\omega}{4\pi}\mathbf{s}\times\widetilde{\delta\epsilon\,\mathbf{E}^{(\mathrm{in})}}(k_0\mathbf{s}) - \frac{k_0^2}{4\pi\mu_0}\mathbf{s}\times\mathbf{s}\times\widetilde{\delta\mu\,\mathbf{H}^{(\mathrm{in})}}(k_0\mathbf{s}), \quad (11.62b)$$

where $\widetilde{\delta\epsilon\,\mathbf{E}^{(\mathrm{in})}}$ and $\widetilde{\delta\mu\,\mathbf{H}^{(\mathrm{in})}}$ denote the spatial Fourier transforms of the products of $\delta\epsilon$ and $\mathbf{E}^{(\mathrm{in})}$ and of $\delta\mu$ and $\mathbf{H}^{(\mathrm{in})}$, respectively.

The "classical" scattering amplitudes result from incident plane waves propagating in a fixed direction. In the EM case there are two such plane waves having orthogonal polarization states for each unit propagation vector $\mathbf{s}_0$, corresponding to so-called transverse electric (TE) or transverse magnetic (TM) polarization states, which we can represent in the form

$$\mathbf{E}^{(\mathrm{in})}_{\mathrm{TE}}(\mathbf{r},\mathbf{s}_0) = \mathbf{A}_0(\mathbf{s}_0)e^{ik_0\mathbf{s}_0\cdot\mathbf{r}}, \qquad \mathbf{H}^{(\mathrm{in})}_{\mathrm{TE}}(\mathbf{r},\mathbf{s}_0) = \frac{k_0}{\omega\mu_0}\mathbf{s}_0\times\mathbf{A}_0(\mathbf{s}_0)e^{ik_0\mathbf{s}_0\cdot\mathbf{r}},$$

$$\mathbf{H}^{(\mathrm{in})}_{\mathrm{TM}}(\mathbf{r},\mathbf{s}_0) = \mathbf{A}_0(\mathbf{s}_0)e^{ik_0\mathbf{s}_0\cdot\mathbf{r}}, \qquad \mathbf{E}^{(\mathrm{in})}_{\mathrm{TM}}(\mathbf{r},\mathbf{s}_0) = -\frac{k_0}{\omega\epsilon_0}\mathbf{s}_0\times\mathbf{A}_0(\mathbf{s}_0)e^{ik_0\mathbf{s}_0\cdot\mathbf{r}},$$

where $\mathbf{A}_0(\mathbf{s}_0)$ is a constant vector orthogonal to the unit propagation vector $\mathbf{s}_0$ ($\mathbf{s}_0\cdot\mathbf{A}_0(\mathbf{s}_0) = 0$) and we have set $\nu = \mathbf{s}_0$. The TE and TM plane waves each separately satisfy the homogeneous Maxwell equations in the background medium.

Since the two scattering amplitudes $\mathbf{f}_e^B$ and $\mathbf{f}_h^B$ are connected via Eqs. (11.18) we need only consider one of the two for any given scattering experiment. Here we will restrict our attention to the electric field scattering amplitude for a TE incident plane wave and the magnetic field scattering amplitude for a TM incident plane wave. On making use of the above incident plane waves in Eqs. (11.62) we then find after a bit of algebra that

$$\begin{aligned}\mathbf{f}_e^{TE}(\mathbf{s}, \mathbf{s}_0) &= -\frac{k_0^2}{4\pi\epsilon_0}[(\mathbf{s}\cdot\mathbf{A}_0(\mathbf{s}_0))\mathbf{s} - \mathbf{A}_0(\mathbf{s}_0)]\widetilde{\delta\epsilon}[k_0(\mathbf{s}-\mathbf{s}_0)] \\ &\quad - \frac{k_0^2}{4\pi\mu_0}[(\mathbf{s}\cdot\mathbf{A}_0(\mathbf{s}_0))\mathbf{s}_0 - (\mathbf{s}\cdot\mathbf{s}_0)\mathbf{A}_0(\mathbf{s}_0)]\widetilde{\delta\mu}[k_0(\mathbf{s}-\mathbf{s}_0)], \qquad (11.63a)\end{aligned}$$

where we have dropped the superscript "B" for notational convenience and the tilde denotes the spatial Fourier transform. In a similar fashion we find the magnetic field scattering amplitude for a TM incident plane wave to be given by

$$\begin{aligned}\mathbf{f}_h^{TM}(\mathbf{s}, \mathbf{s}_0) &= -\frac{k_0^2}{4\pi\epsilon_0}[(\mathbf{s}\cdot\mathbf{A}_0(\mathbf{s}_0))\mathbf{s}_0 - (\mathbf{s}\cdot\mathbf{s}_0)\mathbf{A}_0(\mathbf{s}_0)]\widetilde{\delta\epsilon}[k_0(\mathbf{s}-\mathbf{s}_0)] \\ &\quad - \frac{k_0^2}{4\pi\mu_0}[(\mathbf{s}\cdot\mathbf{A}_0(\mathbf{s}_0))\mathbf{s} - \mathbf{A}_0(\mathbf{s}_0)]\widetilde{\delta\mu}[k_0(\mathbf{s}-\mathbf{s}_0)]. \qquad (11.63b)\end{aligned}$$

11.10.2 Born inverse scattering

We will restrict our attention to the ISCP using far-field data in the form of the Born approximations to the TE electric field scattering amplitude and the TM magnetic field scattering amplitude given in Eqs. (11.63). Scattered field data over spherical surfaces can be converted to these scattering amplitudes using the EM multipole expansion while scattered field data over plane surfaces can be converted to the scattering amplitudes using the EM angular-spectrum expansions.

It can be seen from Eqs. (11.63) that the two scattering amplitudes are linear combinations of the spatial Fourier transforms of the perturbations of the material properties $\delta\epsilon(\mathbf{r})$ and $\delta\mu(\mathbf{r})$ of the medium evaluated over the spatial frequencies $\mathbf{K} = k_0(\mathbf{s} - \mathbf{s}_0)$. These are the same sets of spatial frequencies as those we first encountered in Chapter 6 in our treatment of scalar-wave scattering theory within the Born approximation. For fixed incident-wave direction $\mathbf{s}_0$ and scattered-field vector $\mathbf{s}$ covering the unit sphere the spatial frequency $\mathbf{K} = k_0(\mathbf{s} - \mathbf{s}_0)$ covers the so-called *Ewald sphere* illustrated in Fig. 6.3. If then $\mathbf{s}_0$ is allowed to also span the unit sphere, the resulting set of Ewald spheres will fill Fourier space within a sphere of radius $2k_0$ also illustrated in Fig. 6.3 and called the *Ewald limiting sphere*. By isolating $\widetilde{\delta\epsilon}[k_0(\mathbf{s} - \mathbf{s}_0)]$ and $\widetilde{\delta\mu}[k_0(\mathbf{s} - \mathbf{s}_0)]$ from Eqs. (11.63) we can thus determine these transforms throughout the Ewald limiting sphere from a complete set of scattering experiments and over a finite set of Ewald spheres from a finite set of experiments. The material parameters can then be determined from these transforms using the Born inverse scattering algorithms already developed in Chapters 8 and 9.

There are a number of ways to proceed in isolating the two transforms and we will consider the simplest, which is to take the dot product of both scattering amplitudes with the unit propagation vector $\mathbf{s}_0$ of the incident plane waves. We then obtain the equations

$$\mathbf{s}_0 \cdot \mathbf{f}_{\mathrm{e}}^{\mathrm{TE}}(\mathbf{s}, \mathbf{s}_0) = -\frac{k_0^2}{4\pi}\left[\mathbf{s} \cdot \mathbf{s}_0 \frac{\widetilde{\delta\epsilon}[k_0(\mathbf{s}-\mathbf{s}_0)]}{\epsilon_0} + \frac{\widetilde{\delta\mu}[k_0(\mathbf{s}-\mathbf{s}_0)]}{\mu_0}\right] \mathbf{s} \cdot \mathbf{A}_0(\mathbf{s}_0),$$

$$\mathbf{s}_0 \cdot \mathbf{f}_{\mathrm{h}}^{\mathrm{TM}}(\mathbf{s}, \mathbf{s}_0) = -\frac{k_0^2}{4\pi}\left[\frac{\widetilde{\delta\epsilon}[k_0(\mathbf{s}-\mathbf{s}_0)]}{\epsilon_0} + \mathbf{s} \cdot \mathbf{s}_0 \frac{\widetilde{\delta\mu}[k_0(\mathbf{s}-\mathbf{s}_0)]}{\mu_0}\right] \mathbf{s} \cdot \mathbf{A}_0(\mathbf{s}_0),$$

which are immediately solved to yield

$$\widetilde{\delta\epsilon}[k_0(\mathbf{s}-\mathbf{s}_0)] = -\frac{4\pi\epsilon_0}{k_0^2} \frac{\mathbf{s}_0 \cdot \mathbf{f}_{\mathrm{h}}^{\mathrm{TM}}(\mathbf{s}, \mathbf{s}_0) - (\mathbf{s} \cdot \mathbf{s}_0)\mathbf{s}_0 \cdot \mathbf{f}_{\mathrm{e}}^{\mathrm{TE}}(\mathbf{s}, \mathbf{s}_0)}{[1 - (\mathbf{s} \cdot \mathbf{s}_0)^2]\mathbf{s} \cdot \mathbf{A}_0(\mathbf{s}_0)}, \tag{11.64a}$$

$$\widetilde{\delta\mu}[k_0(\mathbf{s}-\mathbf{s}_0)] = -\frac{4\pi\mu_0}{k_0^2} \frac{\mathbf{s}_0 \cdot \mathbf{f}_{\mathrm{e}}^{\mathrm{TE}}(\mathbf{s}, \mathbf{s}_0) - (\mathbf{s} \cdot \mathbf{s}_0)\mathbf{s}_0 \cdot \mathbf{f}_{\mathrm{h}}^{\mathrm{TM}}(\mathbf{s}, \mathbf{s}_0)}{[1 - (\mathbf{s} \cdot \mathbf{s}_0)^2]\mathbf{s} \cdot \mathbf{A}_0(\mathbf{s}_0)}. \tag{11.64b}$$

Once the transforms of $\delta\epsilon$ and $\delta\mu$ have been determined from the scattering amplitudes via Eqs. (11.64) the material perturbations can be estimated either using the filtered back-propagation algorithm developed in Chapter 8 or, in the limited-data case, via the ISCP algorithm developed in Chapter 9.

Further reading

An excellent modern treatment of both direct and inverse EM problems in inhomogeneous backgrounds is given in Chew's book (Chew, 1990). The inverse source problem for EM sources and fields is treated in Marengo and Devaney (1999) and for so-called EM wavelet sources in Devaney *et al.* (2008). The book by Colton and Kress (Colton and Kress, 1992) also treats inverse EM problems, especially ones involving surface scatterers (see also the references listed at the end of Chapter 7). The classical texts of general EM theory include those by Morse and Feshbach (Morse and Feshbach, 1953), Stratton (Stratton, 1941), Jackson (Jackson, 1998) and Born and Wolf (Born and Wolf, 1999) and the excellent treatise by Muller (Muller, 1969). The little book by Papas (Papas, 1988) has an excellent treatment of multipole expansions and dyadic Green functions.

Problems

11.1 Derive Eqs. (11.15) using Green-function techniques.

11.2 Compute the radiation pattern of the Green-function dyadic.

11.3 Derive Eqs. (11.23).

11.4 Derive Eqs. (11.26) from Eqs. (11.22b) and (11.23b).

11.5 Prove the following identities:

$$\mathbf{L} \cdot [\mathbf{s} \times \mathbf{L}] = [\mathbf{s} \times \mathbf{L}] \cdot \mathbf{L} = 0, \qquad [\mathbf{s} \times \mathbf{L}] \cdot [\mathbf{s} \times \mathbf{L}] = L^2.$$

11.6 Derive Eqs. (11.30) from Eqs. (11.29).

11.7 Prove that an EM non-radiating (NR) source must have zero total charge.

11.8 Derive a general expression for an NR EM source supported within a spherical region.

11.9 Derive the most general form of an EM surface source supported over a plane surface.

11.10 Determine the relationship between the components of the surface source found in the previous problem for it to be NR throughout one of the two half-spaces bounded by the source plane.

11.11 Derive the most general form of an EM surface source supported over a sphere centered at the origin.

11.12 Determine the relationship between the components of the surface source found in the previous problem for it to be NR throughout the interior (exterior) of the sphere.

11.13 Use the results from Example 11.4 to solve the 2D EM ISP for a source compactly supported between two parallel planes in terms of the tangential components of the electric field specified over two bounding parallel planes.

11.14 Fill in the missing steps in the derivation of Eqs. (11.39).

11.15 Derive the expressions for the EM scattering amplitudes given in Eqs. (11.61) from Eqs. (11.60).

11.16 Derive Eqs. (11.62).

11.17 Verify that the EM scattering amplitudes satisfy Eqs. (11.18).

11.18 Compute $\mathbf{f}_{\mathrm{e}}^{\mathrm{TM}}$ and $\mathbf{f}_{\mathrm{h}}^{\mathrm{TE}}$ within the Born approximation in terms of $\widetilde{\delta\epsilon}$ and $\widetilde{\delta\mu}$.

11.19 Express the TE and TM scattering amplitudes within the Born approximation in terms of scattered-field data specified over a spherical surface surrounding the scattering volume.

11.20 Derive a general expression for non-scattering material parameters $\delta\epsilon$ and $\delta\mu$ within the Born approximation for plane-wave incidence.

A Appendix A Proof of the scattering amplitude theorems

A.1 Proof of the reciprocity theorem

The reciprocity theorem states that the scattering amplitude for a compactly supported scattering potential in a homogeneous medium must satisfy the condition

$$f(\mathbf{s},\mathbf{s}_0)=f(-\mathbf{s}_0,-\mathbf{s}). \tag{A.1}$$

This theorem is easily proven using the LS equation Eq. (6.11b) for the total Green function. In particular, we let the source point $\mathbf{r}_0$ recede to infinity along the direction $-\mathbf{s}_0$ in Eq. (6.9b) to obtain

$$\begin{aligned}
G_+(\mathbf{r},-r_0\mathbf{s}_0) &= G_{0+}(\mathbf{r}+r_0\mathbf{s}_0)+\int d^3r'\, G_+(\mathbf{r},\mathbf{r}')V(\mathbf{r}')G_{0+}(\mathbf{r}'+r_0\mathbf{s}_0)\\
&\sim \left\{-\frac{1}{4\pi}\overbrace{\left[e^{ik_0\mathbf{s}_0\cdot\mathbf{r}}+\int d^3r'\, G_+(\mathbf{r},\mathbf{r}')V(\mathbf{r}')e^{ik_0\mathbf{s}_0\cdot\mathbf{r}'}\right]}^{U(\mathbf{r},\mathbf{s}_0)}\right\}\frac{e^{ik_0r_0}}{r_0}\\
&= -\frac{1}{4\pi}U(\mathbf{r},\mathbf{s}_0)\frac{e^{ik_0r_0}}{r_0}, \quad k_0r_0\to\infty,
\end{aligned}$$

where we have made use of Eq. (6.4) and $U(\mathbf{r},\mathbf{s}_0)$ is the sum of the incident and scattered waves for the incident plane wave $\exp(ik_0\mathbf{s}_0\cdot\mathbf{r})$. We now set $\mathbf{r}=r_0\mathbf{s}$ in the above equation to find that

$$\begin{aligned}
G_+(r_0\mathbf{s},-r_0\mathbf{s}_0) &\sim -\frac{1}{4\pi}U(r_0\mathbf{s},\mathbf{s}_0)\frac{e^{ik_0r_0}}{r_0}\\
&= -\frac{1}{4\pi}\left\{e^{ik_0\mathbf{s}\cdot\mathbf{s}_0r_0}+f(\mathbf{s},\mathbf{s}_0)\frac{e^{ik_0r_0}}{r_0}\right\}\frac{e^{ik_0r_0}}{r_0}, \quad k_0r_0\to\infty.
\end{aligned} \tag{A.2}$$

The final step is to use the reciprocity condition for the full Green function. In particular, it follows from this condition that

$$G_+(r_0\mathbf{s},-r_0\mathbf{s}_0)=G_+(-r_0\mathbf{s}_0,r_0\mathbf{s}),$$

which, when used with Eq. (A.2), requires that

$$f(\mathbf{s},\mathbf{s}_0)=f(-\mathbf{s}_0,-\mathbf{s}),$$

which establishes the theorem.

A.2 Proof of the translation theorem

The scattering amplitude from a potential centered at the point $\mathbf{X}$ is found using Eq. (6.22) to be

$$f_{\mathbf{X}}(\mathbf{s};\mathbf{s}_0) = \frac{-1}{4\pi}\int d^3r\, V(\mathbf{r}-\mathbf{X})U_{\mathbf{X}}(\mathbf{r};\mathbf{s}_0)e^{-ik_0\mathbf{s}\cdot\mathbf{r}}$$
$$= \frac{-1}{4\pi}e^{-ik_0\mathbf{s}\cdot\mathbf{X}}\int d^3r\, V(\mathbf{r})U_{\mathbf{X}}(\mathbf{r}+\mathbf{X};\mathbf{s}_0)e^{-ik_0\mathbf{s}\cdot\mathbf{r}}, \qquad \text{(A.3)}$$

where $U_{\mathbf{X}}(\mathbf{r},\mathbf{s}_0)$ denotes the total field (incident plus scattered) generated from the incident plane wave $\exp(ik_0\mathbf{s}_0\cdot\mathbf{r})$ onto the potential $V(\mathbf{r}-\mathbf{X})$. This field satisfies the LS equation Eq. (6.21), which assumes the form

$$U_{\mathbf{X}}(\mathbf{r};\mathbf{s}_0) = e^{ik_0\mathbf{s}_0\cdot\mathbf{r}} + \int d^3r'\, G_{0_+}(\mathbf{r}-\mathbf{r}')V(\mathbf{r}'-\mathbf{X})U_{\mathbf{X}}(\mathbf{r}';\mathbf{s}_0).$$

We now make the simultaneous transformations $\mathbf{r}'\to\mathbf{r}'+\mathbf{X}$ and $\mathbf{r}\to\mathbf{r}+\mathbf{X}$ to obtain

$$U_{\mathbf{X}}(\mathbf{r}+\mathbf{X};\mathbf{s}_0) = e^{ik_0\mathbf{s}_0\cdot(\mathbf{r}+\mathbf{X})} + \int d^3r'\, G_{0_+}(\mathbf{r}-\mathbf{r}')V(\mathbf{r}')U_{\mathbf{X}}(\mathbf{r}'+\mathbf{X};\mathbf{s}_0),$$

which, after minor manipulation, becomes

$$e^{-ik_0\mathbf{s}_0\cdot\mathbf{X}}U_{\mathbf{X}}(\mathbf{r}+\mathbf{X};\mathbf{s}_0) = e^{ik_0\mathbf{s}_0\cdot\mathbf{r}} + \int d^3r'\, G_{0_+}(\mathbf{r}-\mathbf{r}')V(\mathbf{r}')e^{-ik_0\mathbf{s}_0\cdot\mathbf{X}}U_{\mathbf{X}}(\mathbf{r}'+\mathbf{X};\mathbf{s}_0). \qquad \text{(A.4)}$$

We now compare Eq. (A.4) with the LS equation Eq. (6.21) for plane-wave scattering from the unshifted potential to find that

$$U_0(\mathbf{r},\mathbf{s}_0) = e^{-ik_0\mathbf{s}_0\cdot\mathbf{X}}U_{\mathbf{X}}(\mathbf{r}+\mathbf{X};\mathbf{s}_0). \qquad \text{(A.5)}$$

On substituting this result into Eq. (A.3) we conclude that

$$f_{\mathbf{X}}(\mathbf{s};\mathbf{s}_0) = \frac{-1}{4\pi}e^{-ik_0\mathbf{s}\cdot\mathbf{X}}\int d^3r\, V(\mathbf{r})U_{\mathbf{X}}(\mathbf{r}+\mathbf{X};\mathbf{s}_0)e^{-ik_0\mathbf{s}\cdot\mathbf{r}}$$
$$= \frac{-1}{4\pi}e^{-ik_0\mathbf{s}\cdot\mathbf{X}}e^{ik_0\mathbf{s}_0\cdot\mathbf{X}}\int d^3r\, V(\mathbf{r})U_0(\mathbf{r};\mathbf{s}_0)e^{-ik_0\mathbf{s}\cdot\mathbf{r}},$$

which can be written in the form

$$f_{\mathbf{X}}(\mathbf{s};\mathbf{s}_0) = e^{-ik_0(\mathbf{s}-\mathbf{s}_0)\cdot\mathbf{X}}f_0(\mathbf{s},\mathbf{s}_0),$$

which establishes the theorem.

A.3 Proof of the optical theorem

We consider a real-valued scattering potential $V(\mathbf{r})$ embedded in a non-dispersive homogeneous background having real-valued wavenumber k_0. The total field resulting from an incident plane wave $\exp(ik_0\mathbf{s}_0\cdot\mathbf{r})$ then satisfies the pair of equations

$$[\nabla^2 + k_0^2 - V(\mathbf{r})]U(\mathbf{r};\mathbf{s}_0) = 0, \tag{A.6a}$$

$$[\nabla^2 + k_0^2 - V(\mathbf{r})]U^*(\mathbf{r};\mathbf{s}_0) = 0, \tag{A.6b}$$

and the asymptotic conditions

$$U(\mathbf{r};\mathbf{s}_0) \sim e^{ik_0\mathbf{s}_0\cdot\mathbf{s}r} + f(\mathbf{s},\mathbf{s}_0)\frac{e^{ik_0r}}{r}, \tag{A.7a}$$

$$U^*(\mathbf{r};\mathbf{s}_0) \sim e^{-ik_0\mathbf{s}_0\cdot\mathbf{s}r} + f^*(\mathbf{s},\mathbf{s}_0)\frac{e^{-ik_0r}}{r}. \tag{A.7b}$$

as $k_0r \to \infty$ along the direction defined by the unit vector $\mathbf{s}$. Applying standard manipulations to Eqs. (A.6) we then find that

$$\Im \int_{\partial V} dS\, r^2 U^*(\mathbf{r};\mathbf{s}_0)\frac{\partial}{\partial n}U(\mathbf{r};\mathbf{s}_0) = 0, \tag{A.8}$$

where ∂V is any surface completely surrounding the scattering volume τ_0 and ∂n denotes the outward-directed unit normal to ∂V. If we select ∂V to be the surface of a sphere of radius R centered at the origin and such that $k_0R \to \infty$, Eq. (A.8) becomes

$$0 = R^2\Im \int_{4\pi} d\Omega_s \left[e^{-ik_0\mathbf{s}_0\cdot\mathbf{s}R} + f^*(\mathbf{s},\mathbf{s}_0)\frac{e^{-ik_0R}}{R}\right]\left\{\frac{\partial}{\partial R}\left[e^{ik_0\mathbf{s}_0\cdot\mathbf{s}R} + f(\mathbf{s},\mathbf{s}_0)\frac{e^{ik_0R}}{R}\right]\right\}$$

$$= \overbrace{R^2\Im \int_{4\pi} d\Omega_s\, e^{-ik_0\mathbf{s}_0\cdot\mathbf{s}R}\frac{\partial}{\partial R}e^{ik_0\mathbf{s}_0\cdot\mathbf{s}R}}^{I_1} + \overbrace{R^2\Im \int_{4\pi} d\Omega_s\, e^{-ik_0\mathbf{s}_0\cdot\mathbf{s}R} f(\mathbf{s},\mathbf{s}_0)\frac{\partial}{\partial R}\frac{e^{ik_0R}}{R}}^{I_2}$$

$$+ \overbrace{R^2\Im \int_{4\pi} d\Omega_s f^*(\mathbf{s},\mathbf{s}_0)\frac{e^{-ik_0R}}{R}\frac{\partial}{\partial R}e^{ik_0\mathbf{s}_0\cdot\mathbf{s}R}}^{I_3}$$

$$+ \overbrace{R^2\Im \int_{4\pi} d\Omega_s f^*(\mathbf{s},\mathbf{s}_0)\frac{e^{-ik_0R}}{R} f(\mathbf{s},\mathbf{s}_0)\frac{\partial}{\partial R}\frac{e^{ik_0R}}{R}}^{I_4}. \tag{A.9}$$

We now evaluate each of the integrals in Eq. (A.9). The first integral I_1 is easily shown to be zero while for I_4 we find that

$$I_4 = \int_{4\pi} d\Omega_s |f(\mathbf{s},\mathbf{s}_0)|^2 = \frac{E}{2\kappa\omega},$$

where E is the scattered field energy defined in Eq. (6.32). For I_2 we find that

$$I_2 = R\Im ik_0 e^{ik_0R}\int_{4\pi} d\Omega_s\, e^{-ik_0\mathbf{s}_0\cdot\mathbf{s}R} f(\mathbf{s},\mathbf{s}_0)\left[1 + \left(\frac{1}{R}\right)\right], \tag{A.10}$$

where $O(1/R)$ stands for "of order $1/R$." We now use the multipole expansion of the plane wave $\exp(-ik_0\mathbf{s}_0\cdot\mathbf{r})$ which was derived in Example 3.4 of Chapter 3:

$$e^{-ik\mathbf{s}_0\cdot\mathbf{r}} = 4\pi\sum_{l=0}^{\infty}\sum_{m=-l}^{l}(-i)^l j_l(kr)Y_l^m(\hat{\mathbf{r}})Y_l^{m*}(\mathbf{s}_0). \tag{A.11}$$

On setting $\mathbf{r} = R\mathbf{s}$ and using the asymptotic form of the spherical Bessel functions

$$j_l(k_0R) \sim \frac{1}{2ik_0R}[(-i)^l e^{ik_0R} - i^l e^{-ik_0R}], \quad k_0R \to \infty,$$

we find that Eq. (A.11) yields the result

$$e^{-ik\mathbf{s}_0\cdot\mathbf{s}R} \sim \frac{2\pi}{ik_0R}\sum_{l=0}^{\infty}\sum_{m=-l}^{l}(-i)^l[(-i)^l e^{ik_0R} - i^l e^{-ik_0R}]Y_l^m(\mathbf{s})Y_l^{m*}(\mathbf{s}_0). \tag{A.12}$$

On substituting Eq. (A.12) into Eq. (A.10) we then find that

$$\begin{aligned} I_2 &= 2\pi\Im\int_{4\pi} d\Omega_s\left[f(\mathbf{s},\mathbf{s}_0) + O\left(\frac{1}{R}\right)\right]\sum_{m=-l}^{l}[(-i)^{2l}e^{2ik_0R} - 1]Y_l^m(\mathbf{s})Y_l^{m*}(\mathbf{s}_0) \\ &= 2\pi\Im\int_{4\pi} d\Omega_s f(\mathbf{s},\mathbf{s}_0)\sum_{m=-l}^{l}(-i)^{2l}e^{2ik_0R}Y_l^m(\mathbf{s})Y_l^{m*}(\mathbf{s}_0) - 2\pi\Im f(\mathbf{s}_0,\mathbf{s}_0), k_0R \to \infty, \end{aligned} \tag{A.13}$$

where $f(\mathbf{s}_0, \mathbf{s}_0)$ is the scattering amplitude evaluated in the forward direction (i.e., at $\mathbf{s} = \mathbf{s}_0$) and we have used the completeness relation

$$\sum_{m=-l}^{l} Y_l^m(\mathbf{s})Y_l^{m*}(\mathbf{s}_0) = \delta(\mathbf{s} - \mathbf{s}_0),$$

with $\delta(\mathbf{s} - \mathbf{s}_0)$ being the delta function over the unit sphere.

We now evaluate I_3, which assumes the form

$$\begin{aligned} I_3 &= R\Im e^{-ik_0R}\int_{4\pi} d\Omega_s f^*(\mathbf{s},\mathbf{s}_0)\frac{\partial}{\partial R}e^{ik_0\mathbf{s}_0\cdot\mathbf{s}R} \\ &= R\Im e^{-ik_0R}\int_{4\pi} d\Omega_s f^*(\mathbf{s},\mathbf{s}_0)\frac{2\pi}{R}\sum_{l=0}^{\infty}\sum_{m=-l}^{l} i^l[(-i)^l e^{ik_0R} + i^l e^{-ik_0R}]Y_l^{m*}(\mathbf{s})Y_l^m(\mathbf{s}_0), \end{aligned}$$

where we have made use of the complex conjugate of Eq. (A.12). On simplifying the above result we conclude that

$$\begin{aligned} I_3 &= 2\pi\Im\int_{4\pi} d\Omega_s f^*(\mathbf{s},\mathbf{s}_0)\sum_{l=0}^{\infty}\sum_{m=-l}^{l} i^{2l}e^{-2ik_0R}Y_l^{m*}(\mathbf{s})Y_l^m(\mathbf{s}_0) \\ &\quad + 2\pi\Im\int_{4\pi} d\Omega_s f^*(\mathbf{s},\mathbf{s}_0)\sum_{l=0}^{\infty}\sum_{m=-l}^{l} Y_l^{m*}(\mathbf{s})Y_l^m(\mathbf{s}_0) \\ &= 2\pi\Im\int_{4\pi} d\Omega_s f^*(\mathbf{s},\mathbf{s}_0)\sum_{l=0}^{\infty}\sum_{m=-l}^{l} i^{2l}e^{-2ik_0R}Y_l^{m*}(\mathbf{s})Y_l^m(\mathbf{s}_0) + 2\pi\Im f^*(\mathbf{s}_0,\mathbf{s}_0), \end{aligned} \tag{A.14}$$

where we have again used the completeness relationship for the spherical harmonics. On adding I_1 through I_4 we then obtain

$$
\begin{aligned}
I_1 + I_2 + I_3 + I_3 = 0 = {} & 2\pi \Im \int_{4\pi} d\Omega_s f(\mathbf{s}, \mathbf{s}_0) \sum_{m=-l}^{l} (-i)^{2l} e^{2ik_0 R} Y_l^m(\mathbf{s}) Y^{*m}_l(\mathbf{s}_0) \\
& - 2\pi \Im f(\mathbf{s}_0, \mathbf{s}_0) \\
& + 2\pi \Im \int_{4\pi} d\Omega_s f^*(\mathbf{s}, \mathbf{s}_0) \sum_{l=0}^{\infty} \sum_{m=-l}^{l} i^{2l} e^{-2ik_0 R} Y_l^{m*}(\mathbf{s}) Y_l^m(\mathbf{s}_0) \\
& + 2\pi \Im f^*(\mathbf{s}_0, \mathbf{s}_0) + \frac{E}{2\kappa\omega} \\
= {} & -4\pi \Im f(\mathbf{s}_0, \mathbf{s}_0) + \frac{E}{2\kappa\omega} = 0,
\end{aligned}
$$

which then yields the final result

$$
E = 2\kappa\omega k_0 \int_{4\pi} d\Omega_s |f(\mathbf{s}, \mathbf{s}_0)|^2 = 8\pi\kappa\omega \Im f(\mathbf{s}_0, \mathbf{s}_0). \tag{A.15}
$$

B

Appendix B Derivation of the generalized Weyl expansion

In this appendix we derive the generalized Weyl expansion Eq. (9.25a) for the special case of a lossless background characterized by a real-valued scattering potential $V_0(\mathbf{r})$. This assumption allows us to employ, without major modification, some important results from non-relativistic quantum-mechanical scattering theory (collision theory) that require the Helmholtz (Schrödinger) operator to be Hermitian and, hence, the scattering potential (background wavenumber) to be real-valued.[1] The key to the derivation is the use of so-called "off-shell" Green functions and plane-wave scattering states which are briefly reviewed below.

B.1 Off-shell Green functions and scattering wave states

If we define the scattering potential $V_0(\mathbf{r})$ according to Eq. (9.2a) then the background Green functions $G_{0_\pm}(\mathbf{r},\mathbf{r}')$ are limiting values as $p \to k_0 \pm i\epsilon$ of the *off-shell* Green functions $G_{0_\pm}(\mathbf{r},\mathbf{r}';p)$ that satisfy

$$[\nabla^2 + p^2 - V_0(\mathbf{r})]G_{0_\pm}(\mathbf{r},\mathbf{r}';p) = \delta(\mathbf{r}-\mathbf{r}') \tag{B.1}$$

and the radiation conditions

$$G_{0_\pm}(\mathbf{r},\mathbf{r}';p) \sim g_\pm(\mathbf{s},\mathbf{r}';p)\frac{e^{\pm ipr}}{r} + O\left(\frac{1}{r^2}\right), \quad r \to \infty.$$

It is important to note that the scattering potential V_0 in Eq. (B.1) is *fixed* and does not depend on the parameter p; i.e., p plays the role of k_0 but can vary over the complex-p plane, while the scattering potential V_0 is a fixed real-valued function of position $\mathbf{r}$ that is independent of p. We can consider the "on-shell" or background Green functions to be boundary values of the off-shell Green functions on the "energy sphere" $p = k_0$. The off-shell Green function G_{0_+} approaches this boundary from the half-plane $\Im p > 0$ and the off-shell Green function G_{0_-} approaches this boundary from the half-space $\Im p < 0$ corresponding to the two branches of $\sqrt{p^2}$.

We also define off-shell plane-wave scattering states as the solutions to the homogeneous Helmholtz equation

$$[\nabla^2 + p^2 - V_0(\mathbf{r})]\psi_\pm(\mathbf{r},\mathbf{p}) = 0,$$

[1] However, most, if not all, of the results obtained in the appendix can be generalized to lossy backgrounds using the theory of biorthogonal expansions.

where $p = \sqrt{\mathbf{p} \cdot \mathbf{p}}$. The off-shell plane-wave scattering states satisfy the far-field conditions

$$\psi_{\pm}(\mathbf{r}, \mathbf{p}) = e^{i\mathbf{p}\cdot\mathbf{r}} + f_{\pm}(\mathbf{s}, \hat{\mathbf{p}}; p)\frac{e^{\pm ipr}}{r}, \tag{B.2}$$

where $f_{\pm}(\mathbf{s}, \hat{\mathbf{p}}; p)$ is the scattering amplitude of the potential V_0 embedded in the homogeneous medium having wavenumber p. The on-shell plane-wave scattering states $\psi_{\pm}(\mathbf{r}, k_0\mathbf{s})$, like the on-shell Green functions $G_{0_{\pm}}(\mathbf{r}, \mathbf{r}')$, are boundary values of their off-shell continuations on the energy sphere $p = k_0$, with $\psi_{+}(\mathbf{r}, k_0\mathbf{s})$ obtained when p approaches the sphere from the half-plane $\Im p > 0$ and $\psi_{-}(\mathbf{r}, k_0\mathbf{s})$ when p approaches the sphere from the half-plane $\Im p < 0$. The off-shell plane-wave scattering states satisfy the Lippmann–Schwinger (LS) equation with reference wavenumber p which can be written in the form

$$\psi_{\pm}(\mathbf{r}, \mathbf{p}) = e^{i\mathbf{p}\cdot\mathbf{r}} + \int d^3r'\, G_{0_{\pm}}(\mathbf{r} - \mathbf{r}'; p)V_0(\mathbf{r}')\psi_{\pm}(\mathbf{r}', \mathbf{p}). \tag{B.3}$$

Because the background wavenumber $k_0(\mathbf{r})$ and, hence, the background scattering potential $V_0(\mathbf{r})$ are assumed to be real-valued the set of off-shell plane-wave scattering states $\psi_{\pm}(\mathbf{r}, \mathbf{p})$, $\mathbf{p} \in R^3$ are eigenfunctions of the Hermitian operator $-\nabla^2 + V_0(\mathbf{r})$ with eigenvalue p^2 that satisfy the boundary conditions Eq. (B.2) and, hence, form a complete set of functions for expanding any outgoing wavefield in the background medium having reference wavenumber k_0 and, in particular, can represent the background Green function $G_{0_+}(\mathbf{r}, \mathbf{r}')$. Thus, we can write (cf. the discussion in Section 3.1 of Chapter 3)

$$G_{0_+}(\mathbf{r}, \mathbf{r}') = \frac{1}{(2\pi)^3}\int d^3\mathbf{p}\, \frac{\psi_{+}(\mathbf{r}, \mathbf{p})\psi_{+}^{*}(\mathbf{r}'; \mathbf{p})}{k_{+0}^2 - p^2},$$

where $k_{+0}^2 = k_0^2 + i\epsilon$ to insure that the Green function is outgoing (satisfies the SRC) and the factor $1/(2\pi)^3$ is needed for normalization. We note also that since the scattering potential V_0 is real-valued we have that

$$\psi_{+}^{*}(\mathbf{r}'; \mathbf{p}) = \psi_{-}(\mathbf{r}'; -\mathbf{p}),$$

so that we can write the above expansion for G_{0_+} in the alternative form

$$G_{0_+}(\mathbf{r}, \mathbf{r}') = \frac{1}{(2\pi)^3}\int d^3\mathbf{p}\, \frac{\psi_{+}(\mathbf{r}, \mathbf{p})\psi_{-}(\mathbf{r}'; -\mathbf{p})}{k_{+0}^2 - p^2}. \tag{B.4}$$

B.2 Derivation of the generalized Weyl expansion

We now select a fixed Cartesian coordinate system and set $\mathbf{p} = (p_x, p_y, p_z)$ so that Eq. (B.4) assumes the form

$$G_{0_+}(\mathbf{r}, \mathbf{r}') = \frac{1}{(2\pi)^3}\int dp_x\, dp_y\, dp_z\, \frac{\psi_{+}(\mathbf{r}, \mathbf{p})\psi_{-}(\mathbf{r}', -\mathbf{p})}{k_{+0}^2 - p^2}. \tag{B.5}$$

The wavefunctions $\psi_{+}(\mathbf{r}, \mathbf{p})$ and $\psi_{-}(\mathbf{r}, \mathbf{p})$ are analytic functions of the three Cartesian components of $\mathbf{p}$ over the real environment $\mathbf{p} \in R^3$ and can be continued onto complex values of these three components. We note, however, that in this continuation the multivalued function $p = \sqrt{p_x^2 + p_y^2 + p_z^2}$ will lie on different Riemann sheets for the two functions and, in particular, will have a positive imaginary part for ψ_+ and a negative imaginary part for ψ_-. This requirement will be important in the computations we present below.

If we now allow p_z to tend toward infinity in the upper half of the complex-p_z plane with p_x, p_y real we find using the LS equation Eq. (B.3) that

$$\begin{aligned}
\psi_+(\mathbf{r}, \mathbf{p}) &\sim e^{i(p_x x + p_y y)} e^{-|p_z| z} + \int d^3 r' \frac{e^{-|p_z||\mathbf{r}-\mathbf{r}'|}}{|\mathbf{r}-\mathbf{r}'|} V_0(\mathbf{r}')\psi_+(\mathbf{r}', \mathbf{p}) \\
&\approx e^{i(p_x x + p_y y)} e^{-|p_z| z} + V_0(\mathbf{r})\psi_+(\mathbf{r}, \mathbf{p}) \int d^3 \xi \frac{e^{-|p_z||\xi|}}{|\xi|} \to K_+ e^{-|p_z| z}, \\
\psi_-(\mathbf{r}, -\mathbf{p}) &\sim e^{-i(p_x x + p_y y)} e^{+|p_z| z} + \int d^3 r' \frac{e^{-|p_z||\mathbf{r}-\mathbf{r}'|}}{|\mathbf{r}-\mathbf{r}'|} V_0(\mathbf{r}')\psi_-(\mathbf{r}', \mathbf{p}) \\
&\approx e^{i(p_x x + p_y y)} e^{|p_z| z} + V_0(\mathbf{r})\psi_-(\mathbf{r}, \mathbf{p}) \int d^3 \xi \frac{e^{-|p_z||\xi|}}{|\xi|} \to K_- e^{|p_z| z},
\end{aligned}$$

where $K_\pm$ are constants and we have used the fact that $\psi_+(\mathbf{r}, \mathbf{p})$ is evaluated on the Riemann sheet where $\Im p > 0$ and $\psi_+(\mathbf{r}, -\mathbf{p})$ on the sheet where $\Im p < 0$. Using a similar development, it is easy to show that

$$\psi_+(\mathbf{r}, \mathbf{p}) \to K_- e^{+|p_z| z}, \qquad \psi_-(\mathbf{r}, \mathbf{p}) \to K_+ e^{-|p_z| z}$$

as p_z tends to infinity in the l.h.p.

If we make use of these results in Eq. (B.5) we can close the p_z contour of integration in the u.h.p. as long as $z > z'$ and in the l.h.p. if $z < z'$, and we find using Cauchy's integral formula that

$$\begin{aligned}
G_{0_+}(\mathbf{r}, \mathbf{r}') &= -\frac{1}{(2\pi)^3} \int dp_x\, dp_y \int dp_z \frac{\psi_+(\mathbf{r}, \mathbf{p})\psi_-(\mathbf{r}'; -\mathbf{p})}{(p_z - \gamma)(p_z + \gamma)} \\
&= -\frac{i}{8\pi^2} \int \frac{dk_x\, dk_y}{\gamma} \psi_+(\mathbf{r}, \mathbf{k}_0^\pm)\psi_-(\mathbf{r}', -\mathbf{k}_0^\pm),
\end{aligned} \tag{B.6}$$

where

$$\mathbf{k}_0^\pm = k_x \hat{\mathbf{x}} + k_y \hat{\mathbf{y}} \pm \gamma \hat{\mathbf{z}},$$

with

$$\gamma = \begin{cases} \sqrt{k_0^2 - k_x^2 - k_y^2} & k_x^2 + k_y^2 < k_0^2, \\ i\sqrt{k_x^2 + k_y^2 - k_0^2} & k_x^2 + k_y^2 > k_0^2, \end{cases}$$

where $\mathbf{k}_0^+$ is used when $z > z'$ and $\mathbf{k}_0^-$ when $z < z'$. The plane-wave scattering states $\psi_+(\mathbf{r}, \mathbf{k}_0^\pm)$ and $\psi_-(\mathbf{r}, -\mathbf{k}_0^\pm)$ that enter the expansion Eq. (B.6) are seen to be *homogeneous plane waves* in the region $k_x^2 + k_y^2 < k_0^2$ and *evanescent* plane waves in the region $k_x^2 + k_y^2 > k_0^2$. Since $+\gamma$ is used if $z > z'$ and $-\gamma$ if $z < z'$, it is apparent that the evanescent plane waves in Eq. (B.6) *decay* exponentially fast with increasing $|z - z'|$ in both half-spaces $z > z'$ and $z < z'$.

Equation (B.6) is the Cartesian-variable form of the generalized Weyl expansion. By making a transformation to spherical integration coordinates it is easily shown (cf. Section 4.1) that the generalized Weyl expansion can also be expressed in the angle-variable form given in Eq. (B.2)

$$G_{0_+}(\mathbf{r},\mathbf{r}') = -\frac{ik_0}{8\pi^2}\int_{-\pi}^{\pi} d\beta \int_{C_\pm} \sin\alpha\, d\alpha\, \psi_+(\mathbf{r}, k_0\mathbf{s})\psi_-(\mathbf{r}', -k_0\mathbf{s}),$$

where the contour C_+ is used if $z > z'$ and the contour C_- is used if $z < z'$.

References

Andersen, A. H., and Kak, A. C. 1984. Simultaneous Algebraic Reconstruction Technique (SART): A superior implementation of the ART algorithm. *Ultrasonic Imaging*, **6**, 81–94.

Arfken, G. B., and Weber, H. J. 2001. *Mathematical Methods for Physicists*, 5th edn. San Diego, CA: Academic Press.

Arsac, J. 1984. *Diffraction Physics*. New York: North Holland.

Baker, B., and Copson, E. T. 1950. *The Mathematical Theory of Huygen's Principle*, 2nd edn. Oxford: Clarendon Press.

Balanis, C. A. 1989. *Advanced Engineering Electromagnetics*. New York: Wiley.

Belkebir, K., Chaumet, P. C., and Sentenac, A. 2006. Influence of multiple scattering on three-dimensional imaging with optical diffraction tomography. *J. Opt. Soc. Am. A*, **23**(3), 586–595.

Bender, R., Gordon R., and Herman, G. T. 1970. Algebraic Reconstruction Techniques (ART) for three-dimensional electron microscopy and X-ray photography. *J. Theor. Biol.*, **29**, 471–481.

Bertero, M. 1986. Regularization methods for linear inverse problems, in Talenti, G. (ed.), *Inverse Problems*. Berlin: Springer, pp. 52–112.

Bertero, M. 1989. Linear inverse and ill-posed problems. *Adv. Electron. Electron Phys.*, **75**, 1–120.

Bertero, M., and Poccacci, B. 1998. *Introduction to Inverse Problems in Imaging*. Philadelphia, PA: IOP Publishing.

Bertero, M., De Mol, C., and Pike, E. R. 1985. Linear inverse problems with discrete data: I. General formulation and singular system analysis. *Inverse Problems*, **1**, 301–330.

Bertero, M., Mol, C. De, and Pike, E. R. 1988. Linear inverse problems with discrete data: II – Stability and regularization. *Inverse Problems*, **4**, 573–594.

Beylkin, G. 1983. The fundamental identity for iterated spherical means and the inversion formula for diffraction tomography and inverse scattering. *J. Math. Phys.*, **24**, 1399–1400.

Bleistein, N., and Cohen, J. K. 1977. Nonuniqueness in the inverse source problem in acoustics and electromagnetics. *J. Math. Phys.*, **18**(2), 194–201.

Bohm, D., and Weinstein, M. 1949. Finite relativistic charge-current distributions. *Phys. Rev.*, **76**, 867.

Bojarski, N. 1982a. A survey of the near-field far-field inverse scattering inverse source integral equation. *IEEE Trans. Ant. Propag.*, **30**, 975–979.

Bojarski, N. 1982b. A survey of the physical optics inverse scattering identity. *IEEE Trans. Ant. Propag.*, **30**, 980–989.

Bolomey, J., Lesselier, D., Pichot, C., and Tabbara, W. 1981. Spectral and time domain approaches to some inverse scattering problems. *IEEE Trans. Ant. Propag.*, **29**, 206–212.

Borden, B. 1999. *Radar Imaging of Airborne Targets: A Primer for Applied Mathematicians and Physicists*. Bristol: The Institute of Physics.

Borden, B. 2002. Mathematical problems in radar inverse scattering. *Inverse Problems*, **18**, R1.

Borgiotti, G. 1962. A general way of representing the electromagnetic field. *Alta Frequenza*, **31**, 185.

Born, M., and Wolf, E. 1999. *Principles of Optics: Electromagnetic Theory of Propagation, Interference and Diffraction of Light*, 7th edn. Cambridge: Cambridge University Press.

Brander, O., and DeFacio, B. 1986. The role of filters and the singular-value decomposition for the inverse Born approximation. *Inverse Problems*, **2**, 375.

Burridge, R, and Beylkin, G. 1988. On double integrals over spheres. *Inverse Problems*, **4**, 1.

Byrne, C. L., and Fiddy, M. A. 1987. Estimation of continuous object distributions from limited Fourier magnitude measurements. *J. Opt. Soc. Am. A*, **4**(1), 112–117.

Carney, P., and Schotland, J. 2000. Inverse scattering for near-field microscopy. *Appl. Phys. Lett.*, **77**, 2798–2800.

Cha, S., and Vest, C. M. 1979. Interferometry and reconstruction of strongly refracting asymmetric-refractive-index fields. *Opt. Lett.*, **4**, 311–313.

Chambers, D., and Berryman, J. 2004. Analysis of the time-reversal operator for a small spherical scatterer in an electromagnetic field. *IEEE Trans. Ant. Propag.*, **52**(7), 1729–1738.

Chambers, D. H. 2002. Analysis of the time-reversal operator for scatterers of finite size. *J. Acoust. Soc. Am.*, **112**(2), 411–419.

Chambers, D. H., and Gautesen, A. K. 2001. Time reversal for a single spherical scatterer. *J. Acoust. Soc. America*, **109**(6), 2616–2624.

Cheney, M. 2001. The linear sampling method and the MUSIC algorithm. *Inverse Problems*, **17**, 591–595.

Cheney, M., and Borden, B. 2008. Imaging moving targets from scattered waves. *Inverse Problems*, **24**, 591–595.

Cheney, M., and Borden, B. 2009. *Fundamentals of Radar Imaging*. Philadelphia, PA: SIAM.

Chernov, L. A. 1967. *Wave Propagation in a Random Medium*. New York: Dover Publications.

Chew, W., and Wang, Y. 1990. Reconstruction of two-dimensional permittivity distribution using the distorted Born iterative method. *IEEE Trans. Med. Imaging*, **9**, 218–225.

Chew, W. C. 1990. *Waves and Fields in Inhomogeneous Media.* New York: Van Nostrand Reinholt.

Colton, D. L., and Kress, R. 1991. *Integral Equation Methods in Scattering Theory*. Malabar, FL: Krieger.

Colton, D. L., and Kress, R. 1992. *Inverse Acoustic and Electromagnetic Scattering Theory*. New York: Springer.

Courant, R., and Hilbert, D. 1966. *Methods of Mathematical Physics*, Vol. I. New York: Interscience Publishers.

Cowley, J. 1966. *Fourier Transforms and the Theory of Distributions*. Englewood Cliffs, NJ: Prentice-Hall.

Damelin, S., and Devaney, A. J. 2007. Local Paley–Wiener theorems for functions analytic on the unit sphere. *Inverse Problems*, **23**, 463–474.

Das, Y., and Boerner, W. M. 1978. On radar target shape estimation using algorithms for reconstruction from projections. *IEEE Trans. Ant. Propag.*, **26**, 274–278.

Debye, P. 1909. Der Lichtdruck auf Kugeln von beliebigem Material. *Ann. Phys. Leipzig*, **335**, 57–136.

Devaney, A. J. 1971. *A New Theory of the Debye Representation of Classical and Quantized Electromagnetic Fields*. Rochester, NY: Ph.D. Thesis, University of Rochester.

Devaney, A. J. 1979. The inverse problem for random sources. *J. Math. Phys.*, **20**(8), 1687–1691.

Devaney, A. J. 1982. Inversion formula for inverse scattering within the Born approximation. *Opt. Lett.*, **7**, 111.

Devaney, A. J. 1984. Geophysical diffraction tomography. *IEEE Trans. Geosci. Remote Sensing*, **22**, 3.

Devaney, A. J. 1989. Structure determination from intensity measurements in scattering experiments. *Phys. Rev. Lett.*, **62**, 2385.

Devaney, A. J. 2000. Super-resolution processing of multistatic data using time reversal and MUSIC, Unpublished paper, preprint available on the authors web site.

Devaney, A. J., and Sherman, G. 1973. Plane wave representations for scalar wave fields. *SIAM Rev.*, **15**, 765–786.

Devaney, A. J., and Wolf, E. 1973. Radiating and nonradiating classical current distributions and the fields they generate. *Phys. Rev. D*, **81**(4), 1044.

Devaney, A. J., and Wolf, E. 1974. Multipole expansions and plane wave representations of the electromagnetic field. *J. Math. Phys.*, **15**(2), 234–244.

Devaney, A. J., Marengo, E. A., and Gruber, F. K. 2005. Time-reversal-based imaging and inverse scattering of multiply scattering point targets. *J. Acoust. Soc. Am.*, **118**(11), 3129–3138.

Devaney, A. J., Marengo, E., and Li, M. 2007. Inverse source problem in non-homogeneous background media. *SIAM J. Appl. Math.*, **67**, 1353–1378.

Devaney, A. J., Kaiser, G., Marengo, E., Albanese, R., and Erdmann, G. 2008. The inverse source problem for wavelet fields. *IEEE Trans. Ant. Propag.*, **56**, 3179–3187.

Fiddy, M. A., and Testorf, M. 2006. Inverse scattering method applied to the synthesis of strongly scattering structures. *Opt. Express*, **14**(5), 2037–2046.

Fienup, J. R. 1982. Phase retrieval algorithms: a comparison. *Appl. Opt.*, **21**(15), 2758–2769.

Fischer, D., and Wolf, E. 1997. Theory of diffraction tomography for quasi-homogeneous random objects. *Opt. Commun.*, **133**, 17–21.

Fokas, A. S. 2008. *A Unified Approach to Boundary Value Problems*. Philadelphia, PA: SIAM.

Foldy, L. 1945. The multiple scattering of waves. I. General theory of isotropic scattering by randomly distributed scatterers. *Phys. Rev.*, **67**, 107–119.

Gbur, G. 2003. Nonradiating sources and other "invisible" objects, in Wolf, E. (ed), *Progress in Optics*, vol. 45. Amsterdam: Elsevier, pp. 273–315.

Gbur, G., and Wolf, E. 2002. Hybrid diffraction tomography without phase information. *J. Opt. Soc. Am. A*, **19**(11), 2194–2202.

Gerchberg, R. W., and Saxton, W. O. 1972. A practical algorithm for the determination of the phase from image and diffraction plane pictures. *Optik*, **35**, 237–246.

Gilbert, P. 1972. Iterative methods for the three dimensional reconstruction of an object from projections. *J. Theor. Biol.*, **36**, 105.

Goedecke, G. H. 1964. Classically radiationless motions and possible implications for quantum theory. *Phys. Rev.*, **135**, B281–B288.

Gonsalves, R. A. 1976. Phase retrieval from modulus data. *J. Opt. Soc. Am.*, **66**(9), 961–964.

Goodman, J. W. 1968. *Introduction to Fourier Optics*. San Francisco, CA: McGraw-Hill.

Gori, F. 1981. Fresnel transform and sampling theorem. *Opt. Commun.*, **39**(5), 293–297.

Green, C. D. 1969. *Integral Equation Methods*. New York: Barnes and Noble, Inc.

Greenleaf, J. F. 1980. Computerized transmission tomography, in Edmonds, P. D. (ed.), *Methods of Experimental Physics, Ultrasonics*, Vol. 19. New York: Plenum Press, pp. 563–589.

Greenleaf, J. F. 1983. Computerized tomography with ultrasound. *Proc. IEEE*, **71**, 330.

Greenleaf, J. F., and Bahn, R. C. 1981. Clinical imaging with transmissive ultrasonic computerized tomography. *IEEE Trans. Biomed. Eng.*, **28**, 177.

Gruber, F. K., Marengo, E. A., and Devaney, A. J. 2004. Time-reversal imaging with multiple signal classification considering multiple scattering between the targets. *J. Acoust. Soc. Am.*, **115**(6), 3042–3047.

Guizar-Sicairos, M., and Fienup, J. R. 2006. Complex valued object reconstruction from extrapolated intensity measurements, in *Frontiers in Optics*. Rochester, NY: Optical Society of America, paper FMI6.

Hadamard, J. 1952. *Lectures on Cauchy's Problem in Linear Partial Differential Equations*. New York: Dover.

Haddar, H., Colton, D., and Piana, M. 2003. The linear sampling method in inverse electromagnetic scattering theory. *Inverse Problems*, **19**, S105–S137.

Hamam, H., and de Bougrenet de la Tocnaye, J. L. 1995. Efficient Fresnel-transform algorithm based on fractional Fresnel diffraction. *J. Opt. Soc. Am. A*, **12**(9), 1920–1931.

Hansen, A., Yaghjian T., and Devaney, A. J. 2000. Determination of the minimum convex source region from the electromagnetic far-field pattern. *Radio Sci.*, **35**, 417–425.

Hansen, P. 1988. Computation of the singular value expansion. *Computing*, **40**, 185–199.

Hansen, R. C. 1981. Fundamental limitations in antennas. *Proc. IEEE*, **69**(2), 170–182.

Hansen, T. B., and Yaghjian, A. D. 1999. *Plane Wave Theory of Time Domain Fields*. New York: IEEE Press.

He, P., and Greenleaf, J. F. 1986. Attenuation estimation on phantoms – a stability test. *Ultrasonic Imaging*, **8**, 1.

Herman, G. T. 1980. *Image Reconstruction from Projections: The Fundamentals of Computerized Tomography*. New York: Academic Press.

Heyman, E., and Devaney, A. J. 1996. Time-dependent multipoles and their application for radiation from volume source distributions. *J. Math. Phys.*, **37**(2), 682–692.

Hounsfield, G. N. 1973. Computerized transverse axial scanning tomography: Part I, description of the system. *Br. J. Radiol.*, **46**, 1016.

Imbriale, W., and Mittra, R. 1970. The two-dimensional inverse scattering problem. *IEEE Trans. Ant. Propagag.*, **18**, 633–642.

Isakov, V., and Nachman, A. I. 1995. Global uniqueness for a two-dimensional semilinear elliptic inverse problem. *Trans. Am. Math. Soc.*, **347**, 3375.

Ishimaru, A. 1999. *Wave Propagation and Scattering in Random Media*. New York: Wiley-IEEE Press.

Ishimaru, A., Jaruwatanadilok, S., and Kuga, Y. 2007. Time reversal effects in random scattering media on superresolution, shower curtain effects, and backscattering enhancement. *Radio Sci.*, **42**(42), RS6S28.

Iwata, K., and Nagata, R. 1974. Calculation of refractive index distribution from interferograms using the Born and Rytov's approximation. *Japan. J. Appl. Phys.*, **14**, 379.

Jackson, J. D. 1998. *Classical Electrodynamics*, 3rd edn. New York: Wiley.

James, D., and Agarwal, G. 1996. The generalized Fresnel transform and its application to optics. *Opt. Commun.*, **126**(4–6), 207–212.

Johansen, N., Sponheim I., and Devaney, A. J. 1990. Initial testing of a clinical ultrasound mammograph, in Lee, H., and Wade, G. (eds.), *Acoustical Imaging 18*. New York: Plenum Press, pp. 401–410.

Kaiser, G. 2003. Physical wavelets and their sources: Real physics in complex space-time. Topical review. *J. Phys. A: Math. Gen.*, **36**, R29–R338.

Kaiser, G. 2004. Making electromagnetic wavelets. *J. Phys. A: Math. Gen.*, **38**, 5929–5947.

Kaiser, G. 2005. Making electromagnetic wavelets II. Spheroidal shell antennas. *J. Phys. A: Math. Gen.*, **38**, 495–508.

Kak, A. C., and Slaney, M. 1988. *Principles of Computerized Tomographic Imaging*. New York: IEEE Press.

Keller, J. B. 1969. Accuracy and validity of the Born and Rytov approximation. *J. Opt. Soc. Am.*, **59**, 1003.

Kirsch, A. 1993. The domain derivative and two applications in inverse scattering theory. *Inverse Problems*, **9**, 81.

Kirsch, A. 1998. Characterization of the shape of a scattering obstacle using the spectral data of the far field operator. *Inverse Problems*, **14**(6), 1489–1512.

Kirsch, A., and Ritter, S. 2000. A linear sampling method for inverse scattering from an open arc. *Inverse Problems*, **16**, 89.

Kusiak, S., and Sylvester, J. 2005. The convex scattering support in a background medium. *SIAM J. Math. Anal.*, **36**, 1142–1158.

Langenberg, K. J., Brandfaß, M., Hannemann, R., *et al.* 1999. Inverse scattering with acoustic electromagnetic and elastic waves as applied in nondestructive evaluation, in Wirgin, A. (ed), *Wavefield Inversion*. Vienna: Springer-Verlag, pp. 59–118.

Lauer, V. 2002. New approach to optical diffraction tomography yielding a vector equation of diffraction tomography and a novel tomographic microscope. *J. Microscopy*, **205**(2), 165–176.

Lax, M. 1951. Multiple scattering of waves. *Rev. Mod. Phys.*, **23**, 287–310.

Lehman, S., and Devaney, A. J. 2003. Transmission mode time reversal imaging with MUSIC. *J. Acoust. Soc. Am.*, **113**, 2742–2753.

Lerosey, G., de Rosny, J., Tourin, A., and Fink, M. 2007. Focusing beyond the diffraction limit with far-field time reversal. *Science*, **315**(5815), 1120–1122.

Lev-Ari, H., and Devaney, A. J. 2000. The time-reversal technique reinterpreted: Subspace-based signal processing for multi-static target location, in *IEEE Sensor Array and Multichannel Signal Processing Workshop*. Cambridge, MA: IEEE Press, pp. 509–513.

Levy, T. E., and Witten, A. 1996. Denizens of the desert. *Archaeol. Mag.*, 36–40.

Lewis, R. 1969. Physical optics inverse diffraction. *IEEE Trans. Ant. Propag.*, **17**, 308–314.

Lipson, H., and Cochran, W. 1966. *The Determination of Crystal Structures*. Ithaca, NY: Cornell University Press.

Liu, D., Kang, G., Li, L., *et al.* 2005. Electromagnetic time-reversal imaging of a target in a cluttered environment. *IEEE Trans. Ant. Propag.*, **53**, 3058–3066.

Liu, D., Kang, G., Vasudevan, S., *et al.* 2007. Electromagnetic time-reversal source localization in changing media: Experiment and analysis. *IEEE Trans. Ant. Propag.*, **55**, 344–354.

Louis, A. K. 1981. Ghosts in tomography: The null space of the Radon transform. *Math. Methods Appl. Sci.*, **3**, 1.

Luke, D., and Devaney, A. J. 2007. Identifying scattering obstacles by the construction of non-scattering waves. *SIAM J. Appl. Math.*, **68**, 271–291.

Maleki, M. H., Devaney, A. J., and Schatzberg, A. 1992. Tomographic reconstruction from optical scattered intensities. *J. Opt. Soc. Am. A*, **9**, 1356–1363.

Mandel, L., and Wolf, E. 1995. *Optical Coherence and Quantum Optics*. New York: Cambridge University Press.

Marengo, E. A., and Devaney, A. J. 1998. Time-dependent plane wave and multipole expansions of the electromagnetic field. *J. Math. Phys.*, **39**(7), 3643–3660.

Marengo, E. A., and Devaney, A. J. 1999. The inverse source problem of electromagnetics: Linear inversion formulation and minimum energy solution. *IEEE Trans. Ant. Propag.*, **47**(2), 410–412.

Marengo, E. A., and Ziolkowski, R. W. 2000. Nonradiating and minimum energy sources and their fields: generalized source inversion theory and applications. *IEEE Trans. Ant. Propag.*, **48**(10), 1553–1562.

Marengo, E. A., Devaney, A. J., and Gruber, F. K. 2004. Inverse source problem with reactive power constraint. *IEEE Trans. Ant. Propag.*, **52**(6), 1586–1595.

Marengo, E. A., Gruber, F. K., and Simonetti, F. 2007. Time-reversal MUSIC imaging of extended targets. *IEEE Trans. Image Processing*, **16**(8), 1967–1984.

Markel, V., and Schotland, J. 2001. Inverse problem in optical diffusion tomography. I Fourier–Laplace inversion formulas. *J. Opt. Soc. Am. A*, **18**, 1336–1347.

Melamed, T., and Heyman, E. 1997. Spectral analysis of time-domain diffraction tomography. *Radio Sci.*, **32**, 593–603.

Melamed, T., Ehrlich, Y., and Heyman, E. 1996. Short-pulse inversion of inhomogeneous media: a time-domain diffraction tomography. *Inverse Problems*, **12**, 977–993.

Melamed, T, Heyman, E., and Felsen, L. B. 1999a. Local spectral analysis of short-pulse excited scattering from weakly inhomogeneous media; part I: Forward scattering. *IEEE Trans. Ant. Propag.*, **47**, 1208–1217.

Melamed, T., Heyman, E., and Felsen, L. B. 1999b. Local spectral analysis of short-pulse excited scattering from weakly inhomogeneous media; part II: Inverse scattering. *IEEE Trans. Ant. Propag.*, **47**, 1218–1227.

Miller, D. A. 2006. On perfect cloaking. *Opt. Express*, **14**, 12457.

Morse, P. McC., and Feshbach, H. 1953. *Methods of Theoretical Physics, Parts I & II.* New York: McGraw-Hill.

Moses, H. 1984. The time-dependent inverse source problem for the acoustic and electromagnetic equations in the one- and three-dimensional cases. *J. Math. Phys.*, **25**, 1905–1924.

Mueller, R. K., Kaveh, M., and Wade, G. 1979. Reconstructive tomography and applications to ultrasonics. *Proc. IEEE*, **67**, 567.

Muller, C. 1969. *Foundations of the Mathematical Theory of Electromagnetic Waves*. New York: Springer.

Nachman, A. I. 1996. Global uniqueness for a two-dimensional inverse boundary value problem. *Ann. Math*, **143**, 71–96.

Nachman, T., Mast A. I., and Waag, R. C. 1997. Focusing and imaging using eigenfunctions of the scattering operator. *J. Acoust. Soc. Am.*, **102**, 715–725.

Natterer, F. 1986. *The Mathematics of Computerized Tomography*. New York: Wiley.

Naylor, A. W., and Sell, G. R. 1982. *Linear Operator Theory in Engineering and Science*. New York: Springer.

Newton, R. G. 1980a. Inverse scattering I: One dimension. *J. Math. Phys.*, **21**, 493.

Newton, R. G. 1980b. Inverse scattering II: Three dimensions. *J. Math. Phys.*, **21**, 1698.

Newton, R. G. 1982. *Scattering Theory of Waves and Particles*. Berlin: Springer.

Norton, S. J., and Linzer, M. L. 1981. Ultrasonic reflectivity imaging in three dimensions: exact inverse scattering solutions for plane, cylindrical, and spherical apertures. *IEEE Trans. Bio. Engmed.*, **28**, 202.

Oristaglio, M. L. 1985. Accuracy of the Born and Rytov approximations for the reflection and refraction at a plane interface. *J. Opt. Soc. Am.*, **2**, 1987.

Oughstun, K. 2006. *Electromagnetic and Optical Pulse Propagation I*. New York: Springer.

Oughstun, K., and Sherman, G. C. 1997. *Electromagnetic Pulse Propagation in Causal Dielectrics*. New York: Springer.

Paley, R., and Wiener, N. 1934. *Fourier Transforms in the Complex Domain*. Providence, RI: American Mathematical Society.

Pan, S., and Kak, A. 1983. A computational study of reconstruction algorithms for diffraction tomography: Interpolation versus filtered-backpropagation. *IEEE Trans. Acoust., Speech Signal Processing*, **31**(5), 1262–1275.

Pan, Xiaochuan. 1998. Unified reconstruction theory for diffraction tomography, with consideration of noise control. *J. Opt. Soc. Am. A*, **15**(9), 2312–2326.

Papas, C. H. 1988. *Theory of Electromagnetic Wave Propagation*. New York: Dover Publications.

Pierri, R., Liseno, A., Solimene, R., and Soldovieri, R. 2006. Beyond physical optics SVD shape reconstruction of metallic cylinders. *IEEE Trans. Ant. Propag.*, **54**, 655–665.

Porter, R. P. 1970. Diffraction limited scalar image formation with holograms of arbitrary shape. *J. Opt. Soc. Am.*, **60**, 1051–1059.

Prada, C., and Fink, M. 1994. Eigenstates of the time-reversal operator: A solution to selective focusing in multiple target media. *Wave Motion*, **20**, 151–163.

Prada, C., and Fink, M. 1995. Selective focusing through inhomogeneous media: the DORT method, in *IEEE Ultrasonics Symposium, 1995*. New York: IEEE Press, pp. 1449–1453.

Prosser, R. T. 1969. Formal solutions of inverse scattering problems. *J. Math. Phys.*, **10**, 1819–1822.

Radon, J. 1917. Über die Bestimmung von Functionen durch ihre Integralwerte längs gewisser Mannigfaltigkeiten. (On the determination of functions from their integral values along certain manifolds). *Ber. Sächsische Akad. Wissenschaften Leipzig, Math. Phys. Kl.*, **69**, 262–277 (English translation 1986. *IEEE Trans. Med. Imaging*, **5**(4), 170–176.)

Ramm, A. G. 1990. Completeness of the products of solutions of PDE and inverse problems. *Inverse Problems*, **6**, 643.

Richards, J. I., and Youn, H. K. 1995. *Theory of Distributions*. Cambridge: Cambridge University Press.

Robinson, B. S., and Greenleaf, J. F. 1986. The scattering of ultrasound by cylinders: Implications for diffraction tomography. *J. Acoust. Soc. Am.*, **80**, 40–90.

Rohrlich, F. 1965. *Classical Charged Particles*. New York: Addison-Wesley Publishing Co.

Ross, G., Fiddy, M. A., Moezzi, H., and Nieto-Vesperinas, M. 1979. The inverse problem in light scattering, in *Antennas and Propagation Society International Symposium, 1979*, vol. 17, pp. 232–235.

Sabatier, P. 1983. Theoretical considerations for inverse scattering. *Radio Sci.*, **18**, 1–18.

Schmidt, R. 1981. *A Signal Subspace Approach to Multiple Emitter Location and Spectral Estimation*. Palo Alto, CA: Ph.D. Thesis Stanford University.

Schmidt, R. 1986. Multiple emitter location and signal parameter estimation. *IEEE Trans. Ant. Propag.*, **34**, 276–280.

Schotland, J., and Markel, V. 2001. Inverse scattering with diffusing waves. *J. Opt. Soc. Am. A*, **18**, 2767–2777.

Shepp, L. A., and Logan, B. F. 1974. The Fourier reconstruction of the head section. *IEEE Trans. Nucl. Sci.*, **21**, 21.

Sherman, G. 1967. Integral transform formulation of diffraction theory. *J. Opt. Soc. Am.*, **57**, 1490–1498.

Shewell, J., and Wolf, E. 1968. Inverse diffraction and a new reciprocity theorem. *J. Opt. Soc. Am.*, **59**, 1596.

Shlivinski E., Heyman, A., and Kastner, R. 1997. Antenna characterization in the time domain. *IEEE Trans. Ant. Propag.*, **45**, 1140–1149.

Sitenko, A. 1991. *Scattering Theory*. Berlin: Springer.

Slepian, D., and Pollak, H. O. 1961. Prolate spheroidal wave functions, Fourier analysis and uncertainty I. *Bell Syst. Techn. J.*, **40**, 43–63.

Snieder, R. K., and Scales, J. A. 1998. Time-reversed imaging as a diagnostic of wave and particle chaos. *Phys. Rev. E*, **20**, 5668–5675.

Soldovieri, F., and Pierri, R. 2008. Shape reconstruction of metallic objects from intensity scattered field data only. *Opt. Lett.*, **33**, 246–248.

Solna, S., Hou K., and Zhao, H. 2006. A direct imaging algorithm for extended targets. *Inverse Problems*, **22**, 1151.

Sommerfeld, A. 1967. *Partial Differential Equations in Physics*. New York: Academic Press, p. 189.

Spoliansky, C., Prada S., Manneville D., and Fink, M. 1996. Decomposition of the time reversal operator: Detection and selective focusing on two scatterers. *J. Acoust. Soc. Am.*, **99**, 2067 2076.

Stamnes, J. J. 1986. *Waves in Focal Regions*. Bristol: Adam Hilger.

Stoica, P., and Moses, R. L. 1997. *Introduction to Spectral Analysis*. Upper Saddle River, NJ: Prentice Hall.

Stolt, R. H., and Weglein, A. B. 1985. Migration and inversion of seismic data. *Geophysics*, **50**, 2458.

Stratton, J. 1941. *Electromagnetic Theory*, 1st edn. New York: McGraw-Hill.

Sylvester, J., and Uhlmann, G. 1990. The Dirichlet to Neumann map and applications, in Colton, D. L., Ewing, R. E., and Rundell, W. (eds.), *Inverse Problems in Partial Differential Equations*. Philadelphia, PA: SIAM, pp. 101–139.

Sylvester, J., and Winebrenner, D. P. 1998. Linear and nonlinear inverse scattering. *SIAM J. Appl. Math.*, **59**, 669–699.

Sylvester, J., Winebrenner, D. P., and Gylys-Colwell, F. 1996. Layer stripping for the Helmholtz Equation. *SIAM J. Appl. Math.*, **56**, 736–754.

Tabbara, W., Duchene, B., Pichot, Ch., *et al.* 1988. Diffraction tomography: Contribution to the analysis of some applications in microwaves and ultrasonics. *Inverse Problems*, **4**(2), 305.

Tatarski, V. I. 1961. *Wave Propagation in a Turbulent Medium*. New York: McGraw-Hill.

Taylor, J. R. 1972. *Scattering Theory*. New York: John Wiley & Sons Inc.

Taylor, L. S. 1981. The phase retrieval problem. *IEEE Trans. Ant. Propag.*, **29**, 386–391.

Therrien, C. W. 1992. *Discrete Random Signals and Statistical Signal Processing*. Upper Saddle River, NJ: Prentice Hall.

Thomas, C., Prada J. L., and Fink, M. 1995. The iterative time reversal process: Analysis of the convergence. *J. Acoust. Soc. Am.*, **97**, 62–71.

Tsihrintzis, G. A, and Devaney, A. J. 1993. Stochastic diffraction tomography: Theory and computer simulation. *Signal Processing*, **39**, 49–64.

Tsihrintzis, G. A, and Devaney, A. J. 1994. Stochastic geophysical diffraction tomography. *Int. J. Imag. Syst. Techn.*, **5**, 239–242.

Tsihrintzis, G. A, and Devaney, A. J. 2000a. Higher-order (nonlinear) diffraction tomography: Inversion of the Rytov series. *IEEE Trans. Information Theory; Special Issue on Information Theoretic Imaging*, **46**, 1748–1761.

Tsihrintzis, G. A., and Devaney, A. J. 2000b. Higher-order (nonlinear) diffraction tomography: Reconstruction algorithms and computer simulation. *IEEE Trans. Image Processing*, **9**, 1560–1572.

Tsihrintzis, G. A., and Devaney, A. J. 2000c. A Volterra series approach to nonlinear traveltime tomography. *IEEE Trans. Geophys. Remote Sensing; Special Issue on Inverse Scattering*, **38**, 1733–1742.

Twomey, S. 2002. *Introduction to the Mathematics of Inversion in Remote Sensing*. New York: Dover Publications.

Vainshtein, B. K. 1974. *Diffraction of X-rays by Chain Molecules*. New York: John Wiley.

Vaughn, M. 2007. *Introduction to Mathematical Physics*. Weinheim: Wiley-VCH.

Wedberg, T. C., and Stamnes, J. J. 1996a. Quantitative imaging by optical diffraction tomography. *Opt. Rev.*, **2**(1), 28–31.

Wedberg, T. C., and Stamnes, J. J. 1996b. Recent results in optical diffraction microtomography. *Meas. Sci. Technol.*, **7**, 414.

Weinberger, H. F. 1995. *A First Course in Partial Differential Equations*. New York: Dover Publications.

Weston, V. H. 1985. On the convergence of the Rytov approximation for the reduced wave equation. *J. Math. Phys.*, **26**, 1979.

Weyl, H. 1919. Ausbreitung elektromagnetischer Wellen uber einen ebenen Leiter. *Ann. Phys., Leipzig*, **60**, 481–500.

Whittaker, E. T. 1904. On an expression of the electromagnetic field due to electrons by means of two scalar potentials. *Proc. London Math. Soc.*, **1**, 367–372.

Wilcox, C. H. 1957. Debye potentials. *J. Math. Mech.*, **6**, 167.

Witten, A., Gillette, D., King, W. C., and Sypniewski, J. 1992. Geophysical diffraction tomography at a dinosaur site. *Geophysics*, **57**, 187–195.

Wolf, E. 1969. Three dimensional structure determination of semi transparent objects from holographic data. *Opt. Commun.*, **1**, 153.

Wolf, E. 1996. Principles and development of diffraction tomography, in Consortini, A. (ed.), *Trends in Optics*. San Diego, CA: Academic Press.

Wolf, E., and Marchand, W. 1964. Comparison of the Kirchhoff and the Rayleigh–Sommerfeld theories of diffraction at an aperture. *J. Opt. Soc. Am.*, **54**, 587–594.

Yamaguchi, I., and Zhang, T. 1997. Phase-shifting digital holography. *Opt. Lett.*, **22**, 1268.

Young, N. 1988. *An Introduction to Hilbert Space*. Cambridge: Cambridge University Press.

Zemanian, A. H. 1965. *Distribution Theory and Transform Analysis*. New York: Dover Publications.

Zhao, H. 2004. Analysis of the response matrix for an extended target. *SIAM J. Appl. Math.*, **64**(3), 725–745.

Index

Printed in the United States
by Baker & Taylor Publisher Services

Printed in the United States
by Baker & Taylor Publisher Services